Linear Statistical Inference and its Applications

Linear Statistical Inference and its Applications

Second Editon

C. RADHAKRISHNA RAO
Pennsylvania State University

A Wiley-Interscience Publication
JOHN WILEY & SONS, INC.

This text is printed on acid-free paper. ♾

Paperback edition published 2002.

Published simultaneously in Canada.

For ordering and customer service, call 1-800-CALL-WILEY.

Library of Congress Cataloging in Publication Data is available.

ISBN 978-0-471-21875-3

10 9 8 7 6 5 4

To Bhargavi

Preface

The purpose of this book is to present up-to-date theory and techniques of statistical inference in a logically integrated and practical form. Essentially, it incorporates the important developments in the subject that have taken place in the last three decades. It is written for readers with a background knowledge of mathematics and statistics at the undergraduate level.

Quantitative inference, if it were to retain its scientific character, could not be divested of its logical, mathematical, and probabilistic aspects. The main approach to statistical inference is inductive reasoning, by which we arrive at "statements of uncertainty." The rigorous expression that degrees of uncertainty require are furnished by the mathematical methods and probability concepts which form the foundations of modern statistical theory. It was my awareness that advanced mathematical methods and probability theory are indispensable accompaniments in a self-contained treatment of statistical inference that prompted me to devote the first chapter of this book to a detailed discussion of vector spaces and matrix methods and the second chapter to a measure-theoretic exposition of probability and development of probability tools and techniques.

Statistical inference techniques, if not applied to the real world, will lose their import and appear to be deductive exercises. Furthermore, it is my belief that in a statistical course emphasis should be given to both mathematical theory of statistics and to the application of the theory to practical problems. A detailed discussion on the application of a statistical technique facilitates better understanding of the theory behind the technique. To this end, in this book, live examples have been interwoven with mathematical results. In addition, a large number of problems are given at the end of each chapter. Some are intended to complement main results derived in the body of the chapter, whereas others are meant to serve as exercises for the reader to test his understanding of theory.

The selection and presentation of material to cover the wide field of statistical inference have not been easy. I have been guided by my own experience in teaching undergraduate and graduate students, and in conducting and guiding research in statistics during the last twenty years. I have selected and presented the essential tools of statistics and discussed in detail their theoretical bases to enable the readers to equip themselves for consultation work or for pursuing specialized studies and research in statistics.

Why Chapter 1 provides a rather lengthy treatment of the algebra of vectors and matrices needs some explanation. First, the mathematical treatment of statistical techniques in subsequent chapters depends heavily on vector spaces and matrix methods; and second, vector and matrix algebra constitute a branch of mathematics widely used in modern treatises on natural, biological, and social sciences. The subject matter of the chapter is given a logical and rigorous treatment and is developed gradually to an advanced level. All the important theorems and derived results are presented in a form readily adaptable for use by research workers in different branches of science.

Chapter 2 contains a systematic development of the probability tools and techniques needed for dealing with statistical inference. Starting with the axioms of probability, the chapter proceeds to formulate the concepts of a random variable, distribution function, and conditional expectation and distributions. These are followed by a study of characteristic functions, probability distributions in infinite dimensional product spaces, and all the important limit theorems. Chapter 2 also provides numerous propositions, which find frequent use in some of the other chapters and also serve as good equipment for those who want to specialize in advanced probability theory.

Chapter 3 deals with continuous probability models and the sampling distributions needed for statistical inference. Some of the important distributions frequently used in practice, such as the normal, Gamma, Cauchy, and other distributions, are introduced through appropriate probability models on physical mechanisms generating the observations. A special feature of this chapter is a discussion of problems in statistical mechanics relating to the equilibrium distribution of particles.

Chapter 4 is devoted to inference through the technique of analysis of variance. The Gauss-Markoff linear model and the associated problems of estimation and testing are treated in their wide generality. The problem of variance-components is considered as a special case of the more general problem of estimating intraclass correlation coefficients. A unified treatment is provided of multiclassified data under different sampling schemes for classes within categories.

The different theories and methods of estimation form the subject matter of Chapter 5. Some of the controversies on the topic of estimation are examined; and to remove some of the existing inconsistencies, certain modifications are introduced in the criteria of estimation in large samples.

Problems of specification, and associated tests of homogeneity of parallel samples and estimates, are dealt with in Chapter 6. The choice of a mathematical model from which the observations could be deemed to have arisen is of fundamental importance because subsequent statistical computations will be made on the framework of the chosen model. Appropriate tests have been developed to check the adequacy of proposed models on the basis of available facts.

Chapter 7 provides the theoretical background for the different aspects of statistical inference, such as testing of hypotheses, interval estimation, experimentation, the problem of identification, nonparametric inference, and so on.

Chapter 8, the last chapter, is concerned with inference from multivariate data. A special feature of this chapter is a study of the multivariate normal distribution through a simple characterization, instead of through the density function. The characterization simplifies the multivariate theory and enables suitable generalizations to be made from the univariate theory without further analysis. It also provides the necessary background for studying multivariate normal distributions in more general situations, such as distributions on Hilbert space.

Certain notations have been used throughout the book to indicate sections and other references. The following examples will help in their interpretation. A subsection such as **4f·3** means subsection 3 in section f of Chapter 4. Equation (4f.3.6) is the equation numbered 6 in subsection **4f·3** and Table 4f.3β is the table numbered second in subsection **4f·3**. The main propositions (or theorems) in each subsection are numbered: (i), (ii), etc. A back reference such as [(iii). **5d·2**] indicates proposition (iii) in subsection **5d·2**.

A substantial part of the book was written while I was a visiting professor at the Johns Hopkins University, Baltimore, in 1963–1964, under a Senior Scientist Fellowship scheme of the National Science Foundation, U.S.A. At the Johns Hopkins University, I had the constant advice of G. S. Watson, Professor of Statistics, who read the manuscript at the various stages of its preparation. Comments by Herman Chernoff on Chapters 7 and 8, by Rupert Miller and S. W. Dharmadhikari on Chapter 2, and by Ralph Bradley on Chapters 1 and 3, have been extremely helpful in the preparation of the final manuscript. I wish to express my thanks to all of them. The preparation and revision of the manuscript would not have been an easy task without the help of G. M. Das, who undertook the heavy burden of typing and organizing the manuscript for the press with great care and diligence.

Finally, I wish to express my gratitude to the late Sir Ronald A. Fisher and to Professor P. C. Mahalanobis under whose influence I have come to appreciate statistics as the new technology of the present century.

Calcutta, India
June, 1965

C. R. Rao

Preface to the Second Edition

As in the first edition, the aim has been to provide in a single volume a full discussion of the wide range of statistical methods useful for consulting statisticians and, at same time, to present in a rigorous manner the mathematical and logical tools employed in deriving statistical procedures, with which a research worker should be familiar.

A good deal of new material is added, and the book is brought up to date in several respects, both in theory and applications.

Some of the important additions are different types of generalized inverses, concepts of statistics and subfields, MINQUE theory of variance components, the law of iterated logarithms and sequential tests with power one, analysis of dispersion with structural parameters, discrimination between composite hypotheses, growth models, theorems on characteristic functions, etc.

Special mention may be made of the new material on estimation of parameters in a linear model when the observations have a *possibly singular covariance matrix*. The existing theories and methods due to Gauss (1809) and Aitken (1935) are applicable only when the covariance matrix is known to be nonsingular. The new *unified approaches* discussed in the book (Section 4i) are valid for all situations whether the covariance matrix is singular or not.

A large number of new exercises and complements have been added.

I wish to thank Dr. M. S. Avadhani, Dr. J. K. Ghosh, Dr. A. Maitra, Dr. P. E. Nüesch, Dr. Y. R. Sarma, Dr. H. Toutenberg, and Dr. E. J. Williams for their suggestions while preparing the second edition.

New Delhi C. R. Rao

Contents

CHAPTER 1

Algebra of Vectors and Matrices 1

1a. Vector Spaces 2

1a.1 Definition of Vector Spaces and Subspaces, 1a.2 Basis of a Vector Space, 1a.3 Linear Equations, 1a.4 Vector Spaces with an Inner Product

Complements and Problems 11

1b. Theory of Matrices and Determinants 14

1b.1 Matrix Operations, 1b.2 Elementary Matrices and Diagonal Reduction of a Matrix, 1b.3 Determinants, 1b.4 Transformations 1b.5 Generalized Inverse of a Matrix, 1b.6 Matrix Representation, of Vector Spaces, Bases, etc., 1b.7 Idempotent Matrices, 1b.8 Special Products of Matrices

Complements and Problems 30

1c. Eigenvalues and Reduction of Matrices 34

1c.1 Classification and Transformation of Quadratic Forms, 1c.2 Roots of Determinantal Equations, 1c.3 Canonical Reduction of Matrices, 1c.4 Projection Operator, 1c.5 Further Results on g-Inverse, 1c.6 Restricted Eigenvalue Problem

1d. Convex Sets in Vector Spaces 51

1d.1 Definitions, 1d.2 Separation Theorems for Convex Sets

1e. Inequalities 53

1e.1 Cauchy-Schwarz (C-S) Inequality, 1e.2 Hölder's Inequality, 1e.3 Hadamard's Inequality, 1e.4 Inequalities Involving Moments, 1e.5 Convex Functions and Jensen's Inequality, 1e.6 Inequalities in Information Theory, 1e.7 Stirling's Approximation

1f. Extrema of Quadratic Forms 60

1f.1 General Results, 1f.2 Results Involving Eigenvalues and Vectors 1f.3 Minimum Trace Problems

Complements and Problems 67

CHAPTER 2

Probability Theory, Tools and Techniques 79

2a. Calculus of Probability 80

2a.1 The Space of Elementary Events, 2a.2 The Class of Subsets (Events), 2a.3 Probability as a Set Function, 2a.4 Borel Field (σ-field) and Extension of Probability Measure, 2a.5 Notion of a Random Variable and Distribution Function, 2a.6 Multidimensional Random Variable, 2a.7 Conditional Probability and Statistical Independence, 2a.8 Conditional Distribution of a Random Variable

2b. Mathematical Expectation and Moments of Random Variables 92

2b.1 Properties of Mathematical Expectation, 2b.2 Moments, 2b.3 Conditional Expectation, 2b.4 Characteristic Function (c.f.), 2b.5 Inversion Theorems, 2b.6 Multivariate Moments

2c. Limit Theorems 108

2c.1 Kolmogorov Consistency Theorem, 2c.2 Convergence of a Sequence of Random Variables, 2c.3 Law of Large Numbers, 2c.4 Convergence of a Sequence of Distribution Functions, 2c.5 Central Limit Theorems, 2c.6 Sums of Independent Random Variables

2d. Family of Probability Measures and Problems of Statistics 130

2d.1 Family of Probability Measures, 2d.2 The Concept of a Sufficient Statistic, 2d.3 Characterization of Sufficiency

Appendix 2A Stieltjes and Lebesgue Integrals 132

Appendix 2B. Some Important Theorems in Measure Theory and Integration 134

Appendix 2C. Invariance 138

Appendix 2D. Statistics, Subfields, and Sufficiency 139

Appendix 2E. Non-Negative Definiteness of a Characteristic Function 141

Complements and Problems 142

CHAPTER 3

Continuous Probability Models 155

3a. Univariate Models 158

3a.1 Normal Distribution, 3a.2 Gamma Distribution, 3a.3 Beta Distribution, 3a.4 Cauchy Distribution, 3a.5 Student's t Distribution, 3a.6 Distributions Describing Equilibrium States in Statistical Mechanics, 3a.7 Distribution on a Circle

3b. Sampling Distributions 179

3b.1 Definitions and Results, 3b.2 Sum of Squares of Normal Variables, 3b.3 Joint Distribution of the Sample Mean and Variance, 3b.4 Distribution of Quadratic Forms, 3b.5 Three Fundamental Theorems of the Least Squares Theory, 3b.6 The p-Variate Normal Distribution, 3b.7 The Exponential Family of Distributions

3c. Symmetric Normal Distribution 197

3c.1 Definition, 3c.2 Sampling Distributions

3d. Bivariate Normal Distribution 201

3d.1 General Properties, 3d.2 Sampling Distributions

Complements and Problems 209

CHAPTER 4

The Theory of Least Squares and Analysis of Variance 220

4a. Theory of Least Squares (Linear Estimation) 221

4a.1 Gauss-Markoff Setup $(Y, X\beta, \sigma^2 I)$, 4a.2 Normal Equations and Least Squares (l.s.) Estimators, 4a.3 g-Inverse and a Solution of the Normal Equation, 4a.4 Variances and Covariances of l.s. Estimators, 4a.5 Estimation of σ^2, 4a.6 Other Approaches to the l.s. Theory (Geometric Solution), 4a.7 Explicit Expressions for Correlated Observations, 4a.8 Some Computational Aspects of the l.s. Theory, 4a.9 Least Squares Estimation with Restrictions on Parameters,

4a.10 Simultaneous Estimation of Parametric Functions, 4a.11 Least Squares Theory when the Parameters Are Random Variables, 4a.12 Choice of the Design Matrix

4b. Tests of Hypotheses and Interval Estimation 236

4b.1 Single Parametric Function (Inference), 4b.2 More than One Parametric Function (Inference), 4b.3 Setup with Restrictions

4c. Problems of a Single Sample 243

4c.1 The Test Criterion, 4c.2 Asymmetry of Right and Left Femora (Paired Comparison)

4d. One-Way Classified Data 244

4d.1 The Test Criterion, 4d.2 An Example

4e. Two-Way Classified Data 247

4e.1 Single Observation in Each Cell, 4e.2 Multiple but Equal Numbers in Each Cell, 4e.3 Unequal Numbers in Cells

4f. A General Model for Two-Way Data and Variance Components 258

4f.1 A General Model, 4f.2 Variance Components Model, 4f.3 Treatment of the General Model

4g. The Theory and Application of Statistical Regression 263

4g.1 Concept of Regression (General Theory), 4g.2 Measurement of Additional Association, 4g.3 Prediction of Cranial Capacity (a Practical Example), 4g.4 Test for Equality of the Regression Equations, 4g.5 The Test for an Assigned Regression Function, 4g.6 Restricted Regression

4h. The General Problem of Least Squares with Two Sets of Parameters 288

4h.1 Concomitant Variables, 4h.2 Analysis of Covariance, 4h.3 An Illustrative Example

4i. Unified Theory of Linear Estimation 294

4i.1 A Basic Lemma on Generalized Inverse, 4i.2 The General Gauss-Markoff Model (GGM), 4i.3 The Inverse Partitioned Matrix (IPM) Method, 4i.4 Unified Theory of Least Squares

4j. Estimation of Variance Components 302

4j.1 Variance Components Model, 4j.2 MINQUE Theory, 4j.3 Computation under the Euclidian Norm

4k. Biased Estimation in Linear Models 305

4k.1 Best Linear Estimator (BLE), 4k.2 Best Linear Minimum Bias Estimation (BLIMBE)

Complements and Problems 308

CHAPTER 5

Criteria and Methods of Estimation 314

5a. Minimum Variance Unbiased Estimation 315

5a.1 Minimum Variance Criterion, 5a.2 Some Fundamental Results on Minimum Variance Estimation, 5a.3 The Case of Several Parameters, 5a.4 Fisher's Information Measure, 5a.5 An Improvement of Unbiased Estimators

5b. General Procedures 334

5b.1 Statement of the General Problem (Bayes Theorem), 5b.2 Joint d.f. of (θ, x) Completely Known, 5b.3 The Law of Equal Ignorance, 5b.4 Empirical Bayes Estimation Procedures, 5b.5 Fiducial Probability, 5b.6 Minimax Principle, 5b.7 Principle of Invariance

5c. Criteria of Estimation in Large Samples 344

5c.1, Consistency, 5c.2 Efficiency

5d. Some Methods of Estimation in Large Samples 351

5d.1 Method of Moments, 5d.2 Minimum Chi-Square and Associated Methods, 5d.3 Maximum Likelihood

5e. Estimation of the Multinomial Distribution 355

5e.1 Nonparametric Case, 5e.2 Parametric Case

5f. Estimation of Parameters in the General Case 363

5f.1 Assumptions and Notations, 5f.2 Properties of m.l. Equation Estimators

5g. The Method of Scoring for the Estimation of Parameters 366

Complements and Problems 374

CHAPTER 6

Large Sample Theory and Methods 382

6a. Some Basic Results 382

6a.1 Asymptotic Distribution of Quadratic Functions of Frequencies, 6a.2 Some Convergence Theorems

6b. Chi-Square Tests for the Multinomial Distribution 390

6b.1 Test of Departure from a Simple Hypothesis, 6b.2 Chi-Square Test for Goodness of Fit, 6b.3 Test for Deviation in a Single Cell, 6b.4 Test Whether the Parameters Lie in a Subset, 6b.5 Some Examples, 6b.6 Test for Deviations in a Number of Cells

6c. Tests Relating to Independent Samples from Multinomial Distributions 398

6c.1 General Results, 6c.2 Test of Homogeneity of Parallel Samples, 6c.3 An Example

6d. Contingency Tables 403

6d.1 The Probability of an Observed Configuration and Tests in Large Samples, 6d.2 Tests of Independence in a Contingency Table, 6d.3 Tests of Independence in Small Samples

6e. Some General Classes of Large Sample Tests 415

6e.1 Notations and Basic Results, 6e.2 Test of a Simple Hypothesis, 6e.3 Test of a Composite Hypothesis

6f. Order Statistics 420

6f.1 The Empirical Distribution Function, 6f.2 Asymptotic Distribution of Sample Fractiles

6g. Transformation of Statistics 426

6g.1 A General Formula, 6g.2 Square Root Transformation of the Poisson Variate, 6g.3 Sin^{-1} *Transformation of the Square Root of the Binomial Proportion, 6g.4* Tanh^{-1} *Transformation of the Correlation Coefficient*

6h. Standard Errors of Moments and Related Statistics 436

6h.1 Variances and Covariances of Raw Moments, 6h.2 Asymptotic Variances and Covariances of Central Moments, 6h.3 Exact Expressions for Variances and Covariances of Central Moments

Complements and Problems 439

CHAPTER 7

Theory of Statistical Inference 444

7a. Testing of Statistical Hypotheses 445

7a.1 Statement of the Problem, 7a.2 Neyman-Pearson Fundamental Lemma and Generalizations, 7a.3 Simple H_0 *against Simple* H*,*

7a.4 Locally Most Powerful Tests, 7a.5 Testing a Composite Hypothesis, 7a.6 Fisher-Behrens Problem, 7a.7 Asymptotic Efficiency of Tests

7b. Confidence Intervals 470

7b.1 The General Problem, 7b.2 A General Method of Constructing a Confidence Set, 7b.3 Set Estimators for Functions of θ

7c. Sequential Analysis 474

7c.1 Wald's Sequential Probability Ratio Test, 7c.2 Some Properties of the S.P.R.T., 7c.3 Efficiency of the S.P.R.T., 7c.4 An Example of Economy of Sequential Testing, 7c.5 The Fundamental Identity of Sequential Analysis, 7c.6 Sequential Estimation, 7c.7 Sequential Tests with Power One

7d. Problem of Identification—Decision Theory 491

7d.1 Statement of the Problem, 7d.2 Randomized and Nonrandomized Decision Rules, 7d.3 Bayes Solution, 7d.4 Complete Class of Decision Rules, 7d.5 Minimax Rule

7e. Nonparametric Inference 497

7e.1 Concept of Robustness, 7e.2 Distribution-Free Methods, 7e.3 Some Nonparametric Tests, 7e.4 Principle of Randomization

7f. Ancillary Information 505

Complements and Problems 506

CHAPTER 8

Multivariate Analysis 516

8a. Multivariate Normal Distribution 517

8a.1 Definition, 8a.2 Properties of the Distribution, 8a.3 Some Characterizations of N_p, 8a.4 Density Function of the Multivariate Normal Distribution, 8a.5 Estimation of Parameters, 8a.6 N_p as a Distribution with Maximum Entropy

8b. Wishart Distribution 533

8b.1 Definition and Notation, 8b.2 Some Results on Wishart Distribution

8c. Analysis of Dispersion 543

8c.1 The Gauss-Markoff Setup for Multiple Measurements, 8c.2 Estimation of Parameters, 8c.3 Tests of Linear Hypotheses, Analysis of

Dispersion (A.D.), 8c.4 Test for Additional Information, 8c.5 The Distribution of Λ, 8c.6 Test for Dimensionality (Structural Relationship), 8c.7 Analysis of Dispersion with Structural Parameters (Growth Model)

8d. Some Applications of Multivariate Tests 562

8d.1 Test for Assigned Mean Values, 8d.2 Test for a Given Structure of Mean Values, 8d.3 Test for Differences between Mean Values of Two Populations, 8d.4 Test for Differences in Mean Values between Several Populations, 8d.5 Barnard's Problem of Secular Variations in Skull Characters

8e. Discriminatory Analysis (Identification) 574

8e.1 Discriminant Scores for Decision, 8e.2 Discriminant Analysis in Research Work, 8e.3 Discrimination between Composite Hypotheses

8f. Relation between Sets of Variates 582

8f.1 Canonical Correlations, 8f.2 Properties of Canonical Variables, 8f.3 Effective Number of Common Factors, 8f.4 Factor Analysis

8g. Orthonormal Basis of a Random Variable 587

8g.1 The Gram-Schmidt Basis, 8g.2 Principal Component Analysis

Complements and Problems 593

Publications of the Author 605

Author Index 615

Subject Index 618

Linear Statistical Inference and Its Applications

Chapter 1

ALGEBRA OF VECTORS AND MATRICES

Introduction. The use of matrix theory is now widespread in both pure mathematics and the physical and the social sciences. The theory of vector spaces and transformations (of which matrices are a special case) have not, however, found a prominent place, although they are more fundamental and offer a better understanding of problems. The vector space concepts are essential in the discussion of topics such as the theory of games, economic behavior, prediction in time series, and the modern treatment of univariate and multivariate statistical methods.

The aim of the first chapter is to introduce the reader to the concepts of vector spaces and the basic results. All important theorems are discussed in great detail to enable the beginner to work through the chapter. Numerous illustrations and problems for solution are given as an aid to further understanding of the subject. To introduce wide generality (this is important and should not cause any difficulty in understanding the theory) the elements used in the operations with vectors are considered as belonging to any *Field* in which *addition* and *multiplication* are defined in a consistent way (as in the ordinary number system). Thus, the elements $e_1, e_2, \ldots$ (finite or infinite in number) are said to belong to a field F, if they are closed under the operations of addition ($e_i + e_j$) and multiplication ($e_i e_j$), that is, sums and products of elements of F also belong to F, and satisfy the following conditions:

(A_1) $e_i + e_j = e_j + e_i$ (commutative law)

(A_2) $e_i + (e_j + e_k) = (e_i + e_j) + e_k$ (associative law)

(A_3) For any two elements e_i, e_j, there exists an element e_k such that $e_i + e_k = e_j$.

The condition (A_3) implies that there exists an element e_0 such that $e_i + e_0 = e_i$ for all i. The element e_0 is like 0 (zero) of the number system.

(M_1) $e_i e_j = e_j e_i$ (commutative law)

(M_2) $e_i(e_j e_k) = (e_i e_j)e_k$ (associative law)

(M_3) $e_i(e_j + e_k) = e_i e_j + e_i e_k$ (distributive law)

(M_4) For any two elements e_i and e_j such that $e_i \neq e_0$, the zero element, there exists an element e_k such that $e_i e_k = e_j$.

(M_4) implies that there is an element e_1 such that $e_i e_1 = e_i$ for all i. The element e_1 is like 1 (unity) of the number system.

The study of vector spaces is followed by a discussion of the modern matrix theory and quadratic forms. Besides the basic propositions, a number of results used in mathematical physics, economics, biology and statistics, and numerical computations are brought together and presented in a unified way. This would be useful for those interested in applications of the matrix theory in the physical and the social sciences.

1a VECTOR SPACES

1a.1 Definition of Vector Spaces and Subspaces

Concepts such as force, size of an organism, an individual's health or mental abilities, and price level of commodities cannot be fully represented by a single number. They have to be understood by their manifestations in different directions, each of which may be expressed by a single number. The mental abilities of an individual may be judged by his scores $(x_1, x_2, \ldots, x_k)$ in k specified tests. Such an ordered set of measurements may be simply represented by $\mathbf{x}$, called a row vector. If $\mathbf{y} = (y_1, y_2, \ldots, y_k)$ is the vector of scores for another individual, the total scores for the two individuals in the various tests may be represented by $\mathbf{x} + \mathbf{y}$ with the definition

$$\mathbf{x} + \mathbf{y} = (x_1 + y_1, x_2 + y_2, \ldots, x_k + y_k). \tag{1a.1.1}$$

This rule of combining or adding two vectors is the same as that for obtaining the resultant of two forces in two or three dimensions, known as the *parallelogram law of forces*. Algebraically this law is equivalent to finding a force whose components are the sum of the corresponding components of the individual forces.

Given a force $\mathbf{f} = (f_1, f_2, f_3)$, it is natural to define

$$c\mathbf{f} = (cf_1, cf_2, cf_3) \tag{1a.1.2}$$

as a force c times the first, which introduces a new operation of multiplying a vector by a scalar number such as c. Further, given a force $\mathbf{f}$, we can counterbalance it by adding a force $\mathbf{g} = (-f_1, -f_2, -f_3) = (-1)\mathbf{f}$, or by applying $\mathbf{f}$ in the opposite direction, we have the resulting force $\mathbf{f} + \mathbf{g} = (0, 0, 0)$. Thus we have the concepts of a negative vector such as $-\mathbf{f}$ and a null vector $\mathbf{0} = (0, 0, 0)$, the latter having the property $\mathbf{x} + \mathbf{0} \doteq \mathbf{x}$ for all $\mathbf{x}$.

It is seen that we are able to work with the new quantities, to the extent permitted by the operations defined, in the same way as we do with numbers. As a matter of fact, we need not restrict our algebra to the particular type of new quantities, *viz.*, ordered sets of scalars, but consider a collection of elements $\mathbf{x}$, $\mathbf{y}$, $\mathbf{z}$, ..., finite or infinite, which we choose to call vectors and $c_1, c_2, \ldots$, scalars constituting a *field* (like the ordinary numbers with the operations of addition, subtraction, multiplication, and division suitably defined) and lay down certain rules of combining them.

Vector Addition. The operation of addition indicated by + is defined for any two vectors leading to a vector in the set and is subject to the following rules:

(i) $\quad \mathbf{x} + \mathbf{y} = \mathbf{y} + \mathbf{x}$ (commutative law)

(ii) $\quad \mathbf{x} + (\mathbf{y} + \mathbf{z}) = (\mathbf{x} + \mathbf{y}) + \mathbf{z}$ (associative law)

Null Element. There exists an element in the set denoted by $\mathbf{0}$ such that

(iii) $$\mathbf{x} + \mathbf{0} = \mathbf{x}, \qquad \text{for all} \quad \mathbf{x}.$$

Inverse (Negative) Element. For any given element $\mathbf{x}$, there exists a corresponding element $\boldsymbol{\xi}$ such that

(iv) $$\mathbf{x} + \boldsymbol{\xi} = \mathbf{0}.$$

Scalar Multiplication. The multiplication of a vector $\mathbf{x}$ by a scalar c leads to a vector in the set, represented by $c\mathbf{x}$, and is subject to the following rules.

(v) $\quad c(\mathbf{x} + \mathbf{y}) = c\mathbf{x} + c\mathbf{y}$ (distributive law for vectors)

(vi) $\quad (c_1 + c_2)\mathbf{x} = c_1\mathbf{x} + c_2\mathbf{x}$ (distributive law for scalars)

(vii) $\quad c_1(c_2\mathbf{x}) = (c_1c_2)\mathbf{x}$ (associative law)

(viii) $\quad e\mathbf{x} = \mathbf{x}$ (where e is the unit element in the field of scalars).

A collection of elements (with the associated field of scalars F) satisfying the axioms (i) to (viii) is called a *linear vector space* $\mathscr{V}$ or more explicitly $\mathscr{V}(F)$ or $\mathscr{V}_F$. Note that the conditions (iii) and (iv) can be combined into the single condition that for any two elements $\mathbf{x}$ and $\mathbf{y}$ there is a unique element $\mathbf{z}$ such that $\mathbf{x} + \mathbf{z} = \mathbf{y}$.

The reader may satisfy himself that, for a given k, the collection of all ordered sets $(x_1, \ldots, x_k)$ of real numbers with the addition and the scalar multiplication as defined in (1a.1.1) and (1a.1.2) form a vector space. The same is true of all polynomials of degree not greater than k, with coefficients belonging to any field and addition and multiplication by a constant being defined in the usual way. Although we will be mainly concerned with vectors which are ordered sets of real numbers in the treatment of statistical methods covered in this book, the axiomatic set up will give the reader proper insight into the new algebra and also prepare the ground for a study of more complicated vector spaces, like Hilbert space (see Halmos, 1951), which are being increasingly used in the study of advanced statistical methods.

A *linear subspace, subspace,* or *linear manifold* in a vector space $\mathscr{V}$ is any subset of vectors $\mathscr{M}$ closed under addition and scalar multiplication, that is, if $\mathbf{x}$ and $\mathbf{y} \in \mathscr{M}$, then $(c\mathbf{x} + d\mathbf{y}) \in \mathscr{M}$ for any pair of scalars c and d. Any such subset $\mathscr{M}$ is *itself* a vector space with respect to the same definition of addition and scalar multiplication as in $\mathscr{V}$. The subset containing the null vector alone, as well as that consisting of all the elements in $\mathscr{V}$, are extreme examples of subspaces. They are called improper subspaces whereas others are proper subspaces.

As an example, all linear combinations of a given fixed set S of vectors $\boldsymbol{\alpha}_1, \ldots, \boldsymbol{\alpha}_k$ is a subspace called the linear manifold $\mathscr{M}(S)$ spanned by S. This is the smallest subspace containing S.

Consider k linear equations in n variables $x_1, \ldots, x_n$,

$$a_{i1} x_1 + \cdots + a_{in} x_n = 0, \qquad i = 1, \ldots, k,$$

where a_{ij} belongs to any field F. The reader may verify that the totality of solutions $(x_1, \ldots, x_n)$ considered as vectors constitutes a subspace with the addition and scalar multiplication as defined in (1a.1.1) and (1a.1.2).

1a.2 Basis of a Vector Space

A set of vectors $\mathbf{u}_1, \ldots, \mathbf{u}_k$ is said to be linearly dependent if there exist scalars $c_1, \ldots, c_k$, not all simultaneously zero, such that $c_1\mathbf{u}_1 + \cdots + c_k\mathbf{u}_k = \mathbf{0}$, otherwise it is independent. With such a definition the following are true:

1. The null vector by itself is a dependent set.
2. Any set of vectors containing the null vector is a dependent set.
3. A set of non-zero vectors $\mathbf{u}_1, \ldots, \mathbf{u}_k$ is dependent when and only when a member in the set is a linear combination of its predecessors.

A linearly independent subset of vectors in a vector space $\mathscr{V}$, generating or spanning $\mathscr{V}$ is called a basis (*Hamel* basis) of $\mathscr{V}$.

(i) *Every vector space $\mathscr{V}$ has a basis.*

To demonstrate this, let us choose sequentially non-null vectors $\boldsymbol{\alpha}_1, \boldsymbol{\alpha}_2, \ldots,$ in $\mathscr{V}$ such that no $\boldsymbol{\alpha}_i$ is dependent on its predecessors. In this process it may so happen that after the kth stage no independent vector is left in $\mathscr{V}$, in which case $\boldsymbol{\alpha}_1, \ldots, \boldsymbol{\alpha}_k$ constitute a basis and $\mathscr{V}$ is said to be a finite (k) dimensional vector space. On the other hand, there may be no limit to the process of choosing $\boldsymbol{\alpha}_i$, in which case $\mathscr{V}$ is said to have infinite dimensions. Further argument is needed to show the actual existence of an infinite set of independent vectors which generate all the vectors. This is omitted as our field of study will be limited to finite dimensional vector spaces. The following results concerning finite dimensional spaces are important.

(ii) *If* $\boldsymbol{\alpha}_1, \ldots, \boldsymbol{\alpha}_k$ *and* $\boldsymbol{\beta}_1, \ldots, \boldsymbol{\beta}_s$ *are two alternative choices for a basis, then* $s = k$.

Let, if possible, $s > k$. Consider the dependent set $\boldsymbol{\beta}_1, \boldsymbol{\alpha}_1, \ldots, \boldsymbol{\alpha}_k$. If $\boldsymbol{\alpha}_i$ depends on the predecessors, then $\mathscr{V}$ can also be generated by $\boldsymbol{\beta}_1, \boldsymbol{\alpha}_1, \ldots, \boldsymbol{\alpha}_{i-1}, \boldsymbol{\alpha}_{i+1}, \ldots, \boldsymbol{\alpha}_k$, in which case the set $\boldsymbol{\beta}_2, \boldsymbol{\beta}_1, \boldsymbol{\alpha}_1, \ldots, \boldsymbol{\alpha}_{i-1}, \boldsymbol{\alpha}_{i+1}, \ldots, \boldsymbol{\alpha}_k$ is dependent. One more $\boldsymbol{\alpha}$ can now be omitted. The process of adding a $\boldsymbol{\beta}$ and omitting an $\boldsymbol{\alpha}$ can be continued (observe that no $\boldsymbol{\beta}$ can be eliminated at any stage) till the set $\boldsymbol{\beta}_1, \ldots, \boldsymbol{\beta}_k$ is left, which itself spans $\mathscr{V}$ and hence $(s - k)$ of the $\boldsymbol{\beta}$ vectors are redundant. The *cardinal number* k common to all bases represents the minimal number of vectors needed to span the space or the maximum number of independent vectors in the space. We call this number as the dimension or rank of $\mathscr{V}$ and denote it by $d[\mathscr{V}]$.

(iii) *Every vector in* $\mathscr{V}$ *has a unique representation in terms of a given basis.*

If $\sum a_i \boldsymbol{\alpha}_i$ and $\sum b_i \boldsymbol{\alpha}_i$ represent the same vector, then $\sum (a_i - b_i)\boldsymbol{\alpha}_i = \mathbf{0}$, which is not possible unless $a_i - b_i = 0$ for all i, since $\boldsymbol{\alpha}_i$ are independent.

The Euclidean space $E_k(F)$ of all ordered sets of k elements $(x_1, \ldots, x_k)$, $x_i \in F$ (a field of elements) is of special interest. The vectors may be considered as points of k dimensional Euclidean space. The vectors $\mathbf{e}_1 = (1, 0, \ldots, 0)$, $\mathbf{e}_2 = (0, 1, 0, \ldots, 0), \ldots, \mathbf{e}_k = (0, 0, \ldots, 1)$ are in E_k and are independent, and any vector $\mathbf{x} = (x_1, \ldots, x_k) = x_1\mathbf{e}_1 + \cdots + x_k\mathbf{e}_k$. Therefore $d[E_k] = k$, and the vectors $\mathbf{e}_1, \ldots, \mathbf{e}_k$ constitute a basis and thus any other independent set of k vectors. Any vector in E_k can, therefore, be represented as a unique linear combination of k independent vectors, and, naturally, as a combination (not necessarily unique) of *any set* of vectors containing k independent vectors.

When F is the field of *real numbers* the vector space $E_k(F)$ is denoted simply by R^k. In the study of design of experiments we use *Galois fields* (GF) with a finite number of elements, and consequently the vector space has only a finite number of vectors. The notation $E_k(GF)$ may be used for such spaces. The vector space $E_k(F)$ when F is the field of complex numbers is represented by U^k and is called a k dimensional unitary space. The treatment of Section **1a.3** is valid for any F. Later we shall confine our attention to R^k only. We prove an important result in (iv) which shows that study of finite dimensional vector spaces is equivalent to that of E_k.

(iv) *Any vector space* $\mathscr{V}_{\mathrm{F}}$ *for which* $d[\mathscr{V}] = k$ *is isomorphic to* E_k.

If $\boldsymbol{\alpha}_1, \ldots, \boldsymbol{\alpha}_k$ is a basis of $\mathscr{V}_F$, then an element $\mathbf{u} \in \mathscr{V}_F$ has the representation $\mathbf{u} = a_1\boldsymbol{\alpha}_1 + \cdots + a_k\boldsymbol{\alpha}_k$ where $a_i \in F$, $i = 1, \ldots, k$. The correspondence

$$\mathbf{u} \to \mathbf{u}^* = (a_1, \ldots, a_k), \qquad \mathbf{u}^* \in E_k(F)$$

establishes the isomorphism. For if $\mathbf{u} \to \mathbf{u}^*$, $\mathbf{v} \to \mathbf{v}^*$, then $\mathbf{u} + \mathbf{v} \to \mathbf{u}^* + \mathbf{v}^*$ and $c\mathbf{u} \to c\mathbf{u}^*$. This result also shows that any two vector spaces with the same dimensions are isomorphic.

1a.3 Linear Equations

Let $\boldsymbol{\alpha}_1, \ldots, \boldsymbol{\alpha}_m$ be m *fixed* vectors in an arbitrary vector space $\mathscr{V}_F$ and consider the linear equation in the scalars $x_1, \ldots, x_m \in F$ (associated Field)

$$x_1\boldsymbol{\alpha}_1 + \cdots + x_m\boldsymbol{\alpha}_m = \mathbf{0}. \qquad (1a.3.1)$$

(i) *A necessary and sufficient condition that* (1a.3.1) *has a nontrivial solution, that is, not all x_i simultaneously zero, is that $\boldsymbol{\alpha}_1, \ldots, \boldsymbol{\alpha}_m$ should be dependent.*

(ii) *The solutions considered as (row) vectors $\mathbf{x} = (x_1, \ldots, x_m)$ in $E_m(F)$ constitute a vector space.*

This is true for if $\mathbf{x}$ and $\mathbf{y}$ are solutions then $a\mathbf{x} + b\mathbf{y}$ is also a solution. Note that $\boldsymbol{\alpha}_i$, themselves may belong to any vector space $\mathscr{V}_F$.

(iii) *Let $\mathscr{S}$ be the linear manifold or the subspace of solutions and $\mathscr{M}$, that of the vectors $\boldsymbol{\alpha}_1, \ldots, \boldsymbol{\alpha}_m$. Then $d[\mathscr{S}] = m - d[\mathscr{M}]$ where the symbol d denotes the dimensions of the space.*

Without loss of generality let $\boldsymbol{\alpha}_1, \ldots, \boldsymbol{\alpha}_k$ be independent, that is, $d[\mathscr{M}] = k$, in which case,

$$\boldsymbol{\alpha}_j = a_{j1}\boldsymbol{\alpha}_1 + \cdots + a_{jk}\boldsymbol{\alpha}_k, \qquad j = k+1, \ldots, m.$$

We observe that the vectors

$$\begin{aligned} \boldsymbol{\beta}_1 &= (a_{k+1,1}, \ldots, a_{k+1,k}, -1, 0, \ldots, 0) \\ &\quad \cdots \qquad \cdots \qquad \cdots \\ \boldsymbol{\beta}_{m-k} &= (a_{m,1}, \ldots, a_{m,k}, 0, 0, \ldots, -1) \end{aligned} \qquad (1a.3.2)$$

are independent and satisfy the equation (1a.3.1). The set (1a.3.2) will be a basis of the solution space $\mathscr{S}$ if it spans all the solutions. Let $\mathbf{y} = (y_1, \ldots, y_m)$ be any solution and consider the vector

$$\mathbf{y} + y_{k+1}\boldsymbol{\beta}_1 + \cdots + y_m\boldsymbol{\beta}_{m-k}, \qquad (1a.3.3)$$

which is also a solution. But this is of the form $(z_1, \ldots, z_k, 0, \ldots, 0)$ which means $z_1\boldsymbol{\alpha}_1 + \cdots + z_k\boldsymbol{\alpha}_k = 0$. This is possible only when $z_1 = \cdots = z_k = 0$.

Thus (1a.3.3) is a null vector or any solution $\mathbf{y}$ is expressible in terms of the $m-k$ independent vectors $\boldsymbol{\beta}_1, \ldots, \boldsymbol{\beta}_{m-k}$, where k is the same as the number of independent $\boldsymbol{\alpha}$ vectors, which proves the result.

(iv) *The nonhomogeneous equation*

$$x_1\boldsymbol{\alpha}_1 + \cdots + x_m\boldsymbol{\alpha}_m = \boldsymbol{\alpha}_0 \tag{1a.3.4}$$

admits a solution, if and only if, $\boldsymbol{\alpha}_0$ *is dependent on* $\boldsymbol{\alpha}_1, \ldots, \boldsymbol{\alpha}_k$.

The result follows from the definition of dependence of vectors.

(v) *A general solution of the nonhomogeneous equation* (1a.3.4) *is the sum of any particular solution of* (1a.3.4) *and a general solution of the homogeneous equation* (1a.3.1).

The result follows since the difference of any two solutions of (1a.3.4) is a solution of (1a.3.1).

Note that the set of solutions of (1a.3.4) no doubt belong to E_m, but *do not* constitute a subspace as in the case of the solutions of the homogeneous equation (1a.3.1). From (v), the following result (vi) is deduced.

(vi) *The solution of a nonhomogeneous equation is unique when the corresponding homogeneous equation has the null vector as the only solution, that is,* $d[\mathscr{S}] = m - k = 0$ *or when* $\boldsymbol{\alpha}_1, \ldots, \boldsymbol{\alpha}_m$ *are independent.*

A special and an important case of linear equations is where $\boldsymbol{\alpha}_i \in E_n$. If $\boldsymbol{\alpha}_i = (a_{1i}, \ldots, a_{ni})$, then (1a.3.1) may be written

$$\begin{aligned} a_{11}x_1 + \cdots + a_{1m}x_m &= 0 \\ \cdot \quad \cdots \quad \cdot & \\ a_{n1}x_1 + \cdots + a_{nm}x_m &= 0 \end{aligned} \tag{1a.3.5}$$

which are *n linear equations in m unknowns.* The column vectors of the equations (1a.3.5) are $\boldsymbol{\alpha}_1, \ldots, \boldsymbol{\alpha}_m$. Let the row vectors be represented by $\boldsymbol{\beta}_i = (a_{i1}, \ldots, a_{im})$, $i = 1, \ldots, n$. Let only r rows be independent and let $k = d[\mathscr{M}(\boldsymbol{\alpha})]$, the dimension of the manifold generated by the column vectors. If we retain only the r independent rows, the solution space $\mathscr{S}$ will still be the same. Since $d[\mathscr{S}] = m - k$, some k columns in the reduced set of equations are independent. But now the column vector has only r elements, that is, it belongs to E_r and therefore the number of independent column vectors cannot be greater than r. Therefore $r \geqslant k$. We can reverse the roles of columns and rows in the argument and show that $k \geqslant r$. Hence $r = k$ and we have:

(vii) *Considering only the coefficients* a_{ij} *in* (1a.3.5), *the number of independent rows is equal to the number of independent columns and*

$$d[\mathscr{M}(\boldsymbol{\alpha})] = d[\mathscr{M}(\boldsymbol{\beta})], \qquad d(\mathscr{S}) = m - d[\mathscr{M}(\boldsymbol{\alpha})] = m - d[\mathscr{M}(\boldsymbol{\beta})].$$

Let us consider the nonhomogeneous equations obtained by replacing $(0, \ldots, 0)$ on the R.H.S. of (1a.3.5) by $\boldsymbol{\alpha}_0 = (a_1, \ldots, a_n)$. To verify the existence of a solution we need satisfy ourselves that any one of the following equivalent necessary and sufficient conditions is true.

1. The column vector $\boldsymbol{\alpha}_0$ is dependent on $\boldsymbol{\alpha}_1, \ldots, \boldsymbol{\alpha}_m$ by (iv), or

$$d[\mathcal{M}(\boldsymbol{\alpha}_1, \ldots, \boldsymbol{\alpha}_m)] = d[\mathcal{M}(\boldsymbol{\alpha}_1, \ldots, \boldsymbol{\alpha}_m, \boldsymbol{\alpha}_0)].$$

2. Let $\boldsymbol{\gamma}_i$ be an $(m + 1)$-vector obtained from $\boldsymbol{\beta}_i$ by adding the $(m + 1)$th element a_i. Then $d[\mathcal{M}(\boldsymbol{\gamma}_1, \ldots, \boldsymbol{\gamma}_n)] = d[\mathcal{M}(\boldsymbol{\beta}_1, \ldots, \boldsymbol{\beta}_n)]$. This is true for the former is equal to $d[\mathcal{M}(\boldsymbol{\alpha}_1, \ldots, \boldsymbol{\alpha}_m, \boldsymbol{\alpha}_0)]$ being its column equivalent and the latter to $d[\mathcal{M}(\boldsymbol{\alpha}_1, \ldots, \boldsymbol{\alpha}_m)]$.
3. If $c_1, \ldots, c_n$ are such that $c_1\boldsymbol{\beta}_1 + \cdots + c_n\boldsymbol{\beta}_n = 0$, then $c_1a_1 + \cdots + c_na_n = 0$ or equivalently $c_1\boldsymbol{\gamma}_1 + \cdots + c_n\boldsymbol{\gamma}_n = 0$. To prove this we need only observe that the solutions $(c_1, \ldots, c_n)$ of $\sum c_i\boldsymbol{\beta}_i = 0$ and $\sum c_i\boldsymbol{\gamma}_i = 0$ are the same, which follows from the results that every solution of $\sum c\boldsymbol{\gamma}_i = 0$ is obviously a solution of $\sum c_i\boldsymbol{\beta}_i = 0$ and $d[\mathcal{M}(\boldsymbol{\beta})] = d[\mathcal{M}(\boldsymbol{\gamma})]$.
4. If the row vectors $\boldsymbol{\beta}_1, \ldots, \boldsymbol{\beta}_n$ are independent, the nonhomogeneous equation admits a solution for any arbitrary $\boldsymbol{\alpha}_0$.

1a.4 Vector Spaces with an Inner Product

So far we have studied relationships between elements of a vector space through the notion of independence. It would be fruitful to consider other concepts such as distance and angle between vectors which occur in the study of two- and three-dimensional Euclidean geometries. When the associated scalars belong to real or complex numbers, it appears that the extension of these concepts is immediate for vector spaces, through a complex valued function known as the inner product of two vectors $\mathbf{x}$, $\mathbf{y}$, denoted by $(\mathbf{x}, \mathbf{y})$, satisfying the following conditions

(a) $(\mathbf{x}, \mathbf{y}) = \overline{(\mathbf{y}, \mathbf{x})}$, the complex conjugate of $(\mathbf{y}, \mathbf{x})$
(b) $(\mathbf{x}, \mathbf{x}) > 0$ if $\mathbf{x} \neq \mathbf{0}$, $= 0$ if $\mathbf{x} = \mathbf{0}$
(c) $(t\mathbf{x}, \mathbf{y}) = t(\mathbf{x}, \mathbf{y})$, where t is a scalar
(d) $(\mathbf{x} + \mathbf{y}, \mathbf{z}) = (\mathbf{x}, \mathbf{z}) + (\mathbf{y}, \mathbf{z})$

For R^n, that is, n-dimensional Euclidean space with real coordinates, the real valued function $(\mathbf{x}, \mathbf{y}) = \sum x_i y_i$ satisfies the conditions (a) to (d). We choose this function as the inner product of vectors in R^n. For U^n, the unitary space with complex coordinates, $(\mathbf{x}, \mathbf{y})$ may be defined as $\sum x_i\bar{y}_i$. In fact it can be shown that every inner product in U^n is of the form $\sum\sum a_{ij}x_i\bar{y}_j$ where a_{ij} are suitably chosen (see Example **19.5** at the end of this chapter).

The positive square root of $(\mathbf{x}, \mathbf{x})$ is denoted by $\|\mathbf{x}\|$—the *norm* of $\mathbf{x}$ or length of $\mathbf{x}$. The function $\|\mathbf{x} - \mathbf{y}\|$ satisfies the postulates of a distance in a metric space and may be considered as the distance between $\mathbf{x}$ and $\mathbf{y}$. It would be interesting to verify the triangular inequality

$$\|\mathbf{x} - \mathbf{y}\| + \|\mathbf{y} - \mathbf{z}\| \geqslant \|\mathbf{x} - \mathbf{z}\|.$$

and the Cauchy-Shwarz inequality

$$|(\mathbf{x}, \mathbf{y})| \leq \|\mathbf{x}\| \|\mathbf{y}\|$$

Two vectors, $\mathbf{x}$, $\mathbf{y}$ are said to be orthogonal if their inner product, $(\mathbf{x}, \mathbf{y}) = 0$. The angle θ between non-null vectors $\mathbf{x}, \mathbf{y} \in R^k$ is defined by

$$\cos\theta = \frac{(\mathbf{x}, \mathbf{y})}{\|\mathbf{x}\| \cdot \|\mathbf{y}\|}$$

which lies in the closed interval $[-1, 1]$. We examine the properties of vector spaces furnished with an inner product.

(i) *A set of non-null vectors* $\boldsymbol{\alpha}_1, \ldots, \boldsymbol{\alpha}_k$ *orthogonal in pairs is necessarily independent.*

If this is not so, let $\sum a_i \boldsymbol{\alpha}_i = \mathbf{0}$ and $a_j \neq 0$. Taking the inner product of $\boldsymbol{\alpha}_j$ and $\sum a_i \boldsymbol{\alpha}_i$ and putting $(\boldsymbol{\alpha}_i, \boldsymbol{\alpha}_j) = 0$, $i \neq j$, we find $a_j(\boldsymbol{\alpha}_j, \boldsymbol{\alpha}_j) = 0$, which implies $a_j = 0$ since $(\boldsymbol{\alpha}_j, \boldsymbol{\alpha}_j) > 0$, contrary to assumption.

(ii) GRAM-SCHMIDT ORTHOGONALIZATION. A set of orthogonal vectors spanning a vector space $\mathscr{V}$ is called an orthogonal basis (o.b.). If, further, each vector is of unit norm, it is called an *orthonormal basis* (o.n.b.). *We assert that an o.n.b. always exists.* Let $\boldsymbol{\alpha}_1, \boldsymbol{\alpha}_2, \ldots$ be any basis and construct sequentially

$$\begin{aligned}
\boldsymbol{\xi}_1 &= \boldsymbol{\alpha}_1 \\
\boldsymbol{\xi}_2 &= \boldsymbol{\alpha}_2 - a_{21}\boldsymbol{\xi}_1 \\
\cdots &\quad \cdots \quad \cdots \\
\boldsymbol{\xi}_i &= \boldsymbol{\alpha}_i - a_{i,i-1}\boldsymbol{\xi}_{i-1} - \cdots - a_{i1}\boldsymbol{\xi}_1 \\
\cdot &\quad \cdots \qquad \cdots
\end{aligned} \tag{1a.4.1}$$

where the coefficients in the ith equation are determined by equating the inner product of $\boldsymbol{\xi}_i$, with each of $\boldsymbol{\xi}_1, \ldots, \boldsymbol{\xi}_{i-1}$ to zero. Thus having determined $\boldsymbol{\xi}_1, \ldots, \boldsymbol{\xi}_{i-1}$ the coefficients $a_{i1}, \ldots, a_{i,i-1}$ defining $\boldsymbol{\xi}_i$ are obtained as follows:

$$(\boldsymbol{\alpha}_i, \boldsymbol{\xi}_1) = a_{i1}(\boldsymbol{\xi}_1, \boldsymbol{\xi}_1), \ldots, \qquad (\boldsymbol{\alpha}_i, \boldsymbol{\xi}_{i-1}) = a_{i,i-1}(\boldsymbol{\xi}_{i-1}, \boldsymbol{\xi}_{i-1}).$$

The vectors $\boldsymbol{\xi}_1, \boldsymbol{\xi}_2, \ldots$ provide an o.b. The set defined by $\boldsymbol{\eta}_i = \boldsymbol{\xi}_i / \|\boldsymbol{\xi}_i\|$, $i = 1, 2, \ldots$ is an o.n.b. Observe that $\|\boldsymbol{\xi}_i\| \neq 0$, for otherwise $\boldsymbol{\xi}_i = \mathbf{0}$ which implies dependence of $\boldsymbol{\alpha}_1, \ldots, \boldsymbol{\alpha}_i$. Also $\boldsymbol{\xi}_1, \ldots, \boldsymbol{\xi}_i$ constitutes an o.b. of the linear manifold spanned by $\boldsymbol{\alpha}_1, \ldots, \boldsymbol{\alpha}_i$, that is, $\mathcal{M}(\boldsymbol{\xi}_1, \ldots, \boldsymbol{\xi}_i) = \mathcal{M}(\boldsymbol{\alpha}_1, \ldots, \boldsymbol{\alpha}_i)$, $i = 1, 2, \ldots$. The constructive method of establishing the existence of an o.n.b. is known as Gram-Schmidt orthogonalization process.

(iii) ORTHOGONAL PROJECTION OF A VECTOR. *Let $\mathcal{W}$ be a finite dimensional subspace and $\boldsymbol{\alpha}$, a vector $\notin \mathcal{W}$. Then there exist two vectors $\boldsymbol{\gamma}$, $\boldsymbol{\beta}$, such that $\boldsymbol{\alpha} = \boldsymbol{\beta} + \boldsymbol{\gamma}$, $\boldsymbol{\beta} \in \mathcal{W}$ and $\boldsymbol{\gamma} \neq \mathbf{0}$ is orthogonal to $\mathcal{W}(\boldsymbol{\gamma} \perp^{\text{er}} \mathcal{W})$. Furthermore, $\boldsymbol{\gamma}$ and $\boldsymbol{\beta}$ are unique.*

Let $\boldsymbol{\alpha}_1, \ldots, \boldsymbol{\alpha}_r$ be a basis of $\mathcal{W}$. Then construct $\boldsymbol{\xi}_1, \ldots, \boldsymbol{\xi}_r, \boldsymbol{\xi}_{r+1}$ from $\boldsymbol{\alpha}_1, \ldots, \boldsymbol{\alpha}_r, \boldsymbol{\alpha}_{r+1} (= \boldsymbol{\alpha})$ by the Gram-Schmidt orthogonalization process. We find from the last equation in (1a.4.1)

$$\boldsymbol{\alpha} = \boldsymbol{\xi}_{r+1} + (a_{r+1,r}\boldsymbol{\xi}_r + \cdots + a_{r+1,1}\boldsymbol{\xi}_1) = \boldsymbol{\xi}_{r+1} + \boldsymbol{\beta} = \boldsymbol{\gamma} + \boldsymbol{\beta},$$

which defines $\boldsymbol{\gamma}$ and $\boldsymbol{\beta}$. Since $\mathcal{W} = \mathcal{M}(\boldsymbol{\alpha}_1, \ldots, \boldsymbol{\alpha}_r) = \mathcal{M}(\boldsymbol{\xi}_1, \ldots, \boldsymbol{\xi}_r)$, we find $\boldsymbol{\beta} \in \mathcal{W}$ and $\boldsymbol{\xi}_{r+1} = \boldsymbol{\gamma}$ is, by construction, orthogonal to each of $\boldsymbol{\xi}_1, \ldots, \boldsymbol{\xi}_r$ and therefore to the manifold $\mathcal{M}(\boldsymbol{\xi}_1, \ldots, \boldsymbol{\xi}_r)$ or to $\mathcal{W}$. Let $\boldsymbol{\beta}_1 + \boldsymbol{\gamma}_1$ be another resolution. Then $\mathbf{0} = (\boldsymbol{\gamma} - \boldsymbol{\gamma}_1) + (\boldsymbol{\beta} - \boldsymbol{\beta}_1)$. Taking the inner product of both sides with $(\boldsymbol{\gamma} - \boldsymbol{\gamma}_1)$ we find $(\boldsymbol{\gamma} - \boldsymbol{\gamma}_1, \boldsymbol{\gamma} - \boldsymbol{\gamma}_1) = 0$, which implies $\boldsymbol{\gamma} - \boldsymbol{\gamma}_1 = \mathbf{0}$, and also $\boldsymbol{\beta} - \boldsymbol{\beta}_1 = \mathbf{0}$. The vector $\boldsymbol{\gamma}$ is said to be the $\perp^{\text{er}}$ from $\boldsymbol{\alpha}$ to $\mathcal{W}$ and $\boldsymbol{\beta}$ the projection of $\boldsymbol{\alpha}$ on $\mathcal{W}$. $\boldsymbol{\gamma} = \mathbf{0}$ if $\boldsymbol{\alpha} \in \mathcal{W}$.

(iv) *The $\perp^{\text{er}}$ $\boldsymbol{\gamma}$ from $\boldsymbol{\alpha}$ to $\mathcal{W}$ has the property*

$$\|\boldsymbol{\gamma}\| = \inf \|\boldsymbol{\alpha} - \mathbf{x}\|, \qquad \mathbf{x} \in \mathcal{W}. \tag{1a.4.2}$$

Let $\boldsymbol{\alpha} = \boldsymbol{\beta} + \boldsymbol{\gamma}$, $\boldsymbol{\beta} \in \mathcal{W}$ and $\boldsymbol{\gamma} \perp^{\text{er}} \mathcal{W}$. Then $(\mathbf{x} - \boldsymbol{\beta}, \boldsymbol{\gamma}) = 0$ for $\mathbf{x} \in \mathcal{W}$. Hence verify that

$$\begin{aligned}(\mathbf{x} - \boldsymbol{\alpha}, \mathbf{x} - \boldsymbol{\alpha}) &= (\mathbf{x} - \boldsymbol{\beta} - \boldsymbol{\gamma}, \mathbf{x} - \boldsymbol{\beta} - \boldsymbol{\gamma}) \\ &= (\mathbf{x} - \boldsymbol{\beta}, \mathbf{x} - \boldsymbol{\beta}) + (\boldsymbol{\gamma}, \boldsymbol{\gamma}) \geqslant \|\boldsymbol{\gamma}\|^2 \\ &= \|\boldsymbol{\gamma}\|^2 \text{ when } \mathbf{x} = \boldsymbol{\beta}.\end{aligned}$$

Therefore, the shortest distance between $\boldsymbol{\alpha}$ and an element of $\mathcal{W}$ is the length of the $\perp^{\text{er}}$ from $\boldsymbol{\alpha}$ to $\mathcal{W}$.

(v) BESSEL'S INEQUALITY *Let $\boldsymbol{\alpha}_1, \ldots, \boldsymbol{\alpha}_m$ be an orthonormal set of vectors in an inner product space and $\boldsymbol{\beta}$ be any other vector. Then*

$$|(\boldsymbol{\alpha}_1, \boldsymbol{\beta})|^2 + \cdots + |(\boldsymbol{\alpha}_m, \boldsymbol{\beta})|^2 \leqslant \|\boldsymbol{\beta}\|^2 \tag{1a.4.3}$$

with equality iff $\boldsymbol{\beta}$ belongs to the subspace generated by $\boldsymbol{\alpha}_1, \ldots, \boldsymbol{\alpha}_m$.

Let $\boldsymbol{\delta} = \sum (\boldsymbol{\alpha}_i, \boldsymbol{\beta})\boldsymbol{\alpha}_i$. Then $\boldsymbol{\delta}$ is the projection of $\boldsymbol{\beta}$ on $\mathcal{M}(\boldsymbol{\alpha}_i, \ldots, \boldsymbol{\alpha}_m)$ and by proposition (iii) $\boldsymbol{\beta} = \boldsymbol{\delta} + \mathbf{r}$ where $(\boldsymbol{\delta}, \mathbf{r}) = 0$. Hence

$$\|\boldsymbol{\beta}\|^2 = \|\boldsymbol{\delta}\|^2 + \|\mathbf{r}\|^2 \geqslant \|\boldsymbol{\delta}\|^2 = \sum |(\boldsymbol{\alpha}_i, \boldsymbol{\beta})|^2.$$

(vi) *The set of all vectors orthogonal to a given set of vectors S which is a subspace or not, is a subspace represented by* $S^{\perp}$.

This follows from the definition of a subspace. Let the whole vector space $\mathscr{V}$ be of dimension n (finite) and $\boldsymbol{\alpha}_1, \ldots, \boldsymbol{\alpha}_r$ be a basis of the manifold $\mathscr{M}(S)$ spanned by S. If $\boldsymbol{\beta}_1, \ldots, \boldsymbol{\beta}_s$ is a basis of $S^{\perp}$, the set of vectors $\boldsymbol{\alpha}_1, \ldots, \boldsymbol{\alpha}_r, \boldsymbol{\beta}_1, \ldots, \boldsymbol{\beta}_s$ are independent, since every $\boldsymbol{\beta}_i$ is $\perp^{\text{er}}$ to every $\boldsymbol{\alpha}_j$. Obviously $(r + s) \not> n$. If $(r + s) < n$, there exists at least one vector independent of $\boldsymbol{\alpha}_1, \ldots, \boldsymbol{\alpha}_r, \boldsymbol{\beta}_1, \ldots, \boldsymbol{\beta}_s$ and hence we can find by (iii) a non-null vector orthogonal to all of them. Since such a vector is necessarily orthogonal to each of $\boldsymbol{\alpha}_1, \ldots, \boldsymbol{\alpha}_r$ it must belong to $S^{\perp}$, that is, depend on $\boldsymbol{\beta}_1, \ldots, \boldsymbol{\beta}_s$ which is a contradiction. Hence $(r + s) = n$. We thus obtain an important result

$$d[\mathscr{M}(S)] + d[S^{\perp}] = d[\mathscr{V}]. \tag{1a.4.4}$$

Furthermore, every vector in $\mathscr{V}$ can be written as $\mathbf{x} + \mathbf{y}$, $\mathbf{x} \in \mathscr{M}(S)$ and $\mathbf{y} \in S^{\perp}$.

(vii) Direct Sum of Subspaces. A vector space $\mathscr{V}$ is said to be the direct sum of subspaces $\mathscr{V}_1, \ldots, \mathscr{V}_k$ and written $\mathscr{V} = \mathscr{V}_1 \oplus \cdots \oplus \mathscr{V}_k$ if $\boldsymbol{\alpha} \in \mathscr{V}$ can be uniquely expressed as $\boldsymbol{\alpha}_1 + \cdots + \boldsymbol{\alpha}_k$ where $\boldsymbol{\alpha}_i \in \mathscr{V}_i$. In such a case

$$\mathscr{V}_i \cap \mathscr{V}_j = \{\mathbf{0}\} \quad \text{and} \quad d(\mathscr{V}) = d(\mathscr{V}_1) + \cdots + d(\mathscr{V}_k). \tag{1a.4.5}$$

The resolution of $\mathscr{V} = \mathscr{M}(S) + S^{\perp}$ in (vi) is an example of direct sum.

COMPLEMENTS AND PROBLEMS

1 If $\boldsymbol{\alpha}_1, \ldots, \boldsymbol{\alpha}_k$ are k independent vectors in a vector space $\mathscr{V}$, show that it can always be extended to a basis of $\mathscr{V}$, that is, we can add $\boldsymbol{\alpha}_{k+1}, \ldots, \boldsymbol{\alpha}_r$ such that $\boldsymbol{\alpha}_1, \ldots, \boldsymbol{\alpha}_r$ form a basis of $\mathscr{V}$, where $r = d[\mathscr{V}]$.

2 The dimension of the space generated by the following row vectors is 4.

$$\begin{matrix} 1 & 1 & 0 & 1 & 0 & 0 \\ 1 & 1 & 0 & 0 & 1 & 0 \\ 1 & 1 & 0 & 0 & 0 & 1 \\ 1 & 0 & 1 & 1 & 0 & 0 \\ 1 & 0 & 1 & 0 & 1 & 0 \\ 1 & 0 & 1 & 0 & 0 & 1 \end{matrix}$$

[Hint: The dimensions of the spaces formed by the rows and columns are the same. Consider the last 4 columns which are obviously independent and note that the first two depend on them.]

3 Generalize the result of Example 2 to pq row vectors written in q blocks of p rows each, the jth vector in the ith block having unity in the 1st, $(i + 1)$th, and $(q + j + 1)$th columns and zero in the rest of $(p + q - 2)$ columns. Show that the dimension of the space generated is $(p + q - 1)$.

4 If pq elements a_{ij}, $i = 1, \ldots, p; j = 1, \ldots, q$ are such that the tetra difference $a_{ij} + a_{rs} - a_{is} - a_{rj} = 0$ for all i, j, r, s, then a_{ij} can be expressed in the form $a_{ij} = a_i + b_j$ where $a_1, \ldots, a_p$ and $b_1, \ldots, b_q$ are properly chosen.

[Examples 2, 3, and 4 are useful in the study of two way data by analysis of variance of Section **4e**, Chapter **4**].

5 Find the number of independent vectors in the set $(a, b, \ldots, b), (b, a, b, \ldots, b), \ldots, (b, b, \ldots, a)$ when the sum of elements in each vector is zero.

6 Let $\mathscr{S}$ and $\mathscr{T}$ be subspaces and denote the set of vectors common to $\mathscr{S}$ and $\mathscr{T}$ by $\mathscr{S} \cap \mathscr{T}$, the intersection of $\mathscr{S}$ and $\mathscr{T}$, and the set of vectors $\mathbf{x} + \mathbf{y}$, $\mathbf{x} \in \mathscr{S}$, $\mathbf{y} \in \mathscr{T}$ by $\mathscr{S} + \mathscr{T}$. Show that (i) $\mathscr{S} \cap \mathscr{T}$ and $\mathscr{S} + \mathscr{T}$ are also subspaces, (ii) $\mathscr{S} \cap \mathscr{T}$ is the largest subspace contained in both $\mathscr{S}$ and $\mathscr{T}$, (iii) $\mathscr{S} + \mathscr{T}$ is the smallest subspace containing both $\mathscr{S}$ and $\mathscr{T}$ and, (iv)

$$d[\mathscr{S} + \mathscr{T}] = d[\mathscr{S}] + d[\mathscr{T}] - d[\mathscr{S} \cap \mathscr{T}].$$

7 Show that R^n can be expressed as the direct sum of r subspaces, $r \leqslant n$.

8 Let S be any set of vectors and $S^\perp$ the set of all vectors orthogonal to S. Show that $(S^\perp)^\perp \supset S$ and is equal to S only when S is a subspace.

9 Let S and T be any two sets of vectors containing the null vector. Then $(S + T)^\perp = S^\perp \cap T^\perp$.

10 If $\mathscr{S}$ and $\mathscr{T}$ are subspaces, then $(\mathscr{S} \cap \mathscr{T})^\perp = \mathscr{S}^\perp + \mathscr{T}^\perp$.

11 Let $\boldsymbol{\alpha}_1, \ldots, \boldsymbol{\alpha}_m$ be vectors in an *arbitrary vector space* $\mathscr{V}_F$. Then the collection of vectors $t_1\boldsymbol{\alpha}_1 + \cdots + t_m\boldsymbol{\alpha}_m$, $t_i \in F$ as $(t_1, \ldots, t_m)$ varies over *all* the vectors in E_m is the linear manifold $\mathscr{M}(\boldsymbol{\alpha})$ whose dimension is equal to k, the number of independent vectors in the set $\boldsymbol{\alpha}_1, \ldots, \boldsymbol{\alpha}_m$. What can we say about the collection $t_1\boldsymbol{\alpha}_1 + \cdots + t_m\boldsymbol{\alpha}_m$ when $(t_1, \ldots, t_m)$ is confined to a subspace $\mathscr{S}$ of dimension s in E_m? Show that the restricted collection is also a subspace $\subset \mathscr{M}(\boldsymbol{\alpha})$ and the dimension of this space is $s - d[\mathscr{S} \cap \mathscr{T}]$ where $\mathscr{T}$ is the subspace of E_m representing the solutions $(t_1, \ldots, t_m)$ of the equation $t_1\boldsymbol{\alpha}_1 + \cdots + t_m\boldsymbol{\alpha}_m = \mathbf{0}$.

12 Let $\boldsymbol{\alpha}_i$ in Example 11 belong to E_n,

$$\boldsymbol{\alpha}_i = (a_{i1}, \ldots, a_{in}), \qquad i = 1, \ldots, m$$

and represent the column vectors by

$$\boldsymbol{\beta}_j = (a_{1j}, \ldots, a_{mj}), \qquad j = 1, \ldots, n.$$

Let $\mathbf{t} = (t_1, \ldots, t_m)$ satisfy the equations

$$(\mathbf{t}, \boldsymbol{\gamma}_i) = 0, \qquad i = 1, \ldots, k,$$

where $\boldsymbol{\gamma}_1, \ldots, \boldsymbol{\gamma}_k$ are fixed vectors in E_m. Show that the dimension of the subspace generated by $\sum t_i\boldsymbol{\alpha}_i$ is

$$d[\mathscr{M}(\boldsymbol{\beta}_1, \ldots, \boldsymbol{\beta}_n, \boldsymbol{\gamma}_1, \ldots, \boldsymbol{\gamma}_k)] - d[\mathscr{M}(\boldsymbol{\gamma}_1, \ldots, \boldsymbol{\gamma}_k)].$$

13 Find the value of δ for which the equations admit a solution

$$\begin{aligned} 2x_1 - x_2 + 5x_3 &= 4, \\ 4x_1 \quad\quad + 6x_3 &= 1, \\ - 2x_2 + 4x_3 &= 7 + \delta. \end{aligned}$$

14 The equations

$$\begin{aligned} 4x_1 + 3x_2 + 2x_3 + 8x_4 &= \delta_1 \\ 3x_1 + 4x_2 + x_3 + 7x_4 &= \delta_2 \\ x_1 + x_2 + x_3 + x_4 &= \delta_3 \end{aligned}$$

admit a solution for any arbitrary $\delta_1, \delta_2, \delta_3$. [Note that there are three equations and all the three rows are independent.] Determine the totality of the solutions and find the one with a minimum length.

15 Let 0, 1, 2 represent the residue classes (mod 3). They form a Galois Field with three elements, GF (3), with the usual definition of addition and multiplication of residue classes. In this system find a basis of the solutions of

$$\begin{aligned} x_1 + x_2 + x_3 &= 0 \\ x_1 + 2x_2 \quad\quad &= 0 \end{aligned}$$

and examine whether there is a solution which depends on the row vectors of the equations. [Note that such a thing is not possible if we are considering equations in real numbers.] Also solve the equations

$$\begin{aligned} x_1 + 2x_2 + 0 + x_4 &= 1 \\ 2x_1 + 0 + x_3 + 2x_4 &= 2 \end{aligned}$$

16 Show that the projection of $\mathbf{x}$ on the manifold spanned by $\boldsymbol{\alpha}_1, \ldots, \boldsymbol{\alpha}_m$ is given by $a_1\boldsymbol{\alpha}_1 + \cdots + a_m\boldsymbol{\alpha}_m$ where $a_1, \ldots, a_m$ is *any* solution of

$$\begin{aligned} a_1(\boldsymbol{\alpha}_1, \boldsymbol{\alpha}_1) + \cdots + a_m(\boldsymbol{\alpha}_1, \boldsymbol{\alpha}_m) &= (\mathbf{x}, \boldsymbol{\alpha}_1) \\ \ldots \quad\quad \ldots \quad\quad \ldots \quad &\quad \ldots \\ a_1(\boldsymbol{\alpha}_m, \boldsymbol{\alpha}_1) + \cdots + a_m(\boldsymbol{\alpha}_m, \boldsymbol{\alpha}_m) &= (\mathbf{x}, \boldsymbol{\alpha}_m). \end{aligned}$$

This is obtained by expressing the condition that $(\mathbf{x} - a_1\boldsymbol{\alpha}_1 - \cdots - a_m\boldsymbol{\alpha}_m)$ is $\perp^{\text{er}}$ to each of $\boldsymbol{\alpha}_1, \ldots, \boldsymbol{\alpha}_m$. The equations are soluble because the projection exists (prove otherwise using the conditions for solubility of equations given in **1a.3**). Furthermore, $\sum a_i\boldsymbol{\alpha}_i$ is unique and so also the length of the perpendicular, $a_1(\mathbf{x}, \boldsymbol{\alpha}_1) + \cdots + a_m(\mathbf{x}, \boldsymbol{\alpha}_m)$, whatever may be the solution $(a_1, \ldots, a_m)$.

17 Let $\boldsymbol{\alpha}_1, \ldots, \boldsymbol{\alpha}_m$ be vectors in R^k. Find the minimum value of $(\mathbf{b}, \mathbf{b}) = \sum b_i^2$ subject to the conditions $(\mathbf{b}, \boldsymbol{\alpha}_1) = p_1, \ldots, (\mathbf{b}, \boldsymbol{\alpha}_m) = p_m$, where $p_1, \ldots, p_m$ are given scalars.

18 *Steinitz replacement result.* If $\boldsymbol{\alpha}_1, \ldots, \boldsymbol{\alpha}_k$ is a basis of a vector space $\mathscr{V}$ and $\boldsymbol{\beta}_1, \ldots, \boldsymbol{\beta}_p$ are $p(<k)$ independent vectors in $\mathscr{V}$, we can select $k-p$ vectors $\boldsymbol{\alpha}_{(1)}, \ldots, \boldsymbol{\alpha}_{(k-p)}$ from the $\boldsymbol{\alpha}$ vectors such that $\boldsymbol{\beta}_1, \ldots, \boldsymbol{\beta}_p, \boldsymbol{\alpha}_{(1)}, \ldots, \boldsymbol{\alpha}_{(k-p)}$ is a basis of $\mathscr{V}$.

19 Let $\boldsymbol{\alpha}$ and $\boldsymbol{\beta}$ be normalized vectors. Then $|(\boldsymbol{\alpha}, \boldsymbol{\beta})| = 1 \Rightarrow \boldsymbol{\alpha} = c\boldsymbol{\beta}$ where $|c| = 1$.

1b THEORY OF MATRICES AND DETERMINANTS

1b.1 Matrix Operations

A matrix is an arrangement of pq elements (belonging to any field) in p rows and q columns. Each element is identified by the pair (i, j) representing the numbers of the row and column in which it occurs. Matrices are generally represented by capital letters such as $\mathbf{A}$ or more explicitly, if necessary, by the symbol (a_{ij}), a_{ij} being the (i, j) entry. The following operations are defined on matrices.

Addition. If $\mathbf{A} = (a_{ij})$ and $\mathbf{B} = (b_{ij})$ are two matrices of the same order $p \times q$, that is, each with p rows and q columns, then the sum of $\mathbf{A}$ and $\mathbf{B}$ is defined to be the matrix $\mathbf{A} + \mathbf{B} = (a_{ij} + b_{ij})$, of the same order $p \times q$. It may be easily verified that

$$\mathbf{A} + \mathbf{B} = \mathbf{B} + \mathbf{A}$$
$$\mathbf{A} + (\mathbf{B} + \mathbf{C}) = (\mathbf{A} + \mathbf{B}) + \mathbf{C} = \mathbf{A} + \mathbf{B} + \mathbf{C}.$$

Scalar Multiplication. As for vectors, a matrix may be multiplied by a scalar c, the result being a matrix of the same form

$$c\mathbf{A} = (ca_{ij}),$$

that is, each element is multiplied by the scalar c. Verify that

$$(c + d)\mathbf{A} = c\mathbf{A} + d\mathbf{A}$$
$$c(\mathbf{A} + \mathbf{B}) = c\mathbf{A} + c\mathbf{B}.$$

Matrix Multiplication. The product $\mathbf{AB}$ is defined only when the number of columns in $\mathbf{A}$ is the same as that of the rows in $\mathbf{B}$. The result is a matrix with the (i, j) entry as the product of the ith row vector of $\mathbf{A}$ with the jth column vector of $\mathbf{B}$. If $\mathbf{A} = (a_{ij})$ and $\mathbf{B} = (b_{ij})$, the product is obtained by the formula

$$\mathbf{AB} = (c_{ij}), \qquad c_{ij} = \sum_r a_{ir} b_{rj}.$$

If $\mathbf{A}$ is $p \times q$ and $\mathbf{B}$ is $q \times t$, then $\mathbf{C} = \mathbf{AB}$ is $p \times t$. Note the formula for determining the order of the product matrix, $(p \times q)(q \times t) = (p \times t)$. Verify that

$$\begin{pmatrix} -1 & 2 & 3 \\ 5 & 1 & 2 \end{pmatrix} \begin{pmatrix} 1 & 2 & 0 \\ 4 & 1 & 3 \\ 3 & 1 & 2 \end{pmatrix} = \begin{pmatrix} 16 & 3 & 12 \\ 15 & 13 & 7 \end{pmatrix}.$$

The commutative law does not hold good for matrix multiplication, that is, $\mathbf{AB} \neq \mathbf{BA}$ *in general.* It may also be that $\mathbf{AB}$ is defined but not $\mathbf{BA}$. The associative law is satisfied, however,

$$\mathbf{A(BC)} = \mathbf{(AB)C} = \mathbf{ABC}.$$

For the multiplication to be compatible, the matrices $\mathbf{A}$, $\mathbf{B}$, $\mathbf{C}$ should be of the orders $p \times q$, $q \times r$, $r \times s$, in which case $\mathbf{ABC}$ is of the order $p \times s$. The distributive law also holds:

$$\mathbf{A(B + C)} = \mathbf{AB} + \mathbf{AC}$$
$$\mathbf{(A + B)(C + D)} = \mathbf{A(C + D)} + \mathbf{B(C + D)}$$

Null and Identity Matrices. A null matrix, denoted by $\mathbf{0}$, is one with all its elements zero. A square matrix, that is, of the order $q \times q$ is said to be a unit or identity matrix if all its diagonal elements are unity and the off diagonal elements are zero. Such a matrix is denoted by $\mathbf{I}$ or $\mathbf{I}_q$, when it is necessary to indicate the order. Verify that

$$\mathbf{A} + \mathbf{0} = \mathbf{A},$$
$$\mathbf{IA} = \mathbf{A}, \qquad \mathbf{AI} = \mathbf{A},$$

when the products are defined, so that the matrices $\mathbf{0}$ and $\mathbf{I}$ behave like the zero and unit element of the number system.

Transpose of a Matrix. The matrix $\mathbf{A}'$ (also represented by $\mathbf{A}^T$) obtained by interchanging the rows and columns of $\mathbf{A}$ is called the transpose of $\mathbf{A}$. This means that the (i, j) entry of $\mathbf{A}'$ is the same as the (j, i) entry of $\mathbf{A}$. From the definition,

$$\mathbf{(AB)}' = \mathbf{B}'\mathbf{A}', \qquad \mathbf{(ABC)}' = \mathbf{C}'\mathbf{B}'\mathbf{A}', \qquad \text{etc.}$$

A vector in R^n written as a column is a matrix of order $n \times 1$. If $\mathbf{X}$ and $\mathbf{Y}$ are two column vectors, their inner product $(\mathbf{X}, \mathbf{Y})$ can be written in matrix notation as $\mathbf{X}'\mathbf{Y}$. If $\mathbf{A}$ is $m \times n$ matrix, observe that $\mathbf{AX}$ is a column vector which is a linear combination of the column vectors of $\mathbf{A}$. We shall, in further work, refer to a column vector simply as a vector.

Conjugate Transpose. This is defined in the special case where the elements of $\mathbf{A}$ belong to the field of complex numbers. The matrix $\mathbf{A}^*$ is said to be the conjugate transpose of $\mathbf{A}$, if the (i, j) entry of $\mathbf{A}^*$ is the complex conjugate of the (j, i) entry of $\mathbf{A}$. From the definition,

$$(\mathbf{AB})^* = \mathbf{B}^*\mathbf{A}^*, \qquad (\mathbf{ABC})^* = \mathbf{C}^*\mathbf{B}^*\mathbf{A}^*, \qquad \text{etc.}$$

Rank of a Matrix. A matrix may be considered as a set of row (or column) vectors written in a particular order. The rank of a matrix is defined as the number of independent rows, or which is the same as the number of independent columns it contains [see (vii), **1a.3**].

Inverse of a Matrix. A square matrix $\mathbf{A}$ of order $(m \times m)$ is said to be nonsingular if its rank is m. In such a case there exists a unique matrix $\mathbf{A}^{-1}$, known as the reciprocal of $\mathbf{A}$ such that

$$\mathbf{AA}^{-1} = \mathbf{A}^{-1}\mathbf{A} = \mathbf{I}. \tag{1b.1.1}$$

To establish this let $\mathbf{A}^{-1} = (b_{ij})$ exist. If $\mathbf{A} = (a_{ij})$, then $(a_{ij})(b_{ij}) = \mathbf{I}$. Expanding the left-hand side and equating the elements of the first column on both sides, leads to the linear equations in $b_{11}, b_{21}, \ldots, b_{m1}$ (i.e., the first column of $\mathbf{A}^{-1}$):

$$\begin{aligned} a_{11}b_{11} + \cdots + a_{1m}b_{m1} &= 1 \\ a_{21}b_{11} + \cdots + a_{2m}b_{m1} &= 0 \\ \cdots \qquad \cdots \qquad \cdots & \\ a_{m1}b_{11} + \cdots + a_{mm}b_{m1} &= 0 \end{aligned} \tag{1b.1.2}$$

Since (a_{ij}) has full rank, equations (1b.1.2) have a unique solution [(vi), **1a.3**]. Similarly the other columns of $\mathbf{A}^{-1}$ are determined such that $\mathbf{AA}^{-1} = \mathbf{I}$. The same argument shows that we can find a matrix $\mathbf{C}$ such that $\mathbf{CA} = \mathbf{I}$. Post multiplying by $\mathbf{A}^{-1}$ we find

$$\mathbf{CAA}^{-1} = \mathbf{IA}^{-1} = \mathbf{A}^{-1} \Rightarrow \mathbf{CI} = \mathbf{A}^{-1} \Rightarrow \mathbf{C} = \mathbf{A}^{-1},$$

which proves (1b.1.1). If $\mathbf{A}$ and $\mathbf{B}$ are such that $\mathbf{A}^{-1}$ and $\mathbf{B}^{-1}$ exist, then $(\mathbf{AB})^{-1} = \mathbf{B}^{-1}\mathbf{A}^{-1}$, which is true because $\mathbf{ABB}^{-1}\mathbf{A}^{-1} = \mathbf{I}$. Verify that

$$(\mathbf{ABC})^{-1} = \mathbf{C}^{-1}\mathbf{B}^{-1}\mathbf{A}^{-1}$$

and so on, provided that all the inverses exist. Also observe that $(\mathbf{A}')^{-1} = (\mathbf{A}^{-1})'$ and $(\mathbf{A}^*)^{-1} = (\mathbf{A}^{-1})^*$.

Consider a matrix $\mathbf{B}$ of order $m \times n$. If there exists a matrix $\mathbf{C}$ such that $\mathbf{BC} = \mathbf{I}_m$, then $\mathbf{C}$ is called a right regular inverse of $\mathbf{B}$ and is represented by $\mathbf{B}_R^{-1}$. Similarly, a left regular inverse $\mathbf{B}_L^{-1}$ may exist.

A right inverse exists iff $R(\mathbf{B}) = m \leqslant n$. In such a case we can write equations for columns of $\mathbf{C}$ as in (1b.1.2). Solutions exist in view of [(iv), 1a.3.] Similarly a left inverse exists iff $R(\mathbf{B}) = n \leqslant m$.

Unitary Matrix. A square matrix $\mathbf{A}$ of complex elements is said to be unitary if $\mathbf{A}^*\mathbf{A} = \mathbf{I}$. In such a case, $\mathbf{A}^* = \mathbf{A}^{-1}$ and $\mathbf{A}\mathbf{A}^* = \mathbf{I}$. When the elements of $\mathbf{A}$ are real, the matrix is said to be orthogonal if $\mathbf{A}'\mathbf{A} = \mathbf{A}\mathbf{A}' = \mathbf{I}$.

Partitioned Matrices. Sometimes it is convenient to represent a matrix by the juxtaposition of two or more matrices in a partitioned form. Thus a partitioned matrix is represented by

$$\mathbf{A} = \left(\begin{array}{c|c} \mathbf{P} & \mathbf{Q} \\ \hline \mathbf{R} & \mathbf{S} \end{array}\right) = \begin{pmatrix} \mathbf{P} & \mathbf{Q} \\ \mathbf{R} & \mathbf{S} \end{pmatrix}, \tag{1b.1.3}$$

with or without dotted lines, when there is no chance of confusion, whereas in (1b.1.3) the rows in $\mathbf{P}$ equal in number of those in $\mathbf{Q}$, the columns in $\mathbf{P}$ equal those in $\mathbf{R}$, and so on. From the definition of a transpose it follows that

$$\mathbf{A}' = \begin{pmatrix} \mathbf{P}' & \mathbf{R}' \\ \mathbf{Q}' & \mathbf{S}' \end{pmatrix}.$$

The product of two partitioned matrices is obtained by the rule of row by column multiplication and treating matrices as elements. Thus, for examples

$$\mathbf{AB} = \begin{pmatrix} \mathbf{P} & \mathbf{Q} \\ \mathbf{R} & \mathbf{S} \end{pmatrix}\begin{pmatrix} \mathbf{E} \\ \mathbf{G} \end{pmatrix} = \begin{pmatrix} \mathbf{PE} + \mathbf{QG} \\ \mathbf{RE} + \mathbf{SG} \end{pmatrix}$$

$$\mathbf{AC} = \begin{pmatrix} \mathbf{P} & \mathbf{Q} \\ \mathbf{R} & \mathbf{S} \end{pmatrix}\begin{pmatrix} \mathbf{E} & \mathbf{F} \\ \mathbf{G} & \mathbf{H} \end{pmatrix} = \begin{pmatrix} \mathbf{PE} + \mathbf{QG} & \mathbf{PF} + \mathbf{QH} \\ \mathbf{RE} + \mathbf{SG} & \mathbf{RF} + \mathbf{SH} \end{pmatrix}$$

provided the products $\mathbf{PE}$ etc., exist.

A square matrix $\mathbf{A}$ is said to be *symmetric* if $\mathbf{A} = \mathbf{A}'$ and hermitian if $\mathbf{A} = \mathbf{A}^*$.

1b.2 Elementary Matrices and Diagonal Reduction of a Matrix

Consider the following operations on a matrix $\mathbf{A}$ of order $m \times n$:

(e_1) = Multiplying a row (or a column) by a scalar c.
(e_2) = Replacing the rth row (column) by the rth row (column) + λ times the sth row (column).
(e_3) = Interchanging two rows (columns).

All these operations obviously preserve the rank of a matrix. We shall show that they are equivalent to multiplications by nonsingular matrices. The operation (e_1) of multiplying the rth row by the scalar c is equivalent to premultiplying $\mathbf{A}$ by a square matrix of order $m \times m$, known as Kronecker $\boldsymbol{\Delta}$, i.e., a square matrix with only diagonal elements possibly different from zero and the rest zero, with special values for its elements as follows:

$$\Delta = (\delta_{ij}), \quad \delta_{rr} = c; \qquad \delta_{ii} = 1, \quad i \neq r; \qquad \delta_{ij} = 0, \quad i \neq j.$$

The operation (e_2) on rows is same as premultiplying $\mathbf{A}$ by the square matrix $\mathbf{E}_{rs}(\lambda)$ of order $m \times m$ defined as follows:

$$\mathbf{E}_{rs}(\lambda) = (e_{ij}), \quad e_{ii} = 1 \text{ for all } \quad i, \qquad e_{rs} = \lambda, \quad e_{ij} = 0 \quad \text{otherwise.}$$

For similar operations on the columns we need to postmultiply by $n \times n$ square matrices of the type Δ and $\mathbf{E}_{rs}(\lambda)$. Verify that the operation (e_3) of interchanging two rows, say the rth and sth, is equivalent to premultiplication of $\mathbf{A}$ successively by the matrices $\mathbf{E}_{sr}(-1)$, $\mathbf{E}_{rs}(1)$, $\mathbf{E}_{sr}(-1)$, Δ. Similarly, interchanging columns is equal to a series of postmultiplications. The following results regarding the reduction of a matrix to simpler forms are important.

(i) ECHELON FORM. *Let* $\mathbf{A}$ *be a matrix of order* $m \times n$. *There exists an* $m \times m$ *nonsingular matrix* $\mathbf{B}$ *such that* $\mathbf{BA}$ *is in echelon form.*

A matrix is said to be in an echelon form if and only if (a) each row consists of all zeroes or a unity preceded by zeroes and not necessarily succeeded by zeroes, and (b) any column containing the first unity of a row has the rest of the elements as zeroes.

Choose the first non-zero element (if it exists) in the first row, and let it be in the ith column. Divide the first row by the chosen element known as the first *pivotal* element. If we multiply the row so obtained by a constant and if we subtract from the second, the element in the ith column of the second row can be made zero. By similar operations all the elements in the ith column can be liquidated (or swept out) except the one in the pivotal position. Next we move to the second row and repeat the operations. If a row at any stage consists of all zeroes, we proceed to the next row. It is easy to see that the resulting matrix is in an echelon form, but this is obtained only by operations of the type (e_1), (e_2). Hence the reduction is secured through a series of premultiplications by square matrices of the form Δ or $\mathbf{E}_{rs}(\lambda)$, whose product is precisely $\mathbf{B}$. Any product of matrices of the form Δ, $\mathbf{E}_{rs}(\lambda)$ has the same rank m, since multiplication by any such matrix does not alter the rank and each such matrix has rank m. Hence $\mathbf{B}$ is nonsingular.

(ii) HERMITE CANONICAL FORM. *Let* $\mathbf{A}$ *be a square matrix of order* m. *Then there exists a nonsingular matrix* $\mathbf{C}$ *of order* m *such that* $\mathbf{CA} = \mathbf{H}$ *where* $\mathbf{H}$ *is in Hermite canonical form satisfying the property* $\mathbf{H}^2 = \mathbf{H}$ (*idempotent*).

A matrix is said to be in Hermite canonical form if its principal diagonal consists of only zeroes and unities and all elements below the diagonal are zero, such that when a diagonal element is zero, the entire row is zero and when a diagonal element is unity, the rest of the elements in the column are all zero.

Let **B** be such that **BA** is in echelon form. It needs only interchange of some rows to reduce it to the form **H**. But interchange of rows is secured by operations of type (e_3), that is, premultiplications with elementary matrices. If the product of elementary matrices employed to reduce **BA** to the form **H** is **G**, the required matrix to reduce **A** to the form **H** is $\mathbf{C} = \mathbf{GB}$. Obviously $R(\mathbf{A}) = R(\mathbf{CA}) = R(\mathbf{H})$ which is equal to the number of diagonal unities in **H** [using the notation rank $\mathbf{A} = R(\mathbf{A})$]. It is easy to verify that $\mathbf{H}^2 = \mathbf{H}$.

The Hermite canonical reduction is useful in solving linear equations,

$$\mathbf{AX} = \mathbf{Y}, \tag{1b.2.1}$$

specially when **A** is singular. Multiplying both sides of (1b.2.1) by **C** we obtain the equivalent equations, $\mathbf{HX} = \mathbf{CY}$ which has a solution if and only if the ith element of **CY** is zero when the ith diagonal element of **H** is zero. The solution in such a case is $\mathbf{X} = \mathbf{CY}$. The matrix **C** is said to be a *generalized inverse* of **A** when the rank of **A** is not full and is represented by $\mathbf{A}^-$. A more detailed discussion of generalized inverses is given in **1b.5**.

(iii) DIAGONAL REDUCTION. *Let* **A** *be* $m \times n$ *matrix of rank* r. *Then there exist nonsingular square matrices* **B** *and* **C** *of order* m *and* n *respectively such that*

$$\mathbf{BAC} = \begin{pmatrix} \mathbf{I}_r & \mathbf{0} \\ \mathbf{0} & \mathbf{0} \end{pmatrix} \tag{1b.2.2}$$

where $\mathbf{I}_r$ *is the identity matrix of order* r *and the rest of the elements are zero.*

We start with the *echelon* reduction of **A** and by using the unit element in any row as a pivot sweep out all the succeeding non-zero elements if any. This is accomplished by operations of the type (e_2) on the columns, that is, adding constant times one column to another or postmultiplication by an elementary matrix. In the resulting form we need only interchange rows and columns to bring the unit elements to the principal diagonal. Matrix **C** is simply the product of all elementary matrices used in the postmultiplications and **B** that of matrices used in the premultiplications.

(iv) RANK FACTORIZATION. *Also from* (1b.2.2) *we have for* $m \times n$ *matrix* **A** *of rank* r,

$$\begin{aligned} \mathbf{A} &= \mathbf{B}^{-1} \begin{pmatrix} \mathbf{I}_r & \mathbf{0} \\ \mathbf{0} & \mathbf{0} \end{pmatrix} \mathbf{C}^{-1} \\ &= \mathbf{B}_1 \mathbf{C}_1 \end{aligned} \tag{1b.2.3}$$

where $\mathbf{B}_1$ *and* $\mathbf{C}_1$ *are* $m \times r$ *and* $r \times n$ *matrices each of rank* r.

Let **A** *be* $m \times n$ *matrix of rank* r. *Then there exists a nonsingular* $m \times m$ *matrix* **M** *and an orthogonal* $n \times n$ *matrix* **N** *such that*

$$\mathbf{A} = \mathbf{M}\begin{pmatrix} \mathbf{I}_r & \mathbf{0} \\ \mathbf{0} & \mathbf{0} \end{pmatrix}\mathbf{N}.$$

The result is easily established.

(v) Triangular Reduction. *Let* **A** *be a square matrix of order* m. *Then there exists a nonsingular matrix,* **B**, $m \times m$ *such that* **BA** *is in a triangular form, that is, all elements below the diagonal are zero.*

Let the first diagonal element of **A** be non-zero, in which case by multiplying the first row by constants and subtracting from the other rows, the first column can be swept out. If the first element is zero, add any row with a non-zero first element to the first row and then sweep out the column. If all the elements in the first column are zero, proceed to the second row. Repeat the operations disregarding the first row and column. All the operations are of type (e_2) only. The product of the premultiplying matrices is **B**.

(vi) Reduction of a Symmetric Matrix to a Diagonal Form. *Let* **A** *be a square symmetric matrix. Then there exists a nonsingular matrix* **B** *such that* $\mathbf{BAB}'$ *is in a diagonal form, that is,* **A** *can be expressed as product* $\mathbf{C\Delta C}'$ *where* **C** *is nonsingular and* $\mathbf{\Delta}$ *is a diagonal matrix.*

Consider the first diagonal element of **A**. If it is not zero, then by using it as a pivot the first row and column elements can be swept out, leaving a reduced matrix. This is equivalent to pre- and postmultiplications successively by elementary matrices which are transposes, because of symmetry. The reduced matrix is also symmetrical. The first diagonal element in the reduced matrix is considered and if it is not zero, the non-zero elements in the second row and column are swept out. This process does not alter the already reduced first row and column. The process can be continued unless the first diagonal element is zero in a reduced matrix at any stage. If the corresponding row and column have all their elements zero, we consider the next diagonal element and proceed as before. Otherwise we can, by adding (or subtracting) a scalar multiple of a suitable row and the corresponding column, bring a non-zero element to the first position in the reduced matrix and continue the foregoing process. This is also a symmetric operation on the rows and columns using elementary matrices. The result follows from choosing **B** as the product of the elementary matrices used for premultiplication. Finally $\mathbf{C} = \mathbf{B}^{-1}$.

(vii) Triangular Reduction by Orthogonal Matrix. *Let* **A** *be* $m \times n$ *matrix and* $m \geqslant n$. *Then there exists an orthogonal matrix* **B** *of order*

$m \times m$ such that $\mathbf{BA} = \begin{pmatrix} \mathbf{T} \\ \mathbf{0} \end{pmatrix}$ *where* $\mathbf{T}$ *is an upper triangular matrix of order n and* $\mathbf{0}$ *is a matrix order* $m - n \times n$ *and consists of zeroes only.*

Observe that $(\mathbf{I} - 2\mathbf{uu}')$ for any vector $\mathbf{u}$ of unit length is an orthogonal matrix. We shall show that $\mathbf{u}$ can be determined in such a way that the first column of the matrix $(\mathbf{I} - 2\mathbf{uu}')\mathbf{A}$ contains all zeroes except possibly the first element. For this we need to solve the equation

$$(\mathbf{I} - 2\mathbf{u}_1\mathbf{u}_1')\boldsymbol{\alpha}_1 = \boldsymbol{\beta},$$

where $\boldsymbol{\alpha}_1$ is the first column of $\mathbf{A}$ and $\boldsymbol{\beta}$ is a vector with the first element β_1 possibly different from zero and the rest of the elements as zeroes. It is easy to verify that if $\boldsymbol{\alpha}_1 = \mathbf{0}$, or if only its first element is non-zero then $\mathbf{u}_1 = \mathbf{0}$. Otherwise compute

$$\left.\begin{aligned} \beta_1 &= \sqrt{\boldsymbol{\alpha}_1'\boldsymbol{\alpha}_1} \\ \mathbf{u}_1 &= \frac{\boldsymbol{\alpha}_1 - \boldsymbol{\beta}}{2b} \end{aligned}\right\}, \qquad (1b.2.4)$$

where $4b^2 = (\boldsymbol{\alpha}_1 - \boldsymbol{\beta})'(\boldsymbol{\alpha}_1 - \boldsymbol{\beta})$. Verify that for the choice of $\mathbf{u}_1$ as in (1b.2.4), the result of multiplying $\mathbf{A}$ by $(\mathbf{I} - 2\mathbf{u}_1\mathbf{u}_1')$ is to liquidate (sweep out) the first column except for the first element. Omitting the first column and row we have a reduced matrix. We apply a similar operation on the first column of the reduced matrix by an orthogonal matrix $(\mathbf{I} - 2\mathbf{u}_2\mathbf{u}_2')$ of order $(m-1) \times (m-1)$. This is equivalent to multiplying $(\mathbf{I} - 2\mathbf{u}_1\mathbf{u}_1')\mathbf{A}$ from the left by the partitioned matrix

$$\begin{pmatrix} 1 & \mathbf{0}' \\ \mathbf{0} & \mathbf{I} - 2\mathbf{u}_2\mathbf{u}_2' \end{pmatrix}$$

which is also orthogonal. We repeat the operation until all the columns are swept out, which results in a matrix of the type $\begin{pmatrix} \mathbf{T} \\ \mathbf{0} \end{pmatrix}$. The matrix $\mathbf{B}$ of the theorem is then the product of orthogonal matrices, which is also orthogonal. The method of reduction is due to Householder (1964).

(viii) QR Decomposition. *Let* $\mathbf{A}$ *be* $m \times n$ *real matrix of rank r, with* $m \geqslant n$. *Then there exists an* $m \times (m - n)$ *matrix* $\mathbf{Q}$, *which is suborthogonal, i.e.,* $\mathbf{Q}'\mathbf{Q} = \mathbf{I}$, *and an* $n \times n$ *matrix* $\mathbf{R}$ *which is upper triangular such that*

$$\mathbf{A} = \mathbf{QR} \qquad (1b.2.5)$$

The result follows by choosing $\mathbf{Q}$ as the submatrix formed by the first $m - n$ rows of the matrix $\mathbf{B}$ in (vii) and by writing $\mathbf{T}$ as $\mathbf{R}$.

Note that if $R(\mathbf{A}) = n$ and the diagonal elements of $\mathbf{R}$ are required to be non-negative (always possible), then the decomposition is unique. If $R(\mathbf{A}) = n$, then any two decompositions are related by $\mathbf{Q}_1 = \mathbf{Q}_2\mathbf{S}$ and $\mathbf{R}_1 = \mathbf{S}\mathbf{R}_2$ where $\mathbf{S}$ is a diagonal matrix of $+1$'s and -1's.

(ix) GRAM-SCHMIDT TRIANGULAR REDUCTION. *Let* $\mathbf{A}$ *be* $m \times n$ *matrix. Then there exists a* $n \times n$ *upper triangular matrix* $\mathbf{T}$ *and* $m \times n$ *matrix* $\mathbf{S}$ *with its columns orthonormal such that* $\mathbf{A} = \mathbf{ST}$.

Let $\boldsymbol{\alpha}_1, \ldots, \boldsymbol{\alpha}_n$ be the columns of $\mathbf{A}$, $\boldsymbol{\sigma}_1, \ldots, \boldsymbol{\sigma}_n$ the columns of $\mathbf{S}$, and $t_{1i}, \ldots, t_{ii}$ the upper diagonal elements in the ith column of $\mathbf{T}$. Then we have

$$\boldsymbol{\alpha}_i = t_{1i}\boldsymbol{\sigma}_1 + \cdots + t_{ii}\boldsymbol{\sigma}_i,$$

which gives $t_{ji} = \boldsymbol{\sigma}_j'\boldsymbol{\alpha}_i$, $j = 1, \ldots, i-1$ and

$$t_{ii} = (\boldsymbol{\alpha}_i'\boldsymbol{\alpha}_i - t_{1i}^2 - \cdots - t_{i-1,i}^2)^{1/2}$$

as in the Gram-Schmidt orthogonalization process described in [(ii), **1a.4**]. The coefficients t_{ij} and the vectors $\boldsymbol{\sigma}_1, \boldsymbol{\sigma}_2, \ldots$ are constructed sequentially:

$$\begin{aligned} \boldsymbol{\alpha}_1 &= t_{11}\boldsymbol{\sigma}_1 \\ \boldsymbol{\alpha}_2 &= t_{12}\boldsymbol{\sigma}_1 + t_{22}\boldsymbol{\sigma}_2 \\ &\cdot \quad \cdot \cdot \quad \cdot \cdot \end{aligned}$$

with the convention that if any $t_{ii} = 0$, then $t_{ij} = 0$ for $j > i$ and $\boldsymbol{\sigma}_i$ is chosen arbitrarily.

1b.3 Determinants

The determinant of a square matrix $\mathbf{A} = (a_{ij})$ of order m is a real valued function of the elements a_{ij} defined by

$$|\mathbf{A}| = \sum \pm a_{1i}a_{2j} \ldots a_{mp},$$

where the summation is taken over all permutations $(i, j, \ldots, p)$ of $(1, 2, \ldots, m)$ with a plus sign if $(i, j, \ldots, p)$ is an even permutation and a minus sign if it is an odd permutation. For an axiomatic derivation of the determinant of a matrix, see Levi (1942) and Rao (1952).

The cofactor of the element a_{ij}, denoted by A_{ij}, is defined to be $(-1)^{i+j}$ times the determinant obtained by omitting the ith row and jth column. It is easy to establish the following properties.

(a) $|\mathbf{A}| = 0$ if and only if $R(\mathbf{A}) \neq m$.

(b) $|\mathbf{A}|$ changes sign if any two rows (or columns) are interchanged.

(c) $|\mathbf{A}| = \sum_i a_{ri}A_{ri}$ for any r

$= \sum_r a_{ri}A_{ri}$ for any i.

(d) $\sum_i a_{ri} A_{si} = 0$ if $r \neq s$.

(e) If $|\mathbf{A}| \neq 0$, $\mathbf{A}^{-1} = (A_{ij}/|\mathbf{A}|)'$.

(f) If $\mathbf{A}$ is a diagonal matrix or a triangular matrix, $|\mathbf{A}|$ is the product of diagonal elements.

(g) If $\mathbf{A}$ and $\mathbf{B}$ are square matrices $|\mathbf{AB}| = |\mathbf{A}|\,|\mathbf{B}|$.

1b.4 Transformations

Some authors treat matrix theory through linear transformations. A linear transformation from E_n (Euclidean space of n dimensions) to E_m has the property that if $\mathbf{X} \to \mathbf{Y}$ and $\boldsymbol{\xi} \to \boldsymbol{\eta}$, then $a\mathbf{X} + b\boldsymbol{\xi} \to a\mathbf{Y} + b\boldsymbol{\eta}$ where $\boldsymbol{\xi}$, $\mathbf{X} \in E_n$, $\boldsymbol{\eta}$, $\mathbf{Y} \in E_m$ and $a, b, \in F$, the field of scalars associated with the vector spaces E_n and E_m. We shall show the correspondence between linear transformations and matrices.

Consider a linear transformation (from E_n to E_m) which transforms the elementary vector $\mathbf{e}_1, \ldots, \mathbf{e}_n \in E_n$ as follows:

$$\mathbf{e}_1' = (1, 0, \ldots, 0) \to (a_{11}, \ldots, a_{m1}),$$

$$\ldots \qquad\qquad \ldots$$

$$\mathbf{e}_n' = (0, 0, \ldots, 1) \to (a_{1n}, \ldots, a_{mn})$$

Let $\mathbf{X}' = (x_1, \ldots, x_n) = x_1\mathbf{e}_1' + \cdots + x_n\mathbf{e}_n' \in E_n$. Due to linearity

$$\mathbf{X}' \to \left(\sum a_{1i}x_i, \ldots, \sum a_{mi}x_i\right) = \mathbf{Y}'$$

which means that $\mathbf{X} \to \mathbf{Y}$ is secured by the relation $\mathbf{Y} = \mathbf{AX}$ where $\mathbf{A}$ is the $m \times n$ matrix

$$\mathbf{A} = \begin{pmatrix} a_{11} & \cdots & a_{1n} \\ \cdot & \cdots & \cdot \\ a_{m1} & \cdots & a_{mn} \end{pmatrix}.$$

If $\mathbf{Z} = \mathbf{BY}$ and $\mathbf{Y} = \mathbf{AX}$, then $\mathbf{Z} = (\mathbf{BA})\mathbf{X}$ provides a direct transformation from $\mathbf{X}$ to $\mathbf{Z}$, which involves a product of two matrices.

The transformation from $\mathbf{X}$ to $\mathbf{Y}$ need not be one-to-one; it is so only when $\mathbf{A}$ is nonsingular in which case $\mathbf{A}^{-1}$ exists and the inverse transformation is given by $\mathbf{X} = \mathbf{A}^{-1}\mathbf{Y}$ (showing one-to-one correspondence).

ORTHOGONAL TRANSFORMATION. A linear transformation $\mathbf{Y} = \mathbf{CX}$ is said to be orthogonal if $\mathbf{CC}' = \mathbf{I}$. Since $|\mathbf{C}|^2 = 1$, the transformation is nonsingular. The condition $\mathbf{CC}' = \mathbf{I}$ implies that the row vectors of $\mathbf{C}$ are mutually orthogonal and are of length unity, or in other words they constitute an o.n.b. of the space R^n. It is easy to verify that

$$|\mathbf{C}| = \pm 1, \qquad \mathbf{C}' = \mathbf{C}^{-1}, \qquad \text{and therefore} \qquad \mathbf{C}'\mathbf{C} = \mathbf{I}.$$

The last result implies that columns also form an o.n.b.

(i) *An important property of an orthogonal transformation is that it keeps distances and angles invariant.*

Let $\mathbf{X} \to \mathbf{Y}$ and $\boldsymbol{\xi} \to \boldsymbol{\eta}$ then

$$\begin{aligned}(\mathbf{Y} - \boldsymbol{\eta})'(\mathbf{Y} - \boldsymbol{\eta}) &= (\mathbf{CX} - \mathbf{C}\boldsymbol{\xi})'(\mathbf{CX} - \mathbf{C}\boldsymbol{\xi}) \\ &= (\mathbf{X} - \boldsymbol{\xi})'\mathbf{C}'\mathbf{C}(\mathbf{X} - \boldsymbol{\xi}) = (\mathbf{X} - \boldsymbol{\xi})'(\mathbf{X} - \boldsymbol{\xi}),\end{aligned}$$

so that distance between two points is invariant. Similarly

$$(\mathbf{Y}_1 - \boldsymbol{\eta}_1)'(\mathbf{Y}_2 - \boldsymbol{\eta}_2) = (\mathbf{X}_1 - \boldsymbol{\xi}_1)'(\mathbf{X}_2 - \boldsymbol{\xi}_2),$$

which implies that the angle between vectors is also invariant. In fact a necessary and sufficient condition for distance to be invariant is that the transformation is orthogonal.

PROJECTION. An example of linear transformation in R^n is the projection of vectors on a subspace $\mathcal{M}$, for if $\mathbf{Y}_1$, $\mathbf{Y}_2$ are projections of $\mathbf{X}_1$ and $\mathbf{X}_2$, then $a\mathbf{Y}_1 + b\mathbf{Y}_2$ is the projection of $a\mathbf{X}_1 + b\mathbf{X}_2$. Hence projection is secured through a matrix $\mathbf{P}$, i.e., $\mathbf{Y} = \mathbf{PX}$. An explicit representation of $\mathbf{P}$ is given in **1c.4**.

1b.5 Generalized Inverse of a Matrix

Definition. Consider an $m \times n$ matrix $\mathbf{A}$ of any rank. A generalized inverse (or a g-inverse) of $\mathbf{A}$ is a $n \times m$ matrix, denoted by $\mathbf{A}^-$, such that $\mathbf{X} = \mathbf{A}^-\mathbf{Y}$ is a solution of the equation $\mathbf{AX} = \mathbf{Y}$ for any $\mathbf{Y} \in \mathcal{M}(\mathbf{A})$.

We establish the existence of a g-inverse and investigate its properties.

(i) $\mathbf{A}^-$ *is a g-inverse* $\Leftrightarrow \mathbf{AA}^-\mathbf{A} = \mathbf{A}$.

Choose $\mathbf{Y}$ as the ith column $\mathbf{a}_i$ of $\mathbf{A}$. The equation $\mathbf{AX} = \mathbf{a}_i$ is clearly consistent and hence $\mathbf{X} = \mathbf{A}^-\mathbf{a}_i$ is a solution, that is, $\mathbf{AA}^-\mathbf{a}_i = \mathbf{a}_i$ for all i which is the same as $\mathbf{AA}^-\mathbf{A} = \mathbf{A}$. Thus the existence of $\mathbf{A}^- \Rightarrow \mathbf{AA}^-\mathbf{A} = \mathbf{A}$. Conversely if $\mathbf{A}^-$ is such that $\mathbf{AA}^-\mathbf{A} = \mathbf{A}$ and $\mathbf{AX} = \mathbf{Y}$ is consistent, then $\mathbf{AA}^-\mathbf{AX} = \mathbf{AX}$ or $\mathbf{AA}^-\mathbf{Y} = \mathbf{Y}$. Hence $\mathbf{X} = \mathbf{A}^-\mathbf{Y}$ is a solution. Thus, we have an alternative definition of a generalized inverse as any matrix $\mathbf{A}^-$ such that

$$\mathbf{AA}^-\mathbf{A} = \mathbf{A}. \tag{1b.5.1}$$

A g-inverse as defined by (1b.5.1) is not necessarily unique; so some authors define a g-inverse as a matrix $\mathbf{A}^-$ that satisfies some further conditions besides (1b.5.1) which are unnecessary in many applications (Rao, 1955a, 1962f). We may, however, impose suitable conditions on a g-inverse in particular applications in which case we may refer to it as a g-inverse with the stated properties.

(ii) *Let* $\mathbf{A}^-$ *be any g-inverse of* $\mathbf{A}$ *and* $\mathbf{A}^-\mathbf{A} = \mathbf{H}$. *Then:*

(a) $\mathbf{H}^2 = \mathbf{H}$, *that is,* $\mathbf{H}$ *is idempotent.*
(b) $\mathbf{AH} = \mathbf{A}$ *and rank* $\mathbf{A}$ = *rank* $\mathbf{H}$ = *trace* $\mathbf{H}$.
(c) *A general solution of* $\mathbf{AX} = \mathbf{0}$ *is* $(\mathbf{H} - \mathbf{I})\mathbf{Z}$ *where* $\mathbf{Z}$ *is arbitrary.*
(d) *A general solution of a consistent equation* $\mathbf{AX} = \mathbf{Y}$ *is* $\mathbf{A}^-\mathbf{Y} + (\mathbf{H} - \mathbf{I})\mathbf{Z}$ *where* $\mathbf{Z}$ *is arbitrary.*
(e) $\mathbf{Q}'\mathbf{X}$ *has a unique value for all* $\mathbf{X}$ *satisfying* $\mathbf{AX} = \mathbf{Y}$, *if and only if* $\mathbf{Q}'\mathbf{H} = \mathbf{Q}'$.

Since $\mathbf{AA}^-\mathbf{A} = \mathbf{A}$, $\mathbf{A}^-\mathbf{AA}^-\mathbf{A} = \mathbf{A}^-\mathbf{A}$, that is, $\mathbf{H}^2 = \mathbf{H}$. The rest are easy to prove.

(iii) $\mathbf{A}^-$ *exists and rank* $\mathbf{A}^- \geqslant$ *rank* $\mathbf{A}$.

In [(iii), **1b.2**], it is shown that, given a matrix $\mathbf{A}$, there exist nonsingular square matrices $\mathbf{B}$ and $\mathbf{C}$ such that $\mathbf{BAC} = \mathbf{\Delta}$ or $\mathbf{A} = \mathbf{B}^{-1}\mathbf{\Delta}\mathbf{C}^{-1}$ where $\mathbf{\Delta}$ is a diagonal (not necessarily square) matrix. Let $\mathbf{\Delta}^-$ be a matrix obtained by replacing the nonzero elements of $\mathbf{\Delta}'$ by their reciprocals. Then it may be easily seen that $\mathbf{\Delta\Delta}^-\mathbf{\Delta} = \mathbf{\Delta}$. Consider $\mathbf{A}^- = \mathbf{C\Delta}^-\mathbf{B}$. Verify that

$$\mathbf{AA}^-\mathbf{A} = \mathbf{B}^{-1}\mathbf{\Delta}\mathbf{C}^{-1}\mathbf{C}\mathbf{\Delta}^-\mathbf{B}\mathbf{B}^{-1}\mathbf{\Delta}\mathbf{C}^{-1} = \mathbf{B}^{-1}\mathbf{\Delta}\mathbf{C}^{-1} = \mathbf{A},$$

so that $\mathbf{A}^-$ is a generalized inverse. Obviously rank $\mathbf{A}^- \geqslant$ rank $\mathbf{A}$.

In fact $\mathbf{A}^-$ as constructed has the additional property $\mathbf{A}^-\mathbf{A}\,\mathbf{A}^- = \mathbf{A}^-$, that is, $(\mathbf{A}^-)^- = \mathbf{A}$ and rank $\mathbf{A}^-$ = rank $\mathbf{A}$. Every g-inverse need not satisfy this additional condition as shown in (v).

(iv) $\mathbf{A}^-$ *is a g-inverse* $\Leftrightarrow$ $\mathbf{A}^-\mathbf{A}$ *is idempotent and* $R(\mathbf{A}^-\mathbf{A}) = R(\mathbf{A})$ *or* $\mathbf{AA}^-$ *is idempotent and* $R(\mathbf{AA}^-) = R(\mathbf{A})$.

The result is easy to establish. The propositions (i) and (iv) provide alternative definitions of a g-inverse. (A matrix $\mathbf{B}$ is said to be idempotent if $\mathbf{B}^2 = \mathbf{B}$, see **1b.7**.)

(v) CLASS OF ALL g-INVERSES. *The general solution of* $\mathbf{AXA} = \mathbf{A}$ (*i.e.*, *g-inverse of* $\mathbf{A}$) *can be expressed in three alternative forms*:

(a) $\mathbf{X} = \mathbf{A}^- + \mathbf{U} - \mathbf{A}^-\mathbf{AUAA}^-$ (1b.5.2)

(b) $\mathbf{X} = \mathbf{A}^- + \mathbf{V}(\mathbf{I} - \mathbf{AA}^-) + (\mathbf{I} - \mathbf{A}^-\mathbf{A})\mathbf{W}$ (1b.5.3)

(c) $\mathbf{X} = \mathbf{A}^- + \mathbf{V}(\mathbf{I} - \mathbf{P}_\mathbf{A}) + (\mathbf{I} - \mathbf{P}_{\mathbf{A}^*})\mathbf{U}$ (1b.5.4)

where $\mathbf{A}^-$ *is any particular g-inverse,* $\mathbf{P}_\mathbf{A}$ *and* $\mathbf{P}_{\mathbf{A}^*}$ *are the projection operators onto* $\mathcal{M}(\mathbf{A})$ *and* $\mathcal{M}(\mathbf{A}^*)$ *respectively, and* $\mathbf{U}$, $\mathbf{V}$, $\mathbf{W}$ *are arbitrary matrices.*

Obviously $\mathbf{X}$ as in (1b.5.2) satisfies the equation $\mathbf{AXA} = \mathbf{A}$. Further, any given solution $\mathbf{X}$ can be put in the form (1b.5.2) by choosing $\mathbf{U} = \mathbf{X} - \mathbf{A}^-$.

Similarly $\mathbf{X}$ as in (1b.5.3) satisfies the equation $\mathbf{AXA} = \mathbf{A}$. Further, any given solution $\mathbf{X}$ can be put in the form (1b.5.3) by choosing $\mathbf{V} = \mathbf{G} - \mathbf{A}^-$ and $\mathbf{W} = \mathbf{GAA}^-$. The result (1b.5.4) can be established in the same way.

Definition. A choice $\mathbf{G}$ of g-inverse of $\mathbf{A}$ satisfying the conditions $\mathbf{AGA} = \mathbf{A}$ and $\mathbf{GAG} = \mathbf{G}$ is called a reflexive g-inverse of $\mathbf{A}$ and denoted by $\mathbf{A}_r^-$. The conditions can be equivalently written as $\mathbf{AGA} = \mathbf{A}$ and $R(\mathbf{G}) = R(\mathbf{A})$.

The g-inverse given in (iii) is indeed a reflexive g-inverse. But (1b.5.2) through (1b.5.4) show that not all g-inverses need be reflexive.

(vi) *The following results hold for any choice of g-inverse.*

(a) $\mathbf{A}(\mathbf{A}^*\mathbf{A})^-\mathbf{A}^*\mathbf{A} = \mathbf{A}$ *and* $\mathbf{A}^*\mathbf{A}(\mathbf{A}^*\mathbf{A})^-\mathbf{A}^* = \mathbf{A}^*$ (1b.5.5)

(b) *A necessary and sufficient condition that* $\mathbf{BA}^-\mathbf{A} = \mathbf{B}$ *is that* $\mathcal{M}(\mathbf{B}') \subset \mathcal{M}(\mathbf{A}')$.

(c) *Let* $\mathbf{B}$ *and* $\mathbf{C}$ *be non-null. Then* $\mathbf{BA}^-\mathbf{C}$ *is invariant for any choice of* $\mathbf{A}^-$ *iff* $\mathcal{M}(\mathbf{B}') \subset \mathcal{M}(\mathbf{A}')$ *and* $\mathcal{M}(\mathbf{C}) \subset \mathcal{M}(\mathbf{A})$.

The result (a) is proved by multiplying $\mathbf{A}(\mathbf{A}^*\mathbf{A})^-\mathbf{A}^*\mathbf{A} - \mathbf{A}$ by its conjugate transpose and showing that the product is zero. The results (b) and (c) are similarly proved.

(vii) $\mathbf{A}(\mathbf{A}^*\mathbf{A})^-\mathbf{A}^*$ *is hermitian and invariant for any choice of the g-inverse* $(\mathbf{A}^*\mathbf{A})^-$.

Consider the difference $\mathbf{D} = \mathbf{A}(\mathbf{A}^*\mathbf{A})^-\mathbf{A}^* - \mathbf{A}[(\mathbf{A}^*\mathbf{A})^-]^*\mathbf{A}^*$ and show that $\mathbf{DD}^* = \mathbf{0}$ using (a) of (vi), which establishes the hermitian property. Invariance follows since $\mathbf{A}^* \in \mathcal{M}(\mathbf{A}^*\mathbf{A})$.

(viii) MOORE INVERSE. *Moore* (1935) *and Penrose* (1955) *defined a generalized inverse as a matrix* $\mathbf{A}^+$ *to distinguish it from a general g-inverse* $\mathbf{A}^-$ *satisfying the properties*

$$\text{(a) } \mathbf{AA}^+\mathbf{A} = \mathbf{A}, \qquad \text{(b) } \mathbf{A}^+\mathbf{AA}^+ = \mathbf{A}^+,$$
$$\text{(c) } (\mathbf{AA}^+)^* = \mathbf{AA}^+, \qquad \text{(d) } (\mathbf{A}^+\mathbf{A})^* = \mathbf{A}^+\mathbf{A}. \qquad \text{(1b.5.6)}$$

Such an inverse exists and is unique.

It is shown in (1c.3.8) of **1c.3** that there exists a diagonal reduction of $\mathbf{A}$ such that $\mathbf{BAC} = \mathbf{\Delta}$ where $\mathbf{B}$ and $\mathbf{C}$ are orthogonal and $\mathbf{\Delta}$ is a diagonal (not necessarily square) matrix. It is easy to verify that $\mathbf{A}^+ = \mathbf{C\Delta}^-\mathbf{B}$ satisfies the conditions (a) through (d) and also that if $\mathbf{D}$ is any other matrix satisfying (a) through (d), then $\mathbf{D} = \mathbf{A}^+$.

Further results on g-inverses are given in Section **1c.5**.

Some Computations of a g-inverse. Let $\mathbf{A}$ be a $m \times n$ matrix of rank r. Then there exists a nonsingular submatrix $\mathbf{B}$ of order r, obtained by omitting some rows and columns of $\mathbf{A}$, if necessary. By a suitable interchange of columns and rows we can write $\mathbf{A}$ as the partitioned matrix

$$\mathbf{A} = \begin{pmatrix} \mathbf{B} & \mathbf{C} \\ \mathbf{D} & \mathbf{E} \end{pmatrix}.$$

Consider

$$\mathbf{A}^- = \begin{pmatrix} \mathbf{B}^{-1} & \mathbf{0} \\ \mathbf{0} & \mathbf{0} \end{pmatrix}. \tag{1b.5.7}$$

It is easy to verify (observing that $\mathbf{E} = \mathbf{D}\mathbf{B}^{-1}\mathbf{C}$) that $\mathbf{A}\mathbf{A}^-\mathbf{A} = \mathbf{A}$ so that $\mathbf{A}^-$ defined in (1b.5.7) is a g-inverse with rank $\mathbf{A}^-$ = rank $\mathbf{A}$ and has the additional property $\mathbf{A}^-\mathbf{A}\mathbf{A}^- = \mathbf{A}^-$.

Consider

$$\mathbf{A}^- = \begin{pmatrix} \mathbf{B}^{-1} & -\mathbf{K} \\ \mathbf{0} & \mathbf{I} \end{pmatrix}, \tag{1b.5.8}$$

where $\mathbf{K}$ is such that $\mathbf{BK} = \mathbf{C}$ and $\mathbf{DK} = \mathbf{E}$. Then rank $\mathbf{A}^- = \min(m, n)$ and $\mathbf{A}^-$ is a g-inverse with the maximum rank.

For other methods of computing a g-inverse see Example **5** at the end of Section **1b.6**.

1b.6 Matrix Representation of Vector Spaces, Bases, etc.

We have not made full use of the concepts of vector spaces in the study of matrices except in defining the rank of a matrix as the dimension of the vector space generated by the columns or rows of a matrix. We shall now establish some results on matrices using vector space arguments, which are used repeatedly in the later chapters. Let $\mathbf{X}$ be $n \times m$ matrix. The vector space generated by the columns of $\mathbf{X}$ is denoted by $\mathcal{M}(\mathbf{X})$ which is also called the *range space* of $\mathbf{X}$. The subspace of all vectors $\mathbf{Z}$ in E_m such that $\mathbf{XZ} = \mathbf{0}$ is called the *null space* of $\mathbf{X}$ and denoted by $\mathcal{N}(\mathbf{X})$. The dimension, $d[\mathcal{N}(\mathbf{X})]$ is called the *nullity* of $\mathbf{X}$.

(i) $\mathcal{M}(\mathbf{X}) = \mathcal{M}(\mathbf{XX}')$. *That is, the vector spaces generated by the columns of* $\mathbf{X}$ *and* $\mathbf{XX}'$ *are the same. Hence* $d[\mathcal{M}(\mathbf{X})] = d[\mathcal{M}(\mathbf{XX}')] = R(\mathbf{X}) = R(\mathbf{XX}')$.

Verify that if $\boldsymbol{\alpha}$ is a column vector such that $\boldsymbol{\alpha}'\mathbf{X} = \mathbf{0} \Rightarrow \boldsymbol{\alpha}'\mathbf{XX}' = \mathbf{0}$. Conversely $\boldsymbol{\alpha}'\mathbf{XX}' = \mathbf{0} \Rightarrow \boldsymbol{\alpha}'\mathbf{XX}'\boldsymbol{\alpha} = 0 \Rightarrow \boldsymbol{\alpha}'\mathbf{X} = \mathbf{0}$. Hence every vector orthogonal to $\mathbf{X}$ is also orthogonal to $\mathbf{XX}'$. Therefore $\mathcal{M}(\mathbf{X}) = \mathcal{M}(\mathbf{XX}')$. Furthermore, it is easy to establish that, for any matrix $\mathbf{B}$, $\mathcal{M}(\mathbf{X}) = \mathcal{M}(\mathbf{XB}) = \mathcal{M}(\mathbf{XBB}')$ *if and only if* $\mathcal{M}(\mathbf{B}) \supset \mathcal{M}(\mathbf{X}')$.

(ii) *If* $\boldsymbol{\alpha}$ *and* $\boldsymbol{\gamma}$ *are vectors such that* $\boldsymbol{\alpha} = \mathbf{X}\boldsymbol{\gamma}$, *then there exists a vector* $\boldsymbol{\beta}$ *such that* $\boldsymbol{\alpha} = \mathbf{XX}'\boldsymbol{\beta}$

The result is true because $\boldsymbol{\alpha} \in \mathscr{M}(\mathbf{X}) \Rightarrow \boldsymbol{\alpha} \in \mathscr{M}(\mathbf{XX}')$.

(iii) *The set of vectors* $\{\mathbf{X}\boldsymbol{\gamma} : \boldsymbol{\gamma}$ *subject to the condition* $\mathbf{H}\boldsymbol{\gamma} = \mathbf{0}\}$ *where* $\mathbf{H}$ *is a given* $k \times m$ *matrix is a subspace* $\mathscr{S} \subset \mathscr{M}(\mathbf{X})$. *Furthermore,*

$$d[\mathscr{S}] = \text{rank}\begin{pmatrix}\mathbf{X}\\ \mathbf{H}\end{pmatrix} - \text{rank } \mathbf{H}. \tag{1b.6.1}$$

Observe that the set of all vectors $\mathbf{X}\boldsymbol{\gamma}$, where $\boldsymbol{\gamma}$ is arbitrary is the same as $\mathscr{M}(\mathbf{X})$. But when $\boldsymbol{\gamma}$ is restricted as indicated, only a subspace is spanned, whose dimension is computable by the formula (1b.6.1).

1b.7 Idempotent Matrices

Definition. A square matrix $\mathbf{A}$ is said to be idempotent if $\mathbf{A}^2 = \mathbf{A}$.

The following propositions concerning idempotent matrices are of interest.

(i) *The rank of an idempotent matrix is equal to its trace, i.e., the sum of its diagonal elements.*

Consider the rank factorization (1b.2.3) of $\mathbf{A} = \mathbf{B}_1\mathbf{C}_1$. Then $\mathbf{A}^2 = \mathbf{A} \Rightarrow \mathbf{B}_1\mathbf{C}_1\mathbf{B}_1\mathbf{C}_1 = \mathbf{B}_1\mathbf{C}_1$. But $\mathbf{B}_1$ has a left inverse $\mathbf{L}$ and $\mathbf{C}_1$ a right inverse $\mathbf{R}$. Then

$$\mathbf{LB}_1\mathbf{C}_1\mathbf{B}_1\mathbf{C}_1\mathbf{R} = \mathbf{LB}_1\mathbf{C}_1\mathbf{R} \Rightarrow \mathbf{C}_1\mathbf{B}_1 = \mathbf{I}_r.$$

Trace $\mathbf{A}$ = Trace $\mathbf{B}_1\mathbf{C}_1$ = Trace $\mathbf{C}_1\mathbf{B}_1$ = Trace $\mathbf{I}_r = r = R(\mathbf{A})$.

(ii) $\mathbf{A}^2 = \mathbf{A} \Leftrightarrow R(\mathbf{A}) + R(\mathbf{I} - \mathbf{A}) = m$, *the order of* $\mathbf{A}$.

The proof is simple.

(iii) *Let* $\mathbf{A}_i$, $i = 1, \ldots, k$ *be square matrices of order* m *and* $\mathbf{A} = \mathbf{A}_1 + \cdots + \mathbf{A}_k$. *Consider the following statements:*

(*a*) $\mathbf{A}_i^2 = \mathbf{A}_i$ *for all* i.

(*b*) $\mathbf{A}_i\mathbf{A}_j = \mathbf{0}$ *for all* $i \neq j$, $R(\mathbf{A}_i^2) = R(\mathbf{A}_i)$ *for all* i.

(*c*) $\mathbf{A}^2 = \mathbf{A}$.

(*d*) $R(\mathbf{A}) = R(\mathbf{A}_1) + \cdots + R(\mathbf{A}_k)$.

Then any two of (*a*), (*b*) *and* (*c*) *imply all the four. Further,* (*c*) *and* (*d*) $\Rightarrow$ (*a*) *and* (*b*).

That (*a*) and (*b*) $\Rightarrow$ (*c*) is simple to prove. Since the rank of an idempotent matrix is equal to its trace, (*a*) and (*c*) $\Rightarrow$ (*d*) and hence (*a*) and (*b*) $\Rightarrow$ (*c*) and (*d*).

To prove (b) and $(c) \Rightarrow (a)$, observe that (b) and $(c) \Rightarrow \mathbf{A}_i^2 = \mathbf{A}\mathbf{A}_i = \mathbf{A}^2\mathbf{A}_i = \mathbf{A}_i^3 \Rightarrow \mathbf{A}_i^2(\mathbf{I} - \mathbf{A}_i) = \mathbf{0}$. Also $R(\mathbf{A}_i) = R(\mathbf{A}_i^2) \Rightarrow \mathbf{A}_i = \mathbf{D}\mathbf{A}_i^2$ for some $\mathbf{D}$. Hence $\mathbf{A}_i^2(\mathbf{I} - \mathbf{A}_i) = \mathbf{0} \Rightarrow \mathbf{A}_i(\mathbf{I} - \mathbf{A}_i) = \mathbf{0}$. It is already shown (a) and $(b) \Rightarrow (c)$ and (d). Hence (b) and $(c) \Rightarrow (a)$ and (d).

To prove (c) and $(d) \Rightarrow (a)$ and (b), define $\mathbf{A}_0 = \mathbf{I} - \mathbf{A}$. Observe that (c) and $(d) \Rightarrow \mathbf{A}_0 + \mathbf{A}_1 + \cdots + \mathbf{A}_k = \mathbf{I}$ and $R(\mathbf{A}_0) + R(\mathbf{A}_1) + \cdots + R(\mathbf{A}_k) = m$. Further $R(\mathbf{I} - \mathbf{A}_i) = R(\sum \mathbf{A}_i - \mathbf{A}_i) \leqslant \sum R(\mathbf{A}_i) - R(\mathbf{A}_i) = m - R(\mathbf{A}_i)$. However, $R(\mathbf{A}_i) + R(\mathbf{I} - \mathbf{A}_i) \geqslant m$. Hence $R(\mathbf{A}_i) + R(\mathbf{I} - \mathbf{A}_i) = m$, and using the result of (ii), $\mathbf{A}_i^2 = \mathbf{A}_i$ for all i. Similarly $(\mathbf{A}_i + \mathbf{A}_j)^2 = \mathbf{A}_i + \mathbf{A}_j$. Then

$$(\mathbf{A}_i + \mathbf{A}_j)^2 = \mathbf{A}_i + \mathbf{A}_j, \quad \mathbf{A}_i^2 = \mathbf{A}_i, \quad \mathbf{A}_j^2 = \mathbf{A}_j$$
$$\Rightarrow \mathbf{A}_i\mathbf{A}_j + \mathbf{A}_j\mathbf{A}_i = \mathbf{0}$$
$$\Rightarrow (\mathbf{A}_i\mathbf{A}_j + \mathbf{A}_j\mathbf{A}_i)\mathbf{A}_j = \mathbf{A}_i\mathbf{A}_j + \mathbf{A}_j\mathbf{A}_i\mathbf{A}_j = \mathbf{0} \quad (1\text{b}.7.1)$$
$$\Rightarrow \mathbf{A}_j(\mathbf{A}_i\mathbf{A}_j + \mathbf{A}_j\mathbf{A}_i\mathbf{A}_j) = 2\mathbf{A}_j\mathbf{A}_i\mathbf{A}_j = \mathbf{0}. \quad (1\text{b}.7.2)$$

Using (1b.7.2) in (1b.7.1), $\mathbf{A}_i\mathbf{A}_j = \mathbf{0}$ for all $i \neq j$. The results are due to Khatri (1968).

(iv) *Let* $\mathbf{G}_i$ *be* $m \times p_i$ *matrix of rank* r_i, $i = 1, \ldots, k$. *If* $r_1 + \cdots + r_k = m$, *then the statements*

(a) $\mathbf{G}_i^*\mathbf{G}_j = \mathbf{0}$, $i \neq j$ (1b.7.3)

(b) $\mathbf{I} = \mathbf{G}_1(\mathbf{G}_1^*\mathbf{G}_1)^-\mathbf{G}_1^* + \cdots + \mathbf{G}_k(\mathbf{G}_k^*\mathbf{G}_k)^-\mathbf{G}_k^*$ (1b.7.4)

are equivalent.

The proof follows by using the result, (c) and $(d) \Rightarrow (a)$ and (b) in (iii).

1b.8 Special Products of Matrices

(i) KRONECKER PRODUCT. Let $\mathbf{A} = (a_{ij})$ and $\mathbf{B} = (b_{ij})$ be $m \times n$ and $p \times q$ matrices, respectively. Then the Kronecker product

$$\mathbf{A} \otimes \mathbf{B} = (a_{ij}\mathbf{B}) \quad (1\text{b}.8.1)$$

is an $mp \times nq$ matrix expressible as a partitioned matrix with $a_{ij}\mathbf{B}$ as the (i, j)th partition, $i = 1, \ldots, m$ and $j = 1, \ldots, n$.

The following results are consequences of the definition (1b.8.1).

(a) $\mathbf{0} \otimes \mathbf{A} = \mathbf{A} \otimes \mathbf{0} = \mathbf{0}$

(b) $(\mathbf{A}_1 + \mathbf{A}_2) \otimes \mathbf{B} = (\mathbf{A}_1 \otimes \mathbf{B}) + (\mathbf{A}_2 \otimes \mathbf{B})$

(c) $\mathbf{A} \otimes (\mathbf{B}_1 + \mathbf{B}_2) = (\mathbf{A} \otimes \mathbf{B}_1) + (\mathbf{A} \otimes \mathbf{B}_2)$

(d) $a\mathbf{A} \otimes b\mathbf{B} = ab\mathbf{A} \otimes \mathbf{B}$

(e) $\mathbf{A}_1\mathbf{A}_2 \otimes \mathbf{B}_1\mathbf{B}_2 = (\mathbf{A}_1 \otimes \mathbf{B}_1)(\mathbf{A}_2 \otimes \mathbf{B}_2)$

(f) $(\mathbf{A} \otimes \mathbf{B})^{-1} = \mathbf{A}^{-1} \otimes \mathbf{B}^{-1}$, if the inverses exist

(g) $(\mathbf{A} \otimes \mathbf{B})^- = \mathbf{A}^- \otimes \mathbf{B}^-$ using any g-inverses

(h) $(\mathbf{A} \otimes \mathbf{B})' = \mathbf{A}' \otimes \mathbf{B}'$

(i) $(\mathbf{A} \otimes \mathbf{B})(\mathbf{A}^{-1} \otimes \mathbf{B}^{-1}) = \mathbf{I}$

(ii) HADAMARD PRODUCT. If $\mathbf{A} = (a_{ij})$ and $\mathbf{B} = (b_{ij})$ be each $m \times n$ matrices, their Hadamard product is the $m \times n$ matrix of elementwise products

$$\mathbf{A} * \mathbf{B} = (a_{ij} b_{ij}). \tag{1b.8.2}$$

We have the following results involving Hadamard products.

(a) $\mathbf{A} * \mathbf{0} = \mathbf{0}$

(b) $\mathbf{A} * \mathbf{ee}' = \mathbf{A} = \mathbf{ee}' * \mathbf{A}$ where $\mathbf{e}' = (1, \ldots, 1)$

(c) $\mathbf{A} * \mathbf{B} = \mathbf{B} * \mathbf{A}$

(d) $(\mathbf{A} + \mathbf{B}) * \mathbf{C} = \mathbf{A} * \mathbf{B} + \mathbf{B} * \mathbf{C}$

(e) $\operatorname{tr} \mathbf{AB} = \mathbf{e}'(\mathbf{A} * \mathbf{B}')\mathbf{e}$.

(iii) A NEW PRODUCT. Let $\mathbf{A} = (\mathbf{A}_1 \vdots \cdots \vdots \mathbf{A}_k)$ and $\mathbf{B} = (\mathbf{B}_1 \vdots \cdots \vdots \mathbf{B}_k)$ be two partitioned matrices with the same number of partitions. Khatri and Rao (1968) defined the product

$$\mathbf{A} \odot \mathbf{B} = (\mathbf{A}_1 \otimes \mathbf{B}_1 \vdots \cdots \vdots \mathbf{A}_k \otimes \mathbf{B}_k) \tag{1b.8.3}$$

where $\otimes$ denotes Kronecker product. It is easy to verify that

$$(\mathbf{A} \odot \mathbf{B}) \odot \mathbf{C} = \mathbf{A} \odot (\mathbf{B} \odot \mathbf{C}), \qquad (\mathbf{T}_1 \otimes \mathbf{T}_2)(\mathbf{A} \odot \mathbf{B}) = \mathbf{T}_1\mathbf{A} \odot \mathbf{T}_2\mathbf{B},$$

and so on.

COMPLEMENTS AND PROBLEMS

1 *Rank of matrices.* Let $R(\mathbf{A})$ denote the rank of $\mathbf{A}$.

1.1
$$R(\mathbf{AB}) \leqslant \min[R(\mathbf{A}), R(\mathbf{B})]$$

[The rows of $\mathbf{AB}$ are linear combinations of rows of $\mathbf{B}$, and therefore the number of independent rows of $\mathbf{AB}$ is less than or equal to that of $\mathbf{B}$. Hence $R(\mathbf{AB}) \leqslant R(\mathbf{B})$. Similarly the other inequality follows.]

1.2 (a) $R(\mathbf{AA}') = R(\mathbf{A})$.

(b) $R(\mathbf{AB}) = R(\mathbf{A})$ if $\mathbf{B}$ is square and nonsingular.
[$R(\mathbf{AB}) \leqslant R(\mathbf{A})$ by Example 1.1. $\mathbf{A} = (\mathbf{AB})(\mathbf{B}^{-1})$.
Therefore $R(\mathbf{A}) = R[(\mathbf{AB})(\mathbf{B}^{-1})] \leqslant R(\mathbf{AB})$. Hence $R(\mathbf{AB}) = R(\mathbf{A})$.]

(c) Let $\mathbf{A}$ be $m \times n$ with rank m and $\mathbf{S}$ be $r \times m$ with rank r. Then $R(\mathbf{SA}) = r$.

1.3 If $R(\mathbf{A}) = s$, then all subdeterminants of order greater than s vanish, and conversely, if all subdeterminants of order greater than s vanish and at least one sub-determinant of order s does not vanish, then $R(\mathbf{A}) = s$.

1.4 *Sylvester's law.* If $\mathbf{A}$ and $\mathbf{B}$ are square matrices of order n and $R(\mathbf{A}) = r$, $R(\mathbf{B}) = s$, then $R(\mathbf{AB}) \geqslant r + s - n$.

1.5 Every matrix of rank r can be written as the sum of r matrices each of rank unity.

1.6 *Frobenius inequality.* If $\mathbf{A}$, $\mathbf{B}$, $\mathbf{C}$ are any three matrices such that the products considered are defined, then

$$R(\mathbf{AB}) + R(\mathbf{BC}) \leqslant R(\mathbf{B}) + R(\mathbf{ABC}).$$

By putting $\mathbf{B} = \mathbf{I}$ of order $(n \times n)$ in which case $\mathbf{A}$ is $(m \times n)$ and $\mathbf{B}$ is $(n \times p)$,

$$R(\mathbf{A}) + R(\mathbf{C}) - R(\mathbf{AC}) \leqslant n,$$

and if $\mathbf{AC} = 0$, $R(\mathbf{A}) + R(\mathbf{C}) \leqslant n$.

1.7 If p and q are the ranks of $\mathbf{A} - \mathbf{I}$ and $\mathbf{B} - \mathbf{I}$, then the rank of $\mathbf{AB} - \mathbf{I}$ is $\leqslant (p + q)$.

1.8 For any two matrices $\mathbf{A}$ and $\mathbf{B}$ such that $\mathbf{A} + \mathbf{B}$ is defined, $R(\mathbf{A} + \mathbf{B}) \leqslant R(\mathbf{A}) + R(\mathbf{B})$. $R(\mathbf{A} + \mathbf{B}) = R(\mathbf{A}) + R(\mathbf{B}) \Leftrightarrow \mathcal{M}(\mathbf{A}) \cap \mathcal{M}(\mathbf{B}) = \{\mathbf{0}\}$ and $\mathcal{M}(\mathbf{A}') \cap \mathcal{M}(\mathbf{B}') = \{\mathbf{0}\}$.

1.9 The following statements are equivalent:

(a) $R(\mathbf{A}) = R(\mathbf{A}^2)$.
(b) $\mathcal{M}(\mathbf{A}) \cap \mathcal{N}(\mathbf{A}) = \{\mathbf{0}\}$, where $\mathcal{N}(\mathbf{A})$ is the null space of $\mathbf{A}$.
(c) There exists a nonsingular matrix $\mathbf{P}$, such that $\mathbf{A} = \mathbf{P}\begin{pmatrix} \mathbf{D} & \mathbf{0} \\ \mathbf{0} & \mathbf{0} \end{pmatrix}\mathbf{P}^{-1}$, where $\mathbf{D}$ is nonsingular.

1.10 $R(\mathbf{A}^m) = R(\mathbf{A}^{m+1}) \Rightarrow R(\mathbf{A}^m) = R(\mathbf{A}^n)$ for all $n \geqslant m$.

1.11 For any square matrix $\mathbf{A}$ there exists a positive integer m such that $R(\mathbf{A}^m) = R(\mathbf{A}^{m+1})$. (The smallest such positive integer is called the index of $\mathbf{A}$).

2 *Determinants*

2.1 A square matrix $\mathbf{A} = (a_{ij})$ of order n is said to be skew symmetric if $a_{ii} = 0$ and $a_{ij} = -a_{ji}$. Prove that:

(a) $|\mathbf{A}| = 0$ if n is odd and a perfect square if n is even.
(b) $R(\mathbf{A})$ is always even.
(c) Every matrix can be written as the sum of a symmetric and a skew symmetric matrix.

2.2 If $\mathbf{A}$ is $m \times n$, then:

(a) $|\mathbf{AA}'| \geqslant 0$ if $m \leqslant n$, and equal to 0 if $m > n$ and > 0 if $R(\mathbf{A}) = m$.
(b) $|\mathbf{AA}'| =$ sum of squares of all m columned determinants of $\mathbf{A}$ when $m \leqslant n$.

2.3 Vandermonde determinant defined by

$$\begin{pmatrix} 1 & 1 & \cdots & 1 \\ d_1 & d_2 & \cdots & d_n \\ \cdot & \cdot & \cdots & \cdot \\ d_1^{n-1} & d_2^{n-1} & \cdots & d_n^{n-1} \end{pmatrix}$$

has the value $\prod_{i<j} (d_i - d_j)$.

2.4 Show that

$$\begin{vmatrix} \mathbf{A} & \mathbf{C} \\ \mathbf{B} & \mathbf{D} \end{vmatrix} = |\mathbf{A}||\mathbf{D} - \mathbf{BA}^{-1}\mathbf{C}|,$$

where $\mathbf{A}$ and $\mathbf{D}$ are square matrices (may be of different orders) and $\mathbf{A}$ is non-singular. The result is obtained by premultiplication with a determinant of unit value

$$\begin{vmatrix} \mathbf{I} & \mathbf{0} \\ -\mathbf{BA}^{-1} & \mathbf{I} \end{vmatrix} \begin{vmatrix} \mathbf{A} & \mathbf{C} \\ \mathbf{B} & \mathbf{D} \end{vmatrix} = \begin{vmatrix} \mathbf{A} & \mathbf{C} \\ \mathbf{0} & \mathbf{D} - \mathbf{BA}^{-1}\mathbf{C} \end{vmatrix} = |\mathbf{A}||\mathbf{D} - \mathbf{BA}^{-1}\mathbf{C}|.$$

Satisfy yourself that

$$\begin{vmatrix} \mathbf{E} & \mathbf{F} \\ \mathbf{0} & \mathbf{G} \end{vmatrix} = |\mathbf{E}||\mathbf{G}|.$$

2.5 *Expansion of a bordered determinant.* Let $\mathbf{A}$ be $n \times n$ and $\mathbf{X}$ a $n \times 1$ matrices. Further let adj $\mathbf{A} = (A_{ij})'$ where A_{ij} is the cofactor of a_{ij}. Then

(a) $\begin{vmatrix} \mathbf{A} & \mathbf{X} \\ \mathbf{X}' & c \end{vmatrix} = c|\mathbf{A}| - \mathbf{X}'(\text{adj } \mathbf{A})\mathbf{X}$

(b) $\begin{vmatrix} \mathbf{A} & \mathbf{X} \\ \mathbf{X}' & -1 \end{vmatrix} = -|\mathbf{A} + \mathbf{XX}'| = -|\mathbf{A}| - \mathbf{X}'(\text{adj } \mathbf{A})\mathbf{X}$

$$= -|\mathbf{A}|(1 + \mathbf{X}'\mathbf{A}^{-1}\mathbf{X}) \quad \text{if} \quad |\mathbf{A}| \neq 0$$

(c) If $\mathbf{U}$ is a matrix with all its elements unity and $\mathbf{V}$ is a column vector with all its entries unity then

$$|\mathbf{A} + c\mathbf{U}| = |\mathbf{A}| + c\mathbf{V}'(\text{adj } \mathbf{A})\mathbf{V}$$
$$\mathbf{V}' \text{ adj } (\mathbf{A} + c\mathbf{U}) = \mathbf{V}'(\text{adj } \mathbf{A})$$
$$[\text{adj } (\mathbf{A} + c\mathbf{U})]\mathbf{V} = (\text{adj } \mathbf{A})\mathbf{V}$$

2.6 *Determinant of the product of rectangular matrices.* Let $\mathbf{A}$ be $m \times n$ and $\mathbf{B}$ be $n \times m$ and suppose that $m \leqslant n$. Then

$$|\mathbf{AB}| = \sum |\mathbf{A}_m||\mathbf{B}_m|$$

where the summation is over all submatrices of order m. The $\mathbf{B}_m$ is the square submatrix of order m obtained from the rows of $\mathbf{B}$ which have the same indices as the columns of $\mathbf{A}$ specifying $\mathbf{A}_m$.

2.7 Verify, when $\mathbf{A}$, $\mathbf{D}$ are symmetric matrices such that the inverses which occur in the expressions exist, that

$$\begin{pmatrix} \mathbf{A} & \mathbf{B} \\ \mathbf{B}' & \mathbf{D} \end{pmatrix}^{-1} = \begin{pmatrix} \mathbf{A}^{-1} + \mathbf{FE}^{-1}\mathbf{F}' & -\mathbf{FE}^{-1} \\ -\mathbf{E}^{-1}\mathbf{F}' & \mathbf{E}^{-1} \end{pmatrix}$$

where $\mathbf{E} = \mathbf{D} - \mathbf{B}'\mathbf{A}^{-1}\mathbf{B}$, $\mathbf{F} = \mathbf{A}^{-1}\mathbf{B}$. The example provides a method by which the inverse of a higher order matrix can be obtained by inverting only matrices of lower orders. An interesting case is when $\mathbf{B}$ is a column vector which gives a method of finding the inverse of a matrix of order $(n + 1)$ given the inverse of the principal matrix of order n. Use the formula deduced in the example to find the inverse of

$$\begin{pmatrix} 1 & 0 & 0 & 3 \\ 0 & 1 & 0 & 4 \\ 0 & 0 & 1 & 2 \\ 3 & 4 & 2 & 5 \end{pmatrix} \qquad \text{[Hint: Take } \mathbf{A} = \begin{pmatrix} 1 & 0 & 0 \\ 0 & 1 & 0 \\ 0 & 0 & 1 \end{pmatrix} = \mathbf{A}^{-1}$$

$$\mathbf{B}' = (3, 4, 2), \quad \mathbf{D} = (5)]$$

2.8 Let $\mathbf{A}$ be a nonsingular matrix, and $\mathbf{U}$ and $\mathbf{V}$ be two column vectors. Then

$$(\mathbf{A} + \mathbf{UV}')^{-1} = \mathbf{A}^{-1} - \frac{(\mathbf{A}^{-1}\mathbf{U})(\mathbf{V}'\mathbf{A}^{-1})}{1 + \mathbf{V}'\mathbf{A}^{-1}\mathbf{U}}.$$

which gives a method of computing the inverse of $(\mathbf{A} + \mathbf{UV}')$ knowing the inverse of $\mathbf{A}$.

2.9 Let $\mathbf{A}$ and $\mathbf{D}$ be nonsingular matrices of orders m and n and $\mathbf{B}$ be $m \times n$ matrix. Then

$$\begin{aligned}(\mathbf{A} + \mathbf{BDB}')^{-1} &= \mathbf{A}^{-1} - \mathbf{A}^{-1}\mathbf{B}(\mathbf{B}'\mathbf{A}^{-1}\mathbf{B} + \mathbf{D}^{-1})^{-1}\mathbf{B}'\mathbf{A}^{-1}, \\ &= \mathbf{A}^{-1} - \mathbf{A}^{-1}\mathbf{BEB}'\mathbf{A}^{-1} + \mathbf{A}^{-1}\mathbf{BE}(\mathbf{E} + \mathbf{D})^{-1}\mathbf{EBA}'^{-1}\end{aligned}$$

where $\mathbf{E} = (\mathbf{B}'\mathbf{A}^{-1}\mathbf{B})^{-1}$.

3 *Trace of a matrix.* The trace of a square matrix $\mathbf{A}$ is defined as the sum of its diagonal elements and is denoted by Tr $\mathbf{A}$. Prove the following:

(a) $\text{Tr}(\mathbf{A} + \mathbf{B}) = \text{Tr}\,\mathbf{A} + \text{Tr}\,\mathbf{B}$.
(b) $\text{Tr}\,\mathbf{AB} = \text{Tr}\,\mathbf{BA}$.

(c) $\text{Tr } \mathbf{S}^{-1}\mathbf{AS} = \text{Tr } \mathbf{A}$
(d) $\text{Tr } \mathbf{A} = R(\mathbf{A})$, if $\mathbf{A}$ is idempotent.
(e) $\text{Tr}(\mathbf{AA}^-) = R(\mathbf{A})$, where $\mathbf{A}^-$ is a generalized inverse of $\mathbf{A}$.
(f) $\mathbf{X}'\mathbf{AX} = \text{Tr } \mathbf{AXX}'$, where $\mathbf{X}$ is a column vector.

4 *Construction of transformation matrices useful in mathematical statistics.*

4.1 Let $\mathbf{A}$ be $n \times m$ matrix of rank r. Show that there exists an orthogonal matrix $(\mathbf{B}_1 \vdots \mathbf{B}_2)$ of order $n \times n$ such that the partition $\mathbf{B}_1$ is $n \times r$ and $\mathcal{M}(\mathbf{B}_1) = \mathcal{M}(\mathbf{A})$.

4.2 Let $\mathbf{A}_1$ be $n \times m$ matrix of rank r and $\mathbf{A}_2$ be $n \times k$ matrix of rank s, such that $\mathcal{M}(\mathbf{A}_1) \cap \mathcal{M}(\mathbf{A}_2)$ is of dimension t. Then there exists an orthogonal matrix $(\mathbf{B}_1 \vdots \mathbf{B}_2 \vdots \mathbf{B}_3)$ such that $\mathbf{B}_1$ is $n \times r$ and $\mathcal{M}(\mathbf{B}_1) = \mathcal{M}(\mathbf{A}_1)$, and $\mathbf{B}_2$ is $n \times (s - t)$ and $\mathcal{M}(\mathbf{B}_2) \subset \mathcal{M}(\mathbf{A}_2)$.

5 *Computation of a g-inverse.* Let $\mathbf{A}$ be $n \times m$ matrix of rank r. Then there exists $\mathbf{B}$ of order $s \times m$ and rank $(m - r)$ such that $\mathcal{M}(\mathbf{A}') \cap \mathcal{M}(\mathbf{B}')$ is the null vector. Show that

(a) $\mathbf{A}'\mathbf{A} + \mathbf{B}'\mathbf{B}$ is of rank m, and
(b) $\mathbf{A}'\mathbf{A}(\mathbf{A}'\mathbf{A} + \mathbf{B}'\mathbf{B})^{-1}\mathbf{A}'\mathbf{A} = \mathbf{A}'\mathbf{A}$
so that $(\mathbf{A}'\mathbf{A} + \mathbf{B}'\mathbf{B})^{-1}$ is, in fact, a g-inverse of $\mathbf{A}'\mathbf{A}$.
(c) If $s = m - r$, then
$\begin{pmatrix} \mathbf{A}'\mathbf{A} & \mathbf{B}' \\ \mathbf{B} & \mathbf{0} \end{pmatrix}$ is of full rank, and
(d) $\mathbf{C}_1$ is a g-inverse of $\mathbf{A}'\mathbf{A}$, where $\mathbf{C}_1$ is defined by

$$\begin{pmatrix} \mathbf{A}'\mathbf{A} & \mathbf{B}' \\ \mathbf{B} & \mathbf{0} \end{pmatrix}^{-1} = \begin{pmatrix} \mathbf{C}_1 & \mathbf{C}_2' \\ \mathbf{C}_2 & \mathbf{C}_3 \end{pmatrix}$$

6 Let $\mathbf{A}$ and $\mathbf{B}$ be matrices having the same number of columns, and let $\mathcal{M}(\mathbf{A}') \cap \mathcal{M}(\mathbf{B}') = \{\mathbf{0}\}$. Then

(a) $R(\mathbf{A}'\mathbf{A} + \mathbf{B}'\mathbf{B}) = R(\mathbf{A}) + R(\mathbf{B})$, and
(b) $\mathbf{A}'\mathbf{A}(\mathbf{A}'\mathbf{A} + \mathbf{B}'\mathbf{B})^-\mathbf{A}'\mathbf{A} = \mathbf{A}'\mathbf{A}$ for any choice of g-inverse.

7 $\mathbf{ABB}' = \mathbf{CBB}' \Leftrightarrow \mathbf{AB} = \mathbf{CB}$.

1c EIGENVALUES AND REDUCTION OF MATRICES

1c.1 Classification and Transformation of Quadratic Forms

A quadratic form in n variables $x_1, \ldots, x_n$ is a homogeneous quadratic function of the variables

$$Q = \sum_{i=1}^{n} \sum_{j=1}^{n} a_{ij} x_i x_j = \mathbf{X}'\mathbf{AX},$$

where $\mathbf{X}$ is the column vector of variables and $\mathbf{A} = [(a_{ij} + a_{ji})/2]$, which is symmetric, is called the matrix of the quadratic form.

Classification of Quadratic Forms. A real quadratic form $\mathbf{X'AX}$ is said to be (a) positive definite (p.d.) if $\mathbf{X'AX} > 0$ for all non-null $\mathbf{X}$ and negative definite (n.d.) if $-\mathbf{X'AX}$ is p.d., and (b) positive semidefinite (p.s.d.) if $\mathbf{X'AX} \geqslant 0$ and $= 0$ for at least one non-null $\mathbf{X}$ and negative semidefinite (n.s.d.) if $-\mathbf{X'AX}$ is p.s.d.

It is useful to have a general term non-negative definite, n.n.d., for a quadratic form which is p.d. or p.s.d., and n.p.d. for a form which is n.d. or n.s.d. It is also customary to refer to the matrix $\mathbf{A}$ as p.d., n.d., etc., when the associated quadratic form $\mathbf{X'AX}$ is p.d., n.d., etc.

Transformation of Quadratic Forms. Let us consider a nonsingular linear transformation $\mathbf{Y} = \mathbf{B}^{-1}\mathbf{X}$ or $\mathbf{X} = \mathbf{BY}$. Then

$$Q = \mathbf{X'AX} \rightarrow \mathbf{Y'B'ABY} = \mathbf{Y'CY}$$

which is a quadratic form in the new variables with the matrix $\mathbf{C} = \mathbf{B'AB}$. The following results hold under linear transformations.

(i) *The definiteness of Q is invariant under nonsingular linear transformations.*

If $\mathbf{X'AX} = g$ (a given real number) for some $\mathbf{X}$, then for $\mathbf{Y} = \mathbf{B}^{-1}\mathbf{X}$, where $\mathbf{Y}$ is non-null if $\mathbf{X}$ is non-null,

$$\mathbf{Y'CY} = \mathbf{X'B}^{-1\prime}\mathbf{B'ABB}^{-1}\mathbf{X} = \mathbf{X'AX} = g,$$

and conversely if $\mathbf{Y'CY} = g$ for some $\mathbf{Y}$, there is an $\mathbf{X}$ such that $\mathbf{X'AX} = g$. Thus the values assumed by $\mathbf{X'AX}$ for non-null $\mathbf{X}$ are the same as those of $\mathbf{Y'CY}$ for non-null $\mathbf{Y}$. We have proved more than what is necessary for proving definiteness.

(ii) *Every Q can be reduced to a form containing square terms only by a nonsingular linear transformation.*

In [(vi), **1b.2**] it was shown that there exists a matrix $\mathbf{B}$ such that $|\mathbf{B}| \neq 0$ and $\mathbf{B'AB} = \mathbf{\Delta}$ a diagonal matrix. Choose such a $\mathbf{B}$ and construct the transformation $\mathbf{Y} = \mathbf{B}^{-1}\mathbf{X}$ or $\mathbf{X} = \mathbf{BY}$. Then

$$\mathbf{X'AX} \rightarrow \mathbf{Y'B'ABY} = \mathbf{Y'\Delta Y} = d_1 y_1^2 + \cdots + d_n y_n^2, \tag{1c.1.1}$$

where $d_1, \ldots, d_n$ are the diagonal elements of $\mathbf{\Delta}$.

Let $R(\mathbf{A}) = r$. Since $R(\mathbf{A}) = R(\mathbf{\Delta})$, the number of non-zero d_i in (1c.1.1) is equal to r. Let p and q be the numbers of positive and negative d_i. For any nonsingular transformation which reduces Q to a form with only the square terms, it can be easily shown that the *values of p, q, and hence $p - q$ which is* *called the signature of the quadratic form, are invariant.* Obviously $p + q = r$ which is the same for all reductions.

Let $d_1, \ldots, d_p$ be positive without loss of generality. By a further transformation, $u_i = \sqrt{d_i}\, y_i$, $i = 1, \ldots, p$; $u_i = \sqrt{-d_i}\, y_i$, $i = p + 1, \ldots, r$ and $u_i = y_i$, $i = r + 1, \ldots, n$, (1c.1.1), *and hence the original form* $\mathbf{X}'\mathbf{A}\mathbf{X}$, *by a direct transformation form* $\mathbf{X}$ *to* $\mathbf{U}$, *can be reduced to the form*

$$u_1^2 + \cdots + u_p^2 - u_{p+1}^2 - \cdots - u_r^2, \tag{1c.1.2}$$

where p and r are the invariant numbers associated with the quadratic form $\mathbf{X}'\mathbf{A}\mathbf{X}$.

(iii) *If* $\mathbf{X}'\mathbf{A}\mathbf{X}$ *is p.d., there exists a nonsingular transformation* $\mathbf{X} = \mathbf{B}\mathbf{Y}$ *such that*

$$\mathbf{X}'\mathbf{A}\mathbf{X} \to \mathbf{Y}'\mathbf{Y} = y_1^2 + \cdots + y_n^2. \tag{1c.1.3}$$

Since definiteness is preserved it is clear that all the terms in the reduced form (1c.1.1) have positive coefficients d_i, and therefore a further transformation $\sqrt{d_i}\, y_i = u_i$ gives the form (1c.1.3). Let $\mathbf{X} = \mathbf{B}\mathbf{Y}$ be such a transformation. Then

$$\mathbf{X}'\mathbf{A}\mathbf{X} \to \mathbf{Y}'\mathbf{B}'\mathbf{A}\mathbf{B}\mathbf{Y} = \mathbf{Y}'\mathbf{Y} \Rightarrow \mathbf{B}'\mathbf{A}\mathbf{B} = \mathbf{I} \quad \text{or} \quad \mathbf{A} = \mathbf{C}\mathbf{C}' \tag{1c.1.4}$$

where $\mathbf{C}' = \mathbf{B}^{-1}$. This means that a *p.d. matrix* $\mathbf{A}$ *can be expressed as the product of a matrix and its transpose.* The reader may verify that if $\mathbf{A}$ is n.n.d. and of rank $r < n$, then there exists a nonsingular transformation $\mathbf{X} = \mathbf{B}\mathbf{Y}$ such that $\mathbf{X}'\mathbf{A}\mathbf{X} \to y_1^2 + \cdots + y_r^2$. As in (1c.1.4) a n.n.d. matrix $\mathbf{A}$ of rank r can be written as $\mathbf{C}\mathbf{C}'$ where $\mathbf{C}$ is $n \times r$ matrix of rank r.

(iv) *A necessary and sufficient condition that a real quadratic form* $\mathbf{X}'\mathbf{A}\mathbf{X}$ *is p.d. is that $g_i > 0$ for $i = 1, \ldots, n$ where*

$$g_i = \begin{vmatrix} a_{11} & \cdots & a_{1i} \\ \cdot & \cdots & \cdot \\ a_{i1} & \cdots & a_{ii} \end{vmatrix}.$$

By (iii), there is a nonsingular transformation such that

$$\mathbf{X}'\mathbf{A}\mathbf{X} \Leftrightarrow \mathbf{Y}'\mathbf{B}'\mathbf{A}\mathbf{B}\mathbf{Y} = \mathbf{Y}'\mathbf{Y},$$

that is, $\mathbf{B}'\mathbf{A}\mathbf{B} = \mathbf{I}$, $|\mathbf{B}'\mathbf{A}\mathbf{B}| = |\mathbf{B}|^2|\mathbf{A}| = |\mathbf{I}| = 1$. Hence $|\mathbf{B}|^2 g_n > 0$ or $g_n > 0$. Consider the quadratic form with the last variable x_n as zero, which is also p.d. in $(n - 1)$ variables. Hence, by the preceding argument, the determinant of its matrix $g_{n-1} > 0$, and so on, which establishes the necessity.

Since $a_{11} > 0$, the first row and column of $\mathbf{A}$ can be swept out symmetrically by pre- and postmultiplication with elementary matrices. The resulting matrix is

$$\begin{pmatrix} a_{11} & 0 & \cdots & 0 \\ 0 & b_{22} & \cdots & b_{2n} \\ \cdot & \cdot & \cdots & \cdot \\ 0 & b_{n2} & \cdots & b_{nn} \end{pmatrix},$$

where $a_{11}b_{22} = g_2$, because the value of any subdeterminant including the first row and column is unaltered. Since $g_2 > 0$, $b_{22} = g_2/g_1 > 0$. By using b_{22} as pivot, the second row and column can be swept out, yielding

$$\begin{pmatrix} a_{11} & 0 & 0 & \cdots & 0 \\ 0 & b_{22} & 0 & \cdots & 0 \\ \cdot & 0 & c_{33} & \cdots & c_{3n} \\ \cdot & \cdot & \cdot & \cdots & \cdot \\ 0 & 0 & c_{n3} & \cdots & c_{nn} \end{pmatrix},$$

where $a_{11}b_{22}c_{33} = g_3 > 0$. Hence $c_{33} = g_3/g_2 > 0$. The process can be continued until the matrix is reduced to the form

$$\begin{pmatrix} g_1 & 0 & \cdots & 0 \\ 0 & g_2/g_1 & \cdots & 0 \\ \cdot & \cdot & \cdots & \cdot \\ 0 & 0 & \cdots & g_n/g_{n-1} \end{pmatrix}, \tag{1c.1.5}$$

where each diagonal element is positive. We have demonstrated the existence of a matrix **B**, which is the product of the elementary matrices used in sweeping out the columns, such that $\mathbf{B'AB} = \mathbf{\Delta}$ [in the special form (1c.1.5)]. Hence $\mathbf{X} = \mathbf{BY}$ transforms $\mathbf{X'AX}$ to

$$g_1 {y_1}^2 + \frac{g_2}{g_1} {y_2}^2 + \cdots + \frac{g_n}{g_{n-1}} {y_n}^2, \tag{1c.1.6}$$

which is positive definite. This equation establishes sufficiency. It also follows from a consideration of the quadratic form $-\mathbf{X'AX}$ that *a necessary and sufficient condition for* $\mathbf{X'AX}$ *to be negative definite is that* $g_1 < 0$, $g_2 > 0$, $g_3 < 0, \ldots$.

We have seen in (iv) that a p.d. quadratic form Q can be reduced to a sum of squares by a nonsingular transformation. We now investigate the conditions on the transformation $\mathbf{Y} = \mathbf{CX}$, which transforms a p.d. Q as the sum of p.d. quadratic forms in *exclusive subsets* of the new variables $y_1, \ldots, y_n$. For this purpose let us write **Y** in the partitioned form $\mathbf{Y'} = (\mathbf{Y}_1' \vdots \cdots \vdots \mathbf{Y}_k')$ where $\mathbf{Y}_i$ is the column vector of the ith exclusive subset of the variables $y_1, \cdots, y_n$.

The transformation $\mathbf{Y} = \mathbf{CX}$ can then be written $\mathbf{Y}_1 = \mathbf{C}_1\mathbf{X}, \ldots, \mathbf{Y}_k = \mathbf{C}_k\mathbf{X}$, where $\mathbf{C}_1, \ldots, \mathbf{C}_k$ constitute a partition of $\mathbf{C}$, that is $\mathbf{C}' = (\mathbf{C}_1' \mid \cdots \mid \mathbf{C}_k')$. The following proposition then holds.

(v) *A necessary and sufficient condition that under a nonsingular transformation* $\mathbf{Y}_i = \mathbf{C}_i\mathbf{X}$, $i = 1, \ldots, k$,

$$\mathbf{X}'\mathbf{A}\mathbf{X} \to \mathbf{Y}_1'\mathbf{B}_1\mathbf{Y}_1 + \cdots + \mathbf{Y}_k'\mathbf{B}_k\mathbf{Y}_k \tag{1c.1.7}$$

where $\mathbf{A}$ *is p.d. is that* $\mathbf{C}_i\mathbf{A}^{-1}\mathbf{C}_j' = 0$, $i \neq j$, in which case $\mathbf{B}_i = (\mathbf{C}_i\mathbf{A}^{-1}\mathbf{C}_i')^{-1}$.

Writing $\mathbf{A}^{-1/2}\mathbf{C}_i' = \mathbf{G}_i$ in (1b.7.3) and (1b.7.4) of **1b.7** we observe that the following statements

$$\text{(a)}\quad \mathbf{C}_i\mathbf{A}^{-1}\mathbf{C}_j' = \mathbf{0} \text{ for all } i \neq j \tag{1c.1.8}$$

$$\text{(b)}\quad \mathbf{A} = \sum \mathbf{C}_i'(\mathbf{C}_i\mathbf{A}^{-1}\mathbf{C}_i')^{-1}\mathbf{C}_i \tag{1c.1.9}$$

are equivalent. Then

$$(1c.1.9) \Rightarrow \mathbf{X}'\mathbf{A}\mathbf{X} = \sum \mathbf{Y}_i'(\mathbf{C}_i\mathbf{A}^{-1}\mathbf{C}_i')^{-1}\mathbf{Y}_i$$

which proves sufficiency.

To prove necessity, $(1c.1.7) \Rightarrow \mathbf{A} = \mathbf{C}'\mathbf{B}\mathbf{C}$ where $\mathbf{B} = \text{diag.}\,(\mathbf{B}_1, \ldots, \mathbf{B}_k)$ is nonsingular. Therefore $\mathbf{B}^{-1} = \mathbf{C}\mathbf{A}^{-1}\mathbf{C}'$, from which it clearly follows that $\mathbf{C}_i\mathbf{A}^{-1}\mathbf{C}_j' = \mathbf{0}$ for $i \neq j$ and $(\mathbf{C}_i\mathbf{A}^{-1}\mathbf{C}_i')^{-1} = \mathbf{B}_i$.

1c.2 Roots of Determinantal Equations

Consider the determinantal equation $|\mathbf{A} - \lambda\mathbf{I}| = 0$, also called the characteristic equation of $\mathbf{A}$, which is of degree $\leqslant m$ in λ if $\mathbf{A}$ is $m \times m$. Corresponding to any root λ_i (called a latent or a characteristic root or an eigen value of $\mathbf{A}$) there exists a non-null column vector $\mathbf{P}_i$ (called latent, characteristic or eigen vector) such that $\mathbf{A}\mathbf{P}_i = \lambda_i\mathbf{P}_i$ and $\mathbf{P}_i^*\mathbf{P}_i = 1$. The following results are of interest when $\mathbf{A}$ is a real symmetric matrix.

(i) *If* $R(\mathbf{A}) = r$, *the equation* $|\mathbf{A} - \lambda\mathbf{I}| = 0$ *has zero as a root of multiplicity* $(m - r)$.

(ii) *All the latent roots are real and the latent vectors can be chosen to be real.*

Let $\mathbf{A}(\mathbf{X} + i\mathbf{Y}) = (\lambda + i\mu)(\mathbf{X} + i\mathbf{Y})$ where i is imaginary. Equating real and imaginary parts, premultiplying by $\mathbf{Y}'$ and $\mathbf{X}'$ respectively and subtracting we find $\mu = 0$. $\mathbf{Y}$ can be chosen to be zero.

(iii) *The latent vectors* $\mathbf{P}_i$, $\mathbf{P}_j$ *corresponding to* λ_i, $\lambda_j (\lambda_i \neq \lambda_j)$ are *orthogonal.*

From the definition

$$\mathbf{A}\mathbf{P}_i = \lambda_i\mathbf{P}_i, \qquad \mathbf{A}\mathbf{P}_j = \lambda_j\mathbf{P}_j.$$

Multiplication of the first by $\mathbf{P}_j'$ and the second by $\mathbf{P}_i'$ and subtraction gives $(\lambda_i - \lambda_j)\mathbf{P}_i'\mathbf{P}_j = 0 \Rightarrow \mathbf{P}_i'\mathbf{P}_j = 0$.

(iv) *If* $\mathbf{A}$ *is p.d., all the roots are positive; if A is n.n.d., all the roots are non-negative.*

Consider

$$\mathbf{A}\mathbf{P}_i = \lambda_i \mathbf{P}_i \Rightarrow \mathbf{P}_i'\mathbf{A}\mathbf{P}_i = \lambda_i \mathbf{P}_i'\mathbf{P}_i.$$

If $\mathbf{A}$ is p.d., then $\mathbf{P}_i'\mathbf{A}\mathbf{P}_i$ and $\mathbf{P}_i'\mathbf{P}_i$ are both positive. Hence $\lambda_i > 0$. If $\mathbf{A}$ is n.n.d., then $\mathbf{P}_i'\mathbf{A}\mathbf{P}_i$ is non-negative. Hence $\lambda_i \geqslant 0$.

(v) *If* $\mathbf{X}$ *is an arbitrary non-null vector, there exists a latent vector* $\mathbf{Y}$ *belonging to the linear manifold* $\mathscr{M}(\mathbf{X}, \mathbf{A}\mathbf{X}, \mathbf{A}^2\mathbf{X}, \ldots)$.

The vectors $\mathbf{X}, \mathbf{A}\mathbf{X}, \ldots$ cannot all be independent. Let k be the smallest value such that

$$\mathbf{A}^k\mathbf{X} + b_{k-1}\mathbf{A}^{k-1}\mathbf{X} + \cdots + b_0\mathbf{X} = \mathbf{0}. \tag{1c.2.1}$$

Factorizing (1c.2.1), we see that

$$(\mathbf{A} - \mu_1\mathbf{I})\cdots(\mathbf{A} - \mu_k\mathbf{I})\mathbf{X} = \mathbf{0} \Rightarrow (\mathbf{A} - \mu_1\mathbf{I})\mathbf{Y} = \mathbf{0}$$

where

$$\mathbf{Y} = (\mathbf{A} - \mu_2\mathbf{I})\cdots(\mathbf{A} - \mu_k\mathbf{I})\mathbf{X} \neq \mathbf{0}. \tag{1c.2.2}$$

Furthermore, $(\mathbf{A} - \mu_1\mathbf{I})\mathbf{Y} = \mathbf{0}$, that is, $\mathbf{Y}$ is a latent vector with the associated latent root μ_1. Since $\mathbf{A}$ is real, μ_1 is real using result (ii). Similarly each μ_i is real and the equation (1c.2.2) shows

$$\mathbf{Y} \in \mathscr{M}(\mathbf{A}^{k-1}\mathbf{X}, \ldots, \mathbf{X}).$$

1c.3 Canonical Reduction of Matrices

(i) REAL SYMMETRIC MATRIX. *Let* $\mathbf{A}$ *be a* $m \times m$ *real symmetric matrix. Then there exists an orthogonal matrix* $\mathbf{P}$ *such that* $\mathbf{P}'\mathbf{A}\mathbf{P} = \mathbf{\Lambda}$ *or* $\mathbf{A} = \mathbf{P}\mathbf{\Lambda}\mathbf{P}'$ *where* $\mathbf{\Lambda}$ *is a diagonal matrix.*

Actually, it will be shown that if $\lambda_1 \geqslant \cdots \geqslant \lambda_m$ denote the latent roots of $\mathbf{A}$, including the multiplicities, then the ith diagonal element of $\mathbf{\Lambda}$ is λ_i and the ith column vector of $\mathbf{P}$ is $\mathbf{P}_i$, a latent vector corresponding to λ_i.

Suppose there exist s orthonormal vectors $\mathbf{P}_1, \ldots, \mathbf{P}_s$ such that

$$\mathbf{A}\mathbf{P}_i = \lambda_i\mathbf{P}_i, \qquad i = 1, \ldots, s. \tag{1c.3.1}$$

The result (1c.3.1) $\Rightarrow \mathbf{A}^2\mathbf{P}_i = \lambda_i\mathbf{A}\mathbf{P}_i = \lambda_i^2\mathbf{P}_i, \ldots, \mathbf{A}^r\mathbf{P}_i = \lambda^r\mathbf{P}_i, \ldots$. Choose a vector $\mathbf{X} \perp^{er}$ to $\mathscr{M}(\mathbf{P}_1, \ldots, \mathbf{P}_s)$, then

$$\mathbf{X}'\mathbf{A}^r\mathbf{P}_i = \mathbf{X}'\lambda_i^r\mathbf{P}_i = \lambda_i^r\mathbf{X}'\mathbf{P}_i = 0$$

for all r and $i = 1, \ldots, s$. Hence

$$\mathcal{M}(\mathbf{X}, \mathbf{AX}, \mathbf{A}^2\mathbf{X}, \ldots) \perp^{er} \mathcal{M}(\mathbf{P}_1, \ldots, \mathbf{P}_s). \tag{1c.3.2}$$

From result [(v), **1c.2**], there exists a latent vector $\mathbf{P}_{s+1} \in \mathcal{M}(\mathbf{X}, \mathbf{AX}, \mathbf{A}^2\mathbf{X}, \ldots)$, which in view of (1c.3.2) is orthogonal to $\mathbf{P}_1, \ldots, \mathbf{P}_s$.

Since $\mathbf{P}_1$ can be chosen corresponding to any latent vector to start with, we have established the existence of m mutually orthogonal latent vectors $\mathbf{P}_1, \ldots, \mathbf{P}_m$ such that

$$\mathbf{AP}_i = \lambda_i \mathbf{P}_i, \qquad i = 1, \ldots, m, \tag{1c.3.3}$$

which may be written

$$\mathbf{AP} = \mathbf{P\Lambda}, \qquad \mathbf{PP}' = \mathbf{I} \tag{1c.3.4}$$

where $\mathbf{P}$ is the orthogonal matrix with $\mathbf{P}_i$ as its columns and $\mathbf{\Lambda}$ is the diagonal matrix with λ_i as its ith diagonal element.

The result (1c.3.4) can be written in the following useful form:

$$\begin{aligned} \mathbf{A} = \mathbf{P\Lambda P}' &= \lambda_1 \mathbf{P}_1\mathbf{P}_1' + \cdots + \lambda_m \mathbf{P}_m\mathbf{P}_m' \\ \mathbf{I} = \mathbf{PP}' &= \mathbf{P}_1\mathbf{P}_1' + \cdots + \mathbf{P}_m\mathbf{P}_m', \end{aligned} \tag{1c.3.5}$$

which is known as the spectral decomposition of $\mathbf{A}$. The result (1c.3.4) also implies that *there exists an orthogonal transformation* $\mathbf{Y} = \mathbf{P}'\mathbf{X}$ *which transforms the quadratic form* $\mathbf{X}'\mathbf{AX}$ *into*

$$\mathbf{Y}'\mathbf{\Lambda Y} = \lambda_1 y_1^2 + \cdots + \lambda_m y_m^2. \tag{1c.3.6}$$

It has been established in (1c.1.1) that a quadratic form can always be reduced to a simpler form involving square terms only by a transformation not necessarily orthogonal.

Since $\mathbf{P}'\mathbf{AP} = \mathbf{\Lambda}$, $\mathbf{P}'(\mathbf{A} - \lambda\mathbf{I})\mathbf{P} = \mathbf{\Lambda} - \lambda\mathbf{I}$ which implies that

$$|\mathbf{A} - \lambda\mathbf{I}| = |\mathbf{\Lambda} - \lambda\mathbf{I}| = (\lambda - \lambda_1)\cdots(\lambda - \lambda_m), \tag{1c.3.7}$$

that is, the constants $\lambda_1, \ldots, \lambda_m$ are the same as the roots of $|\mathbf{A} - \lambda\mathbf{I}| = 0$ including the multiplicities, so that we have identified the matrix $\mathbf{\Lambda}$.

There must be r orthogonal vectors satisfying the equation $\mathbf{AX} = \lambda_i \mathbf{X}$ if λ_i is a root of multiplicity r. If there is one more vector, then by [(iii), **1c.2**], it is orthogonal to all $\mathbf{P}_i$ and hence must be null. Thus the rank of $(\mathbf{A} - \lambda_i\mathbf{I})$ is $m - r$. The r latent vectors corresponding to the latent root λ_i of multiplicity r generate a subspace called the latent subspace. The latent subspaces corresponding to two different roots are orthogonal so that $\mathbf{A}$ decomposes the entire vector space E_m as a direct sum of orthogonal subspaces

$$E_m = E_{r_1} \oplus \cdots \oplus E_{r_k}$$

where E_{r_i} is the eigen subspace corresponding to the eigen value of multiplicity r_i. Note that $r_i + \cdots + r_k = m$.

(ii) PAIR OF REAL SYMMETRIC MATRICES. *Let* **A** *and* **B** *be real* $m \times m$ *symmetric matrices of which* **B** *is p.d. Then there exists a matrix* **R** *such that* $\mathbf{A} = \mathbf{R}^{-1\prime}\mathbf{\Lambda}\mathbf{R}^{-1}$ *and* $\mathbf{B} = \mathbf{R}^{-1\prime}\mathbf{R}^{-1}$, *where* $\mathbf{\Lambda}$ *is a diagonal matrix.*

Let $\lambda_1 \geqslant \cdots \geqslant \lambda_m$ be the roots of $|\mathbf{A} - \lambda\mathbf{B}| = 0$. Then it is shown that the ith diagonal element of $\mathbf{\Lambda}$ is λ_i and the ith column vector $\mathbf{R}_i$ of **R** satisfies the equation $\mathbf{AR}_i = \lambda_i\mathbf{BR}_i$.

Since **B** is p.d., there exists a nonsingular matrix **T** such that $\mathbf{B} = \mathbf{T}'\mathbf{T}$. Consider

$$|\mathbf{A} - \lambda\mathbf{B}| = 0 = |\mathbf{A} - \lambda\mathbf{T}'\mathbf{T}| \Leftrightarrow |\mathbf{T}^{-1\prime}\mathbf{A}\mathbf{T}^{-1} - \lambda\mathbf{I}| = 0.$$

If $\mathbf{C} = \mathbf{T}^{-1\prime}\mathbf{AT}$, it is symmetric. Hence using (i), there exists a matrix **P** such that

$$\begin{gathered} \mathbf{P}'\mathbf{CP} = \mathbf{\Lambda}, \ \mathbf{P}'\mathbf{P} = \mathbf{I} \\ \mathbf{P}'\mathbf{T}'^{-1}\mathbf{AT}^{-1}\mathbf{P} = \mathbf{\Lambda}, \ \mathbf{P}'\mathbf{P} = \mathbf{I}. \end{gathered} \tag{1c.3.8}$$

Choosing $\mathbf{R} = \mathbf{T}^{-1}\mathbf{P}$, we can write (1c.3.8) as

$$\mathbf{R}'\mathbf{AR} = \mathbf{\Lambda}, \qquad \mathbf{R}'\mathbf{BR} = \mathbf{I}, \tag{1c.3.9}$$

which is the desired result.

An alternative form of (1c.3.9) is

$$\begin{aligned} \mathbf{A} &= \lambda_1\mathbf{S}_1\mathbf{S}_1' + \cdots + \lambda_m\mathbf{S}_m\mathbf{S}_m' \\ \mathbf{B} &= \quad \mathbf{S}_1\mathbf{S}_1' + \cdots + \quad \mathbf{S}_m\mathbf{S}_m' \end{aligned} \tag{1c.3.10}$$

if we write the ith column of $\mathbf{R}^{-1\prime}$ as $\mathbf{S}_i$, analogous to (1c.3.5).

The result (1c.3.10) also implies that there exists a transformation $\mathbf{Y} = \mathbf{R}^{-1}\mathbf{X}$ such that the two quadratic forms $\mathbf{X}'\mathbf{AX}$ and $\mathbf{X}'\mathbf{BX}$ are simultaneously reduced to simpler forms containing square terms only:

$$\begin{aligned} \mathbf{X}'\mathbf{AX} &\rightarrow \lambda_1 y_1{}^2 + \cdots + \lambda_m y_m{}^2 \\ \mathbf{X}'\mathbf{BX} &\rightarrow \quad y_1{}^2 + \cdots + \quad y_m{}^2. \end{aligned} \tag{1c.3.12}$$

Commuting Matrices. It is shown in (i) that a real quadratic form can be reduced to a simpler form by an orthogonal transformation. A natural question that arises is whether or not two quadratic forms can be reduced to simpler forms by a single *orthogonal transformation*. The answer is true for commuting matrices **A** and **B**, i.e., $\mathbf{AB} = \mathbf{BA}$.

(iii) COMMUTING MATRICES. *Let* **A** *and* **B** *be symmetric matrices. A n.s. condition for simultaneous diagonalization of* **A** *and* **B** *by pre and post multiplication by* $\mathbf{C}'$ *and* **C** *where* **C** *is orthogonal is that* $\mathbf{AB} = \mathbf{BA}$(i.e., **A** *and* **B** *commute*).

Let $\mathbf{Q}_1$ be an eigenvector of **B** and $\mathbf{AB} = \mathbf{BA}$. Then $\mathbf{BQ}_1 = \lambda\mathbf{Q}_1$ for some λ. Multiplying by **A**,

$$\mathbf{ABQ}_1 = \lambda\mathbf{AQ}_1 = \mathbf{BAQ}_1$$

so that $\mathbf{AQ}_1$, and similarly $\mathbf{A}^2\mathbf{Q}_1$, $\mathbf{A}^3\mathbf{Q}_1, \ldots$ belong to the eigen space of $\mathbf{B}$ corresponding to the same eigenvalue λ. But by [(v), **1c.2**], $(\mathbf{Q}_1, \mathbf{AQ}_1, \ldots)$ contain an eigenvector of $\mathbf{A}$. Hence there is a common eigenvector of $\mathbf{A}$ and $\mathbf{B}$ which we designate by $\mathbf{P}_1$. Now choose $\mathbf{Q}_2 \perp^{er}$ to $\mathbf{P}_1$ as an eigenvector $\mathbf{B}$. Applying the same argument we find a common eigenvector $\mathbf{P}_2 \perp^{er} \mathbf{P}_1$ and so on. The matrix with $\mathbf{P}_1, \mathbf{P}_2, \ldots$ as columns is precisely the desired matrix $\mathbf{C}$, which proves sufficiency.

If $\mathbf{C'AC} = \mathbf{M}_1$ and $\mathbf{C'BC} = \mathbf{M}_2$ where $\mathbf{M}_1$ and $\mathbf{M}_2$ are diagonal then $\mathbf{AB} = (\mathbf{CM}_1\mathbf{C}')(\mathbf{CM}_2\mathbf{C}') = \mathbf{CM}_1\mathbf{M}_2\mathbf{C}' = \mathbf{CM}_2\mathbf{M}_1\mathbf{C}' = \mathbf{CM}_2\mathbf{C}' \ \mathbf{CM}_1\mathbf{C}' = \mathbf{BA}$, which proves necessity.

Hermitian Matrix. Matrix $\mathbf{A}$ is said to be a Hermitian matrix if $\mathbf{A} = \mathbf{A}^*$ where $\mathbf{A}^*$ is the conjugate transpose of $\mathbf{A}$. Thus a Hermitian matrix is the counterpart of a symmetric matrix when $\mathbf{A}$ has complex elements. The following results for Hermitian matrixes are proved in exactly the same way as (i) and (ii).

(iv) *Let* $\mathbf{A}$ *be Hermitian of order* m. *Then:*

(a) *The eigenvalues of* $\mathbf{A}$ *are real.*
(b) *There exists a unitary matrix* $\mathbf{U}$ (i.e., $\mathbf{U}^*\mathbf{U} = \mathbf{I}$) *such that* $\mathbf{A} = \mathbf{U\Lambda U}^*$ *where* $\mathbf{\Lambda}$ *is the diagonal matrix of the eigenvalues of* $\mathbf{A}$.
(c) *Let* $\mathbf{B}$ *be Hermitian matrix and p.d.* (i.e., $\mathbf{X}^*\mathbf{BX} > 0$ *for all* $\mathbf{X}$). *Then there exists a nonsingular matrix* $\mathbf{P}$ *such that* $\mathbf{P}^*\mathbf{BP} = \mathbf{I}$ *and* $\mathbf{P}^*\mathbf{AP} = \mathbf{\Lambda}$ *where* $\mathbf{\Lambda}$ *is diagonal with elements as the eigenvalues of the matrix* $\mathbf{B}^{-1}\mathbf{A}$.

(v) SINGULAR VALUE DECOMPOSITION. *Let* $\mathbf{A}$ *be* $m \times n$ *matrix of rank* r. *Then* $\mathbf{A}$ *can be expressed as*

$$\mathbf{A} = \mathbf{U\Lambda V}^* \qquad (1c.3.11)$$

where $\mathbf{\Lambda}$ *is* $r \times r$ *diagonal matrix with positive diagonal elements,* $\mathbf{U}$ *is* $m \times r$ *matrix such that* $\mathbf{U}^*\mathbf{U} = \mathbf{I}$ *and* $\mathbf{V}$ *is* $n \times r$ *matrix such that* $\mathbf{V}^*\mathbf{V} = \mathbf{I}$.

Let $\mathbf{U}_i$, $i = 1, \ldots, r$, be orthonormal eigenvectors corresponding to the non-zero eigenvalues λ_i^2, $i = 1, \ldots, r$, of $\mathbf{AA}^*$. Further let $\mathbf{V}_i = \lambda_i^{-1}\mathbf{A}^*\mathbf{U}_i$. Then $\mathbf{V}_i$, $i = 1, \ldots, r$, are orthonormal.

Let $\mathbf{U}_{r+1}, \ldots, \mathbf{U}_m$ be such that $\mathbf{U}_1, \ldots, \mathbf{U}_m$ is a complete set of orthonormal vectors, i.e., $\mathbf{U}_1\mathbf{U}_1^* + \cdots + \mathbf{U}_m\mathbf{U}_m^* = \mathbf{I}$. Then

$$\begin{aligned}\mathbf{A} &= (\mathbf{U}_1\mathbf{U}_1^* + \cdots + \mathbf{U}_m\mathbf{U}_m^*)\mathbf{A} \\ &= (\mathbf{U}_1\mathbf{U}_1^* + \cdots + \mathbf{U}_r\mathbf{U}_r^*)\,\mathbf{A}, \text{ since } \mathbf{U}_i^*\mathbf{A} = \mathbf{0} \quad \text{for} \quad i > r, \\ &= \lambda_1\mathbf{U}_1\mathbf{V}_1^* + \cdots + \lambda_r\mathbf{U}_r\mathbf{V}_r^* = \mathbf{U}\,\mathbf{\Lambda}\,\mathbf{V}^*\end{aligned}$$

taking $\mathbf{U} = (\mathbf{U}_1 \vdots \cdots \vdots \mathbf{U}_r)$ and $\mathbf{V} = (\mathbf{V}_1 \vdots \cdots \vdots \mathbf{V}_r)$. The diagonal values of $\mathbf{\Lambda}$ are called the singular values of $\mathbf{A}$. It may be seen that $\mathbf{V}_i$, $i = 1, \ldots, r$, are orthonormal eigenvectors of $\mathbf{A}^*\mathbf{A}$ corresponding to the eigenvalues λ_i^2, $i = 1, \ldots, r$. Note the alternative ways of representing $\mathbf{A}$:

$$\begin{aligned} \mathbf{A} &= \mathbf{U\Lambda V}^* \\ &= \lambda_1 \mathbf{U}_1 \mathbf{V}_1^* + \cdots + \lambda_r \mathbf{U}_r \mathbf{V}_r^* \\ &= \mathbf{P}\begin{pmatrix} \mathbf{\Lambda} & \mathbf{0} \\ \mathbf{0} & \mathbf{0} \end{pmatrix} \mathbf{Q}^*, \ \mathbf{P}, \text{ and } \mathbf{Q} \text{ are unitary.} \end{aligned} \tag{1c.3.12}$$

(vi) Decomposition of a Normal Matrix. *Let* $\mathbf{A}$ *be an* $n \times n$ *normal matrix*, i.e., $\mathbf{AA}^* = \mathbf{A}^*\mathbf{A}$. *Then there exists a unitary matrix* $\mathbf{U}$ *such that*

$$\mathbf{A} = \mathbf{U}^*\mathbf{\Lambda U}$$

where $\mathbf{\Lambda}$ *is a diagonal matrix.*

The result follows from the singular value decomposition of $\mathbf{A}$, observing that $\mathbf{U}$ and $\mathbf{V}$ can be chosen to be the same in (1c.3.11) when $\mathbf{AA}^* = \mathbf{A}^*\mathbf{A}$.

(vii) Nonsymmetric Matrix. The determinantal equation $|\mathbf{A} - \lambda\mathbf{I}| = 0$ has m roots, some of which may be complex even if $\mathbf{A}$ is real. Corresponding to a root λ_i, there are two vectors

$$\mathbf{AP}_i = \lambda_i \mathbf{P}_i, \qquad \mathbf{A}'\mathbf{Q}_i = \lambda_i \mathbf{Q}_i, \tag{1c.3.13}$$

called the right and left latent vectors. Let us assume that all the roots are distinct and let

$$\mathbf{P} = (\mathbf{P}_1 \vdots \cdots \vdots \mathbf{P}_m) \quad \text{and} \quad \mathbf{Q} = (\mathbf{Q}_1 \vdots \cdots \vdots \mathbf{Q}_m),$$

which leads to $\mathbf{AP} = \mathbf{P\Lambda}$ and $\mathbf{A}'\mathbf{Q} = \mathbf{Q\Lambda}$.

(a) $\mathbf{P}_i$ *are all independent and so also* $\mathbf{Q}_i$. Let there exist constants such that $c_1\mathbf{P}_1 + \cdots + c_m\mathbf{P}_m = \mathbf{0}$. Using (1c.3.13), $c_1\lambda_1\mathbf{P}_1 + \cdots + c_m\lambda_m\mathbf{P}_m = \mathbf{0}$. By eliminating one of the terms, say the mth from the two linear relations connecting $\mathbf{P}_i$, we obtain $d_1\mathbf{P}_1 + \cdots + d_{m-1}\mathbf{P}_{m-1} = \mathbf{0} \Rightarrow d_1\lambda_1\mathbf{P}_1 + \cdots + d_{m-1}\lambda_{m-1}\mathbf{P}_{m-1} = \mathbf{0}$. We can eliminate one more term and eventually show that one of the $\mathbf{P}_i$ is null, which is a contradiction. Hence $\mathbf{P}$ is nonsingular and so also is $\mathbf{Q}$, which leads to the representations

$$\mathbf{A} = \mathbf{P\Lambda P}^{-1}, \qquad \mathbf{A}' = \mathbf{Q\Lambda Q}^{-1}.$$

(b) $\mathbf{P}_i'\mathbf{Q}_j = 0$, $i \neq j$ *and* $\mathbf{P}_i'\mathbf{Q}_i \neq 0$.

$$\mathbf{AP}_i = \lambda_i \mathbf{P}_i \Rightarrow \mathbf{Q}_j'\mathbf{AP}_i = \lambda_i \mathbf{Q}_j'\mathbf{P}_i = \lambda_i \mathbf{P}_i'\mathbf{Q}_j$$

$$\mathbf{A}'\mathbf{Q}_j = \lambda_j \mathbf{Q}_j \Rightarrow \mathbf{Q}_j'\mathbf{AP}_i = \lambda_j \mathbf{Q}_j'\mathbf{P}_i = \lambda_j \mathbf{P}_i'\mathbf{Q}_j$$

and since $\lambda_i \neq \lambda_j$, $\mathbf{P}_i'\mathbf{Q}_j = 0$. Furthermore, if $\mathbf{P}_i'\mathbf{Q}_j = 0$ for all j, then $\mathbf{P}_i = \mathbf{0}$ which is not true, and hence $\mathbf{Q}_i'\mathbf{P}_i \neq 0$.

(c) *Let* $\mathbf{P}_i'\mathbf{Q}_i = d_i$, $\mathbf{D}$ *be the diagonal matrix containing* d_i, *and* $\mathbf{\Delta}$ *be the diagonal matrix containing* $\delta_i = \lambda_i/d_i$. *Then* $\mathbf{A}$ *has the spectral decomposition*

$$\mathbf{A} = \delta_1 \mathbf{P}_1 \mathbf{Q}_1' + \cdots + \delta_m \mathbf{P}_m \mathbf{Q}_m'. \tag{1c.3.14}$$

Since $\mathbf{Q}'\mathbf{P} = \mathbf{D}$, $\mathbf{P}^{-1}(\mathbf{Q}')^{-1} = \mathbf{D}^{-1}$ or $\mathbf{P}^{-1} = \mathbf{D}^{-1}\mathbf{Q}'$. Substituting for $\mathbf{P}^{-1}$ in $\mathbf{A} = \mathbf{P\Lambda P}^{-1}$, we see that

$$\mathbf{A} = \mathbf{P\Lambda D}^{-1}\mathbf{Q}' = \mathbf{P\Delta Q}' = \Sigma\, \delta_i \mathbf{P}_i \mathbf{Q}_i',$$

analogous to (1c.3.5). The decomposition of a nonsymmetric $\mathbf{A}$ may not be as simple when the latent roots are not distinct. For a study of this problem the reader is referred to books by Aitken (1944), Gantmacher (1959), and Turnbull and Aitken (1932).

One of the applications of the results (1c.3.5, 1c.3.14) is in obtaining simple expressions for powers of $\mathbf{A}$. Considering (1c.3.14) and multiplying by $\mathbf{A}$, we have

$$\mathbf{A}^2 = \Sigma \frac{\lambda_i}{d_i} \mathbf{P}_i \mathbf{Q}_i' \mathbf{A} = \Sigma \frac{\lambda_i^2}{d_i} \mathbf{P}_i \mathbf{Q}_i'. \tag{1c.3.15}$$

Similarly,

$$\mathbf{A}^n = \Sigma \frac{\lambda_i^n}{d_i} \mathbf{P}_i \mathbf{Q}_i'.$$

For a symmetric matrix, using (1c.3.5), we see that

$$\mathbf{A}^n = \Sigma\, \lambda_i^n \mathbf{P}_i \mathbf{P}_i'.$$

An application of (1c.3.15) to genetics is given in a book by Fisher (1949).

(viii) CAYLEY-HAMILTON THEOREM. *Every square matrix* $\mathbf{A}$ *satisfies its own characteristic equation.*

Consider the identity

$$(\mathbf{A} - \lambda \mathbf{I}) \operatorname{adj}(\mathbf{A} - \lambda \mathbf{I}) = |\mathbf{A} - \lambda \mathbf{I}|\mathbf{I} \tag{1c.3.16}$$

(adj $\mathbf{B}$ is the matrix of cofactors of $\mathbf{B}$). By definition

$$\operatorname{adj}(\mathbf{A} - \lambda \mathbf{I}) = \mathbf{C}_0 + \lambda_1 \mathbf{C}_1 + \cdots + \lambda^{n-1}\mathbf{C}_{n-1}$$

where $\mathbf{C}_i$ are suitably chosen matrices and

$$|\mathbf{A} - \lambda \mathbf{I}| = a_0 + a_1\lambda + \cdots + a_n\lambda^n$$

the characteristic polynomial. Comparing the coefficients of λ in (1c.3.16)

$$\mathbf{AC}_0 = a_0\mathbf{I},\ \mathbf{AC}_1 - \mathbf{C}_0 = a_1\mathbf{I}, \ldots, -\mathbf{C}_{n-1} = a_n\mathbf{I}.$$

Multiplying the first equation by $\mathbf{I}$, the second equation by $\mathbf{A}$, the third by $\mathbf{A}^2, \ldots$ and adding

$$a_0\mathbf{I} + a_1\mathbf{A} + \cdots + a_n\mathbf{A}^n = \mathbf{0} \tag{1c.3.17}$$

which proves the desired result.

Note that if $\mathbf{A}^{-1}$ exists, then multiplying (1c.3.17) by $\mathbf{A}^{-1}$ we have

$$-a_0\mathbf{A}^{-1} = a_1\mathbf{I} + a_2\mathbf{A} + \cdots + a_n\mathbf{A}^{n-1} \tag{1c.3.18}$$

which gives an expression for $\mathbf{A}^{-1}$ as a polynomial in $\mathbf{A}$.

(ix) *Let* $\mathbf{A}$ *be a square matrix of order m. Then there exists a unitary matrix* $\mathbf{P}$ *such that* $\mathbf{P}^*\mathbf{A}\mathbf{P}$ *is triangular.*

The result may be proved by induction noting that it is true for a one-by-one matrix.

(x) JORDAN DECOMPOSITION. *Let* $\mathbf{A}$ *be a square matrix of order m. Then there exists a nonsingular matrix* $\mathbf{T}$ *such that*

$$\mathbf{T}^{-1}\mathbf{A}\mathbf{T} = \text{diag}(\Lambda_1, \Lambda_2, \ldots, \Lambda_k)$$

where

$$\Lambda_i = \begin{pmatrix} \lambda_i & 1 & 0 & \cdots & 0 \\ 0 & \lambda_i & 1 & \cdots & 0 \\ \cdot & \cdot & & \cdots & \cdot \\ 0 & 0 & & \cdots & \lambda_i \end{pmatrix}$$

and λ_i *is an eigenvalue of* $\mathbf{A}$.

For a proof see Gantmacher (1959).

NON-NEGATIVE MATRICES (SPECTRAL THEORY). An $m \times n$ matrix $\mathbf{A} = (a_{ij})$ is called non-negative if $a_{ij} \geqslant 0$ and positive if $a_{ij} > 0$ for all i and j.

A non-negative matrix is indicated by $\mathbf{A} \geqslant 0$ and a positive matrix by $\mathbf{A} > 0$. We use the notation $\mathbf{C} \geqslant \mathbf{D}$ if $\mathbf{C} - \mathbf{D}$ is non-negative and $\mathbf{C} > \mathbf{D}$ if $\mathbf{C} - \mathbf{D}$ is positive.

A matrix $\mathbf{A}$ is said to be reducible if by the same permutation of columns and rows it can be written as

$$\begin{pmatrix} \mathbf{B} & \mathbf{0} \\ \mathbf{C} & \mathbf{D} \end{pmatrix}$$

where $\mathbf{0}$ is a null matrix and $\mathbf{B}$, $\mathbf{D}$ are square matrices. Otherwise $\mathbf{A}$ is irreducible.

(xi) PERRON'S THEOREM. *A positive square matrix* $\mathbf{A}$ *has a positive characteristic value* $\lambda_0(\mathbf{A})$ *which is a simple root of the characteristic equation and exceeds the moduli of all the other characteristic values. To this "maximal" root* $\lambda_0(\mathbf{A})$ *there corresponds a positive characteristic vector (i.e., all coordinates positive).*

(xii) FROBENIUS THEOREM. *An irreducible matrix* $\mathbf{A} \geqslant 0$ *always has a positive characteristic value* $\lambda_0(\mathbf{A})$ *which is a simple root of the characteristic equation and not smaller than the moduli of other characteristic values. To the "maximal" root* $\lambda_0(\mathbf{A})$ *there corresponds a positive characteristic vector.*

(xiii) *An irreducible matrix* $\mathbf{A} \geqslant 0$ *cannot have two linearly independent non-negative characteristic vectors.*

(xiv) *A non-negative matrix* $\mathbf{A}$ *always has a non-negative characteristic value* $\lambda_0(\mathbf{A})$ *such that the moduli of all the characteristic values of* $\mathbf{A}$ *do not exceed* $\lambda_0(\mathbf{A})$. *To this "maximal" root there corresponds a non-negative characteristic vector.*

(xv) *If* $\mathbf{A} \geqslant \mathbf{B} \geqslant \mathbf{0}$, *then* $\lambda_0(\mathbf{A}) \geqslant \lambda_0(\mathbf{B})$.

(xvi) *If* $\mathbf{A} \geqslant 0$, $\mathbf{X}_0 > \mathbf{0}$ *and* $\mu_0 = \inf\{\mu : \mathbf{Af} \leqslant \mu\mathbf{f}, \mathbf{f} > \mathbf{0}\}$, *then*

$$\sup_{\substack{\mathbf{X}>\mathbf{0} \\ \Sigma x_i = 1}} \inf_{\substack{\mathbf{f}>\mathbf{0} \\ \Sigma f_i = 1}} \frac{\mathbf{f}'\mathbf{AX}}{\mathbf{f}'\mathbf{X}} = \inf_{\substack{\mathbf{f}>\mathbf{0} \\ \Sigma f_i = 1}} \sup_{\substack{\mathbf{X}>\mathbf{0} \\ \Sigma x_i = 1}} \frac{\mathbf{f}'\mathbf{AX}}{\mathbf{f}'\mathbf{X}}.$$

METZLER MATRIX. A matrix $\mathbf{C} = (c_{ij})$ is called a Metzler matrix if $c_{ij} \geqslant 0$ $(i \neq j)$. It may be seen that by adding $\beta\mathbf{I}$, choosing β large enough, we can achieve

$$\mathbf{A} = \mathbf{C} + \beta\mathbf{I} \geqslant \mathbf{0}$$

so that the study of Metzler or M-matrices is equivalent to that of non-negative matrices. Thus we have the following:

(xvii) (a) *The eigenvalue of* $\mathbf{M}$ *an* M*-matrix with the largest real part is real and has an associated non-negative eigenvector.*

(b) *Let* $\sigma(\mathbf{M})$ *be the maximum of real parts of the eigenvalues of* $\mathbf{M}$. *Then* $\sigma(\mathbf{M}) < 0$ *iff there exists* $\mathbf{X} > \mathbf{0}$ *such that* $\mathbf{MX} < \mathbf{0}$.

The proofs of the propositions can be found in Gantmacher (1959).

1c.4 Projection Operator

Definition. Let $\mathscr{S}_1$ and $\mathscr{S}_2$ be two disjoint subspaces such that $\mathscr{V} = \mathscr{S}_1 \oplus \mathscr{S}_2$, so that any vector $\mathbf{X} \in \mathscr{V}$ has the unique decomposition $\mathbf{X} = \mathbf{X}_1 + \mathbf{X}_2$, $\mathbf{X}_1 \in \mathscr{S}_1$ and $\mathbf{X}_2 \in \mathscr{S}_2$. The mapping $\mathbf{P}, \mathbf{X} \to \mathbf{X}_1$, $(\mathbf{PX} = \mathbf{X}_1)$ is called the projection of $\mathbf{X}$ on $\mathscr{S}_1$ along $\mathscr{S}_2$ and $\mathbf{P}$ is called the projection operator.

We have the following propositions regarding the projection operator $\mathbf{P}$.

(i) $\mathbf{P}$ *is a linear transformation and is unique.*

Obviously $\mathbf{P}(a\mathbf{X} + b\mathbf{Y}) = a\mathbf{PX} + b\mathbf{PY}$, which means that $\mathbf{P}$ can be represented by a matrix which is denoted by the same symbol $\mathbf{P}$. $\mathbf{P}$ is necessarily unique.

(ii) $\mathbf{P}$ *is an idempotent matrix, i.e.,* $\mathbf{P}^2 = \mathbf{P}$.

For an arbitrary vector $\mathbf{X}$

$$\mathbf{P}^2\mathbf{X} = \mathbf{P}(\mathbf{PX}) = \mathbf{PX}_1 = \mathbf{X}_1 = \mathbf{PX}$$

which implies $\mathbf{P}^2 = \mathbf{P}$.

(iii) $\mathbf{I} - \mathbf{P}$ *is a projector on* $\mathscr{S}_2$ *along* $\mathscr{S}_1$.

(iv) *Let* $\mathscr{V} = \mathscr{S}_1 \oplus \mathscr{S}_2 + \cdots \oplus \mathscr{S}_k$. *Then there exist matrices* $\mathbf{P}_1, \ldots, \mathbf{P}_k$ *such that* (a) $\mathbf{P}_i^2 = \mathbf{P}_i$, (b) $\mathbf{P}_i\mathbf{P}_j = \mathbf{0}$, $i \neq j$ *and* (c) $\mathbf{I} = \mathbf{P}_1 + \cdots + \mathbf{P}_k$.

The results follow from the unique decomposition

$$\mathbf{X} = \mathbf{X}_1 + \cdots + \mathbf{X}_k, \qquad \mathbf{X}_i \in \mathscr{S}_i, \quad i = 1, \ldots, k.$$

Definition. Let $\mathscr{S}$ be a subspace of $\mathscr{V}$ and $\mathscr{S}^\perp$ the orthogonal complement of $\mathscr{S}$. Then $\mathscr{V} = \mathscr{S} \oplus \mathscr{S}^\perp$. The operator $\mathbf{P}$ which projects on $\mathscr{S}$ along $\mathscr{S}^\perp$ is called an orthogonal projector.

We have the following propositions regarding orthogonal projectors.

(v) *Let the inner product in* $\mathscr{V}$ *be defined by* $(\mathbf{X}, \mathbf{Y}) = \mathbf{Y}^*\mathbf{\Sigma}\mathbf{X}$ *where* $\mathbf{\Sigma}$ *is a p.d. matrix. Then* $\mathbf{P}$ *is an orthogonal projector iff*

$$\text{(a) } \mathbf{P}^2 = \mathbf{P} \quad \textit{and} \quad \text{(b) } \mathbf{\Sigma P} \textit{ is hermitian.} \tag{1c.4.1}$$

The result (a) is true in view of proposition (ii). Further $\mathbf{PX} \in \mathscr{S}$ and $(\mathbf{I} - \mathbf{P})\mathbf{Y} \in \mathscr{S}^\perp$ for any $\mathbf{X}$ and $\mathbf{Y}$, i.e., $\mathbf{X}^*\mathbf{P}^*\mathbf{\Sigma}(\mathbf{I} - \mathbf{P})\mathbf{Y} = \mathbf{0} \Leftrightarrow \mathbf{P}^*\mathbf{\Sigma}(\mathbf{I} - \mathbf{P}) = \mathbf{0} \Leftrightarrow \mathbf{P}^*\mathbf{\Sigma P} = \mathbf{\Sigma P}$ which establishes (b). It is easily seen that if (1c.4.1) holds, then $\mathbf{P}$ projects vectors onto $\mathscr{M}(\mathbf{P})$ along the orthogonal space $\mathscr{M}(\mathbf{I} - \mathbf{P})$.

Note. If $\mathbf{\Sigma} = \mathbf{I}$, then $\mathbf{P}$ is an orthogonal projector iff $\mathbf{P}$ is idempotent and hermitian.

(vi) Explicit Expression for an Orthogonal Projector. *Consider the subspace* $\mathscr{M}(\mathbf{A})$, *i.e., spanned by the columns of a matrix* $\mathbf{A}$ *and the inner product be as in proposition* (v). *Then the orthogonal projector onto* $\mathscr{M}(\mathbf{A})$ *is*

$$\mathbf{P} = \mathbf{A}(\mathbf{A}^*\mathbf{\Sigma A})^-\mathbf{A}^*\mathbf{\Sigma} \tag{1c.4.2}$$

which is unique for any choice of the g-inverse involved in (1c.4.2).

Let $\mathbf{X} \in \mathscr{V}$ and write $\mathbf{X} = \mathbf{A}\boldsymbol{\alpha} + \mathbf{A}^{\perp}\boldsymbol{\beta}$. Multiplying both sides by $\mathbf{A}^*\boldsymbol{\Sigma}$

$$\mathbf{A}^*\boldsymbol{\Sigma}\mathbf{A}\boldsymbol{\alpha} = \mathbf{A}^*\boldsymbol{\Sigma}\mathbf{X}.$$

Choosing solution $\boldsymbol{\alpha} = (\mathbf{A}^*\boldsymbol{\Sigma}\mathbf{A})^-\mathbf{A}^*\boldsymbol{\Sigma}\mathbf{X}$, we have $\mathbf{A}\boldsymbol{\alpha} = \mathbf{A}(\mathbf{A}^*\boldsymbol{\Sigma}\mathbf{A})^-\mathbf{A}^*\boldsymbol{\Sigma}\mathbf{X} \Rightarrow$ $\mathbf{P} = \mathbf{A}(\mathbf{A}^*\boldsymbol{\Sigma}\mathbf{A})^-\mathbf{A}^*\boldsymbol{\Sigma}$ is the required operator. If $\boldsymbol{\Sigma} = \mathbf{I}$,

$$\mathbf{P} = \mathbf{A}(\mathbf{A}^*\mathbf{A})^-\mathbf{A}^*. \tag{1c.4.3}$$

It is already shown in [(vii), **1b.5**] that $\mathbf{A}(\mathbf{A}^*\mathbf{A})^-\mathbf{A}^*$ is symmetric, hermitian and invariant for any choice of g-inverse.

(v) *Let* $\mathbf{P}_1$ *and* $\mathbf{P}_2$ *be projectors. Then* $\mathbf{P}_1 + \mathbf{P}_2$ *is a projector iff* $\mathbf{P}_1\mathbf{P}_2 = \mathbf{P}_2\mathbf{P}_1 = \mathbf{0}$.

1c.5 Further Results on g-Inverse

A general definition of g-inverse of a matrix $\mathbf{A}$, with elements from any field, is given in **1b.5**. Such an inverse denoted by $\mathbf{A}^-$ is shown to exist but is not unique. In this section we define subclasses of g-inverses which are useful in solving some optimization problems. For this purpose we consider matrices $\mathbf{A}$ with complex elements.

(i) MINIMUM NORM g-INVERSE. *Let* $\mathbf{AX} = \mathbf{Y}$ *be a consistent equation and* $\mathbf{G}$ *be a g-inverse of* $\mathbf{A}$ *such that* $\mathbf{GY}$ *is a solution of* $\mathbf{AX} = \mathbf{Y}$ *with a minimum norm. If the norm of a vector* $\boldsymbol{\alpha}$ *is defined by* $\|\boldsymbol{\alpha}\| = (\boldsymbol{\alpha}^*\mathbf{N}\boldsymbol{\alpha})^{1/2}$ *where* $\mathbf{N}$ *is a p.d. matrix, then* $\mathbf{G}$ *satisfies the conditions*

$$\mathbf{AGA} = \mathbf{A}, \qquad (\mathbf{GA})^*\mathbf{N} = \mathbf{NGA}. \tag{1c.5.1}$$

Such a $\mathbf{G}$ *is denoted by* $\mathbf{A}^-_{m(\mathbf{N})}$ *and by* $\mathbf{A}^-_m$ *when* $\mathbf{N} = \mathbf{I}$.

Using [(ii), **1b.5**], a general solution of $\mathbf{AX} = \mathbf{Y}$ is $\mathbf{GY} + (\mathbf{I} - \mathbf{GA})\mathbf{Z}$ where $\mathbf{Z}$ is an arbitrary vector. If $\mathbf{GY}$ has minimum norm, then

$$\begin{aligned}
\|\mathbf{GY}\| \leqslant \|\mathbf{GY} + (\mathbf{I} - \mathbf{GA})\mathbf{Z}\| &\text{ for all } \mathbf{Y} \in \mathscr{M}(\mathbf{A}), \text{ arbitrary } \mathbf{Z}\\
&\Leftrightarrow (\mathbf{GY}, (\mathbf{I} - \mathbf{GA})\mathbf{Z}) = 0 \text{ for } \mathbf{Y} \in \mathscr{M}(\mathbf{A}), \mathbf{Z}\\
&\Leftrightarrow (\mathbf{GA})^*\mathbf{N}(\mathbf{I} - \mathbf{GA}) = \mathbf{0} \Leftrightarrow (\mathbf{GA})^*\mathbf{N} = \mathbf{NGA}
\end{aligned}$$

which, together with the condition that $\mathbf{G}$ is a g-inverse, i.e., $\mathbf{AGA} = \mathbf{A}$, establishes (1c.5.1).

It may be seen that one choice of

$$\mathbf{A}^-_{m(\mathbf{N})} = \mathbf{N}^{-1}\mathbf{A}^*(\mathbf{A}\mathbf{N}^{-1}\mathbf{A}^*)^-. \tag{1c.5.2}$$

(ii) LEAST SQUARES g-INVERSE. *Let* $\mathbf{AX} = \mathbf{Y}$ *be a possibly inconsistent equation and* $\mathbf{G}$ *be a matrix such that* $\mathbf{X} = \mathbf{GY}$ *minimizes* $(\mathbf{Y} - \mathbf{AX})^*\mathbf{M}(\mathbf{Y} - \mathbf{AX})$ *where* $\mathbf{M}$ *is a p.d. matrix. Then it is n.s. that*

$$\mathbf{AGA} = \mathbf{A}, \qquad (\mathbf{AG})^*\mathbf{M} = \mathbf{MAG}. \tag{1c.5.3}$$

Such a $\mathbf{G}$ *is denoted by* $\mathbf{A}^-_{l(\mathbf{M})}$ *and by* $\mathbf{A}^-_l$ *when* $\mathbf{M} = \mathbf{I}$.

By hypothesis

$(\mathbf{AGY} - \mathbf{Y})^*\mathbf{M}(\mathbf{AGY} - \mathbf{Y}) \leqslant (\mathbf{AX} - \mathbf{Y})^*\mathbf{M}(\mathbf{AX} - \mathbf{Y})$ for all $\mathbf{X}, \mathbf{Y}$, i.e.,

$\|\mathbf{AGY} - \mathbf{Y}\| \leqslant \|\mathbf{AX} - \mathbf{Y}\| = \|\mathbf{AX} - \mathbf{AGY} + \mathbf{AGY} - \mathbf{Y}\|$ for all $\mathbf{X}, \mathbf{Y}$

$$\Leftrightarrow \mathbf{A}^*\mathbf{MAG} = \mathbf{A}^*\mathbf{M} \Leftrightarrow \mathbf{AGA} = \mathbf{A}, \quad (\mathbf{AG})^*\mathbf{M} = \mathbf{MAG}.$$

It may be seen that one choice of

$$\mathbf{A}_{l(\mathbf{M})}^{-} = (\mathbf{A}^*\mathbf{MA})^{-}\mathbf{A}^*\mathbf{M} \tag{1c.5.4}$$

and a minimizing vector $\mathbf{X}$ satisfies the equation

$$\mathbf{A}^*\mathbf{MAX} = \mathbf{A}^*\mathbf{MY},$$

for which a general solution is $\mathbf{GY} + (\mathbf{I} - \mathbf{GA})\mathbf{Z}$, where $\mathbf{G}$ satisfies (1c.5.3).

(iii) MINIMUM NORM LEAST SQUARES g-INVERSE. *Let* $\mathbf{AX} = \mathbf{Y}$ *be a possibly inconsistent equation and* $\mathbf{G}$ *be a matrix such that* $\mathbf{GY}$ *has minimum norm [as defined in* (i)] *in the class of* $\mathbf{X}$ *for which* $(\mathbf{Y} - \mathbf{AX})^*\mathbf{M}(\mathbf{Y} - \mathbf{AX})$ *is a minimum. Then it is n.s. that*

$$\mathbf{AGA} = \mathbf{A}, \quad (\mathbf{AG})^*\mathbf{M} = \mathbf{MAG}, \qquad \mathbf{GAG} = \mathbf{G}, \quad (\mathbf{GA})^*\mathbf{N} = \mathbf{NGA}. \tag{1c.5.5}$$

Such a $\mathbf{G}$ *is denoted by* $\mathbf{A}_{\mathbf{MN}}^{+}$ *and by* $\mathbf{A}^{+}$ *when* $\mathbf{M} = \mathbf{I}$ *and* $\mathbf{N} = \mathbf{I}$.

Since $\mathbf{GY}$ minimizes $(\mathbf{Y} - \mathbf{AX})^*\mathbf{M}(\mathbf{Y} - \mathbf{AX})$, the first two conditions of (1c.5.5) are satisfied. The general solution for $\mathbf{X}$ with this property is, as shown in (ii), $\mathbf{GY} + (\mathbf{I} - \mathbf{GA})\mathbf{Z}$ where $\mathbf{Z}$ is arbitrary. Then

$\|\mathbf{GY}\| \leqslant \|\mathbf{GY} + (\mathbf{I} - \mathbf{GA})\mathbf{Z}\|$ for all $\mathbf{Y}, \mathbf{Z}$

$$\Leftrightarrow (\mathbf{GY}, (\mathbf{I} - \mathbf{GA})\mathbf{Z}) = \mathbf{0} \text{ for all } \mathbf{Y}, \mathbf{Z} \Leftrightarrow \mathbf{G}^*\mathbf{N}(\mathbf{I} - \mathbf{GA}) = \mathbf{0}$$

which is equivalent to the last two conditions of (1c.5.5).

(iv) $\mathbf{A}_{\mathbf{MN}}^{+}$ *as defined in* (iii) *is unique if* $\mathbf{M}$ *and* $\mathbf{N}$ *are p.d. matrices.*

The result is true since the minimum norm solution of a linear equation is unique.

It is easy to verify that

$$\mathbf{A}_{\mathbf{MN}}^{+} = \mathbf{N}^{-1}\mathbf{A}^*\mathbf{MA}(\mathbf{A}^*\mathbf{MAN}^{-1}\mathbf{A}^*\mathbf{MA})^{-}\mathbf{A}^*\mathbf{M} \tag{1c.5.6}$$

satisfies the conditions (1c.5.5). The special case $\mathbf{A}^{+}$ when $\mathbf{M} = \mathbf{I}$, $\mathbf{N} = \mathbf{I}$ is already introduced as Moore-Penrose inverse in [(viii), **1b.5**].

An important result which establishes a duality relationship between a minimum norm and least squares inverses is given in (v).

(v) *Let* $\mathbf{M}$ *be p.d. and* $\mathbf{A}$ *be any* $m \times n$ *matrix. Then*

$$(\mathbf{A}^*)_{m(\mathbf{M})}^{-} = [\mathbf{A}_{l(\mathbf{M}^{-1})}^{-}]^*. \tag{1c.5.7}$$

Let $\mathbf{G}$ denote the left hand side of (1c.5.7). Then by definition

$$\mathbf{A}^*\mathbf{G}\mathbf{A}^* = \mathbf{A}^*, \qquad (\mathbf{G}\mathbf{A}^*)^*\mathbf{M} = \mathbf{M}(\mathbf{G}\mathbf{A}^*)$$

which can be rewritten as $\mathbf{A}\mathbf{G}^*\mathbf{A} = \mathbf{A}$, $(\mathbf{A}\mathbf{G}^*)^*\mathbf{M}^{-1} = \mathbf{M}^{-1}(\mathbf{A}\mathbf{G}^*)$, i.e., $\mathbf{G}^* = \mathbf{A}^-_{l(\mathbf{M}^{-1})}$. Then $\mathbf{G} = [\mathbf{A}^-_{l(\mathbf{M}^{-1})}]^*$.

(vi) *If* $\mathbf{M}$ *and* $\mathbf{N}$ *are p.d. matrices, then*

(a) $(\mathbf{A}^+_{\mathbf{MN}})^+_{\mathbf{NM}} = \mathbf{A}$, (1c.5.8)

(b) $(\mathbf{A}^+_{\mathbf{MN}})^* = (\mathbf{A}^*)^+_{\mathbf{N}^{-1}\mathbf{M}^{-1}}$. (1c.5.9)

The results follow from the definition of $\mathbf{A}^+_{\mathbf{MN}}$.

For further results on g-inverse see the book by Rao and Mitra (1971).

1c.6 Restricted Eigenvalue Problem

In some problems we need the stationary points and values of a quadratic form $\mathbf{X}'\mathbf{A}\mathbf{X}$ when the vector $\mathbf{X}$ is subject to the restrictions

$$\mathbf{X}'\mathbf{B}\mathbf{X} = 1, \qquad \mathbf{C}'\mathbf{X} = \mathbf{0} \tag{1c.6.1}$$

where $\mathbf{B}$ is p.d. and $\mathbf{C}$ is $m \times k$ matrix. Let $\mathbf{P} = \mathbf{C}(\mathbf{C}'\mathbf{B}^{-1}\mathbf{C})^-\mathbf{C}'\mathbf{B}^{-1}$ be the projection operator onto $\mathcal{M}(\mathbf{C})$. We have the following proposition.

(i) *The stationary points and values of* $\mathbf{X}'\mathbf{A}\mathbf{X}$ *when* $\mathbf{X}$ *is subject to the restrictions* (1c.6.1) *are the eigenvectors and values of* $(\mathbf{I} - \mathbf{P})\mathbf{A}$ *with respect to* $\mathbf{B}$.

Introducing Lagrangian multipliers λ and $\boldsymbol{\mu}$ we consider the expression

$$\mathbf{X}'\mathbf{A}\mathbf{X} - \lambda(\mathbf{X}'\mathbf{B}\mathbf{X} - 1) - 2\mathbf{X}'\mathbf{C}\boldsymbol{\mu}$$

and equate its derivatives to zero

$$\begin{aligned} \mathbf{A}\mathbf{X} - \lambda\mathbf{B}\mathbf{X} - \mathbf{C}\boldsymbol{\mu} &= 0 \\ \mathbf{C}'\mathbf{X} &= 0 \\ \mathbf{X}'\mathbf{B}\mathbf{X} &= 1 \end{aligned} \tag{1c.6.2}$$

Multiplying the first equation of (1c.6.2) by $\mathbf{I} - \mathbf{P}$ we have

$$(\mathbf{I} - \mathbf{P})\mathbf{A}\mathbf{X} - \lambda\mathbf{B}\mathbf{X} = 0$$

which proves the desired result. For further results, see Rao (1964).

Note. In the special case $\mathbf{A} = \mathbf{a}\mathbf{a}'$ where $\mathbf{a}$ is m vector, $\mathbf{X}'\mathbf{A}\mathbf{X}$ has the maximum at

$$\mathbf{X} \propto \mathbf{B}^{-1}(\mathbf{I} - \mathbf{P})\mathbf{a} \tag{1c.6.3}$$

which is an important result.

The problem where in (1c.6.1), the condition $\mathbf{C}'\mathbf{X} = \mathbf{0}$ is replaced by $\mathbf{C}'\mathbf{X} \geqslant \mathbf{0}$ (i.e., each element of $\mathbf{C}'\mathbf{X}$ is non-negative) is considered in Rao (1964).

1d CONVEX SETS IN VECTOR SPACES

1d.1 Definitions

A subset C of a vector space is called convex, if for any two points $\mathbf{x}_1, \mathbf{x}_2 \in C$, the line segment joining $\mathbf{x}_1$ and $\mathbf{x}_2$ is contained in C, i.e., $\lambda\mathbf{x}_1 + (1-\lambda)\mathbf{x}_2 \in C$ for $0 \leqslant \lambda \leqslant 1$. Examples of convex sets are points on lines, points inside spheres, half spaces, etc.

Given any set A, the set A^* of points generated from finite sets of points by convex combinations,

$$\lambda_1\mathbf{x}_1 + \cdots + \lambda_k\mathbf{x}_k$$
$$\lambda_i \geqslant 0, \qquad \sum \lambda_i = 1, \qquad \mathbf{x}_i \in A,$$

is clearly a convex set containing A. In fact, A^* is the smallest convex set containing A and is called the *convex hull* of A. The convex hull of two points is the line segment joining the two points, and the convex hull of a sphere is the sphere and all points inside the sphere.

Inner, boundary and exterior points, open and closed sets, and bounded sets are defined as in the theory of sets. Thus $\mathbf{x}$ is an *inner point* of a set A if there is a sphere with a center $\mathbf{x}$ which is a subset of A. The set of all inner points of A is said to be *open*; a set is *closed* if its complementary set is open. A boundary point of A (not necessarily belonging to A) is such that every sphere around it contains points in A as well as in A^c, the complement of A. The closure of a set is obtained by adjoining to it all its boundary points not already in it and is represented by $\bar{A}$. The set $\bar{A}$ is also the smallest closed set containing A. An *exterior* point of A is a point of $\bar{A}^c$, the complement of $\bar{A}$.

The following properties of convex sets are easily established. (1) The intersection of any number of convex sets is convex. (2) The closure of a convex set is a convex set. (3) A convex set C and its closure $\bar{C}$ have the same inner, boundary, and exterior points. (4) Let $\mathbf{x}$ be an inner point and $\mathbf{y}$ a boundary point of C. Then the points $(1-\lambda)\mathbf{x} + \lambda\mathbf{y}$ are inner points of C for $0 \leqslant \lambda < 1$, and exterior points of C for $\lambda > 1$.

1d.2 Separation Theorems for Convex Sets

(i) *Let C be a convex set and* $\mathbf{y}$ *an exterior point. Then there exists a vector* $\mathbf{a}$, *with* $\|\mathbf{a}\| = 1$, *such that*

$$\inf_{\mathbf{x} \in C} (\mathbf{a}, \mathbf{x}) > (\mathbf{a}, \mathbf{y}). \tag{1d.2.1}$$

Let $\mathbf{b}$ be a boundary point such that

$$\inf_{\mathbf{x} \in C} \|\mathbf{y} - \mathbf{x}\| = \|\mathbf{b} - \mathbf{y}\|.$$

If $\mathbf{x} \in C$, the points $(1 - \lambda)\mathbf{x} + \lambda\mathbf{b} \in \bar{C}$ for $0 \leqslant \lambda \leqslant 1$ and therefore

$$\|(1 - \lambda)\mathbf{x} + \lambda\mathbf{b} - \mathbf{y}\| \geqslant \|\mathbf{b} - \mathbf{y}\|. \tag{1d.2.2}$$

Expanding (1d.2.2), we see that

$$(1 - \lambda)^2\|\mathbf{x} - \mathbf{b}\|^2 + 2(1 - \lambda)(\mathbf{x} - \mathbf{b}, \mathbf{b} - \mathbf{y}) \geqslant 0$$

for all $\lambda \leqslant 1$. Hence $(\mathbf{x} - \mathbf{b}, \mathbf{b} - \mathbf{y}) \geqslant 0$, giving

$$(\mathbf{x}, \mathbf{b} - \mathbf{y}) \geqslant (\mathbf{b}, \mathbf{b} - \mathbf{y}) > (\mathbf{y}, \mathbf{b} - \mathbf{y}).$$

The result (1d.2.1) is established by choosing $\mathbf{a} = \mathbf{b} - \mathbf{y}$. Surely $\mathbf{a}$ could be normalized.

(ii) *Let C be a convex set and $\mathbf{y}$ be on the boundary of C. Then there exists a supporting plane through $\mathbf{y}$, i.e., there exists a non-null vector $\mathbf{a}$ such that*

$$\inf_{\mathbf{x} \in C} (\mathbf{a}, \mathbf{x}) = (\mathbf{a}, \mathbf{y}), \qquad \textit{that is,} \quad (\mathbf{a}, \mathbf{x}) \geqslant (\mathbf{a}, \mathbf{y}) \quad \textit{for all} \quad \mathbf{x} \in C. \tag{1d.2.3}$$

Let $\mathbf{y}_n$ be a sequence of exterior points such that $\lim \mathbf{y}_n = \mathbf{y}$. Consider $\mathbf{a}_n$ as determined from $\mathbf{y}_n$ by (1d.2.1). Since $\|\mathbf{a}_n\|$ is bounded, there is a subsequence $\mathbf{a}_\nu$ tending to a limit, say, $\mathbf{a}$. Hence

$$(\mathbf{x}, \mathbf{a}) = \lim_{\nu \to \infty} (\mathbf{x}, \mathbf{a}_\nu) \geqslant \lim_{\nu \to \infty} (\mathbf{y}_\nu, \mathbf{a}_\nu) = (\mathbf{y}, \mathbf{a}).$$

(iii) *Let C and D be convex sets with no inner point common. Then there exists a vector $\mathbf{a}$ and a scalar α such that $(\mathbf{a}, \mathbf{x}) \geqslant \alpha$ for all $\mathbf{x} \in C$ and $(\mathbf{a}, \mathbf{y}) \leqslant \alpha$ for all $\mathbf{y} \in D$, that is, there is a hyperplane separating the two sets.*

Consider the convex set of points $\mathbf{x} - \mathbf{y}$, $\mathbf{x} \in C$, $\mathbf{y} \in D$, which does not contain $\mathbf{0}$ as an interior point. By (1d.2.1) and (1d.2.3) there exists $\mathbf{a}$ such that

$$(\mathbf{a}, \mathbf{x} - \mathbf{y}) \geqslant (\mathbf{a}, \mathbf{0}) = 0 \qquad \text{or} \qquad (\mathbf{a}, \mathbf{x}) \geqslant (\mathbf{a}, \mathbf{y}).$$

Hence there exists an α such that

$$(\mathbf{a}, \mathbf{x}) \geqslant \alpha, \quad \mathbf{x} \in C \qquad \text{and} \qquad (\mathbf{a}, \mathbf{y}) \leqslant \alpha, \quad \mathbf{y} \in D. \tag{1d.2.4}$$

If C and D are closed convex sets with no common point then $\mathbf{0}$ is an exterior point of the closed set of points $\mathbf{x} - \mathbf{y}$, in which case, by an application of (1d.2.1), we have the strict separation

$$(\mathbf{a}, \mathbf{x}) > \alpha, \quad \mathbf{x} \in C \qquad \text{and} \qquad (\mathbf{a}, \mathbf{y}) < \alpha, \quad \mathbf{y} \in D. \tag{1d.2.5}$$

A vector $\mathbf{x} \in E_n$ is said to be $> \mathbf{0}$ if all its coordinates are non-negative and at least one is positive. A *negative vector* is one with all coordinates negative.

(iv) *If C is a convex set in E_n such that none of its points is negative, then there exists a vector $\mathbf{a} > \mathbf{0}$ such that $(\mathbf{a}, \mathbf{x}) \geqslant 0$ for all $\mathbf{x} \in C$.*

Consider the convex sets C and Y the set of all negative vectors. By result (iii) there exist $\mathbf{a}$ and α such that

$$(\mathbf{a}, \mathbf{x}) \geqslant \alpha \quad \text{for} \quad \mathbf{x} \in C \qquad \text{and} \qquad (\mathbf{a}, \mathbf{y}) \leqslant \alpha \quad \text{for} \quad \mathbf{y} \in Y. \quad (1\text{d}.2.6)$$

Since $(\mathbf{a}, \mathbf{y}) \to 0$ as $\mathbf{y} \to \mathbf{0}$, $\alpha \geqslant 0$. Furthermore, $\mathbf{a} > \mathbf{0}$, for otherwise we can find some negative vector $\mathbf{y}$ such that $(\mathbf{a}, \mathbf{y}) \geqslant \alpha$ contrary to (1d.2.6). Hence there exists an $\mathbf{a} > \mathbf{0}$ such that $(\mathbf{a}, \mathbf{x}) > 0$.

A simple treatment of convex sets is contained in a book by Berge (1963). The reader may easily verify the following results.

(v) FARKA'S LEMMA. *Let $\boldsymbol{\beta}$ and $\boldsymbol{\alpha}_1, \ldots, \boldsymbol{\alpha}_k$ be vectors in R^n and let $\mathbf{x}$ exist such that $\boldsymbol{\alpha}_i'\mathbf{x} \geqslant 0$ for all i. Then an n.s. condition that $\boldsymbol{\beta}'\mathbf{x} \geqslant 0$ for all $\mathbf{x}$ such that $\boldsymbol{\alpha}'\mathbf{x} \geqslant 0$ is that $\boldsymbol{\beta}$ can be written as $\boldsymbol{\beta} = \sum a_i \boldsymbol{\alpha}_i$, $a_i \geqslant 0$.*

The proof is omitted.

1. If C^* is the convex hull of a subset C of E_n, every point of C^* can be represented as a convex combination of at most $n + 1$ points of C.
2. Let C be a bounded set in E_n and

$$\mathbf{x} = \sum_{i=1}^{\infty} \alpha_i \mathbf{x}_i, \qquad \mathbf{x}_i \in C, \qquad \alpha_i > 0, \qquad \sum \alpha_i = 1.$$

   Then $\mathbf{x} \in C^*$.

3. A point $\mathbf{x} \in C$ is an extreme point of C if there are no distinct points $\mathbf{x}_1$ and $\mathbf{x}_2 \in C$ such that $\mathbf{x} = \lambda\mathbf{x}_1 + (1 - \lambda)\mathbf{x}_2$ for some λ, $(0 < \lambda < 1)$. If C is a closed, bounded, convex set, it is spanned by its extreme points, that is, every $\mathbf{x} \in C$ can be represented in the form

$$\mathbf{x} = \sum_{1}^{k} \alpha_i \mathbf{x}_i, \qquad \alpha_i \geqslant 0, \qquad \sum \alpha_i = 1.$$

   where $\mathbf{x}_1, \ldots, \mathbf{x}_k$ are extreme points of C.
4. A closed, bounded, convex set has extreme points in every supporting hyperplane.
5. A closed convex set in E_n has a finite number of extreme points iff it is the intersection of a finite number of closed half spaces.

1e INEQUALITIES

Some standard inequalities based on vectors, matrices, and determinants are assembled in this section because of their applications in mathematical statistics.

1e.1 Cauchy-Schwarz (C-S) Inequality

(i) *For any two column vectors* **X**, **Y** *of real elements*

$$(\mathbf{X}'\mathbf{Y})^2 \leqslant (\mathbf{X}'\mathbf{X})(\mathbf{Y}'\mathbf{Y}) \qquad or \qquad (\sum x_i y_i)^2 \leqslant (\sum x_i^2)(\sum y_i^2) \quad (1e.1.1)$$

written in terms of the elements of **X** *and* **Y**, *the equality being attained when and only when* $\lambda\mathbf{X} + \mu\mathbf{Y} = 0$ *for real scalars* λ *and* μ.

The quadratic form in λ, μ,

$$(\lambda\mathbf{X} + \mu\mathbf{Y})'(\lambda\mathbf{X} + \mu\mathbf{Y}) = \lambda^2\mathbf{X}'\mathbf{X} + 2\lambda\mu\mathbf{X}'\mathbf{Y} + \mu^2\mathbf{Y}'\mathbf{Y}$$

is non-negative. Hence if we apply [(v), **1c.1**] or elementary algebra, the discriminant of the quadratic form is greater than or equal to zero,

$$\begin{vmatrix} \mathbf{X}'\mathbf{X} & \mathbf{Y}'\mathbf{X} \\ \mathbf{X}'\mathbf{Y} & \mathbf{Y}'\mathbf{Y} \end{vmatrix} \geqslant 0, \qquad \text{or} \qquad (\mathbf{X}'\mathbf{Y})^2 \leqslant (\mathbf{X}'\mathbf{X})(\mathbf{Y}'\mathbf{Y}). \qquad (1e.1.2)$$

The equality is attained when the determinant is zero, which implies that there exist λ and μ such that λ times the first row plus μ times the second row is zero. Hence $(\lambda\mathbf{X} + \mu\mathbf{Y})'(\lambda\mathbf{X} + \mu\mathbf{Y}) = 0$ or $\lambda\mathbf{X} + \mu\mathbf{Y} = \mathbf{0}$.

Two other forms of C-S inequality which are of interest are as follows.

(ii) *Let* **X**, **Y** *be two column vectors and* **A**, *a Gramian matrix, that is,* $\mathbf{A} = \mathbf{B}'\mathbf{B}$. *Then:*

(a) $(\mathbf{X}'\mathbf{A}\mathbf{Y})^2 \leqslant (\mathbf{X}'\mathbf{A}\mathbf{X})(\mathbf{Y}'\mathbf{A}\mathbf{Y})$, *with equality when* $\mathbf{X} \propto \mathbf{Y}$. (1e.1.3)

(b) $(\mathbf{X}'\mathbf{Y})^2 \leqslant (\mathbf{X}'\mathbf{A}\mathbf{X})(\mathbf{Y}'\mathbf{A}^{-1}\mathbf{Y})$ (1e.1.4)

if $\mathbf{A}^{-1}$ *exists, with equality when* $\mathbf{X} \propto \mathbf{A}^{-1}\mathbf{Y}$.

The result (a) is obtained by choosing $\mathbf{U} = \mathbf{B}\mathbf{X}$, $\mathbf{V} = \mathbf{B}\mathbf{Y}$ and applying the C-S inequality on **U** and **V**. Similarly (b) is obtained by choosing

$$\mathbf{U} = (\mathbf{B}^{-1})'\mathbf{Y}, \qquad \mathbf{V} = \mathbf{B}\mathbf{X}.$$

(iii) *The integral version of the* C-S *inequality is*

$$\left(\int_A fg\, dv\right)^2 \leqslant \int_A f^2\, dv \int_A g^2\, dv \qquad (1e.1.5)$$

where f *and* g *are real functions defined in some set* A *and* f^2, g^2 *are integrable with respect to a measure* v.

To prove the result we consider the non-negative quadratic form in λ, μ,

$$\int_A (\lambda f + \mu g)^2\, dv = \lambda^2 \int_A f^2\, dv + 2\lambda\mu \int_A fg\, dv + \mu^2 \int_A g^2\, dv. \quad (1e.1.6)$$

In particular if v is a probability measure we obtain the result

$$[\text{cov}(xy)]^2 \leqslant V(x)V(y), \tag{1e.1.7}$$

where x and y are random variables, cov and V stand for covariance and variance respectively.

The Lagrange Identity. The inequality (1e.1.1) can be derived immediately from the identity

$$(\sum x_i^2)(\sum y_i^2) - (\sum x_i y_i)^2 = \sum_{i<j} (x_i y_j - x_j y_i)^2, \tag{1e.1.8}$$

where x_i and y_j are the components of vectors **X** and **Y** respectively.

1e.2 Hölder's Inequality

If $x_i, y_i \geqslant 0$, $i = 1, \ldots, n$ *and* $(1/p) + (1/q) = 1, p > 1$, *then*

$$\sum_1^n x_i y_i \leqslant \left(\sum_1^n x_i^p\right)^{1/p} \left(\sum_1^n y_i^q\right)^{1/q} \tag{1e.2.1}$$

with equality when and only when $y_i \propto x_i^{p-1}$. *The integral version is for* $f, g \geqslant 0$

$$\int_R fg\, dv \leqslant \left(\int_R f^p\, dv\right)^{1/p} \left(\int_R g^q\, dv\right)^{1/q}.$$

First we establish by differentiation, or otherwise,

$$\min_{x \geqslant 0} \left\{ t(x) = \frac{x^p}{p} + \frac{x^{-q}}{q} \right\} = 1 \qquad \text{when} \quad x = 1.$$

Substituting $x = u^{1/q} v^{-1/p}$, $(u, v \geqslant 0)$ in $t(x) \geqslant 1$, we find

$$uv \leqslant \frac{u^p}{p} + \frac{v^q}{q}, \tag{1e.2.2}$$

with equality when $v = u^{p-1}$.

Let $u_k = x_k / (\sum x_k^p)^{1/p}$, $v_k = y_k / (\sum y_k^q)^{1/q}$. Substituting in (1e.2.2) and summing over k, we have

$$\sum u_k v_k \leqslant \frac{\sum u_k^p}{p} + \frac{\sum v_k^q}{q} = \frac{1}{p} + \frac{1}{q} = 1$$

and obtain the desired inequality. The integral form is proved similarly. *More generally we have the inequality*

$$\sum x_i y_i z_i \cdots \leqslant (\sum x_i^p)^{1/p} (\sum y_i^q)^{1/q} (\sum z_i^r)^{1/r} \cdots$$

$$\text{if} \quad \frac{1}{p} + \frac{1}{q} + \frac{1}{r} \cdots \leqslant 1. \tag{1e.2.3}$$

Minkowski's Inequality. It is easy to deduce from Hölder's inequality, that if $x_i, y_i \geqslant 0$ and $k \geqslant 1$

$$[\sum (x_i + y_i)^k]^{1/k} \leqslant (\sum x_i^k)^{1/k} + (\sum y_i^k)^{1/k}. \tag{1e.2.4}$$

1e.3 Hadamard's Inequality

For a nonsingular real $(n \times n)$ *matrix,* $\mathbf{B} = (b_{ij})$,

$$|\mathbf{B}|^2 \leqslant \prod_{i=1}^{n} \left(\sum_{k=1}^{n} b_{ik}^2 \right). \tag{1e.3.1}$$

As special cases

(i) $$|\mathbf{B}| \leqslant 1, \quad \text{if } \sum b_{ik}^2 = 1, \quad i = 1, \ldots, n, \tag{1e.3.2}$$

(ii) $$|\mathbf{B}| \leqslant M^n n^{n/2}, \quad \text{if } |b_{ij}| \leqslant M. \tag{1e.3.3}$$

To prove (1e.3.1) we consider the positive definite matrix $\mathbf{A} = \mathbf{BB}'$ and apply the inequality $|\mathbf{A}| \leqslant a_{11}, \ldots, a_{nn}$, the product of diagonal elements. Consider the expansion

$$|\mathbf{A}| = a_{11} \begin{vmatrix} a_{22} & \cdots & a_{2n} \\ \cdot & \cdots & \cdot \\ a_{n2} & \cdots & a_{nn} \end{vmatrix} + \begin{vmatrix} 0 & a_{12} & \cdots & a_{1n} \\ a_{21} & a_{22} & \cdots & a_{2n} \\ \cdot & \cdot & \cdots & \cdot \\ a_{n1} & a_{n2} & \cdots & a_{nn} \end{vmatrix}.$$

Since $\mathbf{A}$ is p.d., the matrix (a_{ij}), $i, j = 2, \ldots, n$ is also p.d., and the second term is negative. Hence

$$|\mathbf{A}| \leqslant a_{11} \begin{vmatrix} a_{22} & \cdots & a_{2n} \\ \cdot & \cdots & \cdot \\ a_{n2} & \cdots & a_{nn} \end{vmatrix}$$

and the result, $|\mathbf{A}| \leqslant a_{11} \cdots a_{nn}$, follows by induction. The result (1e.3.1) follows by observing that $a_{ii} = \sum b_{ik}^2$.

1e.4 Inequalities Involving Moments

Let x be a random variable such that $E(x) = \mu$ and moments

$$E(x - \mu)^r = \mu_r, \qquad r = 1, \ldots, 2N$$

exist. The quadratic form $\sum \sum \mu_{i+j-2} y_i y_j$ in y_i is

$$E[y_1 + y_2(x - \mu) + \cdots + y_{N+1}(x - \mu)^N]^2 \geqslant 0. \tag{1e.4.1}$$

Hence (μ_{i+j-2}) is non-negative definite. This is not, however, a sufficient condition for μ_r to be a moment sequence of a random variable. The condition is necessary whether the moments are of a discrete or continuous distribution, or calculated from a given set of observations. In particular, for $N = 2$,

$$\begin{vmatrix} 1 & 0 & \mu_2 \\ 0 & \mu_2 & \mu_3 \\ \mu_2 & \mu_3 & \mu_4 \end{vmatrix} \geqslant 0 \qquad \text{or} \qquad \mu_2{}^3\left(\frac{\mu_4}{\mu_2{}^2} - \frac{\mu_3{}^2}{\mu_2{}^3} - 1\right) \geqslant 0$$

$$\text{or} \qquad \beta_2 - \beta_1 - 1 \geqslant 0, \quad \text{(1e.4.2)}$$

where $\beta_1 = \mu_3{}^2/\mu_2{}^3$ and $\beta_2 = \mu_4/\mu_2{}^2$ are some measures of asymmetry and kurtosis of a distribution.

1e.5 Convex Functions and Jensen's Inequality

A function $f(x)$ is said to be convex if for $\alpha, \beta > 0$, $\alpha + \beta = 1$

$$f(\alpha x + \beta y) \leqslant \alpha f(x) + \beta f(y) \qquad \text{for all } x, y. \quad \text{(1e.5.1)}$$

For such a function we shall first show that at any point x_0, $f'_+(x_0)$, $f'_-(x_0)$, the right and left derivatives exist. Let $x_0 < x_1 < x_2$. Choosing $\alpha = (x_2 - x_1)/(x_2 - x_0)$, $\beta = (x_1 - x_0)/(x_2 - x_0)$, $x = x_0$, and $y = x_2$ so that $\alpha x + \beta y = x_1$, and using (1e.5.1), we see that

$$(x_2 - x_0)f(x_1) \leqslant (x_2 - x_1)f(x_0) + (x_1 - x_0)f(x_2) \quad \text{(1e.5.2)}$$

after multiplication by $(x_2 - x_0)$. Adding $x_0 f(x_0)$ to both sides and rearranging the terms in (1e.5.2), we have

$$\frac{f(x_1) - f(x_0)}{x_1 - x_0} \leqslant \frac{f(x_2) - f(x_0)}{x_2 - x_0}, \quad \text{(1e.5.3)}$$

which shows that $[f(x) - f(x_0)]/(x - x_0)$ decreases as $x \to x_0$. By adding $x_1 f(x_1)$ to both sides of (1e.5.2) and rearranging the terms,

$$\frac{f(x_0) - f(x_1)}{x_0 - x_1} \leqslant \frac{f(x_2) - f(x_1)}{x_2 - x_1}. \quad \text{(1e.5.4)}$$

Equation (1e.5.4) in terms of $x_{-1} < x_0 < x_1$ becomes

$$\frac{f(x_{-1}) - f(x_0)}{x_{-1} - x_0} \leqslant \frac{f(x_1) - f(x_0)}{x_1 - x_0},$$

which shows that $[f(x) - f(x_0)]/(x - x_0)$ is bounded from below, and since it decreases as $x \to x_0$, the right-hand derivative $f'_+(x_0)$ exists. Similarly, $f'_-(x_0)$ exists and obviously $f'_-(x_0) \leqslant f'_+(x_0)$. Let L be such that $f'_-(x_0) \leqslant L \leqslant f'_+(x_0)$. Then, for all x,

$$f(x) \geqslant f(x_0) + L(x - x_0), \quad \text{(1e.5.5)}$$

for if $x_1 > x_0$,

$$\frac{f(x_1) - f(x_0)}{x_1 - x_0} \geqslant f'_+(x_0) \geqslant L,$$

and the reverse relation is true when $x_1 < x_0$. The inequality (1e.5.5) has important applications; it leads to Jensen's inequality.

Jensen's Inequality. *If x is a random variable such that $E(x) = \mu$ and $f(x)$ is a convex function, then*

$$E[f(x)] \geqslant f[E(x)] \tag{1e.5.6}$$

with equality when and only when x is a degenerate distribution at μ.

Consider the equation (1e.5.5) with μ for x_0,

$$f(x) \geqslant f(\mu) + L(x - \mu), \tag{1e.5.7}$$

and take expectations of both sides. The expectation of the second term on the right-hand side of (1e.5.7) is zero, yielding the inequality (1e.5.6).

1e.6 Inequalities in Information Theory

(i) *Let $\sum a_i$ and $\sum b_i$ be convergent sequences of positive numbers such that $\sum a_i \geqslant \sum b_i$. Then*

$$\sum a_i \log \frac{b_i}{a_i} \leqslant 0 \tag{1e.6.1}$$

the equality being attained when and only when $a_i = b_i$. Further if $a_i \leqslant 1$ and $b_i \leqslant 1$ for all i, then

$$2 \sum a_i \log \frac{a_i}{b_i} \geqslant \sum a_i(a_i - b_i)^2. \tag{1e.6.2}$$

To prove the inequalities, note that for $x > 0$, the expansion of $\log x$ at $x = 1$ yields

$$\log x = (x - 1) - (x - 1)^2(2y^2)^{-1} \quad \text{with} \quad y \in (1, x). \tag{1e.6.3}$$

Using the expansion (1e.6.3) for each term in (1e.6.1), we have

$$\sum a_i \log \frac{b_i}{a_i} = (\sum b_i - \sum a_i) - \sum a_i(b_i - a_i)^2(2y_i^2)^{-1} \leqslant 0, \tag{1e.6.4}$$

thus proving (1e.6.1). In (1e.6.4), $y_i^2 \in (a_i^2, b_i^2)$, and if $a_i \leqslant 1$, $b_i \leqslant 1$, the maximum value of y_i^2 is not greater than unity. Hence

$$-\sum a_i(b_i - a_i)^2(2y_i^2)^{-1} \leqslant \frac{-\sum a_i(b_i - a_i)^2}{2}. \tag{1e.6.5}$$

Combining (1e.6.4) and (1e.6.5), we obtain (1e.6.2).

The results (1e.6.1) and (1e.6.2) are true if a_i and b_i are non-negative and the summations are extended over values of i for which $a_i > 0$. In other words we admit the possibility of some of the b_i being zero (but not the a_i).

(ii) *Let f and g be non-negative and integrable functions with respect to a measure μ and S be the region in which $f > 0$. If $\int_S(f - g)\, d\mu \geqslant 0$, then*

$$\int_S f \log \frac{f}{g}\, d\mu \geqslant 0 \tag{1e.6.6}$$

with equality only when $f = g(a.e.\mu)$.

The proof is the same as that of (1e.6.1) with summations replaced by integrals. The reader may directly deduce the inequalities (1e.6.1) and (1e.6.6) by applying Jensen's inequality (1e.5.6).

1e.7 Stirling's Approximation

This is a convenient approximation to factorials, which is useful for the practical computation of large factorials and also in studying the limiting forms of the Binomial, Poisson, Hypergeometric distributions etc. The approximation states

$$n! \simeq \sqrt{2\pi n}\, n^n e^{-n},$$

which implies that

$$\lim_{n\to\infty} \frac{1 \cdot 2 \cdots n}{\sqrt{2\pi n}\, n^n e^{-n}} = 1.$$

For finite n we have more precisely

$$n! = \sqrt{2\pi n}\, n^n e^{-n} e^{\omega(n)/12},$$

where $(n + \frac{1}{2})^{-1} < \omega(n) < n^{-1}$. It may be noted that the absolute error in the approximation $\sqrt{2\pi n}\, n^n e^{-n}$ increases with n but the relative error tends to zero rapidly. The formula is useful when quotients of large factorials have to be evaluated. For proof the reader may consult any book on numerical mathematics.

1f EXTREMA OF QUADRATIC FORMS

This section contains a number of useful results on the extreme values of quadratic forms under linear and quadratic restrictions on the variables.

1f.1 General Results

(i) *Let* $\mathbf{A}$ *be a p.d.* $m \times m$ *matrix and* $\mathbf{U} \in E_m$, *that is,* $\mathbf{U}$ *is an m-vector. Then*

$$\sup_{\mathbf{X} \in E_m} \frac{(\mathbf{U}'\mathbf{X})^2}{\mathbf{X}'\mathbf{A}\mathbf{X}} = \mathbf{U}'\mathbf{A}^{-1}\mathbf{U}, \tag{1f.1.1}$$

and the supremum is attained at $\mathbf{X}_* = \mathbf{A}^{-1}\mathbf{U}$.

The result (1f.1.1) is only a restatement of one of the versions (1e.1.4) of *C-S* inequality.

As a corollary to (i), *for any given* $m \times m$ *matrix* $\mathbf{T}$

$$\sup_{\mathbf{X} \in E_m} \frac{(\mathbf{U}'\mathbf{T}\mathbf{X})^2}{\mathbf{X}'\mathbf{A}\mathbf{X}} = \mathbf{U}'\mathbf{T}\mathbf{A}^{-1}\mathbf{T}'\mathbf{U}. \tag{1f.1.2}$$

(ii) *Let* $\mathbf{A}$ *be a p.d.* $m \times m$ *matrix,* $\mathbf{B}$ *be a* $m \times k$ *matrix, and* $\mathbf{U}$ *be a k-vector. Denote by* $\mathbf{S}^-$ *any generalized inverse of* $\mathbf{B}'\mathbf{A}^{-1}\mathbf{B}$. *Then*

$$\inf_{\mathbf{B}'\mathbf{X}=\mathbf{U}} \mathbf{X}'\mathbf{A}\mathbf{X} = \mathbf{U}'\mathbf{S}^-\mathbf{U}, \tag{1f.1.3}$$

where $\mathbf{X}$ *is a column vector and the infimum is attained at* $\mathbf{X}_* = \mathbf{A}^{-1}\mathbf{B}\mathbf{S}^-\mathbf{U}$.

Observe that the problem is one of seeking the minimum of a quadratic form when $\mathbf{X}$ is subject to the linear restrictions $\mathbf{B}'\mathbf{X} = \mathbf{U}$, which must be consistent.

If $\mathbf{B}'\mathbf{X} = \mathbf{U}$ is consistent, a $\boldsymbol{\lambda} \in E_k$ exists such that $\mathbf{B}'\mathbf{A}^{-1}\mathbf{B}\boldsymbol{\lambda} = \mathbf{U}$ or $\mathbf{S}\boldsymbol{\lambda} = \mathbf{U}$. As a consequence we have

$$\begin{aligned} \mathbf{B}'\mathbf{X}_* &= \mathbf{B}'\mathbf{A}^{-1}\mathbf{B}\mathbf{S}^-\mathbf{U} = \mathbf{S}\mathbf{S}^-\mathbf{S}\boldsymbol{\lambda} = \mathbf{S}\boldsymbol{\lambda} = \mathbf{U} \\ \mathbf{X}_*'\mathbf{A}\mathbf{X}_* &= (\mathbf{U}'\mathbf{S}^{-\prime}\mathbf{B}'\mathbf{A}^{-1})\mathbf{A}(\mathbf{A}^{-1}\mathbf{B}\mathbf{S}^-\mathbf{U}) = \mathbf{U}'\mathbf{S}^-\mathbf{U}. \end{aligned} \tag{1f.1.4}$$

We need only show that for any $\mathbf{X}$, $\mathbf{X}'\mathbf{A}\mathbf{X} \geqslant \mathbf{U}'\mathbf{S}^-\mathbf{U}$, which is arrived at by applying the *C-S* inequality to the vectors $\mathbf{Y}_1 = \mathbf{A}^{1/2}\mathbf{X}$, $\mathbf{Y}_2 = \mathbf{A}^{1/2}\mathbf{X}_*$ and observing that $\mathbf{Y}_1'\mathbf{Y}_2 = \mathbf{X}'\mathbf{B}\mathbf{S}^-\mathbf{U} = \mathbf{U}'\mathbf{S}^-\mathbf{U}$. As a corollary to (ii):

$$\inf_{\mathbf{B}'\mathbf{X}=\mathbf{U}} (\mathbf{X} - \boldsymbol{\xi})'\mathbf{A}(\mathbf{X} - \boldsymbol{\xi}) = (\mathbf{U} - \mathbf{B}'\boldsymbol{\xi})'\mathbf{S}^-(\mathbf{U} - \mathbf{B}'\boldsymbol{\xi}) \tag{1f.1.5}$$

for given $\boldsymbol{\xi} \in E_m$ and the infimum is attained at $\mathbf{X}_* = \boldsymbol{\xi} + \mathbf{A}^{-1}\mathbf{B}\mathbf{S}^-\mathbf{U} - \mathbf{A}^{-1}\mathbf{B}\mathbf{S}^-\mathbf{B}'\boldsymbol{\xi}$. This is obtained by putting $\mathbf{Y} = \mathbf{X} - \boldsymbol{\xi}$ and applying (1f.1.3).

As a particular case of (1f.1.3) we obtain

$$\inf_{\mathbf{B}'\mathbf{X}=\mathbf{U}} \mathbf{X}'\mathbf{X} = \mathbf{U}'\mathbf{S}^-\mathbf{U}, \tag{1f.1.6}$$

where $\mathbf{S}^-$ is a g-inverse of $\mathbf{B}'\mathbf{B}$ and the infimum is attained at $\mathbf{X}_* = \mathbf{B}\mathbf{S}^-\mathbf{U}$.

Observe that in the results (1f.1.3, 1f.1.5, 1f.1.6), $\mathbf{S}^-$ will be a true inverse $\mathbf{S}^{-1}$ if the rank of $\mathbf{B}$ is full, that is, equal to k, the number of rows in $\mathbf{B}$.

An important generalization of (1f.1.6) is to determine the minimum of $\mathbf{X}_1'\mathbf{X}_1$ where $\mathbf{X}_1$ is a part of $\mathbf{X}$, say the vector of the first $q(<k)$ components of $\mathbf{X}$, subject to the condition $\mathbf{B}'\mathbf{X} = \mathbf{U}$. Then we may write the condition $\mathbf{B}'\mathbf{X} = \mathbf{U}$ in the partitioned form

$$(\mathbf{B}_1' \,\vdots\, \mathbf{B}_2')\begin{pmatrix}\mathbf{X}_1\\ \hdashline \mathbf{X}_2\end{pmatrix} = \mathbf{U}. \tag{1f.1.7}$$

Let

$$\begin{pmatrix}\mathbf{B}_1'\mathbf{B}_1 & \mathbf{B}_2' \\ \mathbf{B}_2 & \mathbf{0}\end{pmatrix}^- = \begin{pmatrix}\mathbf{C}_1 & \mathbf{C}_2 \\ \mathbf{C}_3 & \mathbf{C}_4\end{pmatrix}, \tag{1f.1.8}$$

by choosing any g-inverse.

(iii) *Let* $\mathbf{B}_1$, $\mathbf{B}_2$, *and* $\mathbf{C}_1$ *be as defined by* (1f.1.7) *and* (1f.1.8). *Then*

$$\inf_{\mathbf{B}'\mathbf{X}=\mathbf{U}} \mathbf{X}_1'\mathbf{X}_1 = (\mathbf{X}_{1*})'(\mathbf{X}_{1*}) = \mathbf{U}'\mathbf{C}_1\mathbf{U} \tag{1f.1.9}$$

where $\mathbf{X}_{1*} = \mathbf{B}_1\mathbf{C}_1'\mathbf{U}$.

The condition (1f.1.7) implies that column vectors $\mathbf{E}$ and $\mathbf{F}$ exist such that

$$\mathbf{U} = \mathbf{B}_1'\mathbf{B}_1\mathbf{E} + \mathbf{B}_2'\mathbf{F}$$
$$\mathbf{0} = \mathbf{B}_2\mathbf{E}.$$

Then it is easy to verify that

$$\mathbf{X}_{1*} = \mathbf{B}_1\mathbf{C}_1'\mathbf{U}, \qquad \mathbf{X}_{2*} = \mathbf{C}_2'\mathbf{U}$$

satisfies the condition $\mathbf{B}'\mathbf{X} = \mathbf{U}$ and $(\mathbf{X}_{1*})'(\mathbf{X}_{1*}) = \mathbf{U}'\mathbf{C}_1\mathbf{U}$. Furthermore, $(\mathbf{X}_{1*})'\mathbf{X}_1 = \mathbf{U}'\mathbf{C}_1\mathbf{U}$, where $\mathbf{X}_1$ is any other vector satisfying the condition (1f.1.7). The result (1f.1.9) is obtained by applying the *C-S* inequality to the vectors $\mathbf{X}_1$ and $\mathbf{X}_{1*}$.

(iv) *Let* $\mathbf{A}$ *be an n.n.d. matrix of order* m, $\mathbf{B}$ *be a* $m \times k$ *matrix and* $\mathbf{U}$ *be a* k *vector such that* $\mathbf{U} \in \mathcal{M}(\mathbf{B}')$. *Further, let*

$$\begin{pmatrix}\mathbf{A} & \mathbf{B} \\ \mathbf{B}' & \mathbf{0}\end{pmatrix}^- = \begin{pmatrix}\mathbf{C}_1 & \mathbf{C}_2 \\ \mathbf{C}_3 & -\mathbf{C}_4\end{pmatrix} \tag{1f.1.10}$$

be one choice of g-inverse. Then

$$\inf_{\mathbf{B}'\mathbf{X}=\mathbf{U}} \mathbf{X}'\mathbf{A}\mathbf{X} = \mathbf{U}'\mathbf{C}_4\mathbf{U}. \tag{1f.1.11}$$

1f.2 Results Involving Eigenvalues and Vectors

Let $\mathbf{A}$ be a $m \times m$ symmetric matrix, $\lambda_1 \geqslant \cdots \geqslant \lambda_m$ be its eigenvalues, and $\mathbf{P}_1, \ldots, \mathbf{P}_m$ be the corresponding eigenvectors. It has been shown (1c.3.5), that

$$\mathbf{A} = \lambda_1 \mathbf{P}_1 \mathbf{P}_1' + \cdots + \lambda_m \mathbf{P}_m \mathbf{P}_m'$$
$$\mathbf{I} = \mathbf{P}_1 \mathbf{P}_1' + \cdots + \mathbf{P}_m \mathbf{P}_m'.$$

The following propositions hold.

(i)
$$\sup_{\mathbf{X}} \frac{\mathbf{X}'\mathbf{A}\mathbf{X}}{\mathbf{X}'\mathbf{X}} = \lambda_1, \qquad \inf_{\mathbf{X}} \frac{\mathbf{X}'\mathbf{A}\mathbf{X}}{\mathbf{X}'\mathbf{X}} = \lambda_m. \tag{1f.2.1}$$

Observe that any $\mathbf{X}$ can be written $C_1\mathbf{P}_1 + \cdots + C_m\mathbf{P}_m$. Then

(ii)
$$\frac{\mathbf{X}'\mathbf{A}\mathbf{X}}{\mathbf{X}'\mathbf{X}} = \frac{C_1{}^2\lambda_1 + \cdots + C_m{}^2\lambda_m}{C_1{}^2 + \cdots + C_m{}^2}. \tag{1f.2.2}$$

The sup and inf with respect to the vector $(C_1, \ldots, C_m)$ are obviously λ_1 and λ_m. The sup is attained when $\mathbf{X} = \mathbf{P}_1$ and the inf at $\mathbf{X} = \mathbf{P}_m$.

(ii)
$$\sup_{\substack{\mathbf{P}_i'\mathbf{X}=0\\ i=1,\ldots,k}} \frac{\mathbf{X}'\mathbf{A}\mathbf{X}}{\mathbf{X}'\mathbf{X}} = \lambda_{k+1}, \qquad \inf_{\substack{\mathbf{P}_i'\mathbf{X}=0\\ i=1,\ldots,k}} \frac{\mathbf{X}'\mathbf{A}\mathbf{X}}{\mathbf{X}'\mathbf{X}} = \lambda_m. \tag{1f.2.3}$$

Under the given conditions $\mathbf{X} = C_{k+1}\mathbf{P}_{k+1} + \cdots + C_m\mathbf{P}_m$. Hence we obtain (1f.2.3) with the argument used in (i).

(iii) *Let* $\mathbf{B}$ *be a* $m \times k$ *matrix. Then*

$$\inf_{\mathbf{B}} \sup_{\mathbf{B}'\mathbf{X}=0} \frac{\mathbf{X}'\mathbf{A}\mathbf{X}}{\mathbf{X}'\mathbf{X}} = \lambda_{k+1} \tag{1f.2.4}$$

$$\sup_{\mathbf{B}} \inf_{\mathbf{B}'\mathbf{X}=0} \frac{\mathbf{X}'\mathbf{A}\mathbf{X}}{\mathbf{X}'\mathbf{X}} = \lambda_{m-k}. \tag{1f.2.5}$$

The value λ_{k+1} *is attained in* (1f.2.4) *when* $\mathbf{B}$ *is the matrix with* $\mathbf{P}_1, \ldots, \mathbf{P}_k$ *as its columns, and the value* λ_{m-k} *is attained when* $\mathbf{B}$ *is the matrix with* $\mathbf{P}_{m-k+1}, \ldots, \mathbf{P}_m$ *as its columns.*

As in (i) let $\mathbf{X} = C_1\mathbf{P}_1 + \cdots + C_m\mathbf{P}_m = \mathbf{PC}$ where $\mathbf{P}$ is the $m \times m$ matrix with $\mathbf{P}_i$ as its ith column. Then the restriction $\mathbf{B}'\mathbf{X} = \mathbf{0} \Leftrightarrow \mathbf{B}'\mathbf{PC} = \mathbf{0}$ or $\mathbf{G}'\mathbf{C} = \mathbf{0}$ where $\mathbf{G} = \mathbf{P}'\mathbf{B}$. The problem of (1f.2.4) reduces to the determination of

$$\inf_{\mathbf{G}} \sup_{\mathbf{G}'\mathbf{C}=\mathbf{0}} \frac{C_1{}^2\lambda_1 + \cdots + C_m{}^2\lambda_m}{C_1{}^2 + \cdots + C_m{}^2}. \tag{1f.2.6}$$

But

$$\sup_{\mathbf{G}'\mathbf{C}=\mathbf{0}} \frac{\sum_1^m C_i^2 \lambda_i}{\sum_1^m C_i^2} \geqslant \sup_{\mathbf{G}'\mathbf{C}=\mathbf{0}} \frac{\sum_1^{k+1} C_i^2 \lambda_i}{\sum_1^{k+1} C_i^2} \geqslant \lambda_{k+1}, \tag{1f.2.7}$$

if we choose the subset of $\mathbf{C}$ such that $\mathbf{G}'\mathbf{C} = \mathbf{0}$, $C_{k+2} = \cdots = C_m = 0$. Such a subset of $\mathbf{C}$ is not empty. The result (1f.2.7) is independent of $\mathbf{G}$. Hence (1f.2.6) $\geqslant \lambda_{k+1}$. But, by (ii), there exists a matrix $\mathbf{B}$, viz., the matrix with $\mathbf{P}_1, \ldots, \mathbf{P}_k$ as its columns and an $\mathbf{X}$ such that the value λ_{k+1} is attained which establishes (1f.2.4). Similarly (1f.2.5) is proved.

(iv) *Let* $\mathbf{X}_1, \ldots, \mathbf{X}_k$ *be k mutually orthogonal vectors. Then*

$$\sup_{\mathbf{X}_1, \ldots, \mathbf{X}_k} \sum_1^k \frac{\mathbf{X}_i' \mathbf{A} \mathbf{X}_i}{\mathbf{X}_i' \mathbf{X}_i} = \sum_1^k \lambda_i \tag{1f.2.8}$$

and the supremum is attained when $\mathbf{X}_i \propto \mathbf{P}_i$, $i = 1, \ldots, k$.

Choose $\mathbf{X}_i = C_{i1}\mathbf{P}_1 + \cdots + C_{im}\mathbf{P}_m$, with the additional condition $\mathbf{X}_i'\mathbf{X}_i = 1$ (without loss of generality). Then

$$\sum \frac{\mathbf{X}_i' \mathbf{A} \mathbf{X}_i}{\mathbf{X}_i' \mathbf{X}} = \lambda_1 (C_{11}^2 + \cdots + C_{k1}^2) + \cdots + \lambda_m (C_{1m}^2 + \cdots + C_{km}^2), \tag{1f.2.9}$$

where the coefficient of each λ_i is $\leqslant 1$ and the sum of the coefficients is k. The optimum choice of coefficients, to maximize the left-hand side of (1f.2.9), is unity for $\lambda_1, \ldots, \lambda_k$ and zero for the rest, which is possible by choosing $\mathbf{X}_i = \mathbf{P}_i$, $i = 1, \ldots, k$.

(v) *Let* $\mathbf{B}$ *be any matrix of rank k, then*

$$\inf_{\mathbf{B}} \|\mathbf{A} - \mathbf{B}\| = (\lambda_{k+1}^2 + \cdots + \lambda_m^2)^{1/2}, \tag{1f.2.10}$$

and the infimum is attained when

$$\mathbf{B} = \lambda_1 \mathbf{P}_1 \mathbf{P}_1' + \cdots + \lambda_k \mathbf{P}_k \mathbf{P}_k'. \tag{1f.2.11}$$

The symbol $\|\mathbf{C}\|$ denotes the Euclidean norm of a matrix $\mathbf{C} = (C_{ij})$, defined by $(\sum \sum C_{ij}^2)^{1/2}$. Thus $\|\mathbf{C}\|$ is a measure of smallness or the largeness of a matrix. The result (1f.2.11) shows that the choice of $\mathbf{B}$ as in (1f.2.11) provides the closest fit to $\mathbf{A}$ among all matrices with the given rank k.

Observe that

$$\|\mathbf{A} - \mathbf{B}\|^2 = \|\mathbf{P}'\mathbf{A}\mathbf{P} - \mathbf{P}'\mathbf{B}\mathbf{P}\|^2 = \|\mathbf{\Lambda} - \mathbf{G}\|^2$$

where $\mathbf{G} = (g_{ij}) = \mathbf{P}'\mathbf{B}\mathbf{P}$ and of rank k, and $\mathbf{\Lambda}$ is the diagonal matrix with $\lambda_1, \ldots, \lambda_m$ as the diagonal elements.

Let $\mathbf{G} = \mathbf{C}\mathbf{D}$ be the rank factorization (iv, 1b.2) of $\mathbf{G}$. Then $\min \|\mathbf{\Lambda} - \mathbf{C}\mathbf{D}\|^2$ with respect to $\mathbf{D}$ is $\text{tr}\mathbf{\Lambda}^2 - \text{tr}\,\mathbf{C}'\mathbf{\Lambda}^2\mathbf{C}$, which is attained when $\mathbf{D} = \mathbf{C}'\mathbf{\Lambda}$. From (1f.2.8), sup $\mathbf{C}'\mathbf{\Lambda}^2\mathbf{C}$ with respect to $\mathbf{C}$ is attained when $\mathbf{C}_i$, the i-tr column of $\mathbf{C}$ is e_i, the i-tr elementary vector. Hence

$$\min_{\mathbf{G}} \|\mathbf{\Lambda} - G\|^2 = \lambda_{k+1}^2 + \ldots + \lambda_m{}^2 \tag{1f.2.12}$$

Thus the optimum choice of $\mathbf{B}$ is $\mathbf{P}'\mathbf{G}\mathbf{P}$ where $\mathbf{G}$ is as chosen above, i.e., $\mathbf{B}$ is as defined in (1f.2.11). [Also see Example 10, p.70].

(vi) A STURMIAN SEPARATION THEOREM. *Consider the sequence of matrices $\mathbf{A}_r = (a_{ij})$, $i, j = 1, \ldots, r$ for $r = 1, \ldots, m$. Let $\lambda_1(\mathbf{A}_r) \geqslant \cdots \geqslant \lambda_r(\mathbf{A}_r)$ denote the eigenvalues of $\mathbf{A}_r$. Then*

$$\lambda_{k+1}(\mathbf{A}_{i+1}) \leqslant \lambda_k(\mathbf{A}_i) \leqslant \lambda_k(\mathbf{A}_{i+1}). \tag{1f.2.13}$$

The result (1f.2.13) is easily obtained by using the interpretation (1f.2.1, 1f.2.3) of the eigenvalues of a matrix. As a corollary to (vi), we have the following theorem.

(vii) THE POINCARE SEPARATION THEOREM. *Let $\mathbf{B}$ be $m \times k$ matrix such that $\mathbf{B}'\mathbf{B} = \mathbf{I}_k$ (identity matrix of order k). Then*

$$\begin{aligned} \lambda_i(\mathbf{B}'\mathbf{A}\mathbf{B}) &\leqslant \lambda_i(\mathbf{A}), \qquad & i &= 1, \ldots, k \\ \lambda_{k-j}(\mathbf{B}'\mathbf{A}\mathbf{B}) &\geqslant \lambda_{m-j}(\mathbf{A}), \qquad & j &= 0, 1, \ldots, k-1. \end{aligned} \tag{2.1f.14}$$

(viii) *Let $\mathbf{A}$ be p.d. matrix of order m and $\mathbf{B}$, a $m \times k$ matrix of rank k such that the diagonal elements of $\mathbf{B}'\mathbf{B}$ are given numbers $g_1, \ldots, g_k$. Then*

$$\max_{\mathbf{B}} |\mathbf{B}'\mathbf{A}\mathbf{B}| = g_1 \cdots g_k \lambda_1 \cdots \lambda_k. \tag{1f.2.15}$$

Observe that

$$\max_{\mathbf{B}} |\mathbf{B}'\mathbf{A}\mathbf{B}| = g_1 \cdots g_k \max_{\mathbf{C}} |\mathbf{C}'\mathbf{A}\mathbf{C}|,$$

where the diagonal elements of $\mathbf{C}'\mathbf{C}$ are all unity.

Further, let $\gamma_1, \ldots, \gamma_k$ be the eigenvalues of $\mathbf{C}'\mathbf{C}$. Since $\gamma_1 + \cdots + \gamma_k = k$, it follows that $\gamma_1 \cdots \gamma_k \leqslant 1$. Choose the matrix $\mathbf{G}$ such that its ith column is $\gamma_i^{-1/2}$ times the ith eigenvector of $\mathbf{C}'\mathbf{C}$. Then we have the following:

(a) $\mathbf{D}'\mathbf{D} = \mathbf{I}$, choosing $\mathbf{D} = \mathbf{C}\mathbf{G}$

(b) $$\begin{aligned} |\mathbf{D}'\mathbf{A}\mathbf{D}| &= |\mathbf{G}'\mathbf{C}'\mathbf{A}\mathbf{C}\mathbf{G}| = |\mathbf{C}'\mathbf{A}\mathbf{C}|\,|\mathbf{G}'\mathbf{G}| \\ &= |\mathbf{C}'\mathbf{A}\mathbf{C}|(\gamma_1 \cdots \gamma_k)^{-1} \geqslant |\mathbf{C}'\mathbf{A}\mathbf{C}|. \end{aligned}$$

Hence

$$\max_{\mathbf{C}} |\mathbf{C}'\mathbf{A}\mathbf{C}| = \max_{\mathbf{D}'\mathbf{D}=\mathbf{I}} |\mathbf{D}'\mathbf{A}\mathbf{D}|.$$

Now, an application of (1f.2.14) gives

$$|\mathbf{D}'\mathbf{A}\mathbf{D}| = \prod_{i=1}^{k} \lambda_i(\mathbf{D}'\mathbf{A}\mathbf{D}) \leqslant \prod_{i=1}^{k} \lambda_i(\mathbf{A}).$$

But the equality is attained when the columns of **D** are the first k eigenvectors of **A**, thus proving (1f.2.15). [Also see Examples **20–22**.]

Observe that the result

$$\min_{\mathbf{B}} |\mathbf{B}'\mathbf{A}\mathbf{B}| = g_1 \cdots g_k \lambda_{m-k+1} \cdots \lambda_m$$

is not true unless the columns of **B** are assumed to be orthogonal.

Let $\mathscr{C}_1$ be the class of $m \times n$ matrices **A** such that $\mathbf{AX} = 0$ and $\operatorname{tr} \mathbf{A}\mathbf{V}_i' = p_i$, $i = 1, \ldots, k$, where $\mathbf{X}$, $\mathbf{V}_i$ and p_i are given matrices and scalars, $\mathscr{C}_2$ be the same class as $\mathscr{C}_1$ with square symmetric matrices **A** and $\mathbf{V}_i$. Further let **P** be the orthogonal projection operator onto $\mathscr{M}(\mathbf{X})$, and $\mathbf{Q} = \mathbf{I} - \mathbf{P}$, with the inner product definition, $(\mathbf{x}, \mathbf{y}) = \mathbf{y}^*\mathbf{x}$.

1f.3. Minimum Trace Problems

(i) *Min* tr $\mathbf{AA}'$ *when* $\mathbf{A} \in \mathscr{C}_1$ *is attained at*

$$\mathbf{A}_* = \sum \lambda_i \mathbf{V}_i \mathbf{Q} \tag{1f.3.1}$$

where $\lambda_1, \ldots, \lambda_k$ *are solutions of*

$$\sum_{i=1}^{k} \operatorname{tr} \mathbf{V}_i \mathbf{Q} \mathbf{V}_j' = p_j, \qquad j = 1, \ldots, k. \tag{1f.3.2}$$

Observe that $\mathbf{A}_*\mathbf{X} = \mathbf{0}$. Now consider an alternative $\mathbf{A}_* + \mathbf{D}$ such that $\operatorname{tr} \mathbf{D}\mathbf{V}_i' = 0$ and $\mathbf{DX} = \mathbf{0} \Rightarrow \mathbf{QD}' = \mathbf{D}'$. Check that

$$\operatorname{tr} \mathbf{A}_*\mathbf{D}' = \sum \lambda_i \operatorname{tr} \mathbf{V}_i\mathbf{Q}\mathbf{D}' = \sum \lambda_i \operatorname{tr} \mathbf{V}_i \mathbf{D}' = 0.$$

Then

$$\operatorname{tr}(\mathbf{A}_* + \mathbf{D})(\mathbf{A}_* + \mathbf{D})' = \operatorname{tr} \mathbf{A}_*\mathbf{A}_*' + \operatorname{tr} \mathbf{DD}' \geqslant \operatorname{tr} \mathbf{A}_*\mathbf{A}_*'$$

which proves the desired result. It may be noted that the equations (1f.3.2) are consistent.

(ii) *Min* tr **AA** *when* $\mathbf{A} \in \mathscr{C}_2$ *is attained at*

$$\mathbf{A}_* = \sum \lambda_i \mathbf{Q}\mathbf{V}_i\mathbf{Q} \tag{1f.3.3}$$

where $\lambda_i, \ldots, \lambda_k$ *are solutions of*

$$\sum \lambda_i \operatorname{tr} \mathbf{Q}\mathbf{V}_i\mathbf{Q}\mathbf{V}_j = p_j, \qquad j = 1, \ldots, k. \tag{1f.3.4}$$

The result is proved on the same lines as (i).

(iii) *Let* $\mathbf{V}$ *be a p.d. matrix. Then min* tr $\mathbf{AVAV}$ *when* $\mathbf{A} \in \mathscr{C}_2$ *is attained at*

$$\mathbf{A}_* = \sum \lambda_i \mathbf{Q}_V' \mathbf{V}^{-1} \mathbf{V}_i \mathbf{V}^{-1} \mathbf{Q}_V \tag{1f.3.5}$$

where $\lambda_1, \ldots, \lambda_k$ *are solutions of*

$$\sum \lambda_i \operatorname{tr} \mathbf{Q}_V' \mathbf{V}^{-1} \mathbf{V}_i \mathbf{V}^{-1} \mathbf{Q}_V \mathbf{V}_j = p_j, \qquad j = 1, \ldots, k \tag{1f.3.6}$$

and $\mathbf{Q}_V = \mathbf{I} - \mathbf{P}_V$, $\mathbf{P}_V = \mathbf{x}(\mathbf{x}'\mathbf{v}^{-1}\mathbf{x})^{-}\mathbf{x}\mathbf{v}^{-1}$, *the orthogonal projection operator onto* $\mathscr{M}(\mathbf{X})$ *with the inner product definition* $(\mathbf{x}, \mathbf{y}) = \mathbf{y}^*\mathbf{V}^{-1}\mathbf{x}$.

The result is proved by considering the matrix $\mathbf{B} = \mathbf{V}^{1/2}\mathbf{A}\mathbf{V}^{1/2}$ and applying the result of (ii).

Let $\mathscr{C}_3$ be the class of symmetric matrices $\mathbf{A}$ such that $\mathbf{X}'\mathbf{A}\mathbf{X}0 =$ and $\operatorname{tr} \mathbf{A}\mathbf{V}_i = p_i$, $i = 1, \ldots, k$ where $\mathbf{V}_i$ are symmetric.

(iv) *Min* tr $\mathbf{AA}$ *when* $\mathbf{A} \in \mathscr{C}_3$ *is attained at*

$$\mathbf{A}_* = \sum \lambda_i (\mathbf{V}_i - \mathbf{P}\mathbf{V}_i\mathbf{P}) \tag{1f.3.7}$$

where $\lambda_1, \ldots, \lambda_k$ *are solutions of*

$$\sum \lambda_i \operatorname{tr}(\mathbf{V}_i - \mathbf{P}\mathbf{V}_i\mathbf{P})\mathbf{V}_j = p_j, \qquad j = 1, \ldots, k. \tag{1f.3.8}$$

We note that a general solution of $\mathbf{X}'\mathbf{A}\mathbf{X} = \mathbf{0}$ is of the form $\mathbf{T} - \mathbf{PTP}$ and an alternative to $\mathbf{A}_*$ can be written as $\mathbf{A}_* + \mathbf{T} - \mathbf{PTP}$ with the condition $\operatorname{tr}(\mathbf{T} - \mathbf{PTP})\mathbf{V}_i = 0$. Now

$$\operatorname{tr}(\mathbf{V}_i - \mathbf{P}\mathbf{V}_i\mathbf{P})(\mathbf{T} - \mathbf{PTP}) = \operatorname{tr} \mathbf{V}_i(\mathbf{T} - \mathbf{PTP}) = 0.$$

Then

$$\begin{aligned}\operatorname{tr}(\mathbf{A}_* + \mathbf{T} - \mathbf{PTP})(\mathbf{A}_* + \mathbf{T} - \mathbf{PTP}) &= \operatorname{tr} \mathbf{A}_*\mathbf{A}_* + \operatorname{tr}(\mathbf{T} - \mathbf{PTP})(\mathbf{T} - \mathbf{PTP}) \\ &\geqslant \operatorname{tr} \mathbf{A}_*\mathbf{A}_*\end{aligned}$$

which proves the desired result.

(v) *Let* $\mathbf{V}$ *be a p.d. matrix. Then min* tr $\mathbf{AVAV}$ *when* $\mathbf{A} \in \mathscr{C}_3$ *is attained at*

$$\mathbf{A}_* = \sum \lambda_i \mathbf{V}^{-1}(\mathbf{V}_i - \mathbf{P}_V\mathbf{V}_i\mathbf{P}_V')\mathbf{V}^{-1} \tag{1f.3.9}$$

where $\lambda_1, \ldots, \lambda_k$ *are solutions of*

$$\sum \lambda_i \operatorname{tr} \mathbf{V}^{-1}(\mathbf{V}_i - \mathbf{P}_V\mathbf{V}_i\mathbf{P}_V')\mathbf{V}^{-1}\mathbf{V}_j = p_j, \; j = 1, \ldots, k. \tag{1f.3.10}$$

The result is obtained by applying the result of (iv) after making the transformation $\mathbf{V}^{1/2}\mathbf{A}\mathbf{V}^{1/2} = \mathbf{B}$.

Definitions of special matrices discussed in this Chapter are assembled in Table 1.

TABLE 1

Type of Matrix $\mathbf{A}$	Definition
Symmetric	$\mathbf{A}=\mathbf{A}'$ ($\mathbf{A}'$ is the transpose)
Hermitian	$\mathbf{A}=\mathbf{A}^*$ ($\mathbf{A}^*$ is the conjugate transpose)
Idempotent	$\mathbf{A}^2=\mathbf{A}$
Tripotent	$\mathbf{A}^3=\mathbf{A}$
Nilpotent	$\mathbf{A}^k=\mathbf{0}$ for some integer k
p.d. (positive definite)	$\mathbf{XAX}>0$ for all non-null $\mathbf{X}$
p.s.d. (positive semidefinite)	$\mathbf{XAX}\geqslant 0$ for all $\mathbf{X}$ and $\mathbf{XAX}=0$ for some non-null $\mathbf{X}$
n.n.d. (non-negative definite)	$\mathbf{X}^*\mathbf{AX}\geqslant 0$ for all $\mathbf{X}$
Normal	$\mathbf{AA}^*=\mathbf{A}^*\mathbf{A}$
Semisimple	$\mathbf{A}=\mathbf{P\Lambda P}^{-1}$ for some $\mathbf{P}$ and diagonal $\mathbf{\Lambda}$
Orthogonal	$\mathbf{AA}'=\mathbf{A}'\mathbf{A}=\mathbf{I}$
Unitary	$\mathbf{AA}^*=\mathbf{A}^*\mathbf{A}=\mathbf{I}$
g-Inverse	$\mathbf{A}^-$
Reflexive g-inverse	A_r^-
Minimum N-norm g-inverse	$A_{m(\mathbf{N})}^-$ [see (1c.5.1)]
M-Least squares g-inverse	$A_{l(\mathbf{M})}^-$ [see (1c.5.3)]
Minimum N-norm, M-least squares g-inverse	$A_{M\mathbf{N}}^+$ [see (1c.5.5)]

COMPLEMENTS AND PROBLEMS

1 Latent roots of matrices

1.1 For the $m \times m$ matrix

$$(a_{ij}) = \begin{pmatrix} a & b & \cdots & b \\ b & a & \cdots & b \\ \cdot & \cdot & \cdots & \cdot \\ b & b & \cdots & a \end{pmatrix},$$

establish the following:

(i) $|a_{ij}| = (a-b)^{n-1}(a+\overline{n-1}b)$.

(ii) The inverse is also of the same form

$$a^{ii} = \frac{a+\overline{n-2}b}{(a+\overline{n-1}b)(a-b)}, \qquad a^{ij} = \frac{-b}{(a+\overline{n-1}b)(a-b)}, \qquad i \neq j.$$

(iii) The latent roots of the matrix are $(a+\overline{n-1}b)$ with the latent vector $(1, 1, \ldots, 1)$ and $(a-b)$ of multiplicity $(n-1)$ with $(n-1)$ latent (any set of) vectors orthogonal to $(1, 1, \ldots, 1)$.

1.2 For the circulant

$$(a_{ij}) = \begin{pmatrix} a_1 & a_2 & \cdots & a_n \\ a_2 & a_3 & \cdots & a_1 \\ \cdot & \cdot & \cdots & \cdot \\ a_n & a_1 & \cdots & a_{n-1} \end{pmatrix},$$

prove the following results:

(i) $|a_{ij}| = \prod_{i=1}^{n} [a_1 + a_2 \omega_i + a_3 \omega_i^2 + \cdots + a_n \omega_i^{n-1}]$, where ω_i is the ith root of unity.

(ii) The latent roots are $\lambda_i = a_1 + a_2 \omega_i + \cdots + a_n \omega_i^{n-1}$, $i = 1, \ldots, n$ with the associated vectors $(1, \omega_i, \ldots, \omega_i^{n-1})$, $i = 1, \ldots, n$.

1.3 Generalize the results of Example **1** to a matrix of the form

$$\begin{pmatrix} \mathbf{A} & \mathbf{B} & \cdots & \mathbf{B} \\ \mathbf{B} & \mathbf{A} & \cdots & \mathbf{B} \\ \cdot & \cdot & \cdots & \cdot \\ \mathbf{B} & \mathbf{B} & \cdots & \mathbf{A} \end{pmatrix}$$

where $\mathbf{A}$ and $\mathbf{B}$ are themselves square matrices of order k. Show in particular that the value of the determinant is $|\mathbf{A} - \mathbf{B}|^{n-1} |\mathbf{A} + \overline{n-1}\mathbf{B}|$.

1.4 Let $\mathbf{U}$ be an $m \times n$ matrix. The nonvanishing roots of $|\mathbf{UU}' - \lambda\mathbf{I}| = 0$ and of $|\mathbf{U}'\mathbf{U} - \lambda\mathbf{I}| = 0$ are identical. [If λ^* is a non-zero root of $|\mathbf{UU}' - \lambda\mathbf{I}| = 0$, there exists a vector $\mathbf{L}$ such that $\mathbf{UU}'\mathbf{L} = \lambda^*\mathbf{L}$. By multiplying both sides by $\mathbf{U}'$, $\mathbf{U}'\mathbf{UU}'\mathbf{L} = \lambda^*\mathbf{U}'\mathbf{L}$ or $(\mathbf{U}'\mathbf{U} - \lambda^*\mathbf{I})\mathbf{U}'\mathbf{L} = 0$. Hence λ^* is a latent root of $\mathbf{U}'\mathbf{U}$.]

1.5 Prove that $|\mathbf{UU}' - \mathbf{I}| = (\lambda_1 - 1) \cdots (\lambda_r - 1)(0 - 1)^{m-r}$ and $|\mathbf{U}'\mathbf{U} - \mathbf{I}| = (\lambda_1 - 1) \cdots (\lambda_r - 1)(0 - 1)^{n-r}$ where $\lambda_1, \ldots, \lambda_r$ are the non-zero latent roots of $\mathbf{UU}'$ or $\mathbf{U}'\mathbf{U}$.

1.6 Let $\mathbf{U}$ be an $m \times n$ matrix of rank r. The stationary values of $[\mathbf{C}_1'\mathbf{UC}_2]^2$ for variations of vectors $\mathbf{C}_1$ and $\mathbf{C}_2$ of unit length are the latent roots of $\mathbf{UU}'$.

1.7 Show that for an orthogonal matrix, (a) all the latent roots are of modulus unity, (b) if λ is a latent root, $1/\lambda$ is also a latent root, and (c) if the matrix is of odd order, at least one root is ± 1.

1.8 If $\lambda_1, \lambda_2, \ldots$ are the latent roots of a square matrix $\mathbf{A}$, then (a) Trace $\mathbf{A} = \sum \lambda_i$, (b) trace $\mathbf{A}'\mathbf{A} \geqslant \sum \lambda_i^2$, and $|\mathbf{A}| = \lambda_1 \cdots \lambda_k$.

1.9 Let $\mathbf{A}$ and $\mathbf{B}$ be symmetric matrices of order m. If $\lambda_i(\mathbf{C})$ denotes the ith rank latent root of $\mathbf{C}$,

$$\left.\begin{aligned} &\lambda_i(\mathbf{A} + \mathbf{B}) \leqslant \lambda_j(\mathbf{A}) + \lambda_k(\mathbf{B}) \\ &\lambda_{m+i-1}(\mathbf{AB}) \geqslant \lambda_{m-j+1}(\mathbf{A})\lambda_{m-k+1}(\mathbf{B}) \end{aligned}\right\} \quad j + k \leqslant i + 1$$

(Anderson and Dasgupta, 1963).

2 Let $\mathbf{A}$ be $n \times m$ matrix of rank r, $\lambda_1, \ldots, \lambda_r$ be the non-zero eigenvalues and $\mathbf{P}_1, \ldots, \mathbf{P}_r$ be the eigenvectors of $\mathbf{A}'\mathbf{A}$. Then $\lambda_1^{-1}\mathbf{P}_1\mathbf{P}_1' + \cdots + \lambda_r^{-1}\mathbf{P}_r\mathbf{P}_r'$ is a g-inverse of $\mathbf{A}'\mathbf{A}$.

3 A p.d. or p.s.d. matrix is also called a Gramian and is represented by $\mathbf{G}$. The following results, some of which have been already proved, have important applications.

3.1 If $\mathbf{M}$ is $k \times p$ and of rank r, $\mathbf{MM}'$ is Gramian of rank r.

3.2 A matrix $\mathbf{G}$ of order k and rank r admits the decomposition $\mathbf{G} = \mathbf{FF}'$ where $\mathbf{F}$ is $k \times r$ and of rank r. If $\mathbf{M}$ is $k \times s$ such that $\mathbf{MM}' = \mathbf{FF}' = \mathbf{G}$, an orthogonal matrix $\mathbf{U}$ exists such that $\mathbf{M} = (\mathbf{F} \mid \mathbf{0})\mathbf{U}$.

3.3 If $\mathbf{G}$ is nonsingular, all its principal minors are nonsingular.

3.4 $\mathbf{MGM}'$ is Gramian for any $\mathbf{M}$ provided the product $\mathbf{MGM}'$ exists.

3.5 $\mathbf{G}$ admits a unique Gramian square root.
[There exists an orthogonal matrix $\mathbf{U}$ such that $\mathbf{G} = \mathbf{U\Lambda U}'$, $\mathbf{\Lambda}$ being the diagonal matrix of latent, non-negative roots $\lambda_1, \ldots, \lambda_n$. Verify that $\mathbf{M} = \mathbf{U\Lambda}^{1/2}\mathbf{U}'$ is (a) symmetric and (b) has latent roots $\lambda_1^{1/2}, \ldots, \lambda_n^{1/2}$ and hence Gramian. Furthermore, $\mathbf{MM} = \mathbf{U\Lambda}^{1/2}\mathbf{U}'\mathbf{U\Lambda}^{1/2}\mathbf{U}' = \mathbf{U\Lambda U}' = \mathbf{G}$. $\mathbf{M}$ is, therefore, the required root. This is indeed unique, for all such roots have the same latent roots $\lambda_1^{1/2}, \ldots, \lambda_n^{1/2}$ and vectors.]

3.6 $\mathbf{G}$ has the unique decomposition $\mathbf{G} = \mathbf{TT}'$ where $\mathbf{T}$ is a triangular matrix with diagonal elements non-negative.
[Proof consists in showing one-to-one correspondence between $\mathbf{G}$ and such a matrix $\mathbf{T}$.]

3.7 If $\mathbf{G}_1$and $\mathbf{G}_2$ are n.n.d., then so is $\mathbf{G}_1 + \mathbf{G}_2$ and $\mathcal{M}(\mathbf{G}_1) \subset \mathcal{M}(\mathbf{G}_1 + \mathbf{G}_2)$ and $\mathcal{M}(\mathbf{G}_2) \subset \mathcal{M}(\mathbf{G}_1 + \mathbf{G}_2)$.

4 *Lagrange's theorem.* Let $\mathbf{S}$ be any square matrix of order n and rank $r > 0$, and $\mathbf{X}$, $\mathbf{Y}$ be column vectors such that $\mathbf{X}'\mathbf{SY} \neq 0$. Then the residual matrix

$$\mathbf{S}_1 = \mathbf{S} - \frac{\mathbf{SYX}'\mathbf{S}}{\mathbf{X}'\mathbf{SY}}$$

is exactly of rank $r - 1$.

More generally, if $\mathbf{S}$ is $n \times m$ of rank $r > 0$, $\mathbf{A}$ and $\mathbf{B}$ are of orders $s \times n$ and $s \times m$ respectively, where $s \leqslant r$ and $\mathbf{ASB}'$ is nonsingular, the residual matrix

$$\mathbf{S}_1 = \mathbf{S} - \mathbf{SB}'(\mathbf{ASB}')^{-1}\mathbf{AS}$$

is exactly of rank $(r - s)$ and $\mathbf{S}_1$ is Gramian if $\mathbf{S}$ is Gramian.

5 Consider the matrix (x_{ij}), $i = 1, \ldots, m$; $j = 1, \ldots, n$, and define

$$\sum_j x_{ij} = n\bar{x}_i, \qquad S_{tu} = \sum_j (x_{tj} - \bar{x}_t)(x_{uj} - \bar{x}_u).$$

Then the $m \times m$ matrix (S_{tu}) is

(i) positive definite when the rank of $(x_{ij} - \bar{x}_i)$ is m,

(ii) positive semidefinite when the rank of $(x_{ij} - \bar{x}_i)$ is less than m, in which case there exist constants $p_1, \ldots, p_m$ such that

$$\sum p_i(x_{ij} - \bar{x}_i) = 0, \qquad j = 1, \ldots, n,$$

(iii) necessarily positive semidefinite when $n < p$.

6 Reduction of matrices.

6.1 If $\mathbf{A}$ is a square matrix of order m, there exists an orthogonal matrix $\mathbf{C}$ such that $\mathbf{AC}$ is triangular with all elements below the main diagonal zero and with diagonal elements $\geqslant 0$.

6.2 *Polar reduction.* If $|\mathbf{A}| \neq 0$, then $\mathbf{A} = \mathbf{GH}$ where $\mathbf{G}$ is p.d. and $\mathbf{H}$ is orthogonal. [Let $\mathbf{G}$ be the unique Gramian square root of $\mathbf{A}'\mathbf{A}$. $\mathbf{G}$ is p.d. as $|\mathbf{A}| \neq 0$. $\mathbf{H}$ is then $\mathbf{G}^{-1}\mathbf{A}$.]

6.3 Let $\mathbf{A}$ be symmetric of order m and rank m. There exists a triangular matrix $\mathbf{T}$ such that $\mathbf{A} = \mathbf{T}'\mathbf{\Delta}^{-1}\mathbf{T}$ where $\mathbf{\Delta}$ is a diagonal matrix with diagonal elements the same as those in $\mathbf{T}$.

7 If $\mathbf{A} = \mathbf{B} + \mathbf{C}$, $\mathbf{B}$ is p.d., and $\mathbf{C}$ is skew symmetric, then $|\mathbf{A}| \geqslant |\mathbf{B}|$.

8 If $\mathbf{A}$ and $\mathbf{B}$ are real, p.d., and of order n, then

(i) $|\lambda\mathbf{A} + (1-\lambda)\mathbf{B}| \geqslant |\mathbf{A}|^{\lambda}|\mathbf{B}|^{1-\lambda}, \qquad 0 \leqslant \lambda \leqslant 1,$

(ii) $|\mathbf{A} + \mathbf{B}|^{1/n} \geqslant |\mathbf{A}|^{1/n} + |\mathbf{B}|^{1/n}$.

9 If $\mathbf{B}$ is p.d. and $(\mathbf{A} - \mathbf{B})$ is n.n.d., then

(i) $|\mathbf{A} - \lambda\mathbf{B}| = 0$ has all its roots $\lambda \geqslant 1$ and conversely if all roots $\lambda \geqslant 1$, then $(\mathbf{A} - \mathbf{B})$ is n.n.d.;

(ii) $|\mathbf{A}| \geqslant |\mathbf{B}|$ [as a consequence of (i), $|\mathbf{A}|/|\mathbf{B}| = \lambda_1\lambda_2 \cdots \geqslant 1$. It follows from this result that if $\mathbf{C}$ is p.d. and $\mathbf{D}$ is n.n.d. then $|\mathbf{C}| \leqslant |\mathbf{C}+\mathbf{D}|$];

(iii) $(\mathbf{B}^{-1} - \mathbf{A}^{-1})$ is n.n.d. [this is true if $|\mathbf{B}^{-1} - \mu\mathbf{A}^{-1}| = 0$ has all its roots $\mu \geqslant 1$. But the roots of $|\mathbf{B}^{-1} - \mu\mathbf{A}^{-1}| = 0$ are the same as the roots of $|\mathbf{A} - \mu\mathbf{B}| = 0$, and by (i), $\mu \geqslant 1$];

(iv) If $\mathbf{A}_r$ is a principal minor of $\mathbf{A}$ of order r and $\mathbf{B}_r$, the corresponding minor of $\mathbf{B}$, then $\mathbf{A}_r - \mathbf{B}_r$ is n.n.d.

10 Let $\mathbf{A}$ be an $m \times n$ matrix. Determine an $m \times n$ matrix $\mathbf{B}$ of given rank $\mathbf{r}$ such that $\|\mathbf{A} - \mathbf{B}\|$ is a minimum.

[Hint: Consider the first r eigenvectors of the matrix $\mathbf{AA}'$. Let them be $\mathbf{P}_1, \ldots, \mathbf{P}_r$. If $\boldsymbol{\alpha}_i$ is the ith column vector of $\mathbf{A}$, then the ith column of $\mathbf{B}_*$ for which $\|\mathbf{A} - \mathbf{B}\|$ is a minimum is $(\mathbf{P}_1'\boldsymbol{\alpha}_i)\mathbf{P}_1 + \cdots + (\mathbf{P}_r'\boldsymbol{\alpha}_i)\mathbf{P}_r$. Show that for any other choice of r orthonormal vectors which constitute a basis of $\mathbf{B}$ the Euclidean norm $\|\mathbf{A} - \mathbf{B}\|$ possibly has a higher value.]

11 Let (a_{ij}) be $m \times t$ matrix of non-negative terms and

$$\sum_i a_{ij} = a_{.j}, \quad \sum_j a_{ij} = a_{i.}, \quad \sum_i a_{i.} = \sum_j a_{.j} = a_{..}$$

The following then are true:

(i) $mt \sum \sum a_{ij} a_{i.} a_{.j} \geqslant (a_{..})^3$

(ii) $\sum \sum a_{ij}(p_i \sum_1^t a_{is} q_s)(q_j (\sum_1^m a_{rj} p_r) \geqslant (\sum \sum a_{ij} p_i q_j)$, for any p_i, q_j such that

$$\sum_1^m p_i = \sum_1^t q_j = 1, \qquad p_i, q_j > 0.$$

(iii) As a special case when $m = t$, $a_{ij} = a_{ji}$ and $p_i = q_i$, we have

$$\sum \sum a_{ij} p_i' p_j' \geqslant \sum \sum a_{ij} p_i p_j,$$

where $p_i'(\sum \sum a_{ij} p_i p_j) = \sum a_{is} p_s$. (This inequality arises in a natural way in genetical theory. Let $a_{ij} \geqslant 0$ represent the relative viability of a diploid individual with the genotype $A_i A_j$ where $A_1, \ldots, A_m$ are the alleles at a single locus. If p_i is the frequency of A_i in a particular generation, the mean viability of the population under random mating is $\sum \sum a_{ij} p_i p_j$. The frequency p_i of A_i is changed to $(\sum a_{is} p_s)/(\sum \sum a_{ij} p_i p_j)$, and the mean viability in the next generation is $\sum \sum a_{ij} p_i' p_j'$. The inequality asserts that the mean viability cannot decrease from one generation to the next.) [Atkinson, Watterson, andMoran, 1960.]

12 Let $\mathbf{X}$ be a column vector $k \times 1$ and $\mathbf{A}$ be a square symmetric matrix of order k, with all elements non-negative. Then for any integer n,

$$(\mathbf{X}'\mathbf{A}^n\mathbf{X})(\mathbf{X}'\mathbf{X}) \geqslant (\mathbf{X}'\mathbf{A}\mathbf{X})^n$$

with equality if and only if $\mathbf{X}$ is a latent vector of $\mathbf{A}$. This inequality occurs in genetical theory. (Mulholland and Smith, 1959.)

13 Rules for obtaining simultaneous derivatives.

13.1 Let $f(\mathbf{X})$ be a scalar function of vector variable $\mathbf{X}$ and denote by $\partial f/\partial \mathbf{X}$ the column vector of elements, $\partial f/\partial X_1, \ldots, \partial f/\partial X_n$, where $\mathbf{X}' = (X_1, \ldots, X_n)$. Similarly we define $\partial^2 f/\partial \mathbf{X}^2$ as the matrix $(\partial^2 f/\partial X_i \partial X_j)$. Verify the following:

Scalar Function	Vector Derivative	Matrix Derivative
$f(\mathbf{X})$	$\partial f/\partial \mathbf{X}$	$\partial^2 f/\partial \mathbf{X}^2$
$\mathbf{M}'\mathbf{X}$	$\mathbf{M}$	$\mathbf{0}$
$\mathbf{X}'\mathbf{A}\mathbf{Z}$	$\mathbf{A}\mathbf{Z}$	$\mathbf{0}$
$\mathbf{X}'\mathbf{X}$	$2\mathbf{X}$	$2\mathbf{I}$
$\mathbf{X}'\mathbf{A}\mathbf{X}$	$2\mathbf{A}\mathbf{X}$	$2\mathbf{A}$ (when $\mathbf{A}$ is symmetric)

13.2 Let $f(\mathbf{B})$ be a scalar function of a matrix $\mathbf{B} = (b_{ij})$ and denote by $\partial f(\mathbf{B})/\partial \mathbf{B}$ the matrix $(\partial f(\mathbf{B})/\partial b_{ij})$. Verify that:

(i) $\dfrac{\partial}{\partial \mathbf{B}}(\mathbf{Y}'\mathbf{B}\mathbf{Z}) = \mathbf{Y}\mathbf{Z}'$

(ii) $\dfrac{\partial}{\partial \mathbf{B}} \operatorname{tr} \mathbf{B} = \mathbf{I}$

(iii) $\dfrac{\partial}{\partial \mathbf{B}} \mathbf{X}'\mathbf{B}\mathbf{X} = \mathbf{X}\mathbf{X}'$, if $\mathbf{B}$ is asymmetric

$= 2\mathbf{X}\mathbf{X}' - \operatorname{diag} \mathbf{X}\mathbf{X}'$, if $\mathbf{B}$ is symmetric

(iv) $\dfrac{\partial |\mathbf{B}|}{\partial \mathbf{B}} = |\mathbf{B}|(\mathbf{B}^{-1})'$, if $\mathbf{B}$ is asymmetric

$= |\mathbf{B}|(2\mathbf{B}^{-1} - \operatorname{diag} \mathbf{B}^{-1})$, if $\mathbf{B}$ is symmetric

(v) $\dfrac{\partial \operatorname{tr} \mathbf{BC}}{\partial \mathbf{B}} = \mathbf{C}'$, if $\mathbf{B}$ is asymmetric

$= \mathbf{C} + \mathbf{C}' - \operatorname{diag} \mathbf{C}$, if $\mathbf{B}$ is symmetric

(vi) $\dfrac{\partial \operatorname{tr} \mathbf{BCBD}}{\partial \mathbf{B}} = \mathbf{T}'$, if $\mathbf{B}$ is asymmetric

$= \mathbf{T} + \mathbf{T}' - \operatorname{diag} \mathbf{T}$, if $\mathbf{B}$ is symmetric

where $\mathbf{T} = \mathbf{DBC} + \mathbf{CBD}$, and diag $\mathbf{T}$ is the diagonal matrix with the same diagonal as $\mathbf{T}$.

Apply these rules to minimize $(\mathbf{Y} - \mathbf{AX})'(\mathbf{Y} - \mathbf{AX})$ with respect to $\mathbf{X}$, where $\mathbf{A}$ of order $m \times n$ and $\mathbf{Y}$ of order $m \times 1$ are given. Show that the minimizing equation is $\mathbf{A}'\mathbf{AX} = \mathbf{A}'\mathbf{Y}$.

14 Let $\mathbf{R}$ be a correlation matrix (non-n.d. with unity as diagonal elements) and $\mathbf{D}^2$, a diagonal matrix with $0 \leqslant d_i^2 \leqslant 1$. Prove that $\inf_{\mathbf{D}} \operatorname{rank}(\mathbf{R} - \mathbf{D}^2) \geqslant s$, the number of latent roots of $\mathbf{R}$ greater than unity. (Guttman, 1954.)

15 Let $\mathbf{A}_1, \mathbf{A}_2, \ldots, \mathbf{A}_t$ be symmetric square matrices. A necessary and sufficient condition for an orthogonal matrix $\mathbf{C}$ to exist such that $\mathbf{CA}_1\mathbf{C}', \ldots, \mathbf{CA}_t\mathbf{C}'$ are all diagonal is that $\mathbf{A}_i\mathbf{A}_j$ be symmetric for all i and j, that is, $\mathbf{A}_i$ and $\mathbf{A}_j$ commute.

16 A matrix $\mathbf{A}$ is said to be idempotent if $\mathbf{A}^2 = \mathbf{A}$. Establish the following results for a symmetric idempotent matrix $\mathbf{A}$.

16.1 Every latent root of $\mathbf{A}$ is either 0 or 1.

16.2 $|\mathbf{A}| \neq 0 \Rightarrow \mathbf{A} = \mathbf{I}$ (identity).

16.3 $\mathbf{A}$ is non-negative definite and admits the representation $\mathbf{A} = \mathbf{L}_1\mathbf{L}_1' + \cdots + \mathbf{L}_r\mathbf{L}_r'$, where $\mathbf{L}_1, \ldots, \mathbf{L}_r$ are orthonormal vectors and $r = \operatorname{rank} \mathbf{A}$.

16.4 If the ith diagonal element of $\mathbf{A}$ is zero, all the elements in the ith row and column are zero.

16.5 Trace $\mathbf{A}$ = Rank $\mathbf{A}$.

17 Let $\mathbf{A}_1, \ldots, \mathbf{A}_t$ be symmetric. If $\sum \mathbf{A}_i$ is idempotent and $\mathbf{A}_i \mathbf{A}_j = 0$ for $i \neq j$, then each $\mathbf{A}_i$ is idempotent and hence $\text{rank}(\sum \mathbf{A}_i) = \sum \text{rank } \mathbf{A}_i$.

18 Using the notation of the spectral decomposition of a nonsymmetric matrix (1c.3.14)

$$\mathbf{A} = \delta_1 \mathbf{P}_1 \mathbf{Q}_1' + \cdots + \delta_m \mathbf{P}_m \mathbf{Q}_m'$$

show that $\mathbf{A}$ can be written

$$\mathbf{A} = \delta_1 \mathbf{A}_1 + \cdots + \delta_m \mathbf{A}_m$$

where $\sum \mathbf{A}_i = \mathbf{I}$, $\mathbf{A}_i$ is idempotent, and $\mathbf{A}_i \mathbf{A}_j = 0$, $i \neq j$. Furthermore, if $L_j(z)$ is the Lagrange interpolation polynomial,

$$L_j(z) = \prod_{\substack{i=1 \\ i \neq j}}^{m} \frac{z - \delta_i}{\delta_j - \delta_i},$$

then $\mathbf{A}_i = L_j(\mathbf{A})$. [Hint: Choose $\mathbf{A}_i = \mathbf{P}_i \mathbf{Q}_i'$.]

19 *Relation between Two Inner Products in a Finite Dimensional Vector Space* $\mathscr{S}$. Let $\mathbf{X}, \mathbf{Y}$ be two vectors and define by $(\mathbf{X}, \mathbf{Y})_1$ and $(\mathbf{X}, \mathbf{Y})_2$ two possible inner products.

19.1 Show that a necessary and sufficient condition for a vector $\mathbf{P}_1$ to exist such that

$$\frac{(\mathbf{P}_1, \mathbf{P}_1)_1}{(\mathbf{P}_1, \mathbf{P}_1)_2} = \sup_{\mathbf{X}} \frac{(\mathbf{X}, \mathbf{X})_1}{(\mathbf{X}, \mathbf{X})_2}$$

is that $(\mathbf{P}_1, \mathbf{U})_1 = 0 \Leftrightarrow (\mathbf{P}_1, \mathbf{U})_2 = 0$ for any $\mathbf{U}$, that is, the spaces orthogonal to $\mathbf{P}_1$ according to the two definitions are the same. [Hint: Make use of the properties (**1a**.4) of the inner product only.]

19.2 There exists an orthonormal basis of $\mathscr{S}$ relative to both the definitions of inner product.

19.3 Let $\mathbf{P}_1, \ldots, \mathbf{P}_m$ be the orthonormal basis arranged in order such that $\lambda_1 \geqslant \cdots \geqslant \lambda_m$ where $\lambda_i = (\mathbf{P}_i, \mathbf{P}_i)_1/(\mathbf{P}_i, \mathbf{P}_i)_2$. Then λ_i and $\mathbf{P}_i$ are called eigenvalues and vectors of one inner product relative to the other. Note that if $\mathbf{A}$ and $\mathbf{B}$ are p.d. matrices, then $\mathbf{X}'\mathbf{A}\mathbf{Y}$ and $\mathbf{X}'\mathbf{B}\mathbf{Y}$ are valid inner products and $\lambda_1, \ldots, \lambda_m$ are the roots of the determinantal equation $|\mathbf{A} - \lambda \mathbf{B}| = 0$.

19.4 $\|\mathbf{A}\mathbf{X}\| \leqslant \|\mathbf{A}\mathbf{X} + \mathbf{B}\mathbf{Y}\|$ for all $\mathbf{X}, \mathbf{Y} \Leftrightarrow \mathbf{A}^*\mathbf{\Lambda}\mathbf{B} = \mathbf{0}$, where $(\mathbf{X}, \mathbf{Y}) = \mathbf{Y}^*\mathbf{\Lambda}\mathbf{X}$.

19.5 In U^n, the unitary space of n dimensions, every inner product is of the type $(\mathbf{X}, \mathbf{Y}) = \mathbf{Y}^*\mathbf{\Lambda}\mathbf{X}$ where $\mathbf{\Lambda}$ is a p.d. matrix.

20. Properties of p.d. and p.s.d. matrices.

20.1 If $\mathbf{A} = (a_{ij})$ is p.d. or p.s.d. and $a_{ii} = 0$, then $a_{ij} = 0$ for all j.

20.2 If $\mathbf{A}$ is p.d. and $\mathbf{A}^{-1} = (a^{ij})$, then

(a) $a^{ii} \geqslant 1/a_{ii}$, with equality if and only if $a_{ij} = 0$ for $j = 1, \ldots, i-1, i+1, \ldots, m$;

(b) $a^{ii} = 1/a_{ii}$ for all i implies $a_{ij} = 0$, $i \neq j$;

(c) $|\mathbf{A}| \leqslant a_{11} \cdots a_{pp}$ with equality if and only if $a_{ij} = 0$, $i \neq j$.

20.3 Let $\mathbf{A}$ be p.d. and

$$\mathbf{A}^{-1} = \begin{pmatrix} \mathbf{A}_{11} & \mathbf{A}_{12} \\ \mathbf{A}_{21} & \mathbf{A}_{22} \end{pmatrix}^{-1} = \begin{pmatrix} \mathbf{A}^{11} & \mathbf{A}^{12} \\ \mathbf{A}^{21} & \mathbf{A}^{22} \end{pmatrix}$$

where $\mathbf{A}_{11}$ is square. Then $\mathbf{A}^{11} - \mathbf{A}_{11}^{-1}$ is n.n.d.

21 *Kantorovich Inequality.* If $\mathbf{A}$ is p.d. Hermitian with latent roots $\lambda_1 \geqslant \cdots \geqslant \lambda_m > 0$, then

$$1 \leqslant (\mathbf{X}^*\mathbf{A}\mathbf{X})(\mathbf{X}^*\mathbf{A}^{-1}\mathbf{X}) \leq \frac{1}{4}\left[\left(\frac{\lambda_1}{\lambda_m}\right)^{1/2} + \left(\frac{\lambda_m}{\lambda_1}\right)^{1/2}\right]^2.$$

where $\mathbf{X}$ is normalized, i.e., $\mathbf{X}^*\mathbf{X} = 1$.

22 *Extension of the Results of* **1 f**. Let $\mathbf{A}$ be a symmetric matrix and $\mathbf{C}$ be a p.d. matrix. Denote by $\lambda_1 \geqslant \cdots \geqslant \lambda_m$, the roots of $|\mathbf{A} - \lambda\mathbf{C}| = 0$.

22.1 $\sup_{\mathbf{X}} \dfrac{\mathbf{X}'\mathbf{A}\mathbf{X}}{\mathbf{X}'\mathbf{C}\mathbf{X}} = \lambda_1, \quad \inf_{\mathbf{X}} \dfrac{\mathbf{X}'\mathbf{A}\mathbf{X}}{\mathbf{X}'\mathbf{C}\mathbf{X}} = \lambda_m.$

22.2 $\inf_{\mathbf{B}} \sup_{\mathbf{B}'\mathbf{C}\mathbf{X}=0} \dfrac{\mathbf{X}'\mathbf{A}\mathbf{X}}{\mathbf{X}'\mathbf{C}\mathbf{X}} = \lambda_{k+1}, \quad \sup_{\mathbf{B}} \inf_{\mathbf{B}'\mathbf{C}\mathbf{X}=0} \dfrac{\mathbf{X}'\mathbf{A}\mathbf{X}}{\mathbf{X}'\mathbf{C}\mathbf{X}} = \lambda_{m-k},$

where $\mathbf{B}$ is $m \times k$ matrix.

22.3 Let $\mathbf{X}_1, \ldots, \mathbf{X}_k$ be such that $\mathbf{X}_i'\mathbf{C}\mathbf{X}_j = 0$, $i \neq j$. Then

$$\sup_{\mathbf{X}_1, \ldots, \mathbf{X}_k} \sum_1^k \frac{\mathbf{X}_i'\mathbf{A}\mathbf{X}_i}{\mathbf{X}_i'\mathbf{C}\mathbf{X}_i} = \lambda_1 + \cdots + \lambda_k,$$

$$\inf_{\mathbf{X}_1, \ldots, \mathbf{X}_k} \prod_1^k \frac{\mathbf{X}_i'\mathbf{A}\mathbf{X}_i}{\mathbf{X}_i'\mathbf{C}\mathbf{X}_i} = \lambda_{m-k+1} \cdots \lambda_m.$$

23 *Computation of inverse matrices and certain products.* Let $\mathbf{A}$ be a p.d. matrix of order m and $\mathbf{D}$ be an $m \times k$ matrix. In statistical problems we need the computation of the matrices $\mathbf{A}^{-1}$ and $\mathbf{D}'\mathbf{A}^{-1}\mathbf{D}$.

23.1 Show that there exists a lower triangular matrix $\mathbf{B}$ such that $\mathbf{B}\mathbf{A} = \mathbf{T}$ where $\mathbf{T}$ is an upper triangular matrix and $\mathbf{A} = \mathbf{T}'\mathbf{T}$. Hence show that

$$\mathbf{A}^{-1} = \mathbf{B}'\mathbf{B}, \qquad \mathbf{D}'\mathbf{A}^{-1}\mathbf{D} = (\mathbf{B}\mathbf{D})'(\mathbf{B}\mathbf{D}).$$

23.2 The same formulas are applicable for the computation of $\mathbf{A}^-$ and $\mathbf{D'A^-D}$ when $\mathbf{A}$ is p.s.d.

23.3 Satisfy yourself that $\mathbf{T}$ and $\mathbf{B}$ can be obtained simultaneously by the following computational device.

For purposes of illustration take a fourth-order matrix for $\mathbf{A}$ (the elements below the diagonal are omitted) and set out the matrix $\mathbf{A}$ and a unit matrix as shown in Table 2.

(i) Row (2.1) is obtained by dividing each element of (1.1) by $t_{11} = \sqrt{a_{11}}$. Such an operation is indicated by $(2.1) = (1.1)/t_{11}$.

(ii) The element $t_{22} = \sqrt{a_{22} - t_{12}^2}$. Row (2.2) is obtained by the operation $[(1.2) - t_{12}(2.1)] \div t_{22}$, if $t_{22} \neq 0$, otherwise by $[(1.2) - t_{12}(2.1)]$.

(iii) The element $t_{33} = \sqrt{a_{33} - t_{13}^2 - t_{23}^2}$, and Row $(2.3) = [(1.3) - t_{13}(2.1) - t_{23}(2.2)] \div t_{33}$ if $t_{33} \neq 0$, otherwise by using unity instead of t_{33}. Similarly the last row (2.4) is obtained.

For numerical computations involving matrices see books by Dwyer (1951) and Householder (1964).

TABLE 2. Sweep Out by the Square Root Method

Row No.	A				I			
1.1	a_{11}	a_{12}	a_{13}	a_{14}	1			
1.2		a_{22}	a_{23}	a_{24}	.	1		
1.3			a_{33}	a_{34}	.	.	1	
1.4				a_{44}	.	.	.	1
2.1	t_{11}	t_{12}	t_{13}	t_{14}	b_{11}			
2.2		t_{22}	t_{23}	t_{24}	b_{21}	b_{22}		
2.3			t_{33}	t_{34}	b_{31}	b_{32}	b_{33}	
2.4				t_{44}	b_{41}	b_{42}	b_{43}	b_{44}
	T				B			

24 Computation of $(\mathbf{A'A})^{-1}$, $\mathbf{D'(A'A)^{-1}D}$, etc. Let $\mathbf{A'}$ be $m \times k$ matrix of rank k and $\mathbf{D}$ be $m \times p$ matrix. One method of computing $(\mathbf{A'A})^{-1}$, etc., is to compute first $\mathbf{A'A}$ and apply the method of Example **23**. But these can be done directly without computing $\mathbf{A'A}$.

24.1 Use the triangular reduction $\mathbf{RA} = \binom{\mathbf{T}}{0}$ of [(vi), **1b.2**] where $\mathbf{R}$ is orthogonal and $\mathbf{T}$ is upper triangular. Prove that $\mathbf{A'A} = \mathbf{T'T}$ and $(\mathbf{A'A})^{-1} = \mathbf{T}^{-1}(\mathbf{T}^{-1})'$. [It is easy to compute $\mathbf{T}^{-1}$, the inverse of a triangular matrix. Also $\mathbf{D}'(\mathbf{A'A})^{-1}\mathbf{D} = (\mathbf{D'T}^{-1})(\mathbf{D'T}^{-1})'$.]

24.2 Use the Gram-Schmidt reduction $\mathbf{A} = \mathbf{ST}$ where the columns of $\mathbf{S}$ are orthonormal. Show that $\mathbf{A'A} = \mathbf{T'T}$ and further that $\mathbf{T}$ is the same as in Example **24.1**. Thus we have two ways of computing $\mathbf{T}$, and it is believed that the computation of $\mathbf{T}$ as in (**24.1**) is numerically more stable.

25 A square matrix $\mathbf{A}$ is said to be similar to a matrix $\mathbf{B}$ if there exists a nonsingular matrix $\mathbf{P}$ such that $\mathbf{PAP}^{-1} = \mathbf{B}$.

(a) Similarity is an equivalence relation on $n \times n$ matrices.
(b) Let $\mathbf{A}$ be similar to $\mathbf{B}$. Then $\mathbf{A}$ and $\mathbf{B}$ possess the same characteristic polynomial and hence the same eigenvalues, the same trace and determinant.

26 A square matrix $\mathbf{A}$ is said to be semisimple if there exists a nonsingular matrix $\mathbf{P}$ such that $\mathbf{P}^{-1}\mathbf{AP} = \Lambda$, a diagonal matrix. The diagonal elements of Λ are the eigenvalues and the columns of $\mathbf{P}$ are the corresponding eigenvectors of $\mathbf{A}$.

$\mathbf{A}$ is a semisimple if, and only if, the eigen space is the whole of U^n, the n-dimensional unitary space.

27 Let $\mathbf{A}$ be $m \times n$ matrix. Let $(\mathbf{X}, \boldsymbol{\xi}) = \boldsymbol{\xi}^*\mathbf{NX}$ for all $\mathbf{X}, \boldsymbol{\xi} \in U^n$ and $(\boldsymbol{\eta}, \mathbf{Y}) = \mathbf{Y}^*\mathbf{M}\boldsymbol{\eta}$ for all $\boldsymbol{\eta}, \mathbf{Y} \in U^m$ where $\mathbf{M}$ and $\mathbf{N}$ are p.d. The adjoint $\mathbf{A}^\#$ of $\mathbf{A}$ is defined by the relation $(\mathbf{AX}, \mathbf{Y}) = (\mathbf{X}, \mathbf{A}^\#\mathbf{Y})$ for all $\mathbf{X} \in U^n$ and $\mathbf{Y} \in U^m$. Take $\mathbf{M} = \mathbf{N}$ if $m = n$. Show that

(i) $(\mathbf{A}^\#)^\# = \mathbf{A}$
(ii) $\mathbf{A}^\# = \mathbf{N}^{-1}\mathbf{A}^*\mathbf{M}$
(iii) $(\mathbf{AB})^\# = \mathbf{B}^\#\mathbf{A}^\#$
(iv) $(\mathbf{A}^\#)^{-1} = (\mathbf{A}^{-1})^\#$
(v) $\mathbf{A} = \mathbf{A}^\#$ for some choice of the inner product if, and only if, $\mathbf{A}$ is semisimple with real eigenvalues.

28 Let $\mathbf{A} = \mathbf{U}\begin{pmatrix} \Delta & 0 \\ 0 & 0 \end{pmatrix}\mathbf{V}^*$ be the singular value decomposition of $\mathbf{A}$ where $\mathbf{U}$ and $\mathbf{V}$ are unitary and Δ is a diagonal matrix with positive diagonal elements. Then

(a) $\mathbf{G}$ is a g-inverse of $\mathbf{A}$ if, and only if, $\mathbf{G} = \mathbf{V}\begin{pmatrix} \Delta^{-1} & \mathbf{L} \\ \mathbf{M} & \mathbf{N} \end{pmatrix}\mathbf{U}^*$ where $\mathbf{L}$, $\mathbf{M}$, and $\mathbf{N}$ are arbitrary matrices of proper orders.

(b) $\mathbf{G} = \mathbf{A}_r^- \Leftrightarrow \mathbf{G} = \mathbf{V}\begin{pmatrix} \mathbf{\Delta}^{-1} & \mathbf{L} \\ \mathbf{M} & \mathbf{M\Delta L} \end{pmatrix}\mathbf{U}^*$ where $\mathbf{L}$ and $\mathbf{M}$ are arbitrary.

(c) $\mathbf{G} = \mathbf{A}_m^- \Leftrightarrow \mathbf{G} = \mathbf{V}\begin{pmatrix} \mathbf{\Delta}^{-1} & \mathbf{L} \\ \mathbf{0} & \mathbf{N} \end{pmatrix}\mathbf{U}^*$ where $\mathbf{L}$ and $\mathbf{N}$ are arbitrary.

(d) $\mathbf{G} = \mathbf{A}_l^- \Leftrightarrow \mathbf{G} = \mathbf{V}\begin{pmatrix} \mathbf{\Delta}^{-1} & \mathbf{0} \\ \mathbf{M} & \mathbf{N} \end{pmatrix}\mathbf{U}^*$ where $\mathbf{M}$ and $\mathbf{N}$ are arbitrary.

(e) $\mathbf{G} = \mathbf{A}^+ \Leftrightarrow \mathbf{G} = \mathbf{V}\begin{pmatrix} \mathbf{\Delta}^{-1} & \mathbf{0} \\ \mathbf{0} & \mathbf{0} \end{pmatrix}\mathbf{U}^*$.

29 Let $\mathbf{A}$ be a square matrix and λ be an eigenvalue of $\mathbf{A}$. The set of all eigenvectors of $\mathbf{A}$ corresponding to λ together with the null vector form a subspace. The subspace is called the eigen space of $\mathbf{A}$ corresponding to λ. The dimension of this subspace is called the geometric multiplicity of λ. The multiplicity of λ as a root of the equation $|\mathbf{A} - \lambda\mathbf{I}| = 0$ is called the algebraic multiplicity of λ. Show that:

(a) Algebraic multiplicity $\geqslant$ geometric multiplicity for each eigenvalue.
(b) Algebraic multiplicity $=$ geometric multiplicity for all the eigenvalues of $\mathbf{A} \Leftrightarrow \mathbf{A}$ is semisimple.
(c) If $\mathbf{A}$ is real and symmetric, then for each eigenvalue the algebraic multiplicity is equal to its geometric multiplicity.

30 Let $\mathbf{T} = \mathbf{G} + \mathbf{XUX}'$ where $\mathbf{G}$ is n.n.d., $\mathbf{U}$ is symmetric, such that $\mathcal{M}(\mathbf{G}) \subset \mathcal{M}(\mathbf{T})$ and $\mathcal{M}(\mathbf{X}) \subset \mathcal{M}(\mathbf{T})$. Then $R(\mathbf{X}'\mathbf{T}^-\mathbf{X}) = R(\mathbf{X}')$, where $\mathbf{T}^-$ is any g-inverse.

[Note that $\mathbf{X} = (\mathbf{G} + \mathbf{XUX}')\mathbf{\Lambda}$ for some $\mathbf{\Lambda}$, and hence $\mathbf{X}'\mathbf{T}^-\mathbf{X} = \mathbf{X}'\mathbf{\Lambda} = \mathbf{\Lambda}'\mathbf{X}$. Then $\boldsymbol{\lambda}'\mathbf{X}'\mathbf{T}^-\mathbf{X} = \boldsymbol{\lambda}'\mathbf{X}'\mathbf{\Lambda} = \boldsymbol{\lambda}'\mathbf{\Lambda}'\mathbf{X} = \mathbf{0}$

$$= \boldsymbol{\lambda}'\mathbf{\Lambda}'(\mathbf{G} + \mathbf{XUX}')\mathbf{\Lambda} = \mathbf{0} \Rightarrow \boldsymbol{\lambda}'\mathbf{\Lambda}'\mathbf{G}\mathbf{\Lambda} = \mathbf{0} \Rightarrow \boldsymbol{\lambda}'\mathbf{\Lambda}'\mathbf{G} = \mathbf{0}.$$

Hence $\boldsymbol{\lambda}'\mathbf{\Lambda}'(\mathbf{G} + \mathbf{XUX}') = \mathbf{0} = \boldsymbol{\lambda}'\mathbf{X}'$, i.e., $\boldsymbol{\lambda}'\mathbf{X}'\mathbf{T}^-\mathbf{X} = \mathbf{0} \Rightarrow \boldsymbol{\lambda}'\mathbf{X}' = \mathbf{0}$. Hence $\mathcal{M}(\mathbf{X}') \subset \mathcal{M}(\mathbf{X}'\mathbf{T}^-\mathbf{X})$. The other way is obvious.]

31 If $\mathbf{A} = \begin{pmatrix} \mathbf{A}_{11} & \mathbf{A}_{12} \\ \mathbf{A}_{21} & \mathbf{A}_{22} \end{pmatrix}$ is a p.s.d. matrix, then $\mathcal{M}(\mathbf{A}_{21}) \subset \mathcal{M}(\mathbf{A}_{22})$ and consequently $\mathbf{A}_{22}\mathbf{A}^-_{22}\mathbf{A}_{21} = \mathbf{A}_{21}$ for any choice of the g-inverse.

32 *Schur's lemma.* Let $\mathbf{A} = (a_{ij})$ and $\mathbf{B} = (b_{ij})$ be n.n.d. matrices.. Show that their Hadamard product $\mathbf{C} = (a_{ij}b_{ij})$ is also n.n.d.

33 Let $\mathbf{B}$ be $k \times p$ matrix of rank k and $\mathbf{C}$ be $(p-k) \times p$ matrix of rank $(p-k)$ such that $\mathbf{BC}' = \mathbf{0}$. Then

$$\mathbf{C}'(\mathbf{C\Sigma C}')^{-1}\mathbf{C} + \mathbf{\Sigma}^{-1}\mathbf{B}'(\mathbf{B\Sigma}^{-1}\mathbf{B}')^{-1}\mathbf{B\Sigma}^{-1} = \mathbf{\Sigma}^{-1}$$

(The result follows by multiplying both sides by $\mathbf{B}$ and $\mathbf{C\Sigma}$).

REFERENCES

Aitken, A. C. (1944), *Determinants and Matrices* (third edition), University Mathematical Texts, 1, Oliver and Boyd, Edinburgh and London.

Atkinson, F. A., G. A. Watterson, and P. A. P. Moran (1960), A matrix inequality, *Quart. Jour. Math.* (Oxford series) **11**, 137–140.

Anderson, T. W. and S. Dasgupta (1963), Some inequalities on characteristic roots of matrices, *Biometrika* **50**, 522–524.

Berge, Claude (1963), *Topological Spaces* (English first edition), Oliver and Boyd, Edinburgh and London.

Dwyer, P. S. (1951), *Linear Computations*, John Wiley, New York.

Fisher, R. A. (1949), *The Theory of Inbreeding*, Oliver and Boyd, Edinburgh and London.

Gantmacher, F. R. (1959), *The Theory of Matrices*, Vols. I and II, Chelsea, New York.

Guttman, Louis (1954), Some necessary conditions for common factor analysis, *Psychometrika* **19**, 149–161.

Halmos, P. R. (1951), *Introduction to Hilbert Space and the Theory of Spectral Multiplicity*, Chelsea, New York.

Householder, A. S. (1964), *The Theory of Matrices in Numerical Analysis*, Blaisdell, New York.

Khatri, C. G. (1968), Some results for the singular multivariate regression models, *Sankhyá A*, **30**, 267–280.

Levi, F. W. (1942), *Algebra*, Calcutta University Publication, Calcutta.

Moore, E. H. (1935), *General Analysis*, American Phil. Soc., Philadelphia.

Mulholland, H. P. and C. A. B. Smith (1959), An inequality arising in genetical theory, *Am. Math. Monthly* **66**, 673–683.

Penrose, R. (1955), A generalized inverse for matrices, *Proc. Camb. Phil. Soc.* **51**, 406–413.

Turnbull, H. W. and A. C. Aitken (1932), *An Introduction to the Theory of Canonical Matrices*, Blackie, London.

[References to author's books and papers are given at the end of this book.]

Chapter 2

PROBABILITY THEORY, TOOLS AND TECHNIQUES

Introduction. This chapter is intended to provide the basic concepts of probability theory and the tools used in statistical inference. Familiarity with set algebra and elements of measure theory is sufficient to understand the axiomatic development of probability in Section **2a**.

More advanced results of measure theory used in later sections of the chapter, such as Stieltjes integral, Lebesgue monotone and dominated convergence theorems, and Radon-Nikodym theorem, are quoted in the appendix for ready reference. A proper understanding of these theorems is necessary to follow their applications to probability theory.

This chapter covers most of the basic results needed for a study of mathematical statistics. Detailed proofs are given for most of the propositions stated. Students will find it useful to work through the various steps of the proofs for a better understanding of the mathematical arguments involved.

Advanced concepts such as invariance of a family of probability measures, invariance of a statistical procedure, subfields in the probability space, subfields induced by statistics and conditional expectation with respect to subfields are discussed in Appendices **2C** and **2D** at the end of the chapter.

A number of limit theorems in probability are assembled in this chapter; some are proved and others are quoted in the Complements and Examples. A fairly complete treatment of characteristic functions is given.

The material presented in this chapter should give adequate preparation for those interested in specializing in specific areas of advanced probability and stochastic processes.

For further study, the reader may refer to books by Doob (1953), Feller (1966), Gnedenko and Kolmogorov (1954), Grenander (1963), Kagan, Linnik, and Rao (1972), Kolmogorov (1953, 1950), Levy (1937), Loeve (1955), Linnik (1960), Parthasarathy (1967), Ramachandran (1967), and others.

2a CALCULUS OF PROBABILITY

2a.1 The Space of Elementary Events

Events arising out of an experiment in nature or in a laboratory may not be predicted with certainty; but certain events occur more often than others, so that there is a natural enquiry as to whether such phenomenon can be described by attaching precise measures or indices (to be called probabilities) to events. These quantities, to be useful in practice, should satisfy some consistency conditions based on intuitive notions. The calculus of probability is devoted to the construction and study of such quantities. Their applicability to problems of the real world, that is, an examination of the relevance of hypothetical probabilities in a contingent situation or the estimation of appropriate probabilities and prediction of events based on them constitute the subject matter of statistical theory and methods to which the present book is mostly devoted.

As a first step, we must specify the set of all outcomes of an experiment which are distinguishable in some sense. We shall call them *elementary events* for some practical reasons demanding the definition of an event as a more general statement governing an elementary event (i.e., possibly as a set of elementary events).

Consider 10 tokens numbered 1, 2, ..., 10 from which one is drawn at *random*. Any one of the numbers can occur so that the elementary events are specified by the numbers 1, 2, ..., 10. We may ask the question whether a number drawn is a prime, a statement which refers to a set of elementary events, the occurrence of any one of which will ensure the description of the event, viz., "a number drawn is a prime." We might have argued that when the characteristic, primeness of a number, is considered, there are only two alternatives or events to be recognized (prime or not prime). But such a description would not enable us to characterize an observed number in many other ways in which it could possibly be done and which might be of interest, such as whether it is odd or even. It would then be more general to consider an event as a *set of elementary events*. In the example considered there are

$$\binom{10}{1} + \binom{10}{2} + \cdots + \binom{10}{10} = 2^{10} - 1$$

subsets consisting of single numbers, pairs of numbers, etc., to which we may add a hypothetical event that a token drawn is unnumbered, the corresponding set being called the empty set, to make 2^{10} possible events. We may not be interested in all subsets, but in a theoretical development it is necessary to consider all subsets for which a consistent calculus can be constructed.

As another example, consider heights of individuals drawn from a population. The observations can theoretically assume continuous values on the entire real line, or if height is measured in intervals of 0.5 cm, the elementary events will be confined to a countable set of points on the real line. Events of interest in such cases may be the occurrence of individuals with heights in specified intervals. The number of such subsets is not finite, however, and it would be worthwhile to examine the widest possible collection of subsets for which probabilities can be defined.

2a.2 The Class of Subsets (Events)

Let the set of all elementary events be Ω, also called the *sample space*, with elements or points ω representing elementary events. If an event (subset) A is defined by a property π possessed by an element (like a number being prime in the example of **2a.1**), it is natural to consider the event where π does not hold. The subset corresponding to this is the set of all elements in Ω not belonging to A. Such a set is called the *complement* of A and is represented by A^c. The complement of Ω is the empty set containing no element of Ω and is represented by $\varnothing$. If A_1, A_2 are two events with corresponding properties π_1, π_2 (like prime and even), then two other natural events can be defined: either π_1 or π_2 holds (a number is either prime or even) and both π_1 and π_2 hold (a number is prime as well as even). The set corresponding to the former is the set of all elements belonging to either A_1 or A_2 and is represented by $A_1 \cup A_2$ called set *union*. The latter consists of elements belonging to both A_1 and A_2 and is represented by $A_1 \cap A_2$ called set *intersection*. Our class C of sets on which probabilities are to be defined should at least have the property that if A_1 and $A_2 \in C$, then $A_1{}^c$, $A_1 \cup A_2$ and $A_1 \cap A_2 \in C$ to enable us to make useful statements. Such a class of sets is called a *Boolean* field and is represented by $\mathscr{F}$.

It is useful to define another set operation known as the difference $A_1 - A_2$, the set of elements in A_1 which are not elements of A_2; thus $A_1 - A_2 = A_1 \cap A_2{}^c$. Of course, $A_1 - A_2 \in \mathscr{F}$ if $A_1, A_2 \in \mathscr{F}$.

2a.3 Probability as a Set Function

For each set $A \in \mathscr{F}$ we have to assign a value $P(A)$ to be called the *probability* of A, or in other words define a set function P over the members of $\mathscr{F}$. If the notion of probability is to be related to relative frequencies of occurrence of events, the function P has to satisfy some intuitive requirements. First, its range should be $[0, 1]$, the value 0 corresponding to impossibility and 1 to certainty. Second, let $A_1, \ldots, A_k$, $(A_i \in \mathscr{F})$ be disjoint sets whose union is Ω, which means that any elementary event that occurs has one and only one of k *possible descriptions* $A_1, \ldots, A_k$. The relative frequencies of the events $A_1, \ldots, A_k$ must then add up to unity which suggests the requirement $P(A_1) + \cdots + P(A_k) = 1$.

Let $A_i \in \mathscr{F}$, $i = 1, 2, \ldots$ be a countable number of disjoint sets such that $\bigcup A_i = \Omega$. What can we say about $\sum_1^\infty P(A_i)$? From the second intuitive requirement that $\sum P(A_i) = 1$ for any finite decomposition of Ω, it only follows that $\sum_1^\infty P(A_i) \leqslant 1$. To examine the consequences of $\sum_1^\infty P(A_i) < 1$, let us consider the sequence of events $B_k = \bigcup_k^\infty A_i$. The sets $B_1, B_2, \ldots$ form a decreasing sequence tending in the limit to the empty set $\varnothing$. We may then expect $P(B_k)$ to decrease to zero as k increases. This does not happen if $\sum_1^\infty P(A_i) < 1$. It appears, then, that we need as a convenient condition, $\sum P(A_i) = 1$ for a finite or a countable decomposition of Ω. The intuitive requirements examined can now be stated in the form of two fundamental axioms governing the set function P on the chosen field $\mathscr{F}$.

AXIOM 1. $P(A) \geqslant 0$, $A \in \mathscr{F}$.

AXIOM 2. $\bigcup_{i=1}^\infty A_i = \Omega$, $A_i \cap A_j = \varnothing$ *for all* $i \neq j \Rightarrow \sum_1^\infty P(A_i) = 1$.

A set function P defined for all sets in $\mathscr{F}$ and satisfying the axioms 1 and 2 is called a probability measure. The following results are consequences of these axioms.

(i) $0 \leqslant P(A) \leqslant 1$, $A \in \mathscr{F}$

(ii) $P(\varnothing) = 0$ (*follows from axiom* 2 *observing that* $\varnothing \cup (\bigcup_1^\infty A_i) = \Omega = \bigcup_1^\infty A_i)$. *Then axiom* 2 *holds for a finite decomposition of* Ω.

(iii) $P(\Omega) = 1$ [*since* $\Omega \cup \varnothing \cup \varnothing \cdots = \Omega$]

(iv) $P(\bigcup A_i) = \sum P(A_i)$ *for any countable union of disjoint sets in* $\mathscr{F}$, *whose union also belongs to* $\mathscr{F}$.

$$(\bigcup A_i) \cup (\bigcup A_i)^c = \Omega = (\bigcup A_i)^c \cup A_1 \cup A_2 \cdots$$
$$P(\bigcup A_i) + P[(\bigcup A_i)^c] = 1 = P[(\bigcup A_i)^c] + P(A_1) + P(A_2) + \cdots.$$

Hence the required result follows.

(v) $P(A_1 \cup A_2) = P(A_1) + P(A_2)$, *if* $A_1 \cap A_2 = \varnothing$ (*a special case of* iv)

(vi) *Let* A_i *be a nonincreasing sequence of sets in* $\mathscr{F}$ *such that* $\lim_{i\to\infty} A_i = \bigcap A_i \in \mathscr{F}$. *Then* $\lim_{i\to\infty} P(A_i) = P(\lim_{i\to\infty} A_i)$.

The complements A_i^c form a nondecreasing sequence and

$$\bigcup A_i^c = (\bigcap A_i)^c = A_1^c \cup (A_2^c - A_1^c) \cup (A_3^c - A_2^c) \cdots.$$

Using the consequence (iv) of axiom 2 we see that

$$\begin{aligned} P[(\bigcap A_i)^c] &= P(A_1^c) + P(A_2^c - A_1^c) + \cdots \\ &= P(A_1^c) + [P(A_2^c) - P(A_1^c)] + \cdots \\ &= \lim P(A_i^c) \quad \text{as} \quad i \to \infty, \end{aligned}$$

that is,

$$1 - P(\bigcap A_i) = \lim[1 - P(A_i)] \quad \text{as} \quad i \to \infty.$$

or

$$P(\lim A_i) = P(\bigcap A_i) = \lim P(A_i) \quad \text{as} \quad i \to \infty. \tag{2a.3.1}$$

Similarly, if A_i, $i = 1, 2, \ldots$ is a nondecreasing sequence of sets in $\mathscr{F}$ such that $\lim A_i = \bigcup A_i \in \mathscr{F}$, then

$$\lim_{i\to\infty} P(A_i) = P(\bigcup A_i) = P\left(\lim_{i\to\infty} A_i\right) \tag{2a.3.2}$$

(vii) The reader may easily establish the following:

(a) $P(A^c) = 1 - P(A)$

(b) $A_1 \subset A_2 \Rightarrow P(A_2 - A_1) = P(A_2) - P(A_1) \Rightarrow P(A_1) \leqslant P(A_2)$

(c) $P(A_1 \cup A_2) = P(A_1) + P(A_2) - P(A_1 \cap A_2)$ *giving the inequality,* $P(A_1 \cup A_2) \leqslant P(A_1) + P(A_2)$ *and by induction,* $P(\bigcup A_i) \leqslant \sum P(A_i)$ *for any arbitrary sets* $A_1, A_2, \ldots$.

(viii) *Axioms* 1, 2 $\Leftrightarrow$ *results* (i), (iii), (iv) $\Leftrightarrow$ *results* (i), (iii), (v), (vi).

Kolmogorov (1933) originally stated (i), (iii), (v), (vi) as axioms governing a probability function. The equivalences in (viii) are easy to establish.

2a.4 Borel Field (σ-field) and Extension of Probability Measure

A Boolean field need not contain unions of all countable sequences of sets. A field which contains all such unions (and therefore countable intersections) is called a *Borel field* or a σ-field. Given a field $\mathscr{F}$ (or any collection of sets) there exists a minimal Borel field containing $\mathscr{F}$ which we denote by $\mathscr{B}(\mathscr{F})$ and which may be shown as follows. There is at least one Borel field, viz., the class of all sets in Ω, which contains $\mathscr{F}$. Further arbitrary intersections of Borel fields are also Borel fields. Hence the intersection of all Borel fields containing $\mathscr{F}$ is precisely the minimal Borel field, $\mathscr{B}(\mathscr{F})$. A set function defined on $\mathscr{F}$ and satisfying axioms 1, 2 can be uniquely extended to all sets in $\mathscr{B}(\mathscr{F})$, that is, there exists a unique function P^* such that:

(*a*) $0 \leqslant P^*(A) \leqslant 1$, $A \in \mathscr{B}(\mathscr{F})$,

(*b*) $P^*(\bigcup A_n) = \sum P^*(A_n)$ for a countable sequence A_n of disjoint sets in $\mathscr{B}(\mathscr{F})$, and

(*c*) $P^*(A) = P(A)$, if $A \in \mathscr{F}$.

The function P^* is defined as follows. Consider a set A in $\mathscr{B}(\mathscr{F})$ and a collection of sets A_i in $\mathscr{F}$ such that $A \subset \bigcup_1^\infty A_i$. Then

$$P^*(A) = \inf_{A_i} \sum P(A_i).$$

The extension P^* of P is a standard result in measure theory (see Halmos, 1950; Munroe, 1959).

It may then be of some advantage to consider the wider field $\mathscr{B}(\mathscr{F})$ as our basic class of sets for defining the probability function, if we observe that it is enough to specify the function on a Boolean field generating it. We are now in a position to build up a calculus of probability based on the basic space Ω of elements ω (elementary events), a Borel or a σ-field $\mathscr{B}$ of sets (events) in Ω, and a probability measure P on $\mathscr{B}$. The triplet $(\Omega, \mathscr{B}, P)$ is called a *probability space*, while Ω or $(\Omega, \mathscr{B})$ is called the *sample space*.

2a.5 Notion of a Random Variable and Distribution Function

It is convenient at this stage to introduce some notations which we use throughout this chapter. We have already defined the probability space $(\Omega, \mathscr{B}, P)$. The real line $(-\infty, \infty)$ is represented by R and the n-dimensional real Euclidean space by R^n. The points of R^n can be represented by ordered set of real coordinates $(x_1, \ldots, x_n)$. These are particular cases of Ω.

In R^n, consider the class $\mathscr{G}$ of open sets or closed sets or n-dimensional closed or open rectangles or the field generated by finite unions of sets of the form

$$a_i \leqslant x_i < b_i, \qquad \text{where} \quad -\infty \leqslant a_i \leqslant b_i \leqslant \infty,$$
$$i = 1, \ldots, n.$$

The minimal Borel field containing $\mathscr{G}$ is the same in each case and is denoted by $\mathscr{B}_n$. The probability spaces $(R^n, \mathscr{B}_n, P)$ play an important role in practical applications. The concepts of $\mathscr{B}_\infty$ related to R^∞ and $(R^\infty, \mathscr{B}_\infty, P)$ are developed in **2c.1**.

Random Variable (r.v.). A real valued point function $X(\cdot)$ defined on the space $(\Omega, \mathscr{B}, P)$ is called a *random variable* (or measurable in the language of measure theory) if the set $\{\omega\colon X(\omega) < x\} \in \mathscr{B}$ for every x in R.

The function X may be regarded as a mapping of Ω into R which may be written $\Omega \xrightarrow{X} R$. Consequently, given a set S of points in R, there exists a set of points in Ω mapped into S which is called the pre-image of S and is represented by $X^{-1}(S)$. With this notation X is a random variable if and only if

$$X^{-1}(-\infty, x) \in \mathscr{B} \qquad \text{for every } x.$$

Distribution Function (d.f.). The point function

$$F(x) = P\{\omega\colon X(\omega) < x\} = P[X^{-1}(-\infty, x)]$$

defined on R is called a *distribution function* (d.f.). A d.f. so defined has a number of important properties. We use the notation $P(X < x) = P\{\omega\colon X(\omega) < x\}$.

(i) $F(-\infty) = 0$, $F(\infty) = 1$.

The limit of $X^{-1}(-\infty, x)$ as $x \to -\infty$ is $\emptyset$, the empty set in $\mathscr{B}$. Hence $\lim F(x) = 0$ as $x \to -\infty$, that is, $F(-\infty) = 0$. Otherwise there is a contradiction by (2a.3.1). Similarly, $F(\infty) = 1$.

(ii) *F is nondecreasing.*

Let $x_1 < x_2$. Since $(-\infty, x_2) \supset (-\infty, x_1)$, $X^{-1}(-\infty, x_2) \supset X^{-1}(-\infty, x_1)$ and, therefore, $P[X^{-1}(-\infty, x_2)] \geqslant P[X^{-1}(-\infty, x_1)]$.

(iii) *F is continuous at least from the left.*

Let $x_1, x_2, \ldots$ be an increasing sequence of numbers and $A_1, A_2, \ldots$ the corresponding increasing sequence of sets in $\mathscr{B}$. If $\lim x_n = x$, then $\lim A_n = \bigcup A_n = A$, with $A = X^{-1}[(-\infty, x)]$. Using (2a.3.2), we have

$$\lim F(x_n) = \lim P(A_n) = P(\lim A_n) = P(A) = F(x).$$

Notice that if we are approaching the limit from above the pre-image of x is in every A_i and the limiting set is $X^{-1}((-\infty, x])$ whose probability is $F(x) + P(X = x)$. Thus discontinuity can arise by a single point set having non-zero probability.

Note that left continuity is a consequence of the definition $F(x) = P(X < x)$. Some authors define the d.f. $F(x)$ as $P(X \leqslant x)$ in which case it would be right continuous. The values according to two definitions coincide at all continuity points and differ by $P(X = x)$ at discontinuity point x. We follow the definition $F(x) = P(X < x)$, observing that the two definitions are equivalent and the choice is a matter of convention.

(iv) *The set S of points of discontinuity of F is at most countable.*

Denote by S_n the set of points at which F has a jump $\geqslant 1/n$. Then $S = \bigcup S_n$. For every n, the set S_n is finite, for otherwise F would exceed unity. Hence S is at most countable.

(v) *Two distribution functions are the same if they agree on a dense set in R.*

(vi) Decomposition of a d.f. *$F(x) = F_1(x) + F_2(x)$ where F_2 is everywhere continuous and F_1 is a step function. Such a decomposition is unique.*

Let $D = (x_1, x_2, \ldots)$ be the discontinuity points of $F(x)$ and define

$$F_1(x) = \sum_{x_i < x} [F(x_i + 0) - F(x_i)].$$

Obviously $F_1(x)$ is a step function. Then $F_2(x) = F(x) - F_1(x)$ can be easily seen to be a continuous function.

There is a more general decomposition which expresses the continuous part as

$$F_2(x) = F_{21}(x) + F_{22}(x),$$

where F_{21} is absolutely continuous, that is, can be expressed as an indefinite integral of a function and F_{22} has derivative equal to zero almost everywhere. For a proof see Loève (1955).

(vii) *The random variable* $X(\cdot)$ *on* $(\Omega, \mathscr{B}, P)$ *induces a probability measure on* $(R, \mathscr{B}_1)$. *The probability measure is completely specified by* $F(x)$.

Let $S \in \mathscr{B}_1$; then it is easy to see that $X^{-1}(S) \in \mathscr{B}$. Define $Q(S) = P[X^{-1}(S)]$. Obviously, $Q(S)$ is a probability measure on $\mathscr{B}_1$, as may be verified from the fundamental axioms 1, 2, with the property

$$Q(S) = F(x), \quad \text{if} \quad S = (-\infty, x). \tag{2a.5.1}$$

We shall show that if $Q'(S)$ is another probability measure on $\mathscr{B}_1$ with the property (2a.5.1), then $Q(S) = Q'(S)$ for all $S \in \mathscr{B}_1$, which will prove the required proposition.

Let $S = [a, b)$; since $(-\infty, b) = (-\infty, a) + [a, b)$

$$F(b) = F(a) + Q([a, b)) = F(a) + Q'([a, b)).$$

Hence Q and Q' agree on intervals of the form $[a, b)$.

Let $\mathscr{F}_n$ be the class of finite unions of sets of the form $[a, b)$ in $[-n, n)$. Then $\mathscr{F}_n$ is a field and a set $S \in \mathscr{F}_n$ can always be expressed as union of disjoint sets of the form $S_i = [a_i, b_i)$. Therefore

$$Q(S) = \sum Q([a_i, b_i)) = \sum Q'([a_i, b_i)) = Q'(S),$$

so that Q and Q' agree on $\mathscr{F}_n$ and therefore on $\mathscr{B}(\mathscr{F}_n)$.

Now let $S \in \mathscr{B}_1$. Then $S \cap [-n, n) \in \mathscr{B}(\mathscr{F}_n)$. So $Q[S \cap [-n, n)] = Q'[S \cap [-n, n)$ and letting $n \to \infty$, $Q(S) = Q'(S)$.

(viii) *The random variable* $X(\cdot)$ *defines a* σ*-field of sets* $\mathscr{B}_X \subset \mathscr{B}$, *that is,* $\mathscr{B}_X$ *is a sub* σ*-field of* ω *sets.*

Let $S \in \mathscr{B}_1$ and consider $\mathscr{B}_X = \{X^{-1}(S): S \in \mathscr{B}_1\}$. $\mathscr{B}_X$ is a σ-field since:

(a) $X^{-1}(-\infty, \infty) = \Omega$, *that is,* $\Omega \in \mathscr{B}_X$,

(b) $X^{-1}(S^c) = [X^{-1}(S)]^c$, *that is, the complements of sets in* $\mathscr{B}_X$ *belong to* $\mathscr{B}_X$, *and*

(c) $X^{-1}(\bigcup_1^\infty S_i) = \bigcup_1^\infty X^{-1}(S_i)$, *that it, countable union of sets in* $\mathscr{B}_X$ *also belongs to* $\mathscr{B}_X$. *It is obvious that* $\mathscr{B}_X \subset \mathscr{B}$.

(ix) *A Borel function of a random variable is also a random variable.*

A function f which maps points of R, the real line, into R is said to be a Borel function if

$$S \in \mathscr{B}_1 \Rightarrow f^{-1}(S) \in \mathscr{B}_1.$$

More generally, a function f which maps points of R^n, the n dimensional space, into R is said to be a Borel function if

$$S \in \mathscr{B}_1 \Rightarrow f^{-1}(S) \in \mathscr{B}_n.$$

Let X be a random variable, that is, $X^{-1}(S) \in \mathscr{B}$. Consider $f(X(\omega)) = g(\omega)$. Then it is seen that $g^{-1}(S) = X^{-1}[f^{-1}(S)] \in \mathscr{B}$.

(x) *Every* $F(x)$, *with the following properties:*

(a) $F(x)$ *is nondecreasing,*
(b) $F(-\infty) = 0$, $F(\infty) = 1$, *and*
(c) $F(x)$ *is continuous at least from the left,*

defines a random variable of which F *is the d.f.*

Let $\Omega = [0, 1]$, $\mathscr{B}$ the Borel field generated by intervals in $[0, 1]$, and P, the Lebesgue measure so that $P([0, \omega)) = \omega$, $0 \leqslant \omega \leqslant 1$. Then the random variable (r.v.)

$$X(\omega) = \inf_{F(y) > \omega} y, \qquad (0 \leqslant \omega \leqslant 1)$$

has the d.f.

$$\begin{aligned} P[\omega\colon X(\omega) < x] &= P(\omega\colon \inf_{F(y)>\omega} y < x] \\ &= P[\omega\colon 0 \leqslant \omega < F(x)] = F(x). \end{aligned}$$

We have thus exhibited $F(x)$ as arising out of a function (r.v.) defined on a suitable probability space, $(\Omega, \mathscr{B}, P)$.

In practice, two types of distribution functions are of interest. One is the *step function* corresponding to a random variable assuming discrete values $a_1 < a_2 < \cdots$, finite or countable in number, with probabilities p_1, $p_2, \ldots$ $\left(\sum p_i = 1\right)$ respectively. The d.f. has jumps at points $a_1, a_2, \ldots$ and stays constant in the interval $(a_i, a_{i+1}]$. The saltus at a_i is p_i. The second type called *absolutely continuous* has a representation of the form

$$F(x) = \int_{-\infty}^{x} f(x)\, dx \tag{2a.5.2}$$

in which case $f(x)$ is said to be the probability density (p.d.). If $f(x)$ is continuous at x, $F'(x) = f(x)$. There can be other types of distribution functions which are mixtures of the two types considered and which may not be representable in the form (2a.5.2).

Some examples of discrete distributions with finite and countable alternatives useful in statistics are as follows. In each example the variable is represented by X and $P(X=r)$ by p_r. Other values known as parameters occurring in the specification of p_r are represented by Greek letters. The ranges of the variable and parameters are indicated.

Binomial Distribution

$$p_r = \binom{n}{r}\pi^r(1-\pi)^{n-r}, \qquad 0<\pi<1, \quad n \text{ any integer} \qquad (2a.5.3)$$
$$r=0, 1, \ldots, n.$$

Geometric Distribution

$$p_r = (1-\pi)\pi^r, \qquad 0<\pi<1 \qquad (2a.5.4)$$
$$r=0, 1, \ldots, \infty.$$

Poisson Distribution

$$p_r = e^{-\lambda}\frac{\lambda^r}{r!}, \qquad \lambda>0 \qquad (2a.5.5)$$
$$r=0, 1, \ldots, \infty.$$

Negative Binomial Distribution

$$p_r = \binom{\kappa+r-1}{r}\frac{\rho^r}{(1+\rho)^{\kappa+r}}, \qquad \rho>0, \quad \kappa \geqslant 1 \qquad (2a.5.6)$$
$$r=0, 1, \ldots, \infty.$$

Logarithmic Distribution

$$p_r = \frac{1}{-\log(1-\alpha)}\frac{\alpha^r}{r}, \qquad 0<\alpha<1 \qquad (2a.5.7)$$
$$r=1, 2, \ldots, \infty.$$

Hypergeometric Distribution

$$p_r = \binom{A}{r}\binom{N-A}{n-r} \div \binom{N}{n}, \qquad \begin{array}{l} A \leqslant N \text{ (both integers)} \\ n \leqslant N \text{ (} n \text{ integer)} \end{array} \qquad (2a.5.8)$$
$$r=\max(0, n-N+A), \ldots, \min(A, n).$$

An important example of a continuous distribution is the normal distribution with the density function

$$f(x) = \frac{1}{\sqrt{2\pi}\,\sigma} e^{-(x-\mu)^2/2\sigma}, \qquad -\infty<\mu<\infty, 0<\sigma<\infty \qquad (2a.5.9)$$
$$-\infty<x<\infty.$$

It is easily verified that

$$\int_{-\infty}^{\infty} f(x)\,dx = \frac{1}{\sqrt{2\pi}\,\sigma}\int_{-\infty}^{\infty} e^{-(x-\mu)^2/2\sigma^2}\,dx$$

$$= \frac{2}{\sqrt{2\pi}\,\sigma}\int_{0}^{\infty} e^{-y^2/2\sigma^2}\,dy, \qquad \text{substituting} \quad y = x - \mu$$

$$= \frac{1}{\sqrt{\pi}}\int_{0}^{\infty} z^{-1/2}e^{-z}\,dz = \frac{1}{\sqrt{\pi}}\,\Gamma\left(\frac{1}{2}\right) = 1.$$

Other continuous distributions of interest are given in Chapter 3.

2a.6 Multidimensional Random Variable

The collection of m single-valued real functions $X_1(\cdot), \ldots, X_m(\cdot)$ mapping Ω into R^m is called an m dimensional random variable if the set

$$S(\mathbf{x}) = \{\omega\colon X_1(\omega) < x_1, \ldots, X_m(\omega) < x_m\} \in \mathscr{B} \qquad \text{for all } \mathbf{x},$$

where $\mathbf{x}$ stands for $(x_1, \ldots, x_m)$. As in the case of one dimensional random variable we define the distribution function

$$P[S(\mathbf{x})] = F(x_1, \ldots, x_m).$$

By following the arguments of **2a.5**, it can be shown that F is nondecreasing and continuous at least from the left in each of its arguments and defines a probability measure on $\mathscr{B}_n$.

The d.f. F_i of X_i alone is seen to be

$$F_i(x) = F(\infty, \ldots, \infty, x, \infty, \ldots, \infty),$$

with x in the ith position and is called the marginal distribution of X_i. Similarly, the marginal distribution of any subset of the variables is obtained by substituting ∞ for the other variables in $F(x_1, \ldots, x_m)$.

Two important cases we shall be considering are the step functions and absolutely continuous functions. For step functions the probabilities for different combinations of values taken by $x_1, \ldots, x_m$ are specified. For the absolutely continuous functions a density function $f(x_1, \ldots, x_m)$ is given such that

$$F(x_1, \ldots, x_n) = \int_{-\infty}^{x_1} \cdots \int_{-\infty}^{x_m} f(x_1, \ldots, x_m)\,dx_1 \cdots dx_m.$$

If $f(x_1, \ldots, x_m)$ is continuous in all the variables, then

$$\frac{\partial^m F}{\partial x_1 \cdots \partial x_m} = f(x_1, \ldots, x_m).$$

2a.7 Conditional Probability and Statistical Independence

If, in the example of 10 tokens of **2a.1**, we assign a probability of $r/10$ to any subset of r numbers, the axioms 1, 2 of a probability measure are satisfied. With the measure so defined, the probability of the event A, a number being prime, is 4/10; that of the event C, a number being even, is 5/10; and that of the event AC (even prime) is 1/10.

Suppose a token is drawn and is noted to contain a prime number, that is, the event A has occurred. Can we then ask the question: What is the probability that the number is even (prime) or that the event AC has occurred? Naturally the degree of uncertainty about AC is reduced by our knowledge that A has occurred leading us to consider only the possibilities where A is true. In our example we have to consider only the outcomes 2, 3, 5, 7 which are all prime and assign probabilities to their subsets. A natural way of doing this is to assign probabilities to the elementary events 2, 3, 5, 7 in the same ratio as their original probabilities for their relative occurrences are not altered by the knowledge we have. The new probabilities are then $P^*(i) \propto P(i)$, or

$$P^*(i) = \frac{P(i)}{P(A)}, \qquad i = 2, 3, 5, 7$$

$$\sum P^*(i) = 1$$

where $P(A) = P(2) + P(3) + P(5) + P(7)$ is the probability of a number being prime. Now the probability that a given number is even (event C) knowing that it is a prime is

$$P(C \mid A) = P^*(2) = \frac{P(2)}{P(A)} = \frac{P(AC)}{P(A)}.$$

The ratio $P(AC)/P(A)$ when $P(A) \neq 0$ is called the conditional probability of C given A and is denoted by $P(C \mid A)$. $P(C \mid A) = 0$ when $P(A) = 0$.

In the general case $(\Omega, \mathscr{B}, P)$, consider a set A such that $P(A) \neq 0$ and sets of the form AC, $C \in \mathscr{B}$ and define

$$P(C \mid A) = \frac{P(AC)}{P(A)}. \tag{2a.7.1}$$

The set function $P(\cdot \mid A)$ for fixed A, defined for all sets $C \in \mathscr{B}$ satisfies axioms 1 and 2 and is called the conditional probability measure given A.

Independence of Events. Event C is said to be statistically independent of event A if $P(C|A) = P(C)$, in which case

$$P(AC) = P(A)P(C). \tag{2a.7.2}$$

Then $P(A|C) = P(A)$, i.e., A is independent of C. Thus two events A and C are said to be independent if (2a.7.2) holds, including the case $P(A)$ or $P(C)$ is zero.

Events $A_1, A_2, \ldots$ are independent if for every selection of n events $A_{i_1}, \ldots, A_{i_n}$

$$P(A_{i_1} \cdots A_{i_n}) = P(A_{i_1}) \cdots P(A_{i_n})$$
$$n = 2, 3, \ldots$$

2a.8 Conditional Distribution of a Random Variable

Consider a pair of random variables (X_1, X_2) and a set $S \in \mathscr{B}_1$ such that $P(X_1 \in S) \neq 0$. Following the definition (2a.7.1) of conditional probability, we can interpret the ratio

$$\frac{P[\omega\colon X_1(\omega) \in S,\ X_2(\omega) < x_2]}{P[X_1(\omega) \in S]} = F(x_2 \,|\, X_1 \in S) \tag{2a.8.1}$$

as the conditional probability of $X_2 < x_2$, given that $X_1 \in S$. The expression (2a.8.1) considered as a function of x_2 may be called the conditional d.f. of X_2, given that $X_1 \in S$ and denoted by

$$F_2(\cdot \,|\, X_1 \in S). \tag{2a.8.2}$$

If, however, a function $F_2(\cdot|\cdot)$ exists such that

$$P(X_1 \in S, X_2 < x_2) = \int_S F_2(x_2|x_1)\, dF_1(x_1), \tag{2a.8.3}$$

for all S and x_2 where F_1 is the d.f. of X_1, then $F_2(\cdot|x_1)$ could be called the conditional d.f. of X_2 given $X_1 = x_1$. It is difficult to establish the existence of such a function in the general case (see **2b.3** where the existence is established. We consider R^2 as the sample space of two random variables X_1, X_2 and use the argument employed for R^n).

In most of the applications considered in this book, however, we have d.f.'s which are absolutely continuous or which correspond to a discrete distribution. In such cases it is simple to define a conditional distribution. Let $f(x_1, x_2)$ be joint density of X_1, X_2 and $f_1(x_1)$ that of X_1 alone. (For a discrete distribution, "density" is replaced by "probability" and integration is replaced by summation). By definition,

$$P(X_1 \in S, X_2 < x_2) = \int_S \int_{-\infty}^{x_2} f(x_1, x_2)\, dx_1\, dx_2$$
$$= \int_S f_1(x_1)\, dx_1 \int_{-\infty}^{x_2} \frac{f(x_1, x_2)}{f_1(x_1)}\, dx_2, \qquad f_1(x_1) \neq 0,$$

which shows that

$$F_2(x_2 \mid x_1) = \int_{-\infty}^{x_2} \frac{f(x_1, x_2)}{f_1(x_1)} \, dx_2 \tag{2a.8.4}$$

satisfies the definition of a conditional distribution. Furthermore, the ratio $f(x_1, x_2)/f_1(x_1)$ may be considered as the conditional p.d. of X_2 given $X_1 = x_1$. It is immaterial as to how the conditional d.f. and p.d. are defined when $f_1(x_1) = 0$.

Two random variables X_1, X_2 are said to be independent if

$$F_2(x_2 \mid X_1 \in S) = F_2(x_2) \qquad \text{for all } x_2 \quad \text{and} \quad S \in \mathscr{B}_1. \tag{2a.8.5}$$

The choice of S as an interval implies, by (2a.8.5)

$$F(x_1, x_2) = F_1(x_1)F_2(x_2), \tag{2a.8.6}$$

and it is trivially true that (2a.8.6) implies (2a.8.5). When the densities exist, the condition for independence may be stated as

$$f(x_1, x_2) = f_1(x_1)f_2(x_2).$$

2b MATHEMATICAL EXPECTATION AND MOMENTS OF RANDOM VARIABLES

2b.1 Properties of Mathematical Expectation

The mathematical expectation of a random variable X, also called the mean value, is defined by

$$E(X) = \int x \, dF, \tag{2b.1.1}$$

where $F(x)$ is d.f. of X and the integral is Lebesgue-Stieltjes, explained in Appendix **2A** to this chapter. When X has p.d., $f(x)$, we see that

$$E(X) = \int x \, dF = \int x f(x) \, dx \tag{2b.1.2}$$

reduces to a Riemann integral. For a discrete probability distribution with $F(x)$ as a step function,

$$E(X) = \int x \, dF = \sum x_r p_r, \tag{2b.1.3}$$

which is a summation. In (2b.1.3), each x_r is a discontinuity point of F and p_r is the saltus at x_r. *It is interesting to note that in all the three cases the same symbol* $\int x \, dF$ *could be employed.*

Let $Y = g(X)$ be a Borel function of X and denote by $G(y)$ the d.f. of Y. Then by using the definition of integration it is easy to establish that

$$E(Y) = E[g(X)] = \int y \, dG(y) = \int g(x) \, dF(x), \tag{2b.1.4}$$

so that the expected value of a function of an r.v. can be computed directly without determining first the d.f. of the function of the r.v.

More generally, if we are considering a k-dimensional random variable $(X_1, \ldots, X_k)$ and $Y = h(X_1, \ldots, X_k)$ is a Borel function, it may be shown that

$$E(Y) = \int y \, dG(y) = \int h(x_1, \ldots, x_k) \, dF(x_1, \ldots, x_k).$$

The following results are easily established:

(i) $E(cX) = cE(X)$, *where c is constant.*

(ii) $E(X_1 + X_2) = E(X_1) + E(X_2)$ *whether* X_1 *and* X_2 *are independent or not, and more generally*

$$E(c_1 X_1 + \cdots + c_k X_k) = c_1 E(X_1) + \cdots + c_k E(X_k).$$

(iii) $E(X_1 X_2) = E(X_1) E(X_2)$ *when* X_1 *and* X_2 *are independent.*

2b.2 Moments

The sth moment and absolute moment about a given constant a of a random variable X are defined as

$$\alpha_s = E(X - a)^s = \int (x - a)^s \, dF, \quad \text{in general}$$

$$= \int (x - a)^s f(x) \, dx, \quad \text{when p.d. exists}$$

$$= \sum_r (x_r - a)^s p_r, \quad \text{for a discrete distribution,}$$

$$\gamma_s = E|X - a|^s = \int |x - a|^s \, dF.$$

For some theoretical reasons we say α_s exists only when γ_s exists, which is true when the integral defining α_s is of Lebesgue-Stieltjes type (Appendix **2A**). When $a = E(X)$, the moments are said to be central and are represented by μ_s. The following relationships between central moments and moments about an arbitrary origin may be easily verified.

$$\mu_0 = 1, \quad \mu_1 = 0, \quad \mu_2 = \alpha_2 - \alpha_1^2$$
$$\mu_3 = \alpha_3 - 3\alpha_1\alpha_2 + 2\alpha_1^3$$
$$\mu_4 = \alpha_4 - 4\alpha_1\alpha_3 + 6\alpha_1^2\alpha_2 - 3\alpha_1^4$$

and in general

$$\mu_n = \alpha_n - \binom{n}{1}\alpha_1\alpha_{n-1} + \binom{n}{2}\alpha_1{}^2\alpha_{n-2} - \cdots (-1)^{n-1}(n-1)\alpha_1{}^n$$

which is established as follows:

$$\begin{aligned}
\mu_n &= E[X - E(X)]^n = E[X - a - E(X-a)]^n \\
&= E(X - a - \alpha_1)^n \\
&= E\left\{(X-a)^n - \binom{n}{1}(X-a)^{n-1}\alpha_1 + \binom{n}{2}(X-a)^{n-2}\alpha_1{}^2 - \cdots\right\} \\
&= E(X-a)^n - \binom{n}{1}\alpha_1 E(X-a)^{n-1} + \binom{n}{2}\alpha_1{}^2 E(X-a)^{n-2} - \cdots \\
&= \alpha_n - \binom{n}{1}\alpha_1\alpha_{n-1} + \binom{n}{2}\alpha_1{}^2\alpha_{n-2} - \cdots.
\end{aligned}$$

The mean μ and the second central moment μ_2 (variance), generally represented by the symbol σ^2, play an important role in statistical data analysis. Purely as descriptive measures of the distribution, the mean represents a *central* value of the random variable and the variance represents the scatter around the central value. A few results concerning the mean and variance of a distribution are given.

(i) *The second moment $\alpha_2(a)$ is a minimum when taken about the mean μ.*

$$\begin{aligned}
E(X-a)^2 &= E(X - \mu + \mu - a)^2 \\
&= E(X-\mu)^2 + (\mu - a)^2 + 2(\mu - a)E(X - \mu) \\
&= E(X-\mu)^2 + (\mu - a)^2 \geqslant E(X-\mu)^2.
\end{aligned}$$

(ii) *If X is a non-negative random variable and if $E(X)$ exists, then*

$$E(X) = \int [1 - F(x)]\, dx. \tag{2b.2.1}$$

Integrating by parts, we see that

$$\begin{aligned}
\int_0^T x\, dF(x) &= TF(T) - \int_0^T F(x)\, dx \\
&= -T[1 - F(T)] + \int_0^T [1 - F(x)]\, dx.
\end{aligned} \tag{2b.2.2}$$

But

$$T[1 - F(T)] = T\int_T^\infty dF(x) < \int_T^\infty x\, dF(x) \to 0 \quad \text{as} \quad T \to \infty.$$

Hence the result (2b.2.1) follows by taking limits of both sides of (2b.2.2).

Note that if X is any random variable and $E(X)$ exists, then

$$E(X) = \int_0^{\infty} [1 - F(x) + F(-x)]\, dx \tag{2b.2.3}$$

and that

$$\lim_{x \to -\infty} xF(x) = \lim_{x \to \infty} x[1 - F(x)] = 0$$

is a necessary condition for the existence of $E(x)$.

Furthermore, if G and F are distribution functions with finite mean values g and f, then

$$\int_{-\infty}^{+\infty} [G(x) - F(x)]\, dx = g - f. \tag{2b.2.4}$$

Savage (1970) has shown the more general result

$$\int_{-\infty}^{+\infty} [(G * H)(x) - (F * H)(x)]\, dx = g - f \tag{2b.2.5}$$

where H is any distribution function and $G * H$ denotes the convolution distribution (see 2b.4.12).

(iii) CHEBYSHEV'S INEQUALITY. *For any random variable X with mean μ and variance σ^2*

$$P(|X - \mu| \geqslant \lambda\sigma) \leqslant \frac{1}{\lambda^2}, \qquad \lambda > 0. \tag{2b.2.6}$$

This is an important inequality as it is independent of the exact nature of the distribution of X. By definition,

$$\sigma^2 = \int (x - \mu)^2\, dF(x)$$

$$= \int_{-\infty}^{\mu - \lambda\sigma} (x - \mu)^2\, dF + \int_{\mu - \lambda\sigma}^{\mu + \lambda\sigma} (x - \mu)^2\, dF + \int_{\mu + \lambda\sigma}^{\infty} (x - \mu)^2\, dF.$$

Dropping the middle term and replacing $(x - \mu)^2$ by the smallest value in the first and third terms, we have

$$\sigma^2 \geqslant \lambda^2\sigma^2 \int_{-\infty}^{\mu - \lambda\sigma} dF + \lambda^2\sigma^2 \int_{\mu + \lambda\sigma}^{\infty} dF$$

$$\geqslant \lambda^2\sigma^2 P(|x - \mu| \geqslant \lambda\sigma),$$

which yields the inequality (2b.2.6). The result (2b.2.6) is true when μ is not the mean, provided that σ^2 is replaced by $E(X - \mu)^2$.

(iv) *Let X be a non-negative random variable with $E(X) = \mu$. Then*

$$P(X \geqslant t\mu) \leqslant \frac{1}{t}.$$

(v) *Let X_1, X_2 be two random variables with means μ_1, μ_2, variances σ_1^2, σ_2^2 and d.f.'s F_1, F_2. Then*

$$F_1(x + \mu_1) - F_1(-x + \mu_1) \geqslant F_2(x + \mu_2) - F_2(-x + \mu_2) \quad (2b.2.7)$$

for each x implies that $\sigma_1^2 \leqslant \sigma_2^2$.

Let G_1 and G_2 be the d.f.'s of $|X_1 - \mu_1|$ and $|X_2 - \mu_2|$ respectively. Then integrating by parts, we have

$$\begin{aligned}
\sigma_1^2 - \sigma_2^2 &= \lim_{T\to\infty} \int_0^T y^2\, d(G_1 - G_2) \\
&= \lim_{T\to\infty} T^2[G_1(T) - G_2(T)] - 2 \lim_{T\to\infty} \int_0^T y(G_1 - G_2)\, dy \\
&= 0 - 2 \int_0^\infty y(G_1 - G_2)\, dy \leqslant 0,
\end{aligned}$$

since the condition (2b.2.7) implies that $G_1 \geqslant G_2$. The converse proposition is not true, however. If $\sigma_1^2 \leqslant \sigma_2^2$, then it implies that for at least one value of x the inequality (2b.2.7) is true but not necessarily for all values of x.

2b.3 Conditional Expectation

In **2a.8** the conditional d.f. of X_2 given X_1 was defined in some special cases. We now introduce more general concepts by defining conditional expectation without reference to a conditional distribution. It will be seen that conditional probability is a special case of conditional expectation.

Consider a pair of random variables (Y, T) where T may be n-dimensional. The conditional expectation of Y given T is defined as a function of T with value at $T = t$ denoted by $E(Y|t)$ such that

$$\int_{R_1 \times S} y\, dF(y, t) = \int_S E(Y|t)\, dF(t) \quad (2b.3.1)$$

for all $S \in \mathscr{B}_n$, where $R_1 \times S$ is the cylinder set in the (Y, T)-plane with base S in the T-plane. The left-hand side of (2b.3.1) is a set function $Q(S)$ which is absolutely continuous with respect to $P(S)$,

$$P(S) = \int_S dF(t), \quad (2b.3.2)$$

that is, $P(S) = 0 \Rightarrow Q(S) = 0$. Furthermore, if $E(Y)$ exists, $Q(S)$ is countably additive. Under these conditions it follows from the Radon-Nikodym theorem (see Appendix **2B**) that there exists a function g of T defined a.e. $[dF(t)]$ such that

$$Q(S) = \int_S g(t)\, dP = \int_S g(t)\, dF(t). \tag{2b.3.3}$$

The function $g(t)$ is denoted by $E(Y|t)$. We shall consider some general properties of conditional expectation which have important applications in mathematical statistics.

(i) $$E(Y) = \int E(Y|t)\, dF(t) = \underset{T}{E}\,[E(Y|T)] \tag{2b.3.4}$$

where T under E denotes expectation with respect to the d.f. of T.

It may be noted that when a further expectation is taken, $E(Y|t) = E(Y|T = t)$ is replaced by $E(Y|T)$. We shall follow this convention throughout.

Result (2b.3.4) is an immediate consequence of (2b.3.1) if we take S to be the entire plane. The reader may easily establish the result when densities exist and the conditional expectation is defined as expectation with respect to the conditional distribution (density).

Consider random variables Y and Y^2 and define the *conditional variance of Y given $T = t$*

$$V(Y|t) = E(Y^2|t) - [E(Y|t)]^2. \tag{2b.3.5}$$

(ii) $$V(Y) = \underset{T}{E}\,[V(Y|T)] + \underset{T}{V}\,[E(Y|T)].$$

Integrating both sides of (2b.3.5) with respect to $dF(t)$, we see that

$$\begin{aligned} \underset{T}{E}\,[V(Y|T)] &= E(Y^2) - \underset{T}{E}\,[E(Y|T)]^2 \\ &= V(Y) - \underset{T}{V}\,[E(Y|T)]. \end{aligned}$$

Hence

$$V(Y) = \underset{T}{E}\,[V(Y|T)] + \underset{T}{V}\,[E(Y|T)], \tag{2b.3.6}$$

which provides a decomposition of the total variance of Y as the sum of two components: (a) the average conditional variance, and (b) the variance of conditional average. (See also Example 19 at the end of the chapter.)

(iii) *Let $g(T)$ be a Borel function of T and $E|Y| < \infty$, $E|Yg(T)| < \infty$. Then*

(a) $$E[g(T)Y|t] = g(t)E(Y|t) \tag{2b.3.7}$$

(b) $$E\{g(T)[Y - E(Y|T)]\} = 0 \tag{2b.3.8}$$

The assumption that g is a Borel function of T is to ensure that $g(T)$ is itself a random variable with respect to the probability measure induced by T, so that expectations, etc., may be taken. The result (2b.3.8) is a trivial consequence of (2b.3.7). To prove (2b.3.7), consider a simple function of T, which is the indicator function $C_A(t)$ of a set $A \in \mathscr{B}_n$, in the T plane. Then it is easy to see that, for any set $S \in \mathscr{B}_n$

$$\int_{R_1 \times S} y C_A(t)\, dF(y, t) = \int_S C_A(t) E(Y \mid t)\, dF(t) \tag{2b.3.9}$$

so that (2b.3.7) is true for simple functions of T. The equation (2b.3.9) is true if g is a finite linear combination of simple functions. In general, by approximating g by finite linear combinations of simple functions in bounded sets of T and using the condition $E|g(T)Y| < \infty$, the result (2b.3.7) is deduced.

The conditional probability of a set $A \in \mathscr{B}$, given $T = t$, can be exhibited as a conditional expectation, by choosing the random variable $Y(\cdot)$ as the indicator function of the set A. Thus $P(A \mid t) = E(Y \mid t)$, as may be verified from the definition of conditional probability. It must be noted that the R-N theorem establishes the existence of $P(A \mid t)$ as a function of t for fixed A only a.e. $[dF(t)]$ and the exceptional t-set may depend on A. Hence it may not be possible to define $P(A \mid t)$ for all A over a t-set of probability 1—unless the union of exceptional sets is of probability zero—so that a conditional probability measure over $(\Omega, \mathscr{B})$ given $T = t$ may not always exist.

Kolmogorov (1933) and Doob (1953) have, however, provided a suggestive evaluation of conditional expectation of a random variable Y using only conditional probabilities, not first constructing a conditional probability measure over $(\Omega, \mathscr{B})$. They achieved this by considering the sum

$$S(\lambda) = \sum_{-\infty}^{\infty} (j + 1)\lambda P[j\lambda < Y(\omega) < (j + 1)\lambda \mid t] \tag{2b.3.10}$$

which is absolutely convergent for given λ if $E(Y)$ exists. Then

$$\lim_{\lambda \to 0} S(\lambda) = E(Y \mid t).$$

But it is of some interest to recognize the situations in which a conditional probability measure exists.

Discrete Sample Space. *If the sample space has only a countable number of points, the conditional probability measure is always defined.* The conditional probability of any set $A \in \mathscr{B}$ is in fact defined by $P(A \cap \{T = t\})/P(T = t)$ provided the denominator is not zero.

Euclidean Sample Space. *If the sample space is R^n, the conditional probability measure exists.* The proof depends on the fact that a probability measure over $(R^n, \mathscr{B}_n)$ need only be defined on a countable number of sets, for instance on all rectangles with corners having rational coordinates. The union of exceptional sets is then of zero probability measure. The probability measure so obtained over rectangles can be uniquely extended to all sets in $\mathscr{B}_n$. For further details, the reader is referred to Doob (1953).

Conditional Density on the Sample Space. In many applications the sample space is R^n (n-dimensional Euclidean space) and the probability density at any point $\mathbf{x} = (x_1, \ldots, x_n)$ exists. If $p(\mathbf{x})$ represents the density, the probability of a set $A \in \mathscr{B}_n$ is given by

$$P(A) = \int_A p(\mathbf{x})\, dx_1 \cdots dx_n.$$

Let us consider a measurable function T which may be multidimensional, with a p.d. $q(t)$. We may then define a probability density, $p(\mathbf{x})/q(t)$ *over the surface* $T = t$. The function $p(\mathbf{x})/q(t)$ of $\mathbf{x}$ may be called the conditional density at the point $\mathbf{x}$ of R^n, given $T = t$. If T admits continuous partial derivatives with respect to the variables, it follows from the theory of multiple integration that the volume element

$$dx_1 \cdots dx_n = dt\, dS, \tag{2b.3.11}$$

where dS is the surface element and integration of any function, can be performed by integrating first with dS—keeping t fixed—and then integrating with dt. This property is sufficient to establish all the results considered in this section. If Y is any measurable function of $\mathbf{x}$, then its conditional expectation given $T = t$ is

$$E(Y \mid t) = \int Y \frac{p(\mathbf{x})}{q(t)}\, dS \tag{2b.3.12}$$

which is a function of t.

2b.4 Characteristic Function (c.f.)

The *characteristic function* $f(t)$ of a random variable X with the d.f. F is defined by (the range of integration is $(-\infty, \infty)$ unless otherwise stated)

$$\begin{aligned} f(t) &= \int e^{itx}\, dF(x) \\ &= \int \cos tx\, dF(x) + i \int \sin tx\, dF(x) \end{aligned} \tag{2b.4.1}$$

where $i = \sqrt{-1}$. The integral on the right-hand side of (2b.4.1) exists since $|e^{itx}| = |\cos tx + i \sin tx| = 1$, bounded for all x. We have the following results, which may be proved by using standard results of analysis.

(i) $f(0) = 1$, $|f(t)| \leqslant 1, f(-t) = \overline{f(t)}$, *the complex conjugate of* $f(t)$.
(ii) $f(t)$ *is uniformly continuous on the entire real line.*

The result follows since cos tx and sin tx are continuous and bounded on the entire real line.

(iii) $\overline{f(t)}$ *is the c.f. of* $-X$. $f(t)$ *is real if, and only if,* X *is symmetrical. The c.f. of* $bX + a$ *is* $f(bt)\exp(ita)$.
(iv) *If* $\alpha_r = E(X^r)$ *exists, then the rth derivative of* $f(t)$

$$f^{(r)}(t) = (i)^r \int x^r e^{itx}\, dF(x) \tag{2b.4.2}$$

exists and is uniformly continuous. Further,

$$\alpha_s = E(X^s) = (i)^{-s} f^s(0), \qquad s \leqslant r. \tag{2b.4.3}$$

Since $|x^r e^{itx}| \leqslant |x^r|$ and $E(x^r)$ exists, the integral on the right-hand side of (2b.4.2) is absolutely convergent in which case differentiation and integration can be interchanged (see 2B.14 of Appendix **2B**). Thus

$$\frac{d^r f}{dt^r} = \int \frac{d^r e^{itx}}{dt^r}\, dF(x) = (i)^r \int x^r e^{itx}\, dF(x).$$

Continuity of $f^{(r)}$ follows because of continuity and boundedness of e^{itx}. The result (2b.4.3) follows easily.

(v) *If* $f^{(2n)}(0)$ *exists and is finite, then* $E(X^r) < \infty$ *for* $r \leqslant 2n$.

We observe that

$$\begin{aligned}|f^{(2n)}(0)| &= \left|\lim_{u\to 0} \int \left(\frac{e^{iux} - e^{-iux}}{2u}\right)^{2n} dF(x)\right| \\ &= \lim_{u\to 0} \int \left(\frac{\sin ux}{ux}\right)^{2n} x^{2n}\, dF(x) \geqslant \int x^{2n}\, dF(x)\end{aligned} \tag{2b.4.4}$$

using Fatou's lemma (see 2B.5 in Appendix **2B**) for interchanging limits and integration. Then (v) follows from the inequality (2b.4.4), and the result that the finiteness of $E(X^s)$ implies that of $E(X^k)$ for $k \leqslant s$ in view of the inequality $|x^k| \leqslant 1 + |x|^s$.

(vi) *If* $E(X^r)$ *exists,* $f(t)$ *admits the Taylor expansion*

$$f(t) = \sum_0^r \alpha_j \frac{(it)^j}{j!} + o(t^r) \qquad (\text{as } t \to 0). \tag{2b.4.5}$$

(vii) *Let* $\phi(t) = \log f(t)$ *which is called the second characteristic of* X (*or of* F, *its d.f.*). *If* $E(X^r)$ *exists, then* $\phi(t)$ *admits the expansion*

$$\phi(t) = \sum_0^r \kappa_j \frac{(it)^j}{j!} + o(t^r) \qquad (\text{as } t \to 0). \tag{2b.4.6}$$

The coefficients κ_j in the expansion of $\phi(t)$ are called *cumulants* or *semi-invariants* (semi-invariants because (except for κ_1) they are invariant for translations of the random variable). If we consider the random variable $X + a$, its c.f. is $e^{ita}f(t)$ and the second characteristic is

$$\log e^{ita}f(t) = ita + \log f(t) = ita + \phi(t),$$

so that only the coefficient of t is altered, whereas this property is not true of the moment coefficients α_j.

From the expansions in (vi) and (vii) we have the relations connecting κ_j, α_j, and the central moments μ_j,

$$\begin{aligned}
\kappa_1 &= \mu = \alpha_1 &&= \mu \\
\kappa_2 &= \alpha_2 - \alpha_1^2 &&= \mu_2 \\
\kappa_3 &= \alpha_3 - 3\alpha_1\alpha_2 + 2\alpha_1^3 &&= \mu_3 \\
\kappa_4 &= \alpha_4 - 3\alpha_2^2 - 4\alpha_1\alpha_3 + 12\alpha_1^2\alpha_2 - 6\alpha_1^4 &&= \mu_4 - 3\mu_2^2
\end{aligned}$$

and conversely

$$\begin{aligned}
\alpha_1 &= \kappa_1, & \alpha_3 &= \kappa_3 + 3\kappa_2\kappa_1 + \kappa_1^3 \\
\alpha_2 &= \kappa_2 + \kappa_1^2, & \alpha_4 &= \kappa_4 + 3\kappa_2^2 + 4\kappa_1\kappa_3 + 6\kappa_1^2\kappa_2 + \kappa_1^4.
\end{aligned}$$

The coefficients

$$\sqrt{\beta_1} = \frac{\mu_3}{\mu_2^{3/2}} = \frac{\kappa_3}{\kappa_2^{3/2}}$$

$$\beta_2 = \frac{\mu_4}{\mu_2^2} = \frac{\kappa_4}{\kappa_2^2} + 3$$

are called measures of skewness and kurtosis respectively.

Moment Generating Function (m.g.f.). The function

$$M(t) = \int e^{tx}\, dF(x), \tag{2b.4.7}$$

if it exists for an interval of t containing the origin on the real line, is called the moment generating function.

Probability Generating Function (p.g.f.). The functions $f(t)$, $\phi(t)$ and $M(t)$ are defined for any random variable although $M(t)$ may not exist in all cases.

For a discrete random variable X, taking values 0, 1, ..., we can define still another function

$$P(t) = \sum_0^{\infty} p_r t^r, \qquad -1 \leqslant t \leqslant 1$$

called the p.g.f. The p.g.f. is also useful in determining the moments of X. For this purpose we use the relation

$$\left. \frac{d^s P}{dt^s} \right|_{t=1} = E[X(X-1)\cdots(X-s+1)],$$

where the right-hand side is called the s-th *factorial moment* of X, denoted by ϕ_s. The factorial moments are connected with the raw moments α_s by the relations

$$\phi_1 = \alpha_1, \qquad \phi_2 = \alpha_2 - \alpha_1, \qquad \phi_3 = \alpha_3 - 3\alpha_2 + 2\alpha_1$$
$$\phi_4 = \alpha_4 - 6\alpha_3 + 11\alpha_2 - 6\alpha_1,$$

or conversely

$$\alpha_1 = \phi_1, \qquad \alpha_2 = \phi_2 + \phi_1, \qquad \alpha_3 = \phi_3 + 3\phi_2 + \phi_1$$
$$\alpha_4 = \phi_4 + 6\phi_3 + 7\phi_2 + \phi_1.$$

We will derive the p.g.f., m.g.f., and c.f. [$P(t)$, $M(t)$, and $f(t)$] of certain random variables, which are used in later work.

Binomial Distribution

$$p_r = \binom{n}{r} \pi^r (1-\pi)^{n-r}, \qquad r = 0, 1, \ldots, n$$

$$P(t) = \sum_1^n \binom{n}{r} \pi^r (1-\pi)^{n-r} t^r = (1 - \pi + \pi t)^n, \tag{2b.4.8}$$

$$M(t) = P(e^t) = (1 - \pi + \pi e^t)^n,$$

$$f(t) = M(it) = (1 - \pi + \pi e^{it})^n.$$

Poisson Distribution

$$p_r = e^{-\lambda} \frac{\lambda^r}{r!}, \qquad r = 0, 1, \ldots \infty$$

$$P(t) = \sum e^{-\lambda} \frac{\lambda^r t^r}{r} = e^{-\lambda(1-t)}, \tag{2b.4.9}$$

$$M(t) = P(e^t) = e^{-\lambda(1-e^t)},$$

$$f(t) = M(it) = e^{-\lambda(1-e^{it})}.$$

Normal Distribution

$$p(x) = \frac{1}{\sigma\sqrt{2\pi}} e^{-(x-\mu)^2/2\sigma^2}$$

$$M(t) = \frac{1}{\sigma\sqrt{2\pi}} \int_{-\infty}^{\infty} e^{tx} e^{-(x-\mu)^2/2\sigma^2}\, dx$$

$$= e^{t\mu + t^2\sigma^2/2} \frac{1}{\sigma\sqrt{2\pi}} \int_{-\infty}^{\infty} e^{-(y-t\sigma^2)^2/2\sigma^2}\, dy, \qquad y = x - \mu$$

$$= e^{t\mu + t^2\sigma^2/2}. \tag{2b.4.10}$$

$$f(t) = M(it) = e^{it\mu - t^2\sigma^2/2}. \tag{2b.4.11}$$

The d.f. F is said to be degenerate if it has only one point of increase $a \in R$ with $F(a+0) - F(a-0) = 1$. In such a case the *c.f. of F is* $f(t) = \exp(ita)$.

(viii) *f is the c.f. of a degenerate d.f. if its modulus equals 1 for two values of $t \neq 0$ and $at \neq 0$ where α is irrational. In particular, f is degenerate if $|f(t)| = 1$ in a non-degenerate interval.*

(ix) Convolution of Distribution Functions. A function F on R is said to be the convolution of d.f.'s F_1 and F_2 and written $F = F_1 * F_2$, if

$$F(x) = \int F_1(x - y)\, dF_2(y), \qquad x \in R \tag{2b.4.12}$$

(a) *F as defined in (2b.4.12) is a d.f. and if f, f_1 and f_2 are the c.f.'s of F, F_1 and F_2, respectively, then $f = f_1 f_2$.*

As F_1 is nondecreasing, so also is F. Since $F_1(x - y)$ is bounded by unity, by LDC (Lebesgue dominated convergence theorem, 2B.11 of Appendix **2B**), it follows that $F(-\infty) = 0$ and $F(\infty) = 1$. Further, F_1 is left continuous so that the left continuity F again follows by applying the LDC theorem. The first assertion is proved.

To prove the second part, let $a < x_{n1} < \cdots < x_{n,k_n+1} = b$ with $\sup_k(x_{n,k+1} - x_{nk}) \to 0$ as $n \to \infty$. For every $t \in R$

$$\int_a^b e^{itx}\, dF(x) = \lim \sum_k e^{itx_{nk}} F[x_{nk}, x_{n,k+1})$$

$$= \lim \int \sum_k e^{it(x_{nk}-y)} F_1[x_{nk} - y, x_{n,k+1} - y) e^{ity} dF_2(y)$$

from which we have

$$\int_a^b e^{itx}\, dF(x) = \int \left\{ \int_{a-y}^{b-y} e^{itx}\, dF_1(x) \right\} e^{ity}\, dF_2(y),$$

where $F[c, d) = F(d) - F(c)$. Letting $a \to -\infty$ and $b \to \infty$

$$\int e^{itx}\,dF(x) = \int e^{itx}\,dF_1(x) \int e^{ity}\,dF(y)$$

so that $f = f_1 f_2$.

Note that if $f = f_1 f_2$, then from uniqueness of the c.f. (see **2b.5**), $F = F_1 * F_2$. We observe that the product of c.f.'s is also a c.f., and in particular if f is a c.f., so is $|f|^2$.

(b) *If X and Y are independent r.v.'s with c.f.'s f_1 and f_2, and f is the c.f. of $T = X + Y$, then*

$$\begin{aligned} f &= E\{e^{it(X+Y)}\} = E(e^{itX}) \cdot E(e^{itY}) \\ &= f_1 f_2 . \end{aligned} \tag{2b.4.13}$$

Using result of (a), we note that the convolution $F_1 * F_2$ provides the d.f. of the sum of two independent random variables with d.f.'s F_1 and F_2. $F = F_1 * F_2$ is also called the composition of F_1 and F_2.

2b.5 Inversion Theorems

Given a d.f. there exists a c.f. defined by (2b.4.1). The question naturally arises as to whether a c.f. uniquely determines a d.f. The answer is yes and the precise result is stated in (i).

Suppose a d.f. admits moments $\alpha_1, \alpha_2, \ldots$ of all orders. Does a given moment sequence define a d.f. uniquely? The result is not true in general. A sufficient condition for uniqueness is given in (ii).

(i) *If $f(t)$ is the c.f. corresponding to $F(x)$ and $(a - c, a + c)$ is a continuity interval of $F(x)$, then*

$$F(a+c) - F(a-c) = \lim_{T \to \infty} \frac{1}{\pi} \int_{-T}^{T} \frac{\sin ct}{t} e^{-ita} f(t)\,dt, \tag{2b.5.1}$$

so that if there are two distribution functions, they agree at all continuity points of both the distributions and are therefore identical.

Consider

$$\begin{aligned} J &= \frac{1}{\pi} \int_{-T}^{T} \frac{\sin ct}{t} e^{-ita} f(t)\,dt \\ &= \frac{1}{\pi} \int_{-T}^{T} \frac{\sin ct}{t} e^{-ita}\,dt \int e^{itx}\,dF(x). \end{aligned}$$

Since $|(\sin ct/t)e^{it(x-a)}| < |c|$, the order of integration can be interchanged so that J can be written

$$\frac{1}{\pi}\int dF(x)\int_{-T}^{T}\frac{\sin ct}{t}e^{-it(a-x)}\,dt = \int_{-\infty}^{\infty} g(T, x)\,dF(x)$$

where $g(T, x)$ is equal to

$$\frac{1}{\pi}\int_{-T}^{T}\frac{\sin ct}{t}e^{-it(a-x)}\,dt = \frac{2}{\pi}\int_{0}^{T}\frac{\sin ct}{t}\cos(x-a)t\,dt.$$

Using the formula, $2\sin A\cos B = \sin(A+B) + \sin(A-B)$, we have

$$g(T, x) = \frac{1}{\pi}\int_0^T \frac{\sin(x-a+c)t}{t}\,dt + \frac{1}{\pi}\int_0^T\frac{\sin(c+a-x)t}{t}\,dt \quad (2b.5.2)$$

$$\lim_{T\to\infty} g(T, x) = \frac{1}{\pi}\int_0^\infty \frac{\sin(x-a+c)t}{t}\,dt + \frac{1}{\pi}\int_0^\infty\frac{\sin(c+a-x)t}{t}\,dt.$$

Next we use the result

$$\frac{1}{\pi}\int_0^\infty \frac{\sin \alpha t}{t}\,dt = \begin{cases} \frac{1}{2} & \text{if } \alpha > 0 \\ 0 & \text{if } \alpha = 0 \\ -\frac{1}{2} & \text{if } \alpha < 0 \end{cases}$$

for each term in (2b.5.2). Then we have

$$\lim_{T\to\infty} g(T, x) = \begin{cases} 0 & \text{for } x < a-c \text{ or } x > (a+c) \\ \frac{1}{2} & \text{for } x = a-c \text{ or } x = a+c \\ 1 & \text{for } a-c < x < a+c \end{cases}$$

for in the first case the values of the two terms are $-\frac{1}{2}, \frac{1}{2}$ or $\frac{1}{2}, -\frac{1}{2}$; in the second case, $0, \frac{1}{2}$ or $\frac{1}{2}, 0$; and in the third case $\frac{1}{2}, \frac{1}{2}$. Thus, we have established that $\lim g(T, X)$ exists.

Furthermore $|g(T, x)|$ is bounded since $\int_0^T(\sin bt/t)\,dt$ is uniformly bounded in T. Hence by the Lebesgue dominated convergence theorem (Appendix **2B**), we see that

$$\lim_{T\to\infty}\int_{-\infty}^{\infty} g(T, x)\,dF(x) = \int_{-\infty}^{\infty}\lim_{T\to\infty} g(T, x)\,dF(x)$$
$$= \int_{a-c}^{a+c} dF(x) = F(a+c) - F(a-c),$$

which establishes (2b.5.1).

In particular, if $|f(t)|$ is integrable, by dividing both sides of (2b.5.1) by c and taking limits, we have

$$\lim_{c\to 0} \frac{F(a+c)-F(a-c)}{2c} = \frac{1}{2\pi}\int_{-\infty}^{\infty} \lim_{c\to 0} \frac{\sin ct}{ct}\, e^{-ita}f(t)\,dt \tag{2b.5.3}$$

$$F'(a) = \frac{1}{2\pi}\int_{-\infty}^{\infty} e^{-ita}f(t)\,dt \tag{2b.5.4}$$

so that the derivative of $F(x)$ exists and is also continuous. Notice that the integrand in (2b.5.3) is less than $|f(t)|$ in absolute value and the integrability of $|f(t)|$ validates, by dominated convergence theorem, carrying the limit inside the integral. For this purpose, the integrability of $|f(t)|$ appears to be a strong condition.

(ii) THE MOMENT PROBLEM. *Let $\alpha_1, \alpha_2, \ldots$ be the moment sequence of a d.f. They do not, in general, determine the d.f. uniquely. A sufficient condition for the d.f. to be unique is that the series*

$$\sum_1^{\infty} \frac{\alpha_i}{i!} t^i \tag{2b.5.5}$$

is absolutely convergent for some $t > 0$.

The proof of this proposition, known as *Stieltjes moment problem*, consists in showing that under the condition (2b.5.5) the c.f. is uniquely determined. For details of the proof, refer to original papers by Hamburger (1920, 1921). A sufficient condition in the case of a two-dimensional random variable is given by Cramer and Wold (1936).

That the condition (2b.5.5) is not necessary may be seen by showing that the moments of $Z = X\log(1+Y)$ where X and Y are independent variables with densities e^{-x} and e^{-y} determine the distribution of Z uniquely by (b) below although (2b.5.5) is not satisfied (see Dharmadhikari, 1965).

Any one of the following is a sufficient condition for the moment sequence $\alpha_1, \alpha_2, \ldots$ to define a unique d.f.:

(a) The range of the random variable is finite.

(b) $\sum_0^{\infty} (\alpha_{2j})^{-1/2j} = \infty$, when the range of the variable is $(-\infty, \infty)$.

(c) $\sum_0^{\infty} (\alpha_j)^{-1/2j} = \infty$, when the range of the variable is $(0, \infty)$.

(d) $\lim\sup_{n\to\infty}[(\alpha_{2n})^{1/2n}/2n]$ is finite.

2b.6 Multivariate Moments

If $F(x_1, \ldots, x_k)$ is the joint d.f. of the random variables $X_1, \ldots, X_k$, the multivariate moment $\alpha_{r_1 \cdots r_k}$ about the constants $a_1, \ldots, a_k$ is

$$E[(X_1 - a_1)^{r_1} \cdots (X_k - a_k)^{r_k}] = \int \cdots \int (x_1 - a_1)^{r_1} \cdots (x_k - a_k)^{r_k} \, dF(x_1, \ldots, x_k).$$

When $a_i = E(X_i) = \mu_i$, we have central moments as in the case of a single random variable. The second-order moment

$$E(X_i - \mu_i)(X_j - \mu_j)$$

is called the covariance between X_i, X_j and is represented by $\text{cov}(X_i, X_j)$ or simply $C(X_i, X_j)$. Denoting $C(X_i, X_j)$ by σ_{ij} we have the following results:

1. $\sigma_{ii} = C(X_i, X_i) = \sigma_i^2$ (variance of X_i)
2. $\sigma_{ij}^2 \leqslant \sigma_{ii}\sigma_{jj}$ (by an application of *C-S* inequality)
3. $-1 \leqslant \rho_{ij} \leqslant 1$, where $\rho_{ij} = \sigma_{ij}/\sqrt{\sigma_{ii}\sigma_{jj}}$ is called the product moment correlation coefficient between X_i and X_j [(3) follows from (2)].
4. If $a_1, \ldots, a_k$ are fixed coefficients $V(a_1 X_1 + \cdots + a_k X_k)$, the variance of $a_1 X_1 + \cdots + a_k X_k$, is

$$E[\sum a_i(X_i - \mu_i)]^2 = \sum \sum a_i a_j \sigma_{ij}. \tag{2b.6.1}$$

Similarly the covariance between $\sum a_i X_i$ and $\sum b_i X_i$ is

$$\sum \sum a_i b_j \sigma_{ij}. \tag{2b.6.2}$$

In matrix notation, if $\mathbf{\Sigma} = (\sigma_{ij})$ and $\mathbf{A}$, $\mathbf{B}$, $\mathbf{X}$, $\mathbf{M}$ are the column vectors of the coefficients a_i, b_i, random variables X_i and mean values μ_i,

$$\begin{gathered} E(\mathbf{A}'\mathbf{X}) = \mathbf{A}'E(\mathbf{X}) = \mathbf{A}'\mathbf{M} \\ V(\mathbf{A}'\mathbf{X}) = \mathbf{A}'\mathbf{\Sigma}\mathbf{A}, \qquad V(\mathbf{B}'\mathbf{X}) = \mathbf{B}'\mathbf{\Sigma}\mathbf{B} \\ C(\mathbf{A}'\mathbf{X}, \mathbf{B}'\mathbf{X}) = \mathbf{A}'\mathbf{\Sigma}\mathbf{B} = \mathbf{B}'\mathbf{\Sigma}\mathbf{A} \end{gathered} \tag{2b.6.3}$$

where V stands for variance and C for covariance.

5. The matrix $\mathbf{\Sigma}$ is non-negative definite since the quadratic form $\sum \sum a_i a_j \sigma_{ij}$ is not negative being the variance of a random variable (see 2b.6.1).
6. $\mathbf{\Sigma}$ is positive definite if and only if there is no linear relation among the random variables $X_1, \ldots, X_k$ (with probability 1).
7. Let the dispersion (i.e., variance-covariance) matrix $\mathbf{\Sigma}$ (which is n.n.d.) of k-vector random variable $\mathbf{X}$ be of rank $r < k$. Then $\mathbf{X}$ can be expressed as

$$\mathbf{X} = \mathbf{B}\mathbf{Y} + \mathbf{C} \tag{2b.6.4}$$

where $\mathbf{Y}$ is an r-vector random variable with dispersion matrix $\mathbf{I}$, $\mathbf{B}$ is a $k \times r$ matrix of rank r and $\mathbf{C}$ is a k-vector of constants.

In multivariate problems, the conditional distribution of a subset of the variables, say $X_1, \ldots, X_r$, given the rest $X_{r+1}, \ldots, X_k$, is often studied. In particular the conditional expectation and variance of a variable, given other variables, play an important part. A detailed discussion of the problems based on these conditional moments and their computation is given in **4g.1**, **4g.2** of Chapter 4.

2c LIMIT THEOREMS

2c.1 Kolmogorov Consistency Theorem

Let $(\Omega, \mathscr{B}, P)$ be the basic probability space, $\{X_i(\cdot)\}$, $i = 1, 2, \ldots$ be an infinite sequence of random variables, and F_n be d.f. of $(X_1(\cdot), \ldots, X_n(\cdot))$ defined on R^n. Then we have the following consistency property $F_n(t_1, \ldots, t_n) = F_{n+1}(t_1, \ldots, t_n, \infty)$.

Now we consider the converse problem: Given a sequence of d.f.'s F_n on R^n, $n = 1, 2, \ldots$ which have the consistency property does there exist a $(\Omega, \mathscr{B}, P)$ and a sequence of r.v.'s $X_i(\cdot)$ such that F_n is the d.f. of $(X_1(\cdot), \ldots, X_n(\cdot))$, $n = 1, 2, \ldots,$ A theorem of Kolmogorov says that this is the case, which we shall prove.

In fact we shall choose Ω as R^∞, the Euclidean space of infinite sequences, and the ith random variable as a function whose value at a point is the ith coordinate, and define $\mathscr{B}$ and P suitably.

Take a finite dimensional subsample space of the random variables $X_{i_1}, \ldots, X_{i_n}$ and a Borel set A_{in} in it. By associating with each point $(x_{i_1}, \ldots, x_{i_n})$ in A_{in} arbitrary values of other coordinates, we generate infinite sequences which constitute a set A_i in R^∞. Such a set is called a Borel cylinder set with base A_{in} in some E_n. Let $P_n(A_{in})$ be the probability of the set A_{in} induced by distribution function of $X_{1_1}, \ldots, X_{i_n}$ on $\mathscr{B}_n$ in the appropriate E_n, as considered in **2a.6**. Now define

$$P_\infty(A_i) = P_n(A_{in}). \tag{2c.1.1}$$

Thus, the function P_∞ is defined for all Borel cylinder sets obtained by considering all finite subsets of the random variables $X_1, X_2, \ldots$. The class of cylinder Borel sets form a field, $\mathscr{F}_0$ (say). *P_∞ defined by* (2c.1.1) *constitutes a probability measure over $\mathscr{F}_0$.*

We have only to establish countable additivity of P_∞ to qualify P_∞ as a probability measure, that is, if there is a decreasing sequence of sets $A_1, A_2, \ldots,$ in $\mathscr{F}_0$ tending to the empty set $\varnothing$, then $P_\infty(A_i) \to 0$ as $i \to \infty$.

We can assume without loss of generality that the base of A_1 is a set A_{11} on the X_1 axis, the base of A_2 is a set A_{22} in (X_1, X_2) space, ..., the base of A_n is a set A_{nn} in the $(X_1, \ldots, X_n)$ space, and so on. Given A_{nn}, there exists a bounded closed set B_{nn} in the $(X_1, \ldots, X_n)$ space such that $B_{nn} \subset A_{nn}$ and $P_n(A_{nn} - B_{nn}) < \delta/2^n$. Now let B_n be the cylinder set with base B_{nn}. Then $B_n \subset A_n$ and

$$P_\infty(A_n - B_n) = P_n(A_{nn} - B_{nn}) < \frac{\delta}{2^n}.$$

Consider $A_n = \bigcap_1^n A_j$ and $C_n = \bigcap_1^n B_j$. Now

$$P_\infty(A_n - C_n) \leqslant \sum_1^n P_\infty(A_i - B_i) < \delta\left(\frac{1}{2} + \cdots + \frac{1}{2^n}\right) < \delta. \qquad (2c.1.2)$$

Let us assume that $P_\infty(A_n) > \varepsilon$ for every n so that it does not approach zero. Choosing $\delta = \varepsilon/2$, we find from (2c.1.2) that

$$P_\infty(C_n) > P_\infty(A_n) - \delta > \varepsilon/2 > 0,$$

that is, the set C_n is not empty for any n. Let $\xi_n = (\xi_{n1}, \xi_{n2}, \ldots)$ be a point of R^∞ belonging to C_n. Since $C_n = \bigcap_1^n B_i$, $\xi_n \in B_1$ for every n. The first coordinates of the points $\xi_1, \xi_2, \ldots$ belong to B_{11}, and since B_{11} is closed and bounded there is a subsequence of the points such that the first coordinate tends to a limit, say, b_1. The points $\xi_2, \xi_3, \ldots$ belong to B_2, and therefore the first two coordinates of the points $\xi_2, \xi_3, \ldots$ belong to B_{22}. Consider only the points of the first subsequence which belong to $\xi_2, \xi_3, \ldots$. Since B_{22} is bounded and closed there is a subsequence of the first subsequence for which the second coordinate tends to a limit b_2 and the first coordinate to b_1, and so on. Consider the point $b = (b_1, b_2, \ldots)$. Now the point $(b_1, \ldots, b_k) \in B_{kk}$ and therefore $b \in B_k$, for every k. Hence $b \in \bigcap B_k = \bigcap C_k$, that is, $\bigcap C_k \neq \varnothing$. Now

$$C_k \subset A_k \Rightarrow \bigcap C_k \subset \bigcap A_k \Rightarrow \lim A_k = \bigcap A_k \neq \varnothing,$$

which is a contradiction. Hence $P_\infty(A_i) \to 0$.

The function P_∞ defines a unique probability measure Q on the Borel extension $\mathscr{B}(\mathscr{F}_0)$ of $\mathscr{F}_0$, which we designate by $\mathscr{B}_\infty$. The measure space $(R^\infty, \mathscr{B}_\infty, Q)$, is thus completely specified by all finite dimensional distributions.

In particular we are interested in the distributions of some limit functions of the sequences with values at ω such as

$$\lim X_n(\omega), \quad \sup X_n(\omega), \quad \overline{\lim}\, X_n(\omega), \quad \underline{\lim}\, X_n(\omega), \quad \text{etc.}$$

The reader should satisfy himself that such functions are also random variables (i.e., measurable functions) provided they are finite.

2c.2 Convergence of a Sequence of Random Variables

A sequence of random variables $\{X_n\}$, $n = 1, 2, \ldots$ is said to converge to a constant c:

(a) *weakly or in probability* (written $X_n \xrightarrow{P} c$) if, for every given $\varepsilon > 0$,

$$\lim_{n \to \infty} P(|X_n - c| > \varepsilon) = 0, \tag{2c.2.1}$$

(b) *strongly or almost surely* (written $\lim_{n\to\infty} X_n = c$ with probability 1 or $X_n \xrightarrow{\text{a.s.}} c$) if

$$P\left(\lim_{n \to \infty} X_n = c\right) = 1, \tag{2c.2.2}$$

or equivalently,

$$\lim_{N \to \infty} P\left(\sup_{n \geqslant N} |X_n - c| > \varepsilon\right) = 0 \quad \text{for every } \varepsilon, \tag{2c.2.3}$$

(c) *in quadratic mean* (written $X_n \xrightarrow{\text{q.m.}} c$) if

$$\lim_{n \to \infty} E(X_n - c)^2 = 0. \tag{2c.2.4}$$

Definitions (2c.2.1) and (2c.2.4) offer no difficulty since they depend on the computation of probabilities and expectations with respect to a finite dimensional random variable and the taking of limits as the number of variables increase. The concept involved in the definition (2c.2.2) or (2c.2.3) is somewhat deeper. The probability refers to the set in R^∞ where the sequences (points) satisfy the stated property.

A sequence of random variables $\{X_n(\cdot)\}$, $n = 1, 2, \ldots$ is said to converge to a random variable $X(\cdot)$ in the sense of (a), (b), or (c) according as the sequence $\{X_n(\cdot) - X(\cdot)\}$, $n = 1, 2, \ldots$ converges to zero in the sense of (a), (b), or (c).

The following results establish the relationships among various types of convergence:

(i) (c) $\Rightarrow$ (a), *that is, convergence in q.m.* $\Rightarrow$ *convergence in probability.*

This follows at once from Chebyshev's inequality (2b.2.3)

$$P(|X_n - c| > \varepsilon) \leqslant \frac{1}{\varepsilon^2} E(X_n - c)^2.$$

(ii) (b) $\Rightarrow$ (a), *that is, convergence a.s.* $\Rightarrow$ *convergence in probability.*

Since

$$|X_N - c| > \varepsilon \Rightarrow \sup_{n \geq N} |X_n - c| > \varepsilon,$$

$$P(|X_N - c| > \varepsilon) \leq P\left(\sup_{n \geq N} |X_n - c| > \varepsilon\right),$$

from which the desired result follows. No other relations need be true among (a), (b), or (c) without further conditions.

(iii) *If* $X_n \xrightarrow{\text{q.m.}} c$ *in such a way that* $\sum_1^\infty E(X_n - c)^2 < \infty$, *then* $X_n \xrightarrow{\text{a.s.}} c$.

Since $\sum_1^\infty E(X_n - c)^2 < \infty$, the infinite sum

$$\sum_1^\infty (X_n - c)^2 \tag{2c.2.5}$$

must converge except for a set of sequences of measure zero in R^∞. Suppose that (2c.2.5) diverges on a set of measure p. Let A_N be the set in R^∞ such that $\sum_1^N (X_i - c)^2 > \lambda$. The sequence $\{A_N\}$ is nondecreasing and hence by (2a.3.1)

$$\lim P(A_N) = P(\lim A_N) \geq p,$$

since a divergent series eventually exceeds any given value λ. Consequently,

$$\sum_1^N E(X_i - c)^2 = E\left[\sum_1^N (X_i - c)^2\right] \geq \lambda p,$$

which contradicts the assumption, $\sum_1^\infty E(X_i - c)^2 < \infty$ if λ is chosen sufficiently large.

Hence $P(\sum_1^\infty (X_n - c)^2 \text{ converges}) = 1$. But if an infinite series converges the nth term $\to 0$ as $n \to \infty$. Hence

$$P(|X_n - c| \to 0 \text{ as } n \to \infty) \geq P\left(\sum_1^\infty (X_n - c)^2 \text{ converges}\right) = 1.$$

Incidentally we have proven that *if* Y_n, $n = 1, 2, \ldots$, *is a sequence of random variables, then*

$$E\left(\sum_1^\infty Y_n\right) = \sum_1^\infty E(Y_n) \tag{2c.2.6}$$

provided $\sum_1^\infty E|Y_n| < \infty$, *which ensures the convergence of* $\sum_1^\infty Y_n$ *with probability* 1.

2c.3 Law of Large Numbers

In practice, estimates are made of an unknown quantity (parameter) by taking the average of a number of repeated measurements of the quantity, each of which may be in error. It is, therefore, of interest to study the properties of such an estimate. An initial enquiry is made concerning its behavior as the number of measurements increases ($\rightarrow \infty$). Does the estimate converge in some sense to the true value of the parameter under study? The problem can be formulated in the following way.

Let $\{X_n\}$, $n = 1, 2, \ldots$, be a sequence of observations and $\bar{X}_n$, the average of the first n observations. Under what conditions can we assert that

$$\bar{X}_n \rightarrow \xi \text{ (the unknown quantity)} \tag{2c.3.1}$$

in one or other of the modes of convergence (a), (b), or (c) considered in **2c.2**? We shall generalize the problem further and ask for the conditions under which

$$\bar{X}_n - \bar{\xi}_n \rightarrow 0 \tag{2c.3.2}$$

where $\{\xi_n\}$, $n = 1, 2, \ldots$ is a sequence of constants sought to be measured by the sequence of observations $\{X_n\}$, $n = 1, 2, \ldots$. We shall say that the law of large numbers holds if the convergence such as (2c.3.1) or (2c.3.2) takes place. When the convergence is "in probability" (2c.2.1) we shall say that the *weak law* of large numbers (W.L.L.N) holds, and when it is "with probability 1" or a.s. (2c.2.2), the *strong law* of large numbers (S.L.L.N.) holds.

(i) CHEBYSHEV'S THEOREM (W.L.L.N.). *Let* $E(X_i) = \mu_i$, $V(X_i) = \sigma_i^2$, $\text{cov}(X_i, X_j) = 0$, $i \neq j$. *Then*

$$\lim_{n \rightarrow \infty} \frac{1}{n^2} \sum_1^n \sigma_i^2 = 0 \Rightarrow \bar{X}_n - \bar{\mu}_n \xrightarrow{P} 0. \tag{2c.3.3}$$

By Chebyshev's inequality, we see that

$$P(|\bar{X}_n - \bar{\mu}_n| > \varepsilon) \leqslant \frac{1}{\varepsilon^2} \frac{1}{n^2} \sum_1^n \sigma_i^2 \rightarrow 0.$$

Hence the result (2c.3.3). If $\mu_i = \mu$ and $\sigma_i^2 = \sigma^2$, the condition of the theorem is automatically satisfied and we have the result

$$\bar{X}_n \xrightarrow{P} \mu.$$

(ii) KHINCHINE'S THEOREM (W.L.L.N.). *Let* $\{X_i\}$, $i = 1, 2, \ldots$ *be independent and identically distributed (i.i.d.) and* $E(X_i)$ *exists. Then*

$$E(X_i) = \mu < \infty \Rightarrow \bar{X}_n \xrightarrow{P} \mu. \tag{2c.3.4}$$

It is interesting to note that the existence of the second moment as in Chebyshev's theorem is not assumed, but the observations are considered to be i.i.d. The proof of this theorem based on the convergence of c.f.'s is given in Example 4.3 at the end of the chapter. A stronger version of the theorem based on certain deeper propositions of measure theory is given in (iv). We shall, however, give a simple proof (using only elementary results) by Markoff.

Define a pair of new random variables for $i = 1, \ldots, n$ and a fixed δ

$$\begin{aligned} Y_i &= X_i, \quad Z_i = 0, \qquad \text{if} \quad |X_i| < \delta n, \\ Y_i &= 0, \qquad Z_i = X_i \qquad \text{if} \quad |X_i| \geqslant \delta n, \end{aligned}$$

so that $X_i = Y_i + Z_i$. Let $E(Y_i) = \mu_n$ for $i = 1, \ldots, n$. Since $E(X_i) = \mu$

$$|\mu_n - \mu| < \varepsilon \tag{2c.3.5}$$

for any given ε, if n is chosen sufficiently large. Now

$$V(Y_i) = \int_{-\delta n}^{\delta n} x^2 \, dF(x) - \mu_n^2 \leqslant \delta n \int_{-\delta n}^{\delta n} |x| \, dF(x) \leqslant b \, \delta n,$$

where $b = E|X|$ exists by hypothesis. Using Chebyshev's inequality we see that

$$P\left(\left| \frac{1}{n} \sum_1^n Y_i - \mu_n \right| \geqslant \varepsilon \right) \leqslant \frac{b\delta}{\varepsilon^2},$$

or by using (2c.3.5), we have

$$P\left(\left| \frac{1}{n} \sum_1^n Y_i - \mu \right| \geqslant 2\varepsilon \right) \leqslant \frac{b\delta}{\varepsilon^2}.$$

Since

$$P(Z_i \neq 0) = \int_{|x| \geqslant \delta n} dF(x) \leqslant \frac{1}{\delta n} \int_{|x| \geqslant \delta n} |x| \, dF(x) \leqslant \frac{\delta}{n}$$

by choosing n sufficiently large, we have

$$P\left(\sum_1^n Z_i \neq 0 \right) \leqslant \sum_1^n P(Z_i \neq 0) \leqslant \delta. \tag{2c.3.6}$$

Consider

$$\begin{aligned} P\left(\left| \frac{1}{n} \sum X_i - \mu \right| \geqslant 4\varepsilon \right) &= P\left[\left| \left(\frac{1}{n} \sum Y_i - \mu \right) + \frac{1}{n} \sum Z_i \right| \geqslant 4\varepsilon \right] \\ &\leqslant P\left[\left| \frac{1}{n} \sum Y_i - \mu \right| \geqslant 2\varepsilon \right] + P\left[\left| \frac{1}{n} \sum Z_i \right| \geqslant 2\varepsilon \right] \\ &\leqslant \frac{b\delta}{\varepsilon^2} + P\left(\sum Z_i \neq 0 \right) \leqslant \frac{b\delta}{\varepsilon^2} + \delta. \end{aligned} \tag{2c.3.7}$$

Since δ is arbitrary the last quantity can be made small, thus establishing the required result.

Note that we have used the inequality

$$P(\sum Z_i \neq 0) \leqslant \sum P(Z_i \neq 0)$$

at (2c.3.6) and a similar one at (2c.3.7) which the reader can easily verify.

(iii) KOLMOGOROV THEOREM 1 (S.L.L.N.). *Let $\{X_i\}$, $i = 1, 2, \ldots$ be a sequence of independent random variables such that $E(X_i) = \mu_i$ and $V(X_i) = \sigma_i^2$. Then*

$$\sum_1^\infty \frac{\sigma_i^2}{i_i^2} < \infty \Rightarrow \bar{X}_n - \bar{\mu}_n \xrightarrow{\text{a.s.}} 0, \tag{2c.3.8}$$

that is, the sequence $X_1, X_2, \ldots$ obeys the strong law of large numbers.

To prove this let us consider the random variables $Y_i = X_i - \mu_i$ and apply the Hajek-Renyi inequality proved in Example 3.3 at the end of the chapter. The inequality states that

$$P\left\{\max_{m \leqslant i \leqslant n} c_i |Y_1 + \cdots + Y_i| \geqslant \varepsilon\right\} \leqslant \frac{1}{\varepsilon^2}\left(c_m^2 \sum_1^m \sigma_i^2 + \sum_{m+1}^n c_i^2 \sigma_i^2\right).$$

Choosing $c_i = 1/i$, we have

$$P\left\{\max_{m \leqslant i \leqslant n} |\bar{Y}_i| \geqslant \varepsilon\right\} \leqslant \frac{1}{\varepsilon^2}\left(\frac{1}{m^2} \sum_1^m \sigma_i^2 + \sum_{m+1}^n \frac{\sigma_i^2}{i^2}\right).$$

Letting $n \to \infty$, we see that

$$P\left\{\max_{m \leqslant i} |\bar{Y}_i| \geqslant \varepsilon\right\} \leqslant \frac{1}{\varepsilon^2}\left(\frac{1}{m^2} \sum_1^m \sigma_i^2 + \sum_{m+1}^\infty \frac{\sigma_i^2}{i^2}\right). \tag{2c.3.9}$$

Since $\sum \sigma_i^2/i^2$ converges, it follows from (2c.3.9) that

$$\lim_{m \to \infty} P\left\{\max_{m \leqslant i} |\bar{Y}_i| \geqslant \varepsilon\right\} = 0.$$

But this means that

$$P\left\{\lim_{m \to \infty} \bar{Y}_m = 0\right\} = 1,$$

which proves the required result.

If in the statement of the theorem we replace independence of random variables by orthogonality of the centered variables, that is, $\text{cov}(X_i, X_j) = 0$ for all $i \neq j$, then (we need a stronger condition)

$$\sum_1^\infty \frac{\sigma_i^2 (\log i)^2}{i^2} < \infty \Rightarrow \bar{X}_n - \bar{\mu}_n \xrightarrow{\text{a.s.}} 0. \tag{2c.3.10}$$

For a proof of this and related results concerning orthogonal variables, the reader is referred to Doob (1953).

(iv) KOLMOGOROV THEOREM 2 (S.L.L.N.). *Let $X_1, X_2, \ldots$ be a sequence of i.i.d. variables. Then a necessary and sufficient condition that $\bar{X}_n \xrightarrow{\text{a.s.}} \mu$ is that $E(X_i)$ exists and is equal to μ.*

Necessity: Let E_n be the event that $|X_n| \geqslant n$. Then

$$\frac{X_n}{n} = \bar{X}_n - \frac{n-1}{n}\bar{X}_{n-1} \xrightarrow{\text{a.s.}} 0, \tag{2c.3.11}$$

since $\bar{X}_n \xrightarrow{\text{a.s.}} \mu$. The result (2c.3.11) implies that the probability of infinitely many events E_n occurring is zero. Furthermore, the independence of X_n implies the independence of E_n, and by the Borel-Cantelli lemma (proved in Appendix **2B**), we see that

$$\sum_{n=1}^{\infty} P(|X_n| \geqslant n) = \sum_{n=1}^{\infty} P(E_n) < \infty. \tag{2c.3.12}$$

Now, let $p_j = P(|X| \geqslant j)$. Then

$$\begin{aligned} E(|X|) &\leqslant 1(1-p_1) + 2(p_1 - p_2) + \cdots \\ &= 1 + p_1 + p_2 + \cdots = 1 + \sum_{n=1}^{\infty} P(E_n). \end{aligned} \tag{2c.3.13}$$

From (2c.3.12), the last expression in (2c.3.13) is less than ∞. Hence $E(|X|)$ exists, and from the sufficiency condition it follows that $E(X) = \mu$.

Sufficiency: Consider the sequence of truncated variables

$$X_n^* = \begin{cases} X_n & \text{for } |X_n| < n \\ 0 & \text{for } |X_n| \geqslant n. \end{cases}$$

We then obtain

$$V(X_n^*) \leqslant E(X_n^*)^2 = \int_{-n}^{+n} x^2\, dF(x) \leqslant \sum_{k=0}^{n-1} (k+1)^2 P(k \leqslant |X| < k+1)$$

and

$$\begin{aligned} \sum_{n=1}^{\infty} \frac{V(X_n^*)}{n^2} &\leqslant \sum_{n=1}^{\infty} \sum_{k=0}^{n-1} \frac{(k+1)^2}{n^2} P(k \leqslant |X| < k+1) \\ &\leqslant \sum_{k=1}^{\infty} P(k-1 \leqslant |X| < k)\, k^2 \sum_{n=k}^{\infty} \frac{1}{n^2} \\ &\leqslant 2 \sum_{k=1}^{\infty} kP(k-1 \leqslant |X| < k), \qquad \text{since } \sum_{n=k}^{\infty} \frac{1}{n^2} < \frac{1}{k^2} + \frac{1}{k} < \frac{2}{k} \\ &\leqslant 2[1 + E(|X|)] < \infty, \quad \text{since } E(X) \text{ exists.} \end{aligned}$$

Hence by Kolmogorov theorem proved in (iii), the sequence X_n^* obeys the law of large numbers, that is,

$$\bar{X}_n^* - \frac{1}{n}\sum_1^n E(X_i^*) \xrightarrow{\text{a.s.}} 0. \tag{2c.3.14}$$

Next, $E(X_n^*) \to E(X_n) = \mu$ as $n \to \infty$ and hence

$$\frac{1}{n}\sum_1^n E(X_i^*) \to \mu \quad \text{as} \quad n \to \infty. \tag{2c.3.15}$$

Combining (2c.3.14) and (2c.3.15), we have

$$\bar{X}_n^* \xrightarrow{\text{a.s.}} \mu.$$

We shall now show that X_n and X_n^* are equivalent sequences, that is

$$P\{X_n \neq X_n^*, \text{ for some values of } n \geqslant N\} \to 0 \quad \text{as} \quad N \to \infty,$$

which implies that X_n obeys the S.L.L.N. if X_n^* does and that the limits are the same. Consider

$$\begin{aligned} &P\{X_n \neq X_n^* \text{ for some } n \geqslant N\} \\ &\quad \leqslant \sum_{n \geqslant N} P\{X_n \neq X_n^*\} = \sum_{n=N}^{\infty} P\{|X_n| \geqslant n\} \\ &\quad \leqslant \sum_{n=N}^{\infty} (n - N + 1) P\{n \leqslant |X_n| < (n+1)\} \\ &\quad \leqslant \sum_{n=N}^{\infty} nP\{n \leqslant |X| < (n+1)\}, \quad \text{since all } X_n \text{ have the same d.f.} \\ &\quad \leqslant \int_{|x| \geqslant N} |x| \, dF(x) \to 0 \quad \text{as} \quad N \to \infty. \end{aligned}$$

2c.4 Convergence of a Sequence of Distribution Functions

Let us denote by $\{F_{X_n}\}$ or simply by $\{F_n\}$, $n = 1, 2, \ldots$ the sequence of distribution functions of the random variables $\{X_n\}$, $n = 1, 2, \ldots$.

Definition. The sequence of random variables $\{X_n\}$ is said to converge *in distribution* (or in law) to a random variable X with d.f. F_X (simply denoted by F) if $F_n \to F$ as $n \to \infty$ at all continuity points of F. Such a convergence is expressed as $X_n \xrightarrow{L} X$.

The reader may satisfy that it is enough to specify that $F_n \to F$ for a set S of points which is everywhere dense on the real line $(-\infty, \infty)$.

The approximating distribution F is called the limiting or asymptotic distribution of X_n. In statistical applications, limiting distributions play an important role. The random variable X_n stands for a statistic computed from a sample of size n, whose actual distribution is difficult to find. In such a case it may be approximated by the limiting distribution, at least for large values of n. The following results are important in studying limit distributions.

(i) HELLY LEMMA. *Every sequence of distribution functions is weakly compact, that is, there is a subsequence which tends to a function (not necessarily a d.f.) at all continuity points of the latter.*

Let $D = \{r_k\}$ be the set of all rationals. Since $F_n(r_1)$ is bounded, there exists, by Bolzano-Weierstrass compactness theorem, a convergent subsequence. Consider the sequence $\{F_{n_1}(x)\}$ which converges for the particular value $x = r_1$. From the sequence $\{F_{n_1}(x)\}$ we can extract another subsequence $\{F_{n_2}(x)\}$, in a similar way, which converges at $x = r_2$, and of course such a sequence will converge at $x = r_1$ also, and so on. Let us consider a sequence formed by the first member of $\{F_{n_1}\}$, the second member of $\{F_{n_2}\}$, ... Such a sequence of functions $\{F_s\}$ necessarily converges for all $x \in D$, and the limiting function F_D defined for all $x \in D$ is bounded and nondecreasing. Let, for any x

$$F(x) = \underset{r_i < x}{\text{upper bound}}\ F_D(r_i). \tag{2c.4.1}$$

By definition F is continuous from the left, bounded, and nondecreasing.

We shall show that the subsequence determined actually converges to $F(x)$ as defined in (2c.4.1) at all continuity points of F. Let x be such a point. Then we can find a sequence of rational values (x_i', x_i'') such that $x_i' < x < x_i''$ and $F(x_i'') - F(x_i') \to 0$ as $i \to \infty$. For any pair (x_i', x_i'') we have the obvious relationship

$$F_s(x_i') \leqslant F_s(x) \leqslant F_s(x_i'') \tag{2c.4.2}$$

for each s, where $\{F_s(x)\}$ is the sequence tending towards $F_D(x)$. Taking limits of functions in (2c.4.2), we have

$$F_D(x_i') \leqslant \underline{\lim}\ F_s(x) \leqslant \overline{\lim}\ F_s(x) \leqslant F_D(x_i'')$$

for each i. Since the difference $F_D(x_i'') - F_D(x_i')$ can be made arbitrarily small, $\lim F_s(x)$ exists and is equal to $F(x)$ defined in (2c.4.1). Observe that $F(x)$ is also in the interval $[F_D(x_i'), F_D(x_i'')]$.

(ii) HELLY-BRAY THEOREM. *$F_n \to F \Rightarrow \int g\, dF_n \to \int g\, dF$ for every bounded continuous function g.*

Choose two continuity points a, b $(a < b)$ of F and write

$$\int_{-\infty}^{\infty} g\, dF_n - \int_{-\infty}^{\infty} g\, dF$$
$$= \int_{-\infty}^{a} g(dF_n - dF) + \int_{a}^{b} g(dF_n - dF) + \int_{b}^{\infty} g(dF_n - dF). \tag{2c.4.3}$$

Let $|g| < c$. Then the modulus of the first integral in (2c.4.3) gives

$$\left|\int_{-\infty}^{a} g\, dF_n - \int_{-\infty}^{a} g\, dF\right| < c\int_{-\infty}^{a} dF_n + c\int_{-\infty}^{a} dF = c[F_n(a) + F(a)].$$

If a is sufficiently small, $F(a)$ is small and so also is $F_n(a)$ for all $n > n_0$. Hence $c[F_n(a) + F(a)] < \varepsilon/5$ for a suitable choice of a and n_0. Similarly the modulus third integral in (2c.4.3) is $<\varepsilon/5$ for a suitable choice of b and n_0.

In the finite interval (a, b), g is uniformly continuous. Let us divide (a, b) into m intervals

$$x_0 = a < x_1 < \cdots < x_{m-1} < b = x_m$$

where $x_1, \ldots, x_{m-1}$ are continuity points of F and such that

$$[g(x) - g(x_i)] < \varepsilon/5$$

for $(x_i < x < x_{i+1})$ uniformly for all i. Define the function

$$g_m(x) = g(x_i), \qquad x_i \leqslant x < x_{i+1}.$$

Then

$$\int_a^b g_m(x)\, dF_n = \sum g(x_i)[F_n(x_{i+1}) - F_n(x_i)]$$
$$\to \sum g(x_i)[F(x_{i+1}) - F(x_i)] = \int_a^b g_m\, dF$$

as $n \to \infty$, so that for any given m

$$\left|\int_a^b g_m(dF_n - dF)\right| < \frac{\varepsilon}{5}$$

for sufficiently large n. But

$$\left|\int_a^b g\, dF_n - \int_a^b g\, dF\right| = \left|\int_a^b (g - g_m)\, dF_n + \int_a^b g_m(dF_n - dF) + \int_a^b (g - g_m)\, dF\right|$$
$$< \int_a^b \frac{\varepsilon}{5}\, dF_n + \frac{\varepsilon}{5} + \int_a^b \frac{\varepsilon}{5}\, dF < \frac{3}{5}\varepsilon$$

for sufficiently large n. Hence the modulus of the difference (2c.4.3) is $<\varepsilon$ which proves the desired result.

(iii) Extended Helly-Bray Theorem. *$F_n \to G$, not necessarily a d.f. as assumed in* (ii), *and g a continuous function such that $g(\pm\infty) = 0 \Rightarrow \int g\, dF_n \to \int dg\, G$.*

We consider the equation (2c.4.3) and note that the second term on the right-hand side $\to 0$ by the same argument used in (ii) whether G is d.f. or not. Since g is bounded $\int g\,dF_n$ and $\int g\,dG$ exist. By choosing a sufficiently small the first term is bounded by 2 sup $g(x)$ for x in $(-\infty, a)$, which is small since $g(-\infty) = 0$ and g is continuous irrespective of the value of n. Similarly, by choosing b sufficiently large, the third term can be made small. Hence the difference (2c.4.3) can be made as small as we please by choosing sufficiently large n.

(iv) CONTINUITY THEOREM. *Let $f_n(t)$ be c.f. of X_n. If $X_n \xrightarrow{L} X$ then $f_n(t) \to f(t)$, where $f(t)$ is the c.f. of X. If $f_n(t) \to f(t)$ and the limit function is continuous at $t = 0$, then $X_n \xrightarrow{L} X$ and $f(t)$ is the c.f. of X.*

The first assertion is an immediate consequence of Helly-Bray theorem. We prove the second assertion by using the continuity of $f(t)$ at $t = 0$.

Let F_n be the d.f. of X_n and $\{F_n\}$, a subsequence tending to G, a bounded nondecreasing left-continuous function as in (i), Helly lemma. Consider

$$\int_0^v f_m(t)\,dt = \int_0^v \int e^{itx}\,dF_m\,dt$$

$$= \int dF_m \int_0^v e^{itx}\,dt = \int \frac{e^{ivx} - 1}{ix}\,dF_m. \qquad (2c.4.4)$$

The function $(e^{ivx} - 1)/ix \to 0$ as $x \to \mp\infty$. Hence using (iii), Extended Helly-Bray Theorem, we obtain by taking limits on both sides of (2c.4.4)

$$\int_0^v f(t)\,dt = \int \frac{e^{ivx} - 1}{ix}\,dG. \qquad (2c.4.5)$$

Dividing both sides by v and taking limits as $v \to 0$, we have

$$f(0) = \int \lim_{v \to 0} \frac{e^{ivx} - 1}{ix}\,dG = \int dG = G(\infty) - G(-\infty). \qquad (2c.4.6)$$

But $\lim f_m(0) = 1 = f(0)$ and therefore G is a distribution function. Since $F_m \to G$, we have by (ii), Helly-Bray Theorem,

$$f(t) = \lim_{m \to \infty} f_m(t) = \lim_{m \to \infty} \int e^{itx}\,dF_m = \int e^{itx}\,dG.$$

Thus $f(t)$ is a c.f. and G is, of course, unique for all convergent subsequences of F_n. Then $F_n \to G$.

(v) *By combining* (ii) *and* (iii), *the following results are equivalent,*

$$\text{(a) } F_n \to F, \qquad \text{(b) } f_n(t) \to f(t), \qquad \text{(c) } \int g\,dF_n \to \int g\,dF, \qquad (2c.4.7)$$

where g and $f(t)$ are as defined in (ii) *and* (iii) *respectively.*

From (v) it easily follows that *if* $X_n \xrightarrow{L} X$, *then* $aX_n + b \xrightarrow{L} aX + b$ when a and b are constants. Further, $X_n \xrightarrow{P} c\,(constant) \Leftrightarrow F_n \to F_c$, where F_c is the d.f. of the random variable X which takes the value c with probability 1, that is, the step function, $F_c(x) = 0$, if $x \leqslant c$, $F_c(x) = 1$, if $x > c$.

(vi) POLYA'S THEOREM. *If* $F_n \to F$ *and* F *is continuous then the convergence is uniform, that is,*

$$\lim_{n\to\infty} \sup_x |F_n(x) - F(x)| = 0. \tag{2c.4.8}$$

The proof is easy.

(vii) *Let* g *be a continuous function and* $F_n \to F$. *Then*

(a) $\liminf_{n\to\infty} \int |g|\, dF_n \geqslant \int |g|\, dF$,

(b) $\int |g|\, dF_n$ *is uniformly integrable* $\Rightarrow \int g\, dF_n \to \int g\, dF$.

Since

$$\int |g|\, dF_n \geqslant \int_{-c}^{c} |g|\, dF_n$$

$$\liminf_{n\to\infty} \int |g|\, dF_n \geqslant \lim_{n\to\infty} \int_{-c}^{c} |g|\, dF_n$$

$$= \int_{-c}^{c} |g|\, dF \quad \text{for any } c,$$

which proves (a). Because of uniform integrability there exists an ε independent of n such that

$$\varepsilon + \int_c^d |g|\, dF_n \geqslant \int |g|\, dF_n \geqslant \int_{c'}^{d'} |g|\, dF_n, \qquad c' < c, \quad d' > d.$$

Taking limits, since c, d, c', d' are finite, we see that

$$\varepsilon + \int_c^d |g|\, dF \geqslant \int_{c'}^{d'} |g|\, dF.$$

But c' and d' are arbitrary. Hence $|g|$ and therefore g is integrable wi1h respect to dF. Now

$$\int g\, dF_n - \int g\, dF = \int_{-c}^{c} g(dF_n - dF) + \int_{|x|>c} g\, dF_n - \int_{|x|>c} g\, dF. \tag{2c.4.9}$$

All the terms on the right-hand side of (2c.4.9) can be made small by choosing c and n sufficiently large. Hence (b) is proved.

(viii-A) *Let γ_{rn} be the rth absolute moment of X_n. If for some $r > 0$, γ_{rn} is bounded by a constant for all large n, then*

(a) *there is a subsequence F_m converging to a distribution function G,*

(b) *$\alpha_{sm} \to \alpha_s$ for $s < r$ where α_{ms} and α_s are the sth moments of F_m and G. (Both s and r may not be integers.)*

By the Helly lemma, $F_m \to G$, a nondecreasing bounded function. If c is a continuity point of G, then

$$\int_{|x| \geqslant c} dF_m \leqslant c^{-r} \int_{|x| \geqslant c} |x|^r \, dF_m \leqslant c^{-r}k \qquad \text{for large } m. \qquad (2c.4.10)$$

The first term in (2c.4.10) is

$$1 - [F_m(c) - F_m(-c)] \leqslant c^{-r}k.$$

Taking limits as $m \to \infty$, we have

$$1 - [G(c) - G(-c)] \leqslant c^{-r}k,$$

and by letting $c \to \infty$, we see that

$$1 - [G(\infty) - G(-\infty)] \leqslant 0$$

or $G(\infty) - G(-\infty) \geqslant 1$, which shows that G is a d.f., which proves (a).

Since

$$\int_{|x| \geqslant c} |x|^s \, dF_m \leqslant c^{s-r} \int_{|x| \geqslant c} |x|^r \, dF_m < c^{s-r}k,$$

$|x|^s$ is uniformly integrable if $s < r$. So, (b) follows by using (vii).

(viii-B) *Let α_{sn} be the sth moment of F_n and $\alpha_{sn} \to \alpha_s$, where α_s and α_{sn} are finite for all s and all large n. Then*

(a) *$F_n \to F \Rightarrow \alpha_s$ is the sth moment of F,*

(b) *α_s define a unique d.f. $F \Rightarrow F_n \to F$.* (2c.4.11)

The result (a) follows directly from (vii). Again by (vii) there is a subsequence tending to a d.f. G, and α_s are the moments of G. If α_s determine a unique d.f., every subsequence has the same limit and hence $F_n \to F$.

A sufficient condition for uniqueness of F given the moments is that

$$\sum_1^\infty (\alpha_{2s})^{-1/2s} = \infty,$$

so that the result (b) could be stated

$$\sum_1^\infty (\alpha_{2s})^{-1/2s} = \infty \Rightarrow F_n \to F.$$

(ix) *Let* $\{X_n, Y_n\}$, $n = 1, 2, \ldots$ *be a sequence of pairs of variables. Then*

$$|X_n - Y_n| \xrightarrow{P} 0, \; Y_n \xrightarrow{L} Y \Rightarrow X_n \xrightarrow{L} Y, \tag{2c.4.12}$$

that is, the limiting distribution of X_n *exists and is the same as that of* Y.

Denote by F_{X_n} the d.f. of X_n and by F_Y that of Y. Let $Y_n - X_n = Z_n$ and x be a continuity point of F_Y. Then

$$\begin{aligned} F_{X_n}(x) &= P(X_n < x) = P(Y_n < x + Z_n) \\ &= P(Y_n < x + Z_n, Z_n < \varepsilon) + P(Y_n < x + Z_n, Z_n \geqslant \varepsilon) \\ &\leqslant P(Y_n < x + \varepsilon) + P(Z_n \geqslant \varepsilon). \end{aligned}$$

Taking limits, we have

$$\limsup_{n \to \infty} F_{X_n} \leqslant F_Y(x + \varepsilon).$$

Similarly,

$$\liminf_{n \to \infty} F_{X_n} \geqslant F_Y(x - \varepsilon).$$

Since ε is arbitrary and x is a continuity point of F_Y, by letting $\varepsilon \to 0$,

$$\lim_{n \to \infty} F_{X_n}(x) = F_Y(x),$$

which proves (2c.4.12). As a particular case we have

$$X_n \xrightarrow{P} X \Rightarrow X_n \xrightarrow{L} X.$$

(x) *Let* $\{X_n, Y_n\}$, $n = 1, 2, \ldots$ *be a sequence of pairs of random variables. Then*:

(a) $X_n \xrightarrow{L} X, \; Y_n \xrightarrow{P} 0 \Rightarrow X_n Y_n \xrightarrow{P} 0.$

(b) $$\begin{aligned} X_n \xrightarrow{L} X, \; Y_n \xrightarrow{P} c &\Rightarrow X_n + Y_n \xrightarrow{L} X + c \\ &\Rightarrow X_n Y_n \xrightarrow{L} cX \\ &\Rightarrow X_n / Y_n \xrightarrow{L} X/c, \qquad \text{if} \quad c \neq 0. \end{aligned}$$

(c) $X_n \xrightarrow{L} X, \; Y_n \xrightarrow{P} c \Rightarrow$ *that the limit of the joint distribution of* (X_n, Y_n) *exists and is equal to that of* (X, c).

(d) $X_n - Y_n \xrightarrow{P} 0, \; X_n \xrightarrow{L} X \Rightarrow Y_n \xrightarrow{L} X$

To prove (a), consider

$$\begin{aligned} P(|X_n Y_n| > \varepsilon) &= P\left(|X_n Y_n| > \varepsilon, \; |Y_n| \leqslant \frac{\varepsilon}{k}\right) + P\left(|X_n Y_n| > \varepsilon, \; |Y_n| > \frac{\varepsilon}{k}\right) \\ &\leqslant P(|X_n| > k) + P\left(|Y_n| > \frac{\varepsilon}{k}\right) \end{aligned}$$

$$\Rightarrow \limsup_{n \to \infty} P(|X_n Y_n| > \varepsilon) \leqslant P(|X| > k) \quad \text{for any fixed } k. \tag{2c.4.13}$$

But k is arbitrary and hence the R.H.S. of (2c.4.13) can be made small by choosing k large. Hence $P(|X_nY_n| > \varepsilon) \to 0$.

To prove (b), $X_n \xrightarrow{L} X \Rightarrow X_n + c \xrightarrow{L} X + c$. Now

$$(X_n + Y_n) - (X_n + c) = Y_n - c \xrightarrow{P} 0.$$

Hence by (ix), $X_n + Y_n \xrightarrow{L} X + c$. Furthermore $X_n \xrightarrow{L} X \Rightarrow cX_n \xrightarrow{L} cX$. Now

$$X_nY_n - cX_n = X_n(Y_n - c) \xrightarrow{P} 0, \qquad \text{by (a), (x).}$$

Hence by (ix), $X_n Y_n \xrightarrow{L} cX$. Similarly $X_n/Y_n \xrightarrow{L} X/c$.

To prove (c) consider a continuity point x of the distribution of X and a number $y > c$. Now

$$\begin{aligned} P(X_n < x) \geqslant P(X_n < x, Y_n \leqslant y) &= P(X_n < x) - P(X_n < x, Y_n > y) \\ &\geqslant P(X_n < x) - P(Y_n > y). \end{aligned}$$

Taking limits, we see that

$$\lim_{n \to \infty} P(X_n < x, Y_n < y) = P(X < x), \qquad \text{if} \quad y > c.$$

Similarly,

$$\lim_{n \to \infty} P(X_n < x, Y_n < y) = 0, \qquad \text{if} \quad y < c.$$

Note that when $y = c$, the limit is indeterminable and (x, c) for any x is a discontinuity point of the joint distribution of (X, c).

(d) follows from (b).

(xi) *Let* $\{X_n^{(1)}, \ldots, X_n^{(k)}\}$, $n = 1, 2, \ldots$ *be a sequence of vector random variables and let, for any real* $\lambda_1, \ldots, \lambda_k$,

$$\lambda_1 X_n^{(1)} + \cdots + \lambda_k X_n^{(k)} \xrightarrow{L} \lambda_1 X^{(1)} + \cdots + \lambda_k X^{(k)},$$

where $X^{(1)}, \ldots, X^{(k)}$ *have a joint distribution,* $F(x_1, \ldots, x_k)$. *Then the limiting joint distribution function of* $X_n^{(1)}, \ldots, X_n^{(k)}$ *exists and is equal to* $F(x_1, \ldots, x_k)$.

The proof is trivial because

$$\lambda_1 X_n^{(1)} + \cdots + \lambda_n X_n^{(k)} \xrightarrow{L} \lambda_1 X^{(1)} + \cdots + \lambda_k X^{(k)}$$

implies that the corresponding c.f.'s converge. If $f_n(t_1, \ldots, t_k)$ is the c.f. of $(X_n^{(1)}, \ldots, X_n^{(k)})$ and $f(t_1, \ldots, t_k)$ that of $(X^{(1)}, \ldots, X^{(k)})$, then

$$f_n(t_1\lambda_1, \ldots, t_k\lambda_k) \to f(t_1\lambda_1, \ldots, t_k\lambda_k).$$

Since $\lambda_1, \ldots, \lambda_k$ are arbitrary,

$$f_n(t_1, \ldots, t_k) \to f(t_1, \ldots, t_k),$$

which establishes the desired result.

(xii) *If g is a continuous function and $X_n \xrightarrow{L} X$, then $g(X_n) \xrightarrow{L} g(X)$.*

$$
\begin{aligned}
E[e^{itg(X_n)}] &= \int e^{itg(x)}\, dF_n(x) \\
&= \int \cos tg(x)\, dF_n(x) + i \int \sin tg(x)\, dF_n(x) \\
&\to \int \cos tg(x)\, dF(x) + i \int \sin tg(x)\, dF(x), \\
&= E[e^{itg(x)}]
\end{aligned}
$$

using (2c.4.7), since $\cos tg(x)$ and $\sin tg(x)$ are bounded continuous functions of x.

(xiii) *If g is a continuous function and $X_n \xrightarrow{P} X$, then*

$$g(X_n) \xrightarrow{P} g(X). \tag{2c.4.14}$$

Let I be a finite interval such that $P(X \in I) = 1 - \eta/2$ and $n \geqslant n_0$ such that $P(|X_n - X| < \delta) > 1 - \eta/2$. Now $|g(X_n) - g(X)| < \varepsilon$, if $|X_n - X| < \delta$ for any $X \in I$. Hence

$$
\begin{aligned}
P(|g(X_n) - g(X)| < \varepsilon) &\geqslant P(|X_n - X| < \delta, X \in I) \\
&\geqslant P(|X_n - X| < \delta) - P(X \notin I) \\
&\geqslant 1 - \eta \qquad \text{for} \quad n \geqslant n_0 .
\end{aligned}
$$

An important application of (2c.4.14) is that when $X_n \xrightarrow{P} c$ (a constant), $g(X_n) \xrightarrow{P} g(c)$, provided g is a continuous function.

(xiv) *Let g be a continuous function. Then*

$$X_n - Y_n \xrightarrow{P} 0, \; Y_n \xrightarrow{L} Y \Rightarrow g(X_n) - g(Y_n) \xrightarrow{P} 0.$$

Note. Mann and Wald (1943) proved the results (xii)–(xiv) under more general conditions assuming g to be Borel measurable but placing a restriction on the set of discontinuities of g.

It may also be mentioned that the results (i) to (xiv) are true even when X_n, Y_n are vector variables. The proofs are exactly the same.

(xv) A CONVERGENCE THEOREM INVOLVING DENSITIES (Scheffe, 1947): *Let the random variable X_n admit the probability density $p_n(x)$ and $p_n(x) \to p(x)$ as $n \to \infty$. Then*

(a) *$p(x)$ is a density function $\Rightarrow \int_R |p_n(x) - p(x)|\, dx \to 0$,*

(b) *$|p_n(x)| < q(x)$ and $\int q(x)\, dx$ exists $\Rightarrow p(x)$ is a density function and hence by (a), $\int |p_n(x) - p(x)|\, dx \to 0$.*

In either (a) or (b), the result implies

$$\int_S [p_n(x) - p(x)]\, dx \to 0 \qquad \text{uniformly for all Borel sets } S$$

so that $X_n \xrightarrow{L} X$, where X is the random variable defined by the probability density $p(x)$.

Let $\delta_n(x) = p_n(x) - p(x)$ and define

$$\delta_n^+ = \tfrac{1}{2}(\delta_n + |\delta_n|) \geqslant 0, \qquad \delta_n^- = \tfrac{1}{2}(\delta_n - |\delta_n|) \leqslant 0,$$

the positive and negative parts of δ_n. Obviously $\delta_n^+ \to 0$ and $\delta_n^- \to 0$ as $\delta_n \to 0$ for almost all x in R.

Since $p_n(x)$ and $p(x)$ are densities and $\delta_n = \delta_n^+ + \delta_n^-$

$$0 = \int_R [p_n(x) - p(x)]\, dx = \int_R \delta_n\, dx = \int_R \delta_n^+\, dx + \int_R \delta_n^-\, dx. \quad (2c.4.15)$$

Now $p_n(x) - p(x) = \delta_n(x) \Rightarrow \delta_n(x) \geqslant -p(x)$ whether $\delta(x)$ is positive or negative. Hence $\delta_n^-(x)$ satisfies the condition $0 \geqslant \delta_n^-(x) \geqslant -p(x)$, that is, $|\delta_n^-(x)| < p(x)$ and $\int p(x)\, dx$ exists. Applying Lebesgue dominated convergence theorem (Appendix **2B**), we have

$$\int_R \delta_n^-\, dx \to \int_R \lim \delta_n^-\, dx = 0,$$

since $\lim \delta_n^-$ is zero and from (2c.4.15),

$$\int_R \delta_n^+ \to 0$$

and hence

$$\int_R |\delta_n|\, dx = \int_R \delta_n^+\, dx - \int_R \delta_n^-\, dx \to 0,$$

which proves (a). The result (a) also follows by a direct application of Lemma **2B.3** of Appendix **2B**. The result (b) follows from the dominated convergence theorem since

$$1 = \int_R p_n\, dx \to \int_R p(x)\, dx.$$

(xvi) *Let X_n be a discrete variable and $P(X_n = r) = p_n(r)$, $r = 0, 1, 2, \ldots$. Consider $Y_n = (X_n - n\mu)/\sqrt{n}$ where μ is a constant. If $p_n(\sqrt{n}y + n\mu) \to p(y)$ and $p(y)$ is a density function of a random variable Y, then $Y_n \xrightarrow{L} Y$.*

The proof depends on Scheffe's theorem given in (xv), and reader may work out the actual details. For applications of this result, see Okomoto (1960).

(xvii) CONVERGENCE OF DISCRETE DISTRIBUTIONS. *Let $P_n(t)$, $0 \leqslant t \leqslant 1$ be a sequence of probability generating functions, so that for any index n, there exists a discrete probability distribution defined by $P_n(t) = \sum p_{xn} t^x$. Further let $P(t) = \sum p_x t^x$. Then*

$$p_{xn} \to p_x \ (\textit{for each } x) \Leftrightarrow P_n(t) \to P(t) \ (\textit{for each } t).$$

The terms p_x define a distribution if and only if $\sum p_x = 1$, in which case

$$p_{xn} \to p_x \Leftrightarrow \sum_{x=0}^{\infty} |p_{xn} - p_x| \to 0.$$

The proof is similar to that of (xv).

2c.5 Central Limit Theorems

For a sequence $\{X_n\}$, $n = 1, 2, \ldots$ of independent and identically distributed random variables, Section **2c.3** showed that $\bar{X}_n \to \mu$ in the weak or strong sense provided $\mu = E(X)$ exists. But this gives us no idea as to how the distribution of $\bar{X}_n$ can be approximated in large samples. To study this we consider the sequence of random variables

$$Y_n = \frac{\sqrt{n}(\bar{X}_n - \mu)}{\sigma}, \tag{2c.5.1}$$

where σ is some constant. Under some conditions the limiting d.f. is

$$\Phi(x) = \int_{-\infty}^{x} \frac{1}{\sqrt{2\pi}} e^{-x^2/2}\, dx, \tag{2c.5.2}$$

which is known as the *normal* distribution function with mean zero and variance unity introduced in (2a.5.8). The p.d. for the more general form of the normal variable with mean μ and variance σ^2 is

$$\frac{1}{\sqrt{2\pi}\sigma} e^{-(x-\mu)^2/2\sigma^2}. \tag{2c.5.3}$$

The c.f. of (2c.5.2) as derived in (2b.4.11) is

$$\frac{1}{\sqrt{2\pi}} \int_{-\infty}^{\infty} e^{itx} e^{-x^2/2}\, dx = e^{-t^2/2}. \tag{2c.5.4}$$

The technique employed to prove the convergence of (2c.5.1) to a normal r.v. is to obtain the c.f. $f_n(t)$ of Y_n and to show that

$$\lim_{n \to \infty} f_n(t) = e^{-t^2/2},$$

from which, by an application of (2c.4.7), the desired result follows.

We also consider more general problems such as the convergence of r.v.'s

$$Y_n = \frac{\sqrt{n}(\bar{X}_n - \mu_n)}{\sigma_n}$$

where $\{\mu_n, \sigma_n\}$, $n = 1, 2, \ldots$ is a sequence of pairs of numbers.

(i) LINDBERG-LEVY THEOREM (C.L.T.). *Let $X_1, X_2, \ldots$ be a sequence of i.i.d. random variables such that $E(X_n) = \mu$ and $V(X_n) = \sigma^2 \neq 0$ exist. Then the d.f. of $Y_n \to \Phi(x)$, where Y_n is as defined in* (2c.5.1) *and Φ, in* (2c.5.2).

Let $f(t)$ be the c.f. of $X_i - \mu$. Since the first two moments exist, then

$$f(t) = 1 - \tfrac{1}{2}\sigma^2 t^2 + o(t^2). \tag{2c.5.5}$$

The c.f. of $Y_n = \sum_1^n (X_i - \mu)/\sqrt{n}\,\sigma$ is

$$f_n(t) = \left[f\left(\frac{t}{\sigma\sqrt{n}}\right)\right]^n = \left[1 - \frac{t^2}{2n} + o\left(\frac{t^2}{n}\right)\right]^n, \quad \text{using (2c.5.5).}$$

$$\log\left[1 - \frac{t^2}{2n} + o\left(\frac{t^2}{n}\right)\right]^n = n \log\left[1 - \frac{t^2}{2n} + o\left(\frac{t^2}{n}\right)\right] \to \frac{-t^2}{2},$$

that is,

$$f_n(t) \to e^{-t^2/2} \quad \text{as} \quad n \to \infty.$$

Since the limiting distribution is continuous, the convergence of the d.f. of Y_n is actually uniform, and we have the more general result

$$\lim_{n\to\infty} [F_{Y_n}(x_n) - \Phi(x_n)] \to 0, \tag{2c.5.6}$$

where x_n may depend on n in any manner.

The result (2c.5.6) implies that the d.f. of $\bar{X}_n$ can be approximated by that of a normal variable with mean μ and variance σ^2/n for sufficiently large n, which is called the central limit theorem (C.L.T.).

(ii) LIAPUNOV THEOREM (C.L.T.). *Let $\{X_n\}$, $n = 1, 2, \ldots$ be a sequence of independent random variables. Let $E(X_n) = \mu_n$, $E(X_n - \mu_n)^2 = \sigma_n^2 \neq 0$, and $E|X_n - \mu_n|^3 = \beta_n$ exist for each n. Furthermore let*

$$B_n = \left(\sum_1^n \beta_i\right)^{1/3}, \qquad C_n = \left(\sum_1^n \sigma_i^2\right)^{1/2}.$$

Then if $\lim(B_n/C_n) = 0$ as $n \to \infty$, the d.f. of

$$Y_n = \frac{\sum_1^n (X_i - \mu_i)}{C_n}$$

tends to $\Phi(x)$.

The proof of this theorem is similar to that of Lindberg-Levy (C.L.T.) theorem and consists in expanding c.f. of Y_n and taking the limit as $n \to \infty$. For details the reader is referred to (Gnedenko, 1962).

(iii) LINDBERG–FELLER THEOREM. *Let $\{X_n\}$ be a sequence of independent random variables and G_n be the d.f. of X_n. Further, let $E(X_n) = \mu_n$ and $V(X_n) = \sigma_n^2 \neq 0$ exist. Define*

$$Y_n = \frac{\sum_1^n (X_i - \mu_i)}{C_n}, \qquad C_n = \left(\sum_1^n \sigma_i^2\right)^{1/2}.$$

Then the relations

$$\lim_{n\to\infty} \max_{1 \leqslant i \leqslant n} \frac{\sigma_i}{C_n} = 0, \qquad F_{Y_n} \to \Phi$$

hold if and only if, for every $\varepsilon > 0$,

$$\lim_{n\to\infty} \frac{1}{C_n^2} \sum_{i=1}^n \int_{|x-\mu_i| > \varepsilon C_n} (x - \mu_i)^2 \, dG_i(x) = 0.$$

For a proof of this theorem and other forms of central limit theorems see Gnedenko (1962) and Gnedenko and Kolmogorov (1954).

(iv) MULTIVARIATE CENTRAL LIMIT THEOREM (Varadarajan, 1958, Wald and Wolfowitz, 1944). *Let F_n denote the joint d.f. of the k-dimensional random variable $(X_n^{(1)}, \ldots, X_n^{(k)})$, $n = 1, 2, \ldots$ and $F_{\lambda n}$, the d.f. of the linear function $\lambda_1 X_n^{(1)} + \cdots + \lambda_k X_n^{(k)}$. A necessary and sufficient condition that F_n tends to a k-variate d.f. F is that $F_{\lambda n}$ converges to a limit for each vector λ.*

For a proof see Varadarajan (1958). For applications considered in this book the weaker version of the theorem proved in [(x), **2c.4**] is sufficient.

Define F_n and $F_{\lambda n}$ as in (iv). *Let F be the joint d.f. of a k-dimensional random variable $X^{(1)}, \ldots, X^{(k)}$. If for each vector λ, $F_{\lambda n} \to F_\lambda$, the d.f. of $\lambda_1 X^{(1)} + \cdots + \lambda_k X^{(k)}$, then $F_n \to F$.*

As an application, consider i.i.d. k-dimensional variables

$$\mathbf{U}_n' = (U_{1n}, \ldots, U_{kn}), \qquad n = 1, 2, \ldots$$

admitting first- and second-order moments, $E(U_n) = \boldsymbol{\mu}$ and $D(U_n) = \boldsymbol{\Sigma}$. Define the sequence of r.v.'s

$$\overline{\mathbf{U}}_n = (\overline{U}_{1n}, \ldots, \overline{U}_{kn}), \qquad n = 1, 2, \ldots, \quad \overline{U}_{in} = \frac{1}{n} \sum_1^n U_{ij}.$$

Then the a.d. of $\sqrt{n}(\overline{\mathbf{U}}_n - \boldsymbol{\mu})$ is $N_k(\mathbf{0}, \boldsymbol{\Sigma})$, that is, k-variate normal with zero mean and dispersion matrix $\boldsymbol{\Sigma}$ with the density function

$$N_k(\mathbf{u} \,|\, \mathbf{0}. \, \boldsymbol{\Sigma}) = (2\pi)^{-k/2} |\boldsymbol{\Sigma}|^{-1/2} \exp -\tfrac{1}{2}[(\mathbf{u} - \boldsymbol{\mu})' \boldsymbol{\Sigma}^{-1} (\mathbf{u} - \boldsymbol{\mu})]. \quad (2c.5.7)$$

To demonstrate this result, it is enough to show that with respect to any fixed vector $\boldsymbol{\lambda}$ the a.d. of $\sqrt{n}\,\boldsymbol{\lambda}'(\bar{\mathbf{U}}_n - \boldsymbol{\mu})$ is $N_1(0, \boldsymbol{\lambda}'\boldsymbol{\Sigma}\boldsymbol{\lambda})$, since $N_1(0, \boldsymbol{\lambda}'\boldsymbol{\Sigma}\boldsymbol{\lambda})$ is the distribution of $\boldsymbol{\lambda}'\mathbf{U}$ if $\mathbf{U}$ is distributed as $N_k(\mathbf{0}, \boldsymbol{\Sigma})$. [The reader may establish the latter result by computing the c.f. of $\boldsymbol{\lambda}'\mathbf{U}$ when $\mathbf{U}$ has the density (2c.5.7).] Now consider the sequence of one-dimensional r.v.'s

$$U_n = \boldsymbol{\lambda}'\mathbf{U}_n = \lambda_1 U_{1n} + \cdots + \lambda_k U_{kn}$$

with $E(U_n) = \boldsymbol{\lambda}'\boldsymbol{\mu}$, $V(U_n) = \boldsymbol{\lambda}'\boldsymbol{\Sigma}\boldsymbol{\lambda}$. Applying the central limit theorem (Lindberg-Lévy) to the sequence U_n, the a.d. of $\sqrt{n}(\bar{U}_n - \boldsymbol{\lambda}'\boldsymbol{\mu})$ is $N_1(0, \boldsymbol{\lambda}'\boldsymbol{\Sigma}\boldsymbol{\lambda})$. The required result follows from the observation that $\bar{U}_n = \boldsymbol{\lambda}'\bar{\mathbf{U}}_n$.

For other limit theorems see Examples 4.1 to 4.17 at the end of the chapter.

2c.6 Sums of Independent Random Variables

Let $X_1, X_2, \ldots$ be a sequence of independent random variables and S_n denote the partial sum $X_1 + \cdots + X_n$. The law of large numbers (see **2c.3**) and the central limit theorem (see **2c.5**) were concerned with the behavior of $(S_n - a_n)/b_n$, (a_n, b_n) being a suitably defined sequence of pairs of numbers. We pursue this investigation further and quote a few general theorems which are useful in statistical applications. Proofs are omitted, and the interested reader may consult Loeve (1955).

(i) KOLMOGOROV THEOREM. *Let $E(X_n) = 0$ and $E(X_n^2) = \sigma_n^2 < \infty$ and $b_n \uparrow \infty$. If $\sum_1^\infty (\sigma_n^2/b_n^2) < \infty$, then*

$$\frac{S_n}{b_n} \xrightarrow{\text{a.s.}} 0. \tag{2c.6.1}$$

For instance, if $X_1, X_2, \ldots$ are i.i.d. such that $E(X_1) = 0$, $V(X_1) = \sigma^2 < \infty$, then $S_n/\sqrt{n}\log n \xrightarrow{\text{a.s.}} 0$.

(ii) LAW OF ITERATED LOGARITHM. *Let $X_1, X_2, \ldots$ be i.i.d. random variables such that $E(X_1) = 0$, $V(X_1) = \sigma^2 < \infty$. Further, let $h(n) = \sigma(2n \log\log n)^{1/2}$. Then*

$$\limsup [S_n/h(n)] = 1 \text{ a.s.} \tag{2c.6.2}$$

$$\liminf [S_n/h(n)] = -1 \text{ a.s.} \tag{2c.6.3}$$

The statements (2c.6.2) and (2c.6.3) imply that for every $\varepsilon > 0$,

$$P(|S_n| \geqslant (1 - \varepsilon)h(n) \text{ infinitely often}) = 1, \tag{2c.6.4}$$

$$P(|S_n| \geqslant (1 + \varepsilon)h(n) \text{ infinitely often}) = 0. \tag{2c.6.5}$$

From the law of iterated logarithm it readily follows that for any k, however large,

$$P(S_n \geqslant k\sqrt{n} \text{ infinitely often}) = 1 \tag{2c.6.6}$$

since $(1 - \varepsilon)h(n) > k\sqrt{n}$ for sufficiently large n.

(iii) *Let* $X_1, X_2, \ldots$ *be independent random variables such that* $E(X_i) = 0$ *and* $V(X_i) = \sigma_i^2 < \infty$. *Further, let* $s_n^2 = \sigma_1^2 + \cdots + \sigma_n^2$. *If* $s_n^2 \to \infty$, $|X_n|/s_n = o\,(s_n^2/\sqrt{\log\log s_n})$ *and* $t_n = (2 \log\log s_n^2)^{1/2}$, *then*

$$\limsup \frac{S_n}{t_n s_n} = 1 \ a.s. \tag{2c.6.7}$$

$$\liminf \frac{S_n}{t_n s_n} = -1 \ a.s. \tag{2c.6.8}$$

For a proof, see Loeve (1955).

2d FAMILY OF PROBABILITY MEASURES AND PROBLEMS OF STATISTICS

2d.1 Family of Probability Measures

In practice the exact probability measure P applicable to the outcomes of an experiment is unknown, but it may be specified as a member of a class (family) of measures, P_θ, $\theta \in \Theta$, indexed by a parameter θ belonging to a set Θ. The problems of statistics are concerned with making useful statements about the unknown parameter θ on the basis of an observed event (corresponding to a point in the sample space of outcomes).

A sample space with a family of probability measures is denoted by $(\Omega, \mathscr{B}, P_\theta, \theta \in \Theta)$.

2d.2 The Concept of a Sufficient Statistic

A statistic T (a random variable) is said to be sufficient for the family of measures P_θ, if and only if any one of the following equivalent conditions holds:

(a) $P(A \mid t)$ is independent of θ for every $A \in \mathscr{B}$.
(b) $E(Y \mid t)$ is independent of θ for every random variable Y such that $E(Y)$ exists.
(c) The conditional distribution of every random variable Y given $T = t$, which always exists, is independent of θ.

Observe that (b) implies (a) by choosing Y as the characteristic function of A. It is obvious that (a) implies (c) and also that (c) implies (b), thus establishing the equivalence of all the three conditions.

For the definition of $P(A \mid t)$ and $E(Y \mid t)$ see **2b.3** and for conditional d.f., see **2a.8.**

2d.3 Characterization of Sufficiency

The criteria given in Section **2d.2** for a sufficient statistic are difficult to apply in particular situations. The form of the probability density function, if it exists, however, enables us to check whether a given statistic is sufficient or not.

(i) *Let the sample space be discrete and let $P_\theta(\omega)$ be the probability of the point set ω. Then a necessary and sufficient condition for T to be sufficient for θ is that there exists a factorization*

$$P_\theta(\omega) = g_\theta[T(\omega)]h(\omega) \tag{2d.3.1}$$

where the first factor may depend on θ but depends on ω only through $T(\omega)$ whereas the second factor is independent of θ.

Suppose (2d.3.1) is true. Then

$$P_\theta(T = t) = \sum_{T(\omega)=t} P_\theta(\omega) = g_\theta(t) \sum_{T(\omega)=t} h(\omega)$$

$$= g_\theta(t)H(t).$$

The conditional probability of ω given $T = t$ is

$$\frac{P_\theta(\omega)}{P_\theta(t)} = \frac{g_\theta(t)h(\omega)}{g_\theta(t)H(t)} = \frac{h(\omega)}{H(t)},$$

which is independent of θ. Hence T is sufficient.

The necessity is obvious for if we let

$$\frac{P_\theta(\omega)}{P_\theta(t)} = k(\omega, t)$$

be independent of θ, then

$$P_\theta(\omega) = P_\theta(t)k(\omega, t).$$

(ii) GENERAL FACTORIZATION THEOREM. *Let P_θ be a family of probability measures and let P_θ admit a probability density $p_\theta = dP_\theta/d\mu$ with respect to a σ-finite measure μ. Then T is sufficient for P_θ (or simply for θ), if and only if there exist non-negative measurable functions $g_\theta[T(\omega)]$ and $h(\omega)$ such that*

$$P_\theta(\omega) = g_\theta[T(\omega)]h(\omega). \tag{2d.3.2}$$

The most general proof of the factorization (2d.3.2) involves some deeper theorems in measure theory (see Halmos and Savage, 1949. Bahadur, 1954). The proof is easy when the sample space is R^n and T admits continuous first-order partial derivatives using the concept of conditional density over surfaces of constant T as developed in (2b.3.12).

Sufficiency and Huzurbazar's Conjecture. Let $X_1, \ldots, X_n$ be i.i.d. r.v. with distribution function F_θ, $\theta \in \Theta$. According to definition, a measurable function $T = f(X_1, \ldots, X_n)$ is a sufficient statistic if the conditional distribution of any other statistic Y given T is independent of θ. Huzurbazar conjectured that to establish sufficiency of T it is enough to examine only marginal sufficiency, i.e., the conditional distribution of X_i given T is in independent of θ for any i. A partial proof of this proposition is given by J. K. Ghosh, and a more complete proof of a somewhat more general statement is given by V. N. Sudakov (1971).

APPENDIX 2A STIELTJES AND LEBESGUE INTEGRALS

Let $F(x)$ be a function of bounded variation and continuous from the left such as a distribution function. Given a finite interval (a, b) and a function $f(x)$ we can form the sum

$$J_n = \sum_{i=1}^{n} f(x_i')[F(x_i) - F(x_{i-1})] \tag{2A.1}$$

for a division of (a, b) by points x_i such that $a < x_1 < \cdots < x_n < b$ and arbitrary $x_i' \in (x_{i-1}, x_i)$. It may be noted that in the Riemann integration a similar sum is considered with the length of the interval $(x_i - x_{i-1})$ instead of $F(x_i) - F(x_{i-1})$. If the sum (2A.1) tends to a limit

$$J = \lim_{n \to \infty} J_n$$

where $n \to \infty$ such that the length of each interval $\to 0$, then J is called the Stieltjes integral of $f(x)$ with respect to $F(x)$ and is denoted by

$$J = \int_a^b f(x)\, dF(x).$$

The improper integral is defined by

$$\lim_{\substack{a \to -\infty \\ b \to \infty}} \int_a^b f(x)\, dF(x) = \int f(x)\, dF(x).$$

One point of departure from Riemann integration is that it is necessary to specify whether the end points are included in the integration or not. From the definition it is easily shown that

$$\int_a^b f(x)\, dF(x) = \int_{a+0}^b f(x)\, dF(x) + f(a)[F(a+0) - F(a)]$$

where $a + 0$ indicates that the end point a is not included. If $F(x)$ jumps at a, then

$$\int_a^b f(x)\, dF(x) - \int_{a+0}^b f(x)\, dF(x) = f(a)[F(a+0) - F(a-0)],$$

so that the integral taken over an interval that reduces to zero need not be zero. We shall follow the convention that the lower end point is always included but not the upper end point. With this convention, we see that

$$\int_a^b dF(x) = F(b) - F(a).$$

If there exists a function $p(x)$ such that

$$F(x) = \int_{-\infty}^x p(x)\, dx,$$

the Stieltjes integral reduces to a Riemann integral

$$\int f(x)\, dF(x) = \int f(x) p(x)\, dx.$$

We shall consider a general definition of integral with respect to a finite measure space.

Let $(\Omega, \mathscr{B}, \mu)$ be a measure space and s be a simple measurable function, i.e., $s = \sum_i^n \alpha_i \chi_{A_i}$, where $\alpha_n, \ldots, \alpha_n$ are the distinct values of s and χ_{A_i} is the indicator function of the set $A_i \in \mathscr{B}$. Then we define the integral of s over Ω as

$$\int_\Omega s\, d\mu = \sum_1^n \alpha_i\, \mu(A_i) = \int s\, d\mu$$

when the range of integration is understood, where the convention $0. \infty = 0$ is used when $\alpha_i = 0$ and $\mu(A_i) = \infty$. Let f be a nonnegative measurable function. Then we define the Lebesgue integral of f as

$$\int f\, d\mu = \sup \int s\, d\mu$$

the supremum being taken over all simple measurable functions s such that $0 \leqslant s \leqslant f$.

We say that a general function f is Lebesgue integrable $[L_1(\mu)]$ if

$$\int |f| \, d\mu < \infty$$

in which case the Lebesgue integral of f is

$$\int f^+ \, d\mu - \int f^- \, d\mu$$

where f^+ and f^- are the positive and negative parts of f.

Occasionally it is desirable to define the integral of a measurable function f with range in $[-\infty, \infty]$ to be

$$\int f \, d\mu = \int f^+ \, d\mu - \int f^- \, d\mu$$

provided that at least one integral is finite. Then the value of the integral is a number in $[-\infty, \infty]$.

In the special case, μ is generated by a distribution function F the Lebesgue integral is called Stieltjes-Lebesgue integral.

APPENDIX 2B SOME IMPORTANT THEOREMS IN MEASURE THEORY AND INTEGRATION

In this appendix we quote a number of results in measure theory and integration, proofs of which can be found in standard books on measure and integration. A knowledge of these results will be useful in the study of advanced theoretical statistics and in research.

We consider measurable functions $g, g_n, n = 1, 2, \ldots$ defined on a measure space $(\Omega, \mathscr{B}, \mu)$ and the following modes of convergence.

(a) $g_n \to g$, a.e. μ (almost everywhere with respect to measure μ) if $g_n \to g$, pointwise outside a null set.

(b) $g_n \xrightarrow{W} g$, (weakly) if for all integrable functions f, $\int g_n f \, d\mu \to \int gf \, d\mu$.

(c) $g_n \xrightarrow{L_p} g$, (in pth mean) if $\int |g_n - g|^p \, d\mu \to 0$. When $p = 1$, we say the convergence is in the mean.

(d) $g_n \xrightarrow{\mu} g$, (in measure) if $\lim_{n\to\infty} \mu\{|g_n - g| \geqslant \varepsilon\} = 0$ for every $\varepsilon > 0$, the definition being restricted to a.e. finite valued functions g_n.

The following lemmas are easy to prove from definitions.

LEMMA 2B.1. (i) *$g_n \xrightarrow{L_p} g \Rightarrow g_n \xrightarrow{L_r} g$ for $0 < r \leqslant p$, when $\mu(\Omega) < \infty$.*

(ii) *$\int |g_n - g| \, d\mu \to 0 \Leftrightarrow \int_A g_n \, d\mu \to \int_A g \, d\mu$ uniformly for all $A \in \mathscr{B}$ as $n \to \infty$, where $\{g_n, g\} \subset L_1$. Further $g_n \xrightarrow{L_1} g \Rightarrow g_n \xrightarrow{\mu} g$.*

LEMMA 2B.2. *The sequence g_n of uniformly bounded (or only integrable) functions converges to a bounded (or only integrable) function g weakly iff*

$$\int_A g_n \, d\mu \to \int_A g \, d\mu \qquad \text{for all } A\text{:} \quad \mu(A) < \infty. \tag{2B.1}$$

LEMMA 2B.3. *Let g_n be a sequence of measurable functions such that $g_n \to g$, a.e. μ, and $g_n \geqslant f$ (or $g_n \leqslant h$) a.e. where f (or h) is integrable. If g is integrable and*

$$\int g_n \, d\mu \to \int g \, d\mu \Rightarrow \int |g_n - g| \, d\mu \to 0. \tag{2B.2}$$

In particular if p, p_n, $n = 1, 2, \ldots$ is a sequence of probability density functions relative to μ, such that $p_n \to p$ (a.e. μ) then $\int |p_n - p| \, d\mu \to 0$.

LEBESGUE MONOTONE CONVERGENCE (LMC) THEOREM. *Let g_n be a sequence of non-negative, nondecreasing, and measurable functions, and let $g_n \to g$, a.e. μ. Then g is measurable and*

$$\int g_n \, d\mu \to \int g \, d\mu. \tag{2B.3}$$

As a consequence of the LMC theorem, we have the following corollaries.

COROLLARY 1. *The integral is σ-additive on the family of non-negative functions, or in other words*

$$\int \left(\sum g_n\right) d\mu = \sum \int g_n \, d\mu \tag{2B.4}$$

if g_n is a sequence of non-negative functions.

COROLLARY 2. (FATOU'S LEMMA). *Let $g_n \geqslant f$ (integrable) be a sequence of integrable functions. Then* lim inf g_n *is integrable and*

$$\int \liminf_{n \to \infty} g_n \, d\mu \leqslant \liminf_{n \to \infty} \int g_n \, d\mu \tag{2B.5}$$

COROLLARY 3. *Let g be non-negative. Then*

(a) $\displaystyle\int_E g \, d\mu \to 0$ *as* $\mu(E) \to 0$. (2B.6)

(b) $\displaystyle\lambda(E) = \int_E g \, d\mu$, $E \in \mathscr{B}$ *is a measure on* $(\Omega, \mathscr{B})$. (2B.7)

(c) $g = 0$ *a.e., iff* $\int g \, d\mu = 0$. (2B.8)

COROLLARY 4. *Let $g_n \leqslant h$ (integrable) be a sequence of integrable functions, then*

$$\limsup_{n\to\infty} \int g_n \, d\mu \leqslant \int \limsup_{n\to\infty} g_n \, d\mu. \tag{2B.9}$$

COROLLARY 5. *Let $f \leqslant g_n \leqslant F$, where f and F are integrable and $g_n \to g$. Then*

$$\int g_n \, d\mu \to \int g \, d\mu. \tag{2B.10}$$

LEBESGUE DOMINATED CONVERGENCE (LDC) THEOREM. *Let g_n be a sequence of measurable functions such that $|g_n(\omega)| \leqslant G(\omega)$ for all n and ω, where G is integrable over a set S of finite or infinite μ measure. If $g_n \to g$, a.e. μ in S, or only in measure, then*

$$\int_S g_n \, d\mu \to \int_S g \, d\mu. \tag{2B.11}$$

In fact, then

$$\int_S |g_n - g| \, d\mu \to 0, \qquad i.e., \quad \int_A g_n \, d\mu \to \int_A g \, d\mu, \tag{2B.12}$$

uniformly in A. (This is equivalent to saying that $\{g_n\} \subset L_1$ is uniformly integrable, and $g_n \to g$ in measure.)

As consequences of LDC theorem, we have the following corollaries.

COROLLARY 1. *Let $g(\omega, \theta)$ be a measurable function of $\omega \in \Omega$, for each $\theta \in R$, and $g(\omega, \theta) \to g(\omega, \theta_0)$ as $\theta \to \theta_0$ for each ω. Furthermore, let $|g(\omega, \theta)| \leqslant G(\omega)$ for all ω and θ in some open interval including θ_0, and let G be integrable over S. Then*

$$\int_S g(\omega, \theta) \, d\mu \to \int_S g(\omega, \theta_0) \, d\mu \qquad as \quad \theta \to \theta_0. \tag{2B.13}$$

COROLLARY 2. *Let $dg(\omega, \theta)/d\theta$ exist in an interval (a, b) of θ, and $|dg(\omega, \theta)/d\theta| < G(\omega)$ integrable, then in (a, b)*

$$\frac{d}{d\theta} \int g(\omega, \theta) \, d\mu = \int \frac{d}{d\theta} g(\omega, \theta) \, d\mu \tag{2B.14}$$

EXTENDED LDC THEOREM (Pratt, 1960). *Let*

(a) $f_n \leqslant g_n \leqslant F_n$,
(b) $f_n \to f$, $g_n \to g$, $F_n \to F$, *all a.e. μ and,*
(c) f_n, F_n, f, *and F are integrable with*

$$\int f_n \, d\mu \to \int f \, d\mu, \qquad \int F_n \, d\mu \to \int F \, d\mu.$$

Then $\int g\, d\mu$ *is finite and* $\int g_n\, d\mu \to \int g\, d\mu$. *Further,*

$$\int |g_n - g|\, d\mu \to 0 \qquad \text{if} \quad \int |f_n - f|\, d\mu \to 0. \tag{2B.15}$$

In particular, if $f_n \geqslant h$ *or* $f_n \leqslant k$ *a.e.* μ, *then* $\int |g_n - g|\, d\mu \to 0$.

As a consequence, we have the following corollary.

COROLLARY 1. *Let* θ *be a real parameter,* D *denote differentiation with respect to* θ *and* $\int$ *integration with respect to* μ. *If* θ_0 *is an interior point of an interval* I *and*

(i) $Dg(\omega, \theta)$, $Df(\omega, \theta)$, $DF(\omega, \theta)$ *exist for all* $\theta \in I$
(ii) $Df(\omega, \theta) \leqslant Dg(\omega, \theta) \leqslant DF(\omega, \theta)$ *for all* $\theta \in I$
(iii) $D \int h(\omega, \theta_0)\, d\mu = \int Dh(\omega, \theta_0)\, d\mu$ *for* $h = f$ *and* F *with all the quantities finite, then*

$$D \int g(\omega, \theta_0)\, d\mu = \int Dg(\omega, \theta_0)\, d\mu. \tag{2B.16}$$

FUBINI THEOREM. *Let* $(\Omega_i, \mathscr{B}_i, \mu_i)$, $i = 1, 2$, *be two finite measure spaces. If the function* $g(\omega_1, \omega_2)$ *defined on* $\Omega_1 \times \Omega_2$ *is measurable with respect to* $\mathscr{B}_1 \times \mathscr{B}_2$ *and is either non-negative or integrable on* $\Omega = \Omega_1 \times \Omega_2$ *with respect to the measure* $\mu = \mu_1 \times \mu_2$, *then*

$$\int_\Omega g\, d\mu = \int_{\Omega_1} d\mu_1 \int_{\Omega_2} g\, d\mu_2 = \int_{\Omega_2} d\mu_2 \int_{\Omega_1} g\, d\mu_1. \tag{2B.17}$$

RADON-NIKODYM THEOREM. *Let* μ *and* ν *be* σ*-finite measures over* $(\Omega, \mathscr{B})$. *Then there exists a measurable function* g *such that*

$$\nu(A) = \int_A g\, d\mu, \qquad \text{for all} \quad A \in \mathscr{B} \tag{2B.18}$$

if and only if ν *is absolutely continuous with respect to* μ *(i.e.,* $\mu(A) = 0 \Rightarrow \nu(A) = 0$ *for all* $A \in \mathscr{B}$, *in which case the relationship is written* $\nu << \mu$*).*

Note that a measure is said to be finite if $\mu(\Omega) < \infty$ and σ-finite if there exist disjoint sets $A_1, A_2, \ldots$ such that $\bigcup A_i = \Omega$ and $\mu(A_i) < \infty$ for each i.

For proofs of these theorems see Halmos (1950) and Munroe (1953). An introduction to measure theory and related topics with special reference to probability is also contained in a book by Cramer (1946).

BOREL-CANTELLI LEMMA. *Let* $\{A_n\}$, $n = 1, 2, \ldots$, *be a sequence of sets,* A *be the set of elements common to an infinity of these sets, and* P *be a probability set function. Then*

(a) $\sum P(A_n) < \infty \Rightarrow P(A) = 0$
(b) $\sum P(A_n) = \infty$ *and the events* A_n *are independant* $\Rightarrow P(A) = 1$.

It is easy to identify the set A as

$$A = \bigcap_{r=1}^{\infty} \bigcup_{n=r}^{\infty} A_n$$

from which it follows that

$$A \subset \bigcup_{n=r}^{\infty} A_n, \qquad \text{for each } r$$

and

$$P(A) \leqslant P\left(\bigcup_{r}^{\infty} A_n\right) \leqslant \sum_{r}^{\infty} P(A_n) < \varepsilon$$

for sufficiently large r since $\sum P(A_n)$ converges, that is, $P(A) = 0$, thus proving (a).

Let A^c be the complement of A, that is, the set of points common only to a finite number of sets of $\{A_n\}$. Hence A^c is the set of points contained in all but a finite number of sets of $\{A_n^c\}$. Thus

$$A^c = \bigcup_{r=1}^{\infty} \bigcap_{n=r}^{\infty} A_n^c$$

$$1 - P(A) = P(A^c) = P\left(\bigcup_{r=1}^{\infty} \bigcap_{n=r}^{\infty} A_n^c\right) \leqslant \sum_{r=1}^{\infty} P\left(\bigcap_{n=r}^{\infty} A_n^c\right)$$

$$= \sum_{r=1}^{\infty} \prod_{n=r}^{\infty} [(1 - P(A_n)] \quad \text{using independence of events.}$$

Since $\sum P(A_n) = \infty$, the infinite product diverges to zero for each r. Hence $P(A) = 1$, which proves (b).

APPENDIX 2C INVARIANCE

Invariance of a Family of Probability Spaces. Consider a family of probability spaces

$$(\Omega, \mathscr{B}, P_\theta, \theta \in \Theta) \tag{2C.1}$$

as in **2d**. Let g be any one-to-one transformation of Ω onto Ω. The sigma field $\{gB: B \in \mathscr{B}\}$ is denoted by $g\mathscr{B}$. Corresponding to a given measure P_θ, we define gP_θ as the induced measure on $g\mathscr{B}$ satisfying the relation $gP_\theta(gB) = P_\theta(B)$ for all $B \in \mathscr{B}$. Any function ϕ on Ω generates a new function $g\phi$ such that $g\phi(g\omega) = \phi(\omega)$, $\omega \in \Omega$. Then g produces an isomorphism between $(\Omega, \mathscr{B}, P_\theta)$ and $(\Omega, g\mathscr{B}, gP_\theta)$.

Let G be a group of transformations. Then the family $(\Omega, \mathscr{B}, P_\theta, \theta \in \Theta)$ is said to be invariant under G if for any $g \in G$

$$g\Omega = \Omega, \qquad g\mathscr{B} = \mathscr{B}, \qquad gP_\theta = P_{\theta'}, \qquad \theta, \theta' \in \Theta. \tag{2C.2}$$

The θ' determined by g by the last condition in (2C.2) is denoted by $\bar{g}\theta$.

It can be shown (see Ferguson, 1967) that $\bar{G} = \{\bar{g} : g \in G\}$ is a group of one-to-one transformations of Θ onto Θ, and $\bar{G}$ is a homomorphic image of G ($\bar{G}$ and G need not be isomorphic).

Invariance of Statistical Procedure. In practice, we have an observed event (a point in Ω) but the true value of θ or the exact underlying probability measure in the class P_θ, $\theta \in \Theta$ is unknown. The *statistical problem* is one of making a statement on unknown θ given the observed event. Let the class of all possible statements (or decisions) be represented by D (called the decision space). Then, what we need is a function δ (called decision function) which maps Ω into D. We represent by $L(\theta, d)$, the loss incurred in making the decision $d \in D$ when θ is the true value. An optimum statistical procedure is one which uses a decision function δ for which the loss is a minimum in some sense.

Let G be a group of transformations for which the family $(\Omega, B, P_\theta, \theta \in \Theta)$ is invariant. For each $g \in G$, let there exist a transformation g^* of the decision space D onto itself such that g^* is a homomorphism, i.e.,

$$(g_1 g_2)^* = g_1^* g_2^*$$

and the loss function is unchanged under the transformation, i.e.,

$$L(\bar{g}\theta, g^* d) = L(\theta, d), \qquad \theta \in \Theta, \quad d \in D. \tag{2C.3}$$

Then the statistical procedure or the decision function δ is said to be invariant under G if

$$\delta(g\omega) = g^* \delta(\omega) \qquad \text{for all} \quad \omega \in \Omega, \quad g \in G. \tag{2C.4}$$

APPENDIX 2D STATISTICS, SUBFIELDS, AND SUFFICIENCY

Let $(\Omega, \mathscr{B})$ be any measurable space and R^k be the Euclidean space of k (finite) dimensions and $\mathscr{X}$ the Borel field in R^k.

Definition 1. In **2a.5**, a statistic T is defined as a function on Ω into R^k such that $T^{-1}(X) \in \mathscr{B}$ for every $X \in \mathscr{X}$.

For any function T on Ω into R^k, denote by $\mathscr{B}_T$ the collection of sets $\{T^{-1}(X) : X \in \mathscr{X}\}$. Then T is a statistic (1), i.e., according to Definition 1, iff $\mathscr{B}_T \subset \mathscr{B}$.

We say that a subsigma field $\mathscr{C} \subset \mathscr{B}$ is induced by a statistic T if $\mathscr{C} = \mathscr{B}_T$. Not all subsigma fields are induced by a statistic (1) as the proposition (i) shows.

(i) *A subsigma field $\mathscr{C} \subset \mathscr{B}$ is induced by a statistic* (1) *if and only if $\mathscr{C}$ is countably generated.*

The only if part follows since $\mathscr{X}$ in R^k is separable, the subfield $\mathscr{B}_T$ induced by T, statistics (1) is separable.

To prove the converse, let $\mathscr{C} \subset \mathscr{B}$ be countably generated, and consider T on Ω into the real line R given by

$$T(\omega) = \sum_{n=1}^{\infty} \frac{2}{3^n} T_{C_n}(\omega)$$

where I is the indicator function and $\{C_n, n \geqslant 1\}$ is a countable family which generates $\mathscr{C}$. It is easy to see that T is a statistic (1) and $\mathscr{B}_T \subseteq \mathscr{C}$. To go the other way, let $n \geqslant 1$ and B_n be the set of reals in [0, 1] in whose ternary expansion 2 occurs in the nth place. It is easy to see that (a) B_n is a Borel set and (b) $T^{-1}(B_n) = C_n$. Consequently, $\{C_n, n \geqslant 1\} \subseteq \mathscr{B}_T$ so that $\mathscr{C} = \mathscr{B}_T$, which proves the proposition.

Definition 2. Bahadur (1954, 1955) defines a statistic T as any function on Ω with the range space arbitrary (not necessarily Euclidean) and with no measurability restrictions. A statistic satisfying Bahadur's definition is denoted as statistic (2).

The notion of a subsigma field induced by T, a statistic (2), is now defined as follows. Denote by $\mathscr{Y}_T$ the collection $\{A \subseteq Y: T^{-1}(A) \in \mathscr{B}\}$, where Y is the exact range space of T, i.e., $Y = T(\Omega)$. It is clear that (a) $\mathscr{Y}_T$ is a sigma field on Y, and that (b) T is a measurable transformation from $(\Omega, \mathscr{B})$ to $(Y, \mathscr{Y}_T)$. Next denote by $\mathscr{B}_T$ the collection $\{T^{-1}(B): B \in \mathscr{Y}_T\}$. Then $\mathscr{B}_T$ is a subsigma field of $\mathscr{B}$.

A subsigma field $\mathscr{C} \subset \mathscr{B}$ is said to be induced by a statistic (2) if there is a function T such that $\mathscr{C} = \mathscr{B}_T$. The question arises whether every subsigma field of $\mathscr{B}$ is induced by statistic (2) or even whether every countably generated subsigma field of $\mathscr{B}$ is induced by a statistic (2). The answer is no as the following result of Blackwell shows.

(ii) *Let $\mathscr{C}$ be a proper subsigma field of $\mathscr{B}$, such that $\{\omega\} \in \mathscr{C}$ for every $\omega \in \Omega$. Then $\mathscr{C}$ is not induced by a statistic* (2).

Suppose $\mathscr{C}$ is induced by a statistic T so that $\mathscr{B}_T = \mathscr{C}$. As $\{\omega\} \in \mathscr{C}$, $\{\omega\} \in \mathscr{B}_T$ and hence there exists $B \in \mathscr{Y}_T$ such that $\{w\} = T^{-1}(B)$. As B is contained in the exact range of T, it follows easily that $B = \{T(\omega)\}$. Consequently, for every $\omega \in \Omega$, $\{\omega\} = T^{-1}\{T(\omega)\}$. In other words, T is a one-to-one function on Ω onto Y so that for any $A \in \Omega$, $A = T^{-1}(T(A))$. Now let $A \in \mathscr{B}$. From the preceding argument, it follows that $A \subset \mathscr{B}_T$, i.e., $A \in \mathscr{C}$. Thus $\mathscr{B} = \mathscr{C}$ which is a contradiction.

Bahadur and Lehmann (1955) provide a characterization of subsigma fields of $\mathscr{B}$ which are induced by statistic (2).

We have two notions of subsigma fields of $\mathscr{B}$ induced by statistics (1) and (2). Now we show that neither notion implies the other.

Let Ω be the real line, $\mathscr{B}$ be the sigma field of countable and co-countable subsets of the real line and let $\mathscr{C} = \mathscr{B}$. Taking T as the identity mapping on Ω, we see that $\mathscr{C}$ is induced by a statistic (2). But $\mathscr{C}$, being not countably generated, is not generated by statistic (1).

Now let Ω be the real line, M be a fixed but arbitrary non-Borel subset of the real line, and $\mathscr{B}$ be the smallest sigma field on Ω containing all Borel sets of the real line as well as M. Take $\mathscr{C}$ as the Borel sigma field of the real line. Then $\mathscr{C}$, being countably generated, is induced by a statistic (1). But as $\mathscr{C}$ contains all singletons and is a proper subsigma field of $\mathscr{B}$, by Blackwell's proposition, $\mathscr{C}$ is not induced by a statistic (2).

Conditional Expectation with Respect to a Subfield. Consider any subsigma field $\mathscr{B}_0 \subset \mathscr{B}$. If ϕ is any $(\mathscr{B}, P)$ integrable function on Ω, the conditional expectation of ϕ given $\mathscr{B}_0$ is defined as any $\mathscr{B}_0$ measurable function $E_P(\phi \mid \mathscr{B}_0)$ which satisfies

$$\int_{B_0} E(\phi \mid \mathscr{B}_0)\, dP = \int_{B_0} \phi\, dP \qquad \text{for every} \quad B_0 \in \mathscr{B}_0. \tag{2D.1}$$

The Radon-Nikodym theorem ensures the existence and, up to an equivalence among all $\mathscr{B}_0$ measurable functions, the uniqueness of $E_P(\phi \mid \mathscr{B}_0)$, i.e., if alternative functions satisfying (2D.1) exist, then they differ only on P-null sets.

A subfield $\mathscr{B}_0$ is said to be sufficient for the family $(\Omega, \mathscr{B}, P_\theta, \theta \in \Theta)$, if for every $(\mathscr{B}, P_\theta)$ integrable function ϕ on Ω there exists a $\mathscr{B}_0$ measurable function ϕ_0 on Ω equivalent to

$$E_\theta(\phi \mid \mathscr{B}_0) \qquad \text{for all} \quad \theta \in \Theta. \tag{2D.2}$$

For further study of subfields, sufficient subfields etc., the reader is referred to Bahadur (1954, 1955) and Bahadur and Lehmann (1955).

APPENDIX 2E NON-NEGATIVE DEFINITENESS OF A CHARACTERISTIC FUNCTION

BOCHNER'S THEOREM. *A function $f(\cdot)$ defined on R is non-negative definite and continuous with $f(0) = 1$ iff it is a c.f.*

[A complex valued function f is n.n.d. if for each integer n, real numbers $t_1, \ldots, t_n$ and a function $h(t)$ on R.

$$\sum_{i=1}^{n} \sum_{j=1}^{n} f(t_i - t_j) h(t_i) \overline{h(t_j)} \geqslant 0.] \tag{2E.1}$$

The necessity is easy to prove. To prove sufficiency observe that (2E.1) $\Rightarrow$

$$\iint f(u - v) \exp\{-i(u - v)x - (u^2 + v^2)\sigma^2\}\, du\, dv \geqslant 0 \tag{2E.2}$$

for each x and $\sigma^2 > 0$ because the integral (2E.2) can be approximated by means of sums of the form (2E.1). Letting $t = u - v$ and $s = u + v$ so that $u^2 + v^2 = (t^2 + s^2)/2$ in (2E.2) and integrating over s, we obtain

$$g(x, \sigma) = \int f(t)\exp(-itx - t^2\sigma^2/2)\,dt/2\pi \geqslant 0.$$

By monotone convergence theorem (2B.3 of Appendix **2B**)

$$\begin{aligned}\int g(x, \sigma)\,dx &= \lim_{\beta\to 0} \int g(x, \sigma)\exp(-x^2\beta^2/2)\,dx \\ &= \lim_{\beta\to 0} \iint f(t)\exp(-itx - t^2\sigma^2/2 - x^2\beta^2/2)\,dx\,dt/2\pi \\ &= \lim_{\beta\to 0} \int f(t)\exp(-t^2/2\beta^2 - t^2\sigma^2/2)\,dt/\beta\sqrt{2\pi} \\ &= \lim_{\beta\to 0} \int f(u\beta)\exp(-u^2/2 - u^2\beta^2\sigma^2/2)\,du/\sqrt{2\pi} \\ &= \int e^{-u^2/2}\,du/\sqrt{2\pi} = 1.\end{aligned}$$

Thus $g(x, \sigma)$ as a function in x is a density function for each σ. A similar argument shows

$$\begin{aligned}\int g(x, \sigma)\exp(itx)\,dx &= \lim_{\beta\to 0} \int g(x, \sigma)\exp(itx - x^2\beta^2/2)\,dx \\ &= \lim_{\beta\to 0} \int f(t + u\beta)\exp[-(t + u\beta^2/2) - u^2/2]\,du \\ &= f(t)\exp(-t^2\sigma^2/2)\end{aligned}$$

by virtue of continuity of $f(t)$. Hence for each σ, $f(t)\exp(-t^2\sigma^2/2)$ is a c.f. Then by continuity theorem, $\lim_{\sigma\to 0} f(t)\exp(-t^2\sigma^2/2) = f(t)$ is in fact a c.f. This proof is due to Pathak (1966).

COMPLEMENTS AND PROBLEMS

1 *Characteristic functions*

1.1 A d.f. is said to be symmetric about zero if $F(x) = 1 - F(-x - 0)$. Show that $F(x)$ is symmetric if, and only if, its c.f. is real and even.

1.2 If $f(t)$ is a c.f., then $e^{f(t)-1}$ and $|f(t)|^2$ are also c.f.'s.

1.3 Find the distribution functions corresponding to the c.f.'s

$$\cos t, \quad (\cosh t)^{-1}, \quad \cos^2 t, \quad (\cosh^2 t)^{-1}.$$

1.4 Every c.f. can be written in the form

$$f(t) = c_1 f_d(t) + c_2 f_{ac}(t) + c_3 f_s(t)$$

with $c_1 + c_2 + c_3 = 1$, $c_1 \geqslant 0$, $c_2 \geqslant 0$, $c_3 \geqslant 0$, where $f_d(t)$, $f_{ac}(t)$, and $f_s(t)$ respectively are c.f.'s of (purely) discrete, absolutely continuous, and singular distributions.

1.5 Show that for any c.f. $f(t)$

$$1 - |f(2t)|^2 \leqslant 4[1 - |f(t)|^2],$$

and for any real c.f. $f(t)$

$$1 - f(2t) \leqslant 4[1 - f(t)], \qquad 1 + f(2t) \geqslant 2[f(t)]^2.$$

1.6 Let X_1, X_2, and X_3 be the independent random variables such that $X_1 + X_2$ and $X_1 + X_3$ have the same distribution. Does it follow that X_2 and X_3 have the same distribution?

2 *Moments of a distribution.* Let α_s and γ_s denote the sth moment and absolute moment of a random variable X.

2.1 Establish the following inequalities

(a) $\gamma_{s-1}^2 \leqslant \gamma_{s-2}\gamma_s$

(b) $\gamma_1 \leqslant \gamma_2^{1/2} \leqslant \cdots \leqslant \gamma_s^{1/s}$.

2.2 Let μ_s be the sth central moment and define $\beta_1 = \mu_3{}^2/\mu_2{}^3$ and $\beta_2 = \mu_4/\mu_2{}^2$. Then $\beta_2 - \beta_1 - 1 \geqslant 0$. [Hint: The determinant

$$\begin{vmatrix} 1 & 0 & \mu_2 \\ 0 & \mu_2 & \mu_3 \\ \mu_2 & \mu_3 & \mu_4 \end{vmatrix} \geqslant 0.]$$

3 *Inequalities for probabilities*

3.1 If X is a random variable such that $E(e^{aX})$ exists for $a > 0$, then $P(X \geqslant \varepsilon) \leqslant E(e^{aX})/e^{a\varepsilon}$.

3.2 *Kolmogorov's inequality.* Let $X_1, \ldots, X_n$ be n independent random variables such that $E(X_i) = 0$ and $V(X_i) = \sigma_i{}^2 < \infty$. Prove that

$$P\left(\max_{1 \leqslant k \leqslant n} \left|\sum_1^k X_i\right| > \varepsilon\right) < \frac{1}{\varepsilon^2} \sum_1^n \sigma_i{}^2.$$

3.3 *Hājek-Rēnyi* (1955) *inequality.* Let $X_1, X_2, \ldots$ be independent random variables such that $E(X_i) = 0$ and $V(X_i) = \sigma_i{}^2 < \infty$. If $c_1, c_2, \ldots$ is a nonincreasing sequence of positive constants, then for any positive integers m, n with $m < n$ and arbitrary $\varepsilon > 0$,

$$P\left(\max_{m \leqslant k \leqslant n} c_k |X_1 + \cdots + X_k| \geqslant \varepsilon\right) \leqslant \frac{1}{\varepsilon^2}\left(c_m{}^2 \sum_1^m \sigma_i{}^2 + \sum_{m+1}^n c_i{}^2 \sigma_i{}^2\right).$$

To prove the inequality, consider the quantity

$$Y = \sum_{k=m}^{n-1} S_k^2(c_k^2 - c_{k+1}^2) + c_n^2 S_n^2,$$

where $S_k = X_1 + \cdots + X_k$. It is easy to show that

$$E(Y) = c_m^2 \sum_{k=1}^{m} \sigma_k^2 + \sum_{k=m+1}^{n} c_k^2 \sigma_k^2.$$

Let E_i $(i = m, m+1, \ldots, n)$ be the event $c_j|S_j| < \varepsilon$ for $m \leqslant j < i$ and $c_i|S_i| \geqslant \varepsilon$. The events E_i are mutually exclusive and

$$P\left\{\max_{m \leqslant k < n} c_k|S_k| \geqslant \varepsilon\right\} = \sum_{k=m}^{n} P(E_i).$$

Let E_0 denote the event that $c_j|S_j| < \varepsilon$ for $m \leqslant j \leqslant n$. Then by the definition of conditional expectation

$$E(Y) = \sum_{i=0}^{n} E(Y|E_i)P\{E_i\} \geqslant \sum_{i=1}^{n} E(Y|E_i)P\{E_i\}.$$

For $k \geqslant i$,

$$\begin{aligned} E(S_k^2|E_i) &= E[\{S_i^2 + (X_{i+1} + \cdots + X_k)^2 + 2S_i(X_{i+1} + \cdots + X_k)\}|E_i] \\ &\geqslant E(S_i^2|E_i) + 2E[S_i(X_{i+1} + \cdots + X_k)|E_i]. \end{aligned}$$

But the occurrence of the event E_i only imposes a restriction on the first i of the variables X, and the following variables, under this condition, remain independent of one another and of S_i. Hence, for $j > i$, $E(S_i X_j|E_i) = 0$, thus giving the inequality $E(S_k^2|E_i) \geqslant E(S_i^2|E_i)$. When E_i is given, $|S_i| \geqslant \varepsilon/c_i$ so that $E(S_i^2|E_i) \geqslant \varepsilon^2/c_i^2$. Then we have $E(S_k^2|E_i) \geqslant \varepsilon^2/c_i^2$. But

$$\begin{aligned} E(Y|E_i) &= \sum_{k=m}^{n-1} E(S_k^2|E_i)(c_k^2 - c_{k+1}^2) + c_n^2 E(S_n^2|E_i) \\ &\geqslant \sum_{k=i}^{n-1} E(S_k^2|E_i)(c_k^2 - c_{k+1}^2) + c_n^2 E(S_n^2|E_i) \\ &\geqslant \frac{\varepsilon^2}{c_i^2}\left[\sum_{k=i}^{n-1}(c_k^2 - c_{k+1}^2) + c_n^2\right] = \varepsilon^2, \end{aligned}$$

and therefore $E(Y) \geqslant \varepsilon^2 \sum_{k=m}^{n} P(E_i)$. The required result follows from the exact expression derived for $E(Y)$.

Kolmogorov's inequality of Example **3.2** follows, as a special case, choosing $m = 1$, $c_1 = c_2 = \cdots = c_n = 1$.

3.4 Let $X_1, \ldots, X_m$ be m independent random variables such that $P(|X_i + \cdots + X_m| > \varepsilon) \leqslant \alpha$, $i = 1, \ldots, m$. Then

$$P\left\{\max_{1 \leqslant i < m} |X_1 + \cdots + X_i| > 2\varepsilon\right\} \leqslant \frac{P(|X_1 + \cdots + X_m| > \varepsilon)}{1 - \alpha}.$$

3.5 *Chebyschev-type inequalities.* Let X be a random variable such that $E(X) = \mu$, $E(X - \mu)^2 = \sigma^2$, $E|X - \mu|^r = \beta_r$, $E(|x - \mu|) = v$ and $\delta = v/\sigma$. Prove the following:

(a) $P\left(\dfrac{|X - \mu|}{\beta_r^{1/r}} \geqslant \lambda\right) \leqslant \dfrac{1}{\lambda^r}$.

(b) $P\left(\dfrac{|X - \mu|}{\sigma} \geqslant \lambda\right) \leqslant \dfrac{\beta_r}{\sigma^r \lambda^r}$.

(c) $P(|X - \mu| \geqslant \lambda) \leqslant \dfrac{\beta_r}{\lambda^r}$.

(d) *Cantelli inequality*

$$P(X - \mu \leqslant \lambda) \leqslant \frac{\sigma^2}{\sigma^2 + \lambda^2}, \qquad \lambda < 0,$$

$$\geqslant 1 - \frac{\sigma^2}{\sigma^2 + \lambda^2}, \qquad \lambda \geqslant 0.$$

(e) *Peek inequality*

$$P(|X - \mu| \geqslant \lambda\sigma) < \frac{1 - \delta^2}{\lambda^2 - 2\lambda\delta + 1}, \qquad \lambda \geqslant \delta.$$

(f) *Camp-Meidell inequality.* Let the distribution of X have a single mode at μ_0. Let $\tau = \sigma^2 + (\mu - \mu_0)^2$, and $s = |\mu - \mu_0|/\sigma$. Then

$$P(|X - \mu_0| \geqslant \lambda\tau) \leqslant \begin{cases} 1 - \dfrac{\lambda}{\sqrt{3}}, & \lambda \leqslant \dfrac{2}{\sqrt{3}} \\[2ex] \dfrac{4\lambda^2}{9}, & \lambda \geqslant \dfrac{2}{\sqrt{3}} \end{cases}$$

$$P(|X - \mu| \geqslant \lambda\sigma) \leqslant \frac{4}{9}\frac{(1 + s^2)}{(\lambda - s)^2}, \qquad \lambda > s.$$

3.6 *Berge inequality.* Let X_1, X_2 be a pair of variables such that $E(X_i) = \mu_i$, $V(X_i) = \sigma_i^2$ and $\mathrm{cov}(X_1, X_2) = \rho\sigma_1\sigma_2$. Then

$$P\left\{\max\left(\frac{|X_1 - \mu_1|}{\sigma_1}, \frac{|X_2 - \mu_2|}{\sigma_2}\right) \geqslant \lambda\right\} \leqslant \frac{1 + \sqrt{1 - \rho^2}}{\lambda^2}.$$

3.7 Show that for $x > 0$

$$\int_x^\infty e^{-y^2/2}\,dy \leqslant \frac{1}{x}\,e^{-x^2/2} = \frac{1}{x}\int_x^\infty ye^{-y^2/2}\,dy.$$

3.8 $X_1, X_2, \ldots$ are independent and the range of each X_i is $[0, 1]$. Then for any $\varepsilon > 0$

$$P\left\{\frac{1}{n}\sum [X_i - E(X_i)] \geqslant \varepsilon\right\} \leqslant e^{-2n\varepsilon^2}.$$

4 *Limit problems*

4.1 Construct an example to show that Scheffe's convergence theorem (xv, **2c.4**) is not true if the limit function $p(x)$ is not a probability density.

4.2 Let $X_n \xrightarrow{L} X$ and μ_n be a median of X_n. Then any limit point of μ_n is a median of X.

4.3 *Khinchin's theorem* (*W.L.L.N.*). Let $X_1, X_2, \ldots$ be a sequence of i.i.d. variables. Then $E(X_i) = \mu$ is finite $\Rightarrow \bar{X}_n \xrightarrow{P} \mu$.
[Hint: Let $\phi(t)$ be the c.f. of X_i. Then the c.f. of $\bar{X}_n$ is $[\phi(t/n)]^n$. Since $E(X)$ exists

$$\phi\left(\frac{t}{n}\right) = 1 + \frac{\mu it}{n} + o\left(\frac{t}{n}\right) \quad \text{and} \quad \lim_{n\to\infty}\left[1 + \frac{\mu it}{n} + o\left(\frac{t}{n}\right)\right]^n = e^{i\mu t},$$

which is the c.f. of a degenerate distribution at μ.]

4.4 *Kolmogorov theorem* (1929). Let $X_1, X_2, X_3, \ldots$ be independent variables all having the same d.f. $F(x)$ and put $\bar{X}_n = (X_1 + \cdots + X_n)/n$. A necessary and sufficient condition for the existence of a sequence of constants $\mu_1, \mu_2, \ldots$ such that $\bar{X}_n - \mu_n \xrightarrow{P} 0$ is that

$$\int_{|x|>t} dF(x) = o(1/t) \quad \text{as} \quad t \to \infty.$$

If this condition is satisfied we can always take $\mu_n = \int_{-n}^{n} x\,dF(x)$. If in addition the generalized mean value $\mu = \lim_{t\to\infty} \int_{-t}^{t} x\,dF(x)$ exists, it follows that $\bar{X}_n \xrightarrow{P} \mu$. Show that in the case of the Cauchy distribution, the necessary and sufficient condition is not satisfied. Hence, although the generalized mean value exists, $\bar{X}_n$ does not converge to any constant in probability.

4.5 Let $X_1, X_2, \ldots$ be a sequence of independent variables such that $E(X_i) = \mu_i$. Put $\bar{X}_n = (X_1 + \cdots + X_n)/n$ and $\bar{\mu}_n = (\mu_1 + \cdots + \mu_n)/n$. Show that $\bar{X}_n - \bar{\mu}_n \xrightarrow{P} 0$, if

$$\frac{1}{n^{1+\delta}}\sum_{k=1}^{n} |X_k|^{1+\delta} \to 0 \quad \text{as} \quad n \to \infty$$

for a positive $\delta \leqslant 1$.

4.6 Let $X_1, X_2, \ldots$ be independent r.v.'s such that $E(X_i) = \mu_i$ and $V(X_i) = \sigma_i^2$. Put $S_n = X_1 + \cdots + X_n$, $M_n = \mu_1 + \cdots + \mu_n$ and $V_n^2 = \sigma_1^2 + \cdots + \sigma_n^2$. Then a sufficient condition for the d.f. of $(S_n - M_n)/V_n$ to tend to $N(0, 1)$ is that

$$V_n^{-(2+\delta)} \sum_{k=1}^{n} |X_k|^{2+\delta} \to 0 \quad \text{as} \quad n \to \infty$$

for a positive δ. [Hint: Show that the sufficient condition implies Lindberg-Feller condition, (iii), **2c.5**.]

4.7 *Multivariate central limit theorem.* Let $\mathbf{X}_1, \mathbf{X}_2, \ldots$ be a sequence of independent k-dimensional random variables such that $E(\mathbf{X}_i) = \mathbf{0}$ and $D(\mathbf{X}_i) = \mathbf{\Sigma}_i$. Suppose that as $n \to \infty$,

$$\frac{1}{n} \sum \mathbf{\Sigma}_i \to \mathbf{\Sigma} \neq \mathbf{0}$$

and for every $\varepsilon > 0$

$$\frac{1}{n} \sum_{i=1}^{n} \int_{\|\mathbf{X}\| > \varepsilon\sqrt{n}} \|\mathbf{X}\|^2 \, dF_i \to 0$$

where F_i is the d.f. of $\mathbf{X}_i$ and $\|\mathbf{X}\|$ is the Euclidean norm of vector $\mathbf{X}$. Then the r.v. $(\mathbf{X}_1 + \cdots + \mathbf{X}_n)/\sqrt{n}$ converges to the k-variate normal distribution with mean zero and dispersion matrix $\mathbf{\Sigma}$.

4.8 Let $X_1, X_2, \ldots$ be independent r.v.'s with d.f.'s $F_1, F_2, \ldots$. Then $\overline{X}_n \xrightarrow{P} \mu$ a constant, if and only if,

(a) $\sum \int_{|x| \geq n} dF_k \to 0,$

(b) $\frac{1}{n} \sum \int_{|x| < n} x^2 \, dF_k \to 0,$

(c) $\frac{1}{n^2} \sum \left\{ \int_{|x| < n} x^2 \, dF_k - \left(\int_{|x| < n} x \, dF_k \right)^2 \right\} \to 0.$

Observe that the existence of the expectations is not required. For a proof see Loève (1955).

4.9 Suppose that the random variables $X, X_1, X_2, \ldots$ are uniformly bounded. Show that $X_n \xrightarrow{P} X$ if and only if $\lim E(X_n - X)^2 = 0$ as $n \to \infty$.

4.10 If for all real t

$$\lim_{n \to \infty} \int e^{tx} \, dF_n(x) = \int e^{tx} \, dF(x) = M(t)$$

and if $M(t)$ is entire (i.e., there exists an entire function $M(z)$ with a complex argument, which reduces to $M(t)$ on the real line), then $F_n \to F$. For a proof see Kac (1959).

4.11 Let $X_n \xrightarrow{L} X$, and suppose there are constants $a_n > 0$, b_n such that $a_n X_n + b_n \xrightarrow{L} Y$, where X and Y are nondegenerate. Then there exist constants a and b such that $aX + b \overset{L}{=} Y$ ($\overset{L}{=}$ indicates that the distributions are same).

4.12 *Invariance Principle.* Show that if there is one sequence $Y_1, Y_2, \ldots$ of i.i.d. variables such that $E(Y_1) = 0$, $0 < E(Y_1^2) = \sigma^2 < \infty$ such that

$$\frac{Y_1 + \cdots + Y_n}{\sqrt{n}} \xrightarrow{L} Y$$

then the same limiting distribution holds for all sequences $X_1, X_2, \ldots$ of i.i.d. variables such that $E(X_1) = 0$, $0 < E(X_1^2) < \infty$.

4.13 Show that if $X_n \xrightarrow{P} X$, then there exists a subsequence X_{n_k} such that $X_{n_k} \to X$, a.s.

4.14 For $X_1, X_2, \ldots$ independent variables

$$\sum_1^n X_i \xrightarrow{P} \Leftrightarrow \sum_1^n X_i \xrightarrow{\text{a.a.}}.$$

4.15 Let $X_1, X_2, \ldots$ be independent such that $E(X_k) = 0$, $V(X_k) < \infty$. Further, let there exist constants $b_1 < b_2 < \cdots \to \infty$ such that $\sum_1^\infty [V(X_k)/b_k^2] < \infty$. Then $\sum b_n^{-1} X_n$ converges and

$$\frac{X_1 + \cdots + X_n}{b_n} \to 0, \quad \text{a.s.}$$

[Hint: Use *Kronecker's lemma.* Let x_k be an arbitrary numerical sequence and $0 < b_1 < b_2 < \cdots \to \infty$. If the series $\sum b_k x_k$ converges then $\{(x_1 + \cdots + x_n)/b_n\} \to 0$].

4.16 $X_n \xrightarrow{\text{r.m.}} X$ (converges in the rth mean) if $E|X_n - X|^r \to 0$. Show that

(a) $X_n \xrightarrow{\text{r.m.}} X \Rightarrow E|X_n|^r \to E|X|^r$,
(b) $X_n \xrightarrow{\text{r.m.}} X \Rightarrow X_n \xrightarrow{\text{s.m.}} X, 0 \leqslant s \leqslant r$,
(c) $X_n \xrightarrow{\text{r.m.}} X \Rightarrow X_n \xrightarrow{P} X$.

4.17 Give an example where $X_n \xrightarrow{P} 0$ but $E(X_n) \to c \neq 0$.

5 *Zero-or-one law.* Let $X_1, X_2, \ldots$ be any random variables and let $g(X_1, X_2, \ldots)$ be a Baire function of the variables such that the conditional probability

$$P(g = 0 \mid X_1 = x_1, \ldots, X_n = x_n)$$

is the same as the absolute probability $P(g = 0)$ for every n and every vector $(x_1, \ldots, x_n)$. Then $P(g = 0)$ is either zero or one.

Show in particular that if $X_1, X_2, \ldots$ are mutually independent the probability of the convergence of the series $\sum_1^\infty X_n$ is either zero or one.

6 Show that if $F_1(x)$ and $F_2(x)$ are monotonic functions of bounded variation and c_1, c_2 are arbitrary constants

$$\int_a^b f(x)\, d[c_1 F_1(x) + c_2 F_2(x)] = c_1 \int_a^b f(x)\, dF_1(x) + c_2 \int_a^b f(x)\, dF_2(x).$$

7 If X_1, X_2 are two independent random variables with d.f.'s, F_1 and F_2, then the d.f.'s of $Z = X_1 + X_2$ and $Y = X_1/X_2$ are

$$F(z) = \int F_1(z - x)\, dF_2(x) = \int F_2(z - x)\, dF_1(x)$$

$$F(y) = \int_0^\infty F_1(yx)\, dF_2(x) + \int_{-\infty}^0 [1 - F_1(yx)]\, dF_2(x).$$

8 Using the inequalities of Section **1e**, verify the following for random variables X, Y.

(a) $E|X + Y|^r \leqslant C_r(E|X|^r + E|Y|^r)$
where $C_r = 1$ for $r \leqslant 1$, $= 2^{r-1}$ for $r > 1$.

(b) *Holder's inequality*

(i) $E|XY| < (E|X^r|)^{1/r}(E|Y^s|)^{1/s}$, $1 < r < \infty$, $\dfrac{1}{r} + \dfrac{1}{s} = 1.$

(ii) $(E|XY|)^2 \leqslant E(X^2)E(Y^2)$, Cauchy-Schwarz

(iii) $E|X| \leqslant (E|X|^r)^{1/r}$, $r > 1$.

(iv) $(E|X|^t)^{1/t} \leqslant (E|X|^r)^{1/r}$, $0 < t \leqslant r$.

(c) *Minkowski inequality*

$$(E|X + Y|^r)^{1/r} \leqslant (E|X|^r)^{1/r} + (E|Y|^r)^{1/r}, \qquad 1 \leqslant r < \infty.$$

(d) $\log E|X|^r$ is a convex function of r.

(e) *Jensen's inequality.* If g is a convex function and $E(X)$ exists, then

$$g[E(X)] \leqslant E[g(X)].$$

9 *Infinitely divisible distributions.* A d.f. F is called infinitely divisible (weakly) if, for every n, it can be written as a convolution distribution of nondegenerate d.f.'s $F_1, \ldots, F_n$. Or in other words, the random variable X with d.f. F can be expressed as the sum of n nondegenerate i.r.v.'s. The d.f. F is called infinitely divisible if, for every n, X can be expressed as the sum of n

independent and identically distributed random variables. Prove the following with respect to infinitely divisible distributions.

9.1 For any given n the c.f. of X is the nth power of some other c.f.

9.2 The c.f. of X never vanishes.

9.3 The distribution functions of normal and Poisson variables are infinitely divisible.

9.4 The d.f. of a sum of independent random variables having infinitely divisible d.f.'s is itself infinitely divisible.

9.5 The (weak) limit d.f. of a sequence of infinitely divisible d.f.'s is itself infinitely divisible.
[Note: For a canonical representation of an infinitely divisible d.f., refer to books on probability by Doob, Gnedenko, Loève, Ramachandran, etc.]

10 *Discrete distributions*

10.1 Let $P_1(t), \ldots, P_k(t)$ be the p.g.f.'s of k independent discrete random variables $X_1, \ldots, X_k$. Show that the p.g.f. of the sum $X_1 + \cdots + X_k$ is the product $P_1(t) \cdots P_k(t)$.

10.2 Let $S_N = X_1 + \cdots + X_N$ be the sum of N independent discrete random variables each having the same p.g.f., $P(t)$. Further let N itself be a random variable with the p.g.f. $G(t)$. Show that the p.g.f. of the random variable S defined as the sum of a random number of independent random variables is $G[P(t)]$.

10.3 Using the result of Example **10.2**, obtain the distribution of S when each X_i has a Poisson distribution with parameter λ and N has a Poisson distribution with parameter μ.

10.4 Also show that when each X_i has a binomial distribution with parameters $(\pi, n = 1)$ and N is a Poisson variable with parameter μ, S is a Poisson variable with parameter $\mu\pi$.

11 Establish the following conditional distributions.

11.1 Let X and Y be independent binomial variables with parameters (π, n_1) and (π, n_2). Then the conditional distribution of X, Y given $X + Y$ is hypergeometric and is independent of π.

11.2 Let X and Y be independent negative binomial variables with parameters (ρ, κ_1) and (ρ, κ_2). Then the conditional distribution of (X, Y) given $X + Y$ is the negative hypergeometric defined by

$$\binom{X+Y}{X} \frac{\beta(\kappa_1 + X, \kappa_2 + Y)}{\beta(\kappa_1, \kappa_2)}$$

where $B(m, n)$ is the beta function.

12 Characteristic functions of special distributions.

	Density Function	*Characteristic Function*
Normal:	$\frac{1}{\sqrt{2\pi}\sigma} e^{-(x-\mu)^2/2\sigma^2}$	$e^{i\mu t - \sigma^2 t^2/2}$
Laplace:	$\frac{1}{2\sigma} e^{-\|x-\mu\|/\sigma}$	$e^{it\mu}(1 + \sigma^2 t^2)^{-1}$
Rectangular:	$\frac{1}{\theta}$, $a < x < a + \theta$	$e^{ita} \frac{e^{it\theta} - 1}{it\theta}$
Gamma:	$\frac{\alpha^p}{\Gamma(p)} x^{p-1} e^{-\alpha x}$	$\left(\frac{\alpha}{\alpha - it}\right)^p$
Beta:	$\frac{\Gamma(p+q)}{\Gamma(p)\Gamma(q)} x^{p-1}(1-x)^{q-1}$	$\frac{\Gamma(p+q)}{\Gamma(p)} \sum_{j=0}^{\infty} \frac{(it)^j \Gamma(p+j)}{\Gamma(p+q+j)\Gamma(j+1)}$
Cauchy:	$\frac{\delta}{\pi} \frac{1}{\delta^2 + (x-\mu)^2}$	$e^{it\mu} e^{-\|t\delta\|}$
Triangular: $\|x\| < a$	$\frac{1}{a}\left(1 - \frac{\|x\|}{a}\right)$	$2 \frac{1 - \cos at}{a^2 t^2}$
Triangular: $-\infty < x < \infty$	$\frac{1}{\pi} \frac{1 - \cos ax}{ax^2}$	$\begin{cases} 1 - \|t\|/a & \text{for } \|t\| \leqslant a \\ 0 & \text{for } \|t\| > a \end{cases}$
Hyperbolic cosine: $-\infty < x < \infty$	$\frac{1}{\pi \cosh x}$	$\frac{1}{\cosh(\pi t/2)}$
Rectangular: $\|x\| < \theta$	$\frac{1}{2\theta}$	$\frac{\sin \theta t}{\theta t}$
Degenerate:	$X = a$ (with Prob. 1)	e^{ita}

13 *The concept of large and small O.* Let $\{r_n\}$ be a sequence of positive numbers and $\{X_n\}$ be a sequence of random variables. We say $X_n = O_p(r_n)$ or X_n/r_n is bounded in probability if for each ε there is an M_ε and N_ε such that $P[|X_n|/r_n > M_\varepsilon] < \varepsilon$ for $n > N_\varepsilon$. We say $X_n = o_p(r_n)$ if $X_n/r_n \xrightarrow{P} 0$.

Let $X_n - a = o_p(r_n)$ where $r_n \to 0$ as $n \to \infty$ and $g(x)$ have a kth order Taylor expansion at a, that is,

$$g(x) = T_k(x, a) + o(|x - a|^k).$$

Show that $g(X_n) - T_k(X_n, a) = o_p(r_n{}^k)$.

14 Let Y be a number uniformly distributed over the unit interval (0, 1). Let $X_1, X_2, \ldots$ be the successive digits in the decimal expansion of Y, that is,

$$Y = \frac{X_1}{10} + \frac{X_2}{10^2} + \cdots + \frac{X_n}{10^n} + \cdots.$$

Prove that $X_1, X_2, \ldots$ are independent discrete valued random variables uniformly distributed over the integers 0 to 9. Consequently, conclude that for any integer k, the set of numbers Y for which the relative frequency of k in the decimal expansion of Y is $\frac{1}{10}$ has probability 1. Does this contradict the fact that only threes occur in the decimal expansion of $\frac{1}{3}$?

15 Let $\{a_n\}$ be a sequence of numbers such that $a_n \to a$ and $f(a_n) \to b$ as $n \to \infty$. Consider a sequence $\{X_n\}$ of r.v.'s such that $X_n \xrightarrow{P} a$. Show that $f(X_n) \xrightarrow{P} b$. Suppose $X_n/n \xrightarrow{P} 0$. Does this imply that $\max(X_1/n, \ldots, X_n/n) \xrightarrow{P} 0$? [The result is true for a nonstochastic sequence of numbers.]

16 Give examples to show:

(a) Two distinct c.f.'s can coincide within a finite interval.

(b) The relation $F * F_1 = F * F_2 \not\Rightarrow F_1 = F_2$.

17 *Moment problem.* Show that the moments for the two densities of a positive random variable X

$$p_1(x) = (2\pi)^{-1/2} x^{-1} \exp[-(\log x)^2/2]$$
$$p_2(x) = p_1(x)[1 + a \sin(2\pi \log x)], \qquad -1 \leqslant a \leqslant 1,$$

are the same. This interesting example, which shows that the moments may not uniquely define a distribution function, is due to C. C. Heyde.

18 If X_1, X_2 are i.i.d.; then for $t > 0$

$$P\{|X_1 - X_2| > t\} \leqslant 2P\{|X_1| > t/2\}.$$

If $X_1, \ldots, X_n$ are independent with a common distribution function F, then

$$2P\{|X_1 + \cdots + X_n| \geqslant t\} \geqslant 1 - \exp\{-n[1 - F(t) + F(-t)]\}.$$

19 If X, Y, and Z are random variables, prove the following results on conditional expectation using the method of **2b.3**.

(a) $\text{Cov}(X, Y) = \text{Cov}[X, E(Y|X)]$

(b) $E(Z) = E\{E[E(Z|Y)|X]\}$,
$V(Z) = E\{E[V(Z|Y)|X]\} + E\{V[E(Z|Y)|X]\} + V\{E[E(Z|Y)|X]\}$,
$\text{Cov}(X, Z) = \text{Cov}\{X, E[E(Z|Y)|X]\}$.

20 If $(C_1, \ldots, C_k)$ is a partition of the sample space, show that

$$P(A \mid B) = \sum_1^k P(AC_i \mid B)$$
$$= \sum_1^k P(C_i \mid B)P(A \mid BC_i).$$

21 Consider the probability space $(\Omega, \mathscr{B}, P)$. Let Y be a space of points y, $\mathscr{A}$ a σ-algebra of sets of Y, and T: $\Omega \to Y$ a function such that $A \in \mathscr{A} \Rightarrow T^{-1}A \in \mathscr{B}$. Let $f(w, y)$ be a real valued $\mathscr{B} \times \mathscr{A}$ measurable function on $\Omega \times Y$. Then show that (see Bahadur and Bickel, 1968)

$$E[f(\omega, T(\omega)) \mid T(\omega) = y] = E[f(\omega, y) \mid T(\omega) = y].$$

22 Let $\mathbf{X}$ be a vector valued random variable whose distribution depends on a vector parameter $\boldsymbol{\theta}$. Further, let $\mathbf{T}$ be a statistic such that the distribution of $\mathbf{T}$ depends only on $\boldsymbol{\phi}$, a function of $\boldsymbol{\theta}$. Then $\mathbf{T}$ is said to be inference sufficient for $\boldsymbol{\phi}$ if the conditional distribution of $\mathbf{X}$ given $\mathbf{T}$ depends only on functions of $\boldsymbol{\theta}$ which are independent of $\boldsymbol{\phi}$ (see Rao, 1965b).

Let $X_1, \ldots, X_n$ be n independent observations from $N(\mu, \sigma^2)$. Show that $\sum(X_i - \bar{X})^2$ is inference sufficient for the parameter σ^2.

23 Give a simple proof to show that if $E(X_n) \to 0$ and $V(X_n) \to 0$, then $X_n \xrightarrow{P} 0$.

REFERENCES

Bahadur, R. R. (1954), Sufficiency and statistical decision functions, *Ann. Math. Statist.* **25**, 423–462.

Bahadur, R. R. (1955), Statistics and Subfields, *Ann. Math. Statist.* **26**, 490–497.

Bahadur, R. R. and E. L. Lehmann (1955), Two comments on "Sufficiency and statistical decision functions," *Ann. Math. Statist.* **26**, 139–142.

Bahadur, R. R. and P. J. Bickel (1968), Substitution in conditional expectation, *Ann. Math. Statist.* **39**, 377–378.

Cramer, H. (1946), *Mathematical Methods of Statistics*, Princeton University Press, Princeton, N.J.

Cramer, H. and H. Wold (1936), Some theorems on distribution functions, *J. London Math. Soc.* **11**, 290–294.

Dharmadhikari, S. W. (1965), An example in the problem of moments, *Am. Math. Monthly*, **72**, 302–303.

Doob, J. L. (1953), *Stochastic Processes*, John Wiley, New York.

Feller, W. (1966), *An Introduction to Probability Theory and Its Applications*, John Wiley, New York.

Ferguson, T. S. (1967), *Mathematical Statistics, A Decision Theoretic Approach*, Academic Press.

Gnedenko, B. V. (1962), *Theory of Probability*, Chelsea, New York.

Gnedenko, B. V. and A. N. Kolmogorov (1954), *Limit Distributions for Sums of Independent Random Variables*, Addison-Wesley, Reading, Mass.

Grenander, V. (1960), *Probabilities on Algebraic Structures*, John Wiley, New York.

Hājek, J. and A. Rēnyi (1950), Generalisation of an inequality of Kolmogorov, *Acta. Math. Acad. Sci. Hung.*, **6**, 281–283.

Halmos, P. R. (1950), *Measure Theory*, D. Van Nostrand, New York.

Halmos, P. R. and L. J. Savage (1949), Application of the Radon-Nikodym theorem to the theory of sufficient statistics, *Ann. Math. Statist.* **20**, 225–241.

Hamburger, H. (1920, 1921), Uber eine Erweiterung des Stieltjesschen Momentenproblems, *Mathemat. Annalen* **81** 235–319, **82**, 120–167, 167–187.

Kac, M. (1959), *Probability and Related Topics in Physical Sciences*, Interscience, London and New York.

Kolmogorov, A. N. (1929), Bemerkungen zu meiner Arbeit "Uber die Summen zufälliger Grössen," *Math. Annalen* **102**, 484–488.

Kolmogorov, A. N. (1950), *Foundations of the Theory of Probability* (German edition, 1933), Chelsea, New York.

Levy, P. (1937), *Theorie de l'addition des variables aleatoires*, Gauthier-Villars, Paris.

Linnik, Yu. V. (1960), *Decomposition of Probability Laws*. Leningrad University.

Loève, M. (1955), *Probability Theory*, D. Van Nostrand, Princeton, N.J.

Mann, H. B. and A. Wald (1943), On stochastic limit and order relationships, *Ann. Math. Statist.* **14**, 217–226.

Munroe, M. E. (1959), *Introduction to Measure and Integration*, Addison-Wesley, Reading, Mass.

Okamoto Masashi (1959), A convergence theorem for discrete probability distributions, *Inst. Stat. Math. Tokyo* **11**, 107–112.

Parthasarathy, K. R. (1967), *Probability Measures on Metric Spaces*, Academic Press.

Pathak, P. K. (1966), Alternative proofs of some theorems on characteristic functions, *Sankhyā A*, **28**, 309–315.

Pratt, J. W. (1960), Extension of dominated convergence theorem, *Ann. Math. Statist.*, 31, 74–77.

Ramachandran, B. (1967), *Advanced Characteristic Functions*, Statistical Publishing Society.

Savage, L. J. (1970), An unpublished note.

Scheffe, H. (1947), A useful convergence theorem for probability distributions, *Ann. Math. Statist.* **18**, 434–458.

Sudakov, V. N. (1971), A note on marginal sufficiency, *Doklady* 198.

Varadarajan, V. S. (1958), A useful convergence theorem, *Sankhya*, **20**, 221–222.

Wald, A. and J. Wolfowitz (1944), statistical tests based on permutations of the observations, *Ann. Math. Statist.* **15**, 358–372.

Chapter 3

CONTINUOUS PROBABILITY MODELS

Introduction. In Chapter 2 we introduced the concept of a random variable (r.v.) specified by a distribution function (d.f.). In practice, the d.f. F is unknown and the problem is one of drawing inferences about F on the basis of concrete realizations of the r.v. But it would be of great help if it is known *a priori* that $F \in C$, a specified class of d.f.'s, for in such a case our enquiry is restricted to a particular class and not to the wide class of all d.f.'s. The information we use that $F \in C$ is called *specification.* Sometimes it is possible to determine the class C from a knowledge of the mechanism generating the observations by a suitable characterization of the randomness involved, called a *model.*

For example consider n urns each of which contains unknown but fixed numbers of white and black balls. One ball is drawn from each urn and its color noted. What is the probability that out of the n balls drawn, r are white? Suppose that all the balls within any urn are distinguishable and that each urn contains a white and b black balls. Then we have $(a + b)^n$ possible combinations of n balls. If each ball is drawn from an urn after *thorough mixing* we may also make the primitive assumption that all the combinations $(a + b)^n$ are equally likely to occur. It is now easy to show that the proportion of the combinations with r white balls is

$$\binom{n}{r}\pi^r(1 - \pi)^{n-r}$$

where $\pi = a/(a + b)$, the unknown proportion. We thus deduce the binomial distribution (2a.5.3) from a simple model.

In the present chapter, we consider a number of models leading to the well-known continuous distributions. The properties of these distributions are also examined. The probability distributions of certain functions of the observations, which are used in statistical inference, have also been obtained.

The reader will find it useful to study in detail the three fundamental theorems of the least squares theory in **3b.5**, which provide the key to all distribution problems in the univariate and multivariate analyses. Some extensions of these theorems given as examples at the end of the chapter are also useful.

The bivariate normal distribution is introduced in **3d** where sampling distributions of some important statistics are obtained by the use of transformations. These methods could be extended to the multivariate case, but they are not since more sophisticated methods are available as shown in Chapter 8 on multivariate analysis.

Transformation of Variables. Let X be a r.v. with p.d. p_X. What is the p.d. of $Y = f(X)$?

By definition

$$P(Y < y_0) = P\{x : f(x) < y_0\} = \int_A p_X(x)\, dx$$

where A is the region $f(x) < y_0$. Suppose (a) the tranformation $y = f(x)$ is one-to-one and the inverse transformation $x = g(y)$ exists and (b) $g'(y)$ the derivative of g exists and is continuous. Then from the theory of change of variable in Riemann integration

$$\int_A p_X(x)\, dx = \int_{y < y_0} p_X[g(y)]\, |g'(y)|\, dy. \tag{3.1}$$

Hence by definition the p.d. of Y is $p_X[g(y)]|g'(y)|$.

Suppose the transformation is not one-to-one but the range of x can be divided into a finite number of exclusive regions $R_1, \ldots, R_k$ with the corresponding regions $R'_1, \ldots, R'_k$ (may be overlapping) in the range of Y such that the transformation from R_i to R'_i is one-to-one. Let $x = g_i(y)$ be the transformation from R'_i to R_i and let $g'_i(y)$ be continuous. Then the density at y is

$$\sum_{i=1}^{k} \pi_i p_X[g_i(y)]|g'_i(y)|, \tag{3.2}$$

where $\pi_i = 1$ if $y \in R'_i$ and $\pi_i = 0$, otherwise.

For instance if $Y = X^2$, there are two regions to be considered: $R_1 = (-\infty, 0)$ and $R_2 = (0, \infty)$. The corresponding regions in the space of Y are $R'_1 = (0, \infty)$ and $R'_2 = (0, \infty)$ which are completely overlapping. The inverse functions are $x = -\sqrt{y}$ corresponding to (R_1, R'_1) and $x = \sqrt{y}$ corresponding to (R_2, R'_2). Every $y \in R'_1$ and R'_2 and therefore the p.d. at y is

$$[p_X(\sqrt{y}) + p_X(-\sqrt{y})](\tfrac{1}{2} \cdot y^{-1/2}).$$

The results can be extended to several variables in an analogous manner. Let $Y_i = f_i(X_1, \ldots, X_n)$, $i = 1, \ldots, n$ be the given transformation and $p(x_1, \ldots, x_n)$ be the p.d. of $X_1, \ldots, X_n$. Consider the regions $(R_1, R_1'), \ldots, (R_k, R_k')$ where R_i is in the space of X's and R_i' is in the space of Y's and a one-to-one correspondence holds for each pair (R_i, R_i'). Let

$$x_r = g_{ri}(y_1, \ldots, y_n), \qquad r = 1, \ldots, n,$$

be the inverse transformation for (R_i, R_i'). If the partial derivatives of g_{ri} are continuous and the Jacobian (J_i) of the transformation does not vanish in R_i', then the p.d. at $y_1, \ldots, y_n$ is

$$\sum \pi_i p(g_{1i}, \ldots, g_{ni}) | J_i | \tag{3.3}$$

where $\pi_i = 1$ if $(y_1, \ldots, y_n) \in R_i'$ and $\pi_i = 0$, otherwise, and $| J_i |$ is the absolute value of the Jacobian.

Notation. In Chapter 2 and in the example of transformations just considered we denoted a random variable (r.v.) by X and its probability density (p.d.) by $p_X(x)$ or simply $p(x)$ when there is no chance of confusion about the underlying r.v. This cannot be strictly adhered to, for in statistics both capital and lower case letters R, r, F, f, T, t, etc. have been used to denote specific functions of r.v.'s. We should therefore be free to use any symbol to denote a r.v. With such a concession it is also convenient to use the same symbol for the r.v. and in the expression for its p.d., observing that the symbol in the p.d. is only of a dummy character. Thus we may say that x is an r.v. with p.d. $f(x)$ instead of $f(x^*)$, using a different symbol like x^*. Such a convention has the advantage insofar as that a given expression for density can be identified as that of a particular r.v. (with a conventional symbol) that is being considered. Furthermore, most of the propositions we consider involve an application of the formulas (3.1), (3.2), and (3.3) for determining the density function of functions of r.v.'s. For this purpose it is immaterial which symbol (same as that used for the r.v. or a matching symbol like x for the r.v. X) we use for expressing the transformations and in the evaluation of the Jacobians. However, in the derivation of the formulas (3.1) to (3.3) from first principles such a distinction is necessary for purposes of clarity.

We frequently refer to a family of distributions by a symbol such as N for the class of all normal distributions and to a particular member of the family by specifying its parameters in addition such as $N(\mu, \sigma^2)$ for a normal distribution with mean μ and variance σ^2. We use the notation $X \sim N(\mu, \sigma^2)$ to indicate that X is a r.v. distributed as $N(\mu, \sigma^2)$. The density function of $N(\mu, \sigma^2)$ is written $N(x | \mu, \sigma^2)$ or $N(X | \mu, \sigma^2)$ using X itself.

3a UNIVARIATE MODELS

3a.1 Normal Distribution

(i) HITTING THE BULL'S EYE (HERSCHEL'S HYPOTHESIS). Consider a distribution of shots fired at a target (ideally a point) and let (X, Y) be the coordinates (*r.v.'s*) representing the deviation (errors) of a shot with respect to two orthogonal axes through the target point.

Let the following hypotheses be true:

(a) *The marginal density functions $p(x)$, $q(y)$ of the errors X and Y are continuous.*
(b) *The probability density at (x, y) depends only on the distance $r = (x^2 + y^2)^{1/2}$ from the origin (radial symmetry).*
(c) *The errors in x and y directions are independent.*

Then the p.d. of deviation Z in any direction is the normal density

$$\frac{1}{\sigma\sqrt{2\pi}} e^{-z^2/2\sigma^2}.$$

Using (b) and (c), the density at (x, y) is

$$p(x)q(y) = s(r), \qquad r^2 = x^2 + y^2. \tag{3a.1.1}$$

Putting $x = 0$, we find that the functions s and q are proportional and putting $y = 0$, that s and p are proportional. Therefore the functional equation (3a.1.1) reduces to, writing $f(x) = \log[p(x)/p(0)]$,

$$f(x) + f(y) = f(r), \qquad r^2 = x^2 + y^2. \tag{3a.1.2}$$

Further, $f(x) = f(-x) = f(|x|)$, obtained by putting $y = 0$, $x = -x$ in (3a.1.2). Hence, if

$$x^2 = x_1{}^2 + x_2{}^2,$$
$$f(r) = f(y) + f(x_1) + f(x_2), \qquad r^2 = y^2 + x_1{}^2 + x_2{}^2,$$

and so in general

$$f(r) = f(x_1) + \cdots + f(x_k), \qquad \sum x_i{}^2 = r^2.$$

Choosing $k = n^2$ and putting $x = x_1 = \cdots = x_k$, we see that

$$f(nx) = n^2 f(x) \qquad \text{or} \qquad f(n) = n^2 f(1) \qquad \text{for} \quad x = 1.$$

For $x = m/n$ where m is an integer,

$$n^2 f\left(\frac{m}{n}\right) = f\left(n\,\frac{m}{n}\right) = f(m) = m^2 f(1) \qquad \text{or} \qquad f\left(\frac{m}{n}\right) = c\left(\frac{m}{n}\right)^2,$$

where $c = f(1)$, so that $f(x) = cx^2$ for all rational x, and because of continuity the relation is true for all x. Hence

$$p(x) = p(0)e^{cx^2}. \tag{3a.1.3}$$

For (3a.1.3) to be a probability density, c must be negative and may be written as $-1/2\sigma^2$. Integrating (3a.1.3) from $-\infty$ to ∞ and equating the result to unity we find $p(0) = 1/\sigma\sqrt{2\pi}$, so that

$$p(x) = \frac{1}{\sigma\sqrt{2\pi}} e^{-x^2/2\sigma^2} \tag{3a.1.4}$$

which is the famous normal distribution, $N(0, \sigma^2)$ with $E(x) = 0$ and $V(x) = \sigma^2$ already introduced in **2c.5**.

The joint p.d. of the errors X, Y is

$$p(x)p(y) = \frac{1}{\sigma^2 2\pi} e^{-(x^2+y^2)/2\sigma^2}. \tag{3a.1.5}$$

The error in any direction $(\cos\theta, \sin\theta)$ is $Z = X\cos\theta + Y\sin\theta$. To find the p.d. of Z, consider the transformation

$$z = x\cos\theta + y\sin\theta,$$
$$u = x\sin\theta - y\cos\theta.$$

The Jacobian of the transformation $D(z, u)/D(x, y) = 1$. The density (3a.1.5) transforms to

$$\frac{1}{\sigma^2 2\pi} e^{-(z^2+u^2)/2\sigma^2},$$

which shows that U and Z are independent and the p.d. of Z is

$$\frac{1}{\sigma\sqrt{2\pi}} e^{-z^2/2\sigma^2},$$

which is the required result.

The normal distribution with $E(X) = \mu$ and variance σ^2 has the density

$$N(x \mid \mu, \sigma^2) = \frac{1}{\sigma\sqrt{2\pi}} e^{-(x-\mu)^2/2\sigma^2}$$

with the c.f. (2b.4.11)

$$e^{it\mu - \sigma^2 t^2/2}. \tag{3a.1.6}$$

(ii) MAXWELL'S HYPOTHESIS. Maxwell arrived at the normal distribution in deriving the distribution of velocities of molecules under the following assumptions:

(a) *The components of velocity u, v, w in three orthogonal directions are independently distributed.*
(b) *The marginal distributions of u, v, w are the same.*
(c) *The phase space is isotropic, that is, the density of molecules with given velocity components is a function of total velocity and not the direction.*

If $f(\cdot)$ denotes the probability density of any component of velocity, the assumptions (a) to (c) lead to the functional equation

$$f(u)f(v)f(w) = g(V), \qquad V^2 = u^2 + v^2 + w^2, \tag{3a.1.7}$$

similar to the equation (3a.1.1). Hence $f(u)$ is of the form (3a.1.4) which is normal distribution for any single component of velocity and

$$g(V) = \text{const. } e^{-\alpha(u^2+v^2+w^2)}. \tag{3a.1.8}$$

It has been pointed out by Mayer and Mayer (1940) that the assumptions (a) and (b) are not necessary to establish the normal law.

Using the assumption (c), let $g(V)$ be the density at the point (u, v, w). From the principles of physics (preservation of kinetic energy under collision of molecules) it can be deduced that

$$g(V_1)g(V_2) = g(V_1')g(V_2'), \tag{3a.1.9}$$

when the velocities V_1, V_2, V_1', V_2' satisfy the relation

$$V_1^2 + V_2^2 = (V_1')^2 + (V_2')^2. \tag{3a.1.10}$$

The only solution of $g(V)$ under the condition (3a.1.9) with the restriction (3a.1.10) is (3a.1.8), that is, each component of velocity is normally distributed and the distributions of the components are independent.

(iii) LIMIT OF THE BINOMIAL (DE MOIVRE'S THEOREM). *Let r be the number of successes in n Bernoulli trials with probability π for success and define the random variable*

$$x = \frac{r - n\pi}{\sqrt{n\pi\phi}}, \qquad \phi = (1 - \pi).$$

Then the limit d.f. of x is normal.

The c.f. of r is, as shown in (2b.4.8),

$$E(e^{itr}) = (\phi + \pi e^{it})^n$$

and the c.f. $f_x(t)$ of x is

$$\begin{aligned}E(e^{itx}) &= E(e^{it(r-n\pi)/\sqrt{n\pi\phi}}) \\ &= e^{-itn\pi/\sqrt{n\pi\phi}}E(e^{itr/\sqrt{n\pi\phi}}) \\ &= e^{-itn\pi/\sqrt{n\pi\phi}}(\phi + \pi e^{it/\sqrt{n\pi\phi}})^n \\ &= (\phi e^{-it\pi/\sqrt{n\pi\phi}} + \pi e^{it\phi/\sqrt{n\pi\phi}})^n. \qquad (3a.1.11)\end{aligned}$$

Observing that

$$e^{iz} = 1 + (iz) + \cdots + (iz)^k/k! + o(z)^k,$$

and expanding the exponentials inside (3a.1.11), we find

$$f_x(t) = \left[1 - \frac{t^2}{2n} + o\left(\frac{t^2}{n}\right)\right]^n \to e^{-t^2/2}, \quad \text{as} \quad n \to \infty \qquad (3a.1.12)$$

which is the c.f. of $N(0, 1)$. Hence by the continuity theorem [(iv), **2c.4**], the limit d.f. of x is normal. In effect the result (3a.1.12) means that for large n, the d.f. of the binomial variable r can be approximated by the d.f. of a normal variable with mean $n\pi$ and variance $n\pi\phi$.

The result (3a.1.12) is, however, a special case of the *central limit theorem* established in **2c.5**, which says that the limiting distribution of the average of n identical and independently distributed random variables with finite mean and variance is normal in some suitably defined sense.

(iv) HAGEN'S HYPOTHESES (THEORY OF ERRORS). Hagen based his proof of the normal law of error under the following assumptions:

(a) *An error is the sum of a large number of infinitesimal errors, all of equal magnitude, due to different causes.*
(b) *The different components of errors are independent.*
(c) *Each component of error has an equal chance of being positive or negative.*

By assumption (c), each component of error takes the values $\pm\varepsilon$ with probability $\frac{1}{2}$ for each, so that the mean is zero and the variance is ε^2. If $x = \varepsilon_1 + \cdots + \varepsilon_n$ is the total error due to n independent components, then

$$\begin{aligned}E(x) &= E(\varepsilon_1) + \cdots + E(\varepsilon_n) = 0, \\ V(x) &= \sum V(\varepsilon_i) = n\varepsilon^2 = \sigma^2 \qquad \text{(say)}.\end{aligned}$$

Let us find the limiting distribution of x as $n \to \infty$ and $\varepsilon \to 0$ in such a way that σ^2 is finite and fixed. The c.f. of ε_j is

$$\tfrac{1}{2}(e^{it\varepsilon} + e^{-it\varepsilon}),$$

and that of $x = \varepsilon_1 + \cdots + \varepsilon_n$ is

$$\left(1 - \frac{t^2}{2!}\varepsilon^2 + \frac{t^4}{4!}\varepsilon^4 \cdots\right)^n = \left[1 - \frac{t^2}{2}\frac{\sigma^2}{n} + o\left(\frac{1}{n}\right)\right]^n \to e^{-t^2\sigma^2/2} \quad \text{as} \quad n \to \infty,$$

which is the c.f. of $N(0, \sigma^2)$.

(v) SUM OF n INDEPENDENT NORMAL VARIABLES. Let $x_1, \ldots, x_n$ be n independent normal variables with the mean and variance of x_j as μ_j and σ_j^2. The c.f. of the sum $x = x_1 + \cdots + x_n$ is the product of the c.f.'s of $x_1, \ldots, x_n$. Using the result (3a.1.6), we have

$$\prod_{j=1}^{n} e^{it\mu_j - \sigma_j^2 t^2/2} = e^{it\mu - \sigma^2 t^2/2}, \qquad \mu = \sum \mu_j, \qquad \sigma^2 = \sum \sigma_j^2,$$

so that *the sum is exactly normally distributed with the mean equal to sum of the means and the variance equal to sum of the variances. Hence the average* $\bar{x} = (x_1 + \cdots + x_n)/n \sim N(\mu, \sigma^2)$, *where*

$$\mu = \frac{\mu_1 + \cdots + \mu_n}{n}, \qquad \sigma^2 = \frac{\sigma_1^2 + \cdots + \sigma_n^2}{n^2}.$$

The central limit theorem establishes the asymptotic normality of the sum of random variables, independently to some extent of the nature of the distributions of the variables. If the sum of n independent variables is exactly normally distributed what can we say about the distribution of each variable? There is a famous theorem due to Cramer (1937) which asserts that *if the sum of n independent variables is normally distributed, then each variable is normally distributed.*

(vi) MAXIMUM ENTROPY FOR GIVEN MEAN AND VARIANCE. Consider a one-dimensional random variable with probability density (p.d.), $p(x)$ and define entropy of a distribution by

$$-\int p(x) \log p(x)\, dx. \tag{3a.1.13}$$

The significance of entropy and its role in the study of distribution of particles in physics are discussed in **3a.6**. Let us determine $p(x)$ defined on the range $(-\infty, \infty)$ such that (3a.1.13) is a maximum subject to the conditions

$$\left.\begin{aligned} &\int p(x)\, dx = 1, \qquad \int xp(x)\, dx = \mu, \\ &\int (x - \mu)^2 p(x)\, dx = \sigma^2, \end{aligned}\right\} \tag{3a.1.14}$$

where μ and σ^2 have given values.

It is shown in (1e.6.6) that for any two alternative densities p and q

$$-\int p \log \frac{p}{q}\, dx \leqslant 0.$$

Choosing $\log q = \alpha + \beta(x - \mu) + \gamma(x - \mu)^2$ where α, β, γ are determined such that (3a.1.14) is satisfied with q in the place of p, we find

$$-\int p \log p\, dx \leqslant -\int p \log q\, dx = -\int p[\alpha + \beta(x - \mu) + \gamma(x - \mu)^2]\, dx$$

$$= -(\alpha + \gamma\sigma^2)$$

by using the condition (3a.1.14) for p. Hence $-(\alpha + \gamma\sigma^2)$ is a fixed upper bound to entropy when $p(x)$ satisfies the conditions (3a.1.14). But this value is attained when p is chosen as

$$\log p = \alpha + \beta(x - \mu) + \gamma(x - \mu)^2$$

or

$$p = \exp[\alpha + \beta(x - \mu) + \gamma(x - \mu)^2]. \tag{3a.1.15}$$

We have to satisfy ourselves that α, β, γ can be determined to satisfy (3a.1.14). From the form of (3a.1.15) it is clear that the choice

$$\alpha = -\log(\sqrt{2\pi}\,\sigma), \qquad \beta = 0, \qquad \gamma = -\tfrac{1}{2}\sigma^2$$

satisfies (3a.1.14). *Thus the normal distribution is characterized by the property of having the maximum entropy for given mean and variance and the range* $(-\infty, \infty)$ *for the variable.* The reader may satisfy himself that the solution (3a.1.15) is essentially unique, since

$$\int p \log \frac{p}{q}\, dx = 0$$

implies that p and q are almost everywhere equal (see 1e.6.6).

(vii) SQUARE OF A NORMAL VARIABLE. Let $x \sim N(\mu, \sigma^2)$. The problem is to determine the distribution of $y = x^2$. For a given y there are two values of x, $(\pm x)$ so that the transformation is not one-to-one. The p.d.'s for x and $-x$ transform to (observing that $dy = 2x\, dx$)

$$(2\sigma\sqrt{2\pi y})^{-1} \exp\left[\frac{-(\sqrt{y} - \mu)^2}{2\sigma^2}\right]$$

and

$$(2\sigma\sqrt{2\pi y})^{-1} \exp\left[\frac{-(-\sqrt{y} - \mu)^2}{2\sigma^2}\right]$$

so that the total density of y is, using (3.2),

$$(2\sigma\sqrt{2\pi y})^{-1}e^{-(y+\mu^2)/2\sigma^2}[e^{\mu\sqrt{y}/\sigma^2}+e^{-\mu\sqrt{y}/\sigma^2}]$$

$$=(\sigma\sqrt{2\pi y})^{-1}e^{-(y+\mu^2)/2\sigma^2}\cosh\frac{\mu\sqrt{y}}{\sigma^2}. \quad (3a.1.16)$$

If $\mu = 0$, the p.d. of y is

$$G\left(y\,\middle|\,\frac{1}{2\sigma^2},\frac{1}{2}\right)=\frac{1}{\sigma\sqrt{2\pi}}e^{-y/2\sigma^2}y^{-1/2}, \qquad y>0$$
$$=0, \qquad y\leqslant 0 \quad (3a.1.17)$$

which is known as *Rayleigh* distribution in theoretical physics.

This is a particular case of a more general density function, $G(y|\alpha,p)=c.\,e^{-\alpha y}y^{p-1}$, which is called gamma distribution, $G(\alpha,p)$, whose properties are discussed in **3a.2.**

3a.2 Gamma Distribution

The general gamma distribution has the p.d.

$$G(x|\alpha,p)=\frac{\alpha^p}{\Gamma(p)}e^{-\alpha x}x^{p-1}, \qquad \alpha>0,\quad p>0$$
$$0<x<\infty. \quad (3a.2.1)$$

(i) *The* rth *raw moment is seen to be* $\Gamma(p+r)/\alpha^r\Gamma(p)$ *so that*

$$E(x)=p/\alpha \qquad \text{and} \qquad V(x)=p/\alpha^2.$$

(ii) *Let* $x_i \sim G(\alpha,p_i)$, $i=1,\ldots,k$ *be all independent. Then*

$$x_1+\cdots+x_k \sim G\left(\alpha, p=\sum p_i\right), \quad (3a.2.2)$$

that is, the gamma distribution has the reproductive property like the normal distribution but not for variations in both the parameters.

The c.f. of x_j is

$$\int e^{itx_j}G(x_j|\alpha,p_j)\,dx_j=\left(\frac{\alpha}{\alpha-it}\right)^{p_j}. \quad (3a.2.3)$$

Then the c.f. of $x_1+\cdots+x_k$ is $[\alpha/(\alpha-it)]^p$, which establishes the result. *The result is not true if the parameter* α *is not the same for all* x_i.

(iii) *Let* $x \sim G(\alpha,p_1)$ *and* $y \sim G(\alpha,p_2)$ *be independent. Then* $r=x+y$ *and* $f=x/y$ *are independently distributed. The distribution of* r *is* $G(\alpha,p_1+p_2)$ *and that of* f *is* $f(2p_1,2p_2)$ *as defined in* (3a.2.6).

The joint distribution of x and y is

$$\text{c. } e^{-\alpha x-\alpha y}x^{p_1-1}y^{p_2-1}\,dx\,dy. \tag{3a.2.4}$$

Make the transformation $x = r\sin^2\theta$, $y = r\cos^2\theta$, $(0 < r < \infty, 0 < \theta < \pi/2)$, so that $dx\,dy = 2r\sin\theta\cos\theta\,dr\,d\theta$. Expression (3a.2.4) transforms to

$$\text{c. } e^{-\alpha r}r^{p_1+p_2-1}(\sin\theta)^{2p_1-1}(\cos\theta)^{2p_2-1}\,dr\,d\theta,$$

which shows that r and θ and *hence r and f are independently distributed.* The distribution of r is $G(\alpha, p_1 + p_2)$ and θ alone is

$$\frac{2\Gamma(p_1+p_2)}{\Gamma(p_1)\Gamma(p_2)}(\sin\theta)^{2p_1-1}(\cos\theta)^{2p_2-1}\,d\theta, \qquad 0<\theta<\frac{\pi}{2}, \tag{3a.2.5}$$

with the constant supplied. Now

$$f = \frac{x}{y} = \frac{\sin^2\theta}{\cos^2\theta} = \tan^2\theta$$

$$df = 2\tan\theta\sec^2\theta\,d\theta = 2\sin\theta/\cos^3\theta.$$

By substituting in (3a.2.5), the p.d. of f which is the ratio of two independent gamma variables is

$$\frac{\Gamma(p_1+p_2)}{\Gamma(p_1)\Gamma(p_2)}\frac{f^{p_1-1}}{(1+f)^{p_1+p_2}}. \tag{3a.2.6}$$

A converse of proposition (iii) characterizing a gamma distribution is given by Khatri and Rao (1968b).

(iv) *Let $x \sim G(\alpha, p_1)$ and $y \sim G(\alpha, p_2)$. Then the p.d. of $g = x/(x+y)$ is*

$$\frac{\Gamma(p_1+p_2)}{\Gamma(p_1)\Gamma(p_2)}g^{p_1-1}(1-g)^{p_2-1}, \qquad 0<g<1, \tag{3a.2.7a}$$

which is called the beta distribution, $B(p_1, p_2)$.

The result (3a.2.7a) is obtained by making the substitution $g = \sin^2\theta$ in (3a.2.5).

In the special case when $p_1 = p_2 = \frac{1}{2}$, the beta distribution has the p.d.

$$\pi^{-1}g^{-1/2}(1-g)^{-1/2} \tag{3a.2.7b}$$

and the d.f.

$$2\pi^{-1}\sin^{-1}g^{1/2} \tag{3a.2.7c}$$

and, for this reason, (3a.2.7b) is called the *arc sine* density.

(v) WAITING TIME DISTRIBUTION. Consider the occurrence of events over a period of time subject to the condition that the occurrences in two non-overlapping intervals of time are independent. Let $q(t)$ be the probability that in a time interval t no event has occurred. If t_1 and t_2 are two consecutive intervals, then by the hypothesis of independence we have

$$q(t_1)q(t_2) = q(t_1 + t_2),$$

which if q is continuous has a solution of the form

$$q(t) = e^{-\lambda t}.$$

Hence the distribution function of waiting time for an event to happen is $1 - e^{-\lambda t}$. The probability density is

$$\frac{d}{dt}(1 - e^{-\lambda t}) = \lambda e^{-\lambda t} = G(t \mid \lambda, 1), \tag{3a.2.8}$$

which is a special case of the Gamma distribution (3a.2.1). It is easy to deduce that the waiting-time distribution for k events to happen is $G(\lambda, k)$.

Suppose the waiting-time distribution for an event to happen is as (3a.2.8). What is the probability $P(n, t)$ that n events happen in given time t? Let $t_1, \ldots, t_n$ be the time intervals at which the n events happen, and let no event happen in the period $\sum t_i$ to t. Then the probability for n events up to t is

$$\begin{aligned}\int_R \lambda^n e^{-\lambda(t_1 + \cdots + t_n)} e^{-\lambda(t - t_1 - \cdots - t_n)}\, dt_1 \cdots dt_n &= \int_R \lambda^n e^{-\lambda t}\, dt_1 \cdots dt_n \\ &= \lambda^n e^{-\lambda t} \int_R dt_1 \cdots dt_n\end{aligned}$$

over the region R: $t_1 + \cdots + t_n \leqslant t$. Integrating over R, we find that $P(n, t) = e^{-\lambda t}(\lambda t)^n/n!$ which is the Poisson probability.

The χ^2 Distribution. The special case of the gamma distribution with $\alpha = \frac{1}{2}$ and $p = k/2$, where k is an integer is called the χ^2 distribution on k *degrees of freedom* (D.F.). The density function is

$$\chi^2(x \mid k) = \frac{1}{2^{k/2}\Gamma(k/2)} e^{-x/2} x^{(k/2)-1}, \tag{3a.2.9}$$

from which we have the following results.

(vi) *Let* $x_i \sim \chi^2(k_i)$, $i = 1, \ldots, m$, *be independent; then*

$$\sum x_i \sim \chi^2\left(\sum k_i\right). \tag{3a.2.10}$$

(vii) *Let $x_1 \sim \chi^2(k_1)$, $x_2 \sim \chi^2(k_2)$ be independent, then the p.d. of $f = x_1/x_2$ is*

$$f(x \mid k_1, k_2) = \frac{\Gamma\left(\frac{k_1 + k_2}{2}\right)}{\Gamma\left(\frac{k_1}{2}\right)\Gamma\left(\frac{k_2}{2}\right)} \frac{x^{(k_1/2)-1}}{(1+x)^{(k_1+k_2)/2}}, \tag{3a.2.11}$$

the p.d. of $F = (x_1/k_1)/(x_2/k_2)$ is

$$F(x \mid k_1, k_2) = \frac{\left(\frac{k_1}{k_2}\right)^{k_1/2}}{\beta\left(\frac{k_1}{2}, \frac{k_2}{2}\right)} \frac{x^{(k_1/2)-1}}{\left(1 + \frac{k_1 x}{k_2}\right)^{(k_1+k_2)/2}}, \tag{3a.2.12}$$

the p.d. of $z = (1/2)\log F$ is

$$z(x \mid k_1, k_2) = \frac{2\left(\frac{k_1}{k_2}\right)^{k_1/2}}{\beta\left(\frac{k_1}{2}, \frac{k_2}{2}\right)} \frac{e^{k_1 x}}{\left(1 + \frac{k_1}{k_2} e^{2x}\right)^{(k_1+k_2)/2}}, \tag{3a.2.13}$$

and the p.d. of $g = x_1/(x_1 + x_2)$ is

$$B\left(g \,\middle|\, \frac{k_1}{2}, \frac{k_2}{2}\right) = \frac{\Gamma\left(\frac{k_1 + k_2}{2}\right)}{\Gamma\left(\frac{k_1}{2}\right)\Gamma\left(\frac{k_2}{2}\right)} g^{(k_1/2)-1}(1-g)^{(k_2/2)-1}. \tag{3a.2.14}$$

The result (3a.2.10) is the same as (3a.2.2), and (3a.2.11) is the same as (3a.2.6) except for change of notation. The result (3a.2.12) is obtained by changing the variable f to $F = fk_2/k_1$ in (3a.2.11). By transforming the variable F to $z = 2^{-1}\log F$ in (3a.2.12), we have (3a.2.13). (3a.2.14) is the same as (3a.2.7a).

The distribution (3a.2.12) is called Fisher's variance ratio distribution and (3a.2.13) is Fisher's z distribution (Fisher, 1924). These distributions are important in tests of significance.

(viii) *The c.f. of $\chi^2(k)$ is $(1 - 2it)^{-k/2}$, by substituting $\alpha = (1/2)$, $p = k/2$ in* (3a.2.3).

The first four moments of $\chi^2(k)$ are

$$\mu_1' = k \text{ (degrees of freedom)}$$

$$\mu_2 = 2k, \qquad \mu_3 = 8k, \qquad \mu_4 = 48k + 12k^2.$$

3a.3 Beta Distribution

The beta distribution involving two parameters γ, δ has the p.d.

$$B(x|\gamma, \delta) = [\beta(\gamma, \delta)]^{-1}x^{\gamma-1}(1-x)^{\delta-1}, \qquad 0 < x < 1, \tag{3a.3.1}$$

where $\beta(\gamma, \delta)$ is the beta function $\Gamma(\gamma)\Gamma(\delta)/\Gamma(\gamma+\delta)$.

(i) *The* sth *moment about the origin is*

$$[\beta(\gamma, \delta)]^{-1}\int x^{s+\gamma-1}(1-x)^{\delta-1}\,dx = \beta(s+\gamma, \delta)/\beta(\gamma, \delta) \tag{3a.3.2}$$

so that $E(x) = \gamma/(\gamma+\delta)$, $V(x) = \gamma\delta/(\gamma+\delta)^2(\gamma+\delta+1)$.

(ii) *If x and y are independent beta variables with parameters (γ_1, δ_1) and (γ_2, δ_2), the distribution of $u = xy$ is also beta with parameters $(\gamma_2, \delta_1+\delta_2)$ if $\gamma_1 = \gamma_2 + \delta_2$.*

To establish this, make the transformation $u = xy$, $v = x$. The region $(0 < x < 1, 0 < y < 1)$ in the (x, y) plane transforms to the region $(u < v < 1, 0 < u < 1)$ in the (u, v) plane. The Jacobian is $D(x, y)/D(u, v) = 1/v$ and

$$\begin{aligned}\text{c. } & x^{\gamma_1-1}(1-x)^{\delta_1-1}y^{\gamma_2-1}(1-y)^{\delta_2-1}\,dx\,dy \\ &= \text{c. } (1-v)^{\delta_1-1}u^{\gamma_2-1}(v-u)^{\delta_2-1}\,du\,dv.\end{aligned}$$

By integrating over v, from u to 1, the distribution of u is

$$\text{c. } u^{\gamma_2-1}(1-u)^{\delta_1+\delta_2-1}\,du. \tag{3a.3.3}$$

(iii) *In general the product of k beta variables with parameters $(\gamma_1, \delta_1), \ldots, (\gamma_k, \delta_k)$ such that $\gamma_i = \gamma_{i+1} + \delta_{i+1}$, $i = 1, \ldots, k-1$, is also a beta variable with parameters $(\gamma_k, \delta_1 + \cdots + \delta_k)$.*

(iv) A special case of the beta distribution is the rectangular distribution $B(x|1, 1)\,dx = dx$, where the density function is unity. This distribution arises in a natural way when we consider any stochastic variable u with the probability differential $f(u)\,du$ and wish to find the distribution of the variable x defined by

$$x = \int_{-\infty}^{u} f(u)\,du, \tag{3a.3.4}$$

which represents the d.f. of u. Since $dx = f(u)\,du$, *the p.d. of x is simply* 1.

(v) Pearson's p_λ Statistic. Let x have a rectangular distribution. If $y = -2\log_e x$, then $dy = (-2/x)\,dx$ and therefore the distribution of y is $2^{-1}e^{-y/2}\,dy$, which is $G(\frac{1}{2}, 1)$ or $\chi^2(2)$ as defined in (3a.2.9). *If* $x_1, \ldots, x_k$

are k probabilities (3a.3.4) *associated with k independent observations* $u_1, \ldots,$ u_k *from k possibly different populations, then the statistic*

$$P_\lambda = -2 \log_e x_1 - \cdots - 2 \log_e x_k, \tag{3a.3.5}$$

being the sum of k independent χ^2 *variates each on* 2 *D.F., is itself distributed as* χ^2 *on* 2*k D.F.*

This distribution is useful in combining several independent significance tests of statistical hypotheses.

3a.4 Cauchy Distribution

Consider a radioactive source from which α-particles are emanating and let us find the distribution of particles hitting a two-dimensional screen. Let the screen be at a perpendicular distance δ from the source. Consider orthogonal axes (x', y') through the source in a plane parallel to the screen. The direction in which an α particle emanates may be specified by two coordinates—the angle ϕ of the direction with the perpendicular to the plane and the angle θ the projection of the direction on (x', y') plane makes with the x' axis. We assume that the statement, "the direction in which an α particle emanates is random" is equivalent to, "that equal areas on a sphere round the source have the same probability of receiving an α particle." Or, in other words, the probability that a particle hits the region defined by $(\theta, \theta + d\theta)$, $(\phi, \phi + d\phi)$ on the sphere is

$$\text{c.} \sin \phi \, d\theta \, d\phi, \tag{3a.4.1}$$

which is the element of area on the surface of the sphere. What we wish to find is the joint distribution of the coordinates (x, y) of hits by particles, in reference to (orthogonal) axes with an arbitrary origin, on the screen. In reference to the (x, y) axes, let the coordinates of the foot of the perpendicular from the source to the screen be (μ_1, μ_2). Then we have the relations (δ being the length of the perpendicular), assuming that the x and y axes are parallel to the x' and y' axes,

$$\begin{aligned} x &= \mu_1 + \delta \tan \phi \cos \theta \\ y &= \mu_2 + \delta \tan \phi \sin \theta \\ dx\, dy &= b^2 \sin \phi \cos^{-3} \phi \, d\theta \, d\phi. \end{aligned} \tag{3a.4.2}$$

The substitution in (3a.4.1) yields the joint distribution of (x, y)

$$\text{c.} \, [\delta^2 + (x - \mu_1)^2 + (y - \mu_2)^2]^{-3/2} \, dx \, dy \tag{3a.4.3}$$

which may be called a bivariate Cauchy distribution. Integrating with respect to y the marginal p.d. of x alone, with range $(-\infty, \infty)$, is

$$C(x\mid\mu, \delta) = \frac{c\,dx}{\delta^2 + (x-\mu)^2}, \qquad c = \frac{\delta}{\pi}, \tag{3a.4.4}$$

where μ is written for μ_1, which is known as the one-dimensional Cauchy distribution with a location parameter μ and scale parameter δ. The same distribution (3a.4.4) arises when we consider a one-dimensional barrier and a point source at a perpendicular distance δ, the variable being the distance measured from an arbitrary origin of a point where the α particle hits the barrier.

The distribution (3a.4.4) is symmetrical with modal value at $x = \mu/\delta$. The expected value of x is

$$E(x) = \frac{\delta}{\pi}\int_{-\infty}^{+\infty} \frac{x\,dx}{\delta^2 + (x-\mu)^2},$$

but the integral does not exist, although it has the principal value

$$\lim_{m\to\infty} \frac{\delta}{\pi}\int_{-m}^{+m} \frac{x\,dx}{\delta^2 + (x-\mu)^2} = \mu.$$

The second moment is infinite. We thus have an example of a continuous distribution for which the mean and variance do not exist.

The c.f. of (3a.4.4) is

$$e^{it\mu}e^{-|t\delta|}. \tag{3a.4.5}$$

The c.f. of the mean of n independent observations from n Cauchy populations with location parameters $\mu_1, \ldots, \mu_n$ and same scale parameter δ is

$$e^{it\mu}e^{-|t\delta|}, \tag{3a.4.6}$$

where $\mu = (\mu_1 + \cdots + \mu_n)/n$. Hence the mean of n observations has again a Cauchy distribution with the location parameter as the average of individual location parameters. *For observations from the same Cauchy population, we see that the mean of any number of observations has the same distribution as any single observation.*

Thus, the law of large numbers of **2c.3** describing the behaviour of the mean as the number of observations increases does not hold in the case of the Cauchy distibution.

3a.5 Student's t Distribution

(i) Let $y \sim N(0, 1)$ and $x \sim \chi^2(k)$ be independent variables. Student (1908) defined the statistic

$$t = \frac{y}{(x/k)^{1/2}} \tag{3a.5.1}$$

which is the ratio of a normal variable to the square root of an independent χ^2 variable divided by the D.F. The joint distribution of y and x is

$$\text{c. } e^{-y^2/2} e^{-x/2} x^{(k/2)-1}\, dy\, dx. \tag{3a.5.2}$$

By making the transformation to polar coordinates $(0 < r < \infty,\ -\pi/2 < \theta < \pi/2)$,

$$y = r \sin\theta, \quad x = r^2 \cos^2\theta, \quad dx\, dy = 2r^2 \cos\theta\, dr\, d\theta \tag{3a.5.3}$$

(3a.5.2) transforms to

$$\text{c. } e^{-r^2/2} r^k (\cos\theta)^{k-1}\, dr\, d\theta. \tag{3a.5.4}$$

The distribution of θ alone, which is seen to be independent of r, is

$$\text{c. } (\cos\theta)^{k-1}\, d\theta = \left[\beta\left(\frac{1}{2}, \frac{k}{2}\right)\right]^{-1} (\cos\theta)^{k-1}\, d\theta, \tag{3a.5.5}$$

thus supplying the constant to make the total integral unity. The statistic whose distribution is to be found is

$$t = \frac{\sqrt{k}\, y}{\sqrt{x}} = \sqrt{k} \tan\theta, \qquad (-\infty < t < \infty),$$

$$dt = \sqrt{k} \sec^2\theta\, d\theta = \sqrt{k}\left(1 + \frac{t^2}{k}\right) d\theta.$$

The expression (3a.5.5) transforms to

$$S(t \mid k)\, dt = \left[\sqrt{k}\, \beta\left(\frac{1}{2}, \frac{k}{2}\right)\right]^{-1} \left(1 + \frac{t^2}{k}\right)^{-(k+1)/2} dt, \tag{3a.5.6}$$

which is called *Student's t distribution on k degrees of freedom and is represented by* $S(k)$.

(ii) *Let* $y \sim N(\mu, \sigma^2)$ *and* $(x/\sigma^2) \sim \chi^2(k)$ *be independent. Then*

$$\frac{y - \mu}{(x/k)^{1/2}} \sim S(k). \tag{3a.5.7}$$

Result (3a.5.7) is obtained by observing that $(y - \mu)/\sigma \sim N(0, 1)$ and applying (3a.5.6) to the ratio of $[(y - \mu)/\sigma]$ to $(x/k\sigma^2)^{1/2}$.

(iii) *Let* $y \sim N(\mu, \sigma^2)$ *and* $(x/\sigma^2) \sim \chi^2(k)$ *be independent. Then the probability density of* $t = y/(x/k)^{1/2}$ *is*

$$S(t \mid k, \delta) = \frac{k^{k/2}}{\Gamma(k/2)} \frac{e^{-\delta^2/2}}{(k + t^2)^{(k+1)/2}} \sum_{s=0}^{\infty} \Gamma\left(\frac{k + s + 1}{2}\right)\left(\frac{\delta^s}{s!}\right)\left(\frac{2t^2}{k + t^2}\right)^{s/2}, \tag{3a.5.8}$$

where $\delta = \mu/\sigma$, *which is called the noncentral t distribution.*

The joint distribution of y and x is

$$c_1 e^{-(y-\mu)^2/2\sigma^2} e^{-x/\sigma^2} x^{(k/2)-1}\, dy\, dx,$$
$$c_1^{-1} = \sqrt{2\pi}\, \sigma\sigma^k 2^{k/2}\Gamma(k/2).$$

If we make the transformation as in (3a.5.3),

$$y = r \sin\theta, \qquad x = r^2 \cos^2\theta,$$

we see that the joint density of r and θ is

$$c_2\, e^{-(r^2 - 2\mu r \sin\theta)/2\sigma^2} r^k (\cos\theta)^{k-1}\, dr\, d\theta, \tag{3a.5.9}$$
$$c_2 = c_1 e^{-\mu^2/2\sigma^2} = c_1 e^{-\delta^2/2},$$

which shows that r and θ are not independent as in (3a.5.4). Let us expand $\exp(\mu r \sin\theta/\sigma^2)$ and write (3a.5.9) in an infinite series

$$c_2\, e^{-r^2/2\sigma^2}(\cos\theta)^{k-1} \sum \frac{(\mu \sin\theta)^s r^{k+s}}{s!\,\sigma^{2s}}. \tag{3a.5.10}$$

Integrating out term by term with respect to r in (3a.5.10) we obtain the density of θ as

$$2^{-1}c_2(\cos\theta)^{k-1} \sum \Gamma\left(\frac{k+s+1}{2}\right) 2^{(s+k+1)/2}\, \frac{(\delta \sin\theta)^s}{s!}. \tag{3a.5.11}$$

But $t = \sqrt{k} \tan\theta$. Transforming from θ to t and using the value of c_2 in (3a.5.9), we finally obtain the distribution of t as (3a.5.8).

The noncentral beta and f distributions are derived in Example 17 at the end of the chapter.

3a.6 Distributions Describing Equilibrium States in Statistical Mechanics

Consider a general space (for instance, phase space in physics) with the coordinate system $x_1, \ldots, x_m$ and define a density function $p(x_1, \ldots, x_m)$ which is the limit of ratio of number of particles (such as molecules) in a small volume Δv around the point $(x_1, \ldots, x_m)$ to Δv. A system may be said to be in equilibrium if the distribution of particles is as close as possible to equidistribution over the space consistent with the kinematic restrictions. For instance, a particle occupying the position $(x_1, \ldots, x_m)$ will have an energy $E(x_1, \ldots, x_m)$. One restriction that may be imposed is that the average energy per particle be a given constant c, which is expressed by the relation

$$\int Ep\, dv = c, \tag{3a.6.1}$$

where dv represents the volume element. Our aim is to determine p which is as close as possible to the uniform distribution subject to the condition (3a.6.1).

For this we need an index of closeness of a given distribution to a uniform distribution. One such measure with large values indicating a high degree of closeness,

$$-\int p \log p \, dv \tag{3a.6.2}$$

is called *entropy* of the system. The problem then, is one of maximizing (3a.6.2) subject to the condition (3a.6.1).

(i) *The optimum choice of* p, *for which* $-\int p \log p \, dv$ *is a maximum subject to* $\int Ep \, dv = c$ *is*

$$p = e^{\lambda E + \mu} = \alpha e^{\lambda E}, \tag{3a.6.3}$$

where α *is chosen such that* $\int p \, dv = 1$.

In (1e.6.6) we proved the inequality

$$\int p \log \frac{p}{q} \, dv \geqslant 0 \tag{3a.6.4}$$

for any two alternative densities, which implies

$$-\int p \log p \, dv \leqslant -\int p \log q \, dv = -\int p(\lambda E + \mu) \, dv = -(\lambda c + \mu), \tag{3a.6.5}$$

if we choose $\log q = \lambda E + \mu$ and make use of (3a.6.1) and the fact that integral of p is unity. Let λ and μ exist such that

$$\int q \, dv = 1 = \int e^{\lambda E + \mu} \, dv,$$

$$\int Eq \, dv = c = \int E e^{\lambda E + \mu} \, dv.$$

Then from (3a.6.5), it follows that the quantity $(-\lambda c - \mu)$ is a fixed upper bound to (3a.6.2), which is attained for the choice

$$p = \exp(\lambda E + \mu) = \alpha \exp(\lambda E).$$

Solution (3a.6.3) is known as the Maxwell-Boltzmann distribution. The constant λ is written $-1/kT$ where T is temperature and k is Boltzmann's constant. We shall consider a few applications.

Helly's Law for the Equilibrium of Sedimentation. Assume that the energy of a particle of mass m depends only on position coordinates. For example, let the particles be in the earth's gravitational field and have a potential energy $E = mgz$ at altitude z. The number of particles between heights z and $z + dz$, in an equilibrium state according to the distribution (3a.6.3), is

$$Ae^{-mgz/kT} \, dz. \tag{3a.6.6}$$

Thus, if the particles are in suspension in a liquid the number of particles must decrease exponentially with height, and this is actually observed.

Angular Distribution of Axis of Elementary Magnets in a Magnetic Field. The potential energy of a small magnet of moment p in a field H is given by

$$E = -pH \cos \theta,$$

where θ is the angle that the axis of diapole makes with the field. The space in this case is the surface of a sphere representing the direction cosines of the axis of diapole. The density at any point of sphere is

$$Ae^{pH \cos \theta / kT} = Ae^{\kappa \cos \theta}, \tag{3a.6.7}$$

writing κ for pH/kT. Observing that the surface element in polar coordinates is $\sin \theta \, d\theta \, d\phi$ we can evaluate the constant A by the relation

$$1 = \int_0^{\pi} \int_0^{2\pi} Ae^{\kappa \cos \theta} \sin \theta \, d\theta \, d\phi = A \frac{4\pi \sinh \kappa}{\kappa}. \tag{3a.6.8}$$

The density (3a.6.7) can then be written

$$\frac{\kappa}{4\pi \sinh \kappa} e^{\kappa \cos \theta}. \tag{3a.6.9}$$

This distribution is used by Fisher (1953), Watson (1956), and Watson and Williams (1956) in analyzing samples of directions of remanent magnetism in rocks.

Maxwell's Distribution of Velocities. Let x, y, z be position coordinates and u, v, w be velocities in three orthogonal directions of particles in ideal gas. The kinetic energy of a particle with coordinates (x, y, z, u, v, w) is $E = u^2 + v^2 + w^2$. Hence the probability differential corresponding to the optimum distribution (3a.6.3) of particles is

$$Ae^{-(u^2+v^2+w^2)/kT} \, du \, dv \, dw \, dx \, dy \, dz.$$

By integrating over x, y, z, the distribution of velocities alone is

$$Ae^{-(u^2+v^2+w^2)/kT} \, du \, dv \, dw, \tag{3a.6.10}$$

which shows that the velocities in the three directions are independently distributed and each component has a normal distribution. For further applications of (3a.6.3) the reader is referred to *Theoretical Physics* by Joos (1951).

We can try other measures of closeness to a uniform distribution such as

$$\frac{1}{1-\alpha} \log \int p^{\alpha} \, dv, \qquad \alpha > 0 \tag{3a.6.11}$$

as an alternative to (3a.6.2) and find p to maximize (3a.6.11) subject to the condition (3a.6.1). Instead of (3a.6.11) we may choose

$$\int p^\alpha \, dv \tag{3a.6.12}$$

and maximize or minimize according as $\alpha < 1$ or > 1. Using the mean value theorem to the integrand, we can write

$$\int (p^\alpha - q^\alpha) \, dv = \int (p - q)\alpha q^{\alpha - 1} \, dv + \int \alpha(\alpha - 1)(p - q)^2 r^{\alpha - 2} \, dv, \tag{3a.6.13}$$

where $r \in (p, q)$. Let us choose $q^{\alpha - 1} = (\lambda E + \mu)$ where λ and μ are such that from the condition (3a.6.1),

$$\int q \, dv = 1 = \int (\lambda E + \mu)^{1/(\alpha - 1)} \, dv$$

$$\int Eq \, dv = c = \int E(\lambda E + \mu)^{1/(\alpha - 1)} \, dv.$$

By substituting for $q^{\alpha - 1}$, the right-hand side of (3a.6.13) reduces to

$$\int \alpha(\alpha - 1)(p - q)^2 r^{\alpha - 2} \, dv \gtrless 0 \qquad \text{according as} \quad \alpha \gtrless 1,$$

which give $\int p^\alpha \, dv \lesseqgtr \int q^\alpha \, dv$ according as $\alpha \lesseqgtr 1$, for the special choice of q. Thus the extremum of $\int p^\alpha \, dv$ is attained when

$$p = (\lambda E + \mu)^{1/(\alpha - 1)}, \tag{3a.6.14}$$

which provides a family of equilibrium distributions alternative to (3a.6.3). The relevance of these distributions in actual physical situations is worth examining.

3a.7 Distribution on a Circle

There are many practical situations where the basic variable under observation is a direction in two dimensions or simply an angle. In such a case, the sample space consists of points on a unit circle, and we have to consider probability measures on Borel sets on the circumference of a circle. Such distributions are called circular distributions (CD). When the distribution is absolutely continuous with respect to Lebesgue measure on the circumference of the unit circle, it can be specified by its density function $f(x)$, where x is the angle measured from a chosen direction, which is a periodic function (with period 2π) satisfying

$$f(x) \geqslant 0, \qquad \int_0^{2\pi} f(x) \, dx = 1, \qquad 0 < x \leqslant 2\pi. \tag{3a.7.1}$$

Before discussing special cases of CD's, we consider some peculiarities of these distributions in general. Let x be the angle measured from a chosen direction and y be the angle measured from another direction at an angle c from the original direction. Then

$$\begin{aligned} x &= c + y, \qquad 0 < y \leqslant 2\pi - c \\ x &= y - 2\pi + c, \qquad 2\pi - c < y \leqslant 2\pi \end{aligned} \tag{3a.7.2}$$

so that $E(x) = c + E(y)$ need not hold as in the case of translation on an infinite straight line. This shows that $E(x)$ does not indicate any "preferred direction" or a central value of the distribution on the circle.

Thus if we have independent observations of directions of flights of birds (in studying bird migrations) the average of the angles may not be a meaningful estimate of the direction of their habitat.

Similarly, it can be shown that the variance of x in (3a.7.1) is not invariant for a shift of the origin. Thus the usual measures of location and dispersion are not meaningful in summarizing data from a CD.

More meaningful measures in such cases can, however, be constructed. Let us represent a point on the circle by the pair of direction cosines, $(\cos x, \sin x)$ corresponding to angle x. It may be seen that if x and y are as defined in (3a.7.2), then $\cos x = \cos(c + y)$ and $\sin x = \sin(c + y)$ giving

$$E(\cos x) = E(\cos \overline{c + y}), \qquad E(\sin x) = E(\sin \overline{c + y}) \tag{3a.7.3}$$

from which it easily follows

$$\cot^{-1} \frac{E(\cos x)}{E(\sin x)} = c + \cot^{-1} \frac{E(\cos y)}{E(\sin y)}. \tag{3a.7.4}$$

Then a meaningful measure of preferred direction (angle) is

$$\cot^{-1} \frac{E(\cos x)}{E(\sin x)}. \tag{3a.7.5}$$

From (3a.7.3), it also follows that

$$R^2 = [E(\cos x)]^2 + [E(\sin x)]^2 = [E(\cos y)]^2 + [E(\sin y)]^2 \tag{3a.7.6}$$

is independent of c. Further, $0 \leqslant R^2 \leqslant 1$, where the value 0 obtains when x is uniformly distributed (the dispersion is a maximum) over the circle and the value 1 when x is concentrated at a single point (dispersion is a minimum). Then we may use $\sqrt{1 - R^2}$ as a measure of dispersion.

Thus, if $x_1, \ldots, x_n$ are n observed angles from any zero direction, the preferred direction may be estimated by the formula

$$x = \cot^{-1} \frac{\sum \cos x_i}{\sum \sin x_i} \tag{3a.7.7}$$

and the dispersion by $\sqrt{1-r^2}$ where

$$r^2 = \left(\frac{\sum \cos x_i}{n}\right)^2 + \left(\frac{\sum \sin x_i}{n}\right)^2 \tag{3a.7.8}$$

Let $\overrightarrow{OP}_1, \ldots, \overrightarrow{OP}_n$ be unit vectors of n observed points on a unit circle (corresponding to the angles $x_1, \ldots, x_n$) and $\overrightarrow{OP}$ the resultant vector obtained by applying the usual parallelogram law on n vectors. Then (3a.7.7) is simply the direction of $\overrightarrow{OP}$, and r as defined by (3a.7.8) is its length divided by n. Thus (3a.7.7) and (3a.7.8) constitute a natural way of summarising directional data.

Uniform Distribution. As a basic CD, we may consider uniform distribution on a circle with the density

$$f(x) = \frac{1}{2\pi}, \qquad 0 < x \leqslant 2\pi \tag{3a.7.9}$$

where all directions are equally likely. There is no preferred direction, and the measure (3a.7.5) is not defined.

Circular Normal Distribution. Among the unimodal CD's, the one that has found many applications is the circular normal distribution, or what is called the von Mises distribution with the density

$$f(x) = [2\pi I_0(\kappa)]^{-1} \exp[\kappa \cos(x-\beta)] \tag{3a.7.10}$$
$$0 < x \leqslant 2\pi, \quad 0 \leqslant \beta \leqslant 2\pi, \quad 0 \leqslant \kappa < \infty,$$

where $I_0(x)$ is Bessel function of purely imaginary argument. The more general function with index n, $I_n(x)$ is defined by the series

$$I_n(x) = \sum_{r=0}^{\infty} \frac{(x/2)^{n+2r}}{r!\,(n+r)!}. \tag{3a.7.11}$$

Using the density function (3a.7.10)

$$E(\cos x) = \rho \cos \beta, \qquad E(\sin x) = \rho \sin \beta, \tag{3a.7.12}$$
$$\rho = I_1(\kappa)/I_0(\kappa).$$

Then

$$\cot^{-1} \frac{E(\cos x)}{E(\sin x)} = \beta \tag{3a.7.13}$$

which shows that β is the parameter representing the preferred direction. The dispersion $\sqrt{1-\rho^2}$ depends only on the other parameter κ, small values of κ representing higher dispersion. Then we may interpret κ as a measure of dispersion.

If we have n independent observations $x_1, \ldots, x_n$, then β and κ are estimated by the formulae

$$\cot \beta = \frac{\sum \cos x_i}{\sum \sin x_i}, \qquad \frac{I_1(\kappa)}{I_0(\kappa)} = r \tag{3a.7.14}$$

where r is as defined in (3a.7.8).

Wrapped-up Distributions. One can obtain a CD by "wrapping a distribution on a straight line around the unit circle." For instance, wrapping an $N(0, \sigma^2)$ we obtain the wrapped normal with the density

$$f(x) = \frac{1}{2\pi} + \frac{1}{\pi} \sum_{n=1}^{\infty} \cos nx \exp(-n^2\sigma^2/2). \tag{3a.7.15}$$

Similarly, choosing the Cauchy density $\delta/\pi(\delta^2 + x^2)$, $-\infty < x < \infty$, we obtain the wrapped Cauchy with the density

$$f(x) = \frac{1 - \sigma^2}{2\pi[1 + \sigma^2 - 2\sigma \cos x]} \tag{3a.7.16}$$

where $0 < x \leqslant 2\pi$ and $0 \leqslant \sigma = e^{-\delta} \leqslant 1$.

Bimodal Distributions. When the circular data have more than one preferred direction, we have to consider multimodal, circular distributions. For instance, the density

$$f(x) = [2\pi I_0(\kappa)]^{-1} \exp[\kappa \cos 2(x - \beta)] \tag{3a.7.17}$$

provides a distribution with an axial symmetry with modes at β and $\pi + \beta$. Another interesting bimodal density is

$$f(x) = \frac{(1 - \rho^2)^{1/2}}{2\pi(1 - \rho \sin 2x)}. \tag{3a.7.18}$$

It is easily shown that if (X_1, X_2) has bivariate normal density with means zero, unit variances, and correlation ρ, then $x = \tan^{-1}(X_1/X_2)$ has the density (3a.7.18).

Similarly, one can consider distributions on a sphere when directions are observed in three dimensions. An example and characterization of such a distribution is already considered in (3a.6.9).

A detailed treatment of directional data is contained in Mardia (1972). Appendix **2.2** of this book contains a table for obtaining κ for given r in (3a.7.14). See also J. S. Rao and Sengupta (1972) for applications.

3b SAMPLING DISTRIBUTIONS

3b.1 Definitions and Results

The method employed to derive sampling distributions is transformation of variables. Usually, a linear transformation $\mathbf{Y} = \mathbf{AX}$ is made from $\mathbf{X} \to \mathbf{Y}$ (column vectors of the same size) by using a nonsingular matrix $\mathbf{A}$. The Jacobian of transformation $D\mathbf{Y}/D\mathbf{X}$ is $|\mathbf{A}|$ with a positive sign, so that the differential elements are connected by the relation

$$dy_1\, dy_2 \cdots dy_n = |\mathbf{A}|\, dx_1\, dx_2 \cdots dx_n \tag{3b.1.1}$$

which we will represent by $d\mathbf{Y} = |\mathbf{A}|\, d\mathbf{X}$. When $\mathbf{A}$ is an orthogonal matrix, $|\mathbf{A}| = 1$.

Two types of transformation matrices are generally used. One is a completely orthogonal matrix. It is shown in **1b.4** that if $\mathbf{A}$ is an orthogonal matrix, $\mathbf{Y} = \mathbf{AX}$ transforms quadratic forms in a simple way

$$\mathbf{X}'\mathbf{X} \to \mathbf{Y}'\mathbf{Y}$$

$$(\mathbf{X} - \boldsymbol{\mu})'(\mathbf{X} - \boldsymbol{\mu}) \to (\mathbf{Y} - \boldsymbol{\eta})'(\mathbf{Y} - \boldsymbol{\eta}), \qquad \boldsymbol{\eta} = \mathbf{A}\boldsymbol{\mu} \tag{3b.1.2}$$

that is, preserving distances.

Another type is the partitioned matrix (where $\mathbf{A}_i$ is $n_i \times n$ and $\sum n_i = n$)

$$\mathbf{A} = \begin{pmatrix} \mathbf{A}_1 \\ \cdot \\ \cdot \\ \mathbf{A}_k \end{pmatrix}$$

with $\mathbf{A}_i \mathbf{A}_j' = \mathbf{0}$, $i \neq j$, so that the submatrices are orthogonal to one another but may not themselves be orthogonal. The transformation $\mathbf{Y} = \mathbf{AX}$ is written using the submatrices

$$\mathbf{Y}_1 = \mathbf{A}_1\mathbf{X}, \ldots, \mathbf{Y}_k = \mathbf{A}_k\mathbf{X},$$

where $\mathbf{Y}_1, \ldots, \mathbf{Y}_k$ are exclusive subsets of new variables. Under such a transformation it is shown in [(vii), **1c.1**], that

$$\mathbf{X}'\mathbf{X} \to \mathbf{Y}_1'\mathbf{B}_1\mathbf{Y}_1 + \cdots + \mathbf{Y}_k'\mathbf{B}_k\mathbf{Y}_k$$

$$(\mathbf{X} - \boldsymbol{\mu})'(\mathbf{X} - \boldsymbol{\mu}) \to (\mathbf{Y}_1 - \boldsymbol{\eta}_1)'\mathbf{B}_1(\mathbf{Y}_1 - \boldsymbol{\eta}_1) + \cdots + (\mathbf{Y}_k - \boldsymbol{\eta}_k)'\mathbf{B}_k(\mathbf{Y}_k - \boldsymbol{\eta}_k), \tag{3b.1.3}$$

where $\mathbf{B}_i = (\mathbf{A}_i\mathbf{A}_i')^{-1}$ and $\boldsymbol{\eta}_i = \mathbf{A}_i\boldsymbol{\mu}$. Result (3b.1.3) shows that the transformed expression splits up into quadratic forms in exclusive subsets of the new variables. When $\mathbf{A}$ is fully orthogonal, that is, when each row is orthogonal to every other row, full splitting occurs, as in (3b.1.2).

We also make use of some basic notions of vector spaces. A linear manifold generated by the columns of a matrix **B** is represented by $\mathcal{M}(\mathbf{B})$. It is easy to see that $\mathcal{M}(\mathbf{B}) = \mathcal{M}(\mathbf{BB}')$.

If **H** is a matrix of rank k and $\mathcal{M}(\mathbf{H}) \subset \mathcal{M}(\mathbf{B}) = \mathcal{M}(\mathbf{BB}')$, then there exists a matrix **C** such that

$$\mathbf{H} = \mathbf{BB}'\mathbf{C} \quad \text{and} \quad k = \text{rank } \mathbf{B}'\mathbf{C}. \tag{3b.1.4}$$

It may also be noted that a set of linear equations $\mathbf{GX} = \mathbf{F}$ admits a solution if rank $\mathbf{G}$ = number of rows in **G**, whatever **F** may be.

We also record a few other results to make matters simple in the rest of our discussions. Let us represent

$$\int_T f(x_1, \ldots, x_n)\, dx_1 \cdots dx_n = h(T)$$

$$T = g(x_1, \ldots, x_n),$$

as the result of transforming the variables $x_1, \ldots, x_n$ to another set involving T as one of the new variables and integrating out with respect all the other new variables and dropping the differential dT. Then $h(T)$ is the probability density function of T.

We also use the following notations of density which have already been introduced in Section **3a**.

$$N(x \mid \mu, \sigma^2) = (\sigma\sqrt{2\pi})^{-1} e^{-(x-\mu)^2/2\sigma^2}, \qquad -\infty < x < \infty$$

$$G(x \mid \alpha, p) = [\alpha^p/\Gamma(p)] e^{-\alpha x} x^{p-1}, \qquad 0 < x < \infty$$

$$B(x \mid p, q) = [\beta(p, q)]^{-1} x^{p-1}(1-x)^{q-1}, \qquad 0 < x < 1$$

$$\chi^2(x \mid k) = [2^{k/2}\Gamma(k/2)]^{-1} e^{-x/2} x^{(k/2)-1}, \qquad 0 < x < \infty$$

$$S(x \mid k) = \left[\sqrt{k}\beta\left(\frac{1}{2}, \frac{k}{2}\right)\right]^{-1} \left(1 + \frac{x^2}{k}\right)^{-(k+1)/2}, \qquad -\infty < x < \infty$$

$$f(x \mid k_1, k_2) = \left[\beta\left(\frac{k_1}{2}, \frac{k_2}{2}\right)\right]^{-1} \frac{x^{(k_1/2)-1}}{(1+x)^{(k_1+k_2)/2}}, \qquad 0 < x < \infty$$

$$F(x \mid k_1, k_2) = \left[\beta\left(\frac{k_1}{2}, \frac{k_2}{2}\right)\right]^{-1} \left(\frac{k_1}{k_2}\right)^{k_1/2} \frac{x^{(k_1/2)-1}}{\left(1 + \frac{k_1 x}{k_2}\right)^{(k_1+k_2)/2}}, \qquad 0 < x < \infty$$

From what have been proved in Section **3a**, we have the following:

$$\text{(a)} \quad \int_{x = x_1 + \cdots + x_n} G(x_1 \mid \alpha, p_1) \cdots G(x_n \mid \alpha, p_n)\, dx_1 \cdots dx_n = G(x \mid \alpha, p_1 + \cdots + p_n) \tag{3b.1.5}$$

(b) $$\int_{g=x_1/(x_1+x_2)} G(x_1|\alpha, p)G(x_2|\alpha, q)\, dx_1\, dx_2 = B(g|p, q) \tag{3b.1.6}$$

(c) $$\int_{x=x_1/x_2} G(x_1|\alpha, p)G(x_2|\alpha, q)\, dx_1\, dx_2 = f(x|2p, 2q) \tag{3b.1.7}$$

(d) $$\int_{t=x_1/\sqrt{x_2/k}} N(x_1|0, 1)\chi^2(x_2|k)\, dx_1\, dx_2 = S(t|k) \tag{3b.1.8}$$

(e) $$\int_{x=x_1+\cdots+x_n} \chi^2(x_1|k_1)\cdots\chi^2(x_n|k_n)\, dx_1\cdots dx_n = \chi^2(x|k_1+\cdots+k_n) \tag{3b.1.9}$$

(f) $$\int_{x=x_1/x_2} \chi^2(x_1|k_1)\chi^2(x_2|k_2)\, dx_1\, dx_2 = f(x|k_1, k_2) \tag{3b.1.10}$$

(g) $$\int_{x=x_1k_2/x_2k_1} \chi^2(x_1|k_1)\chi^2(x_2|k_2)\, dx_1\, dx_2 = F(x|k_1, k_2) \tag{3b.1.11}$$

(h) $$\int_{g=x_1/(x_1+x_2)} \chi^2(x_1|k_1)\chi^2(x_2|k_2)\, dx_1\, dx_2 = B\left(g\,\middle|\,\frac{k_1}{2}, \frac{k_2}{2}\right) \tag{3b.1.12}$$

(i) $$\int_{r=x_1/(x_2+x_1^2)^{1/2}} N(x_1|0, 1)\chi^2(x_2|k)\, dx_1\, dx_2 = \sqrt{\pi}\left[\Gamma\left(\frac{k}{2}\right)\Big/\Gamma\left(\frac{k+1}{2}\right)\right](1-r^2)^{(k/2)-1} \tag{3b.1.13}$$

(j) $$\int_{y=x^2/\sigma^2} N(x|0, \sigma^2)\, dx = \chi^2(y|1) = G(y|\tfrac{1}{2}, \tfrac{1}{2}) \tag{3b.1.14}$$

(k) $$\int_{y=x^2/\sigma^2} N(x|\mu, \sigma^2)\, dx = e^{-\mu^2/2\sigma^2}\sum \frac{1}{r!}\left(\frac{\mu^2}{2\sigma^2}\right)^r G(y|\tfrac{1}{2}, r+\tfrac{1}{2}) \tag{3b.1.15}$$

Results (j) and (k) are the same as (3a.1.17) and (3a.1.16).

3b.2 Sum of Squares of Normal Variables

Let $x_i \sim N(\mu_i, \sigma^2)$, $i = 1, \ldots, k$ be independent variables. What is the distribution of $x = \sum x_i^2/\sigma^2$?

If $\mu_i = 0$ for all i, $x_i^2/\sigma^2 \sim \chi^2(1)$ and by an application of the reproductive property of χ^2 under convolution (3b.1.9), $x = \sum x_i^2/\sigma^2 \sim \chi^2(k)$, or $\sum x_i^2 \sim \sigma^2\chi^2(k)$.

If $\mu_i \neq 0$ for at least one i, make the orthogonal transformation $\mathbf{Y} = \mathbf{BX}$ such that the first row of $\mathbf{B}$ is $(\mu_1/\mu, \ldots, \mu_k/\mu)$ where $\mu^2 = \mu_1^2 + \cdots + \mu_k^2$. Since the transformation is orthogonal, then

$$\sum (x_i - \mu_i)^2 = (y_1 - \mu)^2 + y_2^2 + \cdots + y_k^2,$$

since $(\mu_1, \ldots, \mu_n) \to (\mu, 0, \ldots, 0)$, so that the joint density of $x_1, \ldots, x_k$

$$c_1 \exp[-\sum (x_i - \mu_i)^2/2\sigma^2]$$

transforms to

$$c_1 \exp\{-[(y_1 - \mu)^2 + y_2^2 + \cdots + y_k^2]/2\sigma^2\},$$

which shows that $y_1, \ldots, y_k$ are independent and normal, with $E(y_1) = \mu$ and $E(y_i) = 0$, $i > 1$. Since $\mathbf{X}'\mathbf{X} \to \mathbf{Y}'\mathbf{Y}$, the given statistic reduces to

$$\begin{aligned} x &= \frac{y_1^2}{\sigma^2} + \frac{y_2^2 + \cdots + y_k^2}{\sigma^2}, \\ &= u + v. \end{aligned}$$

Thus $v \sim \chi^2(k-1)$ or $G[\frac{1}{2}, (k-1)/2]$ and u has the distribution (3b.1.15) independently of v. Therefore the joint distribution of v and u is

$$e^{-\mu^2/2\sigma^2} \sum \frac{1}{r!} \left(\frac{\mu^2}{2\sigma^2}\right)^r G\left(u \,\middle|\, \frac{1}{2}, r + \frac{1}{2}\right) du\, G\left(v \,\middle|\, \frac{1}{2}, \frac{k-1}{2}\right) dv. \quad (3b.2.1)$$

Applying (3b.1.5) to each term of the series in the product (3b.2.1), we obtain the distribution of $x = u + v = \sum x_i^2/\sigma^2$ as

$$e^{-\mu^2/2\sigma^2} \sum \frac{1}{r!} \left(\frac{\mu^2}{2\sigma^2}\right)^r G\left(x \,\middle|\, \frac{1}{2}, r + \frac{k}{2}\right) dx, \quad (3b.2.2)$$

which is called noncentral χ^2 on k degrees of freedom with the noncentrality parameter, $\lambda = (\mu^2/\sigma^2)$ and denoted by $\chi^2(k, \lambda)$. Note that the noncentrality parameter is *the value of the quadratic form when expected values are substituted for the variables.*

Since the characteristic function (c.f.) of $G(\alpha, p)$ is known [see (ii), **3a.2**] the determination of the c.f. of $\chi^2(k, \lambda)$ reduces to a summation of c.f.'s of gamma distributions with coefficients as in (3b.2.2). Hence the c.f. of $\chi^2(k, \lambda)$ is

$$e^{-\lambda/2} \sum \frac{1}{r!} \left(\frac{\lambda}{2}\right)^r (1 - 2it)^{-r-k/2} = (1 - 2it)^{-k/2} e^{\lambda it/(1-2it)}. \quad (3b.2.3)$$

From (3b.2.3) we observe, at once, *that the noncentral χ^2 satisfies the reproductive property with respect to k and λ.* Thus if $x_i \sim \chi^2(k_i, \lambda_i)$, $i = 1, \ldots, m$ and are all independent, then

$$x = x_1 + \cdots + x_m \sim \chi^2(\sum k_i, \sum \lambda_i). \quad (3b.2.4)$$

3b.3 Joint Distribution of the Sample Mean and Variance

Let $x_i \sim N(\mu, \sigma^2)$, $i = 1, \ldots, n$ be all independent. To find the joint distribution of $\bar{x} = (x_1 + \cdots + x_n)/n$ and $s^2 = \sum (x_i - \bar{x})^2/(n-1)$.

Make an orthogonal transformation $\mathbf{Y} = \mathbf{BX}$ such that the first row of $\mathbf{B}$ is $(n^{-1/2}, \ldots, n^{-1/2})$. The following then are true.

(a) $y_1 = \sqrt{n}\bar{x}$.
(b) $y_1^2 + \cdots + y_n^2 = x_1^2 + \cdots + x_n^2 = \sum (x_i - \bar{x})^2 + n\bar{x}^2$
$y_2^2 + \cdots + y_n^2 = (n-1)s^2$.
(c) $(y_1 - \sqrt{n}\mu)^2 + y_2^2 + \cdots + y_n^2 = (x_1 - \mu)^2 + \cdots + (x_n - \mu)^2$.

Because of (c), the density

$$\text{c. } e^{-\Sigma(x_i - \mu)^2/2\sigma^2}$$

transforms to

$$\text{c. } e^{-[(y_1 - \mu\sqrt{n})^2 + y_2^2 + \cdots + y_n^2]/2\sigma^2},$$

so that $y_1, \ldots, y_n$ are independently distributed. Hence

$$\sqrt{n}\bar{x} = y_1 \sim N(\sqrt{n}\mu, \sigma^2)$$

$$\frac{(n-1)s^2}{\sigma^2} = \frac{y_2^2 + \cdots + y_n^2}{\sigma^2} \sim \chi^2(n-1),$$

and are independently distributed. Making a change of scale, the joint distribution of $\bar{x}$ and s^2 is

$$N\left(\bar{x}\middle|\mu, \frac{\sigma^2}{n}\right) d\bar{x}\, \chi^2\left(\frac{\overline{n-1}\, s^2}{\sigma^2}\middle| n-1\right) \frac{n-1}{\sigma^2} ds^2. \tag{3b.3.1}$$

Defining $t = \sqrt{n}(\bar{x} - \mu)/s$, we find that it is the ratio of $\sqrt{n}(\bar{x} - \mu)/\sigma \sim N(0, 1)$ and s/σ such that $s^2/\sigma^2 \sim (n-1)^{-1}\chi^2(n-1)$. Hence, by applying Student's distribution (3b.1.8), $t \sim S(n-1)$.

Distribution of Linear Functions of Normal Variables. Let $x_i \sim N(\mu_i, \sigma^2)$ $i = 1, \ldots, n$ be n independent variables. To find the joint distribution of $p < n$ independent linear functions of $x_1, \ldots, x_n$, which can be written in matrix notation, $\mathbf{Z} = \mathbf{BX}$, where $\mathbf{Z}' = (z_1, \ldots, z_p)$ and $\mathbf{B}$ is a $p \times n$ matrix of rank p.

Let $\boldsymbol{\mu}' = (\mu_1, \ldots, \mu_n)$. Transform $\mathbf{X}$ to $\mathbf{U} + \boldsymbol{\mu}$ so that the new variables $\mathbf{U}$ have zero mean. In terms of $\mathbf{U}$, $\mathbf{Z} = \mathbf{B}(\mathbf{U} + \boldsymbol{\mu})$ or $\mathbf{Z} - \mathbf{B}\boldsymbol{\mu} = \mathbf{BU}$. We shall find the distribution of $\mathbf{V} = \mathbf{BU}$ starting from the distribution of $\mathbf{U}$ observing that the distribution of $\mathbf{Z}$ is obtained by replacing $\mathbf{V}$ by $\mathbf{Z} - \mathbf{B}\boldsymbol{\mu}$.

Consider the case where the rows of $\mathbf{B}$ are orthonormal and thus the matrix $\mathbf{B}$ can be completed by $n - p$ orthonormal vectors leading to an orthogonal matrix $\mathbf{C}$. Let us transform from $(u_1, \ldots, u_n)$ to $(v_1, \ldots, v_p, v_{p+1}, \ldots, v_n)$ using $\mathbf{C}$. The density

$$\text{c. } e^{-(\Sigma u_i^2)/2\sigma^2} \to \text{c. } e^{-(\Sigma v_i^2)/2\sigma^2} \tag{3b.3.2}$$

so that v_i are all independent and normal. The joint density of the subset $v_1, \ldots, v_p$ which are precisely the elements of **BU**, is

$$c.\, e^{-(v_1^2+\cdots+v_p^2)/2\sigma^2} = c.\, e^{-\mathbf{V}'\mathbf{V}/2\sigma^2}. \tag{3b.3.3}$$

If the rows of **B** are not orthonormal, there exists a nonsingular square matrix **D** of order p such that **DB** has its rows orthonormal. Consider the variables $\mathbf{W} = \mathbf{DV} = (\mathbf{DB})\mathbf{U}$. Since the rows of **DB** are orthonormal, the distribution of $\mathbf{W} = (\mathbf{DB})\mathbf{U}$ is, using (3b.3.3),

$$c.\, e^{-\mathbf{W}'\mathbf{W}/2\sigma^2}\, d\mathbf{W} = (2\pi\sigma^2)^{-p/2} e^{-\mathbf{W}'\mathbf{W}/2\sigma^2}\, d\mathbf{W}.$$

The distribution of $\mathbf{V} = \mathbf{D}^{-1}\mathbf{W}$ is therefore (if we observe that $d\mathbf{W} = |\mathbf{D}|\, d\mathbf{V}$)

$$(2\pi\sigma^2)^{-p/2} e^{-\mathbf{V}'\mathbf{D}'\mathbf{D}\mathbf{V}/2\sigma^2} |\mathbf{D}|\, d\mathbf{V}.$$

But the rows of **DB** are orthonormal and therefore $\mathbf{DB}\mathbf{B}'\mathbf{D}' = \mathbf{I} \Rightarrow (\mathbf{BB}')^{-1} = \mathbf{D}'\mathbf{D}$ and $|\mathbf{BB}'| = 1/|\mathbf{D}|^2$. By denoting $(\mathbf{BB}') = \mathbf{A}$, the distributions of **V** and hence that of **Z** are

$$\frac{|\mathbf{A}|^{-1/2}}{(2\pi\sigma^2)^{p/2}} e^{-\mathbf{V}'\mathbf{A}^{-1}\mathbf{V}/2\sigma^2}\, d\mathbf{V} = \frac{|\mathbf{A}|^{-1/2}}{(2\pi\sigma^2)^{p/2}} e^{-(\mathbf{Z}-\mathbf{B}\boldsymbol{\mu})'\mathbf{A}^{-1}(\mathbf{Z}-\mathbf{B}\boldsymbol{\mu})/2\sigma^2}\, d\mathbf{Z}. \tag{3b.3.4}$$

The density function in (3b.3.4), with $\mathbf{B}\boldsymbol{\mu}$ replaced by an arbitrary vector and **A** by any positive definite matrix is called a p-variate normal density.

Suppose the number p of functions $\mathbf{Z} = \mathbf{BX}$ is greater than n or $R(\mathbf{B}) < p$. What is the distribution of **Z**? The method outlined above leading to the density (3b.3.4) fails as $\mathbf{BB}'$ is no longer nonsingular. In fact as the components of **Z** are linearly related the p.d. of **X** with respect to Lebesgue measure does not exist. However, the distribution is well defined through its characteristic function

$$\begin{aligned} E(e^{i\mathbf{t}'\mathbf{Z}}) &= E(e^{i\mathbf{t}'\mathbf{BX}}) = E(e^{i\mathbf{T}'\mathbf{X}}), \qquad \mathbf{T} = \mathbf{B}'\mathbf{t} \\ &= e^{i\mathbf{T}'\boldsymbol{\mu} - \sigma^2\mathbf{T}'\mathbf{T}/2}, \qquad \text{using (3a.1.6)} \\ &= e^{i\mathbf{t}'\mathbf{B}\boldsymbol{\mu} - \sigma^2\mathbf{t}'\mathbf{BB}'\mathbf{t}/2} = e^{-i\mathbf{t}'\mathbf{v} - \mathbf{t}'\boldsymbol{\Sigma}\mathbf{t}/2} \end{aligned} \tag{3b.3.5}$$

which shows that the distribution depends on the two parameters,

$$E(\mathbf{Z}) = \mathbf{v} \qquad \text{and} \qquad D(\mathbf{Z}) = \boldsymbol{\Sigma}. \tag{3b.3.6}$$

We call the distribution as determined by the c.f. (3b.3.5), whether density exists or not, a p-variate normal distribution and represent it by $N_p(\mathbf{v}, \boldsymbol{\Sigma})$. The expression for p.d. exists when $|\boldsymbol{\Sigma}| \neq 0$ as shown in (3b.3.4). Some properties of the distribution defined by (3b.3.4) are studied in Section **3b.6**, while the general case where $\boldsymbol{\Sigma}$ may be singular is considered in Chapter 8. A representation of the density function when $\boldsymbol{\Sigma}$ is singular is given in **8a.4.**

3b.4 Distribution of Quadratic Forms

We shall consider n independent normal variables $y_i \sim N(\mu_i, 1)$, $i = 1, \ldots, n$ and discuss the distribution of quadratic forms. Let $\mathbf{Y}' = (y_1, \ldots, y_n)$ and $\boldsymbol{\mu}' = (\mu_1, \ldots, \mu_n)$.

(i) FISHER-COCHRAN THEOREM. *Let $Q_1, \ldots, Q_k$ be k quadratic forms with ranks $n_1, \ldots, n_k$, such that*

$$\mathbf{Y}'\mathbf{Y} = Q_1 + \cdots + Q_k.$$

Then an n.s. condition that $Q_i \sim \chi^2(n_i, \lambda_i)$ and are independent is $n = \sum n_i$, in which case $\lambda_i = \boldsymbol{\mu}'\mathbf{A}_i\boldsymbol{\mu}$ if $Q_i = \mathbf{Y}'\mathbf{A}_i\mathbf{Y}$ and $\sum \mu_j^2 = \sum \lambda_i$.

We use the result proved in (1c.1.2) that any quadratic form $\mathbf{Y}'\mathbf{A}\mathbf{Y}$ in n variables of rank $m \leqslant n$ can be reduced to $\pm x_1^2 \pm \cdots \pm x_m^2$ by a suitable linear transformation from $\mathbf{Y}$ to $\mathbf{X}$, that is, $\mathbf{Y}'\mathbf{A}\mathbf{Y}$ can be expressed as a combination, with coefficients ± 1 of squares of m linear functions of $\mathbf{Y}$.

Let $Q_i = \mathbf{Y}'\mathbf{A}_i\mathbf{Y}$. Since rank $\mathbf{A}_i = n_i$, there exist n_i linear functions of $\mathbf{Y}$ such that

$$Q_i = \pm(b_{11}y_1 + \cdots + b_{1n}y_n)^2 \pm \cdots \pm (b_{n_i1}y_1 + \cdots + b_{n_in}y_n)^2.$$

Consider the n_1 linear functions associated with Q_1, n_2 with Q_2, and so on. If $n = \sum n_i$, there are n linear functions which may be written as $\mathbf{BY}$, where $\mathbf{B}$ is an $n \times n$ matrix. Now $\sum Q_i$ is a combination of squares of n linear functions with coefficients ± 1, which means

$$\sum Q_i = \mathbf{Y}'\mathbf{B}'\boldsymbol{\Delta}\mathbf{BY},$$

where $\boldsymbol{\Delta}$ is a diagonal matrix with each diagonal element equal to $+1$ or -1. But

$$\mathbf{Y}'\mathbf{Y} = \sum Q_i = \mathbf{Y}'\mathbf{B}'\boldsymbol{\Delta}\mathbf{BY} \quad \text{for all} \quad \mathbf{Y} \Rightarrow \mathbf{I} = \mathbf{B}'\boldsymbol{\Delta}\mathbf{B}. \tag{3b.4.1}$$

Rank $\mathbf{B}$ must be n, and then $\boldsymbol{\Delta} = (\mathbf{B}')^{-1}\mathbf{B}^{-1}$, which is positive definite. Therefore $\boldsymbol{\Delta}$ must have all $+1$ in the diagonal, and hence from (3b.4.1), $\mathbf{B}$ is an orthogonal matrix. The transformation $\mathbf{X} = \mathbf{BY}$ is orthogonal; hence the components $(x_1, \ldots, x_n)$ of $\mathbf{X}$ are all independent and normal. But Q_1 is the sum of squares of the first n_1 linear functions in $\mathbf{BY}$ and therefore Q_1 transforms to $x_1^2 + \cdots + x_{n_1}^2$. Similary, Q_2 transforms to $x_{n_1+1}^2 + \cdots + x_{n_1+n_2}^2$, and so on.

Hence Q_i are independent since they depend on exclusive subsets of independent variables x_i. Furthermore, Q_i is the sum of squares of n_i normal variables and therefore by (3b.2.2), $Q_i \sim \chi^2(n_i, \lambda_i)$, noncentral χ^2 with $\lambda_i = \boldsymbol{\mu}'\mathbf{A}_i\boldsymbol{\mu}$, $i = 1, \ldots, k$. But $\mathbf{Y}'\mathbf{Y} \sim \chi^2(n, \sum \mu_j^2)$, and therefore $\sum \mu_j^2 = \sum \lambda_i$, using (3b.2.4). The sufficiency of $n = \sum n_i$ is established; the necessity is obvious.

It may be seen that $Q_i \sim \chi^2(n_i)$, that is, a central χ^2 when $\boldsymbol{\mu}'\mathbf{A}_i\boldsymbol{\mu} = 0$, which is true if $\mu_j = 0, j = 1, \ldots, n$. On the other hand, if all Q_i have central χ^2 distributions, $\mu_j = 0$ for all j.

If the reader is not familiar with noncentral χ^2, the Fisher-Cochran theorem may be read, substituting 0 for all μ_j and λ_i.

In the rest of the discussion we shall take $\mu_j = 0$ for all j for simplicity of notation. But the conditions given on the matrices and the results remain valid if, instead of a central $\chi^2(k)$, a noncentral $\chi^2(k, \lambda)$ is written. Observe that λ is the value of the quadratic form when the variables are substituted by their expected values. The general result established in (i) validates all these results.

(ii) *An n.s. condition that* $\mathbf{Y}'\mathbf{A}\mathbf{Y}$ *has a chi-square distribution is that* $\mathbf{A}$ *is idempotent, that is,* $\mathbf{A}^2 = \mathbf{A}$, *in which case the degrees of freedom of* χ^2 *is* rank $\mathbf{A}$ = trace $\mathbf{A}$.

Sufficiency is immediately established since

$$\mathbf{Y}'\mathbf{Y} = \mathbf{Y}'\mathbf{A}\mathbf{Y} + \mathbf{Y}'(\mathbf{I} - \mathbf{A})\mathbf{Y},$$

and because $\mathbf{A}^2 = \mathbf{A}$, rank $\mathbf{A}$ + rank$(\mathbf{I} - \mathbf{A}) = n$ (see Example 1.7 at the end of Section **1b**), the Fisher-Cochran theorem applies leading to the desired result.

To prove the necessity we observe that there exists an orthogonal matrix $\mathbf{C}$ (see 1c.3.5) such that under the transformation $\mathbf{Y} = \mathbf{C}\mathbf{X}$

$$\begin{aligned} \mathbf{Y}'\mathbf{A}\mathbf{Y} &\to \mathbf{X}'\mathbf{C}'\mathbf{A}\mathbf{C}\mathbf{X} = \lambda_1 x_1^{\,2} + \cdots + \lambda_m x_m^{\,2} \\ \mathbf{Y}'\mathbf{Y} &\to \mathbf{X}'\mathbf{X} = x_1^{\,2} + \cdots + x_n^{\,2} \end{aligned} \tag{3b.4.2}$$

where $\lambda_1, \ldots, \lambda_m$ are the nonzero eigenvalues of $\mathbf{A}$.

Since $x_j^{\,2} \sim \chi^2(1) = G(\frac{1}{2}, \frac{1}{2})$, the c.f. of $x_j^{\,2}$ is $(1 - 2it)^{-1/2}$. Hence the c.f. of $\lambda_1 x_1^{\,2} + \cdots + \lambda_m x_m^{\,2}$ is

$$[(1 - 2i\lambda_1 t)(1 - 2i\lambda_2 t) \cdots (1 - 2i\lambda_m t)]^{-1/2}. \tag{3b.4.3}$$

But $\mathbf{Y}'\mathbf{A}\mathbf{Y}$ is given to be a χ^2, say on p d.f., in which case its c.f. is

$$(1 - 2it)^{-p/2}. \tag{3b.4.4}$$

Comparing (3b.4.3) and (3b.4.4), $p = m$ and $\lambda_i = 1$ for all i. Hence from (3b.4.2), $\mathbf{C}'\mathbf{A}\mathbf{C}$ is a diagonal matrix with diagonal elements zero or unity and is therefore idempotent. This leads to

$$\begin{aligned} \mathbf{C}'\mathbf{A}\mathbf{C} &= \mathbf{C}'\mathbf{A}\mathbf{C}\mathbf{C}'\mathbf{A}\mathbf{C} \\ &= \mathbf{C}'\mathbf{A}^2\mathbf{C} \Rightarrow \mathbf{A} = \mathbf{A}^2 \end{aligned} \tag{3b.4.5}$$

The c.f. of $x_j^{\,2}$ is that of noncentral χ^2 when $\mu_j \neq 0$. This brings in an extra term in (3b.4.3) and (3b.4.4). The reader may then deduce that (3b.4.5) is true.

The D.F. of the χ^2 is equal to rank $\mathbf{A}$ which is the same as trace $\mathbf{A}$ since $\mathbf{A}$ is idempotent [(ii), **1b.4**].

(iii) *Let* $\mathbf{Y}'\mathbf{Y} = Q_1 + Q_2$, *where* $Q_1 \sim \chi^2(a)$. *Then* $Q_2 \sim \chi^2(n - a)$.

If $Q_1 = \mathbf{Y}'\mathbf{A}\mathbf{Y}$, then by (ii), $\mathbf{A}^2 = \mathbf{A}$. But $Q_2 = \mathbf{Y}'(\mathbf{I} - \mathbf{A})\mathbf{Y}$. Since

$$(\mathbf{I} - \mathbf{A})^2 = \mathbf{I} + \mathbf{A}^2 - 2\mathbf{A} = \mathbf{I} - \mathbf{A}$$

$\mathbf{I} - \mathbf{A}$ is idempotent. Hence the result.

(iv) *Let* $Q = Q_1 + Q_2$, *where* $Q \sim \chi^2(a)$, $Q_1 \sim \chi^2(b)$ *and* Q_2 *is non-negative. Then* $Q_2 \sim \chi^2(a - b)$.

Since $Q \sim \chi^2(a)$, by (3b.4.2) there exists an orthogonal transformation such that $Q \to x_1{}^2 + \cdots + x_a{}^2$ and $\mathbf{Y}'\mathbf{Y} \to x_1{}^2 + \cdots + x_n{}^2$. Let $Q_1 \to \mathbf{X}'\mathbf{B}_1\mathbf{X}$ and $Q_2 \to \mathbf{X}'\mathbf{B}_2\mathbf{X}$, both of which are non-negative. Since

$$x_1{}^2 + \cdots + x_a{}^2 = \mathbf{X}'\mathbf{B}_1\mathbf{X} + \mathbf{X}'\mathbf{B}_2\mathbf{X}.$$

it follows that each of the quadratic forms on the right-hand side must involve $x_1, \ldots, x_a$ only, in which case result (iii) can be applied to establish that $Q_2 \sim \chi^2(a - b)$.

It follows from this result that if $\mathbf{A}$, $\mathbf{B}$, $\mathbf{A} - \mathbf{B}$ are matrices of non-negative quadratic forms and $\mathbf{A}$ and $\mathbf{B}$ are idempotent, then $\mathbf{A} - \mathbf{B}$ is also idempotent. A direct proof of this result would imply the result of (iv).

(v) *Let* $\mathbf{Y}'\mathbf{A}_1\mathbf{Y} \sim \chi^2(a)$ *and* $\mathbf{Y}'\mathbf{A}_2\mathbf{Y} \sim \chi^2(b)$. *An n.s. condition that they are independently distributed is* $\mathbf{A}_1\mathbf{A}_2 = 0$.

Since $\mathbf{A}_1{}^2 = \mathbf{A}_1$, $\mathbf{A}_2{}^2 = \mathbf{A}_2$ it follows that

$$\mathbf{A}_1(\mathbf{I} - \mathbf{A}_1 - \mathbf{A}_2) = \mathbf{A}_2(\mathbf{I} - \mathbf{A}_1 - \mathbf{A}_2) = 0$$

if $\mathbf{A}_1\mathbf{A}_2 = 0$. This means

$$\text{rank}(\mathbf{A}_1) + \text{rank}(\mathbf{A}_2) + \text{rank}(\mathbf{I} - \mathbf{A}_1 - \mathbf{A}_2) = n.$$

But $\mathbf{Y}'\mathbf{Y} = \mathbf{Y}'\mathbf{A}_1\mathbf{Y} + \mathbf{Y}'\mathbf{A}_2\mathbf{Y} + \mathbf{Y}'(\mathbf{I} - \mathbf{A}_1 - \mathbf{A}_2)\mathbf{Y}$. Hence an application of the Fisher-Cochran theorem shows that the quadratic forms are independently distributed. The necessity is easily established by expressing the condition that $(\mathbf{A}_1 + \mathbf{A}_2)$ is idempotent since $\mathbf{Y}'(\mathbf{A}_1 + \mathbf{A}_2)\mathbf{Y} \sim \chi^2(a + b)$.

(vi) *Let* $\mathbf{Y}'\mathbf{Y} = \mathbf{Y}'\mathbf{A}_1\mathbf{Y} + \cdots + \mathbf{Y}'\mathbf{A}_k\mathbf{Y}$. *Then either of the following conditions is n.s. for* $\mathbf{Y}'\mathbf{A}_i\mathbf{Y} \sim \chi^2(n_i, \lambda_i)$ *where* n_i *is the rank* $\mathbf{A}_i$, $i = 1, \ldots, k$ *and for the set* $\mathbf{Y}'\mathbf{A}_1\mathbf{Y}, \ldots, \mathbf{Y}'\mathbf{A}_k\mathbf{Y}$ *to be independently distributed.*

(a) $\mathbf{A}_1, \ldots, \mathbf{A}_k$ *are each idempotent matrices.*
(b) $\mathbf{A}_i\mathbf{A}_j = 0$ *for all* $i \neq j$.

Since $\mathbf{A}_i^2 = \mathbf{A}_i$, trace $\mathbf{A}_i$ = rank $\mathbf{A}_i$ and $\mathbf{I} = \mathbf{A}_i + \cdots + \mathbf{A}_k \Rightarrow$ trace $\mathbf{I} = n = \sum$ trace $\mathbf{A}_i = \sum n_i$. Hence the Fisher-Cochran theorem applies, thus establishing the n.s. condition (a). Now

$$\mathbf{I} = \mathbf{A}_1 + \cdots + \mathbf{A}_k \quad \text{and} \quad \mathbf{A}_i \mathbf{A}_j = 0 \Rightarrow \mathbf{A}_i(\mathbf{I} - \mathbf{A}_i) = 0,$$

that is, $\mathbf{A}_i$ is idempotent. Hence condition (b) is equivalent to (a).

We extend some of the above results to the case where $y_1, \ldots, y_n$ are correlated, more precisely when $\mathbf{Y} \sim N_n(\mathbf{v}, \mathbf{\Sigma})$ as defined by the c.f. (3b.3.5). We observe that when $R(\mathbf{\Sigma}) = r$, the vector $\mathbf{Y}$ has the representation

$$\mathbf{Y} = \mathbf{v} + \mathbf{BX} \tag{3b.4.6}$$

where $\mathbf{X}$ is an r-vector of independent $N(0, 1)$ variables and $\mathbf{BB}' = \mathbf{\Sigma}$. A study of a quadratic form in $\mathbf{Y}$ can be reduced by the relation (3b.4.6) to the study of a quadratic function of $\mathbf{X}$, i.e., of independent normal variables.

(vii) Ogasawara and Takahashi (1951). *Let* $\mathbf{Y} \sim N_n(\mathbf{v}, \Sigma)$. *An n.s. condition that* $(\mathbf{Y} - \mathbf{v})'\mathbf{A}(\mathbf{Y} - \mathbf{v})$ *has* χ^2 *distribution is*

$$\mathbf{\Sigma A \Sigma A \Sigma} = \mathbf{\Sigma A \Sigma} \tag{3b.4.7}$$

in which case the degrees of freedom is $R(\mathbf{A\Sigma})$.

Observe that $(\mathbf{Y} - \mathbf{v})'\mathbf{A}(\mathbf{Y} - \mathbf{v}) = \mathbf{X}'\mathbf{B}'\mathbf{ABX}$ using (3b.4.6). Then applying the result (ii), an n.s. condition is that $\mathbf{B}'\mathbf{AB}$ is idempotent

$$\mathbf{B}'\mathbf{AB}\,\mathbf{B}'\mathbf{AB} = \mathbf{B}'\mathbf{AB} \Leftrightarrow \mathbf{BB}'\mathbf{ABB}'\mathbf{ABB}' = \mathbf{BB}'\mathbf{ABB}',$$

or,

$$\mathbf{\Sigma A \Sigma A \Sigma} = \mathbf{\Sigma A \Sigma}.$$

The degrees of freedom is tr $\mathbf{B}'\mathbf{AB}$ = tr $\mathbf{ABB}'$ = tr $\mathbf{A\Sigma}$.

If $|\mathbf{\Sigma}| \neq 0$, then the condition (3b.4.7) reduces to

$$\mathbf{A \Sigma A} = \mathbf{A}. \tag{3b.4.8}$$

(viii) *Let* $\mathbf{Y} \sim N_n(\mathbf{v}, \mathbf{\Sigma})$. *An n.s. condition that* $\mathbf{P}'\mathbf{Y}$ *and* $(\mathbf{Y} - \mathbf{v})'\mathbf{A}(\mathbf{Y} - \mathbf{v})$ *are independently distributed is*

$$\mathbf{\Sigma A \Sigma P} = \mathbf{0}. \tag{3b.4.9}$$

(ix) *Let* $\mathbf{Y} \sim N_n(\mathbf{v}, \Sigma)$. *An n.s. condition that*

$$(\mathbf{Y} - \mathbf{v})'\mathbf{A}(\mathbf{Y} - \mathbf{v}) \quad \textit{and} \quad (\mathbf{Y} - \mathbf{v})'\mathbf{B}(\mathbf{Y} - \mathbf{v})$$

are independently distributed is

$$\mathbf{\Sigma A \Sigma B \Sigma} = \mathbf{0} \tag{3b.4.10}$$

The results (viii) and (ix) are established in the same way as in the case of independent variables.

For other results, the reader is referred to Chapter 9 of Rao and Mitra (1971).

There is considerable literature on the distribution of quadratic forms. Some references on which the results of **3b.4** are based are, Cochran (1934), Craig (1943), Hogg and Craig (1958), Ogasawara and Takahashi (1951), Ogawa (1949), Rao (1951a, 1953b, 1961a) and Sakamoto (1944). For further references, see Graybill (1961).

3b.5 Three Fundamental Theorems of the Least Squares Theory

Consider independent variables

$$y_i \sim N(x_{i1}\beta_1 + \cdots + x_{im}\beta_m, \sigma^2), \qquad i = 1, \ldots, n, \tag{3b.5.1}$$

where x_{ij} are known coefficients and β_i are unknown parameters. In matrix notation, if $\mathbf{Y}$ stands for the column vector of the variables y_i, $\boldsymbol{\beta}$ for the parameters β_i and $\mathbf{X} = (x_{ij})$ for the matrix of coefficients, then

$$\sum (y_i - x_{i1}\beta_1 - \cdots - x_{im}\beta_m)^2 = (\mathbf{Y} - \mathbf{X}\boldsymbol{\beta})'(\mathbf{Y} - \mathbf{X}\boldsymbol{\beta}).$$

Hence the p.d. of $y_1, \ldots, y_n$ can be written

$$\text{c. } e^{-(\mathbf{Y} - \mathbf{X}\boldsymbol{\beta})'(\mathbf{Y} - \mathbf{X}\boldsymbol{\beta})/2\sigma^2}. \tag{3b.5.2}$$

In this section we shall derive distributions of certain statistics which are fundamental to the theory of least squares (discussed in Chapter 4).

(i) THE FIRST FUNDAMENTAL THEOREM. *Let*

$$R_0^2 = \min_{\beta} (\mathbf{Y} - \mathbf{X}\boldsymbol{\beta})'(\mathbf{Y} - \mathbf{X}\boldsymbol{\beta}).$$

Then $R_0^2 \sim \sigma^2\chi^2(n - r)$ *where r is the rank of* $\mathbf{X}$.

Let $\mathbf{F}$ be a matrix of order $(n \times r)$ such that its columns constitute an orthonormal basis of $\mathcal{M}(\mathbf{X})$. Supplement the matrix $\mathbf{F}$ by $\mathbf{G}$ of order $(n \times n-r)$ such that $(\mathbf{F} \mid \mathbf{G})$ is an orthogonal matrix. By construction

$$\mathbf{F}'\mathbf{G} = 0, \qquad \mathbf{X}'\mathbf{G} = 0, \qquad \mathbf{G}'\mathbf{X} = 0,$$

and the transformation from $\mathbf{Y}$ to $\mathbf{Z} = \begin{pmatrix} \mathbf{Z}_1 \\ \mathbf{Z}_2 \end{pmatrix} = \begin{pmatrix} \mathbf{F}' \\ \mathbf{G}' \end{pmatrix} \mathbf{Y}$ giving

$$\mathbf{Z}_1 = \mathbf{F}'\mathbf{Y}, \qquad \mathbf{Z}_2 = \mathbf{G}'\mathbf{Y}$$

is orthogonal. The vector $\mathbf{X}\boldsymbol{\beta}$ transforms to

$$\zeta_1 = \mathbf{F}'\mathbf{X}\boldsymbol{\beta}, \qquad \zeta_2 = \mathbf{G}'\mathbf{X}\boldsymbol{\beta} = \mathbf{0}.$$

Hence by applying (3b.1.2) (invariance of distance under orthogonal transformation),

$$(\mathbf{Y} - \mathbf{X}\boldsymbol{\beta})'(\mathbf{Y} - \mathbf{X}\boldsymbol{\beta}) = (\mathbf{Z}_1 - \zeta_1)'(\mathbf{Z}_1 - \zeta_1) + \mathbf{Z}_2'\mathbf{Z}_2.$$

The p.d. of $\mathbf{Z}$ is therefore

$$\text{c. } e^{-[(\mathbf{Z}_1-\mathbf{F}'\mathbf{X}\boldsymbol{\beta})'(\mathbf{Z}_1-\mathbf{F}'\mathbf{X}\boldsymbol{\beta})+\mathbf{Z}_2'\mathbf{Z}_2]/2\sigma^2},$$

which shows that $\mathbf{Z}_1$ and $\mathbf{Z}_2$ are independently distributed and that of $\mathbf{Z}_2$ alone is

$$\text{c. } e^{-\mathbf{Z}_2'\mathbf{Z}_2/2\sigma^2}. \tag{3b.5.3}$$

But $\mathbf{Z}_2$ is a vector of $(n-r)$ variables, which by (3b.5.3) are all independent, and $N(0, \sigma^2)$. Hence $\mathbf{Z}_2'\mathbf{Z}_2 \sim \sigma^2\chi^2(n-r)$. Now

$$R_0^2 = \min_{\boldsymbol{\beta}} (\mathbf{Y}-\mathbf{X}\boldsymbol{\beta})'(\mathbf{Y}-\mathbf{X}\boldsymbol{\beta}) = \min_{\boldsymbol{\beta}} [(\mathbf{Z}_1-\boldsymbol{\zeta}_1)'(\mathbf{Z}_1-\boldsymbol{\zeta}_1)+\mathbf{Z}_2\mathbf{Z}_2]$$

$$= \left[\min_{\boldsymbol{\beta}} (\mathbf{Z}_1-\mathbf{F}'\mathbf{X}\boldsymbol{\beta})'(\mathbf{Z}_1-\mathbf{F}'\mathbf{X}\boldsymbol{\beta})\right] + \mathbf{Z}_2'\mathbf{Z}_2 = \mathbf{Z}_2'\mathbf{Z}_2, \tag{3b.5.4}$$

if we choose $\boldsymbol{\beta}$ such that $\mathbf{Z}_1 = \mathbf{F}'\mathbf{X}\boldsymbol{\beta}$, which is possible since the number of equations is r equal to the rank of $\mathbf{F}'\mathbf{X}$, as shown below (see condition 4, p. 8).

$$r = \text{rank } \mathbf{X}' = \text{rank } \mathbf{X}'(\mathbf{F} \mid \mathbf{G})$$
$$= \text{rank}(\mathbf{X}'\mathbf{F} \mid \mathbf{X}'\mathbf{G}) = \text{rank}(\mathbf{X}'\mathbf{F} \mid 0) = \text{rank } \mathbf{X}'\mathbf{F}.$$

Note. A more elegant (alternative) approach to the problem of distribution of R_0^2, which uses the knowledge of a projection operator is as follows.

Observe that $(\mathbf{Y}-\mathbf{X}\boldsymbol{\beta})'(\mathbf{Y}-\mathbf{X}\boldsymbol{\beta})$ is a minimum when $\mathbf{X}\boldsymbol{\beta}$ is the projection of $\mathbf{Y}$ on $\mathcal{M}(\mathbf{X})$. But projection of any vector on $\mathcal{M}(\mathbf{X})$ is secured through an operator, a matrix $\mathbf{P}$, which is symmetric, idempotent, and of rank equal to $r = \text{rank } \mathbf{X}$ (see **1c.4**). Thus the projection $\mathbf{Y}$ on $\mathcal{M}(\mathbf{X})$ equals $\mathbf{PY}$ and therefore $\mathbf{Y}-\mathbf{YP}$ is the perpendicular. Hence

$$R_0^2 = (\mathbf{Y}-\mathbf{PY})'(\mathbf{Y}-\mathbf{PY}) = \mathbf{Y}'(\mathbf{I}-\mathbf{P})(\mathbf{I}-\mathbf{P})\mathbf{Y} = \mathbf{Y}'(\mathbf{I}-\mathbf{P})\mathbf{Y}.$$

The matrix $\mathbf{I}-\mathbf{P}$ is also idempotent and therefore

$$\text{rank}(\mathbf{I}-\mathbf{P}) = \text{trace}(\mathbf{I}-\mathbf{P}) = \text{trace } \mathbf{I} - \text{trace } \mathbf{P} = n - r.$$

Hence, by applying the n.s. condition of [(ii), **3b.4**] on the distribution of quadratic forms, we have

$$\mathbf{Y}'(\mathbf{I}-\mathbf{P})\mathbf{Y} \sim \sigma^2\chi^2(n-r, \lambda), \tag{3b.5.5}$$

where λ is the value of the quadratic form $\mathbf{Y}'(\mathbf{I}-\mathbf{P})\mathbf{Y}/\sigma^2$ at the expected value of $\mathbf{Y}$ (see 3b.2.2). Thus

$$\sigma^2\lambda = E(\mathbf{Y}')(\mathbf{I}-\mathbf{P})E(\mathbf{Y})$$
$$= \boldsymbol{\beta}'\mathbf{X}'(\mathbf{I}-\mathbf{P})\mathbf{X}\boldsymbol{\beta} = \boldsymbol{\beta}'(\mathbf{X}'\mathbf{X}-\mathbf{X}'\mathbf{P}\mathbf{X})\boldsymbol{\beta} = 0,$$

since $\mathbf{PX} = \mathbf{X}$ (since vectors of $\mathcal{M}(\mathbf{X})$ remain unaltered by the projection operator $\mathbf{P}$). Hence the distribution (3b.5.5) is in fact central.

(ii) THE SECOND FUNDAMENTAL THEOREM. *Let* $\mathbf{H}$ *be a matrix of order* $(m \times k)$ *and rank* k *such that* $\mathscr{M}(\mathbf{H}) \subset \mathscr{M}(\mathbf{X}')$ *and*

$$R_1^2 = \min_{\beta} (\mathbf{Y} - \mathbf{X}\boldsymbol{\beta})'(\mathbf{Y} - \mathbf{X}\boldsymbol{\beta})$$

subject to the condition $\mathbf{H}'\boldsymbol{\beta} = \boldsymbol{\xi}$ *(given). Then:*

(a) R_0^2 *and* $R_1^2 - R_0^2$ *are independently distributed.*
(b) $R_0^2 \sim \sigma^2\chi^2(n - r)$ *and* $R_1^2 - R_0^2 \sim$ *as a noncentral* χ^2 *on* k *degrees of freedom.*
(c) *If* $\mathbf{H}'\boldsymbol{\beta} = \boldsymbol{\xi}$ *is true, then* $R_1^2 - R_0^2 \sim \sigma^2\chi^2(k)$ *and*

$$\frac{R_1^2 - R_0^2}{k} \div \frac{R_0^2}{n - r} \sim F(k, n - r).$$

Since $\mathscr{M}(\mathbf{X}') = \mathscr{M}(\mathbf{X}'\mathbf{X})$, the condition $\mathscr{M}(\mathbf{H}) \subset \mathscr{M}(\mathbf{X}') \Rightarrow \mathscr{M}(\mathbf{H}) \subset \mathscr{M}(\mathbf{X}'\mathbf{X})$ and therefore there exists by (3b.1.4) a matrix $\mathbf{C}$ such that $\mathbf{H} = \mathbf{X}'\mathbf{X}\mathbf{C}$ and rank $\mathbf{X}\mathbf{C} = k$. Make the transformation

$$\mathbf{Z}_1 = \mathbf{D}'\mathbf{Y}, \qquad \mathbf{Z}_2 = \mathbf{G}'\mathbf{Y}, \qquad \mathbf{Z}_3 = \mathbf{C}'\mathbf{X}'\mathbf{Y}$$

where $\mathbf{G}$ is as in (i) and $\mathbf{D}'$ is chosen such that $\mathbf{D}'\mathbf{D} = \mathbf{I}$ and orthogonal to $\mathbf{G}$ and $\mathbf{X}\mathbf{C}$. We may note that the numbers of variables in $\mathbf{Z}_1$, $\mathbf{Z}_2$, and $\mathbf{Z}_3$ are $r - k$, $n - r$, and k respectively. The vector $\mathbf{X}\boldsymbol{\beta}$ transforms to

$$\zeta_1 = \mathbf{D}'\mathbf{X}\boldsymbol{\beta}, \qquad \zeta_2 = \mathbf{G}'\mathbf{X}\boldsymbol{\beta} = 0, \qquad \zeta_3 = \mathbf{C}'\mathbf{X}'\mathbf{X}\boldsymbol{\beta} = \mathbf{H}'\boldsymbol{\beta}.$$

Since the submatrices in the transformation are orthogonal to each other the result (3b.1.3) applies. Hence

$$\begin{aligned}(\mathbf{Y} - \mathbf{X}\boldsymbol{\beta})'(\mathbf{Y} - \mathbf{X}\boldsymbol{\beta}) &= (\mathbf{Z}_1 - \mathbf{D}'\mathbf{X}\boldsymbol{\beta})'(\mathbf{Z}_1 - \mathbf{D}'\mathbf{X}\boldsymbol{\beta}) + \mathbf{Z}_2'\mathbf{Z}_2 \\ &\quad + (\mathbf{Z}_3 - \mathbf{H}'\boldsymbol{\beta})'(\mathbf{C}'\mathbf{X}'\mathbf{X}\mathbf{C})^{-1}(\mathbf{Z}_3 - \mathbf{H}'\boldsymbol{\beta}). \qquad (3b.5.6)\end{aligned}$$

Writing the p.d. of $\mathbf{Z}_1, \mathbf{Z}_2, \mathbf{Z}_3$ we find that they are independently distributed. In particular $\mathbf{Z}_3$ has the density

$$\text{c.}\ e^{-(\mathbf{Z}_3 - \mathbf{H}'\boldsymbol{\beta})(\mathbf{C}'\mathbf{X}'\mathbf{X}\mathbf{C})^{-1}(\mathbf{Z}_3 - \mathbf{H}'\boldsymbol{\beta})/2\sigma^2},$$

which is the same as that of k linear functions of normal variables.

With the result (3b.4.7), the quadratic form

$$Q = (\mathbf{Z}_3 - \boldsymbol{\xi})'(\mathbf{C}'\mathbf{X}'\mathbf{X}\mathbf{C})^{-1}(\mathbf{Z}_3 - \boldsymbol{\xi}) \sim \sigma^2\chi^2(k, \lambda),$$

where the noncentrality parameter is, by (3b.2.2),

$$\lambda = \sigma^{-2}(\mathbf{H}'\boldsymbol{\beta} - \boldsymbol{\xi})'(\mathbf{C}'\mathbf{X}'\mathbf{X}\mathbf{C})^{-1}(\mathbf{H}'\boldsymbol{\beta} - \boldsymbol{\xi}).$$

Furthermore, $R_0^2 = \mathbf{Z}_2'\mathbf{Z}_2 \sim \sigma^2\chi^2(n - r)$ as shown in (i) and is independent of Q. When $\mathbf{H}'\boldsymbol{\beta} = \boldsymbol{\xi}$ is true, $\lambda = 0$, in which case $Q \sim \sigma^2\chi^2(k)$. Results (a), (b), and consequently (c) will be provod if it can be shown that $Q = R_1^2 - R_0^2$.

Let us evaluate

$$R_1^2 = \min_{\mathbf{H}'\boldsymbol{\beta}=\boldsymbol{\xi}} (\mathbf{Y} - \mathbf{X}\boldsymbol{\beta})'(\mathbf{Y} - \mathbf{X}\boldsymbol{\beta})$$

$$= \left[\min_{\mathbf{H}'\boldsymbol{\beta}=\boldsymbol{\xi}} (\mathbf{Z}_1 - \mathbf{D}'\mathbf{X}\boldsymbol{\beta})'(\mathbf{Z}_1 - \mathbf{D}'\mathbf{X}\boldsymbol{\beta})\right] + R_0^2 + Q, \quad \text{[using (3b.5.6)]}$$

$$= R_0^2 + Q$$

provided $\boldsymbol{\beta}$ exists such that $\mathbf{Z}_1 - \mathbf{D}'\mathbf{X}\boldsymbol{\beta} = 0$ and $\mathbf{H}'\boldsymbol{\beta} = \boldsymbol{\xi}$. This existence is possible since the number of equations is r and also

$$r = \text{rank } \mathbf{X}' = \text{rank } \mathbf{X}'(\mathbf{D} \,\vdots\, \mathbf{G} \,\vdots\, \mathbf{XC}) = \text{rank}(\mathbf{X}'\mathbf{D} \,\vdots\, \mathbf{0} \,\vdots\, \mathbf{H}) = \text{rank}(\mathbf{X}'\mathbf{D} \,\vdots\, \mathbf{H}).$$

Note. An alternative approach using projection operators is as follows. First observe that if $\mathbf{H}'\boldsymbol{\beta} = \boldsymbol{\xi}$, then $\boldsymbol{\beta} = \boldsymbol{\beta}_0 + \boldsymbol{\gamma}$ where $\boldsymbol{\beta}_0$ is a particular solution of $\mathbf{H}'\boldsymbol{\beta} = \boldsymbol{\xi}$ and $\boldsymbol{\gamma}$ is a general solution of $\mathbf{H}'\boldsymbol{\beta} = \mathbf{0}$. Hence

$$\min_{\mathbf{H}'\boldsymbol{\beta}=\boldsymbol{\xi}} (\mathbf{Y} - \mathbf{X}\boldsymbol{\beta})'(\mathbf{Y} - \mathbf{X}\boldsymbol{\beta}) = \min_{\mathbf{H}'\boldsymbol{\gamma}=0} (\mathbf{Y} - \mathbf{X}\boldsymbol{\beta}_0 - \mathbf{X}\boldsymbol{\gamma})'(\mathbf{Y} - \mathbf{X}\boldsymbol{\beta}_0 - \mathbf{X}\boldsymbol{\gamma}). \quad (3b.5.7)$$

But $\mathbf{X}\boldsymbol{\gamma}$ with the restriction $\mathbf{H}'\boldsymbol{\gamma} = 0$ is a subspace $\mathscr{S} \subset \mathscr{M}(\mathbf{X})$ with

$$d[\mathscr{S}] = \text{rank}(\mathbf{X}' \,\vdots\, \mathbf{H}) - \text{rank } \mathbf{H} = s \quad \text{(say)},$$

whether or not $\mathbf{H}$ satisfies the condition $\mathscr{M}(\mathbf{H}) \subset \mathscr{M}(\mathbf{X}')$. Let $\mathbf{P}$ be the operator for projecting a vector on $\mathscr{M}(\mathbf{X})$ and $\mathbf{U}$ the operator for projecting on $\mathscr{S}$. Then rank $\mathbf{P} = r$ and rank $\mathbf{U} = s$. Now

$$R_1^2 = (\mathbf{Y} - \mathbf{X}\boldsymbol{\beta}_0)'(\mathbf{I} - \mathbf{U})(\mathbf{Y} - \mathbf{X}\boldsymbol{\beta}_0)$$
$$R_0^2 = \mathbf{Y}'(\mathbf{I} - \mathbf{P})\mathbf{Y} = (\mathbf{Y} - \mathbf{X}\boldsymbol{\beta}_0)'(\mathbf{I} - \mathbf{P})(\mathbf{Y} - \mathbf{X}\boldsymbol{\beta}_0).$$

Introduction of the factor $\mathbf{X}\boldsymbol{\beta}_0$ in the expression for R_0^2 does not alter its value. Since $(\mathbf{I} - \mathbf{U})$ is idempotent and of rank $(n - s)$,

$$R_1^2 \sim \sigma^2\chi^2(n - s, \lambda)$$

where

$$\sigma^2\lambda = (\mathbf{X}\boldsymbol{\beta} - \mathbf{X}\boldsymbol{\beta}_0)'(\mathbf{I} - \mathbf{U})(\mathbf{X}\boldsymbol{\beta} - \mathbf{X}\boldsymbol{\beta}_0). \quad (3b.5.8)$$

Now $R_0^2 \sim \sigma^2\chi^2(n - r)$ by the first fundamental theorem and $R_1^2 - R_0^2 \geqslant 0$. Hence, applying [(iv), **3b.4**], we have

$$R_1^2 - R_0^2 \sim \sigma^2\chi^2(r - s, \lambda) \quad \text{independently of } R_0^2.$$

If $\mathbf{H}'\boldsymbol{\beta} = \boldsymbol{\xi}$ is true, then $\boldsymbol{\beta} = \boldsymbol{\beta}_0 + \boldsymbol{\gamma}$ where $\mathbf{H}'\boldsymbol{\gamma} = 0$. Hence

$$\sigma^2\lambda = (\mathbf{X}\boldsymbol{\gamma})'(\mathbf{I} - \mathbf{U})(\mathbf{X}\boldsymbol{\gamma}) = (\mathbf{X}\boldsymbol{\gamma})'(\mathbf{X}\boldsymbol{\gamma}) - (\mathbf{X}\boldsymbol{\gamma}')(\mathbf{X}\boldsymbol{\gamma}) = 0,$$

since $\mathbf{U}\mathbf{X}\boldsymbol{\gamma} = \mathbf{X}\boldsymbol{\gamma}$. Thus $R_1^2 - R_0^2 \sim \sigma^2\chi^2(r - s)$, a central χ^2.

In the special case $\mathcal{M}(\mathbf{H}) \subset \mathcal{M}(\mathbf{X}')$ and $\mathbf{H}$ is of rank k,

$$s = \text{rank}(\mathbf{X}' \,\vdots\, \mathbf{H}) - \text{rank}\,\mathbf{H} = r - k \Rightarrow r - s = k. \tag{3b.5.9}$$

Hence $R_1{}^2 - R_0{}^2 \sim \sigma^2\chi^2(k)$.

(iii) THE THIRD FUNDAMENTAL THEOREM. *Let* $\mathbf{Y}' = (\mathbf{Y}_1' \,\vdots\, \mathbf{Y}_2')$ *with the corresponding partition of the expectation vector* $\boldsymbol{\beta}'\mathbf{X}' = \boldsymbol{\beta}'(\mathbf{X}_1' \,\vdots\, \mathbf{X}_2') = (\boldsymbol{\beta}'\mathbf{X}_1' \,\vdots\, \boldsymbol{\beta}'\mathbf{X}_2')$. *Then the statistic*

$$U = \frac{\min_{\boldsymbol{\beta}}(\mathbf{Y}_1 - \mathbf{X}_1\boldsymbol{\beta})'(\mathbf{Y}_1 - \mathbf{X}_1\boldsymbol{\beta})}{\min_{\boldsymbol{\beta}}(\mathbf{Y} - \mathbf{X}\boldsymbol{\beta})'(\mathbf{Y} - \mathbf{X}\boldsymbol{\beta})}$$

$$\sim B\left(\frac{q - r_1}{2}, \frac{n - q - r + r_1}{2}\right), \tag{3b.5.10}$$

where q is the number of columns in $\mathbf{X}_1$, $r_1 = \text{rank}\ \mathbf{X}_1$ and n, r are as defined in (i).

By making a transformation similar to that in (ii) the statistic U can be expressed as

$$\frac{\chi^2(q - r_1)}{\chi^2(q - r_1) + \chi^2(n - r - q + r_1)},$$

where the two χ^2 variables are independent. Hence by (3b.1.6) the ratio has the beta distribution (3b.5.10).

(iv) A GENERAL THEOREM. *Let* $\mathbf{W}$ *be a p-vector and* $\mathbf{Y}$ *be* $p \times k$ *matrix of random variables such that the conditional distribution of* $\mathbf{W}$ *given* $\mathbf{Y}$

$$\begin{gathered} \mathbf{W} \mid \mathbf{Y} \sim N_p(\mathbf{Y}\boldsymbol{\alpha}, \sigma^2\mathbf{G}), \\ R(\mathbf{Y}'\mathbf{G}\mathbf{Y}) = r \quad (\textit{fixed})\ \textit{with probability 1}, \\ R(\mathbf{G}) = s, \quad \mathbf{G}\ \textit{is free of}\ \mathbf{Y}. \end{gathered} \tag{3b.5.11}$$

Define the statistics

$$\begin{gathered} T_1 = \mathbf{d}'\mathbf{Y}(\mathbf{Y}'\mathbf{G}\mathbf{Y})^-\mathbf{Y}'\mathbf{d}, \\ T_2 = \mathbf{d}'\mathbf{G}^-\mathbf{d}, \qquad \mathbf{d} = \mathbf{W} - \mathbf{Y}\boldsymbol{\alpha}, \end{gathered} \tag{3b.5.12}$$

for any choice of the g-inverses. Then $T_1 \sim \sigma^2\chi^2(r)$ and $T_2 - T_1 \sim \sigma^2\chi^2(s - r)$ and are independently distributed so that $(s - r)T_1/r(T_2 - T_1)$ has the $F(r, s - r)$ distribution.

The result is just a restatement of the propositions, (iv)–(ix), in **3b.4**. Of course the theorem is true when $\mathbf{Y}$ is nonstochastic.

See also examples **2.1–2.11** at the end of the Chapter for alternative proofs and further results.

3b.6 The p-Variate Normal Distribution

Following the expression (3b.3.4) we consider a p-variate normal distribution with the probability density

$$c.\ e^{-(\mathbf{X}-\boldsymbol{\mu})'\boldsymbol{\Sigma}^{-1}(\mathbf{X}-\boldsymbol{\mu})/2} \tag{3b.6.1}$$

where $\boldsymbol{\Sigma}$, a p.d. matrix, and $\boldsymbol{\mu}$ a p-vector are parameters of the distribution.

A more general definition is given in Chapter 8 where the density of $\mathbf{X}$ with respect to Lebesgue measure may not exist. However, the expression (3b.6.1) is of classical interest, and it may be of some interest to study its properties.

(i) *The value of the constant* c *in* (3b.6.1) *is*

$$(2\pi)^{-p/2}|\boldsymbol{\Sigma}|^{-1/2}. \tag{3b.6.2}$$

Let us make the transformation $\mathbf{Y} = \boldsymbol{\Sigma}^{-1/2}\mathbf{X}$ where $\boldsymbol{\Sigma}^{1/2}$ is a square root of $\boldsymbol{\Sigma}$. The differential $d\mathbf{Y} = |\boldsymbol{\Sigma}|^{-1/2}d\mathbf{X}$, so that the density of $\mathbf{Y}$ is

$$c.\ |\boldsymbol{\Sigma}|^{1/2}e^{-(\mathbf{Y}-\mathbf{v})'(\mathbf{Y}-\mathbf{v})/2}, \qquad \mathbf{v} = \boldsymbol{\Sigma}^{-1/2}\boldsymbol{\mu}. \tag{3b.6.3}$$

The components of $\mathbf{Y}$ are independently and normally distributed. Hence integrating (3b.6.3) term by term we have

$$c.\ |\boldsymbol{\Sigma}|^{1/2}(2\pi)^{p/2} = 1$$

giving the value of c as in (3b.6.2).

Note the significance of the transformation $\mathbf{Y} = \boldsymbol{\Sigma}^{-1/2}\mathbf{X}$. It shows that the study of a p-variate normal distribution reduces to that of p independent univariate normal variables considered in **3b.2–3b.5**. For instance, to study the behavior of a statistic $f(\mathbf{X})$ based on a p-variate normal variable $\mathbf{X}$ we need only consider the statistic $f(\boldsymbol{\Sigma}^{1/2}\mathbf{Y})$ based on independent univariate normal variables $\mathbf{Y}$. We shall illustrate this in the following propositions.

(ii) $E(\mathbf{X}) = \boldsymbol{\mu}$, *i.e.* $\boldsymbol{\mu}$ *is the vector of mean values of the components of* $\mathbf{X}$.

Since $\mathbf{X} = \boldsymbol{\Sigma}^{1/2}\mathbf{Y}$, $E(\mathbf{X}) = \boldsymbol{\Sigma}^{1/2}E(\mathbf{Y}) = \boldsymbol{\Sigma}^{1/2}\boldsymbol{\Sigma}^{-1/2}\boldsymbol{\mu} = \boldsymbol{\mu}$, using (3b.6.3).

(iii) $D(\mathbf{X})$, *the dispersion* (*the variance-covariance*) *matrix of* $\mathbf{X}$ *is* $\boldsymbol{\Sigma}$.

Again we use the transformation $(\mathbf{X} - \boldsymbol{\mu}) = \boldsymbol{\Sigma}^{1/2}(\mathbf{Y} - \mathbf{v})$ and note that $D(\mathbf{Y}) = \mathbf{I}$, since the components of $(\mathbf{Y} - \mathbf{v})$ are independent $N(0, 1)$ variables. Then

$$\begin{aligned} D(\mathbf{X}) &= D[\boldsymbol{\Sigma}^{1/2}(\mathbf{Y} - \mathbf{v})] = \boldsymbol{\Sigma}^{1/2}D(\mathbf{Y} - \mathbf{v})\boldsymbol{\Sigma}^{1/2} \\ &= \boldsymbol{\Sigma}^{1/2}\boldsymbol{\Sigma}^{1/2} = \boldsymbol{\Sigma}. \end{aligned}$$

The propositions (ii) and (iii) show that a p-variate normal distribution is completely specified by the mean and dispersion of the random vector $\mathbf{X}$. The distribution with the density (3b.6.1) is denoted by $N_p(\boldsymbol{\mu}, \boldsymbol{\Sigma})$.

(iv) *The marginal distribution of any subset of the components of* $\mathbf{X}$, *say the first is k is k-variate normal.*

Let $\mathbf{X}' = (\mathbf{X}_1' \mid \mathbf{X}_2')$ where $\mathbf{X}_1$ is the subvector with the first k components of $\mathbf{X}$. Then

$$\mathbf{X}' = (\mathbf{X}_1' \mid \mathbf{X}_2') = \mathbf{Y}'\mathbf{\Sigma}^{1/2} = \mathbf{Y}'(\mathbf{C}_1 \mid \mathbf{C}_2) \Rightarrow \mathbf{X}_1 = \mathbf{C}_1'\mathbf{Y}.$$

Hence using (3b.3.4), $\mathbf{X}_1$ has k-variate normal distribution, $N_k(\boldsymbol{\mu}_1, \mathbf{\Sigma}_{11})$ where

$$E(\mathbf{X}_1) = \boldsymbol{\mu}_1, \qquad D(\mathbf{X}_1) = \mathbf{\Sigma}_{11}.$$

Then $\boldsymbol{\mu}_1$ and $\mathbf{\Sigma}_{11}$ are appropriate partitions of $\boldsymbol{\mu}$ and $\mathbf{\Sigma}$ respectively.

For further study of the multivariate normal distribution, the reader is referred to Chapter 8.

3b.7 The Exponential Family of Distributions

Let $(\Omega, \mathscr{B}, \mu)$ be a measure space and $T_1, \ldots, T_q$ represent q-real valued measurable functions over Ω. Then a typical exponential family of distributions is characterized by probability density of the form

$$f(\omega, \theta) = C(\theta)h(\omega)\exp\left[\sum \pi_i(\theta)T_i(\omega)\right], \qquad \omega \in \Omega \tag{3b.7.1}$$

with respect to the measure μ, where $\theta \in \Theta$, a specified parameter space and π_i are given functions. The expression (3b.7.1) can also be written in the alternative form

$$f(\omega, \theta) = \exp[\pi(\theta) + T(\omega) + \sum \pi_i(\theta)T_i(\omega)] \tag{3b.7.2}$$

where $\pi(\theta)$ is obtained from the relation, $\sum f(\omega, \theta) = 1$ in the discrete case, and $\int f(\omega, \theta)\, d\mu = 1$ in the general case. We may assume without loss of generality that there is no linear relationship among $1, T_1(\omega), \ldots, T_q(\omega)$ or among $1, \pi_1(\theta), \ldots, \pi_q(\theta)$. The functions $T_1, \ldots, T_q$ may otherwise be functionally related.

Many discrete and continuous distributions belong to the class (3b.7.1). The reader may verify that the discrete distributions binomial, Poisson, negative binomial, etc., and the continuous distributions, gamma, normal, uniform, etc., are particular cases.

For example, if Ω is the space R^p with general coordinates $X_1, \ldots, X_p$ then the family of p-variate normal distributions considered in **3b.6** is determined by

$$(T_1, \ldots, T_q) = (X_1, \ldots, X_p, X_1^2, \ldots, X_p^2, X_1X_2, X_1X_3, \ldots, X_{p-1}X_p) \tag{3b.7.3}$$

where $q = p + p(p+1)/2$. Thus the exponential family includes the multivariate normal as a special case. Note that in (3b.7.3), $T_1, \ldots, T_q$ are related.

Applying the factorization theorem for densities (see **2d.3**), we find that $(T_1, \ldots, T_q)$ constitutes a sufficient statistic for the exponential family. A special class of the exponential family is

$$f(\omega, \theta) = \exp[h(\theta) + T(\omega) + \theta_1 T_1(\omega) + \cdots + \theta_q T_q(\omega)] \qquad (3b.7.4)$$

where $\theta = (\theta_1, \ldots, \theta_q)$ is considered as a vector of free parameters belonging to a set $\Theta \subset R^q$. In such a case, the following propositions are of interest.

(i) *The natural parameter space is convex.*

(ii) *The distribution of a subset $(T_1, \ldots, T_r)$ of statistics also belongs to the exponential family.*

(iii) *The set $(T_1, \ldots, T_q)$ is sufficient for θ and their distribution is complete, i.e.,*

$$Ef(T_1, \ldots, T_q) = 0 \text{ for each } \theta \in \Theta \Rightarrow f(T_1, \ldots, T_q) = 0 \text{ a.e.}$$

provided Θ contains a q-dimensional rectangle.

The model (3b.7.4) can be extended by the introduction of a set of variables, $X_1, \ldots, X_p$, defined over a space S. Each response $\omega \in \Omega$ is paired with an observable vector $(X_1, \ldots, X_p)$ which influences the response. An extended model may be defined by the density

$$f(\omega, \theta | X_1, \ldots, X_p) = \exp[\alpha + \sum \sum \phi_{ij} X_i T_j(\omega)] \qquad (3b.7.5)$$

where α, the normalizing constant, depends on ϕ_{ij}, the parameters, and X_i. The model (3b.7.5) includes as special cases the conditional distribution of a subset of the variables given the rest in a multivariate normal distribution and the multinomial logit considered by Bock (1969). Thus the families considered in (3b.7.1), (3b.7.4), and (3b.7.5) provide a wide variety of models for use in practical work (see Dempster, 1971).

3c SYMMETRIC NORMAL DISTRIBUTION

3c.1 Definition

The variables $x_1, \ldots, x_p$ are said to have a symmetric normal distribution if their distribution is p-variate normal with $E(x_i) = \mu$ (same for all i) and their dispersion matrix is of the form

$$\mathbf{D} = \sigma^2 \begin{pmatrix} 1 & \rho & \cdots & \rho \\ \rho & 1 & \cdots & \rho \\ \cdot & \cdot & \cdots & \cdot \\ \rho & \rho & \cdots & 1 \end{pmatrix}.$$

Make an orthogonal transformation

$$\mathbf{Y} = \mathbf{C}\mathbf{X} \tag{3c.1.1}$$

with the first row of $\mathbf{C}$ as $(1/\sqrt{p}, \ldots, 1/\sqrt{p})$. It is a straightforward computation to show that

$$E(y_1) = \mu\sqrt{p}, \qquad V(y_1) = (1 + \overline{p-1}\rho)\sigma^2 \tag{3c.1.2}$$

$$E(y_i) = 0, \qquad V(y_i) = (1-\rho)\sigma^2, \qquad i = 2, \ldots, p. \tag{3c.1.3}$$

$$\text{cov}(y_i\, y_j) = 0 \qquad \text{for} \quad i \neq j, \tag{3c.1.4}$$

that is, the transformed variables are all uncorrelated. The results (3c.1.2) to (3c.1.4) show that

$$y_1 \sim N(\mu\sqrt{p}, [1 + (p-1)\rho]\sigma^2)$$

$$y_i \sim N[0, (1-\rho)\sigma^2], \qquad i = 2, \ldots, p \tag{3c.1.5}$$

and are all independent. It may be noted that $\mathbf{C}$ is precisely the matrix of eigenvectors of the matrix $\mathbf{D}$.

3c.2 Sampling Distributions

Joint Distribution of Mean and Variance. Let $x_1, \ldots, x_p$ have a symmetric normal distribution. Then the mean $\bar{x} = (x_1 + \cdots + x_p)/p$ and the variance $W = \sum (x_i - \bar{x})^2 \div (p-1)$ are independently distributed;

$$\bar{x} \sim N(\mu, (1 + \overline{p-1}\rho)\sigma^2/p) \qquad \text{and} \qquad W \sim (1-\rho)\sigma^2\chi^2(p-1).$$

Under the orthogonal transformation (3c.1.1),

$$\sum x_i^2 = \sum y_i^2$$

$$y_1 = \sqrt{p\bar{x}} \qquad \text{and} \qquad W = y_2^2 + \cdots + y_p^2,$$

and by the result (3c.1.5), y_1 and W are independently distributed. y_1 is normal as in (3c.1.5) and $W \sim (1-\rho)\sigma^2\chi^2(p-1)$.

As for uncorrelated variables (3b.3.1), *the sample mean and the sample variance of the observations are independently distributed.*

Let us define t as in **3b.3**

$$t = \sqrt{p}(\bar{x} - \mu)/\sqrt{W/(p-1)}.$$

Then it easily follows that

$$t \sim \frac{1 + \overline{p-1}\rho}{1-\rho} S(p-1).$$

or $(1-\rho)t/(1 + \overline{p-1}\rho)$ has Student's distribution on $(p-1)$ D.F., $S(p-1)$. However, no inference can be drawn on μ when ρ is unknown.

Ratio of Sum of Squares between and within Samples. Consider repeated observations on a p-dimensional symmetric normal variable. Let the ith observation be

$$(x_{1i}, \dots, x_{pi}), \qquad i = 1, \dots, n.$$

Define

$$\bar{x}_i = \frac{(x_{1i} + \cdots + x_{pi})}{p}$$

$$W_i = \sum_j (x_{ji} - \bar{x}_i)^2 \tag{3c.2.1}$$

$$B = p \sum (\bar{x}_i - \bar{x})^2, \qquad W = W_1 + \cdots + W_n,$$

where $\bar{x}$ is the grand average. If $T = \sum \sum (x_{ij} - \bar{x})^2$, we have the relation $T = B + W$. We call B and W as sums of squares between and within samples.

It has already been established that for each i, $\bar{x}_i$ and W_i are independent. $W_1, \dots, W_n$ are independent as they arise from different samples. Since $W_i \sim (1 - \rho)\sigma^2\chi^2(p - 1)$,

$$W = W_1 + \cdots + W_n \sim (1 - \rho)\sigma^2\chi^2(np - n). \tag{3c.2.2}$$

Similarly, $\bar{x}_1, \dots, \bar{x}_n$ are independent and each is normal with the same mean and variance $(1 + \overline{p - 1}\rho)\sigma^2/p$. Therefore

$$\frac{B}{p} = \sum (\bar{x}_i - \bar{x})^2 \sim \frac{1 + \overline{p - 1}\rho}{p}\sigma^2\chi^2(n - 1)$$

or

$$B \sim (1 + \overline{p - 1}\rho)\sigma^2\chi^2(n - 1). \tag{3c.2.3}$$

Combining the results (3c.2.2) and (3c.2.3) and applying (3b.1.10), we have

$$\frac{B}{1 + \overline{p - 1}\rho} \div \frac{W}{1 - \rho} \sim f[n - 1, n(p - 1)], \tag{3c.2.4}$$

where f is the distribution of ratio of two chi-squares. When $\rho = 0$,

$$\frac{B}{W} \sim f[n - 1, n(p - 1)], \qquad \frac{W}{B} \sim f[n(p - 1), n - 1]. \tag{3c.2.5}$$

Also by (3b.1.12), when $\rho = 0$,

$$\frac{W}{W + B} = \frac{W}{T} \sim B\left(\frac{n(p - 1)}{2}, \frac{n - 1}{2}\right). \tag{3c.2.6}$$

Intraclass Correlation Coefficient. Let it be required to estimate the correlation coefficient between heights of brothers on the basis of measurements taken on p brothers in each of n families. The p measurements on a family provide $p(p-1)$ pairs of observations (x, y)—x being the height of one brother and y that of another. From the n families we generate a total of $np(p-1)$ pairs from which a correlation coefficient is computed in the ordinary way. Let the observations be

$$(x_{1i}, \ldots, x_{pi}), \qquad i = 1, \ldots, n \quad \text{(families)}. \tag{3c.2.7}$$

It will be a good exercise to show, using the notations (3c.2.1) that the corrected sum of squares S and products P for $np(p-1)$ pairs (x, y) obtained from (3c.2.7) are

$$S = (p-1)T, \qquad P = pB - T,$$

so that the intraclass correlation is

$$r = \frac{pB - T}{(p-1)T},$$

which gives the relation

$$\frac{W}{T} = \frac{(1-r)(p-1)}{p}.$$

When $\rho = 0$, it is shown (3c.2.6) that

$$W/T \sim B\left(\frac{n(p-1)}{2}, \frac{n-1}{2}\right).$$

Hence

$$u = \frac{(1-r)(p-1)}{p} \sim B\left(\frac{n(p-1)}{2}, \frac{n-1}{2}\right).$$

The density of u is (by writing the beta distribution)

$$\text{c. } u^{[n(p-1)/2]-1}(1-u)^{[(n-1)/2]-1}$$

and hence that of r (the intraclass correlation) is

$$\text{c. } (1-r)^{[n(p-1)/2]-1}(1 + r\overline{p-1})^{[(n-1)/2]-1}. \tag{3c.2.8}$$

To find the distribution of r when $\rho \neq 0$, we observe that

$$\frac{B}{W} = \frac{1 + (p-1)r}{(1-r)(p-1)}$$

and using (3c.2.4)

$$\frac{1+\overline{p-1}r}{(1-r)(p-1)} \cdot \frac{1-\rho}{1+(p-1)\rho} \sim f\left(\frac{n-1}{2}, \frac{n(p-1)}{2}\right). \qquad \text{(3c.2.9)}$$

We can then obtain the distribution of r by change of variable from (3c.2.9).

A Test for Symmetry with Respect to Means. Suppose that there is a natural order for the variables $(x_1, \ldots, x_p)$, as when x_i stands for the height of a brother with parity i in the family. On the basis of observations on n families, is it possible to test whether there are differences in average heights of brothers of different parities?

From the np observations (3c.2.7) let us compute the average height of brothers with parity i

$$\bar{x}_{i.} = \frac{(x_{i1} + \cdots + x_{in})}{n}, \qquad i = 1, \ldots, p$$

and the sum of squares arising from them:

$$\sum (\bar{x}_{i.} - \bar{x})^2.$$

Let $R = n\sum (\bar{x}_{i.} - \bar{x})^2$. We find the distribution of R, as a statistic which reflects the differences between parities.

We recall that the transformation (3c.1.1), $\mathbf{Y} = \mathbf{CX}$ on the ith set $(x_{1i}, \ldots, x_{pi})$ leads to new variables

$$y_{1i}, y_{2i}, \ldots, y_{pi}, \qquad i = 1, \ldots, n$$

where y_{ij} are all independent and

$$V(y_{ij}) = (1-\rho)\sigma^2, \qquad i > 1, \quad j = 1, \ldots, n.$$

We observe the identity

$$\sum_2^p \sum_1^n y_{ij}{}^2 = n \sum_2^p \bar{y}_{i.}{}^2 + \sum_2^p \sum_1^n (y_{ij} - \bar{y}_{i.})^2,$$

$$W = Q_1 + Q_2.$$

Since $\sqrt{n}\,\bar{y}_{i.} \sim N(0, (1-\rho)\sigma^2)$, then $Q_1 \sim (1-\rho)\sigma^2\chi^2(p-1)$, and since $y_{ij} \sim N(0, (1-\rho)\sigma^2)$, then $W \sim (1-\rho)\sigma^2\chi^2[n(p-1)]$. Q_2 is obviously positive and hence by an application of [(iv), **3b.4**],

$$Q_2 \sim (1-\rho)\sigma^2\chi^2[(n-1)(p-1)],$$

and furthermore Q_1 and Q_2 are independently distributed.

The orthogonal transformation $\mathbf{Y} = \mathbf{CX}$ on the averages $\bar{x}_{1.}, \ldots, \bar{x}_{p.}$ leads to $\bar{y}_{1.}, \ldots, \bar{y}_{p.}$. Hence

$$\sum \bar{x}_{i.}{}^2 = \sum \bar{y}_{i.}{}^2.$$

But $y_1 = \sqrt{p}\,\bar{x}$ where $\bar{x}$ is the average of all the np observations. Hence

$$\frac{R}{n} = \sum (\bar{x}_{i.} - \bar{x})^2 = \bar{y}_{2.}{}^2 + \cdots + \bar{y}_{p.}{}^2 = \frac{Q_1}{n}.$$

Therefore, *R has the same distribution as Q_1, that is,* $(1-\rho)\sigma^2\chi^2(p-1)$. Since the distribution of R involves the unknown parameters let us define the statistic

$$F = \frac{R}{p-1} \div \frac{Q_2}{(n-1)(p-1)}. \tag{3c.2.10}$$

The statistic F of (3c.2.10) *has Fisher's distribution* $F[p-1, (n-1)(p-1)]$ *which is independent of unknown parameters ρ and σ.* The statistic F reflects the differences due to parities as measured by R since Q_2 does not involve parity differences. Hence it provides a reasonable criterion for the hypothesis under test. Note that $T = B + W = B + R + Q_2$ where T and B are as defined earlier in (3c.2.1). Thus Q_2 can be computed by the formula $T - B - R$ (using the original observations x_{ij}). The expressions for T, B, R in an easily computable form are

$$T = \sum\sum x_{ij}{}^2 - np\bar{x}^2$$

$$B = \frac{1}{p}\sum x_{.j}{}^2 - np\bar{x}^2$$

$$R = \frac{1}{n}\sum x_{i.}{}^2 - np\bar{x}^2$$

where

$$x_{.j} = \sum_i x_{ij} \qquad \text{and} \qquad x_{i.} = \sum_j x_{ij}.$$

3d BIVARIATE NORMAL DISTRIBUTION

3d.1 General Properties

The multivariate normal distribution has already been introduced as providing the sampling distribution of linear functions of independent normal variables. The bivariate distribution is no doubt a special case, but it is of some interest to examine its properties. The general multivariate normal distribution is discussed in detail in Chapter 8, from a different point of view.

A pair of variables (x, y) is said to have a bivariate normal distribution if the density function is

$$c.\, e^{-Q/2} \tag{3d.1.1}$$

where $Q = \lambda_{11}(x - \xi)^2 + 2\lambda_{12}(x - \xi)(y - \eta) + \lambda_{22}(y - \eta)^2$ is positive definite. Let us write

$$Q = \lambda_{22}\left(y - \eta + \frac{\lambda_{12}}{\lambda_{22}}\overline{x - \xi}\right)^2 + \left(\lambda_{11} - \frac{\lambda_{12}{}^2}{\lambda_{22}}\right)(x - \xi)^2$$

$$= \lambda_{22}u^2 + \left(\lambda_{11} - \frac{\lambda_{12}{}^2}{\lambda_{22}}\right)v^2, \tag{3d.1.2}$$

where u and v are defined by the equation (3d.1.2). Substituting for Q in terms of u and v in (3d.1.1) we find that u and v are independent and normal. A number of results follow as a consequence of this.

1. It is seen that $v \sim N(0, \lambda_{22}/\Delta)$ where $\Delta = \lambda_{11}\lambda_{22} - \lambda_{12}{}^2$. Since $v = x - \xi$, it follows that $E(x) = \xi$ and $V(x) = \lambda_{22}/\Delta = \sigma_1{}^2$ (say).

2. Similarly by symmetry, $E(y) = \eta$ and $V(y) = \lambda_{11}/\Delta = \sigma_2{}^2$.

3. Independence of u and v implies $0 = \text{cov}(u, v) = \text{cov}(x, y) + (\lambda_{12}/\lambda_{22}) V(x) = 0$, yielding $\text{cov}(x, y) = -\lambda_{12}/\Delta = \rho\sigma_1\sigma_2$, where ρ is the correlation between x and y.

4. In terms of σ_1, σ_2, ρ,

$$Q = \frac{1}{1 - \rho^2}\left[\frac{(x - \xi)^2}{\sigma_1{}^2} - \frac{2\rho(x - \xi)}{\sigma_1}\frac{(y - \eta)}{\sigma_2} + \frac{(y - \eta)^2}{\sigma_2{}^2}\right].$$

5. The marginal distribution of x is $N(\xi, \sigma_1{}^2)$, being same as that of v and similarly that of y is $N(\eta, \sigma_2{}^2)$.

6. The conditional density of y given x [dividing (3d.1.1) by the marginal density of x] is

$$\text{c. exp}\left\{\frac{-\lambda_{22}}{2}\left[y - \eta - \frac{\lambda_{21}}{\lambda_{22}}(x - \xi)\right]^2\right\}$$

$$= \text{c. exp}\left\{\frac{-1}{2\sigma_2{}^2(1 - \rho^2)}\left[y - \eta - \frac{\rho\sigma_2}{\sigma_1}(x - \xi)\right]^2\right\}$$

which is also normal with variance of $\sigma_2{}^2(1 - \rho^2)$ and mean (regression function)

$$\eta + \frac{\rho\sigma_2}{\sigma_1}(x - \xi) = \alpha + \beta x, \qquad \beta = \frac{\rho\sigma_2}{\sigma_1}, \qquad \alpha = \eta - \beta\xi,$$

which is linear in x.

7. By writing $\sigma_2{}^2(1 - \rho^2) = \sigma_{2.1}^2$, the density at (x, y) can be written

$$(2\pi\sigma_1\sigma_{2.1})^{-1}\exp\left[-\frac{(x - \xi)^2}{2\sigma_1{}^2} - \frac{(y - \alpha - \beta x)^2}{2\sigma_{2.1}^2}\right]. \tag{3d.1.3}$$

8. From (3d.1.3), $(x - \xi)$ and $(y - \alpha - \beta x)$ are independent and their joint c.f. is (using the formula for the univariate case)

$$e^{-(t_1^2\sigma_1^2 + t_2^2\sigma_{2.1}^2)/2},$$

from which the joint c.f. of $(x - \xi)$ and $(y - \eta)$ is obtained as

$$e^{-(t_1^2\sigma_1^2 + 2\rho\sigma_1\sigma_2 t_1 t_2 + t_2^2\sigma_2^2)/2}. \tag{3d.1.4}$$

If the joint c.f. of x and y is needed, the expression (3d.1.4) is multiplied by

$$e^{it_1\xi + it_2\eta}.$$

9. Substituting $ta = t_1$ and $tb = t_2$ in (3d.1.4) we find the c.f. of $a(x - \xi) + b(y - \eta)$ which is seen to be normally distributed with mean zero and variance $a^2\sigma_1^2 + 2\rho ab\sigma_1\sigma_2 + b^2\sigma_2^2$.

Let us suppose that two variables (x, y) are distributed in such a way that every linear function of x and y has a normal distribution. What can be said about the joint distribution of x and y?

Let $E(x) = E(y) = 0$, without loss of generality, and $V(x) = \sigma_1^2$, $V(y) = \sigma_2^2$ and $\text{cov}(x, y) = \rho\sigma_1\sigma_2$. Consider the linear function $ax + by$ with variance $\sigma^2 = a^2\sigma_1^2 + 2ab\rho\sigma_1\sigma_2 + b^2\sigma_2^2$. Its c.f. is

$$E[\exp it(ax + by)] = e^{-\sigma^2 t^2/2} \quad \text{by hypothesis.}$$

Hence

$$E[\exp(it_1 x + it_2 y)] = e^{-(\sigma_1^2 t_1^2 + 2\rho\sigma_1\sigma_2 t_1 t_2 + \sigma_2^2 t_2^2)/2} \tag{3d.1.5}$$

if we write $at = t_1$ and $bt = t_2$. Equation (3d.1.5) holds good for all a and b and hence for all t_1 and t_2. But the right-hand side of (3d.1.5) is the c.f. of a bivariate normal distribution, as in (3d.1.4). We have already seen that if (x, y) has a bivariate normal distribution, every linear function has a normal distribution. The converse is now established.

3d.2 Sampling Distributions

Let $u_i \sim N(0, \sigma^2)$, $i = 1, \ldots, n$, be n independent variables and define

$$\bar{u} = \frac{(u_1 + \cdots + u_n)}{n}$$

$$S_{uv} = u_1(v_1 - \bar{v}) + \cdots + u_n(v_n - \bar{v})$$
$$S_{uu} = (u_1 - \bar{u})^2 + \cdots + (u_n - \bar{u})^2$$
$$S_{vv} = (v_1 - \bar{v})^2 + \cdots + (v_n - \bar{v})^2$$

where $v_1, \ldots, v_n$ are *fixed numbers*. It is easy to see that since $\bar{u}$ and S_{uv} are linear functions of u_i,

$$\bar{u} \sim N\left(0, \frac{\sigma^2}{n}\right) \qquad \text{and hence } \frac{n\bar{u}^2}{\sigma^2} \sim \chi^2(1) \tag{3d.2.1}$$

$$S_{uv} \sim N(0, \sigma^2 S_{vv}) \qquad \text{and hence } \frac{S_{uv}{}^2}{\sigma^2 S_{vv}} \sim \chi^2(1). \tag{3d.2.2}$$

Furthermore, $\text{cov}(\bar{u}, S_{uv}) = \sum (v_i - \bar{v})\sigma^2/n = 0$. Hence they are independent. Let us write

$$S_{uu} = \sum u_i{}^2 - n\bar{u}^2 = \frac{S_{uv}{}^2}{S_{vv}} + \left(S_{uu} - \frac{S_{uv}{}^2}{S_{vv}}\right),$$

which gives

$$\frac{\sum u_i{}^2}{\sigma^2} = \frac{n\bar{u}^2}{\sigma^2} + \frac{S_{uv}{}^2}{\sigma^2 S_{vv}} + \frac{1}{\sigma^2}\left(S_{uu} - \frac{S_{uv}{}^2}{S_{vv}}\right) \tag{3d.2.3}$$

$$\chi^2(n) = \chi^2(1) + \chi^2(1) + \chi^2(n-2).$$

The distribution of the last term is deduced by an application of [(iv), **3b.4**], because it is positive (by the *C-S* inequality or otherwise) and the distributions of the others (3d.2.1, 3d.2.2) are known. The result (3d.2.3) can be established in many other ways, for instance by making an orthogonal transformation on $(u_1, \ldots, u_n)$. From (3d.2.1) to (3d.2.3) we note that the statistics

$$T_1 = \frac{\sqrt{n}\,\bar{u}}{\sigma} \sim N(0, 1)$$

$$T_2 = \frac{S_{uv}}{\sigma\sqrt{S_{vv}}} \sim N(0, 1) \tag{3d.2.4}$$

$$T_3 = \frac{1}{\sigma^2}\left(S_{uu} - \frac{S_{uv}{}^2}{S_{vv}}\right) \sim \chi^2(n-2)$$

are all independent. If $(v_1, \ldots, v_n)$ is now considered as a random variable independent of $(u_1, \ldots, u_n)$, then $(v_1, \ldots, v_n)$ is independent of T_1, T_2, T_3. Let us add two more statistics which are mutually independent under the assumption $v_i \sim N(0, \sigma_0{}^2)$, $i = 1, \ldots, n$:

$$T_4 = \frac{\sqrt{n}\,\bar{v}}{\sigma_0} \sim N(0, 1)$$

$$T_5 = \frac{S_{vv}}{\sigma_0{}^2} \sim \chi^2(n-1). \tag{3d.2.5}$$

The five statistics T_1, T_2, T_3, T_4, and T_5 are all independent. We shall now deduce the sampling distributions of a number of statistics from a bivariate sample.

Joint Distribution of Means from a Bivariate Sample. If (x, y) is an observation from a bivariate normal distribution, then it is shown in (3d.1.3) that

$$\begin{aligned} v &= x - \xi \sim N(0, \sigma_1^2), \\ u &= y - \eta - \beta(x - \xi) \sim N(0, \sigma_{2.1}^2), \end{aligned} \tag{3d.2.6}$$

and are independent. A sample of n pairs (x, y) gives n pairs (u, v) with the relation

$$\bar{v} = \bar{x} - \xi, \qquad \bar{u} = \bar{y} - \eta - \beta(\bar{x} - \xi).$$

Hence by (3d.2.4) and (3d.2.5), we see that

$$\begin{aligned} T_1 &= \frac{[\bar{y} - \eta - \beta(\bar{x} - \xi)]\sqrt{n}}{\sigma_{2.1}} \\ T_4 &= \frac{(\bar{x} - \xi)\sqrt{n}}{\sigma_1} \end{aligned} \tag{3d.2.7}$$

are independent and each is $N(0, 1)$. The density of T_1 and T_4 is

$$\text{c. } e^{-(T_1^2 + T_4^2)/2}.$$

Changing over to variables $\bar{x}$, $\bar{y}$ by the transformation (3d.2.7), we see that the density of $\bar{x}$, $\bar{y}$ is const. $\exp(Q)$ where

$$\begin{aligned} Q &= -n\left\{\frac{(\bar{x} - \xi)^2}{2\sigma_1^2} + \frac{[\bar{y} - \eta - \beta(\bar{x} - \xi)]^2}{2\sigma_{2.1}^2}\right\}, \\ &= \frac{-n}{2(1 - \rho^2)}\left[\frac{(\bar{x} - \xi)^2}{\sigma_1^2} - \frac{2\rho(\bar{x} - \xi)(\bar{y} - \eta)}{\sigma_1\sigma_2} + \frac{(\bar{y} - \eta)^2}{\sigma_2^2}\right], \end{aligned}$$

which is bivariate normal.

Joint Distribution of Variances and Covariances. Using the relationship (3d.2.6) between (u, v) and (x, y), we find

$$\begin{aligned} S_{vv} &= S_{xx} \\ S_{uv} &= S_{xy} - \beta S_{xx} \\ S_{uu} - \frac{S_{uv}^2}{S_{vv}} &= S_{yy} - \frac{S_{xy}^2}{S_{xx}}. \end{aligned}$$

We use the three statistics T_2, T_3, and T_5 as follows:

$$T_2 = \frac{S_{uv}}{\sigma_{2.1}\sqrt{S_{vv}}} = \frac{S_{xy} - \beta S_{xx}}{\sigma_{2.1}\sqrt{S_{xx}}},$$

$$T_3 = \frac{1}{\sigma_{2.1}^2}\left(S_{uu} - \frac{S_{uv}^2}{S_{vv}}\right) = \frac{1}{\sigma_{2.1}^2}\left(S_{yy} - \frac{S_{xy}^2}{S_{xx}}\right), \tag{3d.2.8}$$

$$T_5 = \frac{S_{vv}}{\sigma_1^2} = \frac{S_{xx}}{\sigma_1^2}.$$

The Jacobian of transformation is

$$\frac{D(T_2, T_3, T_5)}{D(S_{xx}, S_{xy}, S_{yy})} = \frac{1}{\sigma_1^2} \cdot \frac{1}{\sigma_{2.1}\sqrt{S_{xx}}} \cdot \frac{1}{\sigma_{2.1}^2}.$$

The joint density of T_2, T_3, T_5 is (from 3d.2.4, 3d.2.5)

$$\text{c. } e^{-2^{-1}(T_2^2 + T_3 + T_5)} T_3^{(n-4)/2} T_5^{(n-3)/2}. \tag{3d.2.9}$$

But

$$T_2^2 + T_3 + T_5 = \frac{1}{1 - \rho^2}\left(\frac{S_{xx}}{\sigma_1^2} - 2\rho\frac{S_{xy}}{\sigma_1\sigma_2} + \frac{S_{yy}}{\sigma_2^2}\right) = Q.$$

Hence by writing (3d.2.9) in terms of S_{xx}, S_{xy}, S_{yy} and introducing the Jacobian the joint density of sample variances and covariance is

$$\text{c. } e^{-Q/2}(S_{xx}S_{yy} - S_{xy}^2)^{(n-4)/2}. \tag{3d.2.10}$$

Distribution of the Correlation Coefficient. The distribution (3d.2.10) may be used to find that of $r = S_{xy} \div \sqrt{S_{xx}S_{yy}}$ by integrating out with respect to two suitably chosen functions as shown by Fisher (1915) in his original memoir. But we shall determine the distribution of r directly by using the fundamental statistics T_2, T_3, T_5.

Introducing r and writing $\sigma_{2.1}^2 = \sigma_2^2(1 - \rho^2)$, we can write the relationships (3d.2.8) as

$$T_2 = \frac{r\sqrt{S_{yy}}}{\sigma_2\sqrt{1 - \rho^2}} - \frac{\rho\sqrt{S_{xx}}}{\sigma_1\sqrt{1 - \rho^2}},$$

$$T_3 = \frac{S_{yy}(1 - r^2)}{\sigma_2^2(1 - \rho^2)},$$

$$T_5 = \frac{S_{xx}}{\sigma_1^2}.$$

Using T_3 and T_5 to eliminate S_{xx} and S_{yy} from T_2, we obtain

$$T_2 = \frac{r}{\sqrt{1-r^2}}\sqrt{T_3} - \frac{\rho}{\sqrt{1-\rho^2}}\sqrt{T_5}, \tag{3d.2.11}$$

which is a fundamental equation due to Fraser and Sprott. The distribution of r is obtained from the joint distribution of T_2, T_3, and T_5 by using equation (3d.2.11). When $\rho = 0$

$$T_2 = \frac{r}{\sqrt{1-r^2}}\sqrt{T_3},$$

and the distribution of $r/\sqrt{1-r^2}$ is therefore that of $T_2/\sqrt{T_3}$. But

$$T_2/\sqrt{T_3 \div (n-2)} \sim S(n-2),$$

Student's distribution (3b.1.8). Hence

$$\frac{r\sqrt{n-2}}{\sqrt{1-r^2}} \sim S(n-2), \tag{3d.2.12}$$

so that the significance of r can be tested by using tables of percentage points of t. From (3d.2.12) we find the distribution of r by change of variable as

$$\frac{\Gamma[(n-1)/2]}{\sqrt{\pi}\Gamma[(n-2)/2]}(1-r^2)^{(n-4)/2}\,dr, \qquad -1 \leqslant r \leqslant 1. \tag{3d.2.13}$$

To determine the non-null distribution we use the joint distribution of T_2, T_3, and T_5. Let $t = r/\sqrt{1-r^2}$ and $\theta = \rho/\sqrt{1-\rho^2}$, in which case (3d.2.11) reduces to

$$t\sqrt{T_3} = T_2 + \theta\sqrt{T_5} \qquad \text{or} \qquad t = \frac{T_2 + \theta\sqrt{T_5}}{\sqrt{T_3}}. \tag{3d.2.14}$$

The joint density of T_2, T_3, T_5 is

$$\text{c. } e^{-(T_2^2 + T_3 + T_5)/2} T_3^{(n-4)/2} T_5^{(n-3)/2}, \tag{3d.2.15}$$

and the problem is reduced to finding the distribution of t which is the function of T_2, T_3, T_5 defined in (3d.2.14). Substituting

$$T_2 = t\sqrt{T_3} - \theta\sqrt{T_5}$$
$$dT_2 = \sqrt{T_3}\,dt$$

in (3d.2.15), we find the joint density of t, T_3, T_5 to be

$$\text{c. } e^{-[T_3(1+t^2) + T_5(1+\theta^2) - 2t\theta\sqrt{T_3 T_5}]/2}(T_3 T_5)^{(n-3)/2}.$$

Let us transform T_3, T_5 to

$$u = T_3(1 + t^2), \qquad v = T_5(1 + \theta^2),$$

which leads to the joint density of t, u, v

$$\text{c.}\left(\frac{1}{1+t^2}\right)^{(n-1)/2} e^{-(u+v-2\rho r\sqrt{uv})/2}(uv)^{(n-3)/2}$$

and to that of r, u, v $(t = r/\sqrt{1-r^2})$, which is

$$\text{c.}\,(1 - r^2)^{(n-4)/2} e^{-(u+v-2\rho r\sqrt{uv})/2}(uv)^{(n-3)/2}.$$

The density of r alone can be expressed as an integral

$$\text{c.}\,(1 - r^2)^{(n-4)/2} \int_0^\infty \int_0^\infty e^{-(u+v-2\rho r\sqrt{uv})/2}(uv)^{(n-3)/2}\, du\, dv.$$

If we expand $e^{\rho r\sqrt{uv}}$ and integrate out term by term, the non-null distribution of r is obtained as

$$\text{c.}\,(1 - r^2)^{(n-4)/2} \sum_{s=0}^{\infty} \Gamma^2\left(\frac{n+s-1}{2}\right)\frac{(2\rho r)^s}{s!}, \tag{3d.2.16}$$

where the constant may be evaluated as

$$\frac{2^{n-3}}{\pi(n-3)!}(1 - \rho^2)^{(n-1)/2}.$$

Distribution of the Regression Coefficient. The regression coefficient of y on x is estimated by

$$b = \frac{S_{xy}}{S_{xx}}.$$

To find its distribution, the relevant statistics are

$$T_2 = \frac{b-\beta}{\sigma_{2.1}}\sqrt{S_{xx}} = \frac{\sigma_1}{\sigma_{2.1}}\sqrt{T_5}(b-\beta),$$

$$T_3 = \frac{1}{\sigma_{2.1}}\left(S_{yy} - \frac{S_{xy}^2}{S_{xx}}\right), \tag{3d.2.17}$$

$$T_5 = \frac{S_{xx}}{\sigma_1^2}.$$

The most useful distribution is of the statistic

$$t = \frac{(b-\beta)\sqrt{S_{xx}}}{\left[\left(S_{yy} - \dfrac{S_{yx}^2}{S_{xx}}\right)\Big/(n-2)\right]^{1/2}} = \frac{T_2}{\sqrt{T_3 \div (n-2)}} \sim S(n-2),$$

which is Student's distribution on $(n-2)$ D.F. The significance of an observed deviation of the estimate from the hypothetical value β is tested by the t statistic. If the distribution of b alone is desired, however, we have to consider the statistics T_2 and T_5 with the joint density

$$\text{c. } e^{-(T_2{}^2+T_5)/2}T_5^{(n-3)/2}.$$

By transforming to b and T_5 by the relation (3d.2.17) the density transforms to

$$\text{c. } e^{-T_5[\sigma_{2.1}{}^2+\sigma_1{}^2(b-\beta)^2]/2}T_5^{(n-2)/2}.$$

If we integrate out for T_5, the density of b alone is

$$\text{c. } [\sigma_{2.1}^2+\sigma_1{}^2(b-\beta)^2]^{-n/2}. \tag{3d.2.18}$$

COMPLEMENTS AND PROBLEMS

1 *Independence of linear and quadratic forms.* Let $y_i \sim N(\mu_i, \sigma^2)$, $i = 1, \ldots, n$ be all independent and denote $\mathbf{Y}' = (y_1, \ldots, y_n)$ and $\boldsymbol{\mu}' = (\mu_1, \ldots, \mu_n)$. Prove the following results.

1.1 A sufficient condition for $\mathbf{Y}'\mathbf{A}\mathbf{Y}$ and $\mathbf{Y}'\mathbf{B}\mathbf{Y}$ to be independently distributed is $\mathbf{AB} = \mathbf{BA} = 0$. [Hint: by the spectral decomposition (1c.3.4, 1c.3.5) of $\mathbf{A}$ and $\mathbf{B}$

$$\mathbf{A} = \lambda_1\mathbf{P}_1\mathbf{P}_1' + \cdots + \lambda_r\mathbf{P}_r\mathbf{P}_r', \qquad \lambda_i \neq 0, \quad i = 1, \ldots, r$$
$$\mathbf{B} = \nu_1\mathbf{Q}_1\mathbf{Q}_1' + \cdots + \nu_s\mathbf{Q}_s\mathbf{Q}_s', \qquad \nu_j \neq 0, \quad j = 1, \ldots, s$$

where $r = \text{rank } \mathbf{A}$ and $s = \text{rank } \mathbf{B}$ and $\mathbf{P}_i$ are mutually orthonormal and so also are $\mathbf{Q}_j$. $\mathbf{AB} = \mathbf{0} = \sum\sum \lambda_i \mathbf{P}_i\mathbf{P}_i'\mathbf{Q}_j\mathbf{Q}_j'$. Pre- and postmultiplying by $\mathbf{P}_r'$ and $\mathbf{Q}_s$, we find $\mathbf{P}_r'\mathbf{Q}_s = 0$ so that $\mathbf{P}_i$ and $\mathbf{Q}_j$ are all orthonormal. Then $\mathbf{P}_i'\mathbf{Y}$ and $\mathbf{Q}_j'\mathbf{Y}$ are all independent and normal. The sufficiency of the condition follows by observing that $\mathbf{Y}'\mathbf{A}\mathbf{Y} = \lambda_1(\mathbf{P}_1'\mathbf{Y})^2 + \cdots + \lambda_r(\mathbf{P}_r'\mathbf{Y})^2$ and $\mathbf{Y}'\mathbf{B}\mathbf{Y} = \nu_1(\mathbf{Q}_1'\mathbf{Y})^2 + \cdots + \nu_s(\mathbf{Q}_s'\mathbf{Y})^2$ depend on exclusive sets of independent variables. Note that $\mathbf{Y}'\mathbf{A}\mathbf{Y}$ and $\mathbf{Y}'\mathbf{B}\mathbf{Y}$ need not have χ^2 distributions.]

1.2 Let $\mathbf{BY}$ be m linear functions of $\mathbf{Y}$, that is, $\mathbf{B}$ is $m \times n$ matrix. A sufficient condition for $\mathbf{BY}$ to be distributed independently of the quadratic form $\mathbf{Y}'\mathbf{A}\mathbf{Y}$ is that $\mathbf{BA} = \mathbf{0}$.

[Use the spectral decomposition of $\mathbf{A}$ as in Example 1.1. The condition $\mathbf{BA} = \mathbf{0}$ implies that $\mathbf{BP}_r = 0$, $i = 1, \ldots, r$. This means that $\mathbf{BY}$ is independent of the linear functions $\mathbf{P}_1'\mathbf{Y}, \ldots, \mathbf{P}_r'\mathbf{Y}$. But $\mathbf{Y}'\mathbf{A}\mathbf{Y} = \sum \lambda_i(\mathbf{P}_i'\mathbf{Y})^2$.]

1.3 The condition $\mathbf{BA} = \mathbf{0}$ is also necessary for the results of Examples 1.1 and 1.2 to hold.

[The proof of the necessity is somewhat complicated and employs characteristic functions. See Ogawa (1949). It may be noted that in **3b.4** the necessity is established under the additional condition that the matrices of the quadratic forms are symmetric and idempotent.]

1.4 Show that the c.f. of $\mathbf{Y'AY}$ is $|\mathbf{I} - 2i\sigma^2 t\mathbf{A}|^{1/2}$, and the joint c.f. of $\mathbf{Y'AY}$ and $\mathbf{Y'BY}$ is $|\mathbf{I} - 2i\sigma^2 t_1\mathbf{A} - 2i\sigma^2 t_2\mathbf{B}|^{1/2}$.

1.5 *Sakamoto-Craig theorem.* If $\mathbf{Y'AY}$ and $\mathbf{Y'BY}$ are independently distributed $|\mathbf{I} - 2i\sigma^2 t_1\mathbf{A} - 2i\sigma^2 t_2\mathbf{B}| = |\mathbf{I} - 2i\sigma^2 t_1\mathbf{A}|\ |\mathbf{I} - 2i\sigma^2 t_2\mathbf{B}|$ for all t_1, t_2. Hence deduce that $\mathbf{AB} = \mathbf{0}$.

2 Consider the setup of **3b.5** of n independent variables

$$y_i \sim N(x_{i1}\beta_1 + \cdots + x_{im}\beta_m, \sigma^2), \quad i = 1, \ldots, n.$$

Let $\mathbf{Y}' = (y_1, \ldots, y_n)$, $\boldsymbol{\beta}' = (\beta_1, \ldots, \beta_m)$, $\mathbf{X} = (x_{ij})$ of rank r, and $(\mathbf{X'X})^-$ be a generalized inverse of $\mathbf{X'X}$ (**1b.5**). Further let $\hat{\boldsymbol{\beta}} = (\mathbf{X'X})^-\mathbf{X'Y}$ so that $\hat{\boldsymbol{\beta}}$ is a solution of the equation $\mathbf{X'X}\boldsymbol{\beta} = \mathbf{X'Y}$.

2.1 Show that (a) $\mathbf{X} = \mathbf{X}(\mathbf{X'X})^-\mathbf{X'X}$ and (b) the matrix $\mathbf{I} - \mathbf{X}(\mathbf{X'X})^-\mathbf{X}'$ is idempotent. [Hint: Let $\mathbf{G} = \mathbf{X}[\mathbf{I} - (\mathbf{X'X})^-\mathbf{X'X}]$. Compute $\mathbf{G'G}$ and show that it is zero, from which it follows that $\mathbf{G} = \mathbf{0}$. To prove (b), take the square of $\mathbf{I} - \mathbf{X}(\mathbf{X'X})^-\mathbf{X}'$ and use result (a).] (c) Also $\mathbf{X}' = \mathbf{X'X}(\mathbf{X'X})^-\mathbf{X}'$.

2.2 Let $\mathbf{H}$ be a $m \times k$ matrix such that $\mathscr{M}(\mathbf{H}) = \mathscr{M}(\mathbf{X}')$. Show that $\mathbf{Z} = \mathbf{H}\hat{\boldsymbol{\beta}}'$ and $R_0^2 = (\mathbf{Y} - \mathbf{X}\hat{\boldsymbol{\beta}})'(\mathbf{Y} - \mathbf{X}\hat{\boldsymbol{\beta}})$ are independently distributed. [Hint: Since $\mathscr{M}(\mathbf{H}) = \mathscr{M}(\mathbf{X}')$, there exists a matrix $\mathbf{C}$ such that $\mathbf{H} = \mathbf{X'C}$. Hence

$$\mathbf{H}'\hat{\boldsymbol{\beta}} = \mathbf{C'X}(\mathbf{X'X})^-\mathbf{X'Y} = \mathbf{BY}.$$

$$R_0^2 = \mathbf{Y'Y} - \mathbf{Y'X}(\mathbf{X'X})^-\mathbf{X'Y} = \mathbf{Y}'[\mathbf{I} - \mathbf{X}(\mathbf{X'X})^-\mathbf{X}']\mathbf{Y} = \mathbf{Y'AY}.$$

We want to establish the independence of a set of linear functions and a quadratic function of $\mathbf{Y}$. We find

$$\mathbf{BA} = [\mathbf{C'X}(\mathbf{X'X})^-\mathbf{X}'][\mathbf{I} - \mathbf{X}(\mathbf{X'X})^-\mathbf{X}'] = 0,$$

since the product of the first two terms is zero by (c) of Example 2.1. The result follows from the condition of sufficiency of Example 1.2.]

2.3 $\mathbf{Z} = \mathbf{H}'\hat{\boldsymbol{\beta}}$ has a k-variate normal distribution with mean $\mathbf{H}'\boldsymbol{\beta}$ and dispersion matrix $\sigma^2\mathbf{D}$, where $\mathbf{D} = \mathbf{H}'(\mathbf{X'X})^-\mathbf{H}$ and $R_0^2 \sim \sigma^2\chi^2(n - r)$. [Hint: $\mathbf{Z} = \mathbf{C'X}(\mathbf{X'X})^-\mathbf{X'Y} = \mathbf{BY}$, that is, linear functions of normal variables. By applying (3b.3.4), $\mathbf{Z}$ has a k-variate normal distribution. Consider

$$R_0^2 = \mathbf{Y}'[\mathbf{I} - \mathbf{X}(\mathbf{X'X})^-\mathbf{X}']\mathbf{Y}.$$

The matrix $\mathbf{I} - \mathbf{X}(\mathbf{X'X})^-\mathbf{X}'$ is idempotent. Hence $R_0^2 \sim \sigma^2\chi^2(n - r)$ where $n - r = \text{trace } \mathbf{I} - \text{trace } \mathbf{X}(\mathbf{X'X})^-\mathbf{X}' = n - \text{trace}(\mathbf{X'X})^-\mathbf{X'X} = n - \text{rank } \mathbf{X'X} = n - \text{rank } \mathbf{X}$. Furthermore, by applying the sufficiency condition of Example 1.2, $\mathbf{Z}$ and R_0^2 are independently distributed.]

2.4 Let $\boldsymbol{\beta}^*$, $\boldsymbol{\lambda}^*$ be a solution of the equations

$$\mathbf{X}'\mathbf{X}\boldsymbol{\beta} + \mathbf{H}\boldsymbol{\lambda} = \mathbf{X}'\mathbf{Y}$$
$$\mathbf{H}'\boldsymbol{\beta} = \boldsymbol{\theta}_0 \quad \text{(a given value)},$$

which are obtained by differentiating $(\mathbf{Y} - \mathbf{X}\boldsymbol{\beta})'(\mathbf{Y} - \mathbf{X}\boldsymbol{\beta})$ with respect to $\boldsymbol{\beta}$ subject to the condition $\mathbf{H}'\boldsymbol{\beta} = \boldsymbol{\theta}_0$. Show that

$$\begin{aligned} R_1{}^2 &= (\mathbf{Y} - \mathbf{X}\boldsymbol{\beta}^*)'(\mathbf{Y} - \mathbf{X}\boldsymbol{\beta}^*) \\ &= (\mathbf{Z} - \boldsymbol{\theta}_0)\mathbf{D}^{-1}(\mathbf{Z} - \boldsymbol{\theta}_0) + \mathbf{R}_0{}^2, \end{aligned}$$

where $\mathbf{Z}$ and $\mathbf{D}$ are as defined in Example 2.3. [Hint: Observe that $R_1{}^2 = (\mathbf{Y} - \mathbf{X}\hat{\boldsymbol{\beta}})'(\mathbf{Y} - \mathbf{X}\hat{\boldsymbol{\beta}}) + (\hat{\boldsymbol{\beta}} - \boldsymbol{\beta}^*)'\mathbf{X}'\mathbf{X}(\hat{\boldsymbol{\beta}} - \boldsymbol{\beta}^*)$, and it is enough to show that the second term is $(\mathbf{Z} - \boldsymbol{\theta}_0)\mathbf{D}^{-1}(\mathbf{Z} - \boldsymbol{\theta}_0)$. Since $\mathcal{M}(\mathbf{H}) \subset \mathcal{M}(\mathbf{X}'\mathbf{X})$, $\mathbf{H} = \mathbf{X}'\mathbf{X}\mathbf{C}$. From the equations for $\boldsymbol{\beta}^*$, $\boldsymbol{\lambda}^*$, $\mathbf{X}'\mathbf{X}(\hat{\boldsymbol{\beta}} - \boldsymbol{\beta}^*) = \mathbf{H}\boldsymbol{\lambda}^* = \mathbf{X}'\mathbf{X}\mathbf{C}\boldsymbol{\lambda}^*$ so that $(\hat{\boldsymbol{\beta}} - \boldsymbol{\beta}^*)'\mathbf{X}'\mathbf{X}(\hat{\boldsymbol{\beta}} - \boldsymbol{\beta}^*) = \boldsymbol{\lambda}^{*\prime}\mathbf{C}'\mathbf{X}'\mathbf{X}\mathbf{C}\boldsymbol{\lambda}^*$. Also from the equations, $\mathbf{C}'\mathbf{X}'\mathbf{X}\boldsymbol{\beta}^* - \mathbf{H}'\boldsymbol{\beta}^* + \mathbf{C}'\mathbf{H}^*\boldsymbol{\lambda} = \mathbf{C}'\mathbf{X}'\mathbf{Y} - \boldsymbol{\theta}_0$ or $\mathbf{C}'\mathbf{H}\boldsymbol{\lambda}^* = \mathbf{Z} - \boldsymbol{\theta}_0$ or $\mathbf{C}'\mathbf{X}'\mathbf{X}\mathbf{C}\boldsymbol{\lambda}^* = \mathbf{Z} - \boldsymbol{\theta}_0 \Rightarrow \boldsymbol{\lambda}^* = \mathbf{D}^{-1}(\mathbf{Z} - \boldsymbol{\theta}_0)$ where $\mathbf{D} = \mathbf{C}'\mathbf{X}'\mathbf{X}\mathbf{C}$ by substituting for $\mathbf{H}$ in the expression for $\mathbf{D}$ in Example 2.3. Hence the result.]

[Note: In Examples 2.1 to 2.4, we have alternative proofs of the first two fundamental theorems of least squares provided we can assert that the expressions $R_0{}^2$ and $R_1{}^2$ correspond to actual minima, when $\boldsymbol{\beta}$ are unrestricted and restricted to $\mathbf{H}'\boldsymbol{\beta} = \boldsymbol{\theta}_0$ respectively.]

2.5 Relax the condition $\mathcal{M}(\mathbf{H}) \subset \mathcal{M}(\mathbf{X}')$ in (ii), **3b.5**. Then $R_1{}^2 - R_0{}^2$ is a noncentral χ^2 on t degrees of freedom, where $t = \text{rank}[\mathcal{M}(\mathbf{H}) \cap \mathcal{M}(\mathbf{X}')]$. Find the noncentrality parameter.

2.6 Let $R_0{}^2$, $R_1{}^2$, and $R_2{}^2$ be the minima of $(\mathbf{Y} - \mathbf{X}\boldsymbol{\beta})'(\mathbf{Y} - \mathbf{X}\boldsymbol{\beta})$ with no restriction on $\boldsymbol{\beta}$, with $\mathbf{H}'\boldsymbol{\beta} = \boldsymbol{\xi}$, and with $\mathbf{H}'\boldsymbol{\beta} = \boldsymbol{\xi}$, $\mathbf{G}'\boldsymbol{\beta} = \boldsymbol{\eta}$. Then $R_0{}^2$, $R_1{}^2 - R_0{}^2$, and $R_2{}^2 - R_1{}^2$, are independently distributed. $R_0{}^2 \sim \sigma^2\chi^2(n - r)$ and $R_1{}^2 - R_0{}^2$ and $R_2{}^2 - R_1{}^2$ are noncentral χ^2's with k and t degrees of freedom respectively, where $r = \text{rank}\,\mathbf{X}$, $k = \text{rank}[\mathcal{M}(\mathbf{H}) \cap \mathcal{M}(\mathbf{X}')]$ and $t + k = \text{rank}[\mathcal{M}(\mathbf{H}) \cup \mathcal{M}(\mathbf{G}) \cap \mathcal{M}(\mathbf{X}')]$. The noncentrality parameters are zero if the restrictions are true.

2.7 Generalize the results of Example 2.6 to $R_0{}^2, R_1{}^2, R_2{}^2, R_3{}^2, \ldots$ suitably defined.

2.8 Consider a vector $\mathbf{Y}$ of n independent normal variables with a common variance σ^2 and a mean vector $\boldsymbol{\mu}$. Let $\omega_1, \omega_2, \ldots, \omega_k$ be subspaces of the n-dimensional Euclidean space and $R_1{}^2, R_2{}^2, \ldots, R_k{}^2$ be the minima of $(\mathbf{Y} - \boldsymbol{\mu})'(\mathbf{Y} - \boldsymbol{\mu})$ subject to the restrictions $\boldsymbol{\mu} \in \omega_1, \boldsymbol{\mu} \in \omega_1 \cap \omega_2, \ldots, \boldsymbol{\mu} \in \omega_1 \cap \omega_2 \cdots \cap \omega_k$ respectively. Show that $R_1{}^2, R_2{}^2 - R_1{}^2, R_3{}^2 - R_2{}^2, \ldots$ are independent noncentral χ^2 variables with degrees of freedom $k_1 = n - d[\omega_1]$, $k_2 = d[\omega_1] - d[\omega_1 \cap \omega_2]$, $k_3 = d[\omega_1 \cap \omega_2] - d[\omega_1 \cap \omega_2 \cap \omega_3]$, ... and so on, where d represents the dimension of the subspace indicated. The problem is a special case of (2.6) with homogeneous restrictions. [Hint: Construct an orthogonal transformation by taking an o.n.b. of $\omega_1 \cap \cdots \cap \omega_k$ and by adding vectors build up an o.n.b. of $\omega_1 \cap \cdots \cap \omega_{k-1}$, and so on until an o.n.b. of the entire space is obtained.]

2.9 In Example 2.8, state the condition under which $R_i^2 - R_{i-1}^2$ is equal to the minimum of $(\mathbf{Y} - \boldsymbol{\mu})'(\mathbf{Y} - \boldsymbol{\mu})$ subject to the restriction $\boldsymbol{\mu} \in \omega_i$.

2.10 Show that the dimensions of the subspace generated by the vectors $\mathbf{X}\boldsymbol{\beta}$ (where $\mathbf{X}$ is $n \times m$ matrix) when the column vector $\boldsymbol{\beta}$ is restricted to the condition $\mathbf{H}'\boldsymbol{\beta} = \mathbf{0}$ (where $\mathbf{H}'$ is $k \times m$ matrix) is $d\{\mathcal{M}(\mathbf{X}')\} - d\{\mathcal{M}(\mathbf{X}') \cap \mathcal{M}(\mathbf{H})\}$.

2.11 Satisfy yourself that the results of **3b.5** and of the Examples 2.5 to 2.9 remain the same if $\mathbf{Y}$ has an n-variate normal distribution with mean $\mathbf{X}\boldsymbol{\beta}$ and dispersion matrix $\boldsymbol{\Lambda}$ and $R_0^2, R_1^2, \ldots$ are defined as the minima of $(\mathbf{Y} - \mathbf{X}\boldsymbol{\beta})'\boldsymbol{\Lambda}^{-1}(\mathbf{Y} - \mathbf{X}\boldsymbol{\beta})$.

3 Let $x_i \sim N(\mu, \sigma^2)$, $i = 1, \ldots, n$, be n independent variables. If $m_1, \ldots, m_n$ are fixed quantities, find the distribution of the correlation coefficient

$$r = \sum x_i(m_i - \bar{m}) \div [\sum (m_i - \bar{m})^2 \cdot \sum (x_i - \bar{x})^2]^{1/2}.$$

[Hint: By a suitable transformation or otherwise show that $r = x/\sqrt{x^2 + y}$ where $x \sim N(0, 1)$ and $y \sim \chi^2(n - 2)$ are independent. Then use (3b.1.13).]

4 *Laplace distribution.* The density of Laplace distribution is of the form

$$L(x \mid \mu, \lambda) = (2\lambda)^{-1} \exp(-|x - \mu|/\lambda).$$

Show that the c.f. is $(1 + \lambda^2 t^2)^{-1} \exp(it\mu)$. It has moments of all orders and is not preserved under convolution.

5 *Pareto distribution.* The random variable has the density

$$\frac{\alpha}{x_0}\left(\frac{x_0}{x}\right)^{\alpha+1}, \qquad x > x_0, \quad \alpha > 0,$$

and zero otherwise. Examine the moments of this distribution. Show that the mean is finite only when $\alpha > 1$, in which case it is equal to $\alpha/x_0(1 - \alpha)$. The distribution function has the simpler form $1 - (x_0/x)^\alpha$. Hence show that the median is $x_0 2^{1/\alpha}$. Distribution of incomes of individuals whose income exceeds a certain limit x_0 (a truncated distribution) sometimes has Pareto's form.

6 *Log normal distribution.* Consider a series of independent random impulses occurring in the order $\xi_1, \xi_2, \ldots$, the effect of ξ_r being to increase the momentary size of an organism from x_r to x_{r+1} by the relation

$$x_{r+1} = x_r + \xi_r x_r \Rightarrow \frac{x_{r+1}}{x_r} = 1 + \xi_r = \eta_r.$$

If $w = \log x - \log x_0$—the logarithmic difference between the sizes at the final stage and initial stage—then

$$w = \log \eta_0 + \cdots + \log \eta_{n-1},$$

where $\log \eta_i$ are independent. Let each impulse give only a slight growth and n be large. As in [(iii), **3a.1**] let $n \to \infty$, such that $V(w)$ is finite. In the limit, $w = \log x - \log x_0$ has a normal distribution or x has the distribution

$$\frac{1}{x\sigma\sqrt{2\pi}} e^{-(\log x - \mu)^2/2\sigma^2}.$$

If x is measured from α, we have a more general form

$$\frac{1}{\sigma(x-\alpha)\sqrt{2\pi}} e^{-[\log(x-\alpha)-\mu]^2/2\sigma^2}, \qquad x > \alpha.$$

The log normal distribution provides, in some cases, a good fit to income distribution. Find the mean and variance of log normal distribution.

7 *Logistic distribution.* The distribution function is given by

$$F(x) = \frac{1}{1 + e^{-(\alpha+\beta x)}} \qquad \text{with} \quad \alpha > 0.$$

Find the c.f. and moments of x. This is used as a tolerance distribution in bioassay problems.

8 *Pearsonian system of frequency curves.* Let $y = f(x)$ denote the density function. From general considerations, Karl Pearson obtained the differential equation

$$\frac{1}{y}\frac{dy}{dx} = \frac{x + a}{b_0 + b_1 x + b_2 x^2}.$$

Almost all the distributions considered in Chapter 3 are special cases of this system and can be deduced by special choices of the constants a, b_0, b_1, and b_2. Show that the constants can be expressed in terms of the first four moments of the distribution, if they exist.

9 *Life testing models.* Let $F(t)$ be the probability that a structure fails by time t and $\lambda(t)\Delta t + 0(\Delta t)$, the probability of failure in a small interval Δt having survived up to t.

9.1 Show that $F(t)$ satisfies the differential equation

$$\frac{f(t)}{1 - F(t)} = \lambda(t), \qquad \text{where} \quad F'(t) = f(t) \quad \text{is the p.d.}$$

leading to the functional form

$$F(t) = 1 - \exp\left[-\int \lambda(t)\,dt\right].$$

9.2 For the special case $\lambda(t) = c$ (constant), deduce the exponential distribution

$$F(t) = 1 - e^{-ct} \quad \text{or} \quad f(t) = ce^{-ct}.$$

9.3 Show that the exponential distribution is characterized by the property that the conditional distribution of further life, given that a structure has survived up to time T, is independent of T, and is the same as the distribution of life from the start. [There are some structures which obey this property such as an electric fuse and a jeweled bearing in a watch.]

9.4 Show that for the special case $\lambda(t) = c\alpha t^{\alpha-1}$, $\alpha > 1$, $c > 0$, the d.f. is $1 - \exp(-ct^{\alpha})$, which is called *Weibull distribution.*

9.5 Derive the expression for $F(t)$ when $\lambda(t) = \gamma + \delta e^{\mu t}$ which is called Makeham's law. The function $F(t)$ so derived is used in the construction of *mortality tables.*

10 *Order statistics.* Let $x_1, \ldots, x_n$ be n independent observations from a continuous d.f. $F(x)$ and let $y_1, \ldots, y_n$ be the order observations (or statistics as they are called). The p.d. if it exists is denoted by $f(x)$.

10.1 The d.f. of y_1 is $1 - [1 - F(y_1)]^n$. [Hint: the probability that the minimum is less than y_1 is equal to one minus the probability that all the observations are not less than y_1.] If $f(x)$ exists, then the p.d. of y_1 is

$$n[1 - F(y_1)]^{n-1} f(y_1).$$

10.2 Similarly the d.f. and p.d. of the maximum y_n are

$$[F(y_n)]^n \quad \text{and} \quad n[F(y_n)]^{n-1} f(y_n).$$

10.3 The joint d.f. of y_1 and y_n is

$$[F(y_n)]^n - [F(y_n) - F(y_1)]^n.$$

Note that the first term represents the probability that the maximum is less than y_n and the second that all the observations are between y_1 and y_n. The p.d. is obtained by differentiation

$$n(n-1)[F(y_n) - F(y_1)]^{n-2} f(y_1) f(y_n).$$

This is the starting point for finding the distribution of the range $(y_n - y_1)$.

10.4 The kth order statistic is less than y_k if the number of observations less than y_k is k or $k+1$ or $k+2 \cdots$ or n, the probability for which is

$$1 - \sum_{r=0}^{k-1} \binom{n}{r} [F(y_k)]^r [1 - F(y_k)]^{n-r} \quad \text{(by binomial theorem)}$$

$$= 1 - \frac{n!}{(k-1)!\,(n-k)!} \int_0^{1-F} x^{n-k}(1-x)^{k-1}\,dx$$

which is an important result. The p.d. is

$$\frac{n!}{(k-1)!\,(n-k)!} [1 - F(y_k)]^{n-k} [F(y_k)]^{k-1} f(y_k).$$

The distribution of the median and other useful fractiles can be obtained as special cases.

10.5 Show that the distribution of $F(y_k)$ is $B(k, n-k+1)$.

10.6 Show that the joint p.d. of $y_k, y_s, \ldots, y_u$ $(k < s \cdots < u)$ order statistics is

$$\frac{\Gamma(n+1)}{\Gamma(k)\Gamma(s-k) \cdots \Gamma(n-u+1)} [F(y_k)]^{k-1} [F(y_s) - F(y_k)]^{s-k-1} \cdots$$

$$[1 - F(y_u)]^{n-u} f(y_k) \cdots f(y_u).$$

11 *Dirichlet distribution.* Let $x_1, \ldots, x_{k+1}$ be independent gamma variates with parameters $(\alpha, p_1), \ldots, (\alpha, p_{k+1})$. Show that

$$y_i = x_i/(x_1 + \cdots + x_i), \quad i = 2, \ldots, k+1$$

are independently distributed. Further, show that

$$z_i = x_i/(x_1 + \cdots + x_{k+1}), \quad i = 1, \ldots, k$$

have the joint p.d.

$$\frac{\Gamma(p_1) \cdots \Gamma(p_{k+1})}{\Gamma(p_1 + \cdots + p_{k+1})} z_1^{p_1 - 1} \cdots z_k^{p_k - 1} (1 - z_1 - \cdots - z_k)^{p_{k+1} - 1}$$

which may be called Dirichlet distribution.

12 If x_1, x_2 are two independent observations from $N(0, 1)$ find the distribution of $z = x_1/x_2$.

13 Let $x = x_1 + \cdots + x_n$ be the sum of n independent observations from a rectangular distribution $R(0, 1)$. Show that (Rao, 1942a)

$$P(x < X) = \frac{X^n}{n!\,0!} - \frac{(X-1)^n}{(n-1)!\,1!} + \frac{(X-2)^n}{(n-2)!\,2!} \cdots + (-1)^r \frac{(X-r)^n}{(n-r)!\,r!}, \quad r = [X].$$

14 If $f(x)$ is p.d. of a random variable x, $\alpha f(\alpha y)$ is the p.d. of $y = x/\alpha$. Consider random variables y_1, y_2 with p.d.'s $\alpha_1 f(\alpha_1 y_1)$ and $\alpha_2 f(\alpha_2 y_2)$. Find the functional form f if the p.d. of $y = y_1 + y_2$ is of the same form $\alpha f(\alpha y)$

with respect to some α. Deduce that if the first two moments of f exist then it is normal [Polya, 1923].

15 On the circumference of a circle of unit length, n arcs each of length x are marked off at random. (a) Show that the probability that every point of the circle is included in at least one arc is

$$1 - \binom{n}{1}(1-x)^{n-1} + \binom{n}{2}(1-2x)^{n-1} - \cdots \pm \binom{n}{k}(1-kx)^{n-1}, \quad k = \left[\frac{1}{x}\right].$$

(b) Show that the probability for i gaps is

$$\binom{n}{i}\left[(1-ix)^{n-1} - (n-i)[1-(i+1)x]^{n-1} + \cdots \pm \frac{(n-i)!}{(k-i)!(n-k)!}(1-kx)^{n-1}\right] \quad \text{[Stevens, 1939].}$$

16 If x_r has χ^2 distribution on 2 d.f., $r = 1, \ldots, n$, and x_r are independent, then the probability that i of the fractions $x_r/\sum_1^n x_r$ exceed x is the same as the expression (b) in Example 15. Similarly, the probability for the largest fraction to exceed x is the same as the expression (a). These distributions are used by Fisher (1929) for tests of significance in Harmonic analysis.

17 *Noncentral f and beta distributions.* Let $\chi_1^2 \sim \chi^2(k, \lambda)$ [noncentral chi-square (3b.2.2)] and $\chi_2^2 \sim \chi^2(m)$ be independent. Define $f = \chi_1^2/\chi_2^2$ and $g = \chi_1^2/(\chi_1^2 + \chi_2^2)$. It was shown that when $\lambda = 0$, f has the distribution (3b.1.10) of the ratio of central χ^2's and g has the beta distribution (3b.1.12). When $\lambda \neq 0$, the joint p.d. of χ_1^2 and χ_2^2 is

$$e^{-\lambda^2/2} \sum \left(\frac{\lambda^2}{2}\right)^r \frac{1}{r!} G\left(\chi_1^2 \,\middle|\, \frac{1}{2}, \frac{k}{2} + r\right) d\chi_1^2 \, G\left(\chi_2^2 \,\middle|\, \frac{1}{2}, \frac{m}{2}\right) d\chi_2^2.$$

17.1 Apply the result (3b.1.7) to each term to obtain the p.d. of $f = \chi_1^2/\chi_2^2$

$$e^{-\lambda^2/2} \sum_{r=0}^{\infty} \left(\frac{\lambda^2}{2}\right)^r \frac{1}{r!} \frac{\Gamma\left(\frac{k+m}{2} + r\right)}{\Gamma\left(\frac{k}{2} + r\right)\Gamma\left(\frac{m}{2}\right)} \frac{f^{r-1+k/2}}{(1+f)^{r+(k+m)/2}},$$

which is called noncentral f (ratio of χ^2's) and denoted by $f(k, m, \lambda)$. This can also be written in the form using the hypergeometric function ${}_1F_1$,

$$e^{-\lambda^2/2} \frac{\Gamma\left(\frac{k+m}{2}\right)}{\Gamma\left(\frac{k}{2}\right)\Gamma\left(\frac{m}{2}\right)} \frac{f^{(k/2)-1}}{(1+f)^{(k+m)/2}} \, {}_1F_1\left(\frac{k+m}{2}, \frac{k}{2}, \frac{\lambda^2}{2} \frac{f}{1+f}\right) = f(f \mid k, m, \lambda),$$

substituting $F = mf/k$ we have the noncentral F distribution, $F(k, m, \lambda)$.

17.2 Obtain by change of variable, the p.d. of $g = \chi_1^2/(\chi_1^2 + \chi_2^2)$ as

$$e^{-\lambda^2/2} \sum_{r=0}^{\infty} \left(\frac{\lambda^2}{2}\right)^r \frac{1}{r!} B\left(g \,\middle|\, \frac{k}{2} + r, \frac{m}{2}\right)$$

which is called the noncentral beta distribution and is denoted by $B(k, m, \lambda)$. This can also be written in the alternative form

$$e^{-\lambda^2/2} B\left(g \,\middle|\, \frac{k}{2}, \frac{m}{2}\right) {}_1F_1\left(\frac{k+m}{2}, \frac{k}{2}, \frac{\lambda^2 g}{2}\right) = B(g \mid k, m, \lambda).$$

18 Let $p(x)$ be p.d. of a random variable and define entropy by $-\int p \log p \, dx$.

18.1 Show that when the random variable is bounded by a finite interval, say $a \leqslant x \leqslant b$, the entropy is a maximum for the uniform distribution between a and b.

18.2 Let X be a random variable which takes only non-negative values and is such that $E(X) = \mu > 0$ (a fixed number). Show that the entropy is a maximum when the p.d. is of the form $\mu^{-1} \exp(-x/\mu)$ that is, when the distribution is exponential. [Hint: Follow the method outlined in the equations (3a.1.13) to (3a.1.15) and also (3a.6.1) to (3a.6.5).]

19 *Bessel function distribution.* The function

$$I_\rho(x) = \sum_{k=0}^{\infty} \frac{1}{k!\,\Gamma(\rho + k + 1)} \left(\frac{x}{2}\right)^{2k+\rho}$$

is called a modified Bessel function (see 3a.7.10).

Consider a random variable X with the gamma distribution $G(\alpha, \rho + k + 1)$ and take k as an integer valued random variable with a Poisson distribution $P(\lambda)$. The compound distribution of X obtained by summing over k has the density

$$\alpha\lambda^{-\rho} e^{-\lambda - \alpha x} \sum_{0}^{\infty} \frac{(x\alpha\lambda)^{\rho+k}}{k!\,\Gamma(\rho + k + 1)} = \alpha e^{-\lambda - \alpha x} (x\alpha/\lambda)^{\rho/2} I_\rho(2\sqrt{\alpha\lambda x})$$

20 A random variable X is said to have a normal component if it can be expressed as $U + V$ where U and V are independent and V has a normal distribution. Then show that there is a unique maximal decomposition, $X = U + V$, where U, V are independent, V is normal and U has no normal component.

When X is a vector variable, give an example to show that a maximal decomposition exists but may not be unique.

21 *Characterization of the normal law.*

21.1 Let X_1, X_2 be independent variables. Show that if $X_1 + X_2$ and $X_1 - X_2$ are also independent then X_1 and X_2 are normally distributed.

21.2 *Darmois-Skitovic theorem.* Let $X_1, \ldots, X_n$ be independent random variables. If $a_1 X_1 + \cdots + a_n X_n$ and $b_1 X_1 + \cdots + b_n X_n$ are independently distributed then X_i must be normal if $a_i b_i \neq 0$. [(Darmois (1951), Skitovic (1954))]

21.3 If X_1, X_2 are i.i.d. variables such that $E(X_1) = 0$, then

$$E(a_1 X_1 + a_2 X_2 | b_1 X_1 + b_2 X_2) = 0$$

implies that X_1 is normal provided $a_1 b_1 + a_2 b_2 = 0$.

21.4 If $X_1, \ldots, X_n$ are i.i.d. variables such that $E(X_1) = 0$, then

$$E(\overline{X} | X_1 - \overline{X}) = 0$$

implies that X_1 is normal.

(For proofs and a variety of other results see the book by Kagan, Linnik, and Rao, 1972. See also Pathak and Pillai, 1968.)

REFERENCES

Bock, R. D. (1969), Estimating multinomial response relation, *Contributions to Statistics and Probability: Essays in Memory of S. N. Roy*, 111–132.

Cochran, W. G. (1934), The distribution of quadratic forms in a normal system, with applications to the analysis of covariance, *Proc. Camb. Phil. Soc.* **30**, 178–191.

Craig, A. T. (1943), Note on the independence of certain quadratic forms, *Ann. Math. Statist.* **14**, 195–197.

Cramer, H. (1937), *Random Variables and Probability Distributions*, Cambridge Tracts in Mathematics and Math. Physics **36**, Cambridge University Press.

Darmois, G. (1951), Sur une propriete caracteristique de la loi de probabilite de Laplace, *C.R.* **232**, 1999–2000.

Dempster, A. P. (1971), An overview of multivariate data analysis, *J. Multivariate Analysis* **1**, 316–346.

Fisher, R. A. (1915), Frequency distribution of the values of the correlation coefficient in samples from an indefinitely large population, *Biometrika* **10**, 507–521.

Fisher, R. A. (1924), On a distribution yielding the error function of several well-known statistics, *Proc. Int. Math. Congress*, 805–813.

Fisher, R. A. (1929), Tests of significance in harmonic analysis, *Proc. Roy. Soc.* **A125**, 54–59.

Fisher, R. A. (1953), Dispersion on a sphere, *Proc. Roy. Soc.* **A217**, 295–305.

Graybill, F. A. (1961), *Introduction to Linear Statistical Models*, McGraw-Hill, New York.

Hogg, R. V. and A. T. Craig (1958), On the decomposition of certain χ^2 variables, *Ann. Math. Statist.* **29**, 608–610.

Joos, Georg (1934), *Theoretical Physics*, Haffner, New York. (Earlier German edition, 1894.)

Linnik, Yu. V. (1952), Linear statistics and the normal law, *Dokl. Akad. Nauk SSSR* **83**, 353–355.

Mardia, K. V. (1972), *Statistics of Directional Data*, Academic Press, New York.

Mayer, J. W. and M. G. Mayer (1940), *Statistical Mechanics*, Wiley, New York.

Ogasawara, T. and M. Takahashi (1951), Independence of quadratic forms in normal system, *J. Sci. Hiroshima University* **15**, 1–9.

Ogawa, J. (1949), On the independence of linear and quadratic forms of a random sample from a normal population, *Ann. Inst. Math. Stat.* **1**, 83–108.

Pathak, P. K. and R. N. Pillai (1968), On a characterization of the normal law, *Sankhyā A* **30**, 141–144.

Polya, G. (1923), Herleitung des Gausschen Gesetzes aus einer Funktionalgleichung, *Math. Zeitschrift* **18**, 96–108.

Rao, J. S. and S. Sengupta (1972), Mathematical techniques for paleocurrent analysis, *J. Int. Assoc. Math., Geology* **4**, 235–248.

Sakamoto, H. (1956), On independence of statistics (in Japanese), *Res. Memoirs Instit. Stat. Math.* **1**, 1–25.

Skitovic, V. P. (1954), Linear combinations of independent random variables and the normal distribution law, *Izv. Akad. Nauk SSSR Ser. Mat.* **18**, 185–200.

Stevens, W. L. (1939), Solution to a geometrical problem in probability, *Annals of Eugenics* **9**, 315–320.

"Student" (W. S. Gosset) (1908), The probable error of a mean, *Biometrika* **6**, 1–25.

Watson, G. S. (1956), Analysis of dispersion on a sphere, *Monthly Notices of the Royal Astronomical Society* **7**, 153–159.

Watson, G. S. and Williams, E. J. (1956), On the construction of significance tests on the circle and the sphere, *Biometrika* **43**, 344–352.

Chapter 4

THE THEORY OF LEAST SQUARES AND ANALYSIS OF VARIANCE

Introduction. This chapter deals with statistical inference based on *linear models* for the expectations and certain specified structures for the variances and covariances of the observations.

The theory of least squares is concerned with the estimation of parameters in a linear model. The foundations of the theory were laid by Gauss (1809) and more recently by Markoff (1900). Certain improvements have been made by recent writers (Aitken, 1935; Bose, 1950–1951; Neyman and David, 1938; Parzen, 1961; and Rao 1945b,c, 1946b, 1962f). However, the problem remains unsolved in its generality, especially when the observations have a singular dispersion matrix. Specific solutions for the latter case have been given by Goldman and Zelen (1964), Mitra and Rao (1968f), Zyskind and Martin (1969), and others.

Unified approaches to the problem in full generality, which do not depend on nonsingularity or otherwise of the dispersion matrix, are given by Rao, 1971e, 1972d, 1973a) and Rao and Mitra (1971h). These are described in Section **4i** of this chapter. In this approach, we use the generalized inverse (g-inverse) of a matrix developed in **1b.5** and **1c.5** as a principal tool.

While **4f** deals with the estimation of intraclass correlations in a linear model for a two-way classified balanced data, the general problem of estimation of variance components is considered in **4j** using what has been called MINQUE (Minimum Norm Quadratic Unbiased Estimation) theory recently developed by the author (Rao, 1970a, 1971c, 1971d, 1972b).

Section **4g** on regression and least square prediction is treated in the general context of a multivariate distribution and can be studied independently of the least square theory. Measures of association such as multiple and partial correlation coefficients and ratios have been developed from a new point of view with special reference to practical problems.

A brief discussion is given of the design of observations by which observations are generated in such a way that maximum precision is attained in the estimation of certain parameters.

4a THEORY OF LEAST SQUARES (LINEAR ESTIMATION)

4a.1 Gauss-Markoff Setup $(\mathbf{Y}, \mathbf{X\beta}, \sigma^2\mathbf{I})$

Consider uncorrelated observations $(y_1, \ldots, y_n)$ such that

$$\left.\begin{aligned} E(y_i) &= x_{i1}\beta_1 + \cdots + x_{im}\beta_m \\ V(y_i) &= \sigma^2 \end{aligned}\right\} \quad i = 1, 2, \ldots, n, \tag{4a.1.1}$$

where $(\beta_1, \ldots, \beta_m)$ and σ^2 are unknown parameters and $(x_{ij}) = \mathbf{X}$ is a matrix of known coefficients. If $\mathbf{Y}$ and $\boldsymbol{\beta}$ stand for column vectors of the variables y_i and the parameters β_j, equation (4a.1.1) can be written in matrix notation

$$\mathbf{Y} = \mathbf{X\beta} + \boldsymbol{\varepsilon},\ E(\boldsymbol{\varepsilon}) = \mathbf{0},\ D(\boldsymbol{\varepsilon}) = \sigma^2\mathbf{I} \Rightarrow E(\mathbf{Y}) = \mathbf{X\beta},\ D(\mathbf{Y}) = \sigma^2\mathbf{I} \tag{4a.1.2}$$

where D stands for dispersion (variances and covariances) and $\mathbf{I}$ for the unit matrix of order n. The problem is that of estimating the unknown parameters β_j on the basis of observations y_i.

A set up slightly more general than (4a.1.2) considered by Aitken (1935) is

$$\begin{aligned} E(\mathbf{Y}) &= \mathbf{X\beta} \\ D(\mathbf{Y}) &= \sigma^2\mathbf{G}, \quad |\mathbf{G}| \neq 0 \end{aligned} \tag{4a.1.3}$$

which introduces correlations among the observations. In (4a.1.3) $\mathbf{G}$ is supposed to be a known matrix and $\boldsymbol{\beta}$, σ^2 are unknown. The model (4a.1.3) can be reduced to the model (4a.1.2) by considering $\mathbf{Z} = \mathbf{G}^{-1/2}\mathbf{Y}$ giving

$$\begin{aligned} E(\mathbf{Z}) &= \mathbf{G}^{-1/2}\mathbf{X\beta} = \mathbf{U\beta} \\ D(\mathbf{Z}) &= \sigma^2\mathbf{I}. \end{aligned} \tag{4a.1.4}$$

A special case of (4a.1.3) is

$$\begin{aligned} E(\mathbf{Y}) &= \mathbf{X\beta} \\ D(\mathbf{Y}) &= \boldsymbol{\Sigma}, \quad |\boldsymbol{\Sigma}| \neq 0, \end{aligned} \tag{4a.1.5}$$

where $\boldsymbol{\Sigma}$ is known. Then under the transformation $\mathbf{Z} = \boldsymbol{\Sigma}^{-1/2}\mathbf{Y}$,

$$\begin{aligned} E(\mathbf{Z}) &= \boldsymbol{\Sigma}^{-1/2}\mathbf{X\beta} = \mathbf{U\beta}, \\ D(\mathbf{Z}) &= \mathbf{I}, \end{aligned} \tag{4a.1.6}$$

so that the setup (4a.1.5) is reduced to that of (4a.1.2). For all these cases we need only consider the model (4a.1.2) in detail, which will be referred to as $(\mathbf{Y}, \mathbf{X\beta}, \sigma^2\mathbf{I})$.

The general case of $(\mathbf{Y}, \mathbf{X\beta}, \sigma^2\mathbf{G})$ where $\mathbf{G}$ may be singular, so that the transformation of (4a.1.4) is not applicable, is considered in Section **4i**.

We shall use the following notations and results which the reader may verify and with which he should be familiar.

(a) $\mathbf{L'Y}, \mathbf{M'Y}, \ldots$ denote linear functions of $\mathbf{Y}$; $\mathbf{P'\beta}, \mathbf{R'\beta}$, denote linear functions of $\mathbf{\beta}$, where $\mathbf{L}, \mathbf{M}, \ldots, \mathbf{P}, \mathbf{R}, \ldots$ denote column vectors.

(b) $E(\mathbf{L'Y}) = \mathbf{L'}E(\mathbf{Y}) = \mathbf{L'X\beta}$.

(c) $V(\mathbf{L'Y}) = \mathbf{L'\Sigma L}$, where $\mathbf{\Sigma}$ is the dispersion matrix of $\mathbf{Y}$
$= \sigma^2\mathbf{L'L}$, when $\mathbf{\Sigma} = \sigma^2\mathbf{I}$
$\text{cov}(\mathbf{L'Y}, \mathbf{M'Y}) = \mathbf{L'\Sigma M}$, for general $\mathbf{\Sigma}$
$= \sigma^2\mathbf{L'M}$, when $\mathbf{\Sigma} = \sigma^2\mathbf{I}$.

(d) More generally, if $\mathbf{B}$ is an $n \times k$ matrix, then $\mathbf{B'Y}$ denotes k linear functions of $\mathbf{Y}$. It is easy to verify that $E(\mathbf{B'Y}) = \mathbf{B'X\beta}$ and $D(\mathbf{B'Y}) = \mathbf{B'\Sigma B}$ which reduces to $\sigma^2\mathbf{B'B}$ when $\mathbf{\Sigma} = \sigma^2\mathbf{I}$.

(e) Expectation of a quadratic form:

$$\begin{aligned} E(\mathbf{Y'GY}) &= \text{trace } \mathbf{G\Sigma} + E(\mathbf{Y'})\mathbf{G}E(\mathbf{Y}), \text{ where } \mathbf{\Sigma} = D(\mathbf{Y}) \\ &= \sigma^2 \text{ trace } \mathbf{G} + E(\mathbf{Y'})\mathbf{G}E(\mathbf{Y}), \text{ when } \mathbf{\Sigma} = \sigma^2\mathbf{I}. \end{aligned} \tag{4a.1.7}$$

(f) If $\mathbf{A}$ is an idempotent matrix, rank $\mathbf{A}$ = trace $\mathbf{A}$.

(g) Given a matrix $\mathbf{S}$ there exists a matrix $\mathbf{S}^-$ called a *g*-inverse, with the property (see **1b.5**)

$$\mathbf{SS^-S} = \mathbf{S}, \qquad \text{trace } \mathbf{S^-S} = \text{rank } \mathbf{S}. \tag{4a.1.8}$$

A general solution of a consistent equation $\mathbf{S\beta} = \mathbf{Q}$ is $\mathbf{S^-Q} + (\mathbf{I} - \mathbf{S^-S})\mathbf{Z}$ where $\mathbf{Z}$ is arbitrary. When $\mathbf{S}$ is square and nonsingular $\mathbf{S}^- = \mathbf{S}^{-1}$, the true inverse.

(h) The linear manifolds generated by the columns of $\mathbf{A'}$ and $\mathbf{A'A}$ are the same, $\mathcal{M}(\mathbf{A'}) = \mathcal{M}(\mathbf{A'A})$, which implies that if there exists a vector $\mathbf{L}$ such that $\mathbf{P} = \mathbf{A'L}$, then there exists a vector $\mathbf{\lambda}$ such that $\mathbf{P} = \mathbf{A'A\lambda}$ Observe that for any $\mathbf{L}$, $\mathbf{A'L} \in \mathcal{M}(\mathbf{A'}) = \mathcal{M}(\mathbf{A'A})$.

4a.2 Normal Equations and Least Square (l.s.) Estimators

The equation in $\mathbf{\beta}$ (see Example 13.2, at the end of Chapter 1)

$$\mathbf{X'X\beta} = \mathbf{X'Y} \tag{4a.2.1}$$

obtained by differentiating (to minimize)

$$(\mathbf{Y} - \mathbf{X\beta})'(\mathbf{Y} - \mathbf{X\beta}) = \sum (y_i - x_{i1}\beta_1 - \cdots - x_{im}\beta_m)^2 \tag{4a.2.2}$$

the sum of squares of differences between the observations and the expectations is called *normal equation*. The observational equation $\mathbf{Y} = \mathbf{X}\boldsymbol{\beta}$ is in general inconsistent. But the normal equation (4a.2.1) always admits a solution, since $\mathbf{X}'\mathbf{Y} \in \mathcal{M}(\mathbf{X}'\mathbf{X})$, which implies that there exists a $\boldsymbol{\beta}$ such that (4a.2.1) is true. Let $\hat{\boldsymbol{\beta}}$ be any solution of (4a.2.1). Then

$$\begin{aligned}(\mathbf{Y} - \mathbf{X}\boldsymbol{\beta})'(\mathbf{Y} - \mathbf{X}\boldsymbol{\beta}) &= [\mathbf{Y} - \mathbf{X}\hat{\boldsymbol{\beta}} + \mathbf{X}(\hat{\boldsymbol{\beta}} - \boldsymbol{\beta})]'[\mathbf{Y} - \mathbf{X}\hat{\boldsymbol{\beta}} + \mathbf{X}(\hat{\boldsymbol{\beta}} - \boldsymbol{\beta})] \\ &= (\mathbf{Y} - \mathbf{X}\hat{\boldsymbol{\beta}})'(\mathbf{Y} - \mathbf{X}\hat{\boldsymbol{\beta}}) + (\hat{\boldsymbol{\beta}} - \boldsymbol{\beta})'\mathbf{X}'\mathbf{X}(\hat{\boldsymbol{\beta}} - \boldsymbol{\beta}) \\ &\geqslant (\mathbf{Y} - \mathbf{X}\hat{\boldsymbol{\beta}})'(\mathbf{Y} - \mathbf{X}\hat{\boldsymbol{\beta}}),\end{aligned} \tag{4a.2.3}$$

which shows that the minimum of $(\mathbf{Y} - \mathbf{X}\boldsymbol{\beta})'(\mathbf{Y} - \mathbf{X}\boldsymbol{\beta})$ is

$$(\mathbf{Y} - \mathbf{X}\hat{\boldsymbol{\beta}})'(\mathbf{Y} - \mathbf{X}\hat{\boldsymbol{\beta}}),$$

and is attained at $\boldsymbol{\beta} = \hat{\boldsymbol{\beta}}$ and is unique for all solutions $\hat{\boldsymbol{\beta}}$ of (4a.2.1).

A least squares estimator of the parametric function $\mathbf{P}'\boldsymbol{\beta}$ is defined to be $\mathbf{P}'\hat{\boldsymbol{\beta}}$ where $\hat{\boldsymbol{\beta}}$ is *any* solution of the normal equation (4a.2.1). What are the properties of the estimator $\mathbf{P}'\hat{\boldsymbol{\beta}}$ and in what sense is it a good estimator of $\mathbf{P}'\boldsymbol{\beta}$?

(i) $\mathbf{P}'\hat{\boldsymbol{\beta}}$ *is linear in* $\mathbf{Y}$ *and unbiased for* $\mathbf{P}'\boldsymbol{\beta}$, *if and only if*, $\mathbf{P}'\hat{\boldsymbol{\beta}}$ *is unique for all solutions* $\hat{\boldsymbol{\beta}}$ *of the normal equation or equivalently* $\mathbf{P} \in \mathcal{M}(\mathbf{X}')$ *which is the same as* $\mathbf{P} \in \mathcal{M}(\mathbf{X}'\mathbf{X})$, *or equivalently there exists a linear function of* $\mathbf{Y}$ *with expectation* $\mathbf{P}'\boldsymbol{\beta}$.

We do not make any assumption about the number m of unknown parameters β_j or the rank of the matrix $\mathbf{X}$.

Let $\mathbf{P} \in \mathcal{M}(\mathbf{X}'\mathbf{X})$, that is, $\mathbf{P} = \mathbf{X}'\mathbf{X}\boldsymbol{\lambda}$ where $\boldsymbol{\lambda}$ is some vector. Then for any solution $\hat{\boldsymbol{\beta}}$,

$$\mathbf{P}'\hat{\boldsymbol{\beta}} = \boldsymbol{\lambda}'\mathbf{X}'\mathbf{X}\hat{\boldsymbol{\beta}} = \boldsymbol{\lambda}'\mathbf{X}'\mathbf{Y}, \qquad \text{since} \quad \mathbf{X}'\mathbf{X}\hat{\boldsymbol{\beta}} = \mathbf{X}'\mathbf{Y}. \tag{4a.2.4}$$

Thus $\mathbf{P}'\hat{\boldsymbol{\beta}}$ is unique, linear in $\mathbf{Y}$, and $E(\boldsymbol{\lambda}'\mathbf{X}'\mathbf{Y}) = \boldsymbol{\lambda}'\mathbf{X}'\mathbf{X}\boldsymbol{\beta} = \mathbf{P}'\boldsymbol{\beta}$, that is unbiased for $\mathbf{P}'\boldsymbol{\beta}$. Further $\boldsymbol{\lambda}'\mathbf{X}'$ is unique although $\boldsymbol{\lambda}'$ may not be.

Let $\mathbf{P}'\hat{\boldsymbol{\beta}}$ be unique. Corresponding to any vector $\boldsymbol{\lambda}$ such that $\mathbf{X}'\mathbf{X}\boldsymbol{\lambda} = \mathbf{0}$, there is a solution $\bar{\boldsymbol{\beta}} = \hat{\boldsymbol{\beta}} - \boldsymbol{\lambda}$ of (4a.2.1). And $0 = \mathbf{P}'(\hat{\boldsymbol{\beta}} - \bar{\boldsymbol{\beta}}) = \mathbf{P}'\boldsymbol{\lambda}$ for all $\boldsymbol{\lambda}$ such that $\mathbf{X}'\mathbf{X}\boldsymbol{\lambda} = \mathbf{0}$. Therefore $\mathbf{P} \in \mathcal{M}(\mathbf{X}'\mathbf{X})$ and the rest follow.

Let $\mathbf{L}'\mathbf{Y}$ be such that $E(\mathbf{L}'\mathbf{Y}) = \mathbf{P}'\boldsymbol{\beta}$. Then $\mathbf{L}'\mathbf{X}\boldsymbol{\beta} = \mathbf{P}'\boldsymbol{\beta}$ or $\mathbf{L}'\mathbf{X} = \mathbf{P}'$, that is, $\mathbf{P} \in \mathcal{M}(\mathbf{X}')$. We shall say $\mathbf{P}'\boldsymbol{\beta}$ is estimable if any of the conditions in (i) holds.

(ii) *If* $\mathbf{P}'\boldsymbol{\beta}$ *is estimable, then* $\mathbf{P}'\hat{\boldsymbol{\beta}}$ *has minimum variance in the class of linear unbiased estimators* (*LUE*) *of* $\mathbf{P}'\boldsymbol{\beta}$.

Since $\mathbf{P}'\boldsymbol{\beta}$ is estimable, $\mathbf{P} \in \mathcal{M}(\mathbf{X}'\mathbf{X})$, that is, $\mathbf{P} = \mathbf{X}'\mathbf{X}\boldsymbol{\lambda}$ for some $\boldsymbol{\lambda}$. Hence

$$\mathbf{P}'\hat{\boldsymbol{\beta}} = \boldsymbol{\lambda}'\mathbf{X}'\mathbf{X}\hat{\boldsymbol{\beta}} = \boldsymbol{\lambda}'\mathbf{X}'\mathbf{Y}. \tag{4a.2.5}$$

If $\mathbf{L}'\mathbf{Y}$ is any unbiased estimator of $\mathbf{P}'\boldsymbol{\beta}$, then $\mathbf{L}'\mathbf{X} = \mathbf{P}'$. Consider

$$\begin{aligned} V(\mathbf{L}'\mathbf{Y}) &= V(\mathbf{L}'\mathbf{Y} - \boldsymbol{\lambda}'\mathbf{X}'\mathbf{Y} + \boldsymbol{\lambda}'\mathbf{X}'\mathbf{Y}) \\ &= V(\mathbf{L}'\mathbf{Y} - \boldsymbol{\lambda}'\mathbf{X}'\mathbf{Y}) + V(\boldsymbol{\lambda}'\mathbf{X}'\mathbf{Y}) \\ &= V(\mathbf{L}'\mathbf{Y} - \boldsymbol{\lambda}'\mathbf{X}'\mathbf{Y}) + V(\mathbf{P}'\hat{\boldsymbol{\beta}}) \end{aligned} \tag{4a.2.6}$$

where in (4a.2.6) the covariance of $(\mathbf{L}' - \boldsymbol{\lambda}'\mathbf{X}')\mathbf{Y}$ and $\boldsymbol{\lambda}'\mathbf{X}'\mathbf{Y}$ is

$$(\mathbf{L}' - \boldsymbol{\lambda}'\mathbf{X}')\mathbf{X}\boldsymbol{\lambda}\sigma^2 = (\mathbf{L}'\mathbf{X} - \boldsymbol{\lambda}'\mathbf{X}'\mathbf{X})\boldsymbol{\lambda}\sigma^2 = (\mathbf{P}' - \mathbf{P}')\boldsymbol{\lambda}\sigma^2 = 0.$$

From (4a.2.6), $V(\mathbf{L}'\mathbf{Y}) \geqslant V(\mathbf{P}'\hat{\boldsymbol{\beta}})$. We call $\mathbf{P}'\hat{\boldsymbol{\beta}}$, the *BLUE* (best *LUE*) of $\mathbf{P}'\hat{\boldsymbol{\beta}}$.

(iii) *All linear parametric functions are estimable, if and only if* $R(\mathbf{X}) = m$, *the number of unknown parameters* β_j, *where* $R(\mathbf{X})$ *denotes the rank of* $\mathbf{X}$.

If $R(\mathbf{X}) = m$, then $R(\mathbf{X}'\mathbf{X}) = m$ and hence the normal equation has a unique solution and naturally $\mathbf{P}'\hat{\boldsymbol{\beta}}$ is unique for any given $\mathbf{P}$.

If $\mathbf{P}'\boldsymbol{\beta}$ is estimable, then $\mathbf{P} \in \mathscr{M}(\mathbf{X}'\mathbf{X})$. If $\mathbf{P}$ is arbitrary, $R(\mathbf{X}'\mathbf{X})$ must be full, thus establishing the converse.

If $R(\mathbf{X}) < m$, then some parametric functions do not admit unbiased estimators, and under our setup nothing can be inferred about such parametric functions, which are said to be nonestimable or *confounded.*

What the results (i) and (ii) tell us then is that if we are seeking for linear unbiased estimators with minimum variance, of parametric functions, the l.s. estimator of Gauss is the answer. The concept of estimability is due to R. C. Bose.

(iv) *Let* $\mathbf{C}$ *be a g-inverse of* $\mathbf{X}'\mathbf{X}$ *and let* $\mathbf{H} = \mathbf{C}\mathbf{X}'\mathbf{X}$. *Then a necessary and sufficient condition that* $\mathbf{P}'\boldsymbol{\beta}$ *is estimable is that* $\mathbf{P}'(\mathbf{I} - \mathbf{H}) = 0$.

It is shown in [(vi), **1b.5**] that a general solution of $\mathbf{X}'\mathbf{X}\boldsymbol{\beta} = \mathbf{X}'\mathbf{Y}$ is $\mathbf{C}\mathbf{X}'\mathbf{Y} + (\mathbf{I} - \mathbf{H})\mathbf{Z}$ where $\mathbf{Z}$ is arbitrary. Using the n.s. condition in (i) for estimability, *viz.*, that $\mathbf{P}'\hat{\boldsymbol{\beta}}$ is unique for all solutions $\hat{\boldsymbol{\beta}}$, we find $\mathbf{P}'(\mathbf{I} - \mathbf{H})\mathbf{Z} = 0$ for all $\mathbf{Z} \Leftrightarrow \mathbf{P}'(\mathbf{I} - \mathbf{H}) = \mathbf{0}$.

4a.3 g-Inverse and a Solution of the Normal Equation

It is already shown that the normal equation $\mathbf{X}'\mathbf{X}\boldsymbol{\beta} = \mathbf{X}'\mathbf{Y}$ is consistent. Hence if a g-inverse $\mathbf{C}$ of $\mathbf{X}'\mathbf{X}$ is found, a particular solution is given by $\mathbf{C}\mathbf{X}'\mathbf{Y}$. In practice the computation of $\mathbf{C}$ can be reduced to that of determining the true inverse of a suitable matrix. We consider some methods of computing $\mathbf{C}$.

1. If rank $\mathbf{X} = m$, the number of unknown parameters, then $R(\mathbf{X}'\mathbf{X}) = m$ and $\mathbf{C} = (\mathbf{X}'\mathbf{X})^{-1}$ is the true inverse.

2. Let $R(\mathbf{X}) = r < m$. Let it be possible to determine a matrix $\mathbf{H}$ of order $(m - r) \times m$ such that $R(\mathbf{X}' \,\vdots\, \mathbf{H}') = m$, that is, we add some rows to $\mathbf{X}$ to

meet the deficiency in the rank. Then $R(\mathbf{X}'\mathbf{X} + \mathbf{H}'\mathbf{H}) = m$, and the true inverse $(\mathbf{X}'\mathbf{X} + \mathbf{H}'\mathbf{H})^{-1}$ exists. It is easily shown that $\mathbf{C} = (\mathbf{X}'\mathbf{X} + \mathbf{H}'\mathbf{H})^{-1}$ is a g-inverse of $\mathbf{X}'\mathbf{X}$, that is, $\mathbf{X}'\mathbf{X}\mathbf{C}\mathbf{X}'\mathbf{X} = \mathbf{X}'\mathbf{X}$ (see Example 5, p.34).

3. Having determined $\mathbf{H}$ as in (2), we have an alternative method of computing a g-inverse. Observe that the rank of

$$\begin{pmatrix} \mathbf{X}'\mathbf{X} & \mathbf{H}' \\ \mathbf{H} & \mathbf{0} \end{pmatrix}$$

is $2m - r$ and, therefore, the true inverse

$$\begin{pmatrix} \mathbf{X}'\mathbf{X} & \mathbf{H}' \\ \mathbf{H} & \mathbf{0} \end{pmatrix}^{-1} = \begin{pmatrix} \mathbf{C}_1 & \mathbf{C}_2' \\ \mathbf{C}_2 & \mathbf{C}_4 \end{pmatrix} \tag{4a.3.1}$$

exists. The matrix $\mathbf{C}_1$ which appears in the partitioned form of the inverse matrix (4a.3.1) is indeed a generalized inverse of $\mathbf{X}'\mathbf{X}$, that is, it satisfies the equation $\mathbf{X}'\mathbf{X}\mathbf{C}_1\mathbf{X}'\mathbf{X} = \mathbf{X}'\mathbf{X}$.

4. A modification of (3) is to determine matrices $\mathbf{C}_1'$, $\mathbf{C}_2'$ such that

$$(\mathbf{C}_1' \mid \mathbf{C}_2')\begin{pmatrix} \mathbf{X}'\mathbf{X} \\ \mathbf{H} \end{pmatrix} = \mathbf{I}.$$

Then $\mathbf{C}_1$ is a g-inverse of $\mathbf{X}'\mathbf{X}$. In many statistical problems the matrices $\mathbf{C}_1'$ and $\mathbf{C}_2'$ can be written after inspection.

5. Suppose it is possible to find the dependent rows in $\mathbf{X}'\mathbf{X}$. By omitting these rows and the corresponding columns we obtain an $r \times r$ matrix with rank r, which admits a true inverse. Now increase the order of this inverse matrix by inserting rows and columns of zeroes from where the dependent rows and columns are removed. The matrix so obtained is a g-inverse. This is a useful method for in practice we may know where deficiency in $\mathbf{X}'\mathbf{X}$ occurs.

6. Let $\lambda_1, \ldots, \lambda_r$ be the non-zero eigenvalues and let $\mathbf{P}_1, \ldots, \mathbf{P}_r$ the corresponding eigenvectors of $\mathbf{X}'\mathbf{X}$. Then

$$\mathbf{X}'\mathbf{X} = \lambda_1 \mathbf{P}_1 \mathbf{P}_1' + \cdots + \lambda_r \mathbf{P}_r \mathbf{P}_r'.$$

It is easy to verify that

$$\mathbf{C} = (\mathbf{X}'\mathbf{X})^- = \frac{1}{\lambda_1}\mathbf{P}_1\mathbf{P}_1' + \cdots + \frac{1}{\lambda_r}\mathbf{P}_r\mathbf{P}_r'$$

is a g-inverse.

For a general discussion of the computation of a g-inverse, the reader is referred to Rao and Mitra (1971, B3).

4a.4 Variances and Covariances of l.s. Estimators

It is convenient at this stage to denote the normal equation $\mathbf{X}'\mathbf{X}\boldsymbol{\beta} = \mathbf{X}'\mathbf{Y}$ as $\mathbf{S}\boldsymbol{\beta} = \mathbf{Q}$, where $\mathbf{S} = \mathbf{X}'\mathbf{X}$, $\mathbf{Q} = \mathbf{X}'\mathbf{Y}$. An intrinsic property of the normal equation is $D(\mathbf{Q}) = \sigma^2\mathbf{S}$. Let $\mathbf{C}$ be a g-inverse of $\mathbf{S}$, that is, $\mathbf{SCS} = \mathbf{S}$, which gives an explicit representation of a solution of the normal equation, $\hat{\boldsymbol{\beta}} = \mathbf{CQ}$.

(i) *Let* $\mathbf{P}'\hat{\boldsymbol{\beta}}$, $\mathbf{R}'\hat{\boldsymbol{\beta}}$ *be the l.s. estimators of the estimable functions* $\mathbf{P}'\boldsymbol{\beta}$, $\mathbf{R}'\boldsymbol{\beta}$ *and* $\mathbf{C}$ *be any g-inverse of* $\mathbf{S}$. *Then*

$$V(\mathbf{P}'\hat{\boldsymbol{\beta}}) = \sigma^2\mathbf{P}'\mathbf{C}\mathbf{P}, \qquad \operatorname{cov}(\mathbf{P}'\hat{\boldsymbol{\beta}}, \mathbf{R}'\hat{\boldsymbol{\beta}}) = \sigma^2\mathbf{P}'\mathbf{C}\mathbf{R} \tag{4a.4.1}$$

so that $\sigma^2\mathbf{C}$ *can be formally considered as the dispersion matrix of* $\hat{\boldsymbol{\beta}}$ *so long as formulas* (4a.4.1) *are applied to estimable parametric functions.*

Since $\mathbf{P}'\boldsymbol{\beta}$ is estimable, $\mathbf{P} = \mathbf{X}'\mathbf{X}\boldsymbol{\lambda} = \mathbf{S}\boldsymbol{\lambda}$. Now

$$\begin{aligned} V(\mathbf{P}'\hat{\boldsymbol{\beta}}) &= V(\mathbf{P}'\mathbf{C}\mathbf{Q}) = \mathbf{P}'\mathbf{C}D(\mathbf{Q})\mathbf{C}'\mathbf{P} \\ &= \sigma^2\mathbf{P}'\mathbf{CSC}'\mathbf{P} = \sigma^2\boldsymbol{\lambda}'\mathbf{SCSC}'\mathbf{P} = \sigma^2\mathbf{P}'\mathbf{CP}. \end{aligned}$$

The result on the covariance is similarly proved.

When $\mathbf{C} = (C_{ij})$ is a true inverse, all parametric functions are estimable; $\hat{\beta}_i$ is the l.s. estimator of β_i and

$$V(\hat{\beta}_i) = \sigma^2 C_{ii}, \qquad \operatorname{cov}(\hat{\beta}_i, \hat{\beta}_j) = \sigma^2 C_{ij}. \tag{4a.4.2}$$

The expressions (4a.4.2) are not meaningful when $\mathbf{C}$ is not a true inverse, but nonetheless can be used to compute the variance of any linear function of $\hat{\beta}_i$ provided the corresponding linear function in β_i is estimable.

It is not always necessary to obtain $\mathbf{C}$ in advance to compute the l.s. estimators and the expressions for the variances and covariances. Suppose from the normal equation $\mathbf{S}\boldsymbol{\beta} = \mathbf{Q}$ we find that on multiplying by a vector $\boldsymbol{\lambda}$, $\boldsymbol{\lambda}'\mathbf{S}\boldsymbol{\beta} = \boldsymbol{\lambda}'\mathbf{Q}$ reduces to $\mathbf{P}'\boldsymbol{\beta} = \boldsymbol{\lambda}'\mathbf{Q}$. Then $\boldsymbol{\lambda}'\mathbf{Q}$ is the l.s. estimator of $\mathbf{P}'\boldsymbol{\beta}$, and

$$V(\boldsymbol{\lambda}'\mathbf{Q}) = \sigma^2\boldsymbol{\lambda}'\mathbf{S}\boldsymbol{\lambda} = \sigma^2\boldsymbol{\lambda}'\mathbf{P}, \tag{4a.4.3}$$

so that *we have an expression for the variance depending only on* $\boldsymbol{\lambda}$ *and* $\mathbf{P}$. Similarly if $\boldsymbol{\lambda}'\mathbf{S} = \mathbf{P}'$ and $\boldsymbol{\mu}'\mathbf{S} = \mathbf{R}$, then $\boldsymbol{\lambda}'\mathbf{Q}$ and $\boldsymbol{\mu}'\mathbf{Q}$ are the l.s. estimators of $\mathbf{P}'\boldsymbol{\beta}$ and $\mathbf{R}'\boldsymbol{\beta}$ and

$$\operatorname{cov}(\boldsymbol{\lambda}'\mathbf{Q}, \boldsymbol{\mu}'\mathbf{Q}) = \sigma^2\boldsymbol{\lambda}'\mathbf{R} = \sigma^2\boldsymbol{\mu}'\mathbf{P}. \tag{4a.4.4}$$

In many practical problems, formulas (4a.4.3) and (4a.4.4) can be applied, thus avoiding the computation of a g-inverse.

4a.5 Estimation of σ^2

The minimum sum of squares $R_0^2 = (\mathbf{Y} - \mathbf{X}\hat{\boldsymbol{\beta}})'(\mathbf{Y} - \mathbf{X}\hat{\boldsymbol{\beta}})$, introduced in (4a.2.3) has the following alternative expressions, which the reader may verify.

$$\begin{aligned} R_0^2 &= \mathbf{Y}'\mathbf{Y} - \mathbf{Y}'\mathbf{X}\hat{\boldsymbol{\beta}} = \mathbf{Y}'\mathbf{Y} - \hat{\boldsymbol{\beta}}'\mathbf{X}'\mathbf{X}\hat{\boldsymbol{\beta}} \\ &= \mathbf{Y}'\mathbf{Y} - \mathbf{Q}'\hat{\boldsymbol{\beta}} = \mathbf{Y}'\mathbf{Y} - \mathbf{Q}'\mathbf{C}\mathbf{Q} \\ &= \mathbf{Y}'\mathbf{Y} - \mathbf{Y}'\mathbf{X}(\mathbf{X}'\mathbf{X})^{-}\mathbf{X}'\mathbf{Y} = \mathbf{Y}'(\mathbf{I} - \mathbf{X}(\mathbf{X}'\mathbf{X})^{-}\mathbf{X}')\mathbf{Y}. \end{aligned} \tag{4a.5.1}$$

When $|\mathbf{X}'\mathbf{X}| \neq 0$, R_0^2 can be expressed as the ratio

$$\begin{vmatrix} \mathbf{Y}'\mathbf{Y} & \mathbf{X}'\mathbf{Y} \\ \mathbf{Y}'\mathbf{X} & \mathbf{X}'\mathbf{X} \end{vmatrix} \div |\mathbf{X}'\mathbf{X}|. \tag{4a.5.2}$$

The vector $(\mathbf{Y} - \mathbf{X}\hat{\boldsymbol{\beta}})$ is called the residual vector. In some problems it is necessary to examine the residuals. We prove the following,

(i) $E(\mathbf{Y} - \mathbf{X}\hat{\boldsymbol{\beta}}) = 0$.

$E(\mathbf{Y} - \mathbf{X}\hat{\boldsymbol{\beta}}) = E(\mathbf{Y}) - E(\mathbf{X}\hat{\boldsymbol{\beta}}) = \mathbf{X}\boldsymbol{\beta} - \mathbf{X}\boldsymbol{\beta} = 0$. Observe that the step $E(\mathbf{X}\hat{\boldsymbol{\beta}}) = \mathbf{X}\boldsymbol{\beta}$ has to be justified, since $E(\hat{\boldsymbol{\beta}}) \neq \boldsymbol{\beta}$ in general.

(ii) $\text{Cov}(\mathbf{P}'\hat{\boldsymbol{\beta}}, \mathbf{Y} - \mathbf{X}\hat{\boldsymbol{\beta}}) = \mathbf{0}$, *where* $\mathbf{P}'\hat{\boldsymbol{\beta}}$ *is the estimator of an estimable parametric function* $\mathbf{P}'\boldsymbol{\beta}$.

This statement follows from the more general result, $\text{cov}(\mathbf{P}'\hat{\boldsymbol{\beta}}, \mathbf{L}'\mathbf{Y}) = 0$ where $\mathbf{L}'\mathbf{Y}$ is a linear function of $\mathbf{Y}$ such that $E(\mathbf{L}'\mathbf{Y}) = 0$. The latter condition implies $\mathbf{L}'\mathbf{X} = \mathbf{0}$. Now $\mathbf{P}'\hat{\boldsymbol{\beta}} = \boldsymbol{\lambda}'\mathbf{X}'\mathbf{Y}$ where $\mathbf{P} = \mathbf{X}'\mathbf{X}\boldsymbol{\lambda}$ and

$$\text{cov}(\mathbf{P}'\hat{\boldsymbol{\beta}}, \mathbf{L}'\mathbf{Y}) = \text{cov}(\boldsymbol{\lambda}'\mathbf{X}'\mathbf{Y}, \mathbf{L}'\mathbf{Y}) = \sigma^2\mathbf{L}'\mathbf{X}\boldsymbol{\lambda} = 0. \tag{4a.5.3}$$

By (i), $\mathbf{Y} - \mathbf{X}\hat{\boldsymbol{\beta}}$ is a set of linear functions with zero expectations. Hence the desired result follows.

(iii) *The variance-covariance (dispersion) matrix of* $\mathbf{Y} - \mathbf{X}\hat{\boldsymbol{\beta}}$ *is the difference of the dispersion matrices,*

$$D(\mathbf{Y} - \mathbf{X}\hat{\boldsymbol{\beta}}) = D(\mathbf{Y}) - D(\mathbf{X}\hat{\boldsymbol{\beta}}) = \sigma^2(\mathbf{I} - \mathbf{X}(\mathbf{X}'\mathbf{X})^{-}\mathbf{X}'). \tag{4a.5.4}$$

From (ii), $\text{cov}(\mathbf{Y} - \mathbf{X}\hat{\boldsymbol{\beta}}, \mathbf{X}\hat{\boldsymbol{\beta}}) = \mathbf{0}$. Hence (4a.5.4) follows.

(iv) $E(R_0^2) = (n - r)\sigma^2$, *so that an unbiased estimator of* σ^2 *is*

$$\hat{\sigma}^2 = \frac{R_0^2}{n - r} \tag{4a.5.5}$$

where $r = R(\mathbf{X})$.

The simplest way to demonstrate this result is to use the orthogonal transformation of [(i), **3b.5**] from **Y** to **Z** and the representation (3b.5.4), $R_0^2 = z_{r+1}^2 + \cdots + z_n^2$, in terms of new variables, with $E(z_i) = 0$, $V(z_i) = \sigma^2$, $i = r+1, \ldots, n$. Hence $E(R_0^2) = (n-r)\sigma^2$.

Or we may use the representation (4a.5.1)

$$\begin{aligned} R_0^2 &= \mathbf{Y'Y} - \mathbf{Y'X(X'X)^-X'Y} \\ &= \mathbf{Y'(I - X(X'X)^-X')Y} \end{aligned}$$

By applying the property (Ex. 2.1, Chapter 3) $\mathbf{X'X(X'X)^-X'} = \mathbf{X'}$, we find that the introduction of $\mathbf{X\beta}$ does not alter the value of R_0^2.

$$\begin{aligned} R_0^2 &= (\mathbf{Y} - \mathbf{X\beta})'(\mathbf{I} - \mathbf{X(X'X)^-X'})(\mathbf{Y} - \mathbf{X\beta}) \\ E(R_0^2) &= \sigma^2 \operatorname{trace}(\mathbf{I} - \mathbf{X(X'X)^-X'}) \\ &= n\sigma^2 - \sigma^2 \operatorname{trace}(\mathbf{X'X})^-(\mathbf{X'X}) = (n-r)\sigma^2 \end{aligned}$$

since $\operatorname{trace}(\mathbf{X'X})^-(\mathbf{X'X}) = R(\mathbf{X'X}) = r$ by (4a.1.8).

(v) In what sense is $R_0^2/(n-r)$ an optimum estimator of σ^2 beyond being unbiased? Hsu (1938) and Rao (1952a, 1971d) investigated this question by considering quadratic estimators of the form $\mathbf{Y'GY}$ unbiased for σ^2 and trying to minimize its variance. Hsu imposed the condition that $V(\mathbf{Y'GY})$ is independent of $\boldsymbol{\beta}$, the unknown parameters, and Rao imposed the condition that $\mathbf{Y'GY}$ is non-negative. Under either condition, it is shown that $R_0^2/(n-r)$ has minimum variance when (a) $\beta_2 = 3$ or (b) the coefficients of $y_1^2, \ldots, y_n^2$ in the expansion of R_0^2 are the same when $\beta_2 \neq 3$, where β_2 is Pearsonian coefficient of kurtosis of the distribution of individual y_i. Further results cannot be established without knowing the exact value of β_2 or the distribution of y_i. In many situations condition (b) is not satisfied when we cannot claim minimum variance property for the estimator of σ^2 based on least squares.

4a.6 Other Approaches to the l.s. Theory (Geometric Solution)

Let **U** be a projection operator (matrix, **1c.4**) which projects vectors onto the column space of **X**. By definition $\mathbf{UX} = \mathbf{X}$ and **U** is symmetric and idempotent ($\mathbf{U}^2 = \mathbf{U}$). Let **L** be such that $E(\mathbf{L'Y}) = \mathbf{P'\beta}$, that is, $\mathbf{L'X} = \mathbf{P'}$. Consider **UL** and the linear function $(\mathbf{UL})'\mathbf{Y} = \mathbf{L'UY}$. Then

$$\begin{aligned} E(\mathbf{L'UY}) &= \mathbf{L'UX\beta} = \mathbf{L'X\beta} = \mathbf{P'\beta} \\ V(\mathbf{L'Y}) &= V[\mathbf{L'(I-U)Y} + \mathbf{L'UY}] \\ &= V[\mathbf{L'(I-U)Y}] + V(\mathbf{L'UY}) \geqslant V(\mathbf{L'UY}), \end{aligned}$$

since $\text{cov}[\mathbf{L}'(\mathbf{I}-\mathbf{U})\mathbf{Y}, \mathbf{L}'\mathbf{U}\mathbf{Y}] = \sigma^2\mathbf{L}'(\mathbf{I}-\mathbf{U})\mathbf{U}\mathbf{L} = 0$. Hence $\mathbf{L}'\mathbf{U}\mathbf{Y}$ is unbiased for $\mathbf{P}'\boldsymbol{\beta}$ and has smaller variance than $\mathbf{L}'\mathbf{Y}$. If $\mathbf{M}'\mathbf{Y}$ is another unbiased estimator of $\mathbf{P}'\boldsymbol{\beta}$, then $(\mathbf{L}'-\mathbf{M}')\mathbf{X} = \mathbf{0} \Rightarrow (\mathbf{L}-\mathbf{M})$ is orthogonal to columns of $\mathbf{X}$ and therefore $\mathbf{U}(\mathbf{L}-\mathbf{M}) = \mathbf{0}$, that is, $\mathbf{M}'\mathbf{U}\mathbf{Y} = \mathbf{L}'\mathbf{U}\mathbf{Y}$. Hence $\mathbf{V}(\mathbf{M}'\mathbf{Y}) \geqslant \mathbf{V}(\mathbf{M}'\mathbf{U}\mathbf{Y}) = \mathbf{V}(\mathbf{L}'\mathbf{U}\mathbf{Y})$, i.e., $\mathbf{L}'\mathbf{U}\mathbf{Y}$ is a l.s. estimator.

If $\mathbf{X}\boldsymbol{\lambda}$ is the projection of $\mathbf{L}$ on the column space of $\mathbf{X}$, then $\mathbf{X}'(\mathbf{L}-\mathbf{X}\boldsymbol{\lambda}) = \mathbf{0}$, that is, $\boldsymbol{\lambda}$ satisfies the equation $\mathbf{X}'\mathbf{X}\boldsymbol{\lambda} = \mathbf{X}'\mathbf{L}$. Hence $\boldsymbol{\lambda} = (\mathbf{X}'\mathbf{X})^-\mathbf{X}'\mathbf{L}$ and $\mathbf{X}\boldsymbol{\lambda} = \mathbf{X}(\mathbf{X}'\mathbf{X})^-\mathbf{X}'\mathbf{L}$ where $(\mathbf{X}'\mathbf{X})^-$ is a generalized inverse of $\mathbf{X}'\mathbf{X}$. Therefore the projection matrix $\mathbf{U}$ can be chosen as $\mathbf{X}(\mathbf{X}'\mathbf{X})^-\mathbf{X}'$. Now

$$\mathbf{U}\mathbf{L} = \mathbf{X}(\mathbf{X}'\mathbf{X})^-\mathbf{X}'\mathbf{L} = \mathbf{X}(\mathbf{X}'\mathbf{X})^-\mathbf{P}, \qquad \text{using} \quad \mathbf{P} = \mathbf{X}'\mathbf{L}$$
$$\mathbf{L}'\mathbf{U}\mathbf{Y} = \mathbf{P}'(\mathbf{X}'\mathbf{X})^-\mathbf{X}'\mathbf{Y} = \mathbf{P}'\hat{\boldsymbol{\beta}}.$$

Thus $\mathbf{P}'(\mathbf{X}'\mathbf{X})^-\mathbf{X}'\mathbf{Y}$ or $\mathbf{P}'\hat{\boldsymbol{\beta}}$ is the l.s.e. of $\mathbf{P}'\boldsymbol{\beta}$. Note that $\mathbf{P}'[(\mathbf{X}'\mathbf{X})^-]'\mathbf{X}' = \mathbf{P}'(\mathbf{X}'\mathbf{X})^-\mathbf{X}'$.

Extremum of a Quadratic Form. The problem of l.s. estimators may be posed as one of determining a linear function $\mathbf{L}'\mathbf{Y}$ unbiased for a given parametric function $\mathbf{P}'\boldsymbol{\beta}$ and as having minimum variance. Now $V(\mathbf{L}'\mathbf{Y}) = \sigma^2\mathbf{L}'\mathbf{L}$ and $E(\mathbf{L}'\mathbf{Y}) = \mathbf{P}'\boldsymbol{\beta} \Rightarrow \mathbf{X}'\mathbf{L} = \mathbf{P}$, so that the problem is one of minimizing $\mathbf{L}'\mathbf{L}$ subject to the condition, $\mathbf{X}'\mathbf{L} = \mathbf{P}$, or of determining a solution $\mathbf{L}$ of $\mathbf{X}'\mathbf{L} = \mathbf{P}$ with minimum length. This problem is a special case of the one considered in [(ii), **1f.1**] (with $\mathbf{A} = \mathbf{1}$), where it is shown that the infimum of $\mathbf{L}'\mathbf{L}$ is attained at

$$(\mathbf{L}_*)' = \mathbf{P}'(\mathbf{X}'\mathbf{X})^-\mathbf{X}',$$

which gives the minimum variance estimator

$$(\mathbf{L}_*)'\mathbf{Y} = \mathbf{P}'(\mathbf{X}'\mathbf{X})^-\mathbf{X}'\mathbf{Y} = \mathbf{P}'\hat{\boldsymbol{\beta}},$$

the l.s. estimator.

4a.7 Explicit Expressions for Correlated Observations

It is shown in **4a.1** that the problems $(\mathbf{Y}, \mathbf{X}\boldsymbol{\beta}, \sigma^2\mathbf{G})$ and $(\mathbf{Y}, \mathbf{X}\boldsymbol{\beta}, \boldsymbol{\Sigma})$ can be reduced to $(\mathbf{Y}, \mathbf{X}\boldsymbol{\beta}, \sigma^2\mathbf{I})$ by a suitable transformation of the variable $\mathbf{Y}$. We shall give explicit expressions to normal equations, l.s. estimators, etc., in terms of original correlated observations. By substituting for $\mathbf{Y}$ and $\mathbf{X}$ in the expressions for the case $(\mathbf{Y}, \mathbf{X}\boldsymbol{\beta}, \sigma^2\mathbf{I})$ the transformed values, explicit expressions in terms of original observations are obtained. Table 4a.7 summarizes the results.

TABLE 4a.7

Problem	$(\mathbf{Y}, \mathbf{XB}, \sigma^2\mathbf{I})$	$(\mathbf{Y}, \mathbf{X}\boldsymbol{\beta}, \sigma^2\mathbf{G})$	$(\mathbf{Y}, \mathbf{X}\boldsymbol{\beta}, \boldsymbol{\Sigma})$
Transformation	$\mathbf{Y} \to \mathbf{Y}$	$\mathbf{Y} \to \mathbf{G}^{-1/2}\mathbf{Y}, \mathbf{X} \to \mathbf{G}^{-1/2}\mathbf{X}$	$\mathbf{Y} \to \boldsymbol{\Sigma}^{-1/2}\mathbf{Y}, \mathbf{X} \to \boldsymbol{\Sigma}^{-1/2}\mathbf{X}$
Quadratic form to be minimized	$(\mathbf{Y} - \mathbf{X}\boldsymbol{\beta})'(\mathbf{Y} - \mathbf{X}\boldsymbol{\beta})$	$(\mathbf{Y} - \mathbf{X}\boldsymbol{\beta})'\mathbf{G}^{-1}(\mathbf{Y} - \mathbf{X}\boldsymbol{\beta})$	$(\mathbf{Y} - \mathbf{X}\boldsymbol{\beta})'\boldsymbol{\Sigma}^{-1}(\mathbf{Y} - \mathbf{X}\boldsymbol{\beta})$
Normal equation $(\mathbf{S}\boldsymbol{\beta} = \mathbf{Q})$	$\mathbf{S} = \mathbf{X}'\mathbf{X}, \mathbf{Q} = \mathbf{X}'\mathbf{Y}$	$\mathbf{S} = \mathbf{X}'\mathbf{G}^{-1}\mathbf{X}, \mathbf{Q} = \mathbf{X}'\mathbf{G}^{-1}\mathbf{Y}$	$\mathbf{S} = \mathbf{X}'\boldsymbol{\Sigma}^{-1}\mathbf{X}, \mathbf{Q} = \mathbf{X}'\boldsymbol{\Sigma}^{-1}\mathbf{Y}$
Solution $\hat{\boldsymbol{\beta}}$ $(\mathbf{S}^- = \mathbf{C})$	$\hat{\boldsymbol{\beta}} = \mathbf{CQ}$	$\hat{\boldsymbol{\beta}} = \mathbf{CQ}$	$\hat{\boldsymbol{\beta}} = \mathbf{CQ}$
Formal dispersion matrix of $\hat{\boldsymbol{\beta}}$	$\sigma^2\mathbf{C}$	$\sigma^2\mathbf{C}$	$\mathbf{C}$
Residual sum of squares, R_0^2	$\mathbf{Y}'\mathbf{Y} - \mathbf{Q}'\hat{\boldsymbol{\beta}}$	$\mathbf{Y}'\mathbf{G}^{-1}\mathbf{Y} - \mathbf{Q}'\hat{\boldsymbol{\beta}}$	$\mathbf{Y}'\boldsymbol{\Sigma}^{-1}\mathbf{Y} - \mathbf{Q}'\hat{\boldsymbol{\beta}}$
$E(R_0^2)$	$(n - r)\sigma^2$	$(n - r)\sigma^2$	$(n - r)$
Condition for estimability of $\mathbf{P}'\mathbf{B}$	$\mathbf{X}'\mathbf{L} = \mathbf{P}$	$\mathbf{X}'\mathbf{L} = \mathbf{P}$	$\mathbf{X}'\mathbf{L} = \mathbf{P}$
Variance of $\mathbf{P}'\hat{\boldsymbol{\beta}}$	$\sigma^2\mathbf{P}'\mathbf{CP}$	$\sigma^2\mathbf{P}'\mathbf{CP}$	$\mathbf{P}'\mathbf{CP}$
Covariance of $\mathbf{P}'\hat{\boldsymbol{\beta}}, \mathbf{R}'\hat{\boldsymbol{\beta}}$	$\sigma^2\mathbf{P}'\mathbf{CR}$	$\sigma^2\mathbf{P}'\mathbf{CR}$	$\mathbf{P}'\mathbf{CR}$

(G and $\boldsymbol{\Sigma}$ are assumed to be nonsingular)

The essential difference is seen to be only in the computation of the normal equation $\mathbf{S}\boldsymbol{\beta} = \mathbf{Q}$. The rest of the formulas depend only on $\mathbf{S}$ and $\mathbf{Q}$.

4a.8 Some Computational Aspects of the l.s. Theory

The application of the least squares technique involves the numerical reduction of $\mathbf{X}'\mathbf{X}$, the $m \times m$ matrix of normal equations, which may offer some difficulty. A method which has been found to be convenient for computations both on desk calculators and high-speed computers is as follows. First, a matrix $\mathbf{T}$ is found such that $\mathbf{X}'\mathbf{X} = \mathbf{T}'\mathbf{T}$ where $\mathbf{T}$ is upper triangular of order m. When $\mathbf{X}'\mathbf{X}$ is singular, $\mathbf{T}$ has the property that if a diagonal entry is zero, the entire row is zero. The number of non-zero diagonal entries is equal to $R(\mathbf{X})$ which is less than m if $\mathbf{X}'\mathbf{X}$ is singular. Given such a matrix, it is easy to find $\mathbf{T}^-$, its g-inverse. This is done by omitting the zero rows, determining the true inverse of the reduced triangular matrix, and finally restoring the order by inserting the zero rows in the proper positions. The rest of the computations involve only matrix multiplications.

A g-inverse of $\mathbf{X}'\mathbf{X}$ is $\mathbf{C} = \mathbf{T}^-(\mathbf{T}^-)'$, and a solution to normal equation is $\hat{\boldsymbol{\beta}} = \mathbf{C}\mathbf{Q}$. The dispersion matrix of $\hat{\boldsymbol{\beta}}$ for computing the variances and co-variances of estimable parametric functions is $\sigma^2\mathbf{C}$. The least sum of squares which provides an estimate of σ^2 is $R_0{}^2 = \mathbf{Y}'\mathbf{Y} - \hat{\boldsymbol{\beta}}'\mathbf{Q}$.

A numerically stable method has to be used to compute $\mathbf{T}$ itself. Ordinarily the square root method will work provided the matrix $\mathbf{X}'\mathbf{X}$ is not badly conditioned. A better alternative is the triangular reduction of $\mathbf{X}$ by premultiplication with Householder orthogonal matrices [see (vii), **1b.2**].

4a.9 Least Squares Estimation with Restrictions on Parameters

Consider the problem $(\mathbf{Y}, \mathbf{X}\boldsymbol{\beta}, \sigma^2\mathbf{I})$ where $\boldsymbol{\beta}$ is subject to consistent linear restrictions $\mathbf{H}'\boldsymbol{\beta} = \boldsymbol{\xi}$ (given) and $\operatorname{rank} \mathbf{H} = m - s$. A general solution of $\mathbf{H}'\boldsymbol{\beta} = \boldsymbol{\xi}$ is $\boldsymbol{\beta}_0 + \mathbf{B}\boldsymbol{\theta}$ where $\boldsymbol{\beta}_0$ is a particular solution and $\mathbf{B}$ is an $m \times s$ matrix of rank s such that $\mathbf{H}'\mathbf{B} = \mathbf{0}$ and $\boldsymbol{\theta}$ is an arbitrary vector of s elements, which are, so to speak, new parameters.

$$E(\mathbf{Y}) = \mathbf{X}\boldsymbol{\beta} = \mathbf{X}(\boldsymbol{\beta}_0 + \mathbf{B}\boldsymbol{\theta}) \qquad \text{or} \qquad E(\mathbf{Y} - \mathbf{X}\boldsymbol{\beta}_0) = \mathbf{X}\mathbf{B}\boldsymbol{\theta}.$$

Let $\mathbf{Z} = \mathbf{Y} - \mathbf{X}\boldsymbol{\beta}_0$. Then $E(\mathbf{Z}) = \mathbf{X}\mathbf{B}\boldsymbol{\theta}$, and the problem is reduced to that of (4a.1.2), by considering $(\mathbf{Z}, \mathbf{X}\mathbf{B}\boldsymbol{\theta}, \sigma^2\mathbf{I})$, with new variables $\mathbf{Z}$ and new parameters $\boldsymbol{\theta}$. To estimate $\mathbf{P}'\boldsymbol{\beta} = \mathbf{P}'(\boldsymbol{\beta}_0 + \mathbf{B}\boldsymbol{\theta}) = \mathbf{P}'\boldsymbol{\beta}_0 + \mathbf{P}'\mathbf{B}\boldsymbol{\theta}$, we need only consider $\mathbf{P}'\mathbf{B}\boldsymbol{\theta} = \mathbf{R}'\boldsymbol{\theta}$. For the estimation of σ^2 as in **4a.4**, the rank of $\mathbf{F} = \mathbf{X}\mathbf{B}$ is needed. It is easy to establish

$$\operatorname{rank}(\mathbf{X} \,\vdots\, \mathbf{H}') = \operatorname{rank} \mathbf{X}\mathbf{B} + \operatorname{rank} \mathbf{H}, \qquad \text{using} \quad \mathbf{H}'\mathbf{B} = \mathbf{0}. \tag{4a.9.1}$$

The rank of $\mathbf{X}\mathbf{B}$ can thus be deduced from (4a.9.1) by studying the matrices $\mathbf{X}$ and $\mathbf{H}$ without actually computing $\mathbf{B}$ and $\mathbf{F}$.

But we can approach the problem without introducing a smaller number of parameters by applying the result of [(iii), **1f.1**] on the extremum of a quadratic form.

The problem may be posed as one of finding a linear function $\mathbf{L}'\mathbf{Y} + d$ such that

$$V(\mathbf{L}'\mathbf{Y} + d) = \sigma^2 \mathbf{L}'\mathbf{L} \quad \text{is a minimum}$$

subject to the condition

$$E(\mathbf{L}'\mathbf{Y} + d) = \mathbf{P}'\boldsymbol{\beta}, \qquad \text{given} \quad \mathbf{H}'\boldsymbol{\beta} = \boldsymbol{\xi}. \tag{4a.9.2}$$

The condition (4a.9.2) implies that the equation

$$\mathbf{L}'\mathbf{X}\boldsymbol{\beta} + d = \mathbf{P}'\boldsymbol{\beta}, \tag{4a.9.3}$$

is satisfied for all $\boldsymbol{\beta}$ such that $\mathbf{H}'\boldsymbol{\beta} - \boldsymbol{\xi} = 0$. Then there exists a vector $\mathbf{M}$ such that

$$\mathbf{L}'\mathbf{X} + \mathbf{M}'\mathbf{H}' = \mathbf{P}' \qquad \text{or} \qquad (\mathbf{X}' \,\vdots\, \mathbf{H})\begin{pmatrix} \mathbf{L} \\ \mathbf{M} \end{pmatrix} = \mathbf{P} \tag{4a.9.4}$$

and $d = \mathbf{M}'\boldsymbol{\xi}$. The problem then reduces to

$$\begin{gathered} \text{minimizing } \mathbf{L}'\mathbf{L} \\ \text{subject to } (\mathbf{X}' \,\vdots\, \mathbf{H})\begin{pmatrix} \mathbf{L} \\ \hdashline \mathbf{M} \end{pmatrix} = \mathbf{P}. \end{gathered} \tag{4a.9.5}$$

By an application of the result of [(iii), **1f.1**], the optimum choice of $\mathbf{L}$ and $\mathbf{M}$ are

$$\mathbf{L}_* = \mathbf{X}\mathbf{C}_1'\mathbf{P}, \qquad \mathbf{M}_* = \mathbf{C}_2'\mathbf{P} \tag{4a.9.6}$$

where $\mathbf{C}_1$ and $\mathbf{C}_2$ are the submatrices of a generalized inverse

$$\begin{pmatrix} \mathbf{X}'\mathbf{X} & \mathbf{H} \\ \mathbf{H}' & \mathbf{0} \end{pmatrix}^- = \begin{pmatrix} \mathbf{C}_1 & \mathbf{C}_2 \\ \mathbf{C}_3 & \mathbf{C}_4 \end{pmatrix}. \tag{4a.9.7}$$

Hence we have the following result:

The minimum variance unbiased estimator of $\mathbf{P}'\boldsymbol{\beta}$ *is* $(\mathbf{L}_*)'\mathbf{Y} + (\mathbf{M}_*)'\boldsymbol{\xi} = \mathbf{P}'(\mathbf{C}_1\mathbf{X}'\mathbf{Y} + \mathbf{C}_2\boldsymbol{\xi})$, *which may be written as* $\mathbf{P}'\hat{\boldsymbol{\beta}}$ *where* $\hat{\boldsymbol{\beta}}$ *is a solution of the equations*

$$\begin{aligned} \mathbf{X}'\mathbf{X}\boldsymbol{\beta} + \mathbf{H}\boldsymbol{\lambda} &= \mathbf{X}'\mathbf{Y}, \\ \mathbf{H}'\boldsymbol{\beta} \qquad &= \boldsymbol{\xi}. \end{aligned} \tag{4a.9.8}$$

It may be observed that equations (4a.9.8) are the same as the normal equations obtained by minimizing $(\mathbf{Y} - \mathbf{X}\boldsymbol{\beta})'(\mathbf{Y} - \mathbf{X}\boldsymbol{\beta})$ subject to the restriction $\mathbf{H}'\boldsymbol{\beta} = \boldsymbol{\xi}$ using a Lagrangian multiplier $\boldsymbol{\lambda}$.

Variances and Covariance of Estimators. Since

$$\mathbf{P}'\hat{\boldsymbol{\beta}} = \mathbf{P}'\mathbf{C}_1\mathbf{X}'\mathbf{Y} + \mathbf{C}_2\boldsymbol{\xi},$$

then

$$V(\mathbf{P}'\hat{\boldsymbol{\beta}}) = \sigma^2\mathbf{P}'\mathbf{C}_1\mathbf{X}'\mathbf{X}\mathbf{C}_1\mathbf{P} = \sigma^2\mathbf{P}'\mathbf{C}_1\mathbf{P}$$

if we use the properties of the generalized inverse (4a.9.7). Similarly,

$$\text{cov}(\mathbf{P}'\hat{\boldsymbol{\beta}}, \mathbf{R}'\hat{\boldsymbol{\beta}}) = \sigma^2\mathbf{P}'\mathbf{C}_1\mathbf{R} = \sigma^2\mathbf{R}'\mathbf{C}_1\mathbf{P}.$$

Suppose the matrix $\mathbf{C}_1$ is not computed but it is known that $\mathbf{P}'\hat{\boldsymbol{\beta}} = \boldsymbol{\gamma}'\mathbf{X}'\mathbf{Y} + \boldsymbol{\delta}'\boldsymbol{\xi}$ and $\mathbf{R}'\hat{\boldsymbol{\beta}} = \boldsymbol{\phi}'\mathbf{X}'\mathbf{Y} + \boldsymbol{\psi}'\boldsymbol{\xi}$ and $\mathbf{H}'\boldsymbol{\gamma} = \mathbf{H}'\boldsymbol{\phi} = \mathbf{0}$. Then the variances and the covariance

$$V(\mathbf{P}'\hat{\boldsymbol{\beta}}) = \sigma^2\mathbf{P}'\boldsymbol{\gamma}, \qquad V(\mathbf{R}'\hat{\boldsymbol{\beta}}) = \sigma^2\mathbf{R}'\boldsymbol{\phi}$$
$$\text{cov}(\mathbf{P}'\hat{\boldsymbol{\beta}}, \mathbf{R}'\hat{\boldsymbol{\beta}}) = \sigma^2\mathbf{P}'\boldsymbol{\phi} = \sigma^2\mathbf{R}'\boldsymbol{\gamma}$$

involve only the vectors $\mathbf{P}$, $\mathbf{R}$ and $\boldsymbol{\gamma}$, $\boldsymbol{\phi}$. If the condition $\mathbf{H}'\boldsymbol{\gamma} = \mathbf{H}'\boldsymbol{\phi} = \mathbf{0}$ is not satisfied, then

$$V(\mathbf{P}'\hat{\boldsymbol{\beta}}) = \sigma^2\boldsymbol{\gamma}'\mathbf{X}'\mathbf{X}\boldsymbol{\gamma} \qquad \text{and} \qquad \text{cov}(\mathbf{P}'\hat{\boldsymbol{\beta}}, \mathbf{R}'\hat{\boldsymbol{\beta}}) = \sigma^2\boldsymbol{\gamma}'\mathbf{X}'\mathbf{X}\boldsymbol{\phi}.$$

Estimation of σ^2. An unbiased estimator of σ^2 is

$$(\mathbf{Y}'\mathbf{Y} - \hat{\boldsymbol{\beta}}'\mathbf{X}'\mathbf{Y} - \hat{\boldsymbol{\lambda}}'\boldsymbol{\xi}) \div (n - s)$$

where $s = \text{rank}(\mathbf{X}' \,\vdots\, \mathbf{H}) - \text{rank}\,\mathbf{H}$ as shown in (4a.9.1) and $(\hat{\boldsymbol{\beta}}, \hat{\boldsymbol{\lambda}})$ is any solution of (4a.9.8). In terms of the submatrices of the g-inverse (4a.9.7) we see that

$$\hat{\boldsymbol{\beta}} = \mathbf{C}_1\mathbf{X}'\mathbf{Y} + \mathbf{C}_2\boldsymbol{\xi}, \qquad \hat{\boldsymbol{\lambda}} = \mathbf{C}_3\mathbf{X}'\mathbf{Y} + \mathbf{C}_4\boldsymbol{\xi}.$$

4a.10 Simultaneous Estimation of Parametric Functions

Consider the general set up $(\mathbf{Y}, \mathbf{X}\boldsymbol{\beta}, \boldsymbol{\Sigma})$ with or without restrictions on the parameter $\boldsymbol{\beta}$. Let $\mathbf{P}_1'\hat{\boldsymbol{\beta}}, \ldots, \mathbf{P}_k'\hat{\boldsymbol{\beta}}$ be the individual least squares estimators of the parametric functions $\mathbf{P}_1'\boldsymbol{\beta}, \ldots, \mathbf{P}_k'\boldsymbol{\beta}$. Furthermore, let $\mathbf{A}$ be the dispersion matrix of the estimators $\mathbf{P}_1'\hat{\boldsymbol{\beta}}, \ldots, \mathbf{P}_k'\hat{\boldsymbol{\beta}}$. We then have the following optimum property of the l.s. estimators.

(i) *Let* $\mathbf{L}_1'\mathbf{Y}, \ldots, \mathbf{L}_k'\mathbf{Y}$ *be any unbiased estimators of* $\mathbf{P}_1'\boldsymbol{\beta}, \ldots, \mathbf{P}_k'\boldsymbol{\beta}$ *and let the dispersion matrix of the estimators be* $\mathbf{B}$. *Then* $\mathbf{B} - \mathbf{A}$ *is nonnegative definite, implying*

(a) Trace $\mathbf{B} \geqslant$ Trace $\mathbf{A}$,
(b) $|\mathbf{B}| \geqslant |\mathbf{A}|$,
(c) Trace $\mathbf{QB} \geqslant$ Trace $\mathbf{QA}$, where $\mathbf{Q}$ is any n.n.d. matrix, and
(d) maximum latent root of $\mathbf{B} \geqslant$ maximum latent root of $\mathbf{A}$.

Consider the linear parametric function

$$a_1\mathbf{P}_1'\boldsymbol{\beta} + \cdots + a_k\mathbf{P}_k'\boldsymbol{\beta}$$

whose l.s. estimator is $a_1\mathbf{P}_1'\hat{\boldsymbol{\beta}} + \cdots + a_k\mathbf{P}_k'\hat{\boldsymbol{\beta}}$ with variance $\mathbf{a}'\mathbf{A}\mathbf{a}$ where $\mathbf{a}' = (a_1, \ldots, a_k)$. The variance of the alternative estimator $a_1\mathbf{L}_1'\mathbf{Y} + \cdots + a_k\mathbf{L}_k'\mathbf{Y}$ is $\mathbf{a}'\mathbf{B}\mathbf{a}$. Then we have $\mathbf{a}'\mathbf{B}\mathbf{a} \geqslant \mathbf{a}'\mathbf{A}\mathbf{a}$ for all $\mathbf{a}$, which proves all the results (a) – (d).

4a.11 Least Squares Theory when the Parameters are Random Variables

Consider $(\mathbf{Y}, \mathbf{X}\boldsymbol{\beta}, \boldsymbol{\Sigma})$ where $\boldsymbol{\beta}$ itself is a random variable with mean $\boldsymbol{\mu}$ and dispersion matrix $\mathbf{T}$. What is the best linear estimator (or predictor) of the random variable $\mathbf{P}'\boldsymbol{\beta}$. Let us observe that $\boldsymbol{\Sigma} = D(\mathbf{Y} \mid \boldsymbol{\beta})$, that is, the conditional dispersion matrix of $\mathbf{Y}$ given $\boldsymbol{\beta}$. Then we have the following formulas connecting the total dispersion and covariance matrices with the conditional matrices.

$$\begin{aligned} D(\mathbf{Y}) &= ED(\mathbf{Y} \mid \boldsymbol{\beta}) + D[E(\mathbf{Y} \mid \boldsymbol{\beta})] \\ &= \boldsymbol{\Sigma} + D(\mathbf{X}\boldsymbol{\beta}) = \boldsymbol{\Sigma} + \mathbf{XTX}' \end{aligned} \tag{4a.11.1}$$

$$\begin{aligned} C(\mathbf{Y}, \mathbf{P}'\boldsymbol{\beta}) &= E[C(\mathbf{Y}, \mathbf{P}'\boldsymbol{\beta}) \mid \boldsymbol{\beta})] + C[E(\mathbf{Y} \mid \boldsymbol{\beta}), \mathbf{P}'\boldsymbol{\beta}] \\ &= \mathbf{0} + \mathbf{XTP}. \end{aligned} \tag{4a.11.2}$$

We determine a linear function $a + \mathbf{L}'\mathbf{Y}$ such that

$$\left.\begin{aligned} &E(\mathbf{P}'\boldsymbol{\beta} - a - \mathbf{L}'\mathbf{Y}) = 0 \\ &V(\mathbf{P}'\boldsymbol{\beta} - a - \mathbf{L}'\mathbf{Y}) \text{ is a minimum} \end{aligned}\right\}. \tag{4a.11.3}$$

(i) CASE 1, $\boldsymbol{\mu}$ KNOWN. *The optimum choice of* $\mathbf{L}$ *and* a *are*

$$\begin{aligned} \mathbf{L}_* &= (\boldsymbol{\Sigma} + \mathbf{XTX}')^{-1}\mathbf{XTP} = \boldsymbol{\Sigma}^{-1}\mathbf{X}(\mathbf{T}^{-1} + \mathbf{X}'\boldsymbol{\Sigma}^{-1}\mathbf{X})^{-1} \\ a_* &= \boldsymbol{\mu}'\mathbf{P} - \boldsymbol{\mu}'\mathbf{X}'\mathbf{L}_*, \end{aligned} \tag{4a.11.4}$$

and the prediction variance is

$$\begin{aligned} V(\mathbf{P}'\boldsymbol{\beta} - a_* - (\mathbf{L}_*)'\mathbf{Y}) &= \mathbf{P}'\mathbf{TP} - \mathbf{P}'\mathbf{TX}'\mathbf{L}_* \\ &= \mathbf{P}'(\mathbf{T}^{-1} + \mathbf{X}'\boldsymbol{\Sigma}^{-1}\mathbf{X})^{-1}\mathbf{P}. \end{aligned} \tag{4a.11.5}$$

Proof consists in verifying that the choice (4a.11.4) satisfies the first condition in (4a.11.3) and showing that, for any other choice of $\mathbf{L}$ and a,

$$\begin{aligned} V(\mathbf{P}'\boldsymbol{\beta} - a - \mathbf{L}'\mathbf{Y}) &= V(\mathbf{P}'\boldsymbol{\beta} - a_* - (\mathbf{L}_*)'\mathbf{Y}) + V[(\mathbf{L} - \mathbf{L}_*)'\mathbf{Y}] \\ &\geqslant V(\mathbf{P}'\boldsymbol{\beta} - a_* - (\mathbf{L}_*)'\mathbf{Y}). \end{aligned} \tag{4a.11.6}$$

(ii) CASE 2, $\boldsymbol{\mu}$ UNKNOWN. *The optimum choice of* $\mathbf{L}$ *and* a *are*

$$\mathbf{L}_* = \boldsymbol{\Sigma}^{-1}\mathbf{X}(\mathbf{X}'\boldsymbol{\Sigma}^{-1}\mathbf{X})^{-1}\mathbf{P}, \qquad a_* = 0 \tag{4a.11.7}$$

provided there exists an $\mathbf{L}$ *such that* $\mathbf{X}'\mathbf{L} = \mathbf{P}$, *and the prediction variance is*

$$\mathbf{P}'(\mathbf{X}'\boldsymbol{\Sigma}^{-1}\mathbf{X})^{-1}\mathbf{P}. \tag{4a.11.8}$$

Observe that the best estimator and the prediction variance are the same as the l.s. estimator and its variance when $\boldsymbol{\beta}$ is considered as a fixed parameter as in (4a.1.2).

In (4a.11.8) it is assumed that $(\mathbf{X}'\boldsymbol{\Sigma}^{-1}\mathbf{X})^{-1}$ exists. Otherwise we can substitute a g-inverse of $\mathbf{X}'\boldsymbol{\Sigma}^{-1}\mathbf{X}$.

It is easy to verify that $\mathbf{X}'\mathbf{L}_* = \mathbf{P}$, thus satisfying the condition (4a.11.3). Consider

$$V[\mathbf{P}'\boldsymbol{\beta} - (\mathbf{L}_*)'\mathbf{Y} - (\mathbf{L} - \mathbf{L}_*)'\mathbf{Y}] = V[\mathbf{P}'\boldsymbol{\beta} - (\mathbf{L}_*)'\mathbf{Y}] + V[(\mathbf{L} - \mathbf{L}_*)'\mathbf{Y}], \tag{4a.11.9}$$

the covariance term being zero as shown below.

$$\begin{aligned} C[(\mathbf{P}'\boldsymbol{\beta} - (\mathbf{L}_*)'\mathbf{Y}), (\mathbf{L} - \mathbf{L}_*)'\mathbf{Y}] &= (\mathbf{L} - \mathbf{L}_*)'\mathbf{XTP} - (\mathbf{L} - \mathbf{L}_*)'\boldsymbol{\Sigma}\mathbf{L}_* \\ &= 0 - (\mathbf{L} - \mathbf{L}_*)'\boldsymbol{\Sigma}\boldsymbol{\Sigma}^{-1}\mathbf{X}(\mathbf{X}'\boldsymbol{\Sigma}^{-1}\mathbf{X})^{-1}\mathbf{P} = 0 - 0, \end{aligned} \tag{4a.11.10}$$

by using the condition $\mathbf{L}'\mathbf{X} = (\mathbf{L}_*)'\mathbf{X} = \mathbf{P}$. Equation (4a.11.9) proves the desired result. The prediction variance is easily shown to be as given in (4a.11.8).

4a.12 Choice of the Design Matrix

We have not discussed how the choice of $\mathbf{X}$ arises in practice. Each row of $\mathbf{X}$ represents a combination of "factors" (or instrumental variables whose values may be arbitrarily assigned and are not the results of an experiment) at which a certain response (the result of an experiment) is observed. The matrix $\mathbf{X}$, which may be predetermined, is called the design matrix. In the setup $(\mathbf{Y}, \mathbf{X}\boldsymbol{\beta}, \boldsymbol{\Sigma})$, the expected response is assumed to be a linear function of the factor variables.

Given $\mathbf{X}$, it is shown that the dispersion matrix of the l.s. estimators $\hat{\beta}_1, \ldots, \hat{\beta}_m$ is $(\mathbf{X}'\boldsymbol{\Sigma}^{-1}\mathbf{X})^{-1}$ (or a g-inverse when the rank is not full). We now raise the question as to how $\mathbf{X}$ may be chosen to make $(\mathbf{X}'\boldsymbol{\Sigma}^{-1}\mathbf{X})^{-1}$ as small as possible.

If we are allowed to choose the factor values in an unlimited range, the entries of $(\mathbf{X}'\boldsymbol{\Sigma}^{-1}\mathbf{X})^{-1}$ can be made arbitrarily small, since by choosing $\mathbf{X}$ as $a\mathbf{X}$, $(\mathbf{X}'\boldsymbol{\Sigma}\mathbf{X})^{-1}$ changes to $a^{-2}(\mathbf{X}'\boldsymbol{\Sigma}^{-1}\mathbf{X})^{-1} \to 0$ as $a \to \infty$. But in practice it may be wrong to do so since the response model may not be valid for a wide range of the factor values. We have to limit the ranges to small regions in which we are interested and within which we want to explore the response function.

Then we may make a restriction of the form

$$\mathbf{X}_i'\boldsymbol{\Sigma}^{-1}\mathbf{X}_i = C_i^2 \text{ (given)}, \qquad i = 1, \ldots, m, \tag{4a.12.1}$$

where $\mathbf{X}_i$ is the column vector of $\mathbf{X}$ representing the n levels of the ith factor used in an experiment. Subject to the conditions (4a.12.1) we wish to choose the combinations of the factor values or the rows of $\mathbf{X}$ which lead to estimators of $\beta_1, \ldots, \beta_m$ with the least possible variance.

(i) *Let* $\mathbf{X}$ *be a design matrix and* $\hat{\beta}_i$ *be the l.s. estimator of* β_i. *Then, under the conditions* (4a.12.1) *on* $\mathbf{X}$,

$$V(\hat{\beta}_i) \geqslant 1/C_i^2 \tag{4a.12.2}$$

and the minimum is attained when $\mathbf{X}_i'\boldsymbol{\Sigma}^{-1}\mathbf{X}_j = 0$ *for* $j = 1, \ldots, i-1, i+1, \ldots, m$.

We shall prove the result for $i = 1$. The matrix $\mathbf{X}'\boldsymbol{\Sigma}^{-1}\mathbf{X}$ can be written in the partitioned form

$$\mathbf{X}'\boldsymbol{\Sigma}^{-1}\mathbf{X} = \begin{pmatrix} \mathbf{X}_1'\boldsymbol{\Sigma}^{-1}\mathbf{X}_1 & \mathbf{B}' \\ \mathbf{B} & \mathbf{F} \end{pmatrix}. \tag{4a.12.3}$$

Hence $|\mathbf{X}'\boldsymbol{\Sigma}^{-1}\mathbf{X}| = |\mathbf{F}|(\mathbf{X}_1'\boldsymbol{\Sigma}^{-1}\mathbf{X}_1 - \mathbf{B}'\mathbf{F}^{-1}\mathbf{B})$, which gives

$$\frac{|\mathbf{F}|}{|\mathbf{X}'\boldsymbol{\Sigma}^{-1}\mathbf{X}|} = \frac{1}{\mathbf{X}_1'\boldsymbol{\Sigma}^{-1}\mathbf{X}_1 - \mathbf{B}'\mathbf{F}^{-1}\mathbf{B}} \geqslant \frac{1}{\mathbf{X}_1'\boldsymbol{\Sigma}^{-1}\mathbf{X}_1} \tag{4a.12.4}$$

since $\mathbf{B}'\mathbf{F}^{-1}\mathbf{B}$ is positive. But, $V(\hat{\beta}_1) = |\mathbf{F}|/|\mathbf{X}'\boldsymbol{\Sigma}^{-1}\mathbf{X}|$. Hence $V(\hat{\beta}_1) \geqslant 1/C_1^2$. The equality in (4a.12.4) is attained when $\mathbf{B} = 0$, that is, $\mathbf{X}_1'\boldsymbol{\Sigma}^{-1}\mathbf{X}_j = 0$, $j \neq 1$. As a special case when $\boldsymbol{\Sigma} = \sigma^2\mathbf{I}$, we have the following.

(ii) *Under the condition* $\mathbf{X}_i'\mathbf{X}_i = C_i^2$, *the optimum choice of combinations* (*rows of* $\mathbf{X}$) *is when* $\mathbf{X}_i'\mathbf{X}_j = 0$, *that is, the columns of* $\mathbf{X}$ *are orthogonal.*

4b TESTS OF HYPOTHESES AND INTERVAL ESTIMATION

The l.s.e. of a parametric function is only a point estimate, and no exact statement of probability of its deviation from the true value of the parametric function can be made without a specific distribution for the variable $\mathbf{Y}$ being considered. We shall assume that $\mathbf{Y} \sim N(\mathbf{X}\boldsymbol{\beta}, \sigma^2\mathbf{I})$, the setup under which the sampling distributions of several statistics were investigated in **3b.5**. Let $\mathbf{H}'$ be a matrix of order $k \times m$ of rank k, such that its rows depend on the rows of $\mathbf{X}$. This implies [see (i), **4a.2**] that the k parametric functions $\mathbf{H}'\boldsymbol{\beta}$ are individually estimable under the setup $(\mathbf{Y}, \mathbf{X}\boldsymbol{\beta}, \sigma^2\mathbf{I})$, in which case, the l.s.

estimators are $\mathbf{H}'\hat{\boldsymbol{\beta}} = \mathbf{Z}$ (say). In terms of $\mathbf{X}$, $\mathbf{Y}$, we have $\mathbf{H}'\hat{\boldsymbol{\beta}} = \mathbf{H}'(\mathbf{X}'\mathbf{X})^{-}\mathbf{X}'\mathbf{Y}$ and $R_0^2 = \mathbf{Y}'(\mathbf{I} - \mathbf{X}(\mathbf{X}'\mathbf{X})^{-}\mathbf{X}')\mathbf{Y}$. Then

$$\mathbf{H}'(\mathbf{X}'\mathbf{X})^{-}\mathbf{X}'(\mathbf{I} - \mathbf{X}(\mathbf{X}'\mathbf{X})^{-}\mathbf{X}') = \mathbf{0}$$

showing that $\mathbf{H}'\hat{\boldsymbol{\beta}}$ and R_0^2 are independently distributed. (See also Examples 2.2 and 2.3 at the end of Chapter 3.)

From (3b.3.4) the distribution of $\mathbf{Z}$ is k variate normal with mean $\mathbf{H}'\boldsymbol{\beta}$ and dispersion matrix $\sigma^2\mathbf{D}$ (say), and of R_0^2 is $\sigma^2\chi^2(n-r)$. We shall use this result in drawing inferences about unknown values of parametric functions.

4b.1 Single Parametric Function (Inference)

Given an estimable parametric function $\theta = \mathbf{P}'\boldsymbol{\beta}$, let u be its l.s.e. with variance (say) $p^2\sigma^2$. Then

$$u \sim N(\theta, p^2\sigma^2) \qquad \text{and} \qquad R_0^2 \sim \sigma^2\chi^2(n-r)$$

are independent, so that if $s^2 = R_0^2/(n-r)$, then

$$t = \frac{u-\theta}{p\sigma} \div \frac{s}{\sigma} = \frac{u-\theta}{ps} \sim S(n-r) \tag{4b.1.1}$$

which is Student's distribution on $(n-r)$ D.F. [using (3b.1.8)]. If t_α is the α probability point of $|t|$, that is, $P(|t| > t_\alpha) = \alpha$, then

$$P\left\{\frac{|u-\theta|}{ps} \leqslant t_\alpha\right\} = 1 - \alpha, \tag{4b.1.2}$$

that is.

$$P(u - pst_\alpha \leqslant \theta \leqslant u + pst_\alpha) = 1 - \alpha. \tag{4b.1.3}$$

The results (4b.1.2) and (4b.1.3) are extremely important for drawing inferences on θ, the value of the given parametric function.

Testing of Hypothesis. Let the null hypothesis be $\mathbf{P}'\boldsymbol{\beta} = \theta_0$ (an assigned value). If the null hypothesis is true, then using (4b.1.2), we have

$$P\left\{\frac{|u-\theta_0|}{ps} \leqslant t_\alpha\right\} = 1 - \alpha. \tag{4b.1.4}$$

The null hypothesis is thus rejected at α level of significance if, for an observed u, we have

$$\frac{|u-\theta_0|}{ps} > t_\alpha. \tag{4b.1.5}$$

Interval Estimation. Equation (4b.1.3) implies that for an observed u the interval

$$(u - pst_\alpha, u + pst_\alpha) \tag{4b.1.6}$$

is a $(1-\alpha)$ confidence interval of θ. (See **7b** of Chapter 7 for detailed treatment of confidence intervals.)

4b.2 More than One Parametric Function (Inference)

Let us consider k independent estimable linear parametric functions

$$\theta_1 = \mathbf{H}_1'\boldsymbol{\beta}, \quad \ldots, \quad \theta_k = \mathbf{H}_k'\boldsymbol{\beta}, \tag{4b.2.1}$$

which, in matrix notation, may be written $\boldsymbol{\theta} = \mathbf{H}'\boldsymbol{\beta}$ where $\mathbf{H}'$ is a $k \times m$ matrix and $\boldsymbol{\theta}$ is the column vector of $\theta_1, \ldots, \theta_k$. The l.s. estimators of $(\theta_1, \ldots, \theta_k)$ are represented by $(z_1, \ldots, z_k) = \mathbf{Z}'$ and its dispersion matrix by $\sigma^2\mathbf{D}$. Then since $E(\mathbf{Z}) = \boldsymbol{\theta}$, $D(\mathbf{Z}) = \sigma^2\mathbf{D}$, by applying (3b.4.7), we see that

$$(\mathbf{Z} - \boldsymbol{\theta})'\mathbf{D}^{-1}(\mathbf{Z} - \boldsymbol{\theta}) \sim \sigma^2\chi^2(k) \quad \text{and} \quad R_0^2 \sim \sigma^2\chi^2(n-r)$$

and are independent. Hence by (3b.1.11),

$$F = \frac{(\mathbf{Z} - \boldsymbol{\theta})'\mathbf{D}^{-1}(\mathbf{Z} - \boldsymbol{\theta})}{k} \div \frac{R_0^2}{n-r} \sim F(k, n-r). \tag{4b.2.2}$$

The statistic F and its distribution (4b.2.2) play a fundamental role in drawing simultaneous inference on a number of parametric functions. We shall study the various aspects of this statistic.

Test of Multiple Hypotheses. Let it be required to test the null hypothesis that k parametric functions have assigned values,

$$\mathbf{H}_1'\boldsymbol{\beta} = \theta_{10}, \quad \ldots, \quad \mathbf{H}_k'\boldsymbol{\beta} = \theta_{k0},$$

which may be written in matrix notation, $\mathbf{H}'\boldsymbol{\beta} = \boldsymbol{\theta}_0$. The vector of deviations between l.s. estimators and assigned values is $(\mathbf{Z} - \boldsymbol{\theta}_0)$. If in fact $\boldsymbol{\theta}_0$ is not true, the deviations $(\mathbf{Z} - \boldsymbol{\theta}_0)$ are likely to be large. Let us consider the compound deviation (a single measure of deviations)

$$(\mathbf{Z} - \boldsymbol{\theta}_0)'\mathbf{D}^{-1}(\mathbf{Z} - \boldsymbol{\theta}_0).$$

By using the formula (4a.1.3), we see that

$$\begin{aligned}
&\frac{E(\mathbf{Z} - \boldsymbol{\theta}_0)'\mathbf{D}^{-1}(\mathbf{Z} - \boldsymbol{\theta}_0)}{k} \\
&= \frac{\sigma^2}{k}\operatorname{trace}(\mathbf{D}^{-1}\mathbf{D}) + \frac{1}{k}E(\mathbf{Z} - \boldsymbol{\theta}_0)\mathbf{D}^{-1}E(\mathbf{Z} - \boldsymbol{\theta}_0) \\
&= \sigma^2 + \frac{1}{k}(\mathbf{H}'\boldsymbol{\beta} - \boldsymbol{\theta}_0)\mathbf{D}^{-1}(\mathbf{H}'\boldsymbol{\beta} - \boldsymbol{\theta}_0) \\
&= \sigma^2, \quad \text{only if} \quad \mathbf{H}'\boldsymbol{\beta} = \boldsymbol{\theta}_0, \quad \text{that is, the hypothesis is true.}
\end{aligned} \tag{4b.2.3}$$

Since $(\mathbf{H}'\boldsymbol{\beta} - \boldsymbol{\theta}_0)\mathbf{D}^{-1}(\mathbf{H}'\boldsymbol{\beta} - \boldsymbol{\theta}_0)$ is positive definite, (4b.2.3) exceeds σ^2 if and only if the hypothesis is not true. On the other hand, the expected value of $s^2 = [R_0^2/(n-r)]$ is σ^2 without any assumption on the unknown parameters $\boldsymbol{\beta}$. Hence in the ratio

$$F = \frac{(\mathbf{Z} - \boldsymbol{\theta}_0)'\mathbf{D}^{-1}(\mathbf{Z} - \boldsymbol{\theta}_0)}{k} \div \frac{R_0^2}{n-r} \tag{4b.2.4}$$

the numerator is, on the average, larger in magnitude than the denominator if the null hypothesis is not true, depending on the non-null quantity $(\mathbf{H}'\boldsymbol{\beta} - \boldsymbol{\theta}_0)\mathbf{D}^{-1}(\mathbf{H}'\boldsymbol{\beta} - \boldsymbol{\theta}_0)/k$. Thus large values of F indicate departure from null hypothesis. Choosing a value F_α such that

$$P(F > F_\alpha) = \alpha, \qquad \text{where} \quad F \sim F(k, n-r),$$

we may set up the test criterion

$$F = \frac{(\mathbf{Z} - \boldsymbol{\theta}_0)'\mathbf{D}^{-1}(\mathbf{Z} - \boldsymbol{\theta}_0)}{ks^2} > F_\alpha \tag{4b.2.5}$$

for rejecting the hypothesis at α-significance level.

For purposes of computing F, we have an alternative procedure which directly provides the numerator without obtaining the l.s. estimates of individual parametric functions. It was shown in [(ii), **3b.5**] and also in Example 2.4 of Chapter 3 that

$$\begin{aligned} R_1^2 &= \min_{\mathbf{H}'\boldsymbol{\beta} = \boldsymbol{\theta}_0} (\mathbf{Y} - \mathbf{X}\boldsymbol{\beta})'(\mathbf{Y} - \mathbf{X}\boldsymbol{\beta}) \\ &= (\mathbf{Z} - \boldsymbol{\theta}_0)'\mathbf{D}^{-1}(\mathbf{Z} - \boldsymbol{\theta}_0) + R_0^2. \end{aligned} \tag{4b.2.6}$$

If R_1^2 can be obtained directly as the restricted minimum with $\boldsymbol{\beta}$ subject to condition $\mathbf{H}'\boldsymbol{\beta} = \boldsymbol{\theta}_0$, then (4b.2.6) gives the formula

$$(\mathbf{Z} - \boldsymbol{\theta}_0)'\mathbf{D}^{-1}(\mathbf{Z} - \boldsymbol{\theta}_0) = R_1^2 - R_0^2, \tag{4b.2.7}$$

so that we may write the test criterion (4b.2.4) as

$$F = \frac{R_1^2 - R_0^2}{k} \div \frac{R_0^2}{n-r}. \tag{4b.2.8}$$

The equation (4b.2.7) is important in the sense that it provides an explicit representation of $R_1^2 - R_0^2$, which appears in the numerator of the F statistic in (4b.2.8), as a compound measure of the deviations between l.s. estimators and hypothetical values.

The computations leading to the F statistic (4b.2.8) may be presented in tabular form, called the *Analysis of Variance*.

TABLE 4b.2. The Analysis of Variance

	D.F.	Sum of Squares (S.S.)	Mean Square (M.S.)
Deviation from hypothesis $\mathbf{H}'\boldsymbol{\beta} = \boldsymbol{\theta}_0$	k	$* = R_1^2 - R_0^2$	$\dfrac{R_1^2 - R_0^2}{k}$
Residual	$n - r$	$\min\limits_{\boldsymbol{\beta}} (\mathbf{Y} - \mathbf{X}\boldsymbol{\beta})'(\mathbf{Y} - \mathbf{X}\boldsymbol{\beta}) = R_0^2$	$\dfrac{R_0^2}{n - r}$
Total	$n - r + k$	$\min\limits_{\mathbf{H}'\boldsymbol{\beta} = \boldsymbol{\theta}_0} (\mathbf{Y} - \mathbf{X}\boldsymbol{\beta})'(\mathbf{Y} - \mathbf{X}\boldsymbol{\beta}) = R_1^2$	

The entry marked by * is obtained by subtraction. The F statistic is the ratio of M.S. values.

Simultaneous Confidence Intervals (Scheffe, 1959). From (4b.2.2) it follows that

$$P\left\{\frac{(\mathbf{Z} - \boldsymbol{\theta})'\mathbf{D}^{-1}(\mathbf{Z} - \boldsymbol{\theta})}{ks^2} \leqslant F_\alpha\right\} = 1 - \alpha, \qquad (4b.2.9)$$

where $\boldsymbol{\theta} = \mathbf{H}'\boldsymbol{\beta}$ stands for k independent estimable parametric functions of $\boldsymbol{\beta}$. The set of all possible $\boldsymbol{\theta}$, satisfying the inequality inside the brackets of (4b.2.9) for given $\mathbf{Z}$ and s^2, is a closed region, which may be called the $(1 - \alpha)$-confidence region of $\boldsymbol{\theta} = \mathbf{H}'\boldsymbol{\beta}$. Let us represent such a set by C. Then simultaneous confidence intervals for functions $g_i(\boldsymbol{\theta})$, $i = 1, 2, \ldots, k$ with confidence coefficient possibly greater than $(1 - \alpha)$, are given by

$$\left[\min_{\boldsymbol{\theta} \in C} g_i(\boldsymbol{\theta}), \max_{\boldsymbol{\theta} \in C} g_i(\boldsymbol{\theta})\right], \qquad i = 1, \ldots, k. \qquad (4b.2.10)$$

It may be noted that for any particular $g(\boldsymbol{\theta})$

$$P\left[\min_{\boldsymbol{\theta} \in C} g(\boldsymbol{\theta}), \max_{\boldsymbol{\theta} \in C} g(\boldsymbol{\theta})\right] \geqslant 1 - \alpha, \qquad (4b.2.11)$$

so that the confidence coefficient for any particular function is greater than $1 - \alpha$. In (4b.2.9) F_α is the upper α probability value of the F statistic.

Another method applicable to simultaneous confidence intervals of linear parametric functions depends on the Cauchy-Schwarz inequality (see 1e.1.4)

$$\mathbf{A}'\mathbf{B}^{-1}\mathbf{A} = \max_{\mathbf{U}} \frac{(\mathbf{U}'\mathbf{A})^2}{\mathbf{U}'\mathbf{B}\mathbf{U}}, \qquad (4b.2.12)$$

where $\mathbf{U}$ and $\mathbf{A}$ are column vectors and $\mathbf{B}$ is a positive definite matrix.

Applying thc result (4b.2.12) with $\mathbf{A} = \mathbf{Z} - \boldsymbol{\theta}$, $\mathbf{B} = \mathbf{D}$, we have

$$\frac{(\mathbf{Z} - \boldsymbol{\theta})'\mathbf{D}^{-1}(\mathbf{Z} - \boldsymbol{\theta})}{ks^2} = \frac{1}{ks^2} \max_{\mathbf{U}} \frac{[\mathbf{U}'(\mathbf{Z} - \boldsymbol{\theta})]^2}{\mathbf{U}'\mathbf{D}\mathbf{U}}. \tag{4b.2.13}$$

Hence from (4b.2.9) and (4b.2.13), we see that

$$P\left\{\max_{\mathbf{U}} \frac{|\mathbf{U}'(\mathbf{Z} - \boldsymbol{\theta})|}{\sqrt{\mathbf{U}'\mathbf{D}\mathbf{U}}} \leqslant s\sqrt{kF_\alpha}\right\} = 1 - \alpha,$$

that is,

$$P(|\mathbf{U}'(\mathbf{Z} - \boldsymbol{\theta})| \leqslant s\sqrt{kF_\alpha \mathbf{U}'\mathbf{D}\mathbf{U}} \text{ for all } \mathbf{U}) = 1 - \alpha$$

or (4b.2.14)

$$P(\mathbf{U}'\boldsymbol{\theta} \in \mathbf{U}'\mathbf{Z} \pm s\sqrt{kF_\alpha \mathbf{U}'\mathbf{D}\mathbf{U}} \text{ for all } \mathbf{U}) = 1 - \alpha.$$

Equation (4b.2.14) provides simultaneous confidence intervals for all functions of the form $\mathbf{U}'\boldsymbol{\theta} = \mathbf{U}'\mathbf{H}'\boldsymbol{\beta}$, that is, *all linear combinations of the given parametric functions* $\mathbf{H}'\mathbf{B}$. It also follows from (4b.2.14), as pointed out by Scheffe (1959), that for any particular function $\mathbf{U}'\boldsymbol{\theta}$

$$P(\mathbf{U}'\boldsymbol{\theta} \in \mathbf{U}'\mathbf{Z} \pm s\sqrt{kF_\alpha \mathbf{U}'\mathbf{D}\mathbf{U}}) \geqslant 1 - \alpha$$

so that the confidence coefficient is greater than $(1 - \alpha)$.

Confidence Interval for the Ratio of Two Linear Parametric Functions. Let $\theta_1 = \mathbf{P}_1'\boldsymbol{\beta}$ and $\theta_2 = \mathbf{P}_2\boldsymbol{\beta}$ be two linear parametric functions. We want a confidence interval for the ratio $\lambda = \theta_1/\theta_2$. Let z_1 and z_2 be the l.s. estimators of $\mathbf{P}_1'\boldsymbol{\beta}$ and $\mathbf{P}_2'\boldsymbol{\beta}$ and $C_{11}\sigma^2$, $C_{22}\sigma^2$ and $C_{12}\sigma^2$ be the variances and covariance of the estimators. The l.s.e. of the function $\theta_1 - \lambda\theta_2$ is $z_1 - \lambda z_2$ with

$$E(z_1 - \lambda z_2) = 0, \qquad \text{if } \lambda \text{ is the true ratio}$$

and

$$V(z_1 - \lambda z_2) = (C_{11} - 2\lambda C_{12} + \lambda^2 C_{22})\sigma^2.$$

Therefore,

$$F = \frac{(z_1 - \lambda z_2)^2}{s^2(C_{11} - 2\lambda C_{12} + \lambda^2 C_{22})} \sim F(1, n - r) \tag{4b.2.15}$$

and

$$P[(z_1 - \lambda z_2)^2 - F_\alpha s^2(C_{11} - 2\lambda C_{12} + \lambda^2 C_{22}) \leqslant 0] = 1 - \alpha. \tag{4b.2.16}$$

The inequality within the brackets of (4b.2.16) provides a $(1-\alpha)$-confidence *region* for λ. Since the expression in (4b.2.16) is quadratic in λ, the confidence region is the interval between the roots of the quadratic equation or outside this interval, depending on the nature of the coefficients.

4b.3 Setup with Restrictions

Suppose, in the original setup $(\mathbf{Y}, \mathbf{X}\boldsymbol{\beta}, \sigma^2\mathbf{I})$, the parameters are subject to some natural restrictions, $\mathbf{R}\boldsymbol{\beta} = \boldsymbol{\eta}$, and we are required to draw inferences on parametric functions whose values are not specified by the restrictions. This situation presents no difficulty in theory, for as shown in **4a.6** we can replace the original parameters with restrictions by new (smaller-number) parameters without restrictions. If $\boldsymbol{\phi}$ represents the new parameters, the problem is one of $(\mathbf{Y}, \mathbf{B}\boldsymbol{\phi}, \sigma^2\mathbf{I})$. Any parametric function $\mathbf{P}'\boldsymbol{\beta}$ estimable in the original setup is equivalent to a parametric function $\mathbf{U}'\boldsymbol{\phi}$ estimable in the new setup and so on, and therefore no further theoretical discussion of the problem with restrictions is necessary. In practice, however, all the computations may be carried out without introducing any new set of parameters, as shown below. The residual sum of squares is

$$R_0^2 = \min_{\boldsymbol{\phi}} (\mathbf{Y} - \mathbf{B}\boldsymbol{\phi})'(\mathbf{Y} - \mathbf{B}\boldsymbol{\phi}) = \min_{\mathbf{R}\boldsymbol{\beta}=\boldsymbol{\eta}} (\mathbf{Y} - \mathbf{X}\boldsymbol{\beta})'(\mathbf{Y} - \mathbf{X}\boldsymbol{\beta}),$$

with D.F. equal to $(n - f)$, where

$$f = \operatorname{rank} \mathbf{B} = \operatorname{rank}(\mathbf{X}' \,\vdots\, \mathbf{R}') - \operatorname{rank} \mathbf{R}'.$$

Let $\mathbf{H}'\boldsymbol{\beta} = \boldsymbol{\theta}_0$ be a multiple hypothesis of rank k to be tested, such that the value of no linear combination of the individual functions in $\mathbf{H}'\boldsymbol{\beta}$ is implied by the restrictions $\mathbf{R}\boldsymbol{\beta} = \boldsymbol{\eta}$. Let the equivalent hypothesis in terms of $\boldsymbol{\Phi}$ be $\mathbf{G}'\boldsymbol{\Phi} = \boldsymbol{\xi}$. Then

$$\begin{aligned} R_1^2 &= \min_{\mathbf{G}'\boldsymbol{\phi}=\boldsymbol{\xi}} (\mathbf{Y} - \mathbf{B}\boldsymbol{\Phi})'(\mathbf{Y} - \mathbf{B}\boldsymbol{\Phi}) \\ &= \min_{\mathbf{H}'\boldsymbol{\beta}=\boldsymbol{\theta}_0,\, \mathbf{R}\boldsymbol{\beta}=\boldsymbol{\eta}} (\mathbf{Y} - \mathbf{X}\boldsymbol{\beta})'(\mathbf{Y} - \mathbf{X}\boldsymbol{\beta}), \end{aligned}$$

and the test criterion is

$$\frac{R_1^2 - R_0^2}{k} \div \frac{R_0^2}{n-f}.$$

If $R_1^2 - R_0^2$ is sought to be computed by first determining l.s. estimates of $\mathbf{H}'\boldsymbol{\beta}$ and the dispersion matrix of estimates, it has already been explained in **4a.9** how to obtain the latter without changing over to new parameters. It would then appear that all the computations can be carried out without introducing new parameters.

4c PROBLEMS OF A SINGLE SAMPLE

4c.1 The Test Criterion

Let $y_1, \ldots, y_n$ be observations drawn from a population with mean μ and variance σ^2. The following computations are made.

$$R_0^2 = \min_{\mu} \sum (y_i - \mu)^2 = \sum y_i^2 - n\bar{y}^2$$

$$s^2 = \frac{R_0^2}{n-1}.$$

The normal equation for μ is $\sum y = n\mu$, so that the l.s. estimate of μ is

$$\hat{\mu} = \frac{\sum y}{n} = \bar{y}, \qquad \text{with} \quad V(\hat{\mu}) = \frac{\sigma^2}{n}.$$

The t-statistic defined in (4b.1.1) for drawing inference on μ is

$$t = \frac{\sqrt{n}(\bar{y} - \mu)}{s} \sim S(n-1).$$

4c.2 Asymmetry of Right and Left Femora (Paired Comparison)

The mean difference (right femur–left femur) in length between the right and left femora of 36 skeletons of a certain series is found to be 2.0234; the corrected sum of squares R_0^2 of these 36 differences is 418.6875. The estimated variance s^2 on 35 degrees of freedom is $418.6875/35 = 11.9625$.

Two-Sided Test. Are the left and right femora of equal length on the average? We use the statistic

$$t = \frac{\sqrt{n}(\bar{x} - \mu)}{s} = \frac{\sqrt{36}(2.0234)}{\sqrt{11.9625}} = 3.5193,$$

putting $\mu = 0$. If there is no *a priori* information as to whether the left measurement exceeds the right or otherwise, a large value of t in either direction would be evidence against the null hypothesis. We compare the observed value of t, disregarding its sign, with chosen significant points of the distribution of $|t|$, which are given in most statistical tables (see the references at the end of the chapter). For 35 D.F., the 5% value of $|t|$ is 2.03 from R.M.M. Tables, which is the same as the 5% value of t as given in F.Y. Tables. The observed t is significant at the 5% level.

One-Sided Test. If it is known *a priori* that the alternative to the null hypothesis is that the right femur is longer than the left or if the purpose of the test is to discriminate only asymmetry due to the right femur being longer than the left, then the sign of t is important. The hypothesis of equality is rejected in favor of the suggested alternative only when t exceeds the upper 5% value of t which is 1.69 for 35 D.F. from R.M.M. Tables (which is the same as 10% value of t in F.Y. Tables). If the alternative hypothesis is that the left femur has a greater length, then $(-t)$ should exceed the upper 5% tabulated value for significance. In the foregoing, t certainly exceeds the upper 5% value of t, showing that the data are in agreement with the suggested alternative that the right femur is longer than the left. In any problem the decision to use a two-sided or a one-sided test should be taken on some logical grounds.

These tests are useful in situations where the mean values of two series are to be compared, but the observations are such that there is a one-to-one correspondence between a member of one series and a member of the other. In the preceding example, two measurements belong to the same skeleton. The 36 pairs of measurements give rise to 36 differences that can now be treated as a single series in which the expected value of the mean is zero.

On the other hand, the two series may not have any correspondence; for instance, no two measurements are made on the same skeleton, in which case the method of analysis of variance applied to groups (see **4d**) has to be used. In the second test the variation due to skeletons has also to be taken into account, and therefore the precision of the comparison decreases and such a small difference as the foregoing may go undetected, even in a large sample. The variance of femur length is about 400, in which case the variance for difference in means of two independent series of 36 is

$$400\left(\tfrac{1}{36} + \tfrac{1}{36}\right) = 22.22,$$

whereas the corresponding variance for the difference in a correlated series is $11.9625 \div 36 = 0.33$, which admits a considerably more precise assessment of asymmetry. This aspect should be kept in view while conducting any investigation. When two measurements are to be compared, the experiment may be designed so as to obtain correlated series of measurements. The higher the correlation, the greater the advantage. The association should be positive, otherwise the test based on correlated pairs becomes less efficient.

4d ONE-WAY CLASSIFIED DATA

4d.1 The Test Criterion

Let there be k samples of sizes $n_1, \ldots, n_k$ from k populations with unknown means $\mu_1, \ldots, \mu_k$ and with a common unknown variance σ^2. The hypothesis which may be desired to be tested is

$$\mu_1 = \mu_2 = \cdots = \mu_k. \tag{4d.1.1}$$

The observational equations, $n_1 + \cdots + n_k = n$ in number, are given below

	First Sample		...	kth Sample	
	Variable	Expectation	...	Variable	Expectation
	y_{11}	μ_1	...	y_{k1}	μ_k
	.	.	...	.	.
	y_{1n_1}	μ_1	...	y_{kn_k}	μ_k
Total	T_1		...	T_k	

The minimum value of $\sum\sum (y_{ij} - \mu_i)^2$ subject to the condition of the hypothesis (4d.1.1), that is, with common μ, is

$$\min_{\mu} \sum\sum (y_{ij} - \mu)^2 = \sum\sum y_{ij}^2 - \frac{(\sum\sum y_{ij})^2}{n} = R_1^2, \tag{4d.1.2}$$

which is the total corrected sum of squares of all the observations. The minimum value of $\sum\sum (y_{ij} - \mu_i)^2$ without any restriction is

$$\sum_i \min_{\mu_i} \sum_j (y_{ij} - \mu_i)^2 = \sum_{i=1}^{k} \left(\sum_j y_{ij}^2 - \frac{T_i^2}{n_i} \right) = R_0^2. \tag{4d.1.3}$$

The sum of squares due to deviation from the hypothesis with $(k - 1)$ D.F. is then

$$R_1^2 - R_0^2 = \frac{T_1^2}{n_1} + \cdots + \frac{T_k^2}{n_k} - \frac{T^2}{n} \tag{4d.1.4}$$

$$T = T_1 + \cdots + T_k,$$

which depends on the totals of the samples only. It is easier to calculate the expressions (4d.1.2) and (4d.1.4) in practice and derive the expression (4d.1.3) for R_0^2 by subtraction. The scheme of computation is set out in Table 4d.1α. The quantities marked by * are obtained by subtraction. The F statistic is constructed using the mean squares derived from Table 4d.1α.

TABLE 4d.1α. Analysis of Variance, One-Way Classification

	D.F.	S.S.
Deviation from hypothesis or between samples	$k - 1$	$\sum \frac{T_i^2}{n_i} - \frac{T^2}{n}$
Residual or within samples	*	*
Total	$n - 1$	$\sum\sum y_{ij}^2 - \frac{T^2}{n}$

4d.2 An Example

The following data relate to head breadth of 142 skulls belonging to three series. Can the true mean head breadth be considered the same in the three series?

Series	Sample Size	Head Breadth Total	Head Breadth Mean
1	83	11,277	135.87
2	51	7,049	138.22
3	8	1,102	137.75
Total	142	19,428	136.817

The sum of squares between series is

$$11{,}277 \times 135.87 + 7049 \times 138.22 + 1102 \times 137.75 - 19{,}428 \times 136.817 = 238.59$$

The total sum of squares is found to be 4616.64. The analysis of variance is set out in Table 4d.2.

TABLE 4d.2. Analysis of Variance

	D.F	S.S.	M.S.	F
Between	2	238.59	119.29	3.79
Within	139	4378.05	31.50	
Total	141	4616.64		

The variance ratio 3.79 on 2 and 139 D.F. is significant at the 5% level, indicating real differences in mean values (see the R.M.M. Tables for significant values of F at the upper end). We can now examine the nature of these differences by writing the mean values in a descending order.

Series	2	3	1
Mean	138.22	137.75	135.87

It appears that the observed inequality between (2) and (3) is in doubt because the values are close. To examine this we may compute the 5% critical difference

$$\left[\left(\frac{1}{n_2} + \frac{1}{n_3}\right)\hat{\sigma}^2\right]^{1/2} t_\alpha,$$

where n_2 and n_3 are sample sizes, $\hat{\sigma}^2$ is the estimate of σ^2, which is 31.50 on 139 D.F. from the Table 4d.2, and t_α is the α-level significant value of $|t|$ on 139 D.F. Inserting numerical values, the critical difference is

$$[(\tfrac{1}{51} + \tfrac{1}{8})31.50]^{1/2}t_\alpha = 2.13t_\alpha = 2.13 \times 1.96 = 4.17.$$

the 5% value of $|t|$ on 139 D.F. being 1.96. The critical value is considerably larger than the observed difference $138.22 - 137.75 = 0.47$ so that the entire difference in the mean values may be attributed to series (1) being different from (2) and (3). In general, we may compute all possible critical differences arising out of all the pairs in the manner explained and compare with the observed differences. On the basis of such detailed comparisons it should be possible to decide on the order of the groups under comparison with respect to the mean values and on the subsets within which the order is somewhat in doubt. In the present example we could conclude that series (1) takes the last place whereas the order between (2) and (3) is in doubt.

In practical situations, besides the order, the actual magnitudes of differences are also important. The analysis of variance of data, the presentation of mean values, and the examination of individual differences enables us to interpret the magnitudes of differences in a useful way.

4e TWO-WAY CLASSIFIED DATA

4e.1 Single Observation in Each Cell

Let there be pq observations, each of which can be specified in terms of the categories of two classes. The observations may be set out as in Table 4e.1α.

The different categories of A may be varieties of wheat, of B, ovens, and the measurement may be quality of baked bread. A may represent different types of machines and B a number of operators, and the measurement may be efficiency of production. Or A may be different levels of an

TABLE 4e.1α

	Class B				
Class A	B_1	B_2	$\cdots$	B_p	Total
A_1	y_{11}	y_{12}	$\cdots$	y_{1p}	$y_{1\cdot}$
A_2	y_{21}	y_{22}	$\cdots$	y_{2p}	$y_{2\cdot}$
$\cdot$	$\cdot$	$\cdot$	$\cdots$	$\cdot$	$\cdot$
A_q	y_{q1}	y_{q2}	$\cdots$	y_{qp}	$y_{q\cdot}$
Total	$y_{\cdot 1}$	$y_{\cdot 2}$		$y_{\cdot p}$	$y_{\cdot\cdot}$

organic manure and B of an inorganic manure applied to crops. In all these situations, the object is to determine an optimum combination of categories of A and B or to study the differences in categories of one factor for suitably defined categories of another.

Let $E(y_{ij}) = \alpha_i + \beta_j$ and $V(y_{ij}) = \sigma^2$. There are $n = pq$ observations and $p + q$ parameters, but the rank of the matrix of observational equations is only $(p + q - 1)$ (see Example 3, p. 11). The normal equations obtained by minimizing $\sum\sum (y_{ij} - \alpha_i - \beta_j)^2$ are

$$p\alpha_i + \sum \beta_j = y_{i.} \qquad i = 1, \ldots, q$$
$$\sum \alpha_i + q\beta_j = y_{.j}, \qquad j = 1, \ldots, p$$

and a solution is seen to be

$$\hat{\alpha}_i = \frac{y_{i.}}{p}, \qquad \hat{\beta}_j = \frac{y_{.j}}{q} - \frac{y_{..}}{pq}.$$

which is not, however, unique. But we need only one solution for the rest of the computations.

Using the formula (4a.5.1) for the residual sum of squares we see that

$$R_0^2 = \sum\sum y_{ij}^2 - \sum \hat{\alpha}_i y_{i.} - \sum \hat{\beta}_j y_{.j}$$
$$= \left(\sum\sum y_{ij}^2 - \frac{y_{..}^2}{pq}\right) - \left(\sum \frac{y_{i.}^2}{p} - \frac{y_{..}^2}{pq}\right) - \left(\sum \frac{y_{.j}^2}{q} - \frac{y_{..}^2}{pq}\right), \tag{4e.1.1}$$

on $n - r = pq - (p + q - 1) = (p - 1)(q - 1)$ D.F. Suppose we want to test the hypothesis

$$\alpha_1 = \cdots = \alpha_q, \tag{4e.1.2}$$

in which case $E(y_{ij}) = \alpha + \beta_j$ with a common α. Writing the normal equations, solving for $\alpha, \beta_1, \ldots, \beta_p$ and applying the formula for residual sum of squares, we find

$$R_1^2 = \min \sum\sum (y_{ij} - \alpha - \beta_j)^2$$
$$= \left(\sum\sum y_{ij}^2 - \frac{y_{..}^2}{pq}\right) - \left(\sum \frac{y_{.j}^2}{q} - \frac{y_{..}^2}{pq}\right),$$

so that the sum of squares for testing (4e.1.2) is

$$R_1^2 - R_0^2 = \sum \frac{y_{i.}^2}{p} - \frac{y_{..}^2}{pq}. \tag{4e.1.3}$$

with $(q - 1)$ D.F. Similarly the sum of squares for testing the hypothesis $\beta_1 = \cdots = \beta_p$ can be determined. The scheme of computation is presented in Table 4e.1β. The variance ratio for testing the hypothesis (4e.1.2) is V_A/V_e.

TABLE 4e.1β. Analysis of Variance, Two-Way Classification

	D.F.	S.S.	M.S.
Between A classes	$q-1$	$S_A = \frac{1}{p}\sum y_{i.}^2 - \frac{1}{pq} y_{..}^2$	V_A
Between B classes	$p-1$	$S_B = \frac{1}{q}\sum y_{.j}^2 - \frac{1}{pq} y_{..}^2$	V_B
Residual (Interaction) $A \times B$	$(p-1)(q-1)$	$S_{AB} =$ * (by subtraction)	V_e
	$pq-1$	$S = \sum\sum y_{ij}^2 - \frac{1}{pq} y_{..}^2$	

The components of analysis of variance in Table 4e.1β are also useful in testing differences between categories of a factor under more general assumptions on the data. Let us take a concrete example where the factor A represents different kinds of machines and B, operators. If the assumption $E(y_{ij}) = \alpha_i + \beta_j$ is true, then differences in machines are the same for each operator and are measured by differences in the parameters α_i. On the other hand, we may ask the question whether there exist differences between machines *on the average* with reference to a wider group (population) of operators of whom the p operators included in the experiment constitute a sample, *although the differences between machines may not be the same for each operator*. Let us, however, consider a wider null hypothesis that the q observations for an operator on q different machines have a symmetrical distribution, that is, with equal means, variances, and the same correlation for all pairs of observations. Or in other words, the machines are basically the same, except possibly for difference in names. Under these conditions it has been shown in (3c.2.10) that V_A/V_e has the same variance ratio distribution [on $q-1$ and $(p-1)(q-1)$ D.F.] provided the mean values for the q machines are the same over the population of operators. The test V_A/V_e is, however, effective only in detecting differences in mean values for the machines and not other departures in the wider null hypothesis considered. Thus the test V_A/V_e admits different interpretations depending on the model assumed.

Tukey's (1949) Test for Nonadditivity. Let us suppose that departure from additivity can be specified by writing the expectation of y_{ij} as

$$E(y_{ij}) = \mu + \alpha_i + \beta_j + \lambda\alpha_i\beta_j$$
$$\sum \alpha_i = 0, \qquad \sum \beta_j = 0, \tag{4e.1.4}$$

by introducing a product term. If we can construct a test for the hypothesis $\lambda = 0$, it will serve as a test for nonadditivity, *specifically for detecting departures of the type* (4e.1.4). Since the expectations are not linear when $\lambda \neq 0$, the least squares theory is not applicable.

But let us observe that whether $\lambda = 0$ or not

$$E(\bar{y}_{i.} - \bar{y}_{..}) = \alpha_i, \qquad E(\bar{y}_{.j} - \bar{y}_{..}) = \beta_j$$
$$E(\bar{y}_{..}) = \mu$$

so that unbiased estimators of α_i, β_j, and μ are given by

$$\alpha_i^* = \bar{y}_{i.} - \bar{y}_{..}, \qquad \beta_j^* = \bar{y}_{.j} - \bar{y}_{..}, \qquad \mu^* = \bar{y}_{..}$$

Furthermore,

$$E(y_{ij} - \bar{y}_{i.} - \bar{y}_{.j} + \bar{y}_{..}) = \lambda\alpha_i\beta_j$$

and an estimate of λ may be written, when α_i and β_j are known, by using the l.s. theory on equations (4e.1.4), as

$$\hat{\lambda} = \frac{\sum\sum \alpha_i\beta_j(y_{ij} - \alpha_i - \beta_j - \bar{y}_{..})}{\sum\sum \alpha_i^2\beta_j^2}.$$

Substituting estimates α_i^* and β_j^*, we obtain an estimate of λ in terms of known quantities

$$\lambda^* = \frac{\sum\sum(\bar{y}_i - .\bar{y}_{..})(\bar{y}_{.j} - \bar{y}_{..})(y_{ij} - \bar{y}_{i.} - \bar{y}_{.j} + \bar{y}_{..})}{\sum(\bar{y}_{i.} - \bar{y}_{..})^2 \sum(\bar{y}_{.j} - \bar{y}_{..})^2},$$

which simplifies to

$$\frac{pq\sum\sum(\bar{y}_{i.} - \bar{y}_{..})(\bar{y}_{.j} - \bar{y}_{..})y_{ij}}{S_A S_B} \tag{4e.1.5}$$

with variance, for given α_i^* and β_j^*, as

$$\frac{p^2q^2\sigma^2\sum\sum(\bar{y}_{i.} - \bar{y}_{..})^2(\bar{y}_{.j} - \bar{y}_{..})^2}{S_A{}^2 S_B{}^2} = \frac{pq\sigma^2}{S_A S_B}. \tag{4e.1.6}$$

Hence, if $\lambda = 0$, $E(\lambda^* \mid \alpha_i^*, \beta_j^* \text{ for all } i, j) = 0$ and the statistic

$$\frac{(4e.1.5)^2}{(4e.1.6)} = \frac{pq[\sum\sum(\bar{y}_{i.} - \bar{y}_{..})(\bar{y}_{.j} - \bar{y}_{..})y_{ij}]^2}{S_A S_B \sigma^2} \sim \chi^2(1). \tag{4e.1.7}$$

But

$$\frac{\sum\sum(y_{ij} - \bar{y}_{i.} - \bar{y}_{.j} + \bar{y}_{..})^2}{\sigma^2} = \frac{S_{AB}{}^2}{\sigma^2} \sim \chi^2(\overline{p-1}\,\overline{q-1}), \tag{4e.1.8}$$

and we observe that [since y_{ij} in (4e.1.7) can in fact be written $(y_{ij} - \bar{y}_{i.} - \bar{y}_{.j} + \bar{y}_{..})$], (4e.1.8)–(4e.1.7) is non-negative. Hence applying [(iv), **3b.4**]

$$\frac{\chi^2(1)}{\chi^2(\overline{p-1}\,\overline{q-1}) - \chi^2(1)} \cdot \frac{(p-1)(q-1)-1}{1} \tag{4e.1.8}$$

has an exact F distribution on 1 and $(p-1)(q-1)-1$ D.F. We can now write the complete analysis of variance, including the component for non-additivity as in Table 4e.1γ, where S_A, S_B, and S are as in Table 4e.1β and S_N is

$$S_N = pq[\sum\sum(\bar{y}_{i.} - \bar{y}_{..})(\bar{y}_{.j} - \bar{y}_{..})y_{ij}]^2/S_A S_B.$$

as defined in (4e.1.7). The variance ratio for nonadditivity is V_N/V_R with 1 and $(p-1)(q-1)-1$ D.F.

TABLE 4e.1γ. Analysis of Variance of Two-Way Data

	D.F.	S.S.	M.S.
A	$q-1$	S_A	V_A
B	$p-1$	S_B	V_B
Nonadditivity	1	S_N	V_N
Residual	$(p-1)(q-1)-1$	* (by subtraction)	V_R
Total	$pq-1$	S	

Our method of derivation and interpretation of Tukey's test (4e.1.8) suggests the following generalization in testing the specification in a Gauss-Markoff model $(\mathbf{Y}, \mathbf{X}\boldsymbol{\beta}, \sigma^2\mathbf{I})$.

Let us suppose that the actual model is

$$\mathbf{Y} = \mathbf{X}\boldsymbol{\beta} + \mathbf{F}\boldsymbol{\lambda} + \boldsymbol{\varepsilon} \tag{4e.1.9}$$

where $\mathbf{F}$ is $n \times k$ matrix whose (i, j)th element is $F_{ij}(\mathbf{X}\boldsymbol{\beta})$, a given function of $\mathbf{X}\boldsymbol{\beta}$. We wish to test the hypothesis $\boldsymbol{\lambda} = \mathbf{0}$.

Let $\hat{\beta} = (\mathbf{X}'\mathbf{X})^-\mathbf{X}'\mathbf{Y}$ be the least squares estimator of $\boldsymbol{\beta}$ from the model $(\mathbf{Y}, \mathbf{X}\boldsymbol{\beta}, \sigma^2\mathbf{I})$. Further, let

$$\begin{aligned} \mathbf{d} &= \mathbf{Y} - \mathbf{X}\hat{\boldsymbol{\beta}} \\ \mathbf{M} &= (\mathbf{I} - \mathbf{X}(\mathbf{X}'\mathbf{X})^-\mathbf{X}')\hat{\mathbf{F}}, \end{aligned} \tag{4e.1.10}$$

where the (i, j)th element of $\hat{\mathbf{F}}$ is $F_{ij}(\mathbf{X}\hat{\boldsymbol{\beta}})$, and consider (formally) the model $(\mathbf{d}, \mathbf{M}\boldsymbol{\lambda}, \sigma^2\mathbf{I})$, and derive the statistic (4b.2.8) to test the hypothesis $\boldsymbol{\lambda} = \mathbf{0}$ using (formally) the least squares theory of **4b**.

Using the notation of **4b.2**,

$$R_1^2 = \mathbf{d}'\mathbf{d}, \qquad R_0^2 = \mathbf{d}'(\mathbf{I} - \mathbf{M}(\mathbf{M}'\mathbf{M})^- \mathbf{M}')\mathbf{d}$$

so that the required statistic is

$$\frac{a}{b} \cdot \frac{R_1^2 - R_0^2}{R_0^2} = \frac{a}{b} \cdot \frac{\mathbf{d}'\mathbf{M}(\mathbf{M}'\mathbf{M})^- \mathbf{M}'\mathbf{d}}{\mathbf{d}'(\mathbf{I} - \mathbf{M}(\mathbf{M}'\mathbf{M})^- \mathbf{M}')\mathbf{d}}$$

$$= \frac{a}{b} \cdot \frac{\mathbf{Y}'\mathbf{M}(\mathbf{M}'\mathbf{M})^- \mathbf{M}'\mathbf{Y}}{\mathbf{Y}'[\mathbf{I} - \mathbf{X}(\mathbf{X}'\mathbf{X})^- \mathbf{X}' - \mathbf{M}(\mathbf{M}'\mathbf{M})^- \mathbf{M}']\mathbf{Y}} \tag{4e.1.11}$$

The statistic (4e.1.11) is distributed as $F(b, a)$ where

$$b = R(\mathbf{M}), \qquad a = n - R(\mathbf{X}) - b \tag{4e.1.12}$$

and b is supposed to be fixed with probability 1. If $b \neq k$, then we will be testing a subhypothesis that certain linear functions of $\boldsymbol{\lambda}$ vanish.

The proof of (4e.1.11) is the same as that used in the formulae (4e.1.7) and (4e.1.8). However, the result (4e.1.11) is derived more simply by applying the general theorem [(iv), **3b.5**], observing that $\hat{\mathbf{F}}$ and $\mathbf{d}$ are independent. Millikin and Graybill (1970) obtained the test (4e.1.11) though the degrees of freedom a and b are defined differently in their case.

4e.2 Multiple but Equal Numbers in Each Cell

In **4e.1** we considered a partial test of the hypothesis $E(y_{ij}) = \alpha_i + \beta_j$, that is, the effects due to the classes of A (say machines) and B (say operators) are additive. This can, however, be fully tested when each cell contains more than one observation. If there are c observations in the cell (i, j) they may be represented by

$$\begin{gathered} y_{ij1} \cdots y_{ijc} \\ E(y_{ijk}) = \mu_{ij}, \qquad V(y_{ijk}) = \sigma^2. \end{gathered} \tag{4e.2.1}$$

The hypothesis we wish to test is

$$\mu_{ij} = \alpha_i + \beta_j. \tag{4e.2.2}$$

When the additive hypothesis (4e.2.2) is not true, there is said to be interaction between the factors A and B. In the presence of interaction, differences between A classes cannot be specified without reference to a particular B class.

Let

$$y_{ij} = \sum_k y_{ijk}, \qquad y_{..} = \sum_i \sum_j y_{ij},$$

$$y_{i.} = \sum_j y_{ij}, \qquad y_{.j} = \sum_i y_{ij}.$$

Then under the setup (4e.2.1), we have

$$R_0^2 = \left(\sum\sum\sum y_{ijk}^2 - \frac{y_{..}^2}{pqc}\right) - \left(\sum\sum \frac{y_{ij}^2}{c} - \frac{y_{..}^2}{pqc}\right),$$

and under the hypothesis (4e.2.2), as in (4e.1.1), we have

$$R_1^2 = \left(\sum\sum\sum y_{ijk}^2 - \frac{y_{.}^2}{pqc}\right) - \left(\sum \frac{y_{i.}^2}{pc} - \frac{y_{..}^2}{pqc}\right) - \left(\sum \frac{y_{.j}^2}{qc} - \frac{y_{..}^2}{pqc}\right).$$

The sum of squares for interaction is $R_1^2 - R_0^2$ on $(p-1)(q-1)$ D.F. The scheme of computation is given in Table 4e.2α.

TABLE 4e.2α. Analysis of Variance, Two-Way Classification

	D.F.	S.S.	M.S.
Between machines (A)	$q-1$	$\frac{1}{c}\frac{1}{p}\sum_1^q y_{i.}^2 - \frac{1}{cpq} y_{..}^2$	V_A
Between operators (B)	$p-1$	$\frac{1}{c}\frac{1}{q}\sum_1^p y_{.j}^2 - \frac{1}{cpq} y_{..}^2$	V_B
Interaction ($A \times B$)	$(p-1)(q-1)$	* (by subtraction)	V_{AB}
Between pq cells	$pq-1$	$\frac{1}{c}\sum\sum y_{ij}^2 - \frac{1}{cpq} y_{..}^2$	V
Residual	$pq(c-1)$	* (by subtraction)	V_e
Total	$pqc-1$	$\sum\sum\sum y_{ijk}^2 - \frac{1}{cpq} y_{..}^2$	

With the analysis of variance of two-way data as in Table 4e.2α, a variety of hypotheses can be tested.

1. Is the interaction significant? The variance ratio for testing this hypothesis is V_{AB}/V_e as already shown.
2. Are there differences between machines (A classes) when averaged over all operators (B classes) *employed* in the experiment although there may be differences between machines for particular operators? If μ_{ij} is the effect of the combination $(A_i B_j)$, the hypothesis to be tested is

$$\mu_{1.} = \mu_{2.} = \cdots = \mu_{q.}$$

The variance ratio appropriate for this is V_A/V_e.

3. Similarly the variance ratio V_B/V_e is appropriate for the hypothesis

$$\mu_{.1} = \mu_{.2} = \cdots = \mu_{.p}.$$

4. Are there differences between machines on the average over a wider set (population) of operators of whom those examined constitute a sample? The situation is similar to what has been discussed in **4e.1**. The appropriate variance ratio is V_A/V_{AB}, that is, the interaction mean square is the valid denominator whether *additivity holds or not*. The ratio V_B/V_{AB} is used for testing a similar hypothesis on B classes.

The difference between the hypotheses considered in (2) and (4) should be understood. In either case, interaction may be present. In (2) we are interested in average differences between particular operators employed in the experiment, whereas in (4) the differences are with respect to a population of operators not all of whom are included in the experiment but only a small sample. If there is no interaction (2) and (4) are the same.

4e.3 Unequal Numbers in Cells

The following notations are used:

y_{ij} = the total of all observations in the (i, j)th cell
$\bar{y}_{ij}$ = mean in the (i, j)th cell
$y_{.j} = \sum_i y_{ij}$ (total for the jth column)
$y_{i.} = \sum_j y_{ij}$ (total for the ith row)
$y_{..}$ = total of all observations
$\bar{y}_{..}$ = mean of all observations

n_{ij}, $n_{.j}$, $n_{i.}$, and $n_{..}$ are the numbers of observations for the (i, j)th cell, jth column, ith row, and all the cells, respectively.

The data in Table 4e.3α refer to the mean values and totals of nasal height of skulls excavated from three different strata by three observers. It is desired to test for stratum and observer differences.

In problems of this nature it is convenient to set up the figures as in Table 4e.3α for the computation of the various sums of squares. The analysis is carried out in three stages. The total (corrected) sum of squares of all the observations, with 309 degrees of freedom is found to be 5398.4206. For further calculations the entries in Table 4e.3α are sufficient.

The Computation of Between-Cell Sum of Squares. The between-cell sum of squares on $pq - 1 = 8$ degrees of freedom is $\sum\sum y_{ij}\bar{y}_{ij} - y_{..}\bar{y}_{..} = 931.5204$.

TABLE 4e.3α. Mean Values and Totals

Observer		Stratum S_1	S_2	S_3	$y_{i.}$	$\bar{y}_{i.}$
	y_{11}	1071.00	1572.48	913.50	3556.98	
O_1	$\bar{y}_{11}$	51.00	49.14	50.75		50.098
	n_{11}	(21)	(32)	(18)	(71)	
		1966.86	2315.40	1721.52	6003.78	
O_2		46.83	45.40	47.82		46.541
		(42)	(51)	(36)	(129)	
		1219.00	2091.60	1849.20	5159.80	
		48.76	46.48	46.23		46.907
O_3		(25)	(45)	(40)	(110)	
$y_{.j}$		4256.86	5979.48	4484.22	14720.56	
$\bar{y}_{.j}$		48.373	46.715	47.704	$=y_{..}$	47.4857
		(88)	(128)	(94)	(310)	$=\bar{y}_{..}$

Computation of Main Effects and Interaction. Let the expected value of an observation in the (i, j)th cell be $\alpha_i + \beta_j$ representing the additive effects of the ith row and jth column. The normal equations obtained by minimizing

$$\sum\sum n_{ij}(\bar{y}_{ij} - \alpha_i - \beta_j)^2$$

are, omitting the entries below the diagonal because of symmetry,

Observers α_1	α_2	α_3	Strata β_1	β_2	β_3		Marginal total
71	.	.	21	32	18	=	3556.98
	129	.	42	51	36	=	6003.78
		110	25	45	40	=	5159.80
			88	.	.	=	4256.86
				128	.	=	5979.48
					94	=	4484.22

The method of writing these equations is simple. Start with any marginal total say 3556.98, based on 71 observations for observer 1 distributed over the strata as 21, 32, 18. This gives the first equation. There are six marginal totals corresponding to A and B classes which give rise to six equations.

We *need not* actually solve these equations for setting the analysis of variance table. Let us eliminate $\alpha_1, \alpha_2, \alpha_3$ from the last set of equations for $\beta_1, \beta_2, \beta_3$. The rule is to write the equations for α_i or β_j first, which ever contain a larger number of alternatives and eliminate the corresponding constants from the latter set of equations. To eliminate $\alpha_1, \alpha_2, \alpha_3$ from the equation for β_1, we have to multiply the first equation by 21/71, the second by 42/129 and the third by 25/110 and subtract their sum from the equation for β_1. Similarly multiply the first, second, and third equations by 32/71, 51/129, and 45/110 respectively and subtract their sum from the equation for β_2 and so on, obtaining the equations

$$
\begin{array}{ccccc}
\beta_1 & \beta_2 & \beta_3 & & \\
62.432498 & -46.296717 & \cdots\cdots\cdots & = & 77.394630 \\
 & 64.844662 & \cdots\cdots\cdots & = & -108.080590
\end{array}
$$

The coefficients of β_3 and the last equation need not be computed. The equations in β_i are solved by taking the last β to be zero. In the present problem there are only two equations giving $\hat{\beta}_1 = 0.559250$, $\hat{\beta}_2 = -1.170335$. The sum of squares for the main effect of B is simply obtained as

$$\hat{\beta}_1(77.394630) + \hat{\beta}_2(-108.080590) = 171.8919$$

where the values 77.493630, etc., are the right-hand side elements in the reduced equations for β_i. We *need not solve for* α_i to set up the analysis of variance table. But, two more expressions have to be computed. If the B classes are

TABLE 4e.3β. Analysis of Variance for Two-Way Data (Unequal Numbers in Cells)

Source	D.F.	S.S.	S.S.	D.F.	Source
Strata B (ignoring A)	2	147.6319†	634.1516†	2	Observers A (ignoring B)
Observers	2	658.4116*	171.8919†	2	Strata
Interaction	4	125.4769 ←	125.4769*	4	Interaction
Between cells	8	931.5204† →	931.5204	8	
Within cells	301	4466.9002*			
Total	309	5398.4206†			

† Calculated directly as explained in the text.
* Obtained by subtraction. → denotes transfer of a figure (directly computed or obtained by subtraction) from one position to another for further computations.

ignored, the data have only a one-way classification with respect to A and the S.S. due to A (ignoring B) on 2 D.F. is obtained from marginal totals of A

$$\sum y_{i.}\bar{y}_{i.} - y_{..}\bar{y}_{..} = 3556.98 \times 50.098 + \cdots - 14720.56 \times 47.4857 = 634.1516.$$

Similarly, the S.S. due to B (ignoring A) is obtained. The rest of the computations are explained in the Analysis of Variance Table 4e.3β.

The test for interaction on 4 and 301 D.F. is

$$F = \frac{125.4769}{4} \div \frac{4466.9002}{301} = 2.11,$$

and although it is less than the 5% value of F on 4 and 301 D.F., it is somewhat high. Presence of interaction in the present problem would have been most disgusting as it would mean that observer differences exist but are not consistent over the strata and hence no meaningful conclusions could be drawn from the data. On the other hand, some interaction would arise due to samples being not strictly random in any particular stratum for any observer, which may be true (under difficult field conditions) if skulls are taken out from different locations of a stratum by different observers. Since the interaction mean square is more than the within-cell mean square the former is used for testing differences between observers and strata.

The variance ratio on 2 and 4 D.F. for observers is

$$\frac{658.4116}{2} \div 31.3692 = 10.49,$$

which is significant at the 5% level, and the variance ratio on 2 and 4 D.F. for strata is

$$\frac{171.8919}{2} \div 31.3692 = 2.74,$$

which is not very high (not significant at the 5% level).

Suppose we ignored the observers and examined stratum differences. The analysis would then be

	D.F.	M.S.	Ratio
Between strata	2	73.8159	4.31
Within strata	307	17.1035	

with a significant ratio of 4.31, leading to the conclusion that stratum differences exist. A closer analysis reveals that observer differences are important, so some caution is necessary in combining the results of the three different investigators. The stratum differences are not very prominent when the differences due to observers are corrected for.

4f A GENERAL MODEL FOR TWO-WAY DATA AND VARIANCE COMPONENTS

4f.1 A General Model

We have already seen in **4e** how the analysis of variance of two-way data enables us to draw a variety of inferences under different models for the observations. We shall now consider a very general model for which analysis of variance of Table 4e.2α provides a useful *reduction of the observations.* Let the pqc observations, with c observations in each of pq cells, be represented by

$$y_{ijk}, \qquad i = 1, \ldots, q; j = 1, \ldots, p; k = 1, \ldots, c. \tag{4f.1.1}$$

and let their expectations, variances, and covariances be as follows:

$$E(y_{ijk}) = \mu_{ij}, \qquad V(y_{ijk}) = \sigma^2, \qquad \text{cov}(y_{ijk}, y_{ijr}) = \rho_0\sigma^2$$
$$\text{cov}(y_{ijk}, y_{imr}) = \rho_1\sigma^2; \qquad \text{cov}(y_{ijk}, y_{mjr}) = \rho_2\sigma^2,$$
$$\text{cov}(y_{ijk}, y_{smr}) = \rho_3\sigma^2, \tag{4f.1.2}$$

so that the correlation (which may be $+ve$ or $-ve$) between any two observations depends only on the nature of the common suffixes.

4f.2 Variance Components Model

We shall consider a special case, known as the variance components model, which leads to observations of the type (4f.1.1) with the variance covariance structure (4f.1.2). Let $f_{\alpha\beta}$ be a function of a pair of random variables (α, β). Define

$$\underset{\alpha}{E}\,\underset{\beta}{E}(f_{\alpha\beta}) = f_{..}, \qquad \underset{\beta}{E}(f_{\alpha\beta}|\alpha) = f_{\alpha.}, \qquad \underset{\alpha}{E}(f_{\alpha\beta}|\beta) = f_{.\beta}. \tag{4f.2.1}$$

Then it is easy to establish

$$\begin{aligned} EE(f_{\alpha\beta} - f_{..})^2 &= E(f_{\alpha.} - f_{..})^2 + E(f_{.\beta} - f_{..})^2 + EE(f_{\alpha\beta} - f_{\alpha.} - f_{.\beta} + f_{..})^2 \\ \sigma_f^2 &= \sigma_A^2 + \sigma_B^2 + \sigma_{AB}^2. \end{aligned} \tag{4f.2.2}$$

The variable α may be thought of as representing the categories of a *factor A* and β those of a *factor B* and $f_{\alpha\beta}$ is the yield resulting from the choice of α and β. Thus σ_A^2 may be interpreted as the variance in yield due to the categories of A and σ_B^2 as that due to B under a *given sampling scheme for the choices of the categories of A and B.* The last component in (4f.2.2), σ_{AB}^2 represents interaction between factors A and B, that is, a situation in which $f_{\alpha\beta}$ cannot be represented as the sum of two functions, one depending on α only and another depending on β only. If such a decomposition $f_{\alpha\beta} = f_{..} + f_{\alpha.} + f_{.\beta}$ is true, then the effects of A and B are said to be additive and the component σ_{AB}^2 vanishes.

Let us generate random variables in the following manner (sampling scheme). First choose independent samples $\alpha_1, \ldots, \alpha_q$ of α and $\beta_1, \ldots, \beta_p$ of β and associate with each combination (α_i, β_j), c variables:

$$y_{ijk} = f_{\alpha_i\beta_i} + e_{ijk}, \qquad k = 1, \ldots, c \tag{4f.2.3}$$

where

$$E(e_{ijk} | \alpha_i, \beta_j) = 0, \qquad V(e_{ijk} | \alpha_i, \beta_j) = \sigma_e^2,$$
$$\operatorname{cov}(e_{ijk}, e_{ijm} | \alpha_i, \beta_j) = 0.$$

The components e_{ijk} may be thought of as independent errors in the repeated measurement of $f_{\alpha_i\beta_j}$. Now it may be seen that with the components σ_A^2, σ_B^2, and σ_{AB}^2 defined in (4f.2.2) with respect to the particular distribution of (α, β) generated by independent choices on α and β,

$$E(y_{ijk}) = f_{..}, \qquad V(y_{ijk}) = \sigma_f^2 + \sigma_e^2 = \sigma^2,$$
$$\operatorname{cov}(y_{ijk}, y_{ijm}) = \sigma_f^2 = \rho_0\sigma^2, \qquad \operatorname{cov}(y_{ijk}, y_{imr}) = \sigma_A^2 = \rho_1\sigma^2,$$
$$\operatorname{cov}(y_{ijk}, y_{mjr}) = \sigma_B^2 = \rho_2\sigma^2, \qquad \operatorname{cov}(y_{ijk}, y_{smr}) = 0 = \rho_3\sigma^2,$$
$$\sigma_f^2 = \sigma_A^2 + \sigma_B^2 + \sigma_{AB}^2. \tag{4f.2.4}$$

We thus have a particular case of the general model (4f.1.2) with the correspondence between the parameters as shown in (4f.2.4). A model for observations (4f.2.3) with the covariance structure (4f.2.4) is known as the *variance components* model. Other formulations of variance components are given in **4j**.

In theory we may admit other types of sampling such as without replacement in the case of finite alternatives and/or with equal or unequal probabilities or in such a way that α and β are correlated. We shall consider the most general model and show how to estimate the parameters (i.e., the covariance structure) by analysis of variance of two-way data as in **4e.2**. To derive the results in particular cases we have to express the parameters σ_e^2, σ_A^2, σ_B^2, and σ_{AB}^2 in terms of the general parameters σ^2, ρ_0, ρ_1, ρ_2, ρ_3, as is done in (4f.2.4) for a special sampling scheme on (α, β).

4f.3 Treatment of the General Model

Our object is to determine the expectations of the entries in the analysis of variance Table 4e.2α under the general covariance structure (4f.1.2) for the observations. To do this, we construct an orthogonal transformation from variables y_{ijk} to a set of uncorrelated variables. This is achieved in three stages.

Stage 1. Make an orthogonal transformation 0_1 on each set of c variables in the cells obtaining new variables (where $y_{ij} = \sum_k y_{ijk}$),

$$u_{ij} = \frac{y_{ij}}{\sqrt{c}}, \qquad g_{ij2}, \ldots, g_{ijc}, \tag{4f.3.1}$$
$$i = 1, \ldots, q, \qquad j = 1, \ldots, p.$$

Verify that

$$E(g_{ijk}) = 0, \qquad V(g_{ijk}) = (1 - \rho_0)\sigma^2, \qquad E(u_{ij}) = \sqrt{c}\,\mu_{ij},$$
$$V(u_{ij}) = (1 + \overline{c - 1}\rho_0)\sigma^2, \qquad \operatorname{cov}(u_{ij}, u_{im}) = c\rho_1\sigma^2$$
$$\operatorname{cov}(u_{ij}, u_{mj}) = c\rho_2\sigma^2, \qquad \operatorname{cov}(u_{ij}, u_{mr}) = c\rho_3\sigma^2.$$

Stage 2. Apply an orthogonal transformation 0_2 on p variables $u_{i1}, \ldots, u_{ip}$ in ith row to obtain new variables

$$z_{i1} = \sum \frac{u_{ij}}{\sqrt{p}}, \qquad z_{i2}, \ldots, z_{ip} \tag{4f.3.2}$$
$$i = 1, \ldots, q.$$

As in (4f.3.1) it is easy to establish

$$V(z_{i1}) = (1 + \overline{c - 1}\rho_0 + c(p - 1)\rho_1)\sigma^2$$
$$V(z_{ij}) = (1 + \overline{c - 1}\rho_0 - c\rho_1)\sigma^2, \qquad j = 2, \ldots, p$$
$$\operatorname{cov}(z_{i1}, z_{m1}) = c(\rho_2 + \overline{p - 1}\rho_3)\sigma^2,$$
$$\operatorname{cov}(z_{ij}, z_{mj}) = c(\rho_2 - \rho_3)\sigma^2, \qquad j \neq 1$$
$$\operatorname{cov}(z_{ij}, z_{mr}) = 0, \qquad j \neq 1, \qquad r \neq 1.$$

Stage 3. If we see a method in this madness, a third and a final orthogonal transformation 0_3 is called for on each set of q variables $z_{1j}, \ldots, z_{qj}$ to obtain new variables

$$v_{1j} = \sum \frac{z_{ij}}{\sqrt{q}}, \qquad v_{2j}, \ldots, v_{qj} \tag{4f.3.3}$$
$$j = 1, \ldots, p,$$

which are all uncorrelated, but with unequal variances as follows:

$$\begin{aligned}
\sigma_{00} &= V(v_{11}) = [1 + \overline{c - 1}\rho_0 + c(p - 1)\rho_1 \\
&\qquad\qquad + c(q - 1)\rho_2 + c(p - 1)(q - 1)\rho_3]\sigma^2 \\
\sigma_{10} &= V(v_{i1}) = [1 + \overline{c - 1}\rho_0 - c\rho_2 + c(p - 1)(\rho_1 - \rho_3)]\sigma^2, \qquad i > 1 \\
\sigma_{01} &= V(v_{1j}) = [1 + \overline{c - 1}\rho_0 - c\rho_1 + c(q - 1)(\rho_2 - \rho_3)]\sigma^2, \qquad j > 1 \\
\sigma_{11} &= V(v_{ij}) = [1 + \overline{c - 1}\rho_0 - c\rho_1 - c\rho_2 - c\rho_3]\sigma^2, \qquad i, j \geqslant 2 \\
\sigma_e^2 &= V(g_{ijk}) = (1 - \rho_0)\sigma^2, \qquad k \geqslant 2.
\end{aligned} \tag{4f.3.4}$$

For a variance components model, substituting for ρ_i and σ^2 from (4f.2.4) in terms of σ_e^2, σ_A^2, σ_B^2 and σ_{AB}^2, we have

$$\begin{aligned}
\sigma_{00} &= \rho_e^2 + c\sigma_{AB}^2 + cp\sigma_A^2 + cq\sigma_B^2 \\
\sigma_{10} &= \sigma_e^2 + c\sigma_{AB}^2 + cp_A^2, \qquad \sigma_{01} = \sigma_e^2 + c\sigma_{AB}^2 + cq\sigma_B^2 \\
\sigma_{11} &= \sigma_e^2 + c\sigma_{AB}^2, \qquad \sigma_e^2 = \sigma_e^2.
\end{aligned} \tag{4f.3.5}$$

Let us now express the sum of squares of the analysis of variance Table 4f.3α in terms of v_{ij} and g_{ijk}, the new uncorrelated variables. As a consequence of the first transformation

$$\sum\sum\sum y_{ijk}^2 = \sum\sum\left(\frac{y_{ij}}{c}\right)^2 + \sum\sum\sum g_{ijk}^2,$$

TABLE 4f.3α. Expected Values of Mean Squares in a Two-Way Analysis of Variance Under the General Setup (4f.1.2)

	D.F.	S.S. (Original) Variables)	S.S. (New Variables)	M.S.	Expected M.S.
A	$q-1$	$\sum \frac{y_{i.}^2}{pc} - \frac{y_{..}^2}{pqc} = S_A$	$\sum_2^p v_{i1}^2$	V_A	$\sigma_{10} + c\delta_A^2$
B	$p-1$	$\sum \frac{y_{.j}^2}{qc} - \frac{y_{..}^2}{pqc} = S_B$	$\sum_2^q v_{1j}^2$	V_B	$\sigma_{01} + c\delta_B^2$
AB	$(p-1)(q-1)$	$* = S_{AB}$	$\sum_2^p\sum_2^q v_{ij}^2$	V_{AB}	$\sigma_{11} + c\delta_{AB}^2$
Between cells	$pq-1$	$\sum\sum \frac{y_{ij}^2}{c} - \frac{y_{..}^2}{pqc}$	$\sum_1^p\sum_1^q v_{ij} - v_{11}^2$	V	$\sigma^2 + c\delta^2$
Within cells	$pq(c-1)$	$* = S_e$	$\sum_2^c\sum_1^p\sum_1^q g_{ijk}$	V_e	σ_e^2

that is, the sum of squares within cells, on $pq(c-1)$ D.F. is

$$S_e = \left(\sum\sum\sum y_{ijk}^2 - \frac{y_{...}^2}{pqc}\right) - \left(\sum\sum \frac{y_{ij}^2}{c} - \frac{y_{...}^2}{pqc}\right) = \sum\sum\sum g_{ijk}^2,$$

and, therefore $E(S_e) = E[pq(c-1)V_e] = \sum\sum\sum E(g_{ijk}^2) = pq(c-1)\sigma_e^2$.

The transformation from y_{ij} to v_{ij} is orthogonal. Therefore

$$\begin{aligned}\sum\sum v_{ij}^2 - v_{11}^2 &= \sum\sum \frac{y_{ij}^2}{c} - \frac{y_{\ldots}^2}{pqc} \\ &= \sum \frac{y_{i.}^2}{pc} - \frac{y_{\ldots}^2}{pqc} + \sum \frac{y_{.j}^2}{qc} - \frac{y_{\ldots}^2}{pqc} \\ &\quad + \sum\sum \frac{(y_{ij} - \bar{y}_{i.} - \bar{y}_{.j} + \bar{y}_{\ldots})^2}{c} \\ &= S_A + S_B + S_{AB}.\end{aligned} \tag{4f.3.6}$$

In virtue of the orthogonal transformations 0_2 and 0_3,

$$\sum_1^q \frac{y_{i.}^2}{pc} - \frac{y_{\ldots}^2}{pqc} = \sum_2^q v_{i1}^2, \qquad \sum_1^p \frac{y_{.j}^2}{qc} - \frac{y_{\ldots}^2}{pqc} = \sum_2^p v_{1j}^2$$

and therefore

$$S_{AB} = \sum_2^p \sum_2^q v_{ij}^2.$$

Using the identity (4f.3.6) on the mean values μ_{ij}, we have

$$\begin{aligned}c(pq-1)\delta^2 &= c\left(\sum\sum \mu_{ij}^2 - \frac{\mu_{..}^2}{pq}\right) \\ &= c\left(\sum \frac{\mu_{i.}^2}{p} - \frac{\mu_{..}^2}{pq}\right) + c\left(\sum \frac{\mu_{.j}^2}{q} - \frac{\mu_{..}^2}{pq}\right) \\ &\quad + c\sum\sum(\mu_{ij} - \bar{\mu}_{i.} - \bar{\mu}_{.j} + \bar{\mu}_{..})^2 \\ &= c(q-1)\delta_A^2 + c(p-1)\delta_B^2 + c(p-1)(q-1)\delta_{AB}^2.\end{aligned} \tag{4f.3.7}$$

Hence

$$E[S_A] = E\left(\sum_2^q v_{i1}^2\right) = \sum_2^q [V(v_{i1}) + \overline{E(v_{i1})^2}] = (q-1)(\sigma_{10} + c\delta_A^2).$$

Similarly

$$E[S_B] = (p-1)(\sigma_{01} + c\delta_B^2), \qquad E[S_{AB}] = (p-1)(q-1)(\sigma_{00} + c\delta_{AB}^2).$$

The results are summarized in the analysis of variance Table 4f.3α, where σ_{ij} and δ_i^2 are as defined in (4f.3.4) and (4f.3.7) respectively.

It may be noted that in the variance components model, δ_A^2, δ_B^2, δ_{AB}^2 are zero, σ_{ij} have the values given in (4f.3.5), and we have the estimates

$$\begin{aligned}&\hat{\sigma}_e^2 = V_e, \qquad \hat{\sigma}_{AB}^2 = \frac{V_{AB} - V_e}{c} \\ &\hat{\sigma}_A^2 = \frac{V_A - V_{AB}}{pc}, \qquad \hat{\sigma}_B^2 = \frac{V_B - V_{AB}}{qc}.\end{aligned} \tag{4f.3.8}$$

In the general case when δ_A^2, δ_B^2, δ_{AB}^2 are not zero, it is not possible to estimate all the unknowns.

Let us now assume that the variables y_{ijk} have a multivariate normal distribution in which case the transformed variables v_{ij} and g_{ijk} are independent and normal. Hence the sum of squares S_A, S_B, S_{AB}, and S_e due to A, B, AB and within cells are all independently distributed in the most general case, depending on exclusive sets of variables v_{ij} and g_{ijk}. Thus

$$\frac{S_e}{\sigma_e^2} \sim \chi^2[pq(c-1)], \text{ a central } \chi^2,$$

$$\frac{S_A}{\sigma_{10}} \sim \chi^2\left(q-1, \frac{c(q-1)\,\delta_A^2}{\sigma_{10}}\right), \text{ a noncentral } \chi^2 \quad \text{if} \quad \delta_A^2 \neq 0,$$

$$\frac{S_B}{\sigma_{01}} \sim \chi^2\left(p-1, \frac{c(p-1)\,\delta_B^2}{\sigma_{01}}\right), \text{ a noncentral } \chi^2 \quad \text{if} \quad \delta_B^2 \neq 0,$$

$$\frac{S_{AB}}{\sigma_{11}} \sim \chi^2\left((p-1)(q-1), \frac{c(p-1)(q-1)\delta_{AB}^2}{\sigma_{10}}\right),$$

$$\text{a noncentral } \chi^2 \quad \text{if} \quad \delta_{AB}^2 \neq 0.$$

These distributions are basic to all inference problems concerning unknown parameters.

For instance in the variance-components model, since $(S_A/\sigma_{10}) \sim \chi^2\,(q-1)$ and $(S_e/\sigma_e^2) \sim \chi^2[pq(c-1)]$, then

$$\frac{V_A}{V_e}\frac{\sigma_e^2}{\sigma_{10}} \sim F[q-1, pq(c-1)], \tag{4f.3.9}$$

and therefore a lower confidence limit for the ratio σ_{10}/σ_e^2 is obtained by equating the left-hand side of (4f.3.9) to a percentage point of F,

$$\frac{V_A}{V_e}\frac{\sigma_e^2}{\sigma_{10}} = F_\alpha.$$

4g THE THEORY AND APPLICATION OF STATISTICAL REGRESSION

4g.1 Concept of Regression (General Theory)

The theory of regression is concerned with the prediction of one or more variables $(y_1, \ldots, y_q)$ on the basis of information provided by other measurements or concomitant variables, $(x_1, \ldots, x_p) = \mathbf{x}'$. It is customary to call the latter independent or *predictor* variables and the former dependent or *criterion* variables.

Prediction is needed in several practical situations. A meterologist wants to forecast weather several hours ahead on the basis of suitable atmospheric measurements taken at a point in time. An educational institution has to assess the abilities of candidates in particular fields by appropriate tests. An establishment wishes to recruit personnel who will be successful on the jobs assigned to them, by using a suitable test battery. A plant or an animal breeder is interested in choosing individuals whose progeny will give him the maximum return or "pay off."

In all these situations the criteria are some variables observable in the future, which are sought to be predicted by the available measurements for taking decisions. How should the predictors be chosen?

Minimum Mean Square Error Predictor. Consider a single criterion variable y and p predictor variables $x_1, \dots, x_p$ denoted by vector $\mathbf{x}$. Let $f(x_1, \dots, x_p)$ be a predictor of y.

(i) *Let $M(\mathbf{x})$ be $E(y|\mathbf{x})$. Then $E[y - f(\mathbf{x})]^2$ is a minimum when $f(\mathbf{x}) = M(\mathbf{x})$.*

From the definition of conditional expectation

$$E[(y - M)(M - f)] = E[(M - f)E(\overline{y - M}|\mathbf{x})] = 0$$

so that

$$\begin{aligned} E(y - f)^2 &= E(y - M + M - f)^2 \\ &= E(y - M)^2 + E(M - f)^2 \geqslant E(y - M)^2. \end{aligned} \tag{4g.1.1}$$

The lower bound of $E(y - f)^2$ is attained when $f = M$, so that the best choice of the predictor which minimizes the mean square error (m.s.e.) is $M(\mathbf{x})$, the conditional expectation of y given $\mathbf{x}$, which is called the regression of y on $x_1, \dots, x_p$. The m.s.e., $E(y - M)^2$ is the average conditional variance of y given $\mathbf{x}$ and is denoted by $\sigma_{y.\mathbf{x}}^2$.

Predictor Having Maximum Correlation with the Criterion. Let us find f to maximize the product moment correlation between y and f, $\rho(f, y) = \text{cov}(y, f)/\sigma_f \sigma_y$.

(ii) *Let $M(x)$ be as in* (i). *Then $\rho(y, M)$ is non-negative and $\rho(y, M) \geqslant |\rho(y, f)|$ for any function f.*

For any f

$$\text{cov}(y, f) = E\{[f - E(f)]E(y|\mathbf{x})\} = E\{[f - E(f)]M\} = \text{cov}(f, M). \tag{4g.1.2}$$

When $f = M$, $\text{cov}(y, M) = \text{cov}(M, M)$ by using (4g.1.2). Hence $\rho(y, M) = \sigma_M/\sigma_y \geqslant 0$ where $\sigma_M{}^2 = \text{cov}(M, M) = V(M)$ and $\sigma_y{}^2 = V(y)$. Now

$$\begin{aligned}\rho^2(f, y) &= \frac{[\text{cov}(y, f)]^2}{\sigma_f{}^2\sigma_y{}^2} = \frac{[\text{cov}(f, M)]^2}{\sigma_f{}^2\sigma_M{}^2} \cdot \frac{\sigma_M{}^2}{\sigma_y{}^2} \\ &= \rho^2(f, M)\rho^2(y, M) \leqslant \rho^2(y, M). \end{aligned} \tag{4g.1.3}$$

The upper bound of $\rho(f, y)$ is attained when $\rho(f, M) = 1$, which implies that f is a linear function of M. *Again the regression function is the answer.* The square of the maximum correlation attained $\sigma_M{}^2/\sigma_y{}^2$ is known as *correlation ratio* and is represented by $\eta_{y\mathbf{x}}{}^2$. By definition, $0 \leqslant \eta_{y\mathbf{x}}{}^2 \leqslant 1$. We observe the decomposition (or analysis of variance)

$$\begin{aligned} E[y - E(y)]^2 &= E(y - M)^2 + E[M - E(M)]^2 \\ \sigma_y{}^2 \quad &= \quad \sigma_{y.\mathbf{x}}^2 \quad + \quad \sigma_M{}^2, \end{aligned} \tag{4g.1.4}$$

so that $\eta_{y\mathbf{x}}{}^2 = 1 - \sigma_{y.\mathbf{x}}^2/\sigma_y{}^2$, which approaches unity as $\sigma_{y.\mathbf{x}}^2$ the error of prediction $\to 0$, and approaches zero when there is no reduction in error due to the use of $\mathbf{x}$. Thus $\eta_{y\mathbf{x}}{}^2$, written more explicitly as $\eta_{y(x_1, \ldots, x_p)}^2$ or $\eta_{0(1 \cdots p)}^2$, provides a measure of association between y and $x_1, \ldots, x_p$ or of accuracy of prediction, which may be useful in comparing different situations or different choices of predictor variables in any given problem.

Maximizing Expected Performance in Selection. Suppose we want to truncate the distribution of $\mathbf{x}$ in such a way that a given proportion α of the population is left and the expected value of y in the retained distribution of $\mathbf{x}$ is maximized. That is, we want to characterize an α proportion of the population on the basis of $\mathbf{x}$, with the maximum possible average value of y. If $F(\mathbf{x}, y)$ is the joint distribution function, then we need a region w of $\mathbf{x}$ such that

$$\iint_{w \times R} dF(\mathbf{x}, y) = \alpha \text{ (given)},$$

$$\iint_{w \times R} y\,dF(\mathbf{x}, y) \text{ is a maximum.}$$

where R is the entire range of y. Integrating with respect to y, the problem reduces to

$$\text{maximizing} \int_w M(\mathbf{x})\, dF(\mathbf{x}), \tag{4g.1.5}$$

$$\text{subject to} \int_w dF(\mathbf{x}) = \alpha. \tag{4g.1.6}$$

(iii) (*Cochran*, 1951). *The region w defined by $M(\mathbf{x}) \geq \lambda$, where λ is chosen to satisfy* (4g.1.6), *maximizes* (4g.1.5).

The result is an immediate consequence of the Neyman-Pearson lemma proved in Section **7a.2**. Again the answer is based on the regression function.

Linear Regression. The special case when the regression function is linear in $\mathbf{x}$ has been studied extensively. Since $E(y - M)^2$ is a minimum when M is the regression function, we can consider an arbitrary linear function $\alpha + \beta_1 x_1 + \cdots + \beta_p x_p$ and determine the coefficients by minimizing,

$$E(y - \alpha - \beta_1 x_1 - \cdots - \beta_p x_p)^2 = b^2 + \boldsymbol{\beta}'\mathbf{C}\boldsymbol{\beta} - 2\boldsymbol{\beta}'\boldsymbol{\sigma}_0 + V(y), \quad (4g.1.7)$$

where $\mathbf{C}$ is the dispersion matrix of $x_1, \ldots, x_p$, $\boldsymbol{\sigma}_0$ is the vector of covariances of y with $x_1, \ldots, x_p$ and $b = \alpha - E(y) + \beta_1 E(x_1) + \cdots + \beta_p E(x_p)$. The optimum choice of b and $\boldsymbol{\beta}$ are

$$b^* = 0, \qquad \mathbf{C}\boldsymbol{\beta}^* = \boldsymbol{\sigma}_0, \quad (4g.1.8)$$

for it is easily shown by writing $\boldsymbol{\beta} = \boldsymbol{\beta}^* + \boldsymbol{\delta}$, that

$$\begin{aligned} b^2 + \boldsymbol{\beta}'\mathbf{C}\boldsymbol{\beta} - 2\boldsymbol{\beta}'\boldsymbol{\sigma}_0 &= (\boldsymbol{\beta}^*)'\mathbf{C}\boldsymbol{\beta}^* - 2\boldsymbol{\sigma}_0'\boldsymbol{\beta}^* + \boldsymbol{\delta}'\mathbf{C}\boldsymbol{\delta} + b^2 \\ &\geq (\boldsymbol{\beta}^*)'\mathbf{C}\boldsymbol{\beta}^* - 2\boldsymbol{\sigma}_0'\boldsymbol{\beta}^* + (b^*)^2. \end{aligned}$$

The optimum value of α is

$$\alpha^* = E(y) - (\boldsymbol{\beta}^*)'E(\mathbf{x}). \quad (4g.1.9)$$

The minimum value of (4g.1.7) is

$$V(y) + (\boldsymbol{\beta}^*)'\mathbf{C}\boldsymbol{\beta}^* - 2(\boldsymbol{\beta}^*)'\boldsymbol{\sigma}_0 = \sigma_{00} - \boldsymbol{\sigma}_0'\mathbf{C}^{-1}\boldsymbol{\sigma}_0, \quad (4g.1.10)$$

where $\sigma_{00} = V(y)$. Thus we see that *if the regression is linear, then it can be completely specified by mean values and variances and covariances of the variables only* and a knowledge of the exact distribution of the variables is not necessary.

The correlation ratio $\eta_{y\mathbf{x}}^2$ in the case of linear regression, is then

$$\rho_{y\mathbf{x}}^2 = 1 - \frac{\sigma_{y.\mathbf{x}}^2}{\sigma_y^2} = \frac{\boldsymbol{\sigma}_0'\mathbf{C}^{-1}\boldsymbol{\sigma}_0}{\sigma_{00}}, \quad (4g.1.11)$$

and the coefficient ρ so defined by using only the variances and covariances of variables is known by the special name of the *multiple correlation coefficient* whether regression is linear or not. If linear regression does not hold, then η^2 and ρ^2 as defined in (4g.1.11) are different and in fact $\eta^2 \geq \rho^2$.

Best Linear Predictors. We have seen that when the regression is linear, the best predictor can be determined in terms of the first- and second-order moments only. Now we ask, what is the best linear predictor of y in terms of $\mathbf{x}$

whether the regression is linear or not? That is, instead of considering the class of all functions of $\mathbf{x}$, we restrict our enquiry to linear functions of $\mathbf{x}$ only.

(iv) *The linear function* $\alpha^* + \mathbf{x}'\boldsymbol{\beta}^*$ *determined by the equations* (4g.1.8) *and* (4g.1.9) *is the minimum m.s.e. linear predictor of y. It is also the linear function having the maximum correlation with y.*

The first part follows since α^* and $\boldsymbol{\beta}^*$ are, in fact, determined to minimize $E(y - \alpha - \boldsymbol{\beta}'\mathbf{x})^2$.

To prove the second part, let us observe that

$$\text{cov}(y, \boldsymbol{\beta}'\mathbf{x}) = \boldsymbol{\beta}'\boldsymbol{\sigma}_0 = \boldsymbol{\beta}'\mathbf{C}\boldsymbol{\beta}^*$$
$$\text{cov}(y, \mathbf{x}'\boldsymbol{\beta}^*) = (\boldsymbol{\beta}^*)'\mathbf{C}\boldsymbol{\beta}^* = V(\mathbf{x}'\boldsymbol{\beta}^*).$$

Hence, we see that

$$\sigma_y{}^2\rho^2(y, \boldsymbol{\beta}'\mathbf{x}) = \frac{(\boldsymbol{\beta}'\mathbf{C}\boldsymbol{\beta}^*)^2}{\boldsymbol{\beta}'\mathbf{C}\boldsymbol{\beta}} \leqslant (\boldsymbol{\beta}^*)'\mathbf{C}\boldsymbol{\beta}^* = \sigma_y{}^2\rho^2(y, \mathbf{x}'\boldsymbol{\beta}^*),$$

by using the C-S inequality $(\boldsymbol{\beta}'\mathbf{C}\boldsymbol{\beta}^*)^2 \leqslant (\boldsymbol{\beta}^{*\prime}\mathbf{C}\boldsymbol{\beta}^*)(\boldsymbol{\beta}'\mathbf{C}\boldsymbol{\beta})$ and noting that

$$\sigma_y\rho(y, \mathbf{x}'\boldsymbol{\beta}^*) = \frac{\text{cov}(y, \mathbf{x}'\boldsymbol{\beta}^*)}{\sqrt{(\boldsymbol{\beta}^*)'\mathbf{C}\boldsymbol{\beta}^*}} = \sqrt{(\boldsymbol{\beta}^*)'\mathbf{C}\boldsymbol{\beta}^*}.$$

The square of the maximum correlation is

$$\rho_{y\mathbf{x}}{}^2 = \frac{\boldsymbol{\beta}^{*\prime}\mathbf{C}\boldsymbol{\beta}^*}{\sigma_{00}} = \frac{\boldsymbol{\sigma}_0'\mathbf{C}^{-1}\boldsymbol{\sigma}_0}{\sigma_{00}},$$

which is the expression (4g.1.11) derived earlier. Since $\eta_{y\mathbf{x}}$ is the maximum correlation between y and any function of $\mathbf{x}$ and $\rho_{y\mathbf{x}}$ is the maximum correlation between y and linear functions of $\mathbf{x}$, we find that $\eta_{y\mathbf{x}}{}^2 \geqslant \rho_{y\mathbf{x}}{}^2$. The coefficient $\rho_{y\mathbf{x}}{}^2$ is written more explicitly as $\rho^2_{y(x_1, \ldots, x_p)}$ or $\rho^2_{0(1 \cdots p)}$.

Alternative Expressions for ρ^2. Let (ρ_{ij}), $i, j = 0, 1, \ldots, p$ be the correlation matrix of the variables $y, x_1, \ldots, x_p$ and (ρ^{ij}) be the reciprocal of (ρ_{ij}). Similarly, let (σ^{ij}) be the reciprocal of (σ_{ij}), the dispersion matrix of all the variables. Writing the equation $\mathbf{C}\boldsymbol{\beta}^* = \boldsymbol{\sigma}_0$ in full, we have

$$\beta_1^*\sigma_{1i} + \cdots + \beta_p^*\sigma_{pi} = \sigma_{0i}, \qquad i = 1, \ldots, p. \tag{4g.1.12}$$

But from the definition, we have

$$\sigma^{10}\sigma_{1i} + \cdots + \sigma^{p0}\sigma_{pi} = -\sigma_{0i}\sigma^{00}, \qquad i = 1, \ldots, p. \tag{4g.1.13}$$

Comparing (4g.1.12) and (4g.1.13), we have

$$\beta_i^* = -\frac{\sigma^{i0}}{\sigma^{00}} = \frac{\sqrt{\sigma_{00}}}{\sqrt{\sigma_{ii}}}\frac{\rho^{i0}}{\rho^{00}}.$$

Furthermore,

$$\rho^2_{0(1\cdots p)} = \frac{\boldsymbol{\sigma}_0' \mathbf{C}^{-1}\boldsymbol{\sigma}_0}{\sigma_{00}} = \frac{\boldsymbol{\sigma}_0' \boldsymbol{\beta}^*}{\sigma_{00}}$$

$$= -\frac{\sigma_{10}\sigma^{10} + \cdots + \sigma_{p0}\sigma^{p0}}{\sigma_{00}\sigma^{00}} = \frac{\sigma_{00}\sigma^{00} - 1}{\sigma_{00}\sigma^{00}}$$

$$= 1 - \frac{1}{\sigma_{00}\sigma^{00}} \tag{4g.1.14}$$

$$= 1 - \frac{1}{\rho^{00}} \text{ (substituting for } \beta^* \text{ in terms of } \rho^{ij}\text{).}$$

The last equation shows that $\rho^2_{0(1\cdots p)}$ is a function of the correlations only.

4g.2 Measurement of Additional Association

In practice, it is necessary to examine the extent to which accuracy of prediction of the criterion y could be improved by bringing in extra predictor variables. The m.s.e. of prediction of the criterion when variables $x_1, \ldots, x_k$ are used is $\sigma_y^2[1 - \eta^2_{0(1\cdots k)}]$ and when $x_1, \ldots, x_p$, $p > k$, are used is $\sigma_y^2[1 - \eta^2_{0(1\cdots p)}]$. The latter is not greater than the former, and the reduction in mean square error is

$$\sigma_y^2[\eta^2_{0(1\cdots p)} - \eta^2_{0(1\cdots k)}],$$

and the proportional reduction due to the use of the extra variables $x_{k+1}, \ldots, x_p$ (*in addition* to $x_1, \ldots, x_k$) is

$$\frac{\eta^2_{0(1\cdots p)} - \eta^2_{0(1\cdots k)}}{1 - \eta^2_{0(1\cdots k)}}, \tag{4g.2.1}$$

which may be called *partial* correlation ratio and denoted by $\eta^2_{0(k+1\cdots p).(1\cdots k)}$. This must be distinguished from the correlation ratio $\eta^2_{0(k+1\cdots p)}$ which is a measure of the direct association between y and $x_{k+1}, \ldots, x_p$.

If we are considering only linear predictors, the corresponding expression for proportional reduction in m.s.e. is

$$\rho^2_{0(k+1\cdots p).(1\cdots k)} = \frac{\rho^2_{0(1\cdots p)} - \rho^2_{0(1\cdots k)}}{1 - \rho^2_{0(1\cdots k)}}, \tag{4g.2.2}$$

which may be called the (square of) partial multiple correlation of y on $x_{k+1}, \ldots, x_p$ eliminating, so to speak, the association by $x_1, \ldots, x_k$.

The measurement of partial association may be also considered in a slightly different way. The error in predicting y on the basis of $x_1, \ldots, x_k$ is $e = y - M(x_1, \ldots, x_k)$, where M stands for the true regression (or a linear predictor).

If this error could be predicted to some extent by considering extra variables $x_{k+1}, \ldots, x_p$ (in addition to the available variables $x_1, \ldots, x_k$), then the inclusion of the latter would be useful. The measure then depends on the m.s.e. in the prediction of the error e, which is the correlation ratio of (or multiple correlation in the case of a linear prediction) e on $x_1, \ldots, x_p$. It is instructive to verify that the predictor of e is

$$E(e \mid x_1, \ldots, x_p) = M(x_1, \ldots, x_p) - M(x_1, \ldots, x_k) \tag{4g.2.3}$$

with the variance

$$\sigma_y^2[\eta_{0(1 \cdots p)}^2 - \eta_{0(1 \cdots k)}^2], \tag{4g.2.4}$$

whereas the total variance of e is

$$\sigma_y^2[1 - \eta_{0(1 \cdots k)}^2]. \tag{4g.2.5}$$

The required correlation ratio is, therefore, the ratio of (4g.2.4) to (4g.2.5), which is the same as the measure defined in (4g.2.1). Replacing η by ρ we obtain the corresponding expressions for linear predictors. The coefficient $\rho_{0(p) \cdot (1 \cdots p-1)}^2$ which measures the partial association between y and x_p eliminating $x_1, \ldots, x_{p-1}$ is of special interest because it enables us to judge the importance of an individual variable added to an existing set. From (4g.2.2), we have

$$1 - \rho_{0(p) \cdot (1 \cdots p-1)}^2 = \frac{1 - \rho_{0(1 \cdots p)}^2}{1 - \rho_{0(1 \cdots p-1)}^2} = \frac{\begin{vmatrix} 1 & \cdots & \rho_{1p-1} \\ \cdot & \cdots & \cdot \\ \rho_{1p-1} & \cdots & 1 \end{vmatrix}}{\rho^{00} |\rho_{ij}|_{p-1}} \tag{4g.2.6}$$

by using the formula (4g.1.14) for $1 - \rho_{0(1 \ldots k)}^2$, where $|\rho_{ij}|_{p-1}$ denotes the determinant of correlations for $y, x_1, \ldots, x_{p-1}$ only. Furthermore, from the properties of the elements of the reciprocal matrix, we have

$$\rho^{00}\rho^{pp} - (\rho^{0p})^2 = \frac{\begin{vmatrix} 1 & \cdots & \rho_{1p-1} \\ \cdot & \cdots & \cdot \\ \rho_{1p-1} & \cdots & 1 \end{vmatrix}}{|\rho_{ij}|_p}, \tag{4g.2.7}$$

so that from the formula (4g.2.6), we have

$$\rho_{0(p) \cdot (1 \cdots p-1)}^2 = \frac{(\rho^{0p})^2}{\rho^{00}\rho^{pp}},$$

which shows that the measure of partial association is symmetrical in the symbols 0 and p and $\rho_{0(p) \cdot (1 \cdots p-1)}^2$ and $\rho_{p(0) \cdot (1 \cdots p-1)}^2$ have the same value.

Dropping the bracket of the second symbol we may write, after taking square root,

$$\rho_{0p\,.\,(1\,\cdots\,p-1)} = \frac{\rho^{0p}}{\sqrt{\rho^{00}\rho^{pp}}} \tag{4g.2.8}$$

and give the coefficient (4g.2.8) the special name of partial correlation between y and x_p eliminating $x_1, \ldots, x_{p-1}$.

The partial correlation $\rho_{0p\,.\,(1\,\cdots\,p-1)}$, is generally defined as the correlation between the residuals

$$\left.\begin{aligned} e_1 &= y - (\alpha + \beta_1 x_1 + \cdots + \beta_{p-1} x_{p-1}) \\ e_2 &= x_p - (\alpha' + \beta_1' x_1 + \cdots + \beta_{p-1}' x_{p-1}) \end{aligned}\right\} \tag{4g.2.9}$$

after subtracting, the corresponding regressions on $x_1, \ldots, x_{p-1}$ from y and x_p. It would be a good exercise to show that the correlation so determined is exactly (4g.2.8). But the approach leading to the definition of (4g.2.8) appears to be more intuitive, natural, and general.

In the example of Section **4g.3** we shall consider tests of significance for multiple and partial correlations on the assumption of normality of the distribution of variables.

4g.3 Prediction of Cranial Capacity (a Practical Example)

The problem is to estimate the capacity of a skull, which may be broken or damaged so that the capacity cannot be directly measured. In such a case the capacity may be predictable with some accuracy, if at least some external measurements are available (Rao and Shaw, 1948e).

The basic data needed are complete records of capacity as well as external measurements on a number of well-preserved skulls, from which we estimate the regression function for prediction in other cases.

Choice of Regression Equation. Three important measurements from which the cranial capacity C may be predicted are the glabella-occipital length L, the maximum parietal breadth B, and the basio-bregmatic height H'. Since the magnitude to be estimated is a volume, it is appropriate to set up a regression formula of the type

$$C = \gamma L^{\beta_1} B^{\beta_2} H'^{\beta_3}, \tag{4g.3.1}$$

where γ, β_1, β_2, β_3 are constants to be determined. By transforming the variables to

$$y = \log_{10} C, \qquad x_1 = \log_{10} L, \qquad x_2 = \log_{10} B, \qquad x_3 = \log_{10} H',$$

we can write the formula (4g.3.1) as

$$y = \alpha + \beta_1 x_1 + \beta_2 x_2 + \beta_3 x_3, \tag{4g.3.2}$$

where $\alpha = \log_{10} \gamma$. Fortunately, formula (4g.3.1) could easily be reduced to the linear form (4g.3.2). We might ask why a linear regression function of the type

$$C = \alpha + \beta_1 L + \beta_2 B + \beta_3 H' \tag{4g.3.3}$$

in terms of L, B, H' was not adopted. To answer this we must examine the performance of the two functions (4g.3.1) and (4g.3.3) in the prediction of capacity. We choose that which gives a closer prediction in some sense. A wrong choice of formula may lead to grossly inaccurate prediction, and we may have to use our experience and try a number of alternatives before arriving at a reasonable formula.

Estimation of Constants. When the formula is *linear in the unknown constants* as in (4g.3.2), the method of least squares may be employed to estimate the unknowns. We rewrite the equation (4g.3.2) in the form

$$\alpha' - \beta_1(x_1 - \bar{x}_1) - \cdots - \beta_3(x_3 - \bar{x}_3),$$

where $\bar{x}_1$, $\bar{x}_2$, $\bar{x}_3$ are the observed averages for the measurements x_1, x_2, x_3 and

$$\alpha' = \alpha - \beta_1 \bar{x}_1 - \beta_2 \bar{x}_2 - \beta_3 \bar{x}_3 .$$

Let us suppose there are n sets $(y_r, x_{1r}, x_{2r}, x_{3r})$ of measurements. The reader may verify that the normal equations obtained by minimizing

$$\sum_{r=1}^{n} [y_r - \alpha' - \beta_1(x_{1r} - \bar{x}_1) - \beta_2(x_{2r} - \bar{x}_2) - \beta_3(x_{3r} - \bar{x}_3)]^2$$

are, by writing a', b_1, b_2, b_3 for estimates of α', β_1, β_2, β_3,

$$\begin{aligned} na' &= \sum y_r \\ b_1 S_{11} + b_2 S_{12} + b_3 S_{13} &= S_{01} \\ b_1 S_{12} + b_2 S_{22} + b_3 S_{23} &= S_{02} \\ b_1 S_{13} + b_2 S_{23} + b_3 S_{33} &= S_{03}, \end{aligned}$$

where S_{ij} is the usual corrected sum of products,

$$\begin{aligned} S_{ij} &= \sum_r (x_{ir} - \bar{x}_i)(x_{jr} - \bar{x}_j) = \sum x_{ir} x_{jr} - n\bar{x}_i \bar{x}_j \\ S_{0i} &= \sum_r (y_r - \bar{y})(x_{ir} - \bar{x}_i) = \sum y_r x_{ir} - n\bar{y}\bar{x}_i . \end{aligned}$$

Using the measurements on $n = 86$ male skulls from Farringdon Street series obtained by Hooke (1926) the following values are computed:

$$\bar{y} = 3.1685, \qquad \bar{x}_1 = 2.2752, \qquad \bar{x}_2 = 2.1523, \qquad \bar{x}_3 = 2.1128$$

$$(S_{ij}) = \begin{pmatrix} 0.01875 & 0.00848 & 0.00684 \\ 0.00848 & 0.02904 & 0.00878 \\ 0.00684 & 0.00878 & 0.02886 \end{pmatrix}$$

$$S_{01} = 0.03030, \qquad S_{02} = 0.04410, \qquad S_{03} = 0.03629$$

$$S_{00} = \sum y_r^2 - n\bar{y}^2 = 0.02781.$$

These are all the preliminary computations necessary for further analysis.

The reciprocal of the matrix (S_{ij}) denoted by (C_{ij}) [the notation should be (S^{ij}), but in statistical literature this is already known as the C matrix with elements (C_{ij})] is

$$(C_{ij}) = \begin{pmatrix} 64.21 & -15.57 & -10.49 \\ -15.57 & 41.71 & -9.00 \\ -10.49 & -9.00 & 39.88 \end{pmatrix}.$$

The estimates of the parameters (solutions of the normal equations) are

$$b_1 = 64.21 S_{01} - 15.57 S_{02} - 10.49 S_{03} = 0.878$$

by using the first row of (C_{ij}), and

$$b_2 = -15.57 S_{01} + 41.72 S_{02} - 9.00 S_{03} = 1.041$$

by using the second row of (C_{ij}). Similarly $b_3 = 0.733$, and finally $a' = \bar{y} = 3.1685$, thus giving the estimate of α as

$$a = \bar{y} - b_1\bar{x}_1 - b_2\bar{x}_2 - b_3\bar{x}_3 = -2.618.$$

The estimated formula in terms of original measurements is then

$$C = 0.00241 L^{0.878} B^{1.041} H'^{0.733}. \tag{4g.3.4}$$

[The capacity of the Farringdon Street skulls was determined by packing them tightly with mustard seeds and then weighing them. The formula (4g.3.4) is strictly applicable only for predicting capacity determined in this way.]

The residual sum of squares is

$$R_0^2 = S_{00} - \sum b_i S_{0i} = 0.12692 - 0.09911 = 0.02781.$$

The multiple correlation, which provides a measure of accuracy of prediction or the association between the dependent and independent variables, is estimated by

$$R^2 = \frac{b_1 S_{01} + \cdots + b_3 S_{03}}{S_{00}} = \frac{0.09911}{0.12692} = 0.7809.$$

The variance σ^2, of y given x_1, x_2, x_3, is estimated by

$$s^2 = \frac{R_0^{\ 2}}{n-p-1} = \frac{0.02781}{82} = 0.0003391,$$

where p is the number of independent variables. The variances and co-variances of b_i are computed by the formulas

$$V(b_i) = C_{ii}\sigma^2 \qquad \text{and} \qquad \text{cov}(b_i b_j) = C_{ij}\sigma^2.$$

This completes estimation of unknown parameters and standard errors by the method of least squares.

Tests of Hypotheses. We now consider a number of problems that arise in the estimation and use of regression equation. These are stated and examined by using the estimates already obtained.

1. Is there a significant association between the dependent and independent variables? In terms of the parameters the hypothesis is $\beta_1 = \beta_2 = \beta_3 = 0$. This is also equivalent to the statement that the true multiple correlation (4g.1.11) is zero.

We use the analysis of variance technique. The unconditional sum of squares $R_0^{\ 2}$ on $n - p - 1 = 82$ D.F. is already obtained as 0.02781. If the hypothesis $\beta_1 = \beta_2 = \beta_3 = 0$ is true, then $R_1^{\ 2}$, the minimum value of $\sum (y_i - \alpha)^2$, is $\sum y_i^{\ 2} - n\bar{y}^2 = S_{00} = 0.12692$, which has $(n-1)$ D.F. The reduction in the residual sum of squares $R_1^{\ 2} - R_0^{\ 2}$ is due to regression and has $p = 3$ D.F. The analysis of sum of squares is shown in Table 4g.3α.

TABLE 4g.3α. Test of the Hypothesis $\beta_1 = \beta_2 = \beta_3 = 0$ (source: Rao and Shaw, 1948e)

	D.F.	S.S.	M.S.	F
Regression	3	0.09911	0.033037	97.41
Residual	82	0.02781	0.0003391	
Total	85	0.12692		

We observe that the variance ratio of Table 4g.3α can be simply obtained from the value of R^2, the square of the multiple correlation, n, the sample size, and p, the number of independent variable, by the formula

$$F = \frac{R^2}{1 - R^2} \cdot \frac{n - p - 1}{p} = \frac{0.7809}{1 - 0.7809} \cdot \frac{86 - 3 - 1}{3} = 97.41,$$

using the computed value of R^2, so that we can carry out the test of the hypothesis under consideration, knowing only the value of the observed multiple correlation. The variance ratio 97.41 on 3 and 82 D.F. is significant at the 1% level, which indicates the usefulness of the variables in prediction.

2. It may now be examined whether the three linear dimensions appear to the same degree in the prediction formula. From the estimates it is seen that the index b_2 for maximum parietal breadth is higher than the others. This means that a given ratio of increase in breadth counts more for capacity than the corresponding increase in length or height.

The hypothesis relevant to examine this problem is

$$\beta_1 = \beta_2 = \beta_3 = \beta.$$

The minimum value of $\sum [y_i - \alpha - \beta(x_{1i} + x_{2i} + x_{3i})]^2$ has to be found. The normal equation giving the estimate of β is

$$[S_{11} + S_{22} + S_{33} + 2(S_{12} + S_{23} + S_{31})]b = Q = (Q_1 + Q_2 + Q_3)$$

$$0.124856 b = 0.11069, \qquad b = 0.8866.$$

The minimum value with $(n - 2)$ D.F. is

$$(\sum y_i^2 - n\bar{y}^2) - bQ = 0.12692 - 0.09814 = 0.02878.$$

The F ratio of Table 4g.3β is small, so there is no evidence to conclude that β_1, β_2, β_3 are different. The differences, if any, are likely to be small, and a large collection of measurements may be necessary before anything definite can be said about this. Evolutionists believe that the breadth is increasing relatively more than any other magnitude on the skull. If this is true, it is of interest to examine how far the cranial capacity is influenced by the breadth.

So far as the problem of prediction is concerned, the formula

$$C = 0.002342(LBH')^{0.8866}$$

obtained by assuming $\beta_1 = \beta_2 = \beta_3$, may be as useful as the formula derived without such assumption. The variance of the estimate b of β is $\sigma'^2/\sum\sum S_{ij}$, where σ'^2 is the estimate based on 84 D.F. with the corresponding sum of squares given in Table 4g.3β.

TABLE 4g.3β. Test of the Hypothesis $\beta_1 = \beta_2 = \beta_3$

	D.F.	S.S.	M.S.	F
Deviation from equality	2	0.00097	0.000485	1.430
Residual	82	0.02781	0.0003391	
Total	84	0.02878		

3. A simple formula of the type $C = \alpha' LBH'$ is sometimes used for predicting the cranial capacity. A test of the adequacy of such a formula is equivalent to testing the hypothesis

$$\beta_1 = \beta_2 = \beta_3 = 1.$$

The minimum value of $\sum (y_i - \alpha - \beta_1 x_{1i} - \beta_2 x_{2i} - \beta_3 x_{3i})^2$, when $\beta_1 = \beta_2 = \beta_3 = 1$ is

$$(\sum y_i^2 - n\bar{y}^2) + S_{11} + S_{22} + S_{33} + 2(S_{12} + S_{23} + S_{31}) - 2(S_{01} + S_{02} + S_{03}) = 0.03039,$$

which has $(n-1)$ D.F. The residual has $(n-4)$ D.F. so that the difference with 3 D.F. is due to deviation from the hypothesis.

The ratio 2.544 on 3 and 82 D.F. is just below the 5% significance level. It would be of interest to examine this point with more adequate material.

In Table 4g.3γ the sum of squares due to deviation from the hypothesis could be directly calculated from the formula, thus providing a compound measure of the deviations of the estimates b_1, b_2, b_3 from the expected values 1, 1, 1, as

$$\sum\sum S_{ij}(b_i - 1)(b_j - 1) = 0.00258,$$

which is the same as that given in Table 4g.3γ.

4. Having found evidence that the β coefficients differ individually from unity, it is of some interest to examine whether the indices add up to 3 while distributing unequally among the three dimensions used. This requires a test of the hypothesis $\beta_1 + \beta_2 + \beta_3 = 3$. The best estimate of the deviation is $b_1 + b_2 + b_3 - 3 = 2.652 - 3 = -0.348$.

$$V(b_1 + b_2 + b_3 - 3) = (\sum\sum C_{ij})\sigma^2 = 75.68\sigma^2.$$

The ratio on 1 and 82 D.F. is

$$\frac{(0.348)^2}{75.68} \times \frac{1}{0.0003391} = 4.72,$$

which is significant at the 5% level. This shows that the number of dimensions of the prediction formula is not 3.

TABLE 4g.3γ. Test of the Hypothesis $\beta_1 = \beta_2 = \beta_3 = 1$

	D.F.	S.S.	M.S.	F
Deviation from $\beta_1 = \beta_2 = \beta_3 = 1$	3	0.00258	0.00086	2.544
Residual	82	0.02781	0.0003391	
Total	85	0.03039		

5. It is often desirable to test whether the inclusion of an extra variable increases the accuracy of prediction. For instance, in the foregoing example we can test whether H' is necessary in addition to L and B. This is equivalent to testing whether $\beta_3 = 0$. The estimate $b_3 = 0.733$ has the variance $C_{33}\sigma^2$. The ratio on 1 and 82 D.F. is

$$\frac{b_3{}^2}{C_{33}\hat{\sigma}^2} = \frac{(0.733)^2}{39.88} \times \frac{1}{0.0003391}$$

$$= \frac{0.01347}{0.0003391} = 39.72,$$

where for $\hat{\sigma}^2$ the estimate based on 82 D.F., is used. This is significant at the 1% level, showing that H' is also relevant. If b_3 were not significant, the sum of squares due to b_3,

$$\frac{b_3{}^2}{C_{33}} = 0.01347,$$

could be added to the residual sum of squares 0.02781 to obtain a sum 0.04128 based on 83 D.F., giving the estimate of σ^2 as 0.0004973.

If b_3 is declared to be zero, the best estimates of b_1 and b_2 have to be revised, starting with the equation

$$y = \alpha + \beta_1 x_1 + \beta_2 x_2 .$$

It is, however, not necessary to start afresh. The C matrix

$$\begin{matrix} 64.21 & -15.57 & -10.49 \\ -15.57 & 41.71 & -9.00 \\ -10.49 & -9.00 & 39.88 \end{matrix}$$

is reduced by the method of pivotal condensation, starting from the last row and using C_{33} as the pivot. Thus

$$\begin{bmatrix} 64.21 - \dfrac{(10.49)^2}{39.88} & -15.57 - \dfrac{(9.00)(10.49)}{39.88} \\ -15.57 - \dfrac{(9.00)(10.49)}{39.88} & 41.71 - \dfrac{(9.00)^2}{39.88} \end{bmatrix} = \begin{bmatrix} 61.45 & -17.94 \\ -17.94 & 39.68 \end{bmatrix}$$

gives the reduced C matrix for the evaluation of b_1 and b_2.

$$b_1 = 61.45S_{01} - 17.94S_{02} = 1.071$$

$$b_2 = -17.94S_{01} + 39.68S_{02} = 1.206$$

$$a = \bar{y} - b_1\bar{x}_1 - b_2\bar{x}_2 = -1.864$$

The residual sum of squares is

$$\sum (y - \bar{y})^2 - b_1 S_{01} - b_2 S_{02} = 0.0004973,$$

which agrees with the value obtained by adding the sum of squares due to b_3 to the residual, thus providing a check on the calculation of b_1 and b_2. The variance-covariance matrix of b_1, b_2 is σ^2 times the new C matrix.

If more variables are omitted, the method of pivotal condensation has to be carried further. The reduced matrix at each stage gives the C matrix appropriate to the retained variables. It would avoid some confusion if the C matrix could be written in the order in which the variables are eliminated before attempting the method of pivotal condensation. Thus if x_1, x_2, x_3, x_4 are the original variables and if x_2 and x_4 are to be eliminated, we may write the C matrix as

$$\begin{matrix} C_{22} & C_{24} & C_{21} & C_{23} \\ C_{42} & C_{44} & C_{41} & C_{43} \\ C_{12} & C_{14} & C_{11} & C_{13} \\ C_{32} & C_{34} & C_{31} & C_{33} \end{matrix}$$

which is obtained by bringing the second and fourth rows and columns to the first two positions. Now this matrix can be reduced in two stages by the method of forward pivotal condensation.

6. *Confidence interval for the value in an individual case.* Let x_1, x_2, x_3 be the observed values of independent variables in an individual case and y the value to be predicted. Consider $y - Y(x_1, x_2, x_3)$ where $Y(x_1, x_2, x_3)$ is the value of estimated regression for given values of x_1, x_2, x_3.

$$\begin{aligned} V(y - Y) &= \sigma^2 + V(Y) \\ V(Y) &= V[\bar{y} + b_1(x_1 - \bar{x}) + \cdots + b_3(x_3 - \bar{x})] \\ &= \sigma^2\left\{\frac{1}{n} + \sum\sum (x_i - \bar{x}_i)(x_j - \bar{x}_j)C_{ij}\right\}. \end{aligned}$$

Hence

$$t = (y - Y) \div s\left\{1 + \frac{1}{n} + \sum\sum (x_i - \bar{x}_i)(x_j - \bar{x}_j)C_{ij}\right\}^{1/2}$$

has a t distribution on $(n - p - 1)$ D.F., corresponding to the estimate s of σ. We substitute for t, the upper and lower $(\alpha/2)$ probability points of t and obtain the corresponding two values of y, which provide a $(1 - \alpha)$ confidence interval for the unknown value. A skull with $L = 198.5$, $B = 147$,

$H' = 131$, that is, $x_1 = 2.298$, $x_2 = 2.167$, $x_3 = 2.117$, has the value of regression

$$Y = \bar{y} + b_1(x_1 - \bar{x}_1) + \cdots + b_3(x_3 - \bar{x}_3) = 3.2069$$

$$s^2\left\{1 + \frac{1}{n} + \sum\sum (x_i - \bar{x}_i)(x_j - \bar{x}_j)C_{ij}\right\} = 0.0003391(1.04187)$$

$$= 0.0003533.$$

To find 0.95 confidence interval, the equations are

$$\frac{y - 3.2069}{(0.0003533)^{1/2}} = \pm 2.00,$$

where 2.00 is approximately the 5% value of $|t|$ on 84 D.F., or the 0.95 confidence interval for the individual's log capacity is

$$3.2069 \pm 2.00(0.01880) = (3.1693,\ 3.2445).$$

It is an important property that from a confidence interval for y, the interval for any one-to-one function of y can be obtained. Hence the 0.95 confidence interval for the capacity (= antilog y) is (1476, 1755). The interval is somewhat wide, indicating that prediction in any individual case may not be good.

7. *Estimation of mean of the dependent variable (log capacity).* If the problem is one of estimation of the mean log capacity of skulls with L, B, H' as chosen previously then the estimate is obtained as

$$Y = \bar{y} + b_1(x_1 - \bar{x}_1) + \cdots + b_3(x_3 - \bar{x}_3) = 3.2069$$

with

$$V(Y) = \sigma^2\left\{\frac{1}{n} + \sum\sum (x_i - \bar{x}_i)(x_j - \bar{x}_j)C_{ij}\right\}$$

$$= 0.0003391(0.04187) = 0.00001420 \quad (\text{using } s \text{ for } \sigma).$$

The 0.95 confidence interval is

$$3.2069 \pm 2.00(0.0000142)^{1/2} = (3.1994,\ 3.2144),$$

where 2.00 is the 5% value of $|t|$ on 84 D.F. Unfortunately, the confidence interval for mean capacity cannot be obtained from the confidence interval for mean of log capacity. This difficulty would not arise, if capacity itself had linear regression on the concomitant measurements.

It is seen that in the preceding formula for variance of the estimated value Y, the precision of the estimate depends on the closeness of x_1, x_2, x_3 to the averages $\bar{x}_1$, $\bar{x}_2$, $\bar{x}_3$ realized in the sample from which the prediction formula was estimated. In fact, the variance is least (equal to σ^2/n) for the estimate when the measurements on the new specimen coincide with the average values in the earlier sample. The accuracy of prediction diminishes as x_1, x_2, x_3 depart more and more from the average values, and the prediction may become completely unreliable if x_1, x_2, x_3 fall outside the range of values observed in the sample.

For prediction with a single variable, the regression equation is

$$Y = a + b(x - \bar{x})$$

and

$$V(Y) = \left\{\frac{1}{n} + \frac{(x - \bar{x})^2}{S_{11}}\right\}\sigma^2,$$

where S_{11} is the corrected sum of squares for x in the observed sample. As before, accuracy is higher for prediction near the mean. The formula also depends on S_{11}, the scatter of x in the sample; the larger the scatter, the higher the precision of the estimate for any x. Therefore, in choosing the sample for the construction of the prediction formula, we should observe the values of x at the extremities of its range if we want to construct a best prediction formula. This is, no doubt, a theoretically sound policy which can be carried out with advantage when it is known for certain that the regression equation is of the linear form. But, data collected in such a manner are not suitable for judging whether the regression is linear or not, and there is no reason to believe that linearity of regression is universal. The biometric experience is that the regressions are very nearly linear, deviations from linearity being detectable only in large samples. If this is so, the regression line fitted to the data is only an approximation to the true regression function, and the data should allow a closest possible fit of the straight line to the ideal curve. The best plan for this is to choose x from all over its range, or, preferably, to choose x at random so that different values of x may occur with their own probability and exert their influence in the determination of the straight-line fit.

Estimation of Mean Capacity of a Population. The prediction formula obtained can also be used to estimate the mean cranial capacity from a sample of skulls on which only measurements of L, B, H' are available. For this purpose, two methods are available. We may estimate the cranial capacity of each individual skull and calculate the mean of these estimates, or we may apply the formula directly to the mean values of L, B, H' for the sample. It is of interest to know whether these two methods give the same results. For this purpose estimates were made of the mean cranial capacity of an additional 29 male skulls of the Farringdon Street series for which measurements of L, B, H' but not of C were available.

For these 29 skulls the mean of L is 191.1, of B is 143.1, and of H' is 129.0. Applying the formula $C = 0.00241\ L^{0.878}\ B^{1.041}\ H'^{0.733}$ to these mean values, we estimate the mean of C to be 1498.3. If we estimate C for the 29 skulls individually and take the mean of the 29 estimates, we have an estimate of the mean value of C equal to 1498.2.

The same estimates were calculated for the 22 male skulls of the Moorfields series (Hooke, 1926), for which all four measurements were available. For these 22 skulls the mean of L is 189.5, of B is 142.5, and of H' is 128.8, giving an estimate of the mean of C equal to 1479.0. If we estimate C for the 22 skulls individually and calculate the mean, we get an estimate of the mean of C equal to 1480.0. Thus it appears that the two methods give very nearly the same estimates.

Are Only Small Skulls Preserved? A point of some interest is that, whereas the mean value of C for Farringdon Street series as calculated from 86 measured values is 1481.3, the mean value of C as estimated by our formula from the 29 skulls for which measurements of L, B, H' but not of C are available is 1498.3.

Again, for the Moorfields series the mean of L is 189.2 based on 44 measurements, of B is 143.0 based on 46 measurements, and of H' is 129.8 based on 34 measurements. Applying our formula to these mean values (as we may do with some confidence as shown earlier), we obtain an estimate of the mean of C equal to 1490.7. The mean of C as calculated from 22 measured values on well-preserved skulls is only 1473.8.

These results suggest that the skulls that are damaged to such an extent that the cranial capacity cannot be measured are on the whole larger than those that remain intact.

This raises a serious issue. Are not the published mean values of cranial capacities gross underestimates? Can a suitable method be suggested to correct these values? One way would be to use the samples that provide observations on $C, L, B,$ and H' for merely constructing the prediction formula. As observed earlier, the prediction formula, provided the nature of the regression function used is appropriate, could be obtained from samples providing observations on all the measurements although the samples are not drawn at random from the population. For instance, if only small skulls are preserved, the measurements obtained are not strictly random from the population of skulls. Such material is being used just to establish a relationship. Having obtained this formula, the mean values of L, B, H' computed from all the available measurements may be substituted to obtain an estimate of the mean capacity. This value will be higher than the average of the available measurements of the cranial capacity but less than the predicted value on the basis of mean L, B, H' from skulls providing these measurements only.

The extent of underestimation depends on the proportion of the disintegration of the large skulls. This may vary from series to series, and hence for a proper comparison of the mean capacities the correction indicated previously may have to be applied.

4g.4 Test for the Equality of Regression Equations

It is very often necessary to test whether regression functions constructed from two series are the same. Thus if the formula for the prediction of the cranial capacity from a different series is

$$\alpha' + \beta_1' x_1 + \beta_2' x_2 + \beta_3' x_3,$$

two types of hypotheses may be tested:

(i) $\alpha = \alpha' \quad \beta_1 = \beta_1' \quad \beta_2 = \beta_2' \quad \beta_3 = \beta_3'$

(ii) $\beta_1 = \beta_1' \quad \beta_2 = \beta_2' \quad \beta_3 = \beta_3'$, irrespective of whether α equals α' or not.

If (i) is true, the whole regression function is the same in both series; if (ii) is true, the regression functions are the same apart from a change in the constant. These two hypotheses are relevant because many problems arise where a prediction formula constructed from one series may have to be used for a specimen from an entirely different series. An extreme and rather ambitious case of such a use is the prediction of stature of prehistoric men from the length of fossil femur by using a formula connecting the stature of modern man with the length of his long bones (K. Pearson, 1898). Some sort of justification for such a procedure will be available if the first hypothesis is proved to be correct in analogous situations. We first deal with the test procedures when the prediction formulas are estimated for both the series.

Let the derived quantities for the second series be:

Sample size: n'

Mean values: $\bar{x}_1', \bar{x}_2', \bar{x}_3'$, and $\bar{y}'$

Corrected sums of products:

$$\sum (x_{ir}' - \bar{x}_i')(x_{jr}' - \bar{x}_j') = S_{ij}'$$

$$\sum (x_{ir}' - \bar{x}_i')(\bar{y}_r' - \bar{y}') = S_{0i}'.$$

$$\sum (y_r' - \bar{y}')^2 = S_{00}'$$

These are sufficient to determine the regression function

$$y = a' + b_1' x_1 + b_2' x_2 + b_3' x_3.$$

The residual sum of squares

$$R_0^2 = S'_{00} - b'_1 S'_{01} - b'_2 S'_{02} - b'_3 S'_{03} \qquad \text{(for the second sample)}$$

$$+ S_{00} - b_1 S_{01} - b_2 S_{02} - b_3 S_{03} \qquad \text{(for the first sample)}$$

has $(n' - 4) + (n - 4) = (n + n' - 8)$ D.F., where 4 represents the number of constants estimated.

We now throw the two samples together and consider them as a single sample of size$(n + n')$ and determine the regression line and residual sum of squares. The necessary quantities can be computed from those already available:

Sample size: $n + n'$

Mean values:

$$\frac{(n\bar{x}_i + n'\bar{x}'_i)}{(n + n')} = \bar{x}''_i$$

Corrected sums of products:

$$S''_{ij} = S_{ij} + S'_{ij} + \frac{nn'}{n + n'}(\bar{x}_i - \bar{x}'_i (\bar{x}_j - \bar{x}'_j)$$

$$S''_{0i} = S_{0i} + S'_{0i} + \frac{nn'}{n + n'}(\bar{x}_i - \bar{x}'_i)(\bar{y} - \bar{y}')$$

$$S''_{00} = S_{00} + S'_{00} + \frac{nn'}{n + n'}(\bar{y} - \bar{y}')^2$$

If b''_1, b''_2, ... are the regression coefficients, then the residual sum of squares R_1^2 on $(n + n' - 4)$ D.F. is

$$S''_{00} - b''_1 S''_{01} - b''_2 S''_{02} - b''_3 S''_{03}.$$

We set up the analysis of variance as in Table 4g.4α.

TABLE 4g.4α. Analysis of Variance for Testing Equality of Regression Coefficients

Due to	D.F.	S.S.
Deviation from hypothesis	4	* (by subtraction)
Separate regressions (residual)	$n + n' - 8$	R_0^2
Common regression (residual)	$n + n' - 4$	R_1^2

The significance of the ratio of mean squares due to deviation from hypothesis to residual due to separate regressions is tested.

If the object is to test for the equality of the b coefficients only, we calculate the quantities

$$S'''_{0i} = S_{0i} + S'_{0i} \qquad S'''_{ij} = S_{ij} + S'_{ij}$$

and obtain the constants b'''_1, b'''_2, b'''_3 from the equations

$$S'''_{0i} = b'''_1 S'''_{1i} + b'''_2 S'''_{2i} + b'''_3 S'''_{3i}, \qquad i = 1, 2, 3,$$

and find the residual sum of squares $R_2{}^2$ on $(n + n' - 5)$ D.F., where 5 represents the number of regression coefficients plus two for different constants in the series.

$$R_2{}^2 = \sum (y_r - \bar{y})^2 + \sum (y'_r - \bar{y}')^2 - b'''_1 S'''_{01} - b'''_2 S'''_{02} - b_3 S'''_{03}.$$

The test depends on the variance ratio

$$\frac{R_2{}^2 - R_0{}^2}{3} \div \frac{R_0{}^2}{n + n' - 8},$$

on 3 and $(n - n' - 8)$ D.F., where 3 represents the number of regression coefficients under comparison.

The extension of analysis for any number of predictor variables and comparing any number of regression functions is fairly clear.

In biological data it is often found that the mutual correlations and variabilities of measurements are approximately the same for all allied series, in which case the coefficients β_1, β_2, β_3 in the regression formula will not differ much. On the other hand, the mean values differ to some extent from series to series, in which case the constant term will not be the same. This leads us to consider a different problem whether $\alpha = \alpha'$ given $\beta_1 = \beta'_1$, $\beta_2 = \beta'_2, \ldots$. A test for this can be immediately obtained from the sums of squares calculated earlier. The suitable statistic is the variance ratio

$$\frac{R_1{}^2 - R_2{}^2}{1} \div \frac{R_2{}^2}{n + n' - 5}$$

on 1 and $(n + n' - 5)$ degrees of freedom. If the foregoing hypothesis is true, the difference in the mean value of y could be completely explained by differences in the other variables x_1, x_2, x_3. It appears that when a sufficient number of measurements is considered, the extra difference contributed by any other measurement independently of the set already considered is negligibly small. In such situations the equality of the dispersion matrix in both series is sufficient to ensure the equality of the regression functions as whole. Much caution is necessary when the prediction formula based on one or two variables is so used. Such a statistical adventure undertaken by Karl Pearson in predicting the stature of prehistoric men is justifiable, however, if we agree with the last statement of his article (K. Pearson, 1898).

"No scientific investigation is final; it merely represents the most probable conclusion which can be drawn from the data at the disposal of the writer. A wider range of facts, or more refined analysis, experiment, and observation will lead to new formulas and new theories. This is the essence of scientific progress."

4g.5 The Test for an Assigned Regression Function

In **4g.3** it was assumed that the regression function for log capacity is linear in the logarithms of length, breadth, and height. If, at least for some given sets of values of the independent variables, multiple observations on the dependent variable have been recorded, the validity of such an assumption can be tested.

Table 4g.5α gives the mean values of nasal index of people living in various parts of India together with the mean annual temperature and relative humidity of the places. The corrected total sum of squares for nasal index has been found to be 11,140.209 on (344-1) D.F.

TABLE 4g.5α. Nasal Index of the Inhabitants and the Temperature and Humidity of the Region

Region	Sample size (n)	Nasel Index: Mean ($\bar{y}$)	Nasel Index: Total	Temperature: Mean Annual ($\bar{t}$)	Temperature: Total ($n\bar{t}$)	Relative Humidity: Mean Annual ($\bar{h}$)	Relative Humidity: Total ($n\bar{h}$)
Assam	36	83.0	2,988.0	72.6	2,613.6	85	3.060
Orissa	40	80.4	3,216.0	80.3	3,212.0	69	2,760
Bihar	30	80.1	2,403.0	74.8	2,244.0	88	2,640
Malabar	45	77.0	3,465.0	80.2	3,609.0	81	3,645
Bombay	26	76.2	1,981.2	77.6	2,017.6	66	1,716
Madras	35	75.9	2,656.5	81.8	2,863.0	76	2,660
Punjab	28	71.4	1,999.2	76.4	2,139.2	63	1,764
United Province	32	80.8	2,585.6	77.2	2,470.4	69	2,208
Andhra	41	76.8	3,148.8	80.3	3,292.3	69	2,829
Ceylon	31	80.3	2,490.0	80.2	2,486.2	82	2,542
Total	344		26,933.3		26,947.3		25,824
Mean		78.294		78.335		75.070	

If no assumption is made about the regression of nasal index on temperature and relative humidity, the analysis of variance between and within groups is obtained as in Table 4g.5β. The mean square between groups is

TABLE 4g.5β. Analysis of Variance for Nasal Index

	D.F.	S.S.		M.S.
Between groups	9	$\sum (\bar{y})(n\bar{y}) - \frac{\sum n\bar{y}}{\sum n} \sum n\bar{y}$	= 3,169.900	352.21
Within groups	334	* (by subtraction)	7,970.309	23.863
Total	343		11,140.209	

very large, indicating real differences in nasal index. Can these differences be explained by a linear regression of nasal index y on temperature t and relative humidity h of the form

$$y = \alpha + \beta_1 t + \beta_2 h?$$

The normal equations leading to the estimates b_1, b_2 of β_1, β_2 are

$$S_{01} = b_1 S_{11} + b_2 S_{12}$$
$$S_{02} = b_1 S_{12} + b_2 S_{22}$$

where

$$S_{01} = \sum (n\bar{y}\bar{t}) - \frac{(\sum n\bar{t})(\sum n\bar{y})}{\sum n} = -1,072.40$$

$$S_{02} = \sum (n\bar{y}\bar{h}) - \frac{(\sum n\bar{h})(\sum n\bar{y})}{\sum n} = -4,586.57$$

$$S_{12} = \sum (n\bar{t}\bar{h}) - \frac{(\sum n\bar{t})(\sum n\bar{h})}{\sum n} = -2,334.91$$

$$S_{11} = \sum (n\bar{t}\bar{t}) - \frac{(\sum n\bar{t})(\sum n\bar{t})}{\sum n} = 2,721.06$$

$$S_{22} = \sum (n\bar{h}\bar{h}) - \frac{(\sum n\bar{h})(\sum n\bar{h})}{\sum n} = 22,042.32$$

Solving the foregoing equations, we obtain b_1 and b_2 as

$$b_1 = -0.237113 \qquad b_2 = 0.182963.$$

With these values the regression analysis can be set up as the Table 4g.5γ. If the hypotheses concerning the regression is true, the mean square obtained from "Within groups" of Table 4g.4β and "Residual about regression" of Table 5g.5γ will be of the same magnitude. A significant difference would disprove the hypothesis.

The ratio 12.4 on 7 and 334 D.F. is significant at the 1% level, so that the regression of nasal index on temperature and relative humidity cannot be considered linear. It is also seen from Table 4g.5γ that the variance ratio 18.55 on 2 and 341 degrees of freedom is significant, but this may not mean

TABLE 4g.5γ. Regression Analysis

	D.F.	S.S.		M.S.	F
Due to regression	2	$b_1 S_{01} + b_2 S_{02} =$	1,093.45	546.72	18.55
Residual about regression	341	* (by subtraction)	10,046.759	29.463	
Total	343		11,140.209		

that nasal index depends entirely on the weather conditions of the place where the individuals live. The observed differences may be more complex than can be explained by weather differences, or the nature of dependence on weather may itself be very complicated.

In some cases, as in the distribution of heights of father and daughter, it may be desired to test whether the regression of one variable on the other is linear. For such a test, the range of the independent variable has to be divided into a suitable number of class intervals and the variance of the dependent variable analyzed between and within classes. To find an estimate of within variation it is necessary that at least some of the classes contain more than one observation. The regression analysis can be done without the use of the class intervals, or if the data are already grouped the midpoint of the class interval is taken as the value of the independent variable for each observation of the dependent variable in that class. The final test can be carried out as in Table 4g.5δ.

For further examples the reader is referred to books by Fisher (1925), Linnik (1961), Plackett (1960), Scheffe (1959), Williams (1959), and Yates (1949). Applications to design of experiments are contained in papers by Box (1954), Rao (1947c, 1956a, 1971B2), and books by Cochran and Cox (1957) and Kempthorne (1952).

TABLE 4g.5δ. Test for the Specified Regression Function

	D.F.	S.S.	M.S.	F
Deviation from specified regression	7	2,076.45*	296.636	12.4
Within groups	334	7,970.309	23.863	
Residual about regression	341	10.046,759		

* Obtained by subtraction

4g.6 Restricted Regression

In some selection problems (see Kempthorne and Nordskog, 1959) we need a linear selection index based on concomitant (or predictor) variables $X_1, \ldots,$ X_p such that in the selected population we have maximum gain in a criterion variable Y_1 and *status quo* is maintained in other criterion variables Y_2, $\ldots$, Y_k. Or, in other words, we need to find a linear function (an index), $I = \mathbf{L}'\mathbf{X} = L_1 X_1 + \cdots + L_p X_p$ such that $\rho_{IY_1}^2$ is a maximum subject to the conditions $\rho_{IY_2} = \cdots = \rho_{IY_k} = 0$, where ρ_{UV} indicates the correlation between the variables U and V.

Let $\mathbf{C}_i$ be the vector of covariances between Y_i and $(X_1, \ldots, X_p)$. Then the algebraic problem is one of maximizing $(\mathbf{L}'\mathbf{C}_1)^2$ subject to $\mathbf{L}'\mathbf{C}_2 = \cdots =$ $\mathbf{L}'\mathbf{C}_k = 0$ and $\mathbf{L}'\mathbf{\Sigma}\mathbf{L} = 1$, where $\mathbf{\Sigma}$ is the dispersion matrix of $\mathbf{X}$. Consider the expression

$$(\mathbf{L}'\mathbf{C}_1)^2 - \lambda_1(\mathbf{L}'\mathbf{\Sigma}\mathbf{L} - 1) - 2\lambda_2 \mathbf{L}'\mathbf{C}_2 + \cdots - 2\lambda_k \mathbf{L}'\mathbf{C}_k \tag{4g.6.1}$$

where $\lambda_1, \ldots, \lambda_k$ are Lagragian multipliers. Differentiating with respect to $\mathbf{L}$ and equating to zero we have

$$\lambda_1 \mathbf{\Sigma}\mathbf{L} = \mu_1 \mathbf{C}_1 - \lambda_2 \mathbf{C}_2 - \cdots - \lambda_k \mathbf{C}_k \tag{4g.6.2}$$

where μ_1 is written for $\mathbf{L}'\mathbf{C}_1$, or the optimum $\mathbf{L}$ is proportional to

$$\mathbf{L} \propto \mathbf{\Sigma}^{-1}(\mathbf{C}_1 - \mu_2 \mathbf{C}_2 - \cdots - \mu_k \mathbf{C}_k). \tag{4g.6.3}$$

Expressing the conditions $\mathbf{L}'\mathbf{C}_i = 0$, we have the equations

$$\mathbf{C}_i' \mathbf{\Sigma}^{-1}\mathbf{C}_2 \mu_2 + \cdots + \mathbf{C}_i'\mathbf{\Sigma}^{-1}\mathbf{C}_k \mu_k = \mathbf{C}_i'\mathbf{\Sigma}^{-1}\mathbf{C}_1 \tag{4g.6.4}$$
$$i = 2, \ldots, k$$

satisfied by $\mu_2, \ldots, \mu_k$. If $\hat{\mu}_2, \ldots, \hat{\mu}_k$ is a solution then the selection index may be written as

$$I = (\mathbf{C}_1' - \hat{\mu}_2 \mathbf{C}_2' - \cdots - \hat{\mu}_k \mathbf{C}_k')\mathbf{\Sigma}^{-1}\mathbf{X} \tag{4g.6.5}$$

which may be standardized to have variance unity, if necessary, by multiplying by a constant.

The more general problem of maximizing $\mathbf{L}'\mathbf{C}_1$ subject to $\mathbf{L}'\mathbf{C}_i \geqslant 0$, $i = 2, \ldots, k$ and $\mathbf{L}'\mathbf{\Sigma}\mathbf{L} = 1$ is solved in Rao (1962d, 1964a).

4h THE GENERAL PROBLEM OF LEAST SQUARES WITH TWO SETS OF PARAMETERS

4h.1 Concomitant Variables

Suppose that the growth rates of groups of animals receiving different diets are to be compared. The observed differences in growth rates can be attributed to diet only if all the animals treated are similar in some observable aspects such as age, initial weight, or parentage, which influence the growth rate. In fact, if the groups of animals receiving different diets differ in these aspects, it is desirable to compare the growth rates after eliminating these differences.

It may be noted, however, that no bias is introduced in the experiment if the animals which might differ in these aspects are assigned at random to the groups to be treated differently. This procedure enhances the residual variation calculable from the differences in the growth rates of animals receiving the same treatment and thus decreases the efficiency of the experiment.

If the magnitudes of these additional variables are known, it is possible to eliminate the differences caused by them independently of the treatments both from within and between groups and to test for the pure effects of the treatments with greater efficiency. The computational technique relating to this process is known as the adjustment for concomitant variation.

For significant reduction in the residual variation, it must be known that the effects under study are influenced by the concomitant variables. This is important in experimental studies where due consideration is to be given to the cost and time involved in recording the concomitant variables, and can be tested as shown in the example considered in **4h.3**.

On the other hand, the concomitant variables chosen must not have been influenced by the treatments under consideration. Sometimes, in assessing the differences in yields of plants treated differently, concomitant variables such as the number of branches or the quantity of straw are chosen. Some caution is necessary in adopting such a procedure.

4h.2 Analysis of Covariance

Assuming that the regression of y on the concomitant variables $c_1, \ldots, c_k$ to be linear, the observational equations containing the parameters $(\beta_1, \ldots, \beta_m) = \boldsymbol{\beta}'$ under consideration and the regression coefficients $(\gamma_1, \ldots, \gamma_k) = \boldsymbol{\gamma}'$ can be written

$$E(y_i) = x_{i1}\beta_1 + \cdots + x_{im}\beta_m + \gamma_1 c_{1i} + \cdots + \gamma_k c_{ki}$$
$$i = 1, \ldots, n,$$

or in the matrix notation

$$E(\mathbf{Y}) = \mathbf{X}\boldsymbol{\beta} + \mathbf{C}\boldsymbol{\gamma}, \tag{4h.2.1}$$

where $\mathbf{X}$ is the design matrix as before and $\mathbf{C}$ is an $n \times k$ matrix with its ith column as $\mathbf{C}_i$ representing the n observations on the ith concomitant variable. If, therefore, we consider the observations on concomitant variables as fixed, the setup (4h.2.1) is the same as that of the theory of least squares with $(m + k)$ unknown parameters $(\boldsymbol{\beta}' \vdots \boldsymbol{\gamma}')$, a new "design matrix" $(\mathbf{X} \vdots \mathbf{C})$, and the observation vector $\mathbf{Y}$.

No new problem arises in adjustment for concomitant variables either in the estimation of parameters or testing of linear hypotheses. We shall present a computational scheme, however, which enables us to carry out tests of a number of hypotheses of interest in practical situations. The normal equations corresponding to the observational equations (4h.2.1) are

$$\left.\begin{aligned} \mathbf{X}'\mathbf{X}\boldsymbol{\beta} + \mathbf{X}'\mathbf{C}\boldsymbol{\gamma} &= \mathbf{X}'\mathbf{Y} \\ \mathbf{C}'\mathbf{X}\boldsymbol{\beta} + \mathbf{C}'\mathbf{C}\boldsymbol{\gamma} &= \mathbf{C}'\mathbf{Y} \end{aligned}\right\}. \tag{4h.2.2}$$

Let $\hat{\boldsymbol{\beta}}_0$ be a solution of the normal equation

$$\mathbf{X}'\mathbf{X}\boldsymbol{\beta}_0 = \mathbf{X}'\mathbf{Y},$$

ignoring the concomitant variables, and let $\hat{\boldsymbol{\beta}}_i$ be a solution of the normal equation

$$\mathbf{X}'\mathbf{X}\boldsymbol{\beta}_i = \mathbf{X}'\mathbf{C}_i,$$

considering the values of the ith concomitant variable instead of $\mathbf{Y}$ (i.e., substituting $\mathbf{C}_i$ for $\mathbf{Y}$). Define the residual sum of squares and products

$$\begin{aligned}
R_{00} &= (\mathbf{Y} - \mathbf{X}\hat{\boldsymbol{\beta}}_0)'(\mathbf{Y} - \mathbf{X}\hat{\boldsymbol{\beta}}_0) = \mathbf{Y}'\mathbf{Y} - \hat{\boldsymbol{\beta}}_0'\mathbf{X}'\mathbf{Y} & & \\
R_{ii} &= (\mathbf{C}_i - \mathbf{X}\hat{\boldsymbol{\beta}}_i)'(\mathbf{C}_i - \mathbf{X}\hat{\boldsymbol{\beta}}_i) = \mathbf{C}_i'\mathbf{C}_i - \hat{\boldsymbol{\beta}}_i\mathbf{X}'\mathbf{C}_i, & i &= 1, 2, \ldots, k \\
R_{0i} &= (\mathbf{Y} - \mathbf{X}\hat{\boldsymbol{\beta}}_0)'(\mathbf{C}_i - \mathbf{X}\hat{\boldsymbol{\beta}}_i) = \mathbf{Y}'\mathbf{C}_i - \hat{\boldsymbol{\beta}}_0\mathbf{X}'\mathbf{C}_i, & i &= 1, 2, \ldots, k \\
R_{ij} &= (\mathbf{C}_i - \mathbf{X}\hat{\boldsymbol{\beta}}_i)'(\mathbf{C}_j - \mathbf{X}\hat{\boldsymbol{\beta}}_j) = \mathbf{C}_i'\mathbf{C}_j - \hat{\boldsymbol{\beta}}_i\mathbf{X}'\mathbf{C}_j, & i, j &= 1, \ldots, k
\end{aligned} \tag{4h.2.3}$$

Multiplying the first equation of (4h.2.2) by $\hat{\boldsymbol{\beta}}_i'$ and subtracting from ith component of the second equation of (4h.2.2) we obtain equations

$$R_{1i}\gamma_1 + \cdots + R_{ki}\gamma_k = R_{0i} \qquad (4h.2.4)$$
$$i = 1, \ldots, k$$

for the regression parameters only. If $\hat{\gamma}$ is a solution of (4h.2.4), the solution $\hat{\boldsymbol{\beta}}$ satisfying the normal equations (4h.2.2) is seen to be

$$\hat{\boldsymbol{\beta}} = \hat{\boldsymbol{\beta}}_0 - \hat{\gamma}_1\hat{\boldsymbol{\beta}}_1 - \cdots - \hat{\gamma}_k\hat{\boldsymbol{\beta}}_k,$$

which gives the adjustment to be made in $\hat{\boldsymbol{\beta}}_0$ due to concomitant variables. The residual sum of squares appropriate to the setup (4h.2.1) is then

$$R_0^2 = R_{00} - \hat{\gamma}_1 R_{01} - \cdots - \hat{\gamma}_k R_{0k} \qquad (4h.2.5)$$

on $(n - r - k)$ D.F., where $r = \text{rank } \mathbf{X}$ (assuming $|\mathbf{C}'\mathbf{C}| \neq 0$).

Test of the Hypothesis $\boldsymbol{\gamma} = \mathbf{0}$. The adjustment for concomitant variables is not profitable if $\boldsymbol{\gamma} = \mathbf{0}$, and it is therefore important to test this hypothesis. Let us recognize that

$$R_1^2 = \min_{\boldsymbol{\gamma}=\mathbf{0},\boldsymbol{\beta}} (\mathbf{Y} - \mathbf{X}\boldsymbol{\beta} - \mathbf{C}\boldsymbol{\gamma})'(\mathbf{Y} - \mathbf{X}\boldsymbol{\beta} - \mathbf{C}\boldsymbol{\gamma}) = R_{00},$$

so that the S.S. due to deviation from hypothesis is

$$R_1^2 - R_0^2 = \hat{\gamma}_1 R_{01} + \cdots + \hat{\gamma}_k R_{0k} \qquad (4h.2.6)$$

on k D.F. The variance ratio for testing $\boldsymbol{\gamma} = \mathbf{0}$ is

$$\frac{R_1^2 - R_0^2}{k} \div \frac{R_0^2}{n - r - k}. \qquad (4h.2.7)$$

Test of the Hypothesis $\mathbf{H}'\boldsymbol{\beta} = \boldsymbol{\xi}$. We have to find

$$R_2^2 = \min_{\mathbf{H}'\boldsymbol{\beta}=\boldsymbol{\xi}} (\mathbf{Y} - \mathbf{X}\boldsymbol{\beta} - \mathbf{C}\boldsymbol{\gamma})'(\mathbf{Y} - \mathbf{X}\boldsymbol{\beta} - \mathbf{C}\boldsymbol{\gamma}).$$

Let R_{00}', R_{0i}', R_{ij}' be the residual sum of squares and products as in (4h.2.3) when $\boldsymbol{\beta}$ is subject to the restriction $\mathbf{H}'\boldsymbol{\beta} = \boldsymbol{\xi}$. Let $\hat{\gamma}_1', \ldots, \hat{\gamma}_k'$ be solutions using R_{ij}' in the equations (4h.2.4). Then

$$R_2^2 = R_{00}' - \hat{\gamma}_1' R_{01}' - \cdots - \hat{\gamma}_k' R_{0k}'. \qquad (4h.2.8)$$

The variance ratio test is

$$\frac{R_2^2 - R_0^2}{s} \div \frac{R_0^2}{n - r - k}, \qquad (4h.2.9)$$

where s is the D.F. of the hypothesis $\mathbf{H}'\boldsymbol{\beta} = \boldsymbol{\xi}$ and R_0^2 is the pure residual sum of squares (4h.2.5) without any hypothesis on $\boldsymbol{\beta}$ or γ. This completes the formal theory. The method is further explained in the illustrative example of **4h.3**.

4h.3 An Illustrative Example

The following data relate to the initial weights and growth rates of 30 pigs classified according to pen, sex, and type of food given.

The problem is to study the effect of food after eliminating the initial weight.

TABLE 4h.3α. Data for Analysis (Wishart, 1938)

Pen	Treatment	Sex	Initial Weight (w)	Growth Rate in Pounds per Week (g)
I	A	G	48	9.94
	B	G	48	10.00
	C	G	48	9.75
	C	H	48	9.11
	B	H	39	8.51
	A	H	38	9.52
II	B	G	32	9.24
	C	G	28	8.66
	A	G	32	9.48
	C	H	37	8.50
	A	H	35	8.21
	B	H	38	9.95
III	C	G	33	7.63
	A	G	35	9.32
	B	G	41	9.34
	B	H	46	8.43
	C	H	42	8.90
	A	H	41	9.32
IV	C	G	50	10.37
	A	H	48	10.56
	B	G	46	9.68
	A	G	46	10.98
	B	H	40	8.86
	C	H	42	9.51
V	B	G	37	9.67
	A	G	32	8.82
	C	G	30	8.57
	B	H	40	9.20
	C	H	40	8.76
	A	H	43	10.42

The first step in the analysis is to analyze the sum of squares of both the dependent and independent variables and also the sum of products. The analysis of the sum of products is done in the same manner as the analysis of the sum of squares by adopting the rule that the square of a variable in the latter is replaced by the product of the variables involved in the former.

The total sums of squares and products are

$$\left.\begin{aligned} \sum g^2 - \bar{g}\sum g &= 16.6068 \\ \sum wg - \bar{w}\sum g &= 78.979 \\ \sum w^2 - \bar{w}\sum w &= 1108.70 \end{aligned}\right\} \quad 29 \text{ D.F.}$$

If w_i and g_i denote the totals of 6 observations for the ith pen, the sums of squares and products for pens are

$$\left.\begin{aligned} \tfrac{1}{6}\sum g_i^2 - \bar{g}\sum g &= 4.8518 \\ \tfrac{1}{6}\sum g_i w_i - \bar{g}\sum w &= 39.905 \\ \tfrac{1}{6}\sum w_i^2 - \bar{w}\sum w &= 605.87 \end{aligned}\right\} \quad 4 \text{ D.F.}$$

Similarly, the sums of squares and products for food and sex can be obtained.

If w_{ij} and g_{ij} denote the total of 5 observations for the ith type of food and jth sex, the sums of squares and products for the joint effects of food and sex are

$$\left.\begin{aligned} \tfrac{1}{5}\sum\sum g_{ij}^2 - \bar{g}\sum g &= 3.2422 \\ \tfrac{1}{5}\sum\sum g_{ij} w_{ij} - \bar{g}\sum w &= -0.885 \\ \tfrac{1}{5}\sum\sum w_{ij}^2 - \bar{w}\sum w &= 59.90 \end{aligned}\right\} \quad 5 \text{ D.F.}$$

If from these the corresponding expressions for food (2 D.F.) and sex (1 D.F.) are subtracted, the expressions for food × sex interaction are obtained.

TABLE 4h.3β. Analysis of Variance and Covariance

	D.F.	g^2	wg	w^2
Pen	4	4.9607	40.324	605.87
Food	2	2.3242	−0.171	5.40
Sex	1	0.4538	−4.813	32.03
Food × Sex	2	0.4642	4.099	22.47
Error	20	$8.4039 = R_{00}$	$39.540 = R_{01}$	$442.93 = R_{11}$
Total	29	16.6068	78.979	1108.70

Table 4h.3β gives the whole analysis. The error line is obtained by subtraction from the total. There is only one regression coefficient to be estimated:

$$R_{11}\gamma = R_{01}$$

$$442.93\gamma = 39.540, \qquad \hat{\gamma} = 0.089269.$$

The ratio is significant at the 1% level so that the comparisons can be made more efficiently by eliminating the concomitant variations.

TABLE 4h.3γ. Test for Regression from Error Line Above

	D.F.	S.S.	M.S.	F
Regression	1	$\hat{\gamma}R_{01} = 3.5279$	3.5297	13.76
Residual	19	$R_{00} - \hat{\gamma}R_{01} = 4.8742$	0.2565	
Total	20	$R_{00} = 8.4039$		

If the hypothesis specifies that there are no differences in food, the residual sums of squares and products are obtained by adding the rows corresponding to food and error:

$$R'_{00} = 10.7281 \qquad R'_{01} = 39.369 \qquad R'_{11} = 448.33.$$

The new regression coefficient is

$$\gamma' R'_{11} = R'_{01}, \qquad \hat{\gamma}' = 39.369/448.83 = 0.087813.$$

The residual sum of squares when the hypothesis is true is

$$R'_{00} - \hat{\gamma}' R'_{01} = 7.2710 \quad \text{on} \quad 21 \text{ D.F.}$$

TABLE 4h.3δ. Test for Differences in Food, Eliminating the Effects of Initial Weight

	D.F.	S.S.	M.S.	F
Food	2	* 2.3968 (by subtraction)	1.1984	4.67
Residual	19	$R_{00} - \hat{\gamma}R_{01} = 4.8742$	0.2565	
Food + Error	21	$R'_{00} - \hat{\gamma}'R'_{01} = 7.2710$		

The ratio is significant at the 5% level. To test for food without adjustment for concomitant variation, we have to construct the ratio on 2 and 20 D.F.

$$F = \left(\frac{2.3242}{2}\right)\left(\frac{20}{8.4039}\right) = 2.76$$

which is not significant at the 5% level. The quantities used to calculate F are taken from the analysis of variance table (Table 4h.3β). The differences caused by food could be detected when the concomitant variation is eliminated. Similarly, any other effect such as sex or interaction can be tested.

4i UNIFIED THEORY OF LINEAR ESTIMATION

4i.1 A Basic Lemma on Generalized Inverse

The problem of estimation of $\boldsymbol{\beta}$ in the model $(\mathbf{Y}, \mathbf{X}\boldsymbol{\beta}, \sigma^2\mathbf{G})$ was considered in **4a** under the assumption that $\mathbf{G}$ is nonsingular. However, the theory of least squares developed for the purpose cannot be used when $\mathbf{G}$ is singular.

In this section, we present two unified approaches to the problem, which are applicable in all situations whether $\mathbf{G}$ is singular or not. One is a direct approach obtained by minimizing the variance of an unbiased estimator. The other is an analogue of the least squares theory

First, we prove a lemma on generalized inverse of a partitioned matrix which plays a basic role in the unified theory.

(i) A BASIC LEMMA. *Let* $\mathbf{G}$ *be an n.n.d. matrix of order* n *and* $\mathbf{X}$ *be* $n \times m$ *matrix. Further, let*

$$\begin{pmatrix} \mathbf{G} & \mathbf{X} \\ \mathbf{X}' & \mathbf{0} \end{pmatrix}^{-} = \begin{pmatrix} \mathbf{C}_1 & \mathbf{C}_2 \\ \mathbf{C}_3 & -\mathbf{C}_4 \end{pmatrix} \tag{4i.1.1}$$

be one choice of g-inverse. Then the following results hold:

(a) $\mathbf{X}\mathbf{C}_2'\mathbf{X} = \mathbf{X}, \ \mathbf{X}\mathbf{C}_3\mathbf{X} = \mathbf{X}$ (4i.1.2)

(b) $\mathbf{X}\mathbf{C}_4\mathbf{X}' = \mathbf{X}\mathbf{C}_4'\mathbf{X}' = \mathbf{G}\mathbf{C}_3'\mathbf{X}' = \mathbf{X}\mathbf{C}_3\mathbf{G} = \mathbf{G}\mathbf{C}_2\mathbf{X}' = \mathbf{X}\mathbf{C}_2'\mathbf{G}$ (4i.1.3)

(c) $\mathbf{X}'\mathbf{C}_1\mathbf{X}, \ \mathbf{X}'\mathbf{C}_1\mathbf{G}, \ \mathbf{G}\mathbf{C}_1\mathbf{X}$ are all null matrices (4i.1.4)

(d) $\mathbf{G}\mathbf{C}_1\mathbf{G}\mathbf{C}_1\mathbf{G} = \mathbf{G}\mathbf{C}_1\mathbf{G} = \mathbf{G}\mathbf{C}_1'\mathbf{G}\mathbf{C}_1\mathbf{G} = \mathbf{G}\mathbf{C}_1'\mathbf{G}$ (4i.1.5)

(e) $\text{Tr } \mathbf{G}\mathbf{C}_1 = R(\mathbf{G}:\mathbf{X}) - R(\mathbf{X}) = \text{Tr } \mathbf{G}\mathbf{C}_1'$ (4i.1.6)

(f) $\mathbf{G}\mathbf{C}_1\mathbf{G}$ and $\mathbf{X}\mathbf{C}_4\mathbf{X}'$ *are invariant for any choice of* $\mathbf{C}_1$ *and* $\mathbf{C}_4$ (4i.1.7)

We observe that the equations

$$\begin{aligned} \mathbf{G}\mathbf{a} + \mathbf{X}\mathbf{b} &= \mathbf{0} \\ \mathbf{X}'\mathbf{a} &= \mathbf{X}'\mathbf{d} \end{aligned} \tag{4i.1.8}$$

are solvable for any $\mathbf{d}$, in which case, using (4i.1.1)

$$\mathbf{a} = \mathbf{C}_2\mathbf{X}'\mathbf{d}, \ \mathbf{b} = -\mathbf{C}_4\mathbf{X}'\mathbf{d} \tag{4i.1.9}$$

is a solution. Substituting (4i.1.9) in (4i.1.8)

$$\mathbf{G}\mathbf{C}_2\mathbf{X}'\mathbf{d} - \mathbf{X}\mathbf{C}_4\mathbf{X}'\mathbf{d} = \mathbf{0},\ \mathbf{X}'\mathbf{C}_2\mathbf{X}'\mathbf{d} = \mathbf{X}'\mathbf{d} \text{ for all } \mathbf{d}$$
$$\Rightarrow \mathbf{G}\mathbf{C}_2\mathbf{X}' = \mathbf{X}\mathbf{C}_4\mathbf{X}' \text{ and } \mathbf{X}'\mathbf{C}_2\mathbf{X}' = \mathbf{X}'. \tag{4i.1.10}$$

Since

$$\begin{pmatrix} \mathbf{G} & \mathbf{X}' \\ \mathbf{X} & \mathbf{0} \end{pmatrix}\begin{pmatrix} \mathbf{C}_1' & \mathbf{C}_3' \\ \mathbf{C}_2' & -\mathbf{C}_4' \end{pmatrix}\begin{pmatrix} \mathbf{G} & \mathbf{X}' \\ \mathbf{X} & \mathbf{0} \end{pmatrix} = \begin{pmatrix} \mathbf{G} & \mathbf{X}' \\ \mathbf{X} & \mathbf{0} \end{pmatrix}$$

the transpose of the matrix on the right-hand side of (4i.1.1) is also a g-inverse. Then replacing $\mathbf{C}_2$ by $\mathbf{C}_3'$ and $\mathbf{C}_4$ by $\mathbf{C}_4'$ in (4i.1.10) we have

$$\mathbf{G}\mathbf{C}_3'\mathbf{X}' = \mathbf{X}\mathbf{C}_4'\mathbf{X}' \quad \text{and} \quad \mathbf{X}'\mathbf{C}_3'\mathbf{X}' = \mathbf{X}'. \tag{4i.1.11}$$

Multiplying both sides of the first equation in (4i.1.11) by $\mathbf{X}\mathbf{C}_3$

$$\mathbf{X}\mathbf{C}_3\mathbf{G}\mathbf{C}_3'\mathbf{X}' = \mathbf{X}\mathbf{C}_3\mathbf{X}\mathbf{C}_4'\mathbf{X}' = \mathbf{X}\mathbf{C}_4'\mathbf{X}' \tag{4i.1.12}$$

so that $\mathbf{X}\mathbf{C}_4'\mathbf{X}'$ is symmetric. Then (4i.1.10–4i.1.12) prove the results (a) and (b).

To prove (c), we observe that the equations

$$\begin{aligned} \mathbf{G}\mathbf{a} + \mathbf{X}\mathbf{b} &= \mathbf{X}\mathbf{d} \\ \mathbf{X}'\mathbf{a} &= \mathbf{0} \end{aligned} \tag{4i.1.13}$$

can be solved for any $\mathbf{d}$. Then

$$\mathbf{a} = \mathbf{C}_1\mathbf{X}\mathbf{d}, \ \mathbf{b} = \mathbf{C}_3\mathbf{X}\mathbf{d} \tag{4i.1.14}$$

is a solution. Substituting (4i.1.14) in (4i.1.13) and omitting $\mathbf{d}$

$$\mathbf{G}\mathbf{C}_1\mathbf{X} + \mathbf{X}\mathbf{C}_3\mathbf{X} = \mathbf{X},\ \mathbf{X}'\mathbf{C}_1\mathbf{X} = \mathbf{0}. \tag{4i.1.15}$$

But $\mathbf{X}\mathbf{C}_3\mathbf{X} = \mathbf{X}$. Thus (c) is proved. Next we observe that the equations

$$\begin{aligned} \mathbf{G}\mathbf{a} + \mathbf{X}\mathbf{b} &= \mathbf{G}\mathbf{d} \\ \mathbf{X}'\mathbf{a} &= \mathbf{0} \end{aligned} \tag{4i.1.16}$$

can be solved for any $\mathbf{d}$. One solution is

$$\mathbf{a} = \mathbf{C}_1\mathbf{G}\mathbf{d} \quad \text{and} \quad \mathbf{b} = \mathbf{C}_3\mathbf{G}\mathbf{d}. \tag{4i.1.17}$$

Substituting (4i.1.17) in (4i.1.16) and omitting $\mathbf{d}$

$$\mathbf{G}\mathbf{C}_1\mathbf{G} + \mathbf{X}\mathbf{C}_3\mathbf{G} = \mathbf{G},\ \mathbf{X}'\mathbf{C}_1\mathbf{G} = \mathbf{0} \tag{4i.1.18}$$

$$\Rightarrow \mathbf{G}\mathbf{C}_1\mathbf{G}\mathbf{C}_1\mathbf{G} + \mathbf{G}\mathbf{C}_1\mathbf{X}\mathbf{C}_3\mathbf{G} = \mathbf{G}\mathbf{C}_1\mathbf{G} = \mathbf{G}\mathbf{C}_1\mathbf{G}\mathbf{C}_1\mathbf{G} \tag{4i.1.19}$$

since $\mathbf{GC_1X} = \mathbf{0}$. (4i.1.18) also $\Rightarrow \mathbf{GC_1'GC_1G} = \mathbf{GC_1G}$ which proves (d). Now we use the result $R(\mathbf{AA^-}) = R(\mathbf{A}) = \text{Tr}(\mathbf{AA^-})$, where $\mathbf{A}^-$ is any g-inverse of $\mathbf{A}$ [(iv), **1b.5**, (i), **1b.7**]. We have

$$\begin{aligned} R\begin{pmatrix} \mathbf{G} & \mathbf{X} \\ \mathbf{X'} & \mathbf{0} \end{pmatrix}\begin{pmatrix} \mathbf{C_1} & \mathbf{C_2} \\ \mathbf{C_3} & -\mathbf{C_4} \end{pmatrix} &= \text{Tr}\begin{pmatrix} \mathbf{GC_1} + \mathbf{XC_3} & \mathbf{GC_2} - \mathbf{XC_4} \\ \mathbf{X'C_1} & \mathbf{X'C_2} \end{pmatrix} \\ &= \text{Tr}(\mathbf{GC_1} + \mathbf{XC_3}) + \text{Tr}(\mathbf{X'C_2}) \\ &= \text{Tr}(\mathbf{GC_1}) + R(\mathbf{XC_3}) + R(\mathbf{X'C_2}) \\ &= \text{Tr}(\mathbf{GC_1}) + R(\mathbf{X}) + R(\mathbf{X'}). \end{aligned} \tag{4i.1.20}$$

But

$$R\begin{pmatrix} \mathbf{G} & \mathbf{X} \\ \mathbf{X'} & \mathbf{0} \end{pmatrix} = R(\mathbf{G} : \mathbf{X}) + R(\mathbf{X}). \tag{4i.1.21}$$

Equating (4i.1.20) and (4i.1.21) we have the result (e).

The result (f) follows from definition by using the results (4i.1.2–4i.1.6).

(ii) $\begin{pmatrix} \mathbf{G} & \mathbf{X} \\ \mathbf{X'} & \mathbf{0} \end{pmatrix}$ *is nonsingular iff* $R\begin{pmatrix} \mathbf{G} \\ \mathbf{X'} \end{pmatrix} = n$ *and* $R(\mathbf{X}) = m$.

(iii) *Let* $\mathbf{U}$ *be a matrix such that* $\mathcal{M}(\mathbf{X}) \subset \mathcal{M}(\mathbf{G} + \mathbf{XUX'})$ *and* $\mathcal{M}(\mathbf{G}) \subset \mathcal{M}(\mathbf{G} + \mathbf{XUX'})$. *Then*

$$\begin{pmatrix} \mathbf{G} & \mathbf{X} \\ \mathbf{X'} & \mathbf{0} \end{pmatrix}^- = \begin{pmatrix} \mathbf{C_1} & \mathbf{C_2} \\ \mathbf{C_3} & -\mathbf{C_4} \end{pmatrix} \Leftrightarrow \begin{pmatrix} \mathbf{G} + \mathbf{XUX'} & \mathbf{X} \\ \mathbf{X'} & \mathbf{0} \end{pmatrix}^- = \begin{pmatrix} \mathbf{C_1} & \mathbf{C_2} \\ \mathbf{C_3} & -\mathbf{U} - \mathbf{C_4} \end{pmatrix}$$

The propositions (ii) and (iii) are easy to establish.

(iv) EXPLICIT EXPRESSIONS FOR $\mathbf{C_1}, \mathbf{C_2}, \mathbf{C_3}, \mathbf{C_4}$ in (4i.1.1)

(*a*) *If* $\mathcal{M}(\mathbf{X}) \subset \mathcal{M}(\mathbf{G})$, *then one choice for* $\mathbf{C_1}, \mathbf{C_2}, \mathbf{C_3}, \mathbf{C_4}$ *is*

$$\mathbf{C_1} = \mathbf{G^-} - \mathbf{G^-XC_3}, \qquad \mathbf{C_3} = (\mathbf{X'G^-X})^-\mathbf{X'G^-}$$
$$\mathbf{C_2} = \mathbf{C_3'} \qquad \mathbf{C_4} = (\mathbf{X'G^-X})^-$$

(*b*) *If* $R(\mathbf{G}) = n$, *then one choice for* $\mathbf{C_1}, \mathbf{C_2}, \mathbf{C_3}, \mathbf{C_4}$ *is*

$$\mathbf{C_1} = \mathbf{G^{-1}} - \mathbf{G^{-1}XC_3}, \qquad \mathbf{C_3} = (\mathbf{X'G^{-1}X})^-\mathbf{X'G^{-1}}$$
$$\mathbf{C_2} = \mathbf{C_3'} \qquad \mathbf{C_4} = (\mathbf{X'G^{-1}X})^-$$

(c) *In general, one choice for* $\mathbf{C_1}, \mathbf{C_2}, \mathbf{C_3}, \mathbf{C_4}$ is

$$\mathbf{C_1} = \mathbf{W} - \mathbf{WXC_3} \qquad \mathbf{C_3} = \mathbf{TX'W},$$
$$\mathbf{C_2} = \mathbf{C_3'} \qquad \mathbf{C_4} = -\mathbf{U} + \mathbf{T}$$

where $\mathbf{T} = (\mathbf{X'}(\mathbf{G} + \mathbf{X'UX})^-\mathbf{X})^-$, $\mathbf{W} = (\mathbf{G} + \mathbf{X'UX})^-$ *and* $\mathbf{U}$ *is any matrix such that* $\mathcal{M}(\mathbf{X}) \subset \mathcal{M}(\mathbf{G} + \mathbf{XUX'})$ *and* $\mathcal{M}(\mathbf{G}) \subset \mathcal{M}(\mathbf{G} + \mathbf{XUX'})$.

The proposition (iv) is established by verification or by solving an equation of the type

$$\mathbf{Ga} + \mathbf{Xb} = \boldsymbol{\alpha}$$
$$\mathbf{X}'\mathbf{a} = \boldsymbol{\beta}$$

by the usual method of elimination.

4i.2 The General Gauss-Markoff Model (GGM)

Now we consider a general Gauss-Markoff (GGM) model

$$(\mathbf{Y}, \mathbf{X}\boldsymbol{\beta}, \sigma^2\mathbf{G}) \tag{4i.2.1}$$

where there may be deficiency in $R(\mathbf{X})$ and $\mathbf{G}$ is possibly singular. We note that the GGM model includes the case of any specified constraints on the parameter $\boldsymbol{\beta}$. Thus if $\mathbf{R}\boldsymbol{\beta} = \mathbf{C}$ is a set of constraints we can consider

$$\mathbf{Y}_e = \begin{pmatrix} \mathbf{Y} \\ \mathbf{C} \end{pmatrix}, \quad \mathbf{X}_e = \begin{pmatrix} \mathbf{X} \\ \mathbf{R} \end{pmatrix} \tag{4i.2.2}$$

in which case

$$\sigma^2\mathbf{G}_e = D(\mathbf{Y}_e) = \sigma^2\begin{pmatrix} \mathbf{G} & \mathbf{0} \\ \mathbf{0} & \mathbf{0} \end{pmatrix} \tag{4i.2.3}$$

and write the extended model

$$(\mathbf{Y}_e, \mathbf{X}_e\boldsymbol{\beta}, \sigma^2\mathbf{G}_e) \tag{4i.2.4}$$

in the same form as (4i.2.1).

When $\mathbf{G}$ is singular, there are some natural restrictions on the dependent variable $\mathbf{Y}$, which may be examined to make sure that there is no obvious inconsistency in the model.

(i) TEST FOR CONSISTENCY. $\mathbf{Y} \in \mathcal{M}(\mathbf{G} \vdots \mathbf{X})$.

Suppose $\boldsymbol{\alpha}$ is a vector such that $\boldsymbol{\alpha}'\mathbf{X} = \mathbf{0}$ and $\boldsymbol{\alpha}'\mathbf{G} = \mathbf{0}$. Then obviously $E(\boldsymbol{\alpha}'\mathbf{Y}) = 0$ and $E(\boldsymbol{\alpha}'\mathbf{Y})^2 = 0 \Rightarrow \boldsymbol{\alpha}'\mathbf{Y} = 0 \Rightarrow \mathbf{Y} \in \mathcal{M}(\mathbf{G} \vdots \mathbf{X})$ with probability.

(ii) $\mathbf{Y}$ *the random n-vector, and the unknown* $\boldsymbol{\beta}$ *are such that* $(\mathbf{Y} - \mathbf{X}\boldsymbol{\beta}) \in \mathbf{G}$.

The result follows since $\boldsymbol{\alpha}'\mathbf{G} = 0 \Rightarrow \boldsymbol{\alpha}'(\mathbf{Y} - \mathbf{X}\boldsymbol{\beta}) = 0$. Thus the singularity of $\mathbf{G}$ imposes some restrictions on $\mathbf{Y}$ and $\boldsymbol{\beta}$.

If $\mathbf{G}$ is singular, there exists a matrix $\mathbf{H}$ such that $\mathbf{HY} = \mathbf{0}$ with probability 1, which implies that $\mathbf{HX}\boldsymbol{\beta} = \mathbf{0}$ is a restriction on $\boldsymbol{\beta}$. Of course, $\mathbf{H}$ is known only when $\mathbf{Y}$ is known. When $\mathbf{H}$ is known, the condition $\mathbf{X}'\mathbf{L} = \mathbf{p}$ is not necessary for $\mathbf{L}'\mathbf{Y}$ to be unbiased for $\mathbf{p}'\boldsymbol{\beta}$.

(iii) *An n.s. condition that* $\mathbf{L}'\mathbf{Y}$ *is unbiased for* $\mathbf{p}'\boldsymbol{\beta}$ *is that there exists a vector* $\boldsymbol{\lambda}$ *such that* $\mathbf{X}'(\mathbf{L} + \mathbf{H}'\boldsymbol{\lambda}) = \mathbf{p}$, *where* $\mathbf{H}$ *is a matrix of maximum rank such that* $\mathbf{HY} = \mathbf{0}$. *However, given* $\mathbf{L}$ *such that* $\mathbf{L'Y}$ *is unbiased for* $\mathbf{p'\beta}$, *we can find* $\mathbf{M}$ *such that* $\mathbf{M'Y} = \mathbf{L'Y}$ *and* $\mathbf{X'M} = \mathbf{p}$. *Note that* $R(\mathbf{H}) = n - R(\mathbf{G})$ *or* $n - R(\mathbf{G}) - 1$.

The result which is easily established is important. It says that in searching for the BLUE of $\mathbf{p'\beta}$ we need only consider linear functions $\mathbf{L'Y}$ such that $\mathbf{X'L} = \mathbf{p}$, or behave as if $\boldsymbol{\beta}$ is a free parameter. Also if $\mathbf{L'Y}$ has zero expectation we can find a vector $\mathbf{K}$ such that $\mathbf{L'Y} = \mathbf{K'Y}$ and $\mathbf{K'X} = \mathbf{0}$.

4i.3 The Inverse Partitioned Martix (IPM) Method

We consider the GGM model (4i.2.1) and show how the problem of inference on unknown parameters can be reduced to the numerical evaluation of a g-inverse (Rao, 1971e, 1972c)

$$\begin{pmatrix} \mathbf{G} & \mathbf{X} \\ \mathbf{X'} & \mathbf{0} \end{pmatrix}^{-} = \begin{pmatrix} \mathbf{C}_1 & \mathbf{C}_2 \\ \mathbf{C}_3 & -\mathbf{C}_4 \end{pmatrix}. \tag{4i.3.1}$$

More precisely, if by a suitable algorithm (computer program) the matrix on the left-hand side of (4i.3.1) is inverted to obtain $\mathbf{C}_1$, $\mathbf{C}_2$, $\mathbf{C}_3$, $\mathbf{C}_4$, then we have all that is necessary to obtain the BLUE's, their variances and covariances, an unbiased estimator of σ^2, and test criteria for testing linear hypotheses. Propositions (i) and (ii) contain the basic results which hold whether $\mathbf{G}$ is singular or not.

(i) *Let* $\mathbf{C}_1$, $\mathbf{C}_2$, $\mathbf{C}_3$ *and* $\mathbf{C}_4$ *be as defined in (4i.3.1). Then the following hold*:

(a) (Use of $\mathbf{C}_2$ or $\mathbf{C}_3$). *The BLUE of an estimable (i.e., admitting an unbiased estimator) parametric function* $\mathbf{p'\beta}$, *is* $\mathbf{p'\hat{\beta}}$ *where*

$$\hat{\boldsymbol{\beta}} = \mathbf{C}_3\mathbf{Y} \quad \text{or} \quad \hat{\boldsymbol{\beta}} = \mathbf{C}_2'\mathbf{Y} \tag{4i.3.2}$$

(b) (Use of $\mathbf{C}_4$) *The dispersion matrix of* $\hat{\boldsymbol{\beta}}$ *is* $\sigma^2\mathbf{C}_4$ *in the sense*

$$\begin{aligned} V(\mathbf{p'\hat{\beta}}) &= \sigma^2\mathbf{p'C_4p} \\ \text{cov}(\mathbf{p'\hat{\beta}}, \mathbf{q'\hat{\beta}}) &= \sigma^2\mathbf{p'C_4q} = \sigma^2\mathbf{q'C_4p} \end{aligned} \tag{4i.3.3}$$

whenever $\mathbf{p'\beta}$ *and* $\mathbf{q'\beta}$ *are estimable.*

(c) (Use of $\mathbf{C}_1$). *An unbiased estimator of* σ^2 *is*

$$\hat{\sigma}^2 = f^{-1}\mathbf{Y'C_1Y} \tag{4i.3.4}$$

where $f = R(\mathbf{G} \vdots \mathbf{X}) - R(\mathbf{X})$.

Proof of (a): If $\mathbf{L'Y}$ is such that $E(\mathbf{L'Y}) = \mathbf{p'\beta}$, then $\mathbf{X'L} = \mathbf{p}$. Subject to this condition $V(\mathbf{L'Y}) = \sigma^2\mathbf{L'GL}$ or $\mathbf{L'GL}$ has to be minimized. Let $\mathbf{L}_*$ be an optimum choice and $\mathbf{L}$ any other vector such that $\mathbf{X'L} = \mathbf{X'L}_*$. Then

$$\begin{aligned} \mathbf{L'GL} &= (\mathbf{L} - \mathbf{L}_* + \mathbf{L}_*)'\mathbf{G}(\mathbf{L} - \mathbf{L}_* + \mathbf{L}_*) \\ &= (\mathbf{L} - \mathbf{L}_*)'\mathbf{G}(\mathbf{L} - \mathbf{L}_*) + \mathbf{L}_*'\mathbf{G}\mathbf{L}_* + 2\mathbf{L}_*'\mathbf{G}(\mathbf{L} - \mathbf{L}_*) \geqslant \mathbf{L}_*'\mathbf{G}\mathbf{L} \end{aligned}$$

iff $\mathbf{L}_*'\mathbf{G}(\mathbf{L}-\mathbf{L}_*) = 0$ whenever $\mathbf{X}'(\mathbf{L}-\mathbf{L}_*) = \mathbf{0}$, i.e., $\mathbf{GL}_* = -\mathbf{XK}_*$ for a suitable $\mathbf{K}_*$. Then there exist $\mathbf{L}_*$ and $\mathbf{K}_*$ to satisfy the equations

$$\begin{aligned} \mathbf{GL}_* + \mathbf{XK}_* &= \mathbf{0}, \\ \mathbf{X}'\mathbf{L}_* &= \mathbf{p}. \end{aligned} \tag{4i.3.5}$$

Since (4i.3.5) is consistent

$$\mathbf{L}_* = \mathbf{C}_2\mathbf{p} \text{ or } \mathbf{C}_3'\mathbf{p}, \mathbf{K}_* = -\mathbf{C}_4\mathbf{p}$$

is a solution. Then the BLUE of $\mathbf{p}'\boldsymbol{\beta}$ is $\mathbf{L}_*'\mathbf{Y} = \mathbf{p}'\mathbf{C}_2'\mathbf{Y} = \mathbf{p}'\mathbf{C}_3\mathbf{Y}$.

Proof of (b): We use the fact $\mathbf{p} = \mathbf{X}'\mathbf{M}$ for some $\mathbf{M}$. Then

$$\begin{aligned} V(\mathbf{p}'\mathbf{C}_2\mathbf{Y}) &= \sigma^2\mathbf{M}'(\mathbf{XC}_2'\mathbf{G})\mathbf{C}_2\mathbf{X}'\mathbf{M} \\ &= \sigma^2\mathbf{M}'\mathbf{XC}_4(\mathbf{X}'\mathbf{C}_2\mathbf{X}')\mathbf{M} \text{ using (4i.1.3)} \\ &= \sigma^2\mathbf{M}'\mathbf{XC}_4\mathbf{X}'\mathbf{M} \text{ using (4i.1.2)} \\ &= \sigma^2\mathbf{p}'\mathbf{C}_4\mathbf{p}. \end{aligned}$$

Similarly,

$$\operatorname{cov}(\mathbf{p}'\mathbf{C}_2'\mathbf{Y}, \mathbf{q}'\mathbf{C}_2'\mathbf{Y}) = \sigma^2\mathbf{p}'\mathbf{C}_4\mathbf{q} = \sigma^2\mathbf{q}'\mathbf{C}_4\mathbf{p}.$$

Proof of (c): Since $\mathbf{X}'\mathbf{C}_1\mathbf{G} = \mathbf{0}$ and $\mathbf{X}'\mathbf{C}_1\mathbf{X} = \mathbf{0}$ using (4i.1.4),

$$\begin{aligned} \mathbf{Y}'\mathbf{C}_1\mathbf{Y} &= (\mathbf{Y}-\mathbf{X}\boldsymbol{\beta})'\mathbf{C}_1(\mathbf{Y}-\mathbf{X}\boldsymbol{\beta}) \\ E[(\mathbf{Y}-\mathbf{X}\boldsymbol{\beta})'\mathbf{C}_1(\mathbf{Y}-\mathbf{X}\boldsymbol{\beta})] &= \sigma^2 \operatorname{Tr}\mathbf{C}_1 E[(\mathbf{Y}-\mathbf{X}\boldsymbol{\beta})(\mathbf{Y}-\mathbf{X}\boldsymbol{\beta})'] \\ &= \sigma^2 \operatorname{Tr}\mathbf{C}_1\mathbf{G} = \sigma^2[R(\mathbf{G} \mid \mathbf{X}) - R(\mathbf{X})] \end{aligned}$$

using (4i.1.6).

(ii) *Let* $\mathbf{P}'\hat{\boldsymbol{\beta}}$ *be the vector of BLUE's of a set of k estimable parametric functions* $\mathbf{P}'\boldsymbol{\beta}$, $R_0^2 = \mathbf{Y}'\mathbf{C}_1\mathbf{Y}$ *and f be as defined in* (i).

If $\mathbf{Y} \sim N_n(\mathbf{X}\boldsymbol{\beta}, \sigma^2\mathbf{G})$, *then*:

(a) $\mathbf{P}'\hat{\boldsymbol{\beta}}$ *and* $\mathbf{Y}'\mathbf{C}_1\mathbf{Y}$ *are independently distributed, with*

$$\mathbf{P}'\hat{\boldsymbol{\beta}} \sim N_k(\mathbf{P}'\boldsymbol{\beta}, \sigma^2\mathbf{D}), \ \mathbf{D} = \mathbf{P}'\mathbf{C}_4\mathbf{P} \tag{4i.3.6}$$

$$R_0^2 = \mathbf{Y}'\mathbf{C}_1\mathbf{Y} \sim \sigma^2\chi_f^2. \tag{4i.3.7}$$

(b) *Let* $\mathbf{P}'\boldsymbol{\beta} = \mathbf{w}$ *be a null hypothesis. The null hypothesis is consistent iff*

$$\mathbf{DD}^-\mathbf{u} = \mathbf{u}, \qquad \text{where} \quad \mathbf{u} = \mathbf{P}'\hat{\boldsymbol{\beta}} - \mathbf{w}. \tag{4i.3.8}$$

If the hypothesis is consistent, then

$$F = \frac{\mathbf{u}'\mathbf{D}^-\mathbf{u}}{h} \div \frac{R_0^2}{f}, \qquad h = R(\mathbf{D}) \tag{4i.3.9}$$

has a central F distribution on h and f degrees of freedom when the hypothesis is true and otherwise noncentral.

Proof of (a): Since $E(\mathbf{P}'\hat{\boldsymbol{\beta}}) = \mathbf{P}'\boldsymbol{\beta}$ and $D(\mathbf{P}'\hat{\boldsymbol{\beta}}) = \sigma^2\mathbf{P}'\mathbf{C}_4\mathbf{P}$, $\mathbf{P}'\hat{\boldsymbol{\beta}} \sim N_k(\mathbf{P}'\boldsymbol{\beta}, \sigma^2\mathbf{P}'\mathbf{C}_4\mathbf{P})$ using (3b.3.5).

Let $\mathbf{F} = (\mathbf{C}_1 + \mathbf{C}_1')/2$. Then $\mathbf{Y}'\mathbf{C}_1\mathbf{Y} = \mathbf{Y}'\mathbf{F}\mathbf{Y}$ and $\mathbf{GFGFG} = \mathbf{GFG}$ using (4i.1.5). Hence $\mathbf{Y}'\mathbf{C}_1\mathbf{Y}$ has $\sigma^2\chi^2$ distribution by (3b.4.7).

$\mathbf{P}'\hat{\boldsymbol{\beta}} = \mathbf{P}'\mathbf{C}_2'\mathbf{Y} = \mathbf{H}'\mathbf{X}\mathbf{C}_2'\mathbf{Y}$ since $\mathcal{M}(\mathbf{P}) \subset \mathcal{M}(\mathbf{X}')$. Then $\mathbf{GFGC}_2\mathbf{X}'\mathbf{H} = \mathbf{0}$ using (4i.1.3), which shows by (3b.4.9) that $\mathbf{Y}'\mathbf{C}_1\mathbf{Y}$ and $\mathbf{P}'\hat{\boldsymbol{\beta}}$ are independently distributed. [Note that $\mathbf{GF}(\mathbf{GC}_2\mathbf{X}')\mathbf{H} = \mathbf{GF}(\mathbf{XC}_2'\mathbf{GH}) = \mathbf{0}$ since $\mathbf{GFX} = \mathbf{0}$.]

Proof of (b): The hypothesis is consistent if $V(\boldsymbol{\alpha}'\mathbf{P}'\hat{\boldsymbol{\beta}}) = 0 \Rightarrow \boldsymbol{\alpha}'(\mathbf{P}'\hat{\boldsymbol{\beta}} - \mathbf{w}) = 0 \Rightarrow \mathbf{P}'\hat{\boldsymbol{\beta}} - \mathbf{w} \in \mathcal{M}(\mathbf{D})$, an n.s. condition for which is $\mathbf{DD}^-\mathbf{u} = \mathbf{u}$, where $\mathbf{u} = \mathbf{P}'\hat{\boldsymbol{\beta}} - \mathbf{w}$. If the hypothesis is true then $\mathbf{u}'\mathbf{D}^-\mathbf{u}$ has central $\sigma^2\chi^2$ distribution since $\mathbf{DD}^-\mathbf{D} = \mathbf{D}$, with degrees of freedom $h = R(\mathbf{D})$. Since $\mathbf{u}$ and R_0^2 are independently distributed, $\mathbf{u}'\mathbf{D}^-\mathbf{u}$ and R_0^2 are independent, and hence the statistic (4i.3.9) has the F distribution on h and f degrees of freedom. If the hypothesis is not true, the distribution will be noncentral.

4i.4 Unified Theory of Least Squares

Let us consider the model $(\mathbf{Y}, \mathbf{X}\boldsymbol{\beta}, \sigma^2\mathbf{G})$. In **4a.7** (see Table 4a.7), it is shown that when $|\mathbf{G}| \neq 0$, the estimating normal equations for $\boldsymbol{\beta}$ are obtained by equating the derivative of

$$(\mathbf{Y} - \mathbf{X}\boldsymbol{\beta})'\mathbf{G}^{-1}(\mathbf{Y} - \mathbf{X}\boldsymbol{\beta})$$

with respect to $\boldsymbol{\beta}$ to the null vector. Such a procedure is not available when $|\mathbf{G}| = 0$. A general method different from least squares applicable *to all situations* is given in **4i.3**. We shall investigate whether an analogue of least squares theory applicable *to all situations* can be developed. More specifically, can we find a matrix $\mathbf{M}$ such that the estimate $\hat{\boldsymbol{\beta}}$ of $\boldsymbol{\beta}$ is a *stationary* value of

$$(\mathbf{Y} - \mathbf{X}\boldsymbol{\beta})'\mathbf{M}(\mathbf{Y} - \mathbf{X}\boldsymbol{\beta})$$

i.e., where the derivative with respect to $\boldsymbol{\beta}$ vanishes, whether $|\mathbf{G}| = 0$ or $\neq 0$, and an estimate of σ^2 is obtained as

$$(\mathbf{Y} - \mathbf{X}\hat{\boldsymbol{\beta}})'\mathbf{M}(\mathbf{Y} - \mathbf{X}\hat{\boldsymbol{\beta}}) \div f$$

where $f = R(\mathbf{G} \vdots \mathbf{X}) - R(\mathbf{X})$ as in (4i.3.4)?

We show that one choice of $\mathbf{M}$ is $(\mathbf{G} + \mathbf{XUX}')^-$ for any g-inverse, where $\mathbf{U}$ is any symmetric matrix such that $R(\mathbf{G} : \mathbf{X}) = R(\mathbf{G} + \mathbf{XUX}')$, the necessity of which was established in (Rao, 1973a). One simple choice of $\mathbf{U} = k^2\mathbf{I}$ for $k \neq 0$ as shown in (Rao and Mitra, 1971h). k can be zero when $\mathcal{M}(\mathbf{X}) \subset \mathcal{M}(\mathbf{G})$. Let us write $\mathbf{T} = \mathbf{G} + \mathbf{XUX}'$ and $\mathbf{T}^-$ any g-inverse (not necessarily symmetric).

(i) *Let* $(\mathbf{Y}, \mathbf{X\beta}, \sigma^2\mathbf{G})$ *be GGM model (i.e.,* $\mathbf{G}$ *is possibly singular and* $\mathbf{X}$ *possibly deficient in rank), and* $\mathbf{T}$, $\mathbf{T}^-$, $\mathbf{U}$ *be as defined above.*

(a) *The BLUE of an estimable function* $\mathbf{p'\beta}$ *is* $\mathbf{p'\hat{\beta}}$ *where* $\hat{\boldsymbol{\beta}}$ *is a stationary value of* $(\mathbf{Y} - \mathbf{X\beta})'\mathbf{T}^-(\mathbf{Y} - \mathbf{X\beta})$.

(b) $V(\mathbf{p'\hat{\beta}}) = \sigma^2\mathbf{p}'[(\mathbf{X'T^-X})^- - \mathbf{U}]\mathbf{p}$,

$$\text{cov}(\mathbf{p'\hat{\beta}}, \mathbf{q'\hat{\beta}}) = \sigma^2\mathbf{p}'[(\mathbf{X'T^-X})^- - \mathbf{U}]\mathbf{q}. \tag{4i.4.1}$$

(c) *An unbiased estimator of* σ^2 *is*

$$\hat{\sigma}^2 = f^{-1}(\mathbf{Y} - \mathbf{X\hat{\beta}})'\mathbf{T}^-(\mathbf{Y} - \mathbf{X\hat{\beta}}) = f^{-1}R_0^2 \tag{4i.4.2}$$

where

$$f = R(\mathbf{G} \vdots \mathbf{X}) - R(\mathbf{X}).$$

Proof of (a): $\hat{\boldsymbol{\beta}}$ satisfies the equation

$$\mathbf{X'T^-X\hat{\beta}} = \mathbf{X'T^-Y} \tag{4i.4.3}$$

observing that $\mathbf{X'T^-X} = \mathbf{X'(T^-)'X}$ and $\mathbf{X'T^-Y} = \mathbf{X'(T^-)'Y}$ since $\mathbf{X}$ and $\mathbf{Y} \in \mathcal{M}(\mathbf{T})$ and $\mathbf{T}$ is symmetric. Let $\hat{\boldsymbol{\beta}} = (\mathbf{X'T^-X})^-\mathbf{X'T^-Y}$ be a solution, which exists since $R(\mathbf{X'T^-X}) = R(\mathbf{X'})$ as shown in Example 30, p. 77 (end of Chapter 1).

Let $\mathbf{p'\beta}$ be an estimable function, i.e., $\mathbf{p} = \mathbf{X'L}$ for some $\mathbf{L}$. Then

$$\mathbf{p'\hat{\beta}} = \mathbf{L'X(X'T^-X)^-X'T^-Y}$$
$$E(\mathbf{p'\hat{\beta}}) = \mathbf{L'X(X'T^-X)^-X'T^-X\beta} = \mathbf{L'X\beta} = \mathbf{p'\beta}.$$

since $\mathbf{X'} \in \mathcal{M}(\mathbf{X'T^-X})$ using [vi(b), **1b.5**]. Further, if $\mathbf{Z}$ is such that $\mathbf{Z'X} = \mathbf{0}$, then

$$\text{cov}(\mathbf{p'\hat{\beta}}, \mathbf{Z'Y}) = \sigma^2\mathbf{L'X(X'T^-X)^-X'T^-GZ} = \mathbf{0}$$

since $\mathbf{X'T^-GZ} = \mathbf{X'T^-(G + XUX')Z} = \mathbf{X'T^-TZ} = \mathbf{X'Z} = \mathbf{0}$, as $\mathbf{X} \in \mathcal{M}(\mathbf{T})$. Thus $\mathbf{p'\hat{\beta}}$ is the BLUE.

Proof of (b):

$$\begin{aligned} V(\mathbf{p'\hat{\beta}}) &= V(\mathbf{L'X(X'T^-X)^-X'T^-Y}) \\ &= V(\mathbf{L'WY}), \ \mathbf{W} = \mathbf{X(X'T^-X)^-X'T^-} \\ &= \sigma^2\mathbf{L'W(G + XUX')W'L} - \mathbf{L'WXUX'W'L} \\ &= \sigma^2[\mathbf{L'X(X'T^-X)^-X'L} - \mathbf{L'XUX'L}] \\ &= \sigma^2\mathbf{p}'[(\mathbf{X'T^-X})^- - \mathbf{U}]\mathbf{p}. \end{aligned}$$

Similarly, the expression for covariance is established.

Proof of (c): It is easily shown, along the lines of (4a.5.4),

$$\begin{aligned} E(\mathbf{Y}-\mathbf{X}\hat{\boldsymbol{\beta}})(\mathbf{Y}-\mathbf{X}\hat{\boldsymbol{\beta}})' &= D(\mathbf{Y}) - D(\mathbf{X}\hat{\boldsymbol{\beta}}) \\ &= \sigma^2\{\mathbf{G} - \mathbf{X}[(\mathbf{X}'\mathbf{T}^-\mathbf{X})^- - \mathbf{U}]\mathbf{X}'\} \\ &= \sigma^2\{\mathbf{T} - \mathbf{X}(\mathbf{X}'\mathbf{T}^-\mathbf{X})^-\mathbf{X}')\} \end{aligned}$$

$$\begin{aligned} \sigma^{-2}E(\mathbf{Y}-\mathbf{X}\hat{\boldsymbol{\beta}})'\mathbf{T}^-(\mathbf{Y}-\mathbf{X}\hat{\boldsymbol{\beta}}) &= \operatorname{Tr}\mathbf{T}^-[\mathbf{T} - \mathbf{X}(\mathbf{X}'\mathbf{T}^-\mathbf{X})^-\mathbf{X}'] \\ &= \operatorname{Tr}\mathbf{T}^-\mathbf{T} - \operatorname{Tr}(\mathbf{X}'\mathbf{T}^-\mathbf{X})(\mathbf{X}'\mathbf{T}^-\mathbf{X})^- \\ &= R(\mathbf{T}) - R(\mathbf{X}) = f, \end{aligned}$$

which proves (c).

Tests of linear hypotheses are carried out in the same way as in proposition (ii) of **4i.3**, by taking $\mathbf{C}_4$ in (4i.3.6) as $\mathbf{X}[(\mathbf{X}'\mathbf{T}^-\mathbf{X})^- - \mathbf{U}]\ \mathbf{X}'$ and $\mathbf{Y}'\mathbf{C}_1\mathbf{Y}$ as R_0^2 defined in (4i.4.2).

Note that $\mathbf{U}$ can be chosen, if necessary, such that $\mathbf{T} = \mathbf{G} + \mathbf{XUX}'$ is n.n.d., in addition to the condition $R(\mathbf{G} \vdots \mathbf{X}) = R(\mathbf{T})$. In such a case $(\mathbf{Y}-\mathbf{X}\boldsymbol{\beta})'\mathbf{T}^-(\mathbf{Y}-\mathbf{X}\boldsymbol{\beta})$ is non-negative for any choice of $\mathbf{T}^-$ and the expression attains a minimum at a stationary point $\hat{\boldsymbol{\beta}}$.

In general $\mathbf{T}^-$ is not a g-inverse of $\mathbf{G}$. However, we can choose $\mathbf{U}$ in such a way that $\mathbf{G}$ and $\mathbf{XUX}'$ generate virtually disjoint spaces, in which case $\mathbf{T}^-$ is a g-inverse of $\mathbf{T}$. Unfortunately for any testable hypothesis $\mathbf{P}'\boldsymbol{\beta} = \mathbf{W}$, the result $\mathbf{u}'\mathbf{D}^-\mathbf{u} = \min_{\mathbf{P}'\boldsymbol{\beta}=\mathbf{W}} (\mathbf{Y}-\mathbf{X}\boldsymbol{\beta})'\mathbf{M}(\mathbf{Y}-\mathbf{X}\boldsymbol{\beta}) - \min_{\boldsymbol{\beta}} (\mathbf{Y}-\mathbf{X}\boldsymbol{\beta})'\mathbf{M}(\mathbf{Y}-\mathbf{X}\boldsymbol{\beta})$ where $\mathbf{u}'\mathbf{D}^-\mathbf{u}$ is as defined in (4i.3.9) is not true when $\mathbf{G}$ is singular however $\mathbf{M}$ is chosen unless $\mathcal{M}(\mathbf{X}) \subset \mathcal{M}(\mathbf{G})$. (See Rao 1972d.)

4j ESTIMATION OF VARIANCE COMPONENTS

4j.1 Variance Components Model

In **4f** we introduced a general two-way classification model and considered a special problem of estimation of variance components. Now we consider the general linear model, $\mathbf{Y} = \mathbf{X}\boldsymbol{\beta} + \boldsymbol{\varepsilon}$, where the error component $\boldsymbol{\varepsilon}$ has the structure

$$\boldsymbol{\varepsilon} = \mathbf{U}_1\boldsymbol{\xi}_1 + \cdots + \mathbf{U}_k\boldsymbol{\xi}_k \tag{4j.1.1}$$

$\boldsymbol{\xi}_i$ being a n_i-vector of hypothetical (unobservable) variables such that

$$\begin{aligned} E(\boldsymbol{\xi}_i) = \mathbf{0},\ D(\boldsymbol{\xi}_i) = \sigma_i^2\mathbf{I}, \qquad & i = 1, \ldots, k, \\ E(\boldsymbol{\xi}_i\boldsymbol{\xi}_j) = \mathbf{0}, \qquad & i \neq j, \end{aligned} \tag{4j.1.2}$$

so that

$$D(\boldsymbol{\varepsilon}) = \sigma_1^2\mathbf{U}_1\mathbf{U}_1' + \cdots + \sigma_k^2\mathbf{U}_k\mathbf{U}_k' = \sigma_1^2\mathbf{V}_1 + \cdots + \sigma_k^2\mathbf{V}_k. \tag{4j.1.3}$$

The design matrices $\mathbf{X}$, $\mathbf{U}_i$ (or $\mathbf{V}_i$), and the observation vector $\mathbf{Y}$ are known while $\boldsymbol{\beta}$ and $\boldsymbol{\sigma} = (\sigma_1^2, \ldots, \sigma_k^2)$ are unknown parameters under estimation. The variances $\sigma_1^2, \ldots, \sigma_k^2$ of the hypothetical variables in $\boldsymbol{\xi}_1, \ldots, \boldsymbol{\xi}_k$ are called variance components.

Let $\alpha_1, \ldots, \alpha_k$ be a priori (or approximate) values of $\sigma_1, \ldots, \sigma_k$. (If no *a priori* information is available, then each α_i may be taken to be unity.) Then we can rewrite (4j.1.1) and (4j.1.3) as

$$\boldsymbol{\varepsilon} = \mathbf{W}_1\boldsymbol{\eta}_1 + \cdots + \mathbf{W}_k\boldsymbol{\eta}_k, \ \mathbf{W}_i = \alpha_i\mathbf{U}_i, \ \boldsymbol{\xi}_i = \alpha_i\boldsymbol{\eta}_i,$$
$$D(\boldsymbol{\varepsilon}) = \gamma_1^2\mathbf{T}_1 + \cdots + \gamma_k^2\mathbf{T}_k, \ \mathbf{T}_i = \mathbf{W}_i\mathbf{W}_i' = \alpha_i^2\mathbf{V}_i \qquad (4j.1.4)$$

where $\gamma_i^2 = \sigma_i^2/\alpha_i^2$, the scaled variance components to be estimated.

4j.2 MINQUE Theory

Let us consider the estimation of a linear function

$$\sum p_i\sigma_i^2 = \sum q_i\gamma_i^2 \qquad (4j.2.1)$$

of the variance components by a quadratic function $\mathbf{Y'AY}$ subject to the following conditions.

(a) *Invariance and Unbiasedness.* If instead of $\boldsymbol{\beta}$, we consider $\boldsymbol{\beta}_d = \boldsymbol{\beta} - \boldsymbol{\beta}_0$, then the model becomes $\mathbf{Y}_d = \mathbf{X}\boldsymbol{\beta}_d + \boldsymbol{\varepsilon}$, where $\mathbf{Y}_d = \mathbf{Y} - \mathbf{X}\boldsymbol{\beta}_0$. In such a case, the estimator is $\mathbf{Y}_d'\mathbf{A}\mathbf{Y}_d$, which is the same as $\mathbf{Y'AY}$ for all $\boldsymbol{\beta}_0$, iff $\mathbf{AX} = \mathbf{0}$, which may be called the condition of invariance for translation in the $\boldsymbol{\beta}$ parameter. Now for such an estimator

$$E(\mathbf{Y'AY}) = \text{Tr } \mathbf{A}D(\boldsymbol{\varepsilon}) = \sum \gamma_i^2 \text{ Tr } \mathbf{AT}_i = \sum \gamma_i^2 q_i$$
$$\Rightarrow \text{Tr } \mathbf{AT}_i = q_i, \qquad i = 1, \ldots, k \qquad (4j.2.2)$$

which are the conditions for unbiasedness under invariance.

(b) *Minimum norm.* If the hypothetical variables $\boldsymbol{\eta}_i$ were known, then a natural estimator of γ_i^2 is $\boldsymbol{\eta}_i'\boldsymbol{\eta}_i/n_i$ and hence that of $\sum q_i\gamma_i^2$ is

$$(q_1/n_1)\boldsymbol{\eta}_1'\boldsymbol{\eta}_1 + \cdots + (q_k/n_k)\boldsymbol{\eta}_k'\boldsymbol{\eta}_k = \boldsymbol{\eta}'\boldsymbol{\Delta}\boldsymbol{\eta} \quad \text{(say)} \qquad (4j.2.3)$$

where $\boldsymbol{\Delta}$ is a suitably defined diagonal matrix and $\boldsymbol{\eta}' = (\boldsymbol{\eta}_1' \,\vdots\, \cdots \,\vdots\, \boldsymbol{\eta}_k')$. But the proposed estimator is

$$\mathbf{Y'AY} = \boldsymbol{\varepsilon}'\mathbf{A}\boldsymbol{\varepsilon} = \boldsymbol{\eta}'\mathbf{W'AW}\boldsymbol{\eta} \qquad (4j.2.4)$$

since $\mathbf{AX} = \mathbf{0}$, where $\mathbf{W} = (\mathbf{W}_1 \,\vdots\, \cdots \,\vdots\, \mathbf{W}_k)$. The difference between (4j.2.3) and (4j.2.4) is $\boldsymbol{\eta}'(\mathbf{W'AW} - \boldsymbol{\Delta})\boldsymbol{\eta}$ which can be made small, in some sense, by minimizing

$$\|\mathbf{W'AW} - \boldsymbol{\Delta}\| \qquad (4j.2.5)$$

where the norm is suitably chosen.

An estimator $\mathbf{Y'AY}$ where $\mathbf{A}$ is such that $\|\mathbf{W'AW} - \mathbf{\Delta}\|$ is a minimum, subject to $\mathbf{AX} = \mathbf{0}$ and $\text{Tr}\, \mathbf{AT}_i = q_i$, $i = 1, \ldots, k$ is called the MINQUE (Minimum Norm Quadratic Unbiased Estimator) of the parametric function $\sum q_i \gamma_i^2$.

In the literature (Rao, 1970a, 1971c, 1971d, 1972b), MINQUE is also obtained with the condition $\mathbf{X'AX} = \mathbf{0}$, instead of $\mathbf{AX} = \mathbf{0}$, to ensure unbiasedness only and not invariance, and a distinction is made between MINQUE's with and without invariance. But for the present discussion we shall impose the invariance condition. The condition $\mathbf{X'AX} = \mathbf{0}$ was first considered by Focke and Dewess (1972) for estimation of variance components.

4j.3 Computations under the Euclidean Norm

If we choose the Euclidean norm (i.e., $\|\mathbf{B}\|^2$ = the sum of squares of all the elements of $\mathbf{B}$), then

$$\begin{aligned}\|\mathbf{W'AW} - \mathbf{\Delta}\|^2 &= \text{Tr}\, \mathbf{W'AWW'AW} - 2\,\text{Tr}\, \mathbf{W'AW\Delta} + \text{Tr}\, \mathbf{\Delta\Delta} \\ &= \text{Tr}\, \mathbf{AWW'AWW'} - \text{Tr}\, \mathbf{\Delta\Delta} = \text{Tr}\, \mathbf{ATAT} - \text{Tr}\, \mathbf{\Delta\Delta}\end{aligned}$$

where $\mathbf{T} = \sum \mathbf{T}_i = \sum \alpha_i^2 \mathbf{V}_i$, observing that

$$\text{Tr}\, \mathbf{W'AW\Delta} = \text{Tr}\, \mathbf{AW\Delta W'} = \text{Tr} \sum (q_i/n_i)\mathbf{AT}_i = \text{Tr}\, \mathbf{\Delta\Delta}.$$

Thus, in the case of the Euclidean norm, the problem reduces to that of determining $\mathbf{A}$ such that $\text{Tr}\, \mathbf{ATAT}$ is a minimum subject to the conditions $\mathbf{AX} = \mathbf{0}$ and $\text{Tr}\, \mathbf{AT}_i = q_i$, $i = 1, \ldots, k$.

The algebraic problem is already solved in [(iii), **1f.3**], where it is shown that the minimum is attained at

$$\mathbf{A} = \sum \lambda_i \mathbf{RT}_i\mathbf{R}, \quad \mathbf{R} = \mathbf{T}^{-1} - \mathbf{T}^{-1}\mathbf{X}(\mathbf{X'T}^{-1}\mathbf{X})^{-}\mathbf{X'T}^{-1} \tag{4j.3.1}$$

where λ_i satisfies the equations

$$\sum_{i=1}^{k} \lambda_i\, \text{Tr}\, \mathbf{RT}_i\mathbf{RT}_j = q_j, \qquad j = 1, \ldots, k. \tag{4j.3.2}$$

Then the MINQUE of $\sum q_i \gamma_i^2$ is $\boldsymbol{\lambda}'\mathbf{Q}$ where $\boldsymbol{\lambda}' = (\lambda_1, \ldots, \lambda_k)$ is a solution of (4j.3.2) and $\mathbf{Q}' = (Q_1, \ldots, Q_k)$, $Q_i = \mathbf{Y'RT}_i\mathbf{RY}$. We may write the equation (4j.3.2) in the form $\mathbf{S}\boldsymbol{\lambda} = \mathbf{q}$, in which case $\boldsymbol{\lambda} = \mathbf{S}^{-}\mathbf{q}$ is a solution and

$$\boldsymbol{\lambda}'\mathbf{Q} = \mathbf{q}'(\mathbf{S}^{-})'\mathbf{Q} = \mathbf{q}'\mathbf{S}^{-}\mathbf{Q} = \mathbf{q}'\hat{\boldsymbol{\gamma}}$$

where $\hat{\boldsymbol{\gamma}}$ is a solution of $\mathbf{S}\boldsymbol{\gamma} = \mathbf{Q}$.

Thus the computational procedure consists in first setting up the equation $\mathbf{S}\boldsymbol{\gamma} = \mathbf{Q}$ where the (i, j)th element of $\mathbf{S}$ is $\text{Tr}\, \mathbf{RT}_i\mathbf{RT}_j$ and the ith element of

$\mathbf{Q}$ is $\mathbf{Y}'\mathbf{RT}_i\mathbf{RY}$ and obtaining a solution $\hat{\gamma} = \mathbf{S}^-\mathbf{Q}$. The solution for the original variance components is

$$\hat{\boldsymbol{\sigma}} = (\hat{\sigma}_1^2, \ldots, \hat{\sigma}_k^2) = (\alpha_1^2\hat{\gamma}_1^2, \ldots, \alpha_k^2\hat{\gamma}_k^2)$$

in the sense that the estimate of any estimable function $\mathbf{p}'\boldsymbol{\sigma}$ is $\mathbf{p}'\hat{\boldsymbol{\sigma}}$. [$\mathbf{p}'\boldsymbol{\sigma}$ or equivalently $\mathbf{q}'\gamma$ is said to be estimable provided a matrix $\mathbf{A}$ exists such that $\mathbf{AX} = \mathbf{0}$ and $\text{Tr}\,\mathbf{AT}_i = q_i$, $i = 1, \ldots, k$].

It may be verified that $\text{Tr}\,\mathbf{ATAT}$ is proportional to $V(\mathbf{Y}'\mathbf{AY})$ when $\mathbf{Y}$ has the n-variate normal distribution (which is equivalent to saying that all the hypothetical variables in (4j.1.1) have univariate normal distributions) and the true variance components are $\alpha_1^2, \ldots, \alpha_k^2$. Thus the MINQUE is the locally minimum variance unbiased quadratic estimator under normal distribution for the hypothetical variables.

4k BIASED ESTIMATION IN LINEAR MODELS

4k.1 Best Linear Estimation (BLE)

Let us consider the GGM model $(\mathbf{Y}, \mathbf{X}\boldsymbol{\beta}, \sigma^2\mathbf{G})$ and let $\mathbf{L}'\mathbf{Y}$ be an estimator of $\mathbf{p}'\boldsymbol{\beta}$, the given linear parametric function. The MSE (mean square error) of $\mathbf{L}'\mathbf{Y}$ is

$$E(\mathbf{L}'\mathbf{Y} - \mathbf{p}'\boldsymbol{\beta})^2 = \sigma^2\mathbf{L}'\mathbf{GL} + (\mathbf{X}'\mathbf{L} - \mathbf{p})'\boldsymbol{\beta}\boldsymbol{\beta}'(\mathbf{X}'\mathbf{L} - \mathbf{P}) \tag{4k.1.1}$$

which involves both the unknown parameters σ^2 and $\boldsymbol{\beta}$ and as it stands is not a suitable criterion for minimizing. Then we have three possibilities.

(a) Choose an *a priori* value of $\sigma^{-1}\boldsymbol{\beta}$, say $\mathbf{b}$, based on our previous knowledge and substitute $\sigma^2\mathbf{W} = \sigma^2\mathbf{bb}'$ for $\boldsymbol{\beta}\boldsymbol{\beta}'$ in (4k.1.1) which then takes the form $\sigma^2\mathbf{S}$ where

$$S = \mathbf{L}'\mathbf{GL} + (\mathbf{X}'\mathbf{L} - \mathbf{p})'\mathbf{W}(\mathbf{X}'\mathbf{L} - \mathbf{p}). \tag{4k.1.2}$$

(b) If $\boldsymbol{\beta}$ is considered as a random variable with an *a priori* mean dispersion $E(\boldsymbol{\beta}\boldsymbol{\beta}') = \sigma^2\mathbf{W}$, then (4k.1.1) reduces to $\sigma^2 S$ when a further expectation of (4k.1.1) with respect to $\boldsymbol{\beta}$ is taken.

(c) We observe that (4k.1.1) consists of two parts, one representing the variance and the other bias. In such a case, the choice of $\mathbf{W}$ in S represents the relative weight we attach to bias compared to variance. Then $\mathbf{W}$ may be suitably chosen depending on the relative importance of bias and variance in the estimator.

In the cases (b) and (c), the matrix $\mathbf{W}$ is n.n.d. and of rank possibly greater than one. The result (4k.1.3) of (i) is true for any choice of $\mathbf{W}$ whatever its rank may be provided $\mathbf{Y} \in \mathcal{M}(\mathbf{V} + \mathbf{XWX}')$.

Thus we are led to choose S as the criterion and the proposition (i) provides the answer to BLE.

(i) *The BLE of* $\mathbf{p}'\boldsymbol{\beta}$ *is* $\mathbf{p}'\tilde{\boldsymbol{\beta}}$ *where*

$$\tilde{\boldsymbol{\beta}} = \mathbf{WX}'(\mathbf{G} + \mathbf{XWX}')^{-}\mathbf{Y}. \tag{4k.1.3}$$

The minimum value of S is attained when

$$(\mathbf{G} + \mathbf{XWX}')\mathbf{L} = \mathbf{XWp} \quad \text{or} \quad \mathbf{ML} = \mathbf{XWp},\ \mathbf{M} = \mathbf{G} + \mathbf{XWX}',$$

so that $\mathbf{L} = \mathbf{M}^{-}\mathbf{XWp}$ is a solution and the BLE of $\mathbf{p}'\boldsymbol{\beta}$ is

$$\mathbf{L}'\mathbf{Y} = \mathbf{p}'\mathbf{WX}'\mathbf{M}^{-}\mathbf{Y} = \mathbf{p}'\tilde{\boldsymbol{\beta}}. \tag{4k.1.4}$$

BLE always exists whether $\mathbf{p}'\boldsymbol{\beta}$ admits a linear unbiased estimator or not.

In order to compare BLE with BLUE, let us assume that $\mathbf{X}$ is of full rank so that $\boldsymbol{\beta}$ admits unbiased estimation and $R(\mathbf{M}) = R(\mathbf{G} + \mathbf{XWX}') = R(\mathbf{G}:\mathbf{X})$. Then $\hat{\boldsymbol{\beta}}$, the BLUE of $\boldsymbol{\beta}$, is

$$\hat{\boldsymbol{\beta}} = (\mathbf{X}'\mathbf{M}^{-}\mathbf{X})^{-1}\mathbf{X}'\mathbf{M}^{-}\mathbf{Y}, \text{ using (4i.4.3)} \tag{4k.1.5}$$

whereas the BLE of $\boldsymbol{\beta}$ is

$$\tilde{\boldsymbol{\beta}} = \mathbf{WX}'\mathbf{M}^{-}\mathbf{Y} = \mathbf{WX}'\mathbf{M}^{-}\mathbf{X}\hat{\boldsymbol{\beta}} = \mathbf{T}\hat{\boldsymbol{\beta}} \text{ say} \tag{4k.1.6}$$

which establishes the relationship between $\hat{\boldsymbol{\beta}}$ and $\tilde{\boldsymbol{\beta}}$. The mean dispersion error of $\tilde{\boldsymbol{\beta}}$ is

$$E(\tilde{\boldsymbol{\beta}} - \boldsymbol{\beta})(\tilde{\boldsymbol{\beta}} - \boldsymbol{\beta})' = \mathbf{T}E(\hat{\boldsymbol{\beta}} - \boldsymbol{\beta})(\hat{\boldsymbol{\beta}} - \boldsymbol{\beta})'\mathbf{T}' + (\mathbf{T} - \mathbf{I})\boldsymbol{\beta}\boldsymbol{\beta}'(\mathbf{T} - \mathbf{I})'$$

$$\mathbf{F} = \mathbf{TDT}' + (\mathbf{T} - \mathbf{I})\boldsymbol{\beta}\boldsymbol{\beta}'(\mathbf{T} - \mathbf{I})' \tag{4k.1.7}$$

where $\mathbf{D}$ is the dispersion matrix of $\hat{\boldsymbol{\beta}}$ and $\boldsymbol{\beta}$ is the true value.

The matrix $\mathbf{D} - \mathbf{TDT}$ is n.n.d. so that $\mathbf{D} - \mathbf{F}$ will be n.n.d. for a certain range of $\boldsymbol{\beta}$. Thus, if we have some knowledge about the domain in which $\boldsymbol{\beta}$ is expected to lie then we may chose $\mathbf{W}$ suitably to ensure that BLE has uniformly smaller mean square error than the BLUE.

The reader may refer to related work by Hoerl and Kennard (1970a, 1970b). Note that $\mathbf{T}$ in (4k.1.6) is a sort shrinkage factor bringing $\hat{\boldsymbol{\beta}}$ towards zero, which is the usual technique employed in improving unbiased estimators at least in some range of the parameter space.

4k.2 Best Linear Minimum Bias Estimation (BLIMBE)

Let us consider the GGM model $(\mathbf{Y}, \mathbf{X}\boldsymbol{\beta}, \sigma^2\mathbf{G})$ which includes the case of constraints on the $\boldsymbol{\beta}$ parameter. When $R(\mathbf{X})$ is not full, not all linear parametric functions admit linear unbiased estimators (LUE). In fact, an n.s. condition for $\mathbf{p}'\boldsymbol{\beta}$ to have a LUE is that $\mathbf{p} \in \mathcal{M}(\mathbf{X}')$. We raise two questions.

(a) What is the minimum restriction to be put on $\boldsymbol{\beta}$ such that every parametric function admits a LUE and hence the BLUE?
(b) In what sense can we find a BLIMBE of $\mathbf{p}'\boldsymbol{\beta}$ if it does not admit a LUE?

The answer to question (a) is contained in proposition (i) and to (b) in (iii).

(i) *Let $R(\mathbf{X}) = r < m$, the number of parameters. The minimum restriction on $\boldsymbol{\beta}$ can be expressed in two alternative forms:*

(a) $\mathbf{R}\boldsymbol{\beta} = \mathbf{c}$, $R(\mathbf{R}) = m - r$ *and* $\mathcal{M}(\mathbf{R}') \cap \mathcal{M}(\mathbf{X}') = \{\mathbf{0}\}$.
(b) $\boldsymbol{\beta} = \mathbf{d} + \mathbf{T}\mathbf{X}'\boldsymbol{\gamma}$ *where* $\mathbf{T}$ *is any matrix such that* $R(\mathbf{XTX}') = R(\mathbf{X})$ *and* $\boldsymbol{\gamma}$ *is arbitrary parameter.*

The first restriction is obvious, and the second can be deduced from the first. Both the restrictions imply that $\boldsymbol{\beta}$ is confined to a hyperplane of dimension r.

To answer the second question, we follow the method due to Chipman, (1964). The bias in $\mathbf{L}'\mathbf{Y}$ as an estimator of $\mathbf{p}'\boldsymbol{\beta}$ is $(\mathbf{X}'\mathbf{L} - \mathbf{p})'\ \boldsymbol{\beta}$ which may be made small by finding $\mathbf{L}$ such that $\|\mathbf{X}'\mathbf{L} - \mathbf{p}\|$ is a minimum, choosing a suitable norm.

(ii) *Let* $\|\mathbf{X}'\mathbf{L} - \mathbf{p}\|^2 = (\mathbf{X}'\mathbf{L} - \mathbf{p})'\mathbf{B}(\mathbf{X}'\mathbf{L} - \mathbf{p})$ *where* $\mathbf{B}$ *is p.d. Then a LIMBE (Linear minimum bias estimator) of* $\mathbf{p}'\boldsymbol{\beta}$ *is*

$$\mathbf{p}'[(\mathbf{X}')^-_{l(B)}]'\mathbf{Y} = \mathbf{p}'\mathbf{X}^-_{m(B^{-1})}\mathbf{Y}. \tag{4k.2.2}$$

The result (4k.2.2) follows from the definition of a least squares g-inverse and its equivalent as a minimum norm g-inverse (see 1c.5.7).

LIMBE may not be unique. In such a case, we choose one with the smallest variance, which may be called BLIMBE (Best linear minimum bias estimator).

(iii) *The BLIMBE of* $\mathbf{p}'\boldsymbol{\beta}$ *i.e., a linear function* $\mathbf{L}'\mathbf{Y}$ *such that* $\mathbf{L}'\mathbf{G}\mathbf{L}$ *is a minimum in the class of* $\mathbf{L}$ *which minimizes* $(\mathbf{X}'\mathbf{L} - \mathbf{p})'\mathbf{B}(\mathbf{X}'\mathbf{L} - \mathbf{p})$ *is* $\mathbf{p}'\hat{\boldsymbol{\beta}}$ *where*

$$\hat{\boldsymbol{\beta}} = [(\mathbf{X}')_{\mathbf{BG}}{}^+]'\mathbf{Y} \tag{4k.2.3}$$

and $(\mathbf{X}')_{\mathbf{BG}}{}^+$ *is the minimum* **G**-*norm* **B**-*least squares g-inverse of* $\mathbf{X}'$.

The result (4k.2.3) follows from the definition of a minimum norm least squares g-inverse [(iii), **1c.5**].

If $\mathbf{G}$ is of full rank then

$$\hat{\boldsymbol{\beta}} = [(\mathbf{X}')_{\mathbf{BG}}{}^+]'\mathbf{Y} = \mathbf{X}^+_{\mathbf{G}^{-1}\mathbf{B}^{-1}}\mathbf{Y} \tag{4k.2.4}$$

using the result (1c.5.9).

COMPLEMENTS AND PROBLEMS

1 *Linear estimation.* Consider the setup $(\mathbf{Y}, \mathbf{X\beta}, \sigma^2\mathbf{I})$ of **4a** and a parametric function $\mathbf{P}'\boldsymbol{\beta}$. Let $\mathbf{P}$, $\boldsymbol{\beta}$, $\mathbf{L}$, $\mathbf{D}$, $\boldsymbol{\lambda}$ denote column vectors.

1.1 Show that if $\mathbf{P} = \mathbf{X}'\mathbf{L}$ admits a solution for $\mathbf{L}$, then $\mathbf{L}'\mathbf{Y}$ is unbiased for $\mathbf{P}'\boldsymbol{\beta}$. Let $\mathbf{L}_0$ be a solution $\in \mathscr{M}(\mathbf{X})$ (which exists and is unique). Then $V(\mathbf{L}_0'\mathbf{Y}) \leqslant V(\mathbf{L}'\mathbf{Y})$ for any other solution. [See examples **16** and **17**, p. 13.]

1.2 Find $\mathbf{P}$ such that $\mathbf{P}'\boldsymbol{\beta}$ is estimable and the ratio of the variance of its l.s.e. to $\mathbf{P}'\mathbf{P}$ is a maximum (or minimum).

1.3 Show that in general a linear function $\mathbf{L}'\mathbf{Y}$ has minimum variance as an unbiased estimate of $E(\mathbf{L}'\mathbf{Y})$ if and only if $\text{cov}(\mathbf{L}'\mathbf{Y}, \mathbf{D}'\mathbf{Y}) = 0$, that is, $\mathbf{L}'\mathbf{D} = 0$ for all $\mathbf{D}$ such that $E(\mathbf{D}'\mathbf{Y}) = 0$ that is $\mathbf{D}'\mathbf{X} = 0$.

1.4 Show that the number of independent linear functions of $\mathbf{Y}$ with zero expectation is $n - r$ where $r = R(\mathbf{X})$.

1.5 From the results of Examples **1.3** and **1.4** deduce that minimum variance estimators are of the form $\boldsymbol{\lambda}'\mathbf{X}'\mathbf{Y}$, and conversely.

1.6 Show that in the setup $(\mathbf{Y}, \mathbf{X\beta}, \sigma^2\mathbf{I})$ of the least squares theory the conditions of linearity and unbias of an estimator are equivalent to the conditions of linearity and bounded mean square error, assuming that the unknown parameters are unbounded in range. (See Example **4** for other conditions leading to least square estimators.)

2 *Combination of estimators*

2.1 If $t_1, \ldots, t_k$ are unbiased estimators of a (single) parameter θ and $\text{cov}(t_i, t_j) = \sigma_{ij}$, find the linear function of $t_1, \ldots, t_k$ unbiased for θ and having minimum variance.

2.2 Let $t_1, \ldots, t_k$ be uncorrelated estimators of θ, with variances $\sigma_1^2, \ldots, \sigma_k^2$. Show that $\sum c_i t_i$, $(\sum c_i = 1)$ has minimum variance when $c_i = \sigma_i^{-2}/\sum \sigma_i^{-2}$. Find the minimum variance attained.

2.3 Show that if $t_1, \ldots, t_k$ are unbiased minimum variance estimators of the parameters $\theta_1, \ldots, \theta_k$, then $c_1 t_1 + \cdots + c_k t_k$ is the unbiased minimum variance estimator of $c_1\theta_1 + \cdots + c_k\theta_k$.

3 Consider the observations $(y_1, \ldots, y_n) = \mathbf{Y}'$ such that

$$E(y_i) = x_{i1}\beta_1 + \cdots + x_{im}\beta_m$$

and $\text{cov}(y_i, y_j) = \rho\sigma^2$, $V(y_i) = \sigma^2$. Make an orthogonal transformation from $\mathbf{Y}$ to $\mathbf{Z}$ such that $z_1 = (y_1 + \cdots + y_n)/\sqrt{n}$. Show that $\mathbf{Z}_2' = (z_2, \ldots, z_n)$ are uncorrelated and have the same variance $\sigma^2(1 - \rho)$ and $E(\mathbf{Z}_2) = \mathbf{U}\boldsymbol{\beta}$, where $\mathbf{U}$

depends on $\mathbf{X}$ and the particular matrix of transformation. Show that when ρ and σ^2 are unknown, unbiased minimum variance linear estimators of parametric functions are l.s. estimators under the setup $[\mathbf{Z}_2, \mathbf{U\beta}, \sigma^2(1-\rho)\mathbf{I}]$.

4 In the set up $(\mathbf{Y}, \mathbf{X\beta}, \sigma^2\mathbf{I})$ we have constructed an orthogonal transformation $\mathbf{Z}_1 = \mathbf{BY}$, $\mathbf{Z}_2 = \mathbf{DY}$ [see (i), **3b.5**] such that $E(\mathbf{Z}_2) \equiv 0$, $R(\mathbf{D}) = n - r$, $E(\mathbf{Z}_1) = \mathbf{BX\beta} \neq \mathbf{0}$, and all components of $\mathbf{Z}_1$, $\mathbf{Z}_2$ are uncorrelated. We may agree that $\mathbf{Z}_2$ is superfluous for $\boldsymbol{\beta}$ since $E(\mathbf{Z}_2) \equiv 0$ and $C(\mathbf{Z}_1, \mathbf{Z}_2) = 0$. Let $f(\mathbf{Z}_1)$ be any general function of $\mathbf{Z}_1$ which provides a consistent estimator of a parametric function $\mathbf{P}'\boldsymbol{\beta}$, that is, $f(\mathbf{BX\beta}) \equiv \mathbf{P}'\boldsymbol{\beta}$, where $\mathbf{BX\beta}$ is the expected value of $\mathbf{Z}_1$ or the value of $\mathbf{Z}_1$ when there are no errors in the observations. Show that $f(\mathbf{Z}_1)$ is linear in $\mathbf{Z}_1$ and is the l.s. estimator. [We thus have a logical justification of l.s. estimation without appealing to unbias, linearity of estimator, or minimum variance. Superfluity in the observations $\mathbf{Y}$ is, in general, deliberately introduced to obtain an estimate of precision. Thus, the components of $\mathbf{Z}_2$ enable us to estimate the precision of estimators of the unknown parameters $\boldsymbol{\beta}$, although, by themselves they do not provide information on $\boldsymbol{\beta}$.]

5 *Weighing designs.* Suppose that there are m objects whose individual weights have to be ascertained. One method is to weigh each object r times and take the average value as the estimate of its weight. Such a procedure needs a total of mr weighings and the precision of each estimated weight is σ^2/r, where σ^2 is the variance of error in an individual observation. Another method is to weigh the objects in combinations. Each operation consists in placing some of the objects in one balance pan and others in the other pan and placing weights to achieve equilibrium. This results in an observational equation of the type

$$x_1\omega_1 + \cdots + x_m\omega_m = y,$$

where $\omega_1, \ldots, \omega_m$ are hypothetical weights of the objects, $x_i = 0$, 1, or -1 according as the ith object is not used, placed in the left pan or in the right pan and y is the weight required for equilibrium. The n operations described as using different combinations of the objects for the left and right pans yield n observational equations, from which the unknown weights $\omega_1, \ldots, \omega_m$ may be estimated by the method of least squares. The $n \times m$ design matrix of the observational equations $\mathbf{X}$ is a special type where each entry is 0, 1, or -1. The problem is to choose $\mathbf{X}$ in such a way that the precision of the individual estimates is as high as possible.

5.1 Using the result of **4a.12**, show that maximum precision is attained when each entry of $\mathbf{X}$ is either 1 or -1 and the columns of $\mathbf{X}$ are orthogonal.

5.2 For the optimum design in (**5.1**) the precision of each estimate is σ^2/n for n weighing operations. Show that to achieve the same precision by weighing each object individually the number of weighing operations required is mn.

5.3 Estimate the weights of 4 objects from the following observational equations (ignoring bias).

ω_1	ω_2	ω_3	ω_4	Weight
1	1	1	1	20.2
1	−1	1	−1	8.1
1	1	−1	−1	9.7
1	−1	−1	1	1.9
1	1	1	1	19.9
1	−1	1	−1	8.3
1	1	−1	−1	10.2
1	−1	−1	1	1.8

Find the dispersion matrix of the estimates and also the least square estimate of σ^2, the variance of each measurement.

5.4 For the following weighing design, show that the individual estimates of ω_2 and ω_3 cannot be found, but that their sum is estimable. Find the *g*-inverse of the matrix of normal equations and estimate the linear function $\omega_2 + \omega_3$ and the standard error of the estimate.

ω_1	ω_2	ω_3	Weight
1	.	.	5.1
.	1	1	8.2
1	−1	−1	4.9
−1	1	1	3.1

6 Let $a_1 + b_1(x - \bar{x}_1)$ and $a_2 + b_2(x - \bar{x}_2)$ be two estimated regression functions from independent samples of sizes n_1 and n_2. Further let $S_{yy}^{(1)}$, $S_{xy}^{(1)}$, $S_{xx}^{(1)}$ be the corrected sum of squares and products for the first sample and with index (2) for the second sample. Find the confidence interval for the value ξ of x at which the true regression functions cut.

Hint: let $z = a_1 + b_1(\xi - \bar{x}_1) - a_2 - b_2(\xi - \bar{x}_2)$, then $E(z) = 0$ and $V(z) = c\sigma^2$, where

$$c = \frac{1}{n_1} + \frac{(\xi - \bar{x}_1)^2}{S_{xx}^{(1)}} + \frac{1}{n_2} + \frac{(\xi - \bar{x}_2)^2}{S_{xx}^{(2)}}.$$

The $(1 - \alpha)$ confidence region is obtained by solving the quadratic in ξ

$$\frac{z^2}{c\hat{\sigma}^2} = F_\alpha(1, n_1 + n_2 - 4),$$

where

$$\hat{\sigma}^2(n_1 + n_2 - 4) = S_{yy}^{(1)} - \frac{[S_{xy}^{(1)}]^2}{S_{xx}^{(1)}} + S_{yy}^{(2)} - \frac{[S_{xy}^{(2)}]^2}{S_{xx}^{(2)}}.$$

7 Show that a random variable can be written as the sum of two variables, one of which has multiple correlation unity (perfectly determined) with a given set of variables and another uncorrelated with any of the given variables.

8 Let (ρ_{ij}) be the correlation matrix of p variables $x_1, \ldots, x_p$. From the definitions of partial and multiple correlation coefficients given in **4g.1**, show that

(a) $1 - \rho_{1(2\cdots p)}^2 = (1 - \rho_{12}{}^2)(1 - \rho_{13.2}^2)\cdots(1 - \rho_{1p.(2\cdots p-1)}^2)$.

(b) $\rho_{12.3} = (\rho_{12} - \rho_{13}\rho_{23})/(1 - \rho_{13}{}^2)^{1/2}(1 - \rho_{23}{}^2)^{1/2}$.

(c) $\rho_{12} = (\rho_{12.3} - \rho_{13.2}\rho_{23.1})/(1 - \rho_{13.2}^2)^{1/2}(1 - \rho_{23.1}^2)^{1/2}$

(d) $\rho_{12.(34\cdots p-1)} = \dfrac{\rho_{12.(3\cdots p)} + \rho_{1p.(2\cdots p-1)}\rho_{2p.(13\cdots p-1)}}{(1 - \rho_{1p.(2\cdots p-1)}^2)^{1/2}(1 - \rho_{2p.(13\cdots p-1)}^2)^{1/2}}$.

(e) $\rho_{12.(34\cdots p)} = \dfrac{\rho_{12.(3\cdots p-1)} - \rho_{1p.(3\cdots p-1)}\rho_{2p.(3\cdots p-1)}}{(1 - \rho_{1p.(3\cdots p-1)}^2)^{1/2}(1 - \rho_{2p.(3\cdots p-1)}^2)^{1/2}}$.

9 Let us define $\rho_{1(2.34\cdots p)}$ as the correlation between x_1 and the residual of x_2 removing the regression on $x_3, \ldots, x_p$. Such a coefficient is called *part* correlation. Unlike partial correlation it is not symmetrical in symbols 1 and 2. Show that

(a) $\rho_{1(2.34\cdots p)}^2 \leqslant \rho_{12.(3\cdots p)}^2$.

(b) $\rho_{1(234\cdots p)}^2 = \rho_{1p}^2 + \rho_{1(p-1.p)}^2 + \cdots + \rho_{1(2.34\cdots p)}^2$.

10 *Estimation of horizontal distance between two parallel regression lines (Fieller's theorem).* Let $\bar{y}^{(1)}, \bar{x}^{(1)}$ be the mean values and $S_{yy}^{(1)}, S_{xy}^{(1)}, S_{xx}^{(1)}$, the sum of squares and products from a sample of size n_1 on a pair of variables (y, x), and with index (2) for an independent sample of size n_2. If the true regressions are given to be parallel, show that the estimate of the common regression coefficient is

$$b = \frac{S_{xy}^{(1)} + S_{xy}^{(2)}}{S_{xx}^{(1)} + S_{xx}^{(2)}} = \frac{S_{xy}}{S_{xx}}.$$

If η is the horizontal distance between the two regression lines then the statistic $d = a_1 - a_2 + b\eta$ where $a_1 = \bar{y}^{(1)} - b\bar{x}^{(1)}$ and $a_2 = y^{(2)} - b\bar{x}^{(2)}$ has zero expectation and variance

$$\sigma^2\left\{\frac{1}{n_2} + \frac{1}{n_2} + \frac{[\bar{x}^{(1)} - \bar{x}^{(2)} - \eta]^2}{S_{xx}}\right\} = \sigma^2 C.$$

An estimate of σ^2 is

$$\hat{\sigma}^2 = S_{yy}^{(1)} + S_{yy}^{(2)} - \frac{S_{xy}{}^2}{S_{xx}}$$

on $(n_1 + n_2 - 3)$ D.F. Hence show that the confidence interval for η is given by solving the quadratic in η

$$\frac{d^2}{\hat{\sigma}^2 C} = F_\alpha(1, n_1 + n_2 - 3).$$

11 *Robustness of estimators in linear models.* Consider $(\mathbf{Y}, \mathbf{X\beta}, \sigma^2\mathbf{G})$ and let $\mathbf{P}'\hat{\boldsymbol{\beta}}$ be the BLUE of $\mathbf{P}'\boldsymbol{\beta}$ assuming $\mathbf{G} = \mathbf{I}$. If $\mathbf{P}'\hat{\boldsymbol{\beta}}$ is also the BLUE under $(\mathbf{Y}, \mathbf{X\beta}, \boldsymbol{\sigma}^2\mathbf{G})$ for all estimable $\mathbf{P}'\boldsymbol{\beta}$, then it is n.s. that

$$\mathbf{G} = \mathbf{XAX}' + \mathbf{ZBZ}'$$

where $\mathbf{A}$ and $\mathbf{B}$ are arbitrary symmetric matrices and $\mathbf{Z}$ is a matrix of maximum rank such that $\mathbf{X}'\mathbf{Z} = \mathbf{0}$. [For a proof and a further discussion of the problem, see Rao (1967c, 1968a, 1971e).]

REFERENCES

Aitken, A. C. (1935), On least squares and linear combination of observations, *Proc. Roy. Soc. Edin.* **A**, 55, 42–48.

Bose, R. C. (1950–51), *Least Square Aspects of the Analysis of Variance*, Mimeographed series, no. 9. North Carolina University.

Box, G. E. P. (1954), The exploration and exploitation of response surfaces. Some general considerations and examples, *Biometrics* **10**, 16–60.

Chipman, J. S. (1964), On least squares with insufficient observations, *J. Amer. Statist. Assoc.* **59**, 1078–1111.

Cochran, W. G. (1951), Improvement by means of selection, *Proc.* (second) *Berkeley Symp. on Math. Statist. Prob.*, 449–470.

Cochran, W. G. and G. M. Cox, (1957), *Experimental Designs* (second edition), Wiley, New York.

Fisher, R. A. (1925), *Statistical Methods for Research Workers* (first edition 1925, eleventh edition 1950), Oliver and Boyd, Edinburgh.

Focke, J. and Dewess, G. (1972), Über die Shätzmetnode MINQUE von C. R. Rao und ihre verallgemeinerung, *Math. Operationsforsch. u. Statist.* **3**, 129–144.

Gauss, K. F. (1809), *Werke* **4**, 1–93, Gottingen.

Goldman, A. J. and Zelen, M. (1964), Weak generalized inverses and minimum variance linear unbiased estimation. *J. of Research Nat. Bureau of Standards* **68B**, 151–172.

Hoerl, A. E. and Kennard, R. W. (1970*a*), Ridge regression: Biased estimation for nonorthogonal problems, *Technometrics* **12**, 55–68.

Hoerl, A. E. and Kennard, R. W. (1970b), Ridge regression: Application to nonorthogonal problems, *Technometrics* **12**, 69–82.

Hooke, B. G. E. (1926), A third study of the English skull with special reference to the Farrington Street crania, *Biometrika* **18**, 1–55.

Hsu, P. L. (1958), On the best unbiased quadratic estimate of variance, *Statist. Res. Mem.* **2**, 91–104.

Kempthorne, O. (1952), *The Design and Analysis of Experiments.* Wiley, New York.

Kempthorne, O. and Nordskog, A. W. (1959), Restricted selection indices, *Biometrics*, **15**, 10–19.

Linnik, Yu. V. (1961), *Method of Least Squares and Principles of the Theory of Observations* (Translated from Russian), Pergamon Press, London.

Markoff, A. A. (1900), *Wahrscheinlichkeitsrechnung*, Tebner, Leipzig.

Milliken, George A. and Graybill, Franklin, A. (1970), Extensions of the general linear hypothesis model. *J. Amer. Statist. Assoc.* **65**, 797–807.

Neyman, J. and F. N. David (1938), Extension of the Markoff theorem on least squares, *Stst. Res. Mem.* **2**, 105–116.

Parzen, E. (1961), An approach to time series analysis, *Ann. Math. Statist.* **32**, 951–989.

Pearson, K. (1899), Mathematical contributions to the theory of evolution V. On the reconstruction of the stature of prehistoric races, *Philos. Trans. Roy. Soc.* **A192**, 169–244.

Plackett, R. L. (1960), *Principles of Regression Analysis*, Clarendon Press, Oxford.

Scheffe, Henry (1959), *The Analysis of Variance*, Wiley, New York.

Tukey, John. (1949), One degree of freedom for non-additivity, *Biometrics* **5**, 232–242.

Williams, E. J. (1959), *Regression Analysis*, Wiley, New York.

Wishart, J. (1938), Growth rate determinations in nutrition studies with the bacon pig and their analysis. *Biometrika* **30**, 16–28.

Yates, F. (1949), *Sampling Methods for Censuses and Surveys*, Charles Griffin, London.

Zyskind, G. and Martin, F. B. (1969), On best linear estimation and a general Gauss-Markoff theorem in linear models with arbitrary non-negative covariance structure, *SIAM, J. Appl. Math.* **17**, 1190–1202.

STATISTICAL TABLES

Biometrika tables (1956, 4th edition), *Biometrika Tables for Statisticians* Vol. 1. By E. S. Pearson and H. O. Hartley, Cambridge University Press, Cambridge.

R. M. M. tables (1965), *Tables and Formulae in Statistical Work*. By C. R. Rao, A. Matthai, and S. K. Mitra, Statistical Publishing Society, Calcutta.

F. Y. tables (1963) (6th edition). *Statistical Tables For Biological, Agricultural and Medical Research.* By R. A. Fisher and F. Yates, Oliver and Boyd, London.

Chapter 5

CRITERIA AND METHODS OF ESTIMATION

Introduction. In Chapter 4 we considered the estimation of parameters using only the specified structure of the first and second moments of the observations. In the present chapter we consider more general situations where the distribution function (d.f.) of the observations is specified to be a member of a family $\mathscr{F}$ of d.f.'s. Then the wider problem of estimation is one of choosing a d.f. belonging to $\mathscr{F}$ on the basis of observed data or a given realization of the observations. In some problems our interest may lie only in a particular characteristic (parameter or a function) of the d.f., in which case we need only estimate the parameter and not necessarily the entire d.f.

In either case what is needed is a function of the *observations*, to be called an *estimator* whose value at a given realization of the observations is the *estimate* of the unknown quantity (which may be a d.f. or a parameter). To determine an estimator we need a set of criteria by which its performance can be judged. Obviously any set of criteria depends on the purpose for which an estimate is obtained. Consequently there need not be a single set of criteria by which all estimators can be judged or a single estimator of any given parameter, which is appropriate for all situations. Our attempt in the present chapter is to discuss various criteria of estimation and lay down appropriate procedures for obtaining estimates.

The subject is full of controversies both with respect to the methods of estimation and the expression of uncertainty in the estimates. Some discussion is devoted to these controversies.

The examples at the end of the chapter contain numerous results on estimation. They may be read as additional material on theories and methods of estimation. A numerical example is given at the end of the chapter to illustrate the scoring method for maximum likelihood estimation.

5a MINIMUM VARIANCE UNBIASED ESTIMATION

5a.1 Minimum Variance Criterion

In Chapter 4 we have introduced the method of least squares for estimating parametric functions. It was shown that, for any given parametric function, the least squares estimator possesses the least variance in the class of *linear* unbiased estimators. The status of the least squares estimator in the entire class of unbiased estimators could not be studied as the specification of the distribution of the observations was confined to the first two moments only. It is, therefore, of some interest to examine some general methods of obtaining *minimum variance estimators by considering the entire class of unbiased estimators* in cases where the distribution of the observations is restricted to a smaller class.

Intuitively, by an estimator of a parameter θ we mean a function T of the observations $(x_1, \ldots, x_n)$ which is closest to the true value in some sense. In laying down criteria of estimation one attempts to provide a measure of closeness of an estimator to the true value of a parameter and to impose suitable restrictions on the class of estimators. An optimum estimator in the restricted class is determined by minimizing the measure of closeness. Some restrictions on the class of estimators seem to be necessary to avoid trivial estimators such as $T = c$ (a constant), which is best in any sense when the unknown value of the parameter is c.

There need not be a single set of criteria of estimation useful in all situations. We shall examine some sets of criteria available in statistical literature and discuss the merits and demerits of each.

It would, of course, be ideal if there exists a function T such that, compared to any other statistic T', the probabilities satisfy the relationship

$$P(\theta - \lambda_1 < T < \theta + \lambda_2) \geqslant P(\theta - \lambda_1 < T' < \theta + \lambda_2) \qquad (5a.1.1)$$

for all possible λ_1 and λ_2 in a chosen interval $(0, \lambda)$ and for all θ. If the condition (5a.1.1) is satisfied for all λ, a necessary condition is

$$E(T - \theta)^2 \leqslant E(T' - \theta)^2, \qquad (5a.1.2)$$

where E stands for expectation, that is, the mean square error (m.s.e) of T is a minimum. If, further, it is assumed that the estimator should be unbiased, it follows that

$$V(T) \leqslant V(T'), \qquad (5a.1.3)$$

where V stands for variance.

Estimators satisfying the criterion of *highest concentration* (5a.1.1) or the criterion of *minimum mean square error* (5a.1.2) do not generally exist. The restriction that the estimator should be unbiased, however, provides us with optimum estimators, in the sense of having minimum variance, in some problems. The condition of unbiasedness may be particularly unattractive in that many biased estimators with possibly smaller m.s.e. lose their claim as estimating functions. There are also situations in which unbiased estimators do not exist and in which no minimum variance estimator exists although unbiased estimators exist. [See Example **12** at the end of the chapter.]

Consider the statistic

$$s^2 = \frac{\sum (x_i - \bar{x})^2}{n - 1}$$

based on n observations $x_1, \ldots, x_n$ from $N(\mu, \sigma^2)$. It is known that $E(s^2) = \sigma^2$, so that s^2 is unbiased for σ^2. Since $(n - 1)s^2/\sigma^2 \sim \chi^2(n - 1)$,

$$V\left[\frac{(n - 1)s^2}{\sigma^2}\right] = V[\chi^2(n - 1)] = 2(n - 1)$$

giving $V(s^2) = 2\sigma^4/(n - 1)$. Let us consider

$$E(cs^2 - \sigma^2)^2 = E[c(s^2 - \sigma^2) - \sigma^2(1 - c)]^2$$

$$= \left\{\frac{2c^2}{n - 1} + (1 - c)^2\right\}\sigma^4,$$

which attains the minimum when $c = (n - 1)/(n + 1)$, the minimum value being

$$\frac{2\sigma^4}{n + 1} < \frac{2\sigma^4}{n - 1}.$$

Therefore, for all n and σ

$$E\left\{\frac{\sum (x_i - \bar{x})^2}{n + 1} - \sigma^2\right\}^2 < E\left\{\frac{\sum (x_i - \bar{x})^2}{n - 1} - \sigma^2\right\}^2.$$

If we use the criterion of m.s.e. the estimator $\sum (x_i - \bar{x})^2/(n + 1)$ which is biased for σ^2 is better than s^2. Which then should be preferred? The answer obviously depends on the purpose for which an estimate is obtained.

Let r be the number of successes in n Bernoulli trials with an unknown probability of success π. But there is no function f of r such that $E(f) = 1/\pi$, a situation in which unbiased estimation is not possible.

The most important role of unbiased minimum variance estimation is in the pooling of information supplied by independent estimators of a parameter. Let $T_1, T_2, \ldots$ be estimators of θ such that

$$E(T_i) = \theta, \qquad V(T_i) = \sigma_i^2 < \sigma^2,$$

and suppose that nothing is specified about the actual distribution of T_i. If $\bar{T}_k = (T_1 + \cdots + T_k)/k$, then

$$E(\bar{T}_k) = \theta, \qquad V(\bar{T}_k) = \frac{\sum \sigma_i^2}{k^2} < \frac{\sigma^2}{k},$$

so that as $k \to \infty$, $V(\bar{T}_k) \to 0$, which implies that $\bar{T}_k$ gets closer and closer to the true value as $k \to \infty$. On the other hand if T_i is biased, and has an unknown bias β, then $\bar{T}_k$ approaches the wrong value $\theta + \beta$, instead of the true value θ.

A bias in any individual estimator may not be serious if it is small compared to its (standard) error. But when several such estimators are pooled in the manner indicated, the bias remains the same while the variance $\to 0$ as $k \to \infty$. The bias becomes large compared to the standard error after some stage.

Or suppose that we wish to examine the relationship between a response y and a factor variable x by obtaining independent estimates of y at some values of x. There is a possibility of making systematic errors in the estimated relationship if the estimators of y for different values of x are not unbiased. It is clear that if it is desirable to be unbiased in these situations, estimators with smaller variances are preferable to ensure rapid convergence to the true value.

5a.2 Some Fundamental Results on Minimum Variance Estimation

Let $\mathscr{P}_\theta$ be a class of probability measures over the sample space of observations X, indexed by a parameter θ (which may be vector valued) with values in a specified parameter space Θ. In most applications X is generally an n-dimensional vector, but it could be more general for purposes of the present discussion. Let U_g be the class of all unbiased estimators (functions of X) of g, a specified function of θ, and U_0 be the class of all functions with zero expectation. Thus $T \in U_g$, iff $E(T \mid \theta) = g(\theta)$ for each $\theta \in \Theta$, and $f \in U_0$, iff $E(f \mid \theta) = 0$ for each θ. We shall prove a number of results concerning minimum variance unbiased estimators (m.v.u.e.).

(i) *A necessary and sufficient condition that an estimator $T \in U_g$ has minimum variance at the value $\theta = \theta_0$ is that* $\text{cov}(T, f \mid \theta_0) = 0$ for every $f \in U_0$ *such that* $V(f \mid \theta_0) < \infty$ *provided* $V(T \mid \theta_0) < \infty$.

The necessity is easily proved by considering $(T + \lambda f) \in U_g$ for arbitrary λ and showing that, for any λ in the interval $[0, -2\,\text{cov}(T, f)/V(f)]$

$$V(T + \lambda f) = V(T) + 2\lambda\,\text{cov}(T, f) + \lambda^2 V(f) \leqslant V(T) \qquad (5a.2.1)$$

unless $\text{cov}(T, f) = 0$. To prove sufficiency, let $T' \in U_g$ be another estimator such that $V(T') < \infty$ at θ_0. Then $(T - T') \in U_0$, and by the condition given

$$E[T(T - T') \mid \theta_0] = 0,$$

that is,

$$V(T) = \text{cov}(T, T') \quad \text{or} \quad \sqrt{V(T)} = \rho\sqrt{V(T')} \leqslant \sqrt{V(T')}, \tag{5a.2.2}$$

where ρ is the correlation between T and T'. Hence (i) is proved. [In the proof all variances and covariances are computed at θ_0].

Note. (a) The correlation between a m.v.u.e., and any unbiased estimator is non-negative (5a.2.2).

(b) If there are two unbiased estimators with the same minimum variance, their correlation $\rho(\theta_0)$ is unity, that is, they are the same except for a set of samples of probability measure zero for θ_0 [follows from (5a.2.2)].

(c) If we restrict estimators to a particular class of functions G, such as the class of linear or continuous functions etc., then besides being unbiased the necessary and sufficient condition is $E(Tf \mid \theta_0) = 0$ for all $f \in U_0 \cap G$, and $V(f \mid \theta_0) < \infty$.

(d) Let us restrict T to any class H. Then a necessary and sufficient condition that the m.s.e., $E(T - \theta)^2$, is a minimum at θ_0 is that

$$E[(T - \theta_0)(T - W) \mid \theta_0] = 0$$

for any $W \in H$.

(e) If T_1 and T_2 are m.v.u. estimators of $g_1(\theta)$, $g_2(\theta)$, then $b_1 T_1 + b_2 T_2$ is a m.v.u. estimator of $b_1 g_1(\theta) + b_2 g_2(\theta)$ when b_1, b_2 are fixed constants. [This is an immediate consequence of result (i).]

(f) If T has minimum variance for each θ, it is said to be a uniformly m.v.u. estimator.

For example, consider the observational setup $(\mathbf{Y}, \mathbf{X}\boldsymbol{\beta}, \sigma^2\mathbf{I})$ of the theory of least squares discussed in Chapter 4. Let $\mathbf{Y} \sim N(\mathbf{X}\boldsymbol{\beta}, \sigma^2\mathbf{I})$ and $f(\mathbf{Y}) \in U_0$, so that

$$\int f(\mathbf{Y}) e^{-(\mathbf{Y} - \mathbf{X}\boldsymbol{\beta})'(\mathbf{Y} - \mathbf{X}\boldsymbol{\beta})/2\sigma^2}\, dv \equiv 0,$$

or

$$\int f(\mathbf{Y})\, e^{-\mathbf{Y}'\mathbf{Y}/2\sigma^2 + \boldsymbol{\beta}'\mathbf{X}'\mathbf{Y}/\sigma^2} dv \equiv 0. \tag{5a.2.3}$$

Differentiating (5a.2.3) with respect to β_i, the ith component of $\boldsymbol{\beta}$, we see that

$$\int f(\mathbf{Y})(x_{1i} y_1 + \cdots + x_{ni} y_n) e^{-(\mathbf{Y} - \mathbf{X}\boldsymbol{\beta})'(\mathbf{Y} - \mathbf{X}\boldsymbol{\beta})/2\sigma^2}\, dv = 0, \tag{5a.2.4}$$

which shows that

$$Q_i = x_{1i} y_1 + \cdots + x_{ni} y_n,$$

and, therefore, any linear function of Q_i, $\lambda_1 Q_1 + \cdots + \lambda_m Q_m = \boldsymbol{\lambda}'\mathbf{Q} = \boldsymbol{\lambda}'\mathbf{X}'\mathbf{Y}$ (in matrix notation) is a uniformly m.v.e. of its expected value. But for any vector $\boldsymbol{\lambda}$, $\boldsymbol{\lambda}'\mathbf{X}'\mathbf{Y}$ is the least squares estimator (l.s.e.) of a parametric function as derived in (4a.2.5) (where it was only shown that it has minimum variance in the class of linear unbiased estimators). *Thus the assumption of normality enables us to prove the stronger result that the l.s.e. has minimum variance in the entire class of unbiased estimators.*

It was shown by the author (Rao, 1959c) that if each component of $(\mathbf{Y} - \mathbf{X}\boldsymbol{\beta})$ has the same distribution with moments of all orders finite and the l.s. estimators have minimum variance in the class of all unbiased estimators, then each component of $(\mathbf{Y} - \mathbf{X}\boldsymbol{\beta})$ has a normal distribution. We thus have a new type of characterization of the normal distribution.

By differentiating (5a.2.4) with respect to β_j we find $E(fQ_iQ_j) = 0$ so that Q_iQ_j is an m.v.e. Differentiating (5a.2.3) with respect to σ^2 we find that $\mathbf{Y}'\mathbf{Y}$ is an m.v.e. Hence for any arbitrary constants b_{ij}, $\mathbf{Y}'\mathbf{Y} - \sum\sum b_{ij} Q_i Q_j$ is an m.v.e. But for a suitable choice of b_{ij},

$$\mathbf{Y}'\mathbf{Y} - \sum\sum b_{ij} Q_i Q_j = \mathbf{Y}'\mathbf{Y} - \mathbf{Q}'\mathbf{B}\mathbf{Q} = \mathbf{Y}'\mathbf{Y} - \mathbf{Y}'\mathbf{X}\mathbf{B}\mathbf{X}'\mathbf{Y} = R_0{}^2,$$

[where $\mathbf{B} = (b_{ij})$] is the residual sum of squares in the least squares theory and is unbiased for $(n - r)\sigma^2$. *Thus, under the normality assumption, $R_0{}^2/(n - r)$ is the m.v.u.e. of σ^2 in the entire class of unbiased estimators.*

As examples of estimators with restrictions on the functional form, let us consider a p.d. of the type $P(x_1 - \theta, \ldots, x_n - \theta)$ of n observations $x_1, \ldots, x_n$. Let us restrict the class of estimating functions to those which satisfy the translation property,

$$f(x_1 + \alpha, \ldots, x_n + \alpha) = \alpha + f(x_1, \ldots, x_n) \quad \text{for any } \alpha. \tag{5a.2.5}$$

The function $\bar{x} = (x_1 + \cdots + x_n)/n$ satisfies the property (5a.2.5). Any other function differs from $\bar{x}$ by h, which is a function of the differences $y_1 = x_2 - x_1, \ldots, y_{n-1} = x_n - x_{n-1}$ only. Let y represent the set $y_1, \ldots, y_{n-1}$. Then it is easily seen that

$$E(\bar{x} \mid y) = \theta + e(y)$$

where $e(y)$ does not involve θ and

$$\begin{aligned} E[\bar{x} + h(y) - \theta]^2 &= \underset{y}{E}\,\underset{\bar{x}}{E}\,[\bar{x} - \theta - e(y) + e(y) + h(y)]^2 \\ &= E[v(y)] + E[e(y) + h(y)]^2, \end{aligned} \tag{5a.2.6}$$

where $v(y) = E([\bar{x} - \theta - e(y)]^2 \mid y)$ does not involve $h(y)$. The minimum of (5a.2.6) is, therefore, attained when $h(y) = -e(y)$. For such a choice the statistic $\bar{x} - e(y)$ satisfies (5a.2.5) and uniformly has the least mean square error, as well as being unbiased. For other examples see (Rao, 1952a, 1952b).

(ii) *If the sample consists of n independent observations from the same distribution, then minimum variance unbiased estimators are symmetric in the observations.*

It is easy to establish that if $T(x_1, \ldots, x_n)$ is an unbiased estimator of a parameter, the symmetric function $\sum T(x_a, x_b, \ldots)/n!$ where the summation is over all possible permutations of $x_1, \ldots, x_n$ is also unbiased and has no larger variance than that of T.

Let $V[T(x_1, \ldots, x_n)] = \sigma^2$. Then by symmetry $V[T(x_a, x_b, \ldots)] = \sigma^2$ for any permutation of $x_1, \ldots, x_n$. Furthermore, for any two permutations $x_a, x_b, \ldots$ and $x_{a'}, x_{b'}, \ldots$, $\operatorname{cov}[T(x_a, x_b, \ldots), T(x_{a'}, x_{b'}, \ldots)] \leqslant \sigma^2$. Now

$$\begin{aligned} V[\textstyle\sum T(x_a, x_b, \ldots)/n!] &= \frac{1}{(n!)^2} \sum V[T(x_a, x_b, \ldots)] \\ &\quad + \frac{1}{(n!)^2} \sum\sum \operatorname{cov}[T(x_a, x_b, \ldots), T(x_{a'}, x_{b'}, \ldots)] \\ &\leqslant \sigma^2 \frac{n! + n!(n! - 1)}{(n!)^2} = \sigma^2. \end{aligned}$$

The importance of result (ii) is realized in the estimation of parameters when the exact form of the underlying distribution is not known. For instance $\bar{x} = (x_1 + \cdots + x_n)/n$ is symmetric and unbiased for the mean of the distribution. This is unique as an unbiased estimator when nothing is known about the form of the distribution except that its first moment exists. For, if an alternative to $\bar{x}$ is $\bar{x} + \delta$ where δ is symmetric in x_i and $E[\delta] = 0$, then it can be shown by taking particular distributions that $\delta \equiv 0$. Similarly if $x_1, \ldots, x_n$ are i.i.d. variables, then

$$s^2 = \frac{x_1{}^2 + \cdots + x_n{}^2 - n\bar{x}^2}{n - 1}$$

is the m.v. unbiased estimator of the population variance when nothing is specified about the distribution of x_i.

(iii) *Let T be a sufficient statistic for θ where both T and θ may be vector valued and T_1 any other statistic. If g is any function of θ, then*

$$E[T_1 - g(\theta)]^2 \geqslant E[h(T) - g(\theta)]^2$$

where $h(t) = E(T_1 \mid T = t)$ *is independent of* θ. $h(T)$ *is unbiased for* $g(\theta)$ *if* T_1 *is unbiased.*

Since T is sufficient for θ, the conditional expectation of T_1 given $T = t$ is independent of θ. Thus $h(t)$ is independent of θ. Using the result (2b.3.4) for conditional expectation, we seen that

$$E(T_1) = \underset{T}{E}[E(T_1 \mid T)] = E[h(T)] \tag{5a.2.7}$$

which shows that T_1 and $h(T)$ have the same expectation. Furthermore, by using (2b.3.7),

$$\begin{aligned} E\{h(T)[T_1 - h(T)]\} &= E\{h(T)[E(T_1 - h(T) \mid T)]\} \\ &= 0, \qquad \text{since} \qquad E(T_1 \mid T = t) = h(t). \end{aligned} \tag{5a.2.8}$$

Consider

$$\begin{aligned} E[T_1 - g(\theta)]^2 &= E[T_1 - h(T) + h(T) - g(\theta)]^2 \\ &= E[T_1 - h(T)]^2 + E[h(T) - g(\theta)]^2 \\ &\geqslant E[h(T) - g(\theta)]^2, \end{aligned} \tag{5a.2.9}$$

since the product term

$$\begin{aligned} E[T_1 - h(T)][h(T) - g(\theta)] &= E\{[T_1 - h(T)]h(T)\} - g(\theta)E[T_1 - h(T)] \\ &= 0 - 0 = 0, \end{aligned}$$

if we use (5a.2.8) and (5a.2.7). The equality in (5a.2.9) is attained only when $T_1 = h(T)$ almost everywhere. From (5a.2.7), $E[h(T)] = g(\theta)$ if $E(T_1) = g(\theta)$.

The result (iii) proved by the author (Rao, 1945d) and also by Blackwell (1947) establishes an important property of the sufficient statistic. Given any statistic, we can find a function of the sufficient statistic which is uniformly better in the sense of mean square error or minimum variance (if no bias is imposed).

Note 1. A sufficient statistic is said to be complete if no function of it has zero expectation unless it is zero almost everywhere with respect to each of the measures $\mathscr{P}_\theta$. *If a complete sufficient statistic exists, then every function of it is a uniformly m.v.u.e. of its expected value.* In view of (iii) we have to search for m.v. estimators only in the class of functions of a sufficient statistic. The condition of completeness implies that there is a unique function of the sufficient statistic unbiased for any given parametric function. In such a case, to find the m.v.e. it is enough to start with any unbiased estimator and take its conditional expectation, given the sufficient statistic Rao (1946h, 1948d).

Note 2. It is seen that in (5a.2.7), sufficiency of T is used only to ensure that $h(T)$ is a statistic independent of the unknown parameter. Then the proposition (iii) remains valid for any two particular statistics (T_1, T) which satisfy the property that $E(T_1 \mid T = t)$ is independent of θ for all t, without demanding that T is sufficient, as pointed out by Arnold and Katti (1972). The proof remains the same.

(iv) *Instead of mean square error, let us consider any convex loss function* $W(T_1, \theta)$. *By using Jensen's inequality* (1e.5.6) *we have* $E[W(T_1, \theta) \mid T] \geqslant W(h(T), \theta)$. *Hence, taking expectations over T, we see that*

$$E[W(T_1, \theta)] \geqslant E[W(h(T), \theta)], \tag{5a.2.10}$$

which gives a generalization of (5a.2.9) *to more general functions that measure the deviation between the estimator and the parameter.*

Example 1. Consider the statistic r, the observed number of successes in n Bernoullian trials with probability π of success. Let $f(r)$ be a function such that $E[f(r)] = 0$ for all π, that is,

$$\sum f(r) \binom{n}{r} \pi^r (1 - \pi)^{n-r} = 0, \qquad \text{for } 0 < \pi < 1$$

$$\Rightarrow \sum_0^n \binom{n}{r} f(r) x^r \qquad = 0, \qquad \text{where } x = \frac{\pi}{1 - \pi}. \tag{5a.2.11}$$

The vanishing of the polynomial (5a.2.11) in x for $0 < x < \infty$ implies that all its coefficients are zero, or, $f(r) = 0$ for all r. Hence r is a complete sufficient statistic and, therefore, any function of r is an m.v.e. It may be verified that

$$E\left(\frac{r}{n}\right) = \pi, \qquad E\left(\frac{r}{n} \frac{r-1}{n-1}\right) = \pi^2, \qquad E\left(\frac{r}{n} \frac{n-r}{n-1}\right) = \pi(1 - \pi),$$

so that the functions inside the expectation are m.v. estimators of the corresponding expected values. We see that not only π but $\pi(1 - \pi)$, which occurs in the variance of the estimator of π, also admits m.v.u. estimation. For other examples see (Lehmann and Scheffe, 1950; Rao, 1945d, 1946h, 1948d, 1949c).

Example 2. An extremely interesting application of (iii) by Kolmogorov (1950) is in the unbiased estimation of the proportion π of the normal population above a given value c on the basis of a sample $x_1, \ldots, x_n$. In this application,

$$\bar{x} = \frac{x_1 + \cdots + x_n}{n} \qquad \text{and} \qquad S^2 = (x_1 - \bar{x})^2 + \cdots + (x_n - \bar{x})^2$$

are sufficient for the unknown mean μ and variance σ^2 of the population. It may be verified that they are also complete. Let us consider the statistic

$$\left.\begin{aligned} T_1 &= 0, \quad \text{if } x_1 \leqslant c \\ &= 1, \quad \text{if } x_1 > c \end{aligned}\right\} \qquad \text{where } x_1 \text{ is the first observation,}$$

which is clearly unbiased for π. An application of result (iii) shows that the minimum variance unbiased estimator of π is $E(T_1 | \bar{x}, S)$. To find an explicit value of this statistic, consider

$$\begin{aligned} E(T_1 | \bar{x}, S) &= P[x_1 > c | \bar{x}, S] \\ &= P\left\{\frac{\sqrt{n}(x_1 - \bar{x})}{\sqrt{n-1}\, S} > \frac{\sqrt{n}(c - \bar{x})}{\sqrt{n-1}\, S} \,|\, \bar{x}, S\right\} \\ &= P[u > u_0 | \bar{x}, s], \end{aligned}$$

where $u = \sqrt{n}(x_1 - \bar{x})/\sqrt{n-1}\, S$ and $u_0 = \sqrt{n}(c - \bar{x})/\sqrt{n-1}\, S$. But the random variable u is distributed independently of $\bar{x}$, S and has the density, const. $(1 - u^2)^{(n-4)/2}$ as shown in Example 3, Chapter 3 if we choose $m_1 = 1 - n^{-1}, m_2 = \cdots = m_n = n^{-1}$. Hence

$$P[u > u_0 | \bar{x}, S] = \int_{u_0}^{1} \text{const.}\, (1 - u^2)^{(n-4)/2}\, du.$$

To obtain the numerical value of this integral corresponding to an observed value of u_0, that is, the value of the estimator $E(T_1 | \bar{x}, S)$ for observed values of $\bar{x}$, S, we use a table of the beta distribution $B[1, (n-2)/2]$ and determine the probability p of a beta variable exceeding u_0^2. If u_0 is positive, the required estimate is $p/2$, and if u is negative, the estimate is $1 - p/2$.

(v) *Let $\mathscr{P}_\theta$ admit a probability density (or probability in the case of a discrete variable) $P(\cdot, \theta)$ for each θ with respect to a σ-finite measure v; let T be any unbiased estimator of $g(\theta)$ and $A(\phi, \theta) = V[P(X, \phi)/P(X, \theta) | \theta]$. Then*

$$V(T | \theta) \geqslant \sup_{\phi \in C} \frac{[g(\phi) - g(\theta)]^2}{A(\phi, \theta)}. \tag{5a.2.12}$$

Where $S_\theta = \{X : P(X,\theta) > 0\}$ and $C = \{\phi : S_\phi \subset S_\theta\}$. Then

$$\int_{S_\theta} TP(X, \phi)\, dv = g(\phi), \qquad \text{for all } \phi.$$

Hence

$$\int_{S_\theta} T \frac{P(X, \phi) - P(X, \theta)}{P(X, \theta)} P(X, \theta)\, dv = g(\phi) - g(\theta),$$

which gives

$$\text{cov}\left\{T, \frac{P(X, \phi)}{P(X, \theta)} - 1 \,|\, \theta\right\} = g(\phi) - g(\theta).$$

Therefore, by using the *C-S* inequality $[\text{cov}(x, y)]^2 \leqslant V(x)V(y)]$, we see that

$$V(T\,|\,\theta) \geqslant \frac{[g(\phi) - g(\theta)]^2}{V[P(X, \phi)/P(X, \theta)\,|\,\theta]}, \qquad \text{for all } \phi, \tag{5a.2.13}$$

and hence (5a.2.12) of (v) is established without any regularity conditions on $P(X, \theta)$ or $g(\theta)$.

(a) Let $g(\theta)$ admit the first derivative $g'(\theta)$ and $A(\phi, \theta)/(\phi - \theta)^2 \to J(\theta)$ as $\phi \to \theta$. Then $V(T) \geqslant [g'(\theta)]^2/J(\theta)$.

The result is obtained by dividing the numerator and denominator of the right-hand side of (5a.2.13) by $(\phi - \theta)^2$ and taking the limit as $\phi \to \theta$.

(b) Let $g(\theta)$ and $P(X, \theta)$ admit the first derivative and for a fixed θ and $|\phi - \theta| < \varepsilon$ (given), let

$$\left| \frac{P(X, \phi) - P(X, \theta)}{(\phi - \theta)P(X, \theta)} \right| < G(X, \theta), \tag{5a.2.14}$$

and $E\{[G(X, \theta)]^2\,|\,\theta\}$ exist. Then

$$V(T) \geqslant [g'(\theta)]^2/\mathscr{I}(\theta), \tag{5a.2.15}$$

where

$$\mathscr{I}(\theta) = \int \frac{1}{P(X, \theta)} \left(\frac{dP(X, \theta)}{d\theta} \right)^2 dv$$

is Fisher's information measure on the parameter θ in an observation from the sample space of X [see **5a.5** for further discussion on $\mathscr{I}(\theta)$].

Under the conditions assumed,

$$\begin{aligned} J(\theta) &= \lim_{\phi \to \theta} \frac{A(\phi, \theta)}{(\phi - \theta)^2} = \lim_{\phi \to \theta} \int \frac{[P(X, \phi) - P(X, \theta)]^2}{[(\phi - \theta)P(X, \theta)]^2} P(X, \theta)\, dv \\ &= \int \left(\frac{P'(X, \theta)}{P(X, \theta)} \right)^2 P(X, \theta)\, dv = \mathscr{I}(\theta). \end{aligned} \tag{5a.2.16}$$

if we take the limit in (5a.2.16) inside the integral, by Lebesgue dominated convergence theorem using (5a.2.14). Instead of this we may have any other sufficient condition for taking the limit inside integration.

Hence if we substitute $\mathscr{I}(\theta)$ for $J(\theta)$ in the result $V(T) \geqslant [g'(\theta)]^2/J(\theta)$, we get the result (5a.2.15), which may be called the information limit (see Cramer, 1946; Rao, 1945d). The inequality in the form (5a.2.12) is due to Chapman and Robbins (1951). A number of other inequalities are now available (see Bhattacharya, 1946; Gart, 1959; Hodges and Lehmann, 1951; and Stein, 1950).

A direct proof of (5a.2.15) may be given using (5a.2.14). Since T is unbiased for $g(\theta)$, then

$$\int T \frac{P(X, \phi) - P(X, \theta)}{(\phi - \theta)P(X, \theta)} P(X, \theta)\, dv = \frac{g(\phi) - g(\theta)}{\phi - \theta}. \tag{5a.2.17}$$

Let $V(T)$ be finite, in which case

$$\left| T \frac{P(X, \phi) - P(X, \theta)}{(\phi - \theta)P(X, \theta)} \right| < |TG(X, \theta)|,$$

where $E[|TG(X, \theta)|]$ exists since $E(T^2)$ and $E[G(X, \theta)]^2$ exist. Hence by taking limits of both sides of (5a.2.17) as $\phi \to \theta$, we see that

$$\int T \frac{P'(X, \theta)}{P(X, \theta)} P(X, \theta)\, dv = g'(\theta),$$

that is, $\text{cov}[T, P'(X, \theta)/P(X, \theta)] = g'(\theta)$. Hence $V(T) \geqslant [g'(\theta)]^2/\mathscr{I}(\theta)$.

When the equality in (5a.2.15) is attained, the correlation between T and $P'(\theta)/P(\theta)$ is unity, in which case

$$T = \lambda \frac{P'}{P} + g(\theta), \qquad (\lambda \text{ being independent of } X) \tag{5a.2.18}$$

except for a set of X of probability measure zero with respect to $P(\cdot, \theta)$. Integrating (5a.2.18) with respect to θ, we see that $P(X, \theta)$ has the form

$$\exp[g_1(\theta)T(X) + g_2(\theta) + U(X)]$$

and shows that T is sufficient for θ.

Example 1. Consider n independent observations $x_1, \ldots, x_n$ from a population with probability density $p(x, \theta)$. In such a case $P(X, \theta) = p(x_1, \theta) \cdots p(x_n, \theta)$ and $\mathscr{I}(\theta) = ni(\theta)$ where

$$i(\theta) = \int \left[\frac{p'(x, \theta)}{p(x, \theta)}\right]^2 p(x, \theta)\, dx$$

If $p(x, \theta) = N(x | \theta, \sigma^2)$, it may be easily verified that $i(\theta) = 1/\sigma^2$, and hence a lower bound to the variance of an unbiased estimator of θ is $1/ni = \sigma^2/n$. The statistic $\bar{x}$ is unbiased for θ and

$$V(\bar{x}) = \frac{\sigma^2}{n} = \frac{1}{\mathscr{I}(\theta)},$$

and therefore $\bar{x}$ is an m.v.u.e. of θ.

Example 2. If $p(x, \theta)$ has the gamma form

$$\frac{\theta^k}{\Gamma(k)} e^{-\theta x} x^{k-1},$$

then $i(\theta) = k/\theta^2$. The statistic $(\bar{x}/k)$ is unbiased for $1/\theta$ and

$$V\left(\frac{\bar{x}}{k}\right) = \frac{1}{k\, n\, \theta^2} = \left[\frac{d(1/\theta)}{d\theta}\right]^2 \frac{1}{\mathscr{I}(\theta)}$$

so that $(\bar{x}/k)$ is an m.v.u.e. of $1/\theta$.

5a.3 The Case of Several Parameters

Suppose that θ in $P(X, \theta)$ is vector valued and let

$$P_{ij} = -\frac{\partial^2 \log P}{\partial\theta_i\, \partial\theta_j} \quad \text{and} \quad E(P_{ij}) = \mathscr{I}_{ij}.$$

The matrix $\mathscr{I} = (\mathscr{I}_{ij})$, $i, j = 1, \ldots, k$ is called the information matrix. Let T be a sufficient statistic for θ. The following main result may now be proved.

(i) *Let* $f_1, \ldots, f_r$ *be* r *statistics such that*

$$\left.\begin{aligned} &\text{(a)}\ \ E(f_i) = g_i(\theta_1, \ldots, \theta_k) \\ &\text{(b)}\ \ E[(f_i - g_i)(f_j - g_j)] = V_{ij} \end{aligned}\right\} \qquad i, j = 1, \ldots, r.$$

Then the following are true.

(A) *There exist functions* $m_1, \ldots, m_r$ *of the sufficient statistic* T *such that*
- (A.1) $E(m_i) = g_i(\theta_1, \ldots, \theta_k)$,
- (A.2) *if* $\mathbf{U} = (U_{ij})$ *where* $U_{ij} = E[(m_i - g_i)(m_j - g_j)]$ *and* $\mathbf{V} = (V_{ij})$, *then the matrix* $(\mathbf{V} - \mathbf{U})$ *is non-negative definite.*

(B) *Suppose*

$$\frac{\partial}{\partial\theta_j}\int f_i P(X, \theta)\, dv = \int f_i \frac{\partial P(X, \theta)}{\partial\theta_j}\, dv = \frac{\partial g_i}{\partial\theta_j},$$

and let $\mathbf{\Delta}$ *be the matrix* $(\partial g_i/\partial\theta_j)$, $i = 1, \ldots, r$; $j = 1, \ldots, k$. *Then* $\mathbf{V} - \mathbf{\Delta}\mathscr{I}^{-1}\mathbf{\Delta}'$ *is non-negative definite, where* $\mathscr{I}^{-1}$ *is the inverse of* $\mathscr{I}$, *the information matrix.* [$\mathscr{I}^{-1}$ *is replaced by a g-inverse* $\mathscr{I}^{-}$ *if* $\mathscr{I}$ *is singular.*]

To prove (A), consider a linear function $\mathbf{L}'\mathbf{F}$ where $\mathbf{F}' = (f_1, \ldots, f_r)$ and $\mathbf{L}$ is a fixed vector and let $E(f_i | T) = m_i$. Then $E(\mathbf{L}'\mathbf{F} | T) = \mathbf{L}'\mathbf{M}$ where $\mathbf{M}' = (m_1, \ldots, m_r)$. Applying result [(iii), **5a.2**] to the single function $\mathbf{L}'\mathbf{F}$, $V(\mathbf{L}'\mathbf{F}) \geqslant V(\mathbf{L}'\mathbf{M})$, gives

$$\mathbf{L}'\mathbf{V}\mathbf{L} \geqslant \mathbf{L}'\mathbf{U}\mathbf{L} \qquad \text{or} \qquad \mathbf{L}'(\mathbf{V} - \mathbf{U})\mathbf{L} \geqslant 0$$

for any fixed $\mathbf{L}$. Hence $\mathbf{V} - \mathbf{U}$ is non-negative definite, which proves (A).

To prove (B), consider the relation

$$\int f_i \frac{\partial P}{\partial \theta_j} dv = \frac{\partial g_i}{\partial \theta_j}$$

and the dispersion matrix of $f_1, \ldots, f_r, (1/P)(\partial P/\partial \theta_1), \ldots, (1/P)(\partial P/\partial \theta_k)$, which may be written as the partitioned matrix

$$\begin{pmatrix} \mathbf{V} & \mathbf{\Delta} \\ \mathbf{\Delta}' & \mathscr{I} \end{pmatrix}.$$

The determinants (denoting a unit matrix by $\mathbf{I}$)

$$\begin{vmatrix} \mathbf{I} & -\mathbf{\Delta}\mathscr{I}^{-1} \\ \mathbf{0} & \mathscr{I}^{-1} \end{vmatrix} \quad \text{and} \quad \begin{vmatrix} \mathbf{V} & \mathbf{\Delta} \\ \mathbf{\Delta}' & \mathscr{I} \end{vmatrix}$$

are non-negative and so also their product

$$\begin{vmatrix} \mathbf{V} - \mathbf{\Delta}\mathscr{I}^{-1}\mathbf{\Delta}' & \mathbf{0} \\ \mathscr{I}^{-1}\mathbf{\Delta}' & \mathbf{I} \end{vmatrix} = |\mathbf{V} - \mathbf{\Delta}\mathscr{I}^{-1}\mathbf{\Delta}'|.$$

This result holds true even for a subset of the statistics $f_1, \ldots, f_r$ which means that $\mathbf{V} - \mathbf{\Delta}\mathscr{I}^{-1}\mathbf{\Delta}'$ is non-negative definite. This proves (B).

A series of results can be obtained from the main results (A) and (B).

(a) *By considering only the diagonal elements in* $\mathbf{V} - \mathbf{\Delta}\mathscr{I}^{-1}\mathbf{\Delta}'$

$$V_{ii} \geqslant \sum \sum \mathscr{I}^{mn} \frac{\partial g_i}{\partial \theta_m} \frac{\partial g_i}{\partial \theta_n} \tag{5a.3.1}$$

where $\mathscr{I}^{mn}$ *are the elements of the matrix reciprocal to the information matrix* $(\mathscr{I}_{mn})$.

This shows that the variance of the estimator of g_i is not less than a quantity which is defined independently of any method of estimation. This is the generalization to many parameters of the expression derived in (5a.2.15). If $g_i = \theta_i$ $(i = 1, \ldots, k)$, the relationship (5a.3.1) reduces to

$$V_{ii} \geqslant \mathscr{I}^{ii}. \tag{5a.3.2}$$

Observe that $\mathscr{I}^{ii}$ is not less than $1/\mathscr{I}_{ii}$ which is the limit obtained in (5a.2.15) for the estimate of θ_i. When the values of $\theta_1, \ldots, \theta_{i-1}, \theta_{i+1}, \ldots, \theta_k$ are known, the limit (5a.2.15) is applicable. If not, the estimate of θ_i has to be independent of the other parameters and for this reason the limit is possibly increased.

(b) *Since the matrix* $(\mathbf{V} - \mathbf{U})$ *is non-negative definite, it follows that* $V_{ii} \geqslant U_{ii}$, $(i = 1, \ldots, r)$, *which shows that estimators with the minimum attainable variances are explicit functions of the sufficient statistic.*

(c) *Since the matrix* $\mathbf{V} - \mathbf{\Delta}\mathscr{I}^{-1}\mathbf{\Delta}'$ *is non-negative definite, it follows that*

$$|\mathbf{V}| \geqslant |\mathbf{\Delta}\mathscr{I}^{-1}\mathbf{\Delta}'|. \tag{5a.3.3}$$

The quantity $|\mathbf{V}|$ is called the generalized variance of the estimators. The result (5a.3.3) shows that this is not less than a quantity which is defined independently of any method of estimation.

(d) *Since* $(\mathbf{V} - \mathbf{U})$ *is non-negative definite, it follows that* $|\mathbf{V}| \geqslant |\mathbf{U}|$.

This shows that the estimates with the minimum possible generalized variance are functions of the sufficient statistic.

(e) *If*

$$V_{ii} = \sum \sum \mathscr{I}^{mn} \frac{\partial g_i}{\partial \theta_m} \frac{\partial g_i}{\partial \theta_n} \tag{5a.3.4}$$

in which case the estimate of g_i has the minimum variance, then

$$V_{ij} = \sum \sum \mathscr{I}^{mn} \frac{\partial g_i}{\partial \theta_m} \frac{\partial g_j}{\partial \theta_n} \tag{5a.3.5}$$

so that the covariance of an m.v.u.e. with any unbiased estimator of any other parametric function has a fixed value defined independently of any method of estimation.

This follows from the fact that the determinant

$$\begin{vmatrix} V_{ii} - \sum \sum \mathscr{I}^{mn} \dfrac{\partial g_i \partial g_i}{\partial \theta_m \partial \theta_n} & V_{ij} - \sum \sum \mathscr{I}^{mn} \dfrac{\partial g_i \partial g_j}{\partial \theta_m \partial \theta_n} \\ V_{ij} - \sum \sum \mathscr{I}^{mn} \dfrac{\partial g_i \partial g_j}{\partial \theta_m \partial \theta_n} & V_{jj} - \sum \sum \mathscr{I}^{mn} \dfrac{\partial g_j \partial g_j}{\partial \theta_m \partial \theta_n} \end{vmatrix},$$

which is a subdeterminant of $|\mathbf{V} - \mathbf{\Delta} \mathscr{I}^{-1} \mathbf{\Delta}'|$, is not less than zero.

Example. Consider n independent observations from the normal population $N(\theta, \sigma^2)$. It is easy to verify that the information matrix for θ and σ^2 is

$$\mathscr{I} = \begin{pmatrix} \dfrac{n}{\sigma^2} & 0 \\ 0 & \dfrac{n}{2\sigma^4} \end{pmatrix}$$

with its reciprocal

$$\mathscr{I}^{-1} = \begin{pmatrix} \dfrac{\sigma^2}{n} & 0 \\ 0 & \dfrac{2\sigma^4}{n} \end{pmatrix}$$

Since $\bar{x}$, as an estimate of θ, has the minimum possible variance, it follows that any unbiased estimator of σ^2 has zero correlation with $\bar{x}$, since $\mathscr{I}^{\theta\sigma^2} = 0$. This result can be extended to the case of a multivariate normal population where it can be shown that the means are uncorrelated with all possible unbiased estimators of the variances and covariances.

For the estimation of σ^2, let us consider $s^2 = \sum(x_i - \bar{x})^2/(n-1)$ with

$$E(s^2) = \sigma^2, \qquad V(s^2) = \frac{2\sigma^4}{n-1} \tag{5a.3.6}$$

showing that the lower limit $2\sigma^4/n$ is not attained. But this is the minimum attainable, as shown below. If any estimator of σ^2 differs from s^2 by $f(s, \bar{x})$, then

$$\int f(s, \bar{x}) \exp\{-[n(\bar{x} - \theta)^2 + \overline{n-1}\, s^2]/2\sigma^2\}\, dv = 0.$$

Twice differentiation with respect to θ leads to the result

$$E[(\bar{x} - \theta)^2 f(s, \bar{x})] = 0.$$

Differentiation with respect to σ^2 gives

$$E[n(\bar{x} - \theta)^2 f(s, \bar{x}) + (n-1)s^2 f(s, \bar{x})] = 0$$

or $\text{cov}[s^2, f(s, \bar{x})] = 0$, giving

$$V[s^2 + f(\bar{x}, s)] = V(s^2) + V[f(s, \bar{x})]$$

which means that $V(s^2)$ is the least possible variance for an unbiased estimator of σ^2. Thus $\bar{x}$ and s^2 are the m.v. unbiased estimators of θ and σ^2.

5a.4 Fisher's Information Measure

Let X be a vector valued or any general random variable (r.v.) with values in space S and having probability density $P(\cdot, \theta)$ with respect to σ-finite measure v. Further let $P(\cdot, \theta)$ be differentiable with respect to θ and for any measurable set $C \subset S$

$$\frac{d}{d\theta} \int_C P(X, \theta)\, dv = \int_C \frac{dP(X, \theta)}{d\theta}\, dv. \tag{5a.4.1}$$

Fisher's information measure on θ contained in the r.v. X is defined by

$$\mathscr{I}(\theta) = E\left(\frac{d \log P}{d\theta}\right)^2 = V\left(\frac{d \log P}{d\theta}\right), \qquad \text{since} \quad E\left(\frac{d \log P}{d\theta}\right) = 0. \tag{5a.4.2}$$

We shall study some properties of the information measure (5a.4.2).

(i) *Let $\mathscr{I}_1$ and $\mathscr{I}_2$ be informations contained in two independent variables X_1 and X_2 respectively and $\mathscr{I}$ contained in (X_1, X_2) jointly. Then $\mathscr{I} = \mathscr{I}_1 + \mathscr{I}_2$.*

If $P_1(\cdot, \theta)$ and $P_2(\cdot, \theta)$ are the densities of X_1, X_2, the joint density of (X_1, X_2) is $P_1(\cdot, \theta)P_2(\cdot, \theta)$. By definition

$$\begin{aligned}
\mathscr{I} &= E\left\{\frac{d}{d\theta}[\log P_1(X_1, \theta)P_2(X_2, \theta)]\right\}^2 \\
&= E\left(\frac{d \log P_1}{d\theta}\right)^2 + E\left(\frac{d \log P_2}{d\theta}\right)^2 + 2E\left(\frac{d \log P_1}{d\theta} \cdot \frac{d \log P_2}{d\theta}\right) \\
&= \mathscr{I}_1 + \mathscr{I}_2,
\end{aligned}$$

since

$$E\left(\frac{d \log P_1}{d\theta} \cdot \frac{d \log P_2}{d\theta}\right) = E\left(\frac{d \log P_1}{d\theta}\right)E\left(\frac{d \log P_2}{d\theta}\right) = 0$$

(ii) *Let $X_1, \ldots, X_k$ be independent and identically distributed random variables and $\mathscr{I}$ be the information in each. The information in $(X_1, \ldots, X_k)$ is then $k\mathscr{I}$.*

(iii) *Let T be a measurable function of X, with a density function $\phi(\cdot, \theta)$ with respect to a σ-finite measure ν, satisfying a condition similar to* (5a.4.1). *Further let $\mathscr{I}_T$ be the information contained in T, that is,*

$$\mathscr{I}_T = E[\phi'(T, \theta)/\phi(T, \theta)]^2.$$

Then

$$\text{(a)} \qquad E\left(\frac{P'(X, \theta)}{P(X, \theta)} \,\middle|\, T = t\right) = \frac{\phi'(t, \theta)}{\phi(t, \theta)},$$

$$\text{(b)} \qquad \mathscr{I} \geqslant \mathscr{I}_T. \qquad (5a.4.3)$$

Let A be a measurable set in the T space and A' be the corresponding set in the X space. Then (using the same symbols X and T in the densities for convenience)

$$\begin{aligned}
\int_{A'} \frac{P'(X, \theta)}{P(X, \theta)} \cdot P(X, \theta)\, dv &= \frac{d}{d\theta}\int_{A'} P(X, \theta)\, dv \\
&= \frac{d}{d\theta}\int_A \phi(T, \theta)\, d\nu = \int_A \phi'(T, \theta)\, d\nu \\
&= \int_A \frac{\phi'(T, \theta)}{\phi(T, \theta)} \cdot \phi(T, \theta)\, d\nu
\end{aligned}$$

for any A. It then follows from the definition of conditional expectation that

$$E\left(\frac{P'(X, \theta)}{P(X, \theta)} \,\middle|\, T = t\right) = \frac{\phi'(t, \theta)}{\phi(t, \theta)}.$$

Consider

$$E\left\{\frac{\phi'(T,\theta)}{\phi(T,\theta)}\cdot\frac{P'(X,\theta)}{P(X,\theta)}\right\} = \underset{T}{E}\left\{\frac{\phi'(T,\theta)}{\phi(T,\theta)}\, E\left[\frac{P'(X,\theta)}{P(X,\theta)}\,\middle|\, T\right]\right\} \quad \text{using (2b.3.7)}$$

$$= \underset{T}{E}\left[\frac{\phi'(T,\theta)}{\phi(T,\theta)}\right]^2 = \mathscr{I}_T.$$

Now

$$E\left(\frac{P'}{P}-\frac{\phi'}{\phi}\right)^2 = E\left(\frac{P'}{P}\right)^2 + E\left(\frac{\phi'}{\phi}\right)^2 - 2E\left(\frac{P'}{P}\cdot\frac{\phi'}{\phi}\right)$$

$$= \mathscr{I} + \mathscr{I}_T - 2\mathscr{I}_T = \mathscr{I} - \mathscr{I}_T \geqslant 0.$$

If X stands for an entire sample and T for a statistic, the result (ii, b) shows that there is generally loss of information in replacing X by T. When there is no loss of information, that is, $\mathscr{I} = \mathscr{I}_T$,

$$E\left(\frac{P'}{P}-\frac{\phi'}{\phi}\right)^2 = 0$$

$$\Rightarrow \frac{P'}{P} = \frac{\phi'}{\phi} \quad \text{a.e. } (v),$$

or $P(X, \theta) = \phi(T, \theta)\Psi(X, T)$, where Ψ is independent of θ, a situation in which T is sufficient for θ.

(iv) In the multiparameter case, let

$$\mathscr{I}_{rs} = E\left(\frac{\partial \log P}{\partial \theta_r}\cdot\frac{\partial \log P}{\partial \theta_s}\right), \qquad r, s = 1, \ldots, k.$$

Then the matrix $\mathscr{I} = (\mathscr{I}_{rs})$ is defined to be the information matrix.

It is easy to prove as in the case of a single parameter that

(a) $\mathscr{I} = \mathscr{I}_1 + \mathscr{I}_2$, *where* $\mathscr{I}_1$ *and* $\mathscr{I}_2$ *are information matrices due to two independent random variables* X_1 *and* X_2 *and* $\mathscr{I}$, *due to* (X_1, X_2) *jointly.*

(b) *The matrix* $\mathscr{I} - \mathscr{I}_T$ *is non-n.d. (positive or semi-positive definite), where* $\mathscr{I}_T$ *is the information matrix in a function* T *of* X.

Information as a Measure of the Intrinsic Accuracy of a Distribution. By information on an unknown parameter θ contained in a random variable or in its distribution, we mean the extent to which uncertainty regarding the unknown value of θ is reduced as a consequence of an observed value of the random variable. If there is a unique observation with probability 1 corresponding to each value of the parameter, we have a situation where the random variable has the maximum information. On the other hand if the random variable has the *same* distribution for all values of the parameter, there is no basis for making statements about θ on the basis of an observed value of such a random variable. The sensitiveness of a random variable with respect to a parameter may then be judged by the extent to which its distribution is altered by a change in the value of the parameter. If $P(\cdot, \theta)$ and $P(\cdot, \theta')$ denote the probability densities of X for two values θ, θ' of the parameter, the difference in the distributions could be measured by some distance function, such as Hellinger distance,

$$\cos^{-1} \int \sqrt{P(X, \theta)P(X, \theta')}\, dv. \tag{5a.4.4}$$

If we let $\theta' = \theta + \delta\theta$ and expand $P(X, \theta')$ by Taylor series, (5a.4.4) reduces to

$$\cos^{-1} \int P(\theta)\left\{1 - \frac{1}{8}\left[\frac{P'(\theta)}{P(\theta)}\right]^2 (\delta\theta)^2\right\} dv = \cos^{-1}[1 - \tfrac{1}{8}\mathscr{I}(\theta)(\delta\theta)^2], \tag{5a.4.5}$$

where $\mathscr{I}(\theta)$ is Fisher's information, neglecting higher powers of $\delta\theta$. Since $\mathscr{I}(\theta)$ is positive, the distance (5a.4.5) increases with increase in the value of $\mathscr{I}(\theta)$ and is thus a measure of sensitivity of the random variable with respect to an infinitesimal change in the value of the parameter.

The reader may verify that a variety of distance measures yield Fisher's information as an index of sensitivity for small changes in the value of the parameter (Rao, 1962e,g). In the multiparameter case we obtain the quadratic differential metric

$$\sum \sum \mathscr{I}_{ij}\, \delta\theta_i\, \delta\theta_j$$

in the place of $\mathscr{I}(\theta)(\delta\theta)^2$ in (5a.4.5), as the distance between the distributions of the random variable for the parameters $(\theta_1, \theta_2, \ldots)$ and $(\theta_1 + \delta\theta_1, \theta_2 + \delta\theta_2, \ldots)$. The sensitivity of the random variable with respect to changes in the parameter may then be judged by examining the information matrix as a whole (Rao, 1945d).

5a.5 An Improvement of Unbiased Estimators

In Section **4k**, it is shown that biased estimators of parameters in a linear model may be found with a smaller mean square error than the BLUE's in a certain region of the parameter space around an *a priori* value. Indeed improvements can always be made over unbiased estimators if we have some *a priori* information about the unknown parameters.

Suppose we are estimating the unknown mean μ of a normal distribution for which the sample average $\bar{X}$ is sufficient and also the minimum variance unbiased estimator. If it is known that $\mu \in (a, b)$, then we can construct the alternative estimator

$$T = \begin{cases} a & \text{if } \bar{X} < a \\ \bar{X} & \text{if } a \leqslant \bar{X} \leqslant b \\ b & \text{if } \bar{X} > a. \end{cases} \tag{5a.5.1}$$

It is readily seen that T is not unbiased for $\bar{X}$, but

$$E(T - \mu)^2 < E(\bar{X} - \mu)^2 \qquad \text{when} \quad a < \mu < b$$

so that T is uniformly better than $\bar{X}$ as an estimator of μ if the *a priori* information $\mu \in (a, b)$ is correct.

Suppose $\mathbf{T}' = (T_1, \ldots, T_k)$ is an unbiased estimator of $\boldsymbol{\tau}'_{\boldsymbol{\theta}} = (\tau_1(\boldsymbol{\theta}), \ldots, \tau_k(\boldsymbol{\theta}))$ with the nonsingular dispersion matrix $\mathbf{V}_{\boldsymbol{\theta}}$. Sometimes it may be possible to find a constant b independent of θ such that

$$\mathbf{V}_{\boldsymbol{\theta}} - E_{\boldsymbol{\theta}}[(b\mathbf{T} - \boldsymbol{\tau}_{\boldsymbol{\theta}})(b\mathbf{T} - \boldsymbol{\tau}_{\boldsymbol{\theta}})'], \tag{5a.5.2}$$

where the second term in (5a.5.2) is the mean dispersion error of $b\mathbf{T}$, is n.n.d. for the entire range of the unknown parameter $\boldsymbol{\theta}$. If so, then (5a.5.2) implies that $b\mathbf{T}$ is a uniformly better estimator of $\boldsymbol{\tau}$ than $\mathbf{T}$ under any quadratic loss function. Examples of such improvement have been known, and the following proposition due to Perlman (1972) provides a general result.

An n.s. condition that a constant b, $0 < b < 1$ exists such that (5a.5.2) *holds is that the quantity $\boldsymbol{\tau}'_{\boldsymbol{\theta}}\mathbf{V}_{\boldsymbol{\theta}}^{-1}\boldsymbol{\tau}_{\boldsymbol{\theta}}$ is bounded over the range of* $\boldsymbol{\theta}$.

We rewrite (5a.5.2) as $\mathbf{V}_{\boldsymbol{\theta}} - b^2\mathbf{V}_{\boldsymbol{\theta}} + (1 - b)^2\boldsymbol{\tau}_{\boldsymbol{\theta}}\boldsymbol{\tau}'_{\boldsymbol{\theta}}$ which is n.n.d. iff $m\mathbf{V}_{\boldsymbol{\theta}} - \boldsymbol{\tau}_{\boldsymbol{\theta}}\boldsymbol{\tau}'_{\boldsymbol{\theta}}$ is n.n.d., where $m = (1 + b)/(1 - b)$. The latter implies

$$m \geqslant \sup_{\mathbf{x}} \frac{\mathbf{x}'\boldsymbol{\tau}_{\boldsymbol{\theta}}\boldsymbol{\tau}'_{\boldsymbol{\theta}}\mathbf{x}}{\mathbf{x}'\mathbf{V}_{\boldsymbol{\theta}}\mathbf{x}} = \boldsymbol{\tau}'_{\boldsymbol{\theta}}\mathbf{V}_{\boldsymbol{\theta}}^{-1}\boldsymbol{\tau}_{\boldsymbol{\theta}} \tag{5a.5.3}$$

using C.S. inequality (P.54). Thus the existence of b, $0 < b < 1$ implies that $\boldsymbol{\tau}'_{\boldsymbol{\theta}}\mathbf{V}_{\boldsymbol{\theta}}^{-1}\boldsymbol{\tau}_{\boldsymbol{\theta}}$ is bounded for all $\boldsymbol{\theta}$. Conversely, if $\sup \boldsymbol{\tau}'_{\boldsymbol{\theta}}\mathbf{V}_{\boldsymbol{\theta}}^{-1}\boldsymbol{\tau}_{\boldsymbol{\theta}} = m_0 < \infty$, then for any b such that $0 < b < 1$ and $b > (m_0 - 1)/(m_0 + 1)$, (5a.5.2) is satisfied.

Note that (5a.5.2) cannot be a null matrix for that would imply $m\mathbf{I} - \mathbf{V}_{\boldsymbol{\theta}}^{-1/2}\boldsymbol{\tau}_{\boldsymbol{\theta}}\boldsymbol{\tau}'_{\boldsymbol{\theta}}\mathbf{V}_{\boldsymbol{\theta}}^{-1/2}$ is null, which is not possible since the rank of $\mathbf{V}_{\boldsymbol{\theta}}^{-1/2}\boldsymbol{\tau}_{\boldsymbol{\theta}}\boldsymbol{\tau}'_{\boldsymbol{\theta}}\mathbf{V}_{\boldsymbol{\theta}}^{-1/2}$ is unity.

Thus if we have the *a priori* information (5a.5.3), we can uniformly improve on unbiased estimators.

The reader may verify that if $E(y) = \mu$, $V(y) = \sigma^2$ and $\theta = \sigma/\mu$, then

$$E(y - \mu)^2 \geqslant E[(1 + c^2)^{-1}y - \mu]^2 \text{ for all } \mu \tag{5a.5.4}$$

if $c^2(1 - \theta^2) \leqslant \theta^2$. An approximate knowledge of the coefficient of variation θ, may enable us to improve upon the unbiased estimator y by a suitable choice of c. The best advantage is obtained when c equals θ.

5b GENERAL PROCEDURES

5b.1 Statement of a General Problem (Bayes' Theorem)

Consider a probability space $(\Omega, \mathscr{B})$ and random variables [ω functions]

$$\theta_1, \ldots, \theta_m; \quad X_1, \ldots, X_n$$

of which θ_i are hypothetical (unobservable) and X_j are observable. The problem of inference which is of practical interest is the prediction of certain unobservable variables in terms of the observables. It is clear that the general problem requires the study of the conditional distribution of $\boldsymbol{\theta}$, the vector of θ_i variables, given $\mathbf{X}$, the vector of X_j variables as in the case of classical regression problem. We consider different situations depending on the uncertainty of the specification of the joint d.f. of $(\boldsymbol{\theta}, \mathbf{X})$.

To simplify the problem and to avoid certain difficulties associated with conditional distributions let us suppose that the conditional p.d. of $\mathbf{X}$, given $\boldsymbol{\theta}$, exists and represent it by $P(\cdot|\cdot)$. If F is the marginal d.f. of $\boldsymbol{\theta}$, then the marginal p.d. of $\mathbf{X}$ at $\mathbf{X}_0$ is

$$G(\mathbf{X}_0) = \int P(\mathbf{X}_0|\boldsymbol{\theta})\, dF(\boldsymbol{\theta}) \tag{5b.1.1}$$

and the conditional probability distribution of $\boldsymbol{\theta}$ given $\mathbf{X} = \mathbf{X}_0$ is

$$dF(\cdot|\mathbf{X}_0) = \frac{P(\mathbf{X}_0|\cdot)}{G(\mathbf{X}_0)}\, dF(\cdot) \tag{5b.1.2}$$

The d.f. F may be called *a prior* d.f. of $\boldsymbol{\theta}$ and $F(\cdot|\mathbf{X}_0)$ *a posterior* d.f. of $\boldsymbol{\theta}$ given $\mathbf{X} = \mathbf{X}_0$—terms which may not be of particular significance in the context of the general problem posed. We consider some examples.

The result (5b.1.2) is known as Bayes theorem.

5b.2 Joint d.f. of $(\boldsymbol{\theta}, \mathbf{X})$ Completely Known

When the joint d.f. of $(\boldsymbol{\theta}, \mathbf{X})$ is completely known the expression (5b.1.2) provides the most satisfactory form of expressing the uncertainty about $\boldsymbol{\theta}$ based on the knowledge of $\mathbf{X}$. Given $\mathbf{X} = \mathbf{X}_0$, we can specify the probability of $\boldsymbol{\theta}$ lying in any given region A by computing the value of

$$\int_A dF(\boldsymbol{\theta}|\mathbf{X}_0) = \int_A \frac{P(\mathbf{X}_0|\boldsymbol{\theta})}{G\mathbf{X}(_0)}\, dF(\boldsymbol{\theta}). \tag{5b.2.1}$$

If a $(1-\alpha)$-confidence interval is desired for a function $g(\boldsymbol{\theta})$ of $\boldsymbol{\theta}$, we may determine (a, b) such that $(b-a)$ is of minimum length subject to the condition

$$\int_{a<g(\boldsymbol{\theta})<b} \frac{P(\mathbf{X}_0|\boldsymbol{\theta})}{G(\mathbf{X}_0)}\, dF(\boldsymbol{\theta}) = 1 - \alpha. \tag{5b.2.2}$$

A point estimate of $g(\boldsymbol{\theta})$ is provided by its expected value with respect to the conditional distribution

$$\int g(\boldsymbol{\theta})\, dF(\boldsymbol{\theta} \mid \mathbf{X}_0) = \int g(\boldsymbol{\theta}) \frac{P(X_0 \mid \boldsymbol{\theta})}{G(X_0)}\, dF(\boldsymbol{\theta}), \tag{5b.2.3}$$

which is the regression of $g(\boldsymbol{\theta})$ on $\mathbf{X}$ (at $\mathbf{X}_0$) and *which minimizes the mean square error* (see 4g.1.1), and so on.

As an illustration, let us consider the prediction of the unknown binomial probability on the basis of an observed number of successes r out of n trials. If the *prior* (marginal) p.d. of π is

$$\frac{\pi^{\alpha-1}(1-\pi)^{\gamma-1}}{\beta(\alpha, \gamma)}, \qquad 0 < \pi < 1, \tag{5b.2.4}$$

the joint probability distribution of r and π is

$$\binom{n}{r} \pi^r (1-\pi)^{n-r} \frac{\pi^{\alpha-1}(1-\pi)^{\gamma-1}}{\beta(\alpha, \gamma)}\, d\pi \tag{5b.2.5}$$

and that of r alone, integrating with respect to π, is

$$\binom{n}{r} \frac{\beta(\alpha + r, \gamma + n - r)}{\beta(\alpha, \gamma)} \tag{5b.2.6}$$

where $\beta(p, q)$ is the beta function. The posterior distribution of π given r is (5b.2.5)/(5b.2.6), which is

$$\frac{\pi^{\alpha+r-1}(1-\pi)^{\gamma+n-r-1}}{\beta(\alpha + r, \gamma + n - r)}\, d\pi. \tag{5b.2.7}$$

The consequences of selecting any particular value of π as an estimate can be completely studied by using the distribution (5b.2.7).

If a point estimate of π is desired we may take the mean or the mode of the distribution (5b.2.7). The mean, which minimizes the mean square error for given r, is

$$\frac{\alpha + r}{\alpha + \gamma + n}, \tag{5b.2.8}$$

which may be called a Bayes estimator with respect to the minimum mean square error criterion and the prior d.f. (5b.2.4). The estimator (5b.2.8) is seen to be different from the traditional estimator r/n.

We can use the distribution (5b.2.7) in determining the probability of s successes in a further sequence of m trials. For a given π, the probability of s successes is $[m!/s!(m-s)!]\pi^s(1-\pi)^{m-s}$. Integrating this with respect to the posterior distribution (5b.2.7), we obtain the probability distribution of s given r as

$$\frac{m!}{(m-s)!\,s!}\;\frac{\beta(\alpha+r+s,\gamma+m+n-r-s)}{\beta(\alpha+r,\gamma+n-r)}. \tag{5b.2.9}$$

5b.3 The Law of Equal Ignorance

Unfortunately, the joint d.f. of $(\boldsymbol{\theta}, \mathbf{X})$ is seldom completely known in practical situations. The functional form of $P(\cdot\,|\,\cdot)$, the conditional density of $\mathbf{X}$ given $\boldsymbol{\theta}$, can be specified with at least a reasonable degree of approximation, but the specification of prior F is difficult. It is sometimes suggested that in the absence of any knowledge about the prior distribution of $\boldsymbol{\theta}$ some consistent rules may be given for their choice. Some attempts made by Jeffreys (1939) and others in this direction do not seem to have yielded any useful results.

In the problem of the binomial parameter discussed in **5b.2**, let us use the *law of equal ignorance* which gives a uniform distribution for π, which is a special case of (5b.2.4) with $\alpha = 1$ and $\gamma = 1$. In such a case the posterior distribution of π given r successes is

$$\frac{\pi^r(1-\pi)^{n-r}}{\beta(r+1,n-r+1)}\,d\pi. \tag{5b.3.1}$$

The probability of a further observation being a failure, using the formula (5b.2.9) with $\alpha = \gamma = 1$, $m = 1$, $s = 0$, is $(n+1-r)/(n+2)$. In particular the probability that n successive failures will be followed by failure is $(n+1)/(n+2)$ which is the celebrated *law of succession.*

But the procedure leading to (5b.3.1) is quite arbitrary, for if instead of the parameter π, we consider the parameter θ such that $\sin\theta = 2\pi - 1$ and if uniform distribution of θ is considered, the posterior distribution of θ given r is seen to be proportional to

$$(1+\sin\theta)^r(1-\sin\theta)^{n-r}\,d\theta. \tag{5b.2.3}$$

The posterior distribution of the original parameter π derived from (5b.2.3) is then proportional to

$$\pi^{r-1/2}(1-\pi)^{n-r-1/2}\,d\pi \tag{5b.3.3}$$

which is different from (5b.3.1). Such arbitrariness or inconsistency noted in (5b.3.3) is inherent in any rule of assigning prior distributions.

5b.4 Empirical Bayes Estimation Procedures

It is clear that when F is not completely known the conditional distribution (5b.1.2) cannot be used to make probability statements concerning the unknown variables. But in some situations it may be possible to compute the regression of g on $\mathbf{X}$ without the complete knowledge of the d.f. of $\boldsymbol{\theta}$. But the estimate may involve unknown parameters which are characteristics of the marginal distribution of $\mathbf{X}$ alone. If past data on $\mathbf{X}$ are available, the marginal distribution of $\mathbf{X}$ can be estimated empirically and consequently an estimate of the regression estimate can be made available. We shall illustrate the procedure by some examples.

Let the joint p.d. of a single parameter θ and random variables $X_1, \ldots, X_n$ (to be denoted by the symbol $\mathbf{X}$) be

$$(2\pi\sigma_1^2)^{-1/2}e^{-(\theta-\mu)^2/2\sigma_1^2}(2\pi\sigma_2^2)^{-n/2}e^{-\Sigma(X_i-\theta)^2/2\sigma_2^2} \tag{5b.4.1}$$

where μ, σ_1^2, σ_2^2 are unknown. The marginal density of $X_1, \ldots, X_n$ [obtained by integrating (5b.4.1) with respect to θ], is proportional to

$$\exp\left\{-\frac{n(\bar{X}-\mu)^2}{2(n\sigma_1^2+\sigma_2^2)} - \frac{S^2}{2\sigma_2^2}\right\} \tag{5b.4.2}$$

where $\bar{X} = (X_1 + \cdots + X_n)/n$ and $S^2 = \sum (X_i - \bar{X})^2$. The conditional density of θ given $\mathbf{X}$ is then proportional to

$$e^{-(\theta-\alpha)^2/2\beta^2} \tag{5b.4.3}$$

where

$$\alpha = \frac{(\mu/\sigma_1^2) + (n\bar{X}/\sigma_2^2)}{(1/\sigma_1^2) + (n/\sigma_2^2)}, \qquad \beta^2 = \frac{\sigma_1^2\sigma_2^2}{n\sigma_1^2 + \sigma_2^2}.$$

From (5b.4.3), $E(\theta \mid \mathbf{X}) = \alpha$, that is, the regression of θ on $\mathbf{X}$ is α which involves the unknown constants μ, σ_1^2, and σ_2^2, *which are the parameters of the marginal probability density* (5b.4.2) *of* $\mathbf{X}$ *only*. If we have past observations on $\mathbf{X}$, all the unknown parameters can be estimated as shown below.

Let the past data consist of p independent sets of n observations

$$\begin{array}{ccc} X_{11} & \cdots & X_{n1} \\ \cdot & \cdots & \cdot \\ X_{1p} & \cdots & X_{np}. \end{array}$$

An analysis of variance between and within sets yields the following table with expectations involving σ_1^2, σ_2^2, which is a special case of the variance components estimation studied in **4f.1** to **4f.3**.

	S.S.	Expectation
Between sets	B	$(p-1)(\sigma_2^2 + n\sigma_1^2)$
Within sets	W	$p(n-1)\sigma_2^2$

The estimates of σ_1^2 and σ_2^2 are

$$\hat{\sigma}_2^2 = \frac{W}{p(n-1)}, \qquad \hat{\sigma}_1^2 = \frac{B}{n(p-1)} - \frac{W}{pn(n-1)}.$$

An estimate of μ is provided by the grand mean of the observations

$$\hat{\mu} = \frac{\sum \sum X_{ij}}{pn}.$$

Thus, an estimate of the regression of θ on any future observation $\mathbf{X}' = (X_1, \ldots, X_n)$ is

$$\left(\frac{\hat{\mu}}{\hat{\sigma}_1^{\,2}} + \frac{n\bar{X}}{\hat{\sigma}_2^{\,2}}\right) \div \left(\frac{1}{\hat{\sigma}_1^{\,2}} + \frac{n}{\hat{\sigma}_2^{\,2}}\right),$$

involving only $\bar{X}$, the average of the observations, and estimates of $\hat{\mu}$, $\hat{\sigma}_1^{\,2}$, $\hat{\sigma}_2^{\,2}$ based on past data.

The reader may note that the present problem differs from the classical regression problem which requires past data *simultaneously* on θ and $\mathbf{X}$. This is because the parameters occurring in the regression are *estimable from the observations on* $\mathbf{X}$ *alone* which may not be always possible.

An interesting example by Robbins (1955) arises in the estimation of the Poisson parameter μ given the observation, without making any assumption on the prior distribution of μ. The situation we consider is (μ, X) where the distribution X given μ is Poisson.

Let x be an observation on the random variable X. The conditional distribution of μ given $X = x$ is, using the formula (5b.1.2),

$$e^{-\mu}\frac{\mu^x}{x!}\,dF(\mu) \div G(x),$$

where

$$G(x) = \int e^{-\mu}\frac{\mu^x}{x!}\,dF(\mu)$$

is the marginal probability of x. The regression of μ on X at $X = x$ is

$$\int \mu e^{-\mu}\frac{\mu^x}{x!}\,dF(\mu) \div G(x) = (x + 1)G(x + 1) \div G(x).$$

We do not know $G(x)$ and $G(x + 1)$. But if we have past observations

$$x_1, x_2, \ldots, x_N$$

on the random variable X, an estimate of $G(x)$ is

$$\hat{G}(x) = \frac{\text{No of past observations equal to } x}{N},$$

and similarly an estimate $\hat{G}(x + 1)$ of $G(x + 1)$ is obtained. Hence an estimate of the regression estimate of μ is $(x + 1)\hat{G}(x + 1)/\hat{G}(x)$.

The reader may note that the regression of μ on X is estimable although simultaneous observations on μ and X are not possible.

We have seen in **5b.2** to **5b.4**, how various degrees of knowledge concerning F, the prior distribution of $\boldsymbol{\theta}$, affect the procedure of estimation. A particularly difficult case of the general problem of section **5b.1** is when F is taken to be a single point distribution with an unknown point of concentration. This situation leads to the classical theory of estimation where the parameter is considered as fixed but unknown.

5b.5 Fiducial Probability

A question may be raised as to whether statements of probability concerning an unknown parameter can be made independently of the prior distribution. Such an approach was developed by Fisher using what is known as *Fiducial argument*. Without going into the complications of the theory, the method is briefly indicated. Let T be a sufficient statistic for θ and $g(T, \theta)$ a function of θ and T such that the distribution of $g(T, \theta)$ is independent of θ. Let

$$P[g(T, \theta) < \lambda \mid \theta] = F(\lambda) \tag{5b.5.1}$$

and $g(T, \theta)$ be such that

$$g(T, \theta) < \lambda \Leftrightarrow \theta > h(T, \lambda), \tag{5b.5.2}$$

in which case the equation (5b.5.1) may be written

$$P[\theta > h(T, \lambda)] = F(\lambda). \tag{5b.5.3}$$

The argument leading to (5b.5.3) involving two variables θ and T is perfectly valid. The next step, which is controversial, consists of asserting that

$$P[\theta > h(T_0, \lambda)] = F(\lambda) \tag{5b.5.4}$$

where T_0 is an observed value of the random variable T.

For instance, if $\bar{x}$ is the average of a sample of n observations from a normal population $N(\theta, 1)$, the function $y = (\bar{x} - \theta) \sim N(0, 1/\sqrt{n})$ and

$$P\left(\bar{x} - \theta < \frac{\lambda}{\sqrt{n}}\right) = \int_{-\infty}^{\lambda/\sqrt{n}} N(y \mid 0, 1/\sqrt{n})\, dy = \Phi(\lambda).$$

Suppose the value of $\bar{x}$ in a particular sample is 5.832, the fiducial argument leads to the statement

$$P\left(\theta > 5.832 - \frac{\lambda}{\sqrt{n}}\right) = \Phi(\lambda). \tag{5b.5.5}$$

Several objections may be raised against such a statement. First, that "θ need not be a random variable and has a fixed but unknown value, such as the distance between the sun and the earth at a particular point of time," which is sought to be estimated by the observations. Second, that even if θ is a random variable with an unknown prior distribution, a statement such as (5b.5.5) cannot be true independently of the prior distribution. For we can always find a prior distribution with respect to which the posterior distribution given $\bar{x}$ is different from (5b.5.5). The interpretation of (5b.5.5) as a probability statement has to be sought in a different way. Some attempts have been made in this direction by Dempster (1963), Fraser (1961), and Sprott (1961).

5b.6 Minimax Principle

Let us consider the $(\boldsymbol{\theta}, \mathbf{X})$ situation of **5b.1** and an estimator $T(\mathbf{X})$ of $g(\boldsymbol{\theta})$. Denote by

$$M_T(\boldsymbol{\theta}) = \underset{\mathbf{X}}{E}\{[T(\mathbf{X}) - g(\boldsymbol{\theta})]^2 \,|\, \boldsymbol{\theta}\} \tag{5b.6.1}$$

the expected mean square error for given $\boldsymbol{\theta}$, which depends only on the conditional distribution of $\mathbf{X}$ given $\boldsymbol{\theta}$. Instead of the squared error we could have chosen any other measure of deviation between T and $g(\boldsymbol{\theta})$ and denoted its expectation by $M_T(\boldsymbol{\theta})$ which may be called loss function. The particular choice of the squared error is made for illustrating the arguments. If the prior d.f. F of $\boldsymbol{\theta}$ is known, the overall mean square error is

$$M_T = \underset{\boldsymbol{\theta}}{E}\,[M_T(\boldsymbol{\theta})] = \int M_T(\boldsymbol{\theta})\, dF(\boldsymbol{\theta}). \tag{5b.6.2}$$

It has already been shown in (5b.2.3) that when $M_T(\boldsymbol{\theta})$ is as defined in (5b.6.1), M_T attains the minimum value when T is chosen as the regression of $g(\boldsymbol{\theta})$ on $\mathbf{X}$, that is, $T = E_{\boldsymbol{\theta}}[g(\boldsymbol{\theta})\,|\,\mathbf{X}]$. We thus have a satisfactory solution when F is known.

When F is unknown we have no integrated measure such as (5b.6.2) for judging the goodness of an estimator. The choice between any two estimators T, T' then poses a serious problem when *a comparison of two functions* $M_T(\boldsymbol{\theta})$, $M_{T'}(\boldsymbol{\theta})$ *are involved.* If there exists a T such that $M_T(\boldsymbol{\theta}) \leqslant M_{T'}(\boldsymbol{\theta})$ for all $\boldsymbol{\theta}$ and T', then T is indeed the best choice. But since this does not generally happen we are forced to bring in other considerations.

We say that T is better than T' iff $M_T(\boldsymbol{\theta}) \leqslant M_{T'}(\boldsymbol{\theta})$ for all $\boldsymbol{\theta}$, and for at least one value of $\boldsymbol{\theta}$ the strict inequality holds. *An estimator T is said to be admissible if there is no other estimator better than T.* The class of admissible estimators is generally very wide, and for any two estimators T and T' in this class there exist non-empty sets S and S' of $\boldsymbol{\theta}$ such that

$$M_T(\boldsymbol{\theta}) \leqslant M_{T'}(\boldsymbol{\theta}), \quad \boldsymbol{\theta} \in S \qquad \text{and} \qquad M_{T'}(\boldsymbol{\theta}) \leqslant M_T(\boldsymbol{\theta}) \quad \text{for} \quad \boldsymbol{\theta} \in S'.$$

Consider the problem of estimating the binomial proportion π on the basis of n Bernoullian trials. If r denotes the number of successes the mean square error of the estimator $(r + \alpha)/(n + \beta)$ considered in (5b.2.8) is

$$E\left(\frac{r+\alpha}{n+\beta} - \pi\right)^2 = \frac{n\pi(1-\pi) + (\alpha - \beta\pi)^2}{(n+\beta)^2}. \tag{5b.6.3}$$

For the special choice $\alpha = \sqrt{n}/2$, $\beta = \sqrt{n}$, the estimator

$$T' = (r + \sqrt{n}/2)/(n + \sqrt{n}) \tag{5b.6.4}$$

has the mean square error $n/4(n + \sqrt{n})^2$ which is independent of π. The mean square error of the traditional estimator $T = r/n$ is $\pi(1 - \pi)/n$. Now T' is better than T for values of π such that

$$\frac{n}{4(n+\sqrt{n})^2} \leqslant \frac{\pi(1-\pi)}{n}$$

or in the interval

$$\pi \in (\tfrac{1}{2} \pm a), \qquad \text{where} \quad a = \frac{1}{2}\left[1 - \left(1 + \frac{1}{\sqrt{n}}\right)^{-2}\right]^{1/2}.$$

In the complementary intervals $(0, \frac{1}{2} - a)$, $(\frac{1}{2} + a, 1)$, T is better than T'. The interval in which T' is better than T, however, tends to zero as $n \to \infty$. But for a finite n the choice between T and T' cannot be made without other considerations. We have already seen that a restriction such as an unbiased estimator selects T as one with minimum mean square error. We shall now consider another principle which prefers T' to T.

Minimax Principle. The philosophy behind this principle is to judge the goodness of an estimator T by $\sup_{\boldsymbol{\theta}} M_T(\boldsymbol{\theta})$, that is by the worst that can happen. *The minimax principle suggests the choice of an estimator for which* $\sup_{\boldsymbol{\theta}} M_T(\boldsymbol{\theta})$ *is a minimum.*

With respect to the single measure $\sup_{\boldsymbol{\theta}} M_T(\boldsymbol{\theta})$, T_1 is better than T_2 if

$$\sup_{\boldsymbol{\theta}} M_{T_1}(\boldsymbol{\theta}) < \sup_{\boldsymbol{\theta}} M_{T_2}(\boldsymbol{\theta}),$$

and if there exists a T^* such that

$$\sup_{\boldsymbol{\theta}} M_{T^*}(\boldsymbol{\theta}) \leqslant \sup_{\boldsymbol{\theta}} M_T(\boldsymbol{\theta}) \qquad \text{for all } T,$$

then T^* is said to be a minimax estimator.

There is, however, no simple method of computing minimax estimators. But the following result is important in that it enables us to recognize minimax estimators in certain situations.

Consider the $(\boldsymbol{\theta}, \mathbf{X})$ situation of **5b.1** and let the loss in estimating $g(\boldsymbol{\theta})$ by $T(\mathbf{X})$ be measured by the squared error $[T(\mathbf{X}) - g(\boldsymbol{\theta})]^2$. Suppose that there exists a prior d.f. F^* of $\boldsymbol{\theta}$, such that for the associated regression estimator T^* of g given $\mathbf{X}$ (see 5b.2.3), the loss function

$$M_{T^*}(\boldsymbol{\theta}) = \underset{\mathbf{X}}{E}\{[T^*(\mathbf{X}) - g(\boldsymbol{\theta})]^2 \mid \boldsymbol{\theta}\}$$

is independent of $\boldsymbol{\theta}$. Then T^* is a minimax estimator with respect to the specified loss function.

Since T^* is a regression estimator corresponding to the prior d.f. F^* of $\boldsymbol{\theta}$, the overall mean square error

$$\int M_{T^*}(\boldsymbol{\theta})\, dF^*(\boldsymbol{\theta}).$$

is a minimum. That is, if T is any other estimator,

$$\int M_{T^*}(\boldsymbol{\theta})\, dF^*(\boldsymbol{\theta}) \leqslant \int M_T(\boldsymbol{\theta})\, dF^*(\boldsymbol{\theta}).$$

But it is given that the left-hand side quantity is a constant, say c, hence

$$c \leqslant \int M_T(\boldsymbol{\theta})\, dF^*(\boldsymbol{\theta}),$$

which implies $c \leqslant \sup M_T(\boldsymbol{\theta})$ for any T, that is, T^* is a minimax estimator.

In the problem of estimating the binomial proportion we find that the estimator T' considered in (5b.6.4) is in fact the Bayes estimator (5b.2.8) with the choice $\alpha = \sqrt{n}/2$ and $\gamma = \sqrt{n}/2$ for the prior probability density (5b.2.4). Furthermore, the associated loss function is $n/4(n + \sqrt{n})^2$ which is independent of π. Hence

$$T^* = \frac{r + \sqrt{n}/2}{n + \sqrt{n}}$$

is the minimax estimator of π with respect to the minimum mean square error criterion.

In a general loss function $W[T(\mathbf{X}), g(\boldsymbol{\theta})]$, there is first the problem of determining the Bayes estimator, that is, of determining a function $T(\mathbf{X})$ such that

$$\int W[T(\mathbf{X}), g(\boldsymbol{\theta})] \frac{P(\mathbf{X} \mid \boldsymbol{\theta})}{G(\mathbf{X})}\, dF(\boldsymbol{\theta})$$

is a minimum. If we can find a prior distribution F^* such that for the corresponding Bayes estimator T^*, the expected loss function

$$M_{T^*}(\boldsymbol{\theta}) = \int W[T^*(\mathbf{X}), g(\boldsymbol{\theta})] P(\mathbf{X} \mid \boldsymbol{\theta})\, dv$$

is independent of $\boldsymbol{\theta}$, then T^* can be recognized to be a minimax estimator.

The reader should not be surprised to find that different criteria lead to different solutions (estimators). The choice of a criterion in any given situation depends on the purpose for which an estimate is obtained. For instance, a minimax estimator may give higher losses than another estimator in a wide range of the parameter space, but may indeed be useful if we want some insurance against undue loss resulting from our estimate in certain situations.

5b.7 Principle of Invariance

In (5a.2.5), we restricted the class of estimators to those having the translation property,

$$f(x_1 + \alpha, \dots, x_n + \alpha) = \alpha + f(x_1, \dots, x_n) \tag{5b.7.1}$$

without explaining the rationale of such a procedure. The p.d. of $x_1, \dots, x_n$ was of the form $p(x_1 - \theta, \dots, x_n - \theta)$, which is invariant under translation, i.e., if instead of x_i we consider $y_i = x_i + \alpha$ as our observations, where α is a known constant, then y_i would have the density $p(y_1 - \theta', \dots, y_n - \theta')$ where $\theta' = \theta + \alpha$. Since the p.d.'s are of the same form, the estimating function of θ in terms of x_i or θ' in terms of y_i should be of the same form. Thus if $f(x_1, \dots, x_n)$ estimates θ, then $f(y_1, \dots, y_n)$ estimates $\theta' = \theta + \alpha$ and consistency demands that (5b.7.1) should hold. We have already formulated, in Appendix **2C** of Chapter 2, the principle of invariance in a more general setup applicable to the general decision problem (to cover estimation, testing of hypotheses, etc.).

Although the principle of invariance is very much discussed in literature, there is some controversy about its application. [See comments by Barnard in Rao (1962d) and Stein (1962).]

Let us consider independent random variables $X_i \sim N(\mu_i, 1)$, $i = 1, \dots, n$ and the problem of simultaneous estimation of $\mu_1, \dots, \mu_n$ using the loss function

$$(t_1 - \mu_1)^2 + \cdots + (t_n - \mu_n)^2. \tag{5b.7.2}$$

It is seen that the decision problem is invariant under translations of the random variables. In such a case, the estimators t_i should have the translation property (See Appendix **2C** of Chapter 2)

$$t_i(x_1 + \alpha_1, \dots, x_n + \alpha_n) = \alpha_i + t_i(x_1, \dots, x_n), \qquad i = 1, \dots, n. \tag{5b.7.3}$$

Subject to (5b.7.3), the expected value of the loss (5b.7.2) is a minimum when $t_i = x_i$, the observed value of the r.v. X_i. However, let us consider an alternative estimator

$$t_i^* = \left(1 - \frac{n-2}{\sum x_i^2}\right)x_i, \qquad i = 1, 2, \dots, n. \tag{5b.7.4}$$

Then it is shown by Stein (1962) that

$$E[\sum(t_i^* - \mu_i)^2] = n - (n-2)^2 E[(2K + n - 2)^{-1}] \tag{5b.7.5}$$

where K is a Poisson variable with mean $(\sum \mu_i^2)/2$. But $E[\sum(x_i - \mu_i)^2] = n$ which is larger than (5b.7.5) for $n \geqslant 3$. In fact, (5b.7.5) has the value 2 when $\sum \mu_i^2 = 0$ and approaches n as $\sum \mu_i^2 \to \infty$. Thus the gain is appreciable but the estimators t_i^* do not satisfy the principle of invariance.

This example only illustrates that any principle, however attractive it may appear, may not have universal validity and its relevance in any particular problem has to be judged by the reasonableness of the solution it provides.

5c CRITERIA OF ESTIMATION IN LARGE SAMPLES

5c.1 Consistency

Let $\{x_n\}$, $n = 1, 2, \ldots$ be a sequence of observations (random variables) providing information on a parameter θ and let T_n be an estimator of θ based on the first n variables. Our object is to examine the properties of the sequence $\{T_n\}$ as $n \to \infty$.

The estimator T_n is said to be consistent for θ if $T_n \to \theta$ in probability or with probability 1. The former may be called weak consistency (W.C.) and the latter, strong consistency (S.C.).

Consistency as defined refers only to a limiting property of the estimator T_n as $n \to \infty$. Therefore, a certain amount of caution is necessary in applying it as a criterion of estimation in practical situations. For if T_n is a consistent estimator, we can replace it by another defined as

$$T_n' = \begin{cases} 0 & \text{for } n \leqslant 10^{10} \\ T_n & \text{for } n > 10^{10} \end{cases} \tag{5c.1.1}$$

which is also consistent, but would indeed be rejected as an estimate in any practical situation where the sample size may be large but not indefinitely large.

Thus the criterion of consistency is not valid with reference to a particular sample size. One of the desirable properties to be satisfied, however, is that, with increase in the observations, the estimator tends to a definite quantity which is the true value to be estimated.

For instance, let $x_1, x_2, \ldots$ be observations from a Cauchy population with the density function proportional to $[1 + (x - \mu)^2]^{-1}$. It is known that $\bar{x}_n = (x_1 + \cdots + x_n)/n$ has the same distribution as that of a single observation (see **3a.4**). Hence the probability of $|\bar{x}_n - \mu| > \delta$ remains the same for all n, so that $\bar{x}_n$ is not consistent for μ. The median of the observations is known, however, to be a consistent estimator.

By the strong law of large numbers it follows that $\bar{x}_n$ is a consistent estimator of the mean provided $\{x_n\}$ is a sequence of i.i.d. variables and $E(x_i)$ exists. The latter condition is not satisfied for the Cauchy distribution.

By using Tschebyshev's inequality [(iii), **2b.2**], it is easy to establish *that if $E(T_n) = \theta_n \to \theta$ and $V(T_n) \to 0$, then T_n is weakly consistent for θ.* Consider the estimator $T_n = \sum (x_i - \bar{x})^2/n$ of σ^2, where $x_1, \ldots, x_n$ are n independent observations from $N(\theta, \sigma^2)$.

$$E(T_n) = \frac{n-1}{n}\sigma^2 \to \sigma^2$$

$$V(T_n) = \frac{2\sigma^4(n-1)}{n^2} \to 0.$$

Hence T_n is consistent for σ^2, although biased.

Fisher Consistency (F.C.). (Fisher, 1922, 1956) Fisher introduced another definition of consistency which seems to place a restriction on the estimating function and thus eliminate the possibility of statistics such as (5c.1.1) being considered. The definition is applicable to any sample size and may, therefore, be more meaningful in practical applications.

We shall first consider samples from a finite multinomial distribution with the cell probabilities $\pi_1(\boldsymbol{\theta}), \ldots, \pi_k(\boldsymbol{\theta})$ depending on a vector parameter $\boldsymbol{\theta}$. Let n be the total sample size, n_i be the observed frequency in the ith cell, $p_i = n_i/n$ be the observed proportion in the ith cell, and $g(\boldsymbol{\theta})$ be the parametric function to be estimated.

An estimator T is said to be F.C. for $g(\boldsymbol{\theta})$ iff, (i) T is defined as a continuous function over the set of vectors $(x_1, \ldots, x_k)$ such that $x_i \geqslant 0$ and $x_1 + \cdots + x_k = 1$ with the value of T at $x_i = p_i$, $i = 1, \ldots, k$, as the estimate of $g(\boldsymbol{\theta})$ based on the sample, and (ii) the value of T at $x_i = \pi_i(\boldsymbol{\theta})$, $i = 1, \ldots, k$, is $g(\boldsymbol{\theta})$ for all admissible values of $\boldsymbol{\theta}$, that is, $T[\pi_1(\boldsymbol{\theta}), \ldots, \pi_k(\boldsymbol{\theta})] \equiv g(\boldsymbol{\theta})$.

What, in effect, the definition demands is that the estimator should be an explicit function of the observed proportions only, which may be written as $T(p_1, \ldots, p_k)$, and that it should have the true value of $g(\boldsymbol{\theta})$ when the observed proportions happen to coincide with the true proportions (probabilities). It would be, indeed, anomalous if we happen to realize the true proportions but the method of estimation does not automatically lead to the true value of $g(\boldsymbol{\theta})$. In this sense, F.C. may also be called *method consistency* (M.C.).

It is easy to see that an *F.C. estimator as defined above is also S.C.* since $p_i \xrightarrow{\text{a.s.}} \pi_i(\boldsymbol{\theta})$ and the continuity of T implies that [(xiii), **2c.4**]

$$T(p_1, \ldots, p_k) \xrightarrow{\text{a.s.}} T[\pi_1(\boldsymbol{\theta}), \ldots, \pi_k(\boldsymbol{\theta})] = g(\boldsymbol{\theta}).$$

In the general case let $x_1, \ldots, x_n$ be i.i.d. observations on a random variable with a distribution function $F(\cdot, \boldsymbol{\theta})$ and S_n be the empirical d.f. based on the observations. An estimator T is said to be F.C. iff:

(a) T is defined as a weakly continuous functional on the space of d.f.'s with the value of T for given S_n, as the estimate of $g(\boldsymbol{\theta})$ based on the observations, and
(b) $T[F(\cdot, \boldsymbol{\theta})] \equiv g(\boldsymbol{\theta})$ identically in $\boldsymbol{\theta}$.

As for the multinomial distribution, F.C. $\Rightarrow$ S.C. since, as shown in (6f.1.3) of Chapter 6,

$$\sup_x |S_n(x) - F(x, \boldsymbol{\theta})| \xrightarrow{\text{a.s.}} 0,$$

which implies that

$$T[S_n] \xrightarrow{\text{a.s.}} T[F(\cdot, \boldsymbol{\theta})] = g(\boldsymbol{\theta})$$

It may be observed that the application of F.C. as stated is limited to i.i.d. observations only.

5c.2 Efficiency

There is some amount of confusion in statistical literature regarding the concept of efficiency of an estimator.

It is shown in (5a.2.15) that if T is an unbiased estimator of $g(\theta)$, a function of the parameter θ, then under some conditions

$$V(T) \geqslant \frac{[g'(\theta)]^2}{\mathscr{I}(\theta)} \tag{5c.2.1}$$

where $\mathscr{I}(\theta)$ is Fisher's information on θ contained in the sample. From the inequality (5c.2.1), some authors define T to be an efficient (unbiased) estimator of $g(\theta)$ if

$$V(T) = \frac{[g'(\theta)]^2}{\mathscr{I}(\theta)},$$

that is, when the lower bound in (5c.2.1) is attained.

Logically, this does not appear to be a good definition, for the expression $[g'(\theta)]^2/\mathscr{I}(\theta)$ is only one possible lower bound and there do exist a large number of sharper lower bounds any one of which could have been equally chosen to define efficiency. Further in many cases it is known that the *minimum attainable* variance is larger than the lower bound (5c.2.1). For instance the unbiased estimator of σ^2 based on n observations from a normal population, $s^2 = \sum (x_i - \bar{x})^2/(n-1)$, has the least possible variance $2\sigma^4/(n-1)$, whereas the lower bound (5c.2.1) has the value $2\sigma^4/n$ (see 5a.3.6).

If it is desired to link efficiency with variance, an unbiased estimator of a parameter may be defined to be efficient if it has the least possible variance and not necessarily a particular value which may not be attainable for any estimator in a given situation. In such a case, however, the term, efficiency, would be synonymous with minimum variance and would not refer to any intrinsic property of the estimator such as concentration around the true value, etc.

Sometimes efficiency is defined as a limiting property.. An estimator T_n (based on a sample of size n) is said to be asymptotically efficient for $g(\theta)$ if

$$\left.\begin{array}{l} E(T_n) \to g(\theta) \\ |V(T_n) - [g'(\theta)]^2/\mathscr{I}| \to 0 \end{array}\right\} \quad \text{as} \quad n \to \infty. \tag{5c.2.2}$$

Efficiency so defined is not applicable to situations where $E(T_n)$ and $V(T_n)$ do not exist but the distribution of T_n becomes concentrated at $g(\theta)$ as $n \to \infty$.

CAN Estimators. A slightly different concept of asymptotic efficiency which has been in use for a long time is based on the asymptotic variance of an estimator. The class of estimators is restricted to what are called consistent asymptotically normal (CAN) estimators. An estimator T_n is said to be a CAN estimator of $g(\theta)$ if the asymptotic distribution of $\sqrt{n}[T_n - g(\theta)]$ is normal. A CAN estimator T_n is said to be the best or efficient if the variance of the limiting distribution of $\sqrt{n}[T_n - g(\theta)]$ has the least possible value. It was thought that when i.i.d. observations are considered, the variance of the limiting distribution of $\sqrt{n}[T_n - g(\theta)]$ has the lower bound $[g'(\theta)]^2/i$, where i is Fisher's information on θ in a single observation. An estimator T_n for which the stated lower bound is attained for the asymptotic distribution is taken to be efficient. But unfortunately the result concerning the lower limit to asymptotic variance is not true without further conditions on the estimator (Kallianpur and Rao, 1955d; Rao, 1963a). In fact, the asymptotic variance of a consistent estimator can be arbitrarily small so that the concept of efficiency based on asymptotic variance is void.

An example to this effect is provided by Hodges (see Le Cam, 1953). Let T_n be any consistent estimator such that the asymptotic variance of $\sqrt{n}(T_n - \theta)$ is $v(\theta)$. Consider the estimator

$$T_n' = \begin{cases} \alpha T_n, & \text{if } |T_n| < n^{-1/4} \\ T_n, & \text{if } |T_n| \geqslant n^{-1/4} \end{cases} \tag{5c.2.3}$$

where α is a constant. It is easy to see that $\sqrt{n}(T_n' - \theta)$ is also asymptotically normally distributed but with variance $\alpha^2 v(0)$ at $\theta = 0$ and $v(\theta)$ elsewhere. Since α is arbitrary, the asymptotic variance of T' can be made smaller than that of T at $\theta = 0$ and equal elsewhere. Therefore, there is no lower bound to the asymptotic variance of a CAN estimator, so there does not exist any *best* CAN estimator, without any further conditions on the estimator. In statistical literature a best CAN estimator is referred to as a best asymptotically normal (BAN) estimator. What we have demonstrated is that BAN estimators do not exist unless possibly when suitable restrictions are placed on the class of estimators. We shall examine some of the restrictions in later sections.

Fisher's (1925) Definition of Asymptotic Efficiency. Let us consider i.i.d. observations and let $i(\theta)$ be the information on θ in a single observation. It is shown in (5a.4.3) that if T_n is an estimator of θ and ni_{T_n} is the information on θ contained in T_n, then

$$ni_{T_n} \leqslant ni \qquad \text{or} \qquad i_{T_n} \leqslant i.$$

Fisher defines the ratio i_{T_n}/i (which is $\leqslant 1$) as the efficiency of an estimator T_n for finite n and the limit of i_{T_n}/i as $n \to \infty$ as the (asymptotic) efficiency of an estimator in large samples.

It may be noted that efficiency so defined does not refer to properties such as bias or consistency with respect to any function of the parameter. In fact, if T_n is efficient in the sense $(i_{T_n}/i) = 1$ or $\to 1$ as $n \to \infty$, then any one-to-one function of T_n is also equally efficient for θ.

We have seen in **5a.5** that Fisher's information measures an intrinsic character of the distribution of a random variable with respect to an unknown parameter. The intrinsic character considered is the discriminatory power between close alternative values of the parameter. The information in a sample of size n is ni and that in the statistic is $ni_{T_n} \leqslant ni$. When equality obtains the discriminatory powers of the sample and the statistic are the same. But a statistic involves reduction of data and consequently its discriminatory power is not greater than that of the entire sample. Hence if a choice has to be made among several statistics it is preferable to choose one for which ni_{T_n}/ni is a maximum.

It has already been shown that when T_n is a sufficient estimator of θ, $ni_{T_n} = ni$, so that there is no loss of information for any given n. But a sufficient estimator (one-dimensional measurable random variable) does not always exist It is possible, however, to find a large class of estimators (as shown in Section **5d**) with the property, $\lim i_{T_n}/i = 1$.

A New Definition of Asymptotic Efficiency. Let $P(\cdot, \ldots, \cdot \,|\, \theta)$ be the p.d. of the r.v.'s $x_1, \ldots, x_n$ and define the r.v.

$$z_n = \frac{1}{n} \frac{d \log P(x_1, \ldots, x_n \,|\, \theta)}{d\theta}. \tag{5c.2.4}$$

A consistent estimator T_n of θ is said to be first order efficient (f.o.e.) if

$$\sqrt{n}\,|T_n - \theta - \beta(\theta) z_n| \to 0 \tag{5c.2.5}$$

in probability or with probability 1, *where* β *does not involve the observations* (Rao, 1961a, 1961b, 1962d,g, 1963a, 1965a).

The condition (5c.2.5) also implies that the asymptotic correlation between T_n and z_n is unity, so that the f.o.e. of any estimator may be measured by the square of the (asymptotic) correlation between T_n and z_n.

In the multiparameter case we have a vector of derivatives

$$\mathbf{Z}_n = (z_n^1, \ldots, z_n^q)$$

$$z_n^i = \frac{1}{n}\frac{\partial \log P}{\partial \theta_i}, \qquad i = 1, \ldots, q$$

and a vector of deviations of the estimators from the true values

$$\mathbf{D}_n' = (\mathbf{T}_n - \boldsymbol{\theta})' = (T_n^1 - \theta_1, \ldots, T_n^q - \theta_q)$$

and $\mathbf{T}_n$ *is said to be f.o.e. if*

$$\sqrt{n}\,|\mathbf{D}_n - \mathbf{B}\mathbf{Z}_n| \to 0 \tag{5c.2.6}$$

in probability or with probability 1, *where* $\mathbf{B}$ *is a matrix of constants which may depend on* θ.

We have the following results as a consequence of the definitions (5c.2.5) and (5c.2.6) of asymptotic efficiency.

(i) *Let* $x_1, x_2, \ldots$ *be a sequence of i.i.d. random variables with probability density* $p(x, \theta)$. *Further let*

$$\int p'(x, \theta)\, dx = 0 \qquad \text{and} \qquad \int \frac{p'^2}{p}\, dx = i(\theta) > 0. \tag{5c.2.7}$$

Then the condition (5c.2.5) *implies that the asymptotic distribution of* $\sqrt{n}(T_n - \theta)$ *is* $N(0, \beta^2 i)$.

Under the conditions (5c.2.7), by the central limit theorem, the asymptotic distribution of

$$\sqrt{n}\, z_n = \left[\frac{p'(x_1, \theta)}{p(x_1, \theta)} + \cdots + \frac{p'(x_n, \theta)}{p(x_n, \theta)}\right] \div \sqrt{n}$$

is $N[0, i(\theta)]$. By (5c.2.5), the asymptotic distribution $\sqrt{n}(T_n - \theta)$ is the same as that of $\beta\sqrt{n}\, z_n$ by [(ix), 2c.4]. Hence the result.

(ii) *If* T_n' *and* T_n' *are both f.o.e., then they are equivalent in the sense that there exist* $\gamma(\theta)$ *and* $\alpha(\theta) = \theta[1 - \gamma(\theta)]$ *such that*

$$\sqrt{n}\,|T_n - \alpha(\theta) - \gamma(\theta)T_n'| \to 0 \tag{5c.2.8}$$

in probability or with probability 1. *That is, asymptotically one is a linear function of the other.*

(iii) *In the multiparameter case, under conditions similar to those in* (i), *the asymptotic distribution of* $\sqrt{n}\,\mathbf{Z}_n$ *is q-variate normal with mean zero and dispersion matrix* $\mathscr{I} = (i_{rs})$, *which is the information matrix. The limit condition* (5c.2.6) *implies that* $\sqrt{n}(\mathbf{T}_n - \boldsymbol{\theta})$ *is also asymptotically q-variate normal with the dispersion matrix* $\mathbf{B}\mathscr{I}\mathbf{B}'$.

The asymptotic normality of $\sqrt{n}\,\mathbf{Z}_n$ follows by an application of the multivariate central limit theorem [(iv) **2c.5**]. Hence the rest follow.

(iv) $\sqrt{n}\,|T_n - \theta - \beta z_n| \to 0$ *in probability implies that* $\lim(i_{T_n}/i) \to 1$ *as* $n \to \infty$.

The result (iv) shows that the definition (5c.2.5) of efficiency implies Fisher's efficiency. The proof of the result is somewhat involved (see Doob, 1934, 1936; Rao, 1961a).

It is shown in later sections that the class of estimators satisfying the definitions (5c.2.5) or (5c.2.6) is not empty. Thus the definitions (5c.2.5) and (5c.2.6) of asymptotic efficiency appear to be satisfactory.

Best CUAN Estimators. One of the advantages in considering CAN estimators is that inferences on θ based on the estimator T_n can be drawn using the normal distribution. Thus if the asymptotic distribution of $\sqrt{n}(T_n - \theta)$ is $N[0, v(\theta)]$, then to test the hypothesis $\theta = \theta_0$, we can use the critical region

$$\frac{\sqrt{n}\,|(T_n - \theta_0)|}{\sqrt{v(\theta_0)}} > d_\alpha \tag{5c.2.9}$$

where d_α is the two-sided α point of $N(0, 1)$. If $v(\theta)$ is continuous in θ, then the test (5c.2.9) is equivalent to, by (*xb*), **2c.4**,

$$\frac{\sqrt{n}\,|(T_n - \theta_0)|}{\sqrt{v(T_n)}} > d_\alpha. \tag{5c.2.10}$$

From the form of the test, we may be tempted to construct a $(1 - \alpha)$ confidence interval of the type

$$T_n \pm \frac{d_\alpha \sqrt{v(T_n)}}{\sqrt{n}} \tag{5c.2.11}$$

for the parameter θ. The procedure (5c.2.11) is not justified, however, unless the convergence to normality of $\sqrt{n}(T_n - \theta)$ is uniform in compact intervals of θ. If, therefore, we intend to use an estimator for inferences of the type (5c.2.9) and (5c.2.11), that is, both for testing simple hypotheses and constructing confidence intervals, it is necessary that the estimator be consistent and uniformly asymptotically normal (CUAN). It has been shown (Rao, 1963a), under suitable regularity conditions on the probability density of i.i.d. observations, that if T_n *is CUAN, then the asymptotic variance* $v(\theta)$ *has the lower bound* $1/i(\theta)$. Thus the best CUAN estimator is one for which the asymptotic variance $v(\theta)$ has the value $1/i(\theta)$. Therefore the concept of minimum variance is not void when the further condition of uniform convergence to normality is imposed on the CAN estimator (Rao, 1965a).

5d SOME METHODS OF ESTIMATION IN LARGE SAMPLES

5d.1 Method of Moments

Let $x_1, \ldots, x_n$ be i.i.d. observations from a distribution involving unknown parameters $\theta_1, \ldots, \theta_q$. Further, let the first q raw moments of the distribution exist as explicit functions $\alpha_r(\theta_1, \ldots, \theta_q)$, $r = 1, \ldots, q$ of the unknown parameters. If

$$a_r = (\sum x_i^r) \div n$$

denote the moment functions, then the method of moments consists of equating the realized values a_{r0} in a sample and the hypothetical moments

$$\alpha_r(\theta_1, \ldots, \theta_q) = a_{r0}, \qquad r = 1, \ldots, q \tag{5d.1.1}$$

and solving for $\theta_1, \ldots, \theta_q$.

We observe that a_r is the mean of n random variables, and if $E(x_i^r)$, the rth raw moment, exists, then by the law of large numbers $a_r \to \alpha_r(\theta_1, \ldots, \theta_q)$ with probability 1, so that a_r is a consistent (besides being unbiased) estimator of α_r. It is easy to see that if the correspondence between $\theta_1, \ldots, \theta_q$ and $\alpha_1, \ldots, \alpha_q$ is one-to-one and inverse functions

$$\theta_i = f_i(\alpha_1, \ldots, \alpha_q), \qquad i = 1, \ldots, q$$

are continuous in $\alpha_1, \ldots, \alpha_q$, then

$$\hat{\theta}_i = f_i(a_{10}, \ldots, a_{q0}), \qquad i = 1, \ldots, q$$

are solutions of (5d.1.1) and $f_i(a_1, \ldots, a_q)$ is a consistent estimator of θ_i, $i = 1, \ldots, q$.

Thus the method of moments provides, under suitable conditions, consistent estimators and the estimating equations are simple in many situations. The method is not applicable when the theoretical moments do not exist as in the case of a Cauchy distribution.

But the estimators obtained by this method are not generally efficient. For some numerical computations of the loss of efficiency the reader is referred to Fisher (1922).

5d.2 Minimum Chi-Square and Associated Methods

These methods are applicable in situations where the observations are continuous measurements but are grouped in suitable class intervals, or the observations themselves are frequencies of a finite number of mutually exclusive events. In either case we have a set of observed frequencies $n_1, \ldots, n_k$ in k cells with the hypothetical probabilities $\pi_1, \ldots, \pi_k$ as functions of q unknown parameters. We first define a measure of discrepancy between the observed frequencies $n_1, \ldots, n_k$ and the hypothetical expectations $n\pi_1(\boldsymbol{\theta}), \ldots, n\pi_k(\boldsymbol{\theta})$, where $n = \sum n_i$. The estimates are obtained by minimizing such a measure with respect to $\boldsymbol{\theta}$. Some of the measures used for the estimation of parameters are as follows:

(a) *Chi-square*

$$\chi^2 = \sum \frac{[n_i - n\pi_i(\boldsymbol{\theta})]^2}{n\pi_i(\boldsymbol{\theta})}.$$

(b) *Modified chi-square*

$$\text{mod } \chi^2 = \sum \frac{[n_i - n\pi_i(\boldsymbol{\theta})]^2}{n_i}.$$

with n_i replaced by unity, if its value is zero.

(c) *Hellinger Distance*

$$\text{H.D.} = \cos^{-1} \sum \sqrt{(n_i/n)\pi_i(\boldsymbol{\theta})}$$

(d) *Kullback-Leibler separator*

$$\text{K.L.S.} = \sum \pi_i(\boldsymbol{\theta}) \log \frac{\pi_i(\boldsymbol{\theta})}{n_i/n}.$$

(e) *Haldane's discrepancy*

$$D_k = \frac{(n+k)!}{n!} \sum \frac{n_i!\,\pi_i^{k+1}(\boldsymbol{\theta})}{(n_i+k)!}, \qquad k \neq -1$$

$$D_{-1} = -\frac{1}{n} \sum n_i \log \pi_i(\boldsymbol{\theta}).$$

All these methods provide reasonably good estimators in the sense that, under suitable regularity conditions, they are consistent and f.o.e. according to the criteria (5c.2.5) and (5c.2.6). For a detailed study of these methods the reader is referred to papers by Neyman (1949) and Rao (1955c, 1961a, b, 1963a).

There are, however, differences between these methods brought out by considering what has been termed as the second-order efficiency. On the basis of second-order efficiency, it is shown that there exists, under some regularity conditions, a method of estimation superior to all those considered in this section (see Rao, 1961a, b, 1962d). This is the method of maximum likelihood which will be discussed in greater detail in the rest of the chapter.

5d.3 Maximum Likelihood

Let X denote the realized value of a set of observations and $P(X, \boldsymbol{\theta})$ denote the joint density, where $\boldsymbol{\theta} = (\theta_1, \ldots, \theta_q)$, the vector of parameters belongs to a set $\Theta \subset E_q$. The likelihood of $\boldsymbol{\theta}$ given the observations is defined to be a function of $\boldsymbol{\theta}$:

$$L(\boldsymbol{\theta} \mid X) \propto P(X, \boldsymbol{\theta}). \tag{5d.3.1}$$

The principle of maximum likelihood (m.l.) consists of accepting $\hat{\boldsymbol{\theta}} = (\hat{\theta}_1, \ldots, \hat{\theta}_q)$ as the estimate of $\boldsymbol{\theta}$, where

$$L(\hat{\boldsymbol{\theta}} \mid X) = \sup_{\boldsymbol{\theta} \in \Theta} L(\boldsymbol{\theta} \mid X). \tag{5d.3.2}$$

The supremum may not always be attained, in which case it may be possible to obtain a "near m.l." estimate

$$L(\boldsymbol{\theta}^* \mid X) \geqslant c \sup_{\boldsymbol{\theta} \in \Theta} L(\boldsymbol{\theta} \mid X),$$

where c is a fixed number such that $0 < c < 1$. There may be a set of samples for which $\hat{\theta}$ or θ^* does not exist. Under regularity conditions on $P(X, \boldsymbol{\theta})$, the frequency of such samples will be shown to be negligible.

In practice it is convenient to work with $l(\boldsymbol{\theta} \mid X) = \log L(\boldsymbol{\theta} \mid X)$, in which case $\hat{\boldsymbol{\theta}}$ of (5d.3.2) satisfies the equation

$$l(\hat{\boldsymbol{\theta}} \mid X) = \sup_{\boldsymbol{\theta} \in \Theta} l(\boldsymbol{\theta} \mid X). \tag{5d.3.3}$$

When the supremum in (5d.3.3) is attained at an interior point of Θ and $l(\boldsymbol{\theta} \mid X)$ is a differentiable function of $\boldsymbol{\theta}$, then the partial derivatives vanish at that point, so that $\hat{\boldsymbol{\theta}}$ is a solution of the equations

$$\frac{\partial l(\boldsymbol{\theta} \mid X)}{\partial \theta_i} = 0, \qquad i = 1, \ldots, q. \tag{5d.3.4}$$

Equations (5d.3.4) are called m.l. equations and any solution of them an m.l. equation estimate.

The function $\hat{\theta}$ defined by (5d.3.3) over the sample space of observations X is called an m.l. estimator. Important results concerning m.l. estimators are given in **5e** and **5f**.

We shall consider some simple examples to illustrate the wide applicability of the m.l. method before discussing general results. Let $x_1, \ldots, x_n$ be observations from $N(\mu, \sigma^2)$, $-\infty < \mu < \infty$, $0 < \sigma^2 < \infty$. It may be seen that

$$l(\mu, \sigma^2 | X) = -n \log \sigma - \frac{\sum (x_i - \mu)^2}{2\sigma^2}$$

$$\frac{\partial l}{\partial \mu} = \sum x_i - n\mu = 0, \qquad \frac{\partial l}{\partial \sigma^2} = \frac{\sum (x_i - \mu)^2}{2\sigma^4} - \frac{n}{2\sigma^2} = 0,$$

which give the estimates

$$\hat{\mu} = \bar{x} \qquad \text{and} \qquad \hat{\sigma}^2 = \frac{\sum (x_i - \bar{x})^2}{n} = s^2. \tag{5d.3.5}$$

Now

$$l(\hat{\mu}, \hat{\sigma} | X) = -n \log s - \frac{n}{2},$$

$$l(\mu, \sigma | X) = -n \log \sigma - \frac{ns^2 + n(\bar{x} - \mu)^2}{2\sigma^2}.$$

To show that $\hat{\mu}, \hat{\sigma}$ actually provide a supremum of the likelihood we examine the inequality $l(\mu, \sigma | X) \leqslant l(\hat{\mu}, \hat{\sigma} | X)$, that is,

$$-n \log \sigma - \frac{ns^2 + n(\bar{x} - \mu)^2}{2\sigma^2} \leqslant -n \log s - \frac{n}{2}$$

$$0 \leqslant \left\{\left(\frac{s^2}{2\sigma^2} - \frac{1}{2}\right) - \log \frac{s}{\sigma}\right\} + \frac{(\bar{x} - \mu)^2}{2\sigma^2}. \tag{5d.3.6}$$

But for any $x > 0$, $\log x < (x^2 - 1)/2$ so that the first term in brackets of (5d.3.6) is not negative. The second term is also not negative so that the inequality (5d.3.6) is true.

Let $x_1, \ldots, x_n$ be observations from a Poisson distribution with parameter μ. It is seen that

$$l(\mu | x_1, \ldots, x_k) = -n\mu + (x_1 + \cdots + x_n) \log \mu,$$

giving the estimate $\hat{\mu} = (x_1 + \cdots + x_n)/n$, which may be shown to provide the supremum of the likelihood.

In the two examples we find that the m.l. estimators are functions of the minimal sufficient statistic. This is true in general because of the factorization of the density function into two parts—one involving the sufficient statistic and the parameter and another independent of $\boldsymbol{\theta}$. Maximizing the likelihood is equivalent to maximizing the first factor so that $\boldsymbol{\theta}$ is an explicit function of the sufficient statistic.

Limitation of the Likelihood Principle. Let us suppose that there are N counters numbered 1, ..., N with unknown values $X_1, \ldots, X_N$ written on them. We draw n counters at random without replacement and record for each counter drawn the pair of values

$$(a = \text{number of the counter}, \quad b = \text{the value of } X \text{ on it}).$$

We may represent the sample outcome as

$$S : [(a_1, b_1), \ldots, (a_n, b_n)].$$

Given S, how do we estimate the unknown parameter $\tau = X_1 + \cdots + X_N$?

It is seen that the probability of S is $n!(N-n)!/N!$ irrespective of what the observations are, and consequently the likelihood of the unknown parameters $X_1, \ldots, X_N$ given S is zero if the values not consistent with those observed on the counters drawn, and constant otherwise irrespective of the values on unobserved counters. Thus the likelihood principle, as pointed out by Godambe, does not enable us to prefer any one set of parametric values for the unobserved counters against others (or in other words no estimation of the unknown parameters is possible by this principle).

While the likelihood is uninformative on unknown values, we cannot say that the sample does not contain information on τ. For instance the estimator

$$N \frac{b_1 + \cdots + b_n}{n}$$

is unbiased for τ, which provides evidence that the sample has information on τ. Thus, the likelihood approach seems to have some limitations.

For a discussion on the subject, the reader is referred to a paper by Rao (1971b) and others in Godambe and Sprott (1971).

5e ESTIMATION OF THE MULTINOMIAL DISTRIBUTION

5e.1 Nonparametric Case

A multinomial distribution is specified by a vector $\boldsymbol{\pi} = (\pi_1, \ldots, \pi_k)$ of probabilities of k mutually exclusive events. If n independent trials yield n_1 events of the first kind, n_2 of the second kind, and so on, the probability of $n_1, \ldots, n_k$ given $\pi_1, \ldots, \pi_k$ is

$$\frac{n!}{n_1! \ldots n_k!} \pi_1^{n_1} \ldots \pi_k^{n_k}. \tag{5e.1.1}$$

Let us denote the expression (5e.1.1) by $L(\boldsymbol{\pi} | n_1, \ldots, n_k)$ or simply by $L(\boldsymbol{\pi})$ as a function of $\boldsymbol{\pi}$. The true distribution $\boldsymbol{\pi}$ applicable to any particular situation is unknown but is given to be a member of a set A (admissible set). The object is to estimate the true distribution on the basis of n independent observations.

In practice, a distribution is usually estimated by first estimating some parameters which may be used as an index to identify the distribution. In theory, we need not make any distinction between the estimation of a distribution or of its parameters identifying it; however, in some situations estimation of parameters may be of prime interest. We shall consider such problems in **5e.2**. In the present section we prove some important results concerning the m.l. estimator of $\boldsymbol{\pi}$, that is, of the distribution as a whole. The same results are true of some other methods of estimation such as minimum Chi-square, minimum Hellinger distance, etc. The reader may easily supply the proofs for these methods on similar lines.

We define $\boldsymbol{\pi}^*$ to be an approximate m.l. estimator of $\boldsymbol{\pi}$ if

$$L(\boldsymbol{\pi}^*) \geqslant c \sup_{\pi \in A} L(\boldsymbol{\pi}), \qquad 0 < c < 1. \tag{5e.1.2}$$

The reason for defining an approximate m.l. estimator (which always exists) is to cover the situations where an m.l. estimator $\hat{\boldsymbol{\pi}}$ such that

$$L(\hat{\boldsymbol{\pi}}) = \sup_{\pi} L(\boldsymbol{\pi}) \tag{5e.1.3}$$

may not exist. We have the following results concerning an estimator of the distribution.

(i) *Let $\boldsymbol{\pi}^*$ be an approximate m.l. estimator of $\boldsymbol{\pi}$ and $\boldsymbol{\pi}^0$ be the true distribution. Then $\boldsymbol{\pi}^* \to \boldsymbol{\pi}^0$ with probability* 1, *without any assumption (restrictions) on the admissible set A.*

There may be a multiplicity of estimators $\boldsymbol{\pi}^*$ satisfying (5e.1.2) including the m.l., estimator when it exists. The result (i) asserts the convergence of $\boldsymbol{\pi}^*$ for any choice (and, in fact, uniformly for all choices). We denote by $\mathbf{p}$, the vector with elements $p_1, \cdots, p_k$.

Let $p_i = n_i/n$ be the observed proportion of events of the ith kind. By the strong law of large numbers $p_i \to \pi_i^0$, $i = 1, \ldots, k$ with probability 1, and to prove the desired result it is enough to show that $\boldsymbol{\pi}^*$ as a function of $\mathbf{p}$ and n tends to $\boldsymbol{\pi}^0$ as $\mathbf{p} \to \boldsymbol{\pi}^0$ and $n \to \infty$. Without loss of generality let $\pi_i^0 > 0$, $i = 1, \ldots, k$. Taking logarithms of (5e.1.2) and dividing by n, we have

$$\begin{aligned}\sum p_i \log \pi_i^* &\geqslant \frac{\log c}{n} + \sup_{\pi \in A} \sum p_i \log \pi_i \\ &\geqslant \frac{\log c}{n} + \sum p_i \log \pi_i^0. \end{aligned} \tag{5e.1.4}$$

We also have, however, the obvious inequality (1e.6.1) in information theory

$$\sum p_i \log \pi_i^* \leqslant \sum p_i \log p_i \tag{5e.1.5}$$

Combining (5e.1.4) and (5e.1.5), we see that

$$\sum p_i \log p_i \geqslant \sum p_i \log \pi_i^* \geqslant \frac{\log c}{n} + \sum p_i \log \pi_i^0$$

$$0 \geqslant \sum p_i \log \frac{\pi_i^*}{p_i} \geqslant \frac{\log c}{n} + \sum p_i \log \frac{\pi_i^0}{p_i}.$$

As $\mathbf{p} \to \boldsymbol{\pi}^0$ and $n \to \infty$, the last term $\to 0$. Hence, as a function of the r.v.'s $p_1, \ldots, p_k$,

$$\lim_{n\to\infty} \sum p_i \log \frac{\pi_i^*}{p_i} = 0 \qquad \text{with probability 1.} \tag{5e.1.6}$$

By Taylor expansion of the logarithm in each term of (5e.1.6) at the value of the argument equal to 1,

$$\begin{aligned} -\sum p_i \log \frac{\pi_i^*}{p_i} &= \sum p_i \left(\frac{\pi_i^*}{p_i} - 1\right)^2 \frac{1}{2y_i^2}, \qquad y_i \in \left(1, \frac{\pi_i^*}{p_i}\right) \\ &= \sum p_i \frac{(\pi_i^* - p_i)^2}{2z_i^2}, \qquad z_i \in (p_i, \pi_i^*) \\ &\geqslant \sum p_i (\pi_i^* - p_i)^2, \qquad \text{since } z_i \leqslant 1. \end{aligned}$$

By using (5e.1.6), $\sum p_i(\pi_i^* - p_i)^2 \to 0$ with probability 1. Then $p_i \to \pi_i^0 \Rightarrow \pi_i^* \to \pi_i^0$ with probability 1, for otherwise there is a contradiction.

(ii) *Let* $A_\varepsilon = \{\boldsymbol{\pi} : \sum \pi_i^0 \log(\pi_i/\pi_i^0) \geqslant -\varepsilon\}$ *and for all sufficiently small* ε, *let* A_ε *be in the interior of* A. *Then an m.l. estimator exists with probability* 1.

Consider the set $A_\varepsilon^c = A - A_\varepsilon$ where

$$\sum \pi_i^0 \log \pi_i < -\varepsilon + \sum \pi_i^0 \log \pi_i^0.$$

Let S be the set of $\boldsymbol{\pi}$ in A_ε^c such that

$$\sum (\pi_i^0 - \varepsilon_1) \log \pi_i < -\varepsilon + \sum \pi_i^0 \log \pi_i^0$$

where $\varepsilon_1 > 0$. As $p_i \to \pi_i^0$, $p_i > \pi_i^0 - \varepsilon_1$, in which case

$$\begin{aligned} \sum p_i \log \pi_i \leqslant \sum (\pi_i^0 - \varepsilon_1) \log \pi_i &< -\varepsilon + \sum \pi_i^0 \log \pi_i^0 \\ &\leqslant -\varepsilon_2 + \sum p_i \log \pi_i^0, \end{aligned}$$

where $\varepsilon_2 \leqslant \varepsilon$. Outside S and within A_ε^c, $\log \pi_i$ are bounded, and therefore as $p_i \to \pi_i^0$,

$$\sum p_i \log \frac{\pi_i}{\pi_i^0} \leqslant -\varepsilon_2.$$

Hence in A_ε^c, when $\mathbf{p}$ is sufficiently close to $\boldsymbol{\pi}^0$,

$$n \sum p_i \log \frac{\pi_i}{\pi_i^0} \leqslant -n\varepsilon_2 ;$$

or by taking exponentials we see that

$$L(\boldsymbol{\pi}^0) \geqslant e^{n\varepsilon_2} L(\boldsymbol{\pi}), \tag{5e.1.7}$$

which implies that

$$L(\boldsymbol{\pi}^0) > \sup_{\boldsymbol{\pi} \in A_\varepsilon^c} L(\boldsymbol{\pi}),$$

and hence, with probability 1,

$$\sup_{\boldsymbol{\pi} \in A} L(\boldsymbol{\pi}) = \sup_{\boldsymbol{\pi} \in A_\varepsilon} L(\boldsymbol{\pi}), \qquad \text{since} \quad \boldsymbol{\pi}^0 \in A_\varepsilon . \tag{5e.1.8}$$

But $L(\boldsymbol{\pi})$ is a continuous function of $\boldsymbol{\pi}$, and for sufficiently small ε, A_ε belongs to the interior of A. Hence the supremum is attained in A_ε.

Thus as $n \to \infty$, an m.l. estimator of $\boldsymbol{\pi}$ exists, and by result (i) it converges to the true distribution.

(iii) *Let λ be an a priori probability measure on A and let A_ε be the set as defined in* (ii). *Further let $P(A_\varepsilon | n_i)$ be the posterior probability that $\boldsymbol{\pi} \in A_\varepsilon$ given $n_1, \ldots, n_k$, that is,*

$$P(A_\varepsilon | n_i) = \frac{\int_{A_\varepsilon} L(\boldsymbol{\pi})\, d\lambda}{\int_A L(\boldsymbol{\pi})\, d\lambda} .$$

Then $P(A_\varepsilon | n_i) \to 1$ with probability 1 as $n \to \infty$ with respect to the true distribution $\boldsymbol{\pi}^0$, if there exists an $\varepsilon' < \varepsilon$ such that

$$P(A_{\varepsilon'}) = \int_{A_{\varepsilon'}} d\lambda > 0.$$

The result $P(A_\varepsilon | n_i) \to 1$ implies that the posterior probability distribution is concentrated at $\boldsymbol{\pi}^0$ with probability 1 as $n \to \infty$. As in (i) and (ii), we observe that $P(A_\varepsilon | n_i)$ is a function of $\mathbf{p}$ and n and show that $P(A_\varepsilon | n_i) \to 1$ as $\mathbf{p} \to \boldsymbol{\pi}^0$ and $n \to \infty$, which would imply the desired convergence in probability.

Within $A_{\varepsilon'}$

$$\sum \pi_i^0 \log \frac{\pi_i}{\pi_i^0} \geqslant -\varepsilon',$$

and $\log \pi_i$ are bounded. Hence as $\mathbf{p} \to \boldsymbol{\pi}^0$,

$$\sum p_i \log \frac{\pi_i}{\pi_i^0} \geqslant -\varepsilon_1, \qquad \varepsilon' < \varepsilon_1 < \varepsilon.$$

If we multiply by n and take exponentials, we see that

$$L(\pi) \geqslant e^{-n\varepsilon_1} L(\pi_0), \quad \text{in} \quad A_{\varepsilon'}.$$

In equation (5e.1.7) of (ii) we can choose ε_2 such that $\varepsilon_1 < \varepsilon_2 < \varepsilon$, so that in $A_\varepsilon{}^c$

$$L(\pi) \leqslant e^{-n\varepsilon_2} L(\pi_0),$$

and hence

$$\int_{A_\varepsilon{}^c} L(\pi)\, d\lambda \leqslant e^{-n\varepsilon_2} L(\pi_0)[1 - P(A_{\varepsilon'})]. \tag{5e.1.9}$$

But

$$\begin{aligned} \int L(\pi)\, d\lambda &\geqslant \int_{A_{\varepsilon'}} L(\pi)\, d\lambda \\ &\geqslant e^{-n\varepsilon_1} L(\pi_0) P(A_{\varepsilon'}). \end{aligned} \tag{5e.1.10}$$

Taking the ratio of (5e.1.9) to (5e.1.10), we see that

$$\begin{aligned} 1 - P(A_\varepsilon | n_i) &\leqslant e^{-n(\varepsilon_2 - \varepsilon_1)} \frac{1 - P(A_{\varepsilon'})}{P(A_{\varepsilon'})} \\ &\to 0 \quad \text{as} \quad n \to \infty, \end{aligned}$$

which is the required result.

5e.2 Parametric Case

The results of **5e.1** in terms of the estimators and the true values of parameters specifying the distribution need not be true without further assumptions. The difficulty arises when the parameters identifying the distribution are not well-behaved functionals of the distribution. Let us consider π_i as a function of $\boldsymbol{\theta} = (\theta_1, \ldots, \theta_q)$, a vector valued parameter belonging to an admissible set Θ. The true value $\boldsymbol{\theta}^0$ is supposed to be an interior point of Θ. We therefore make the following assumptions.

ASSUMPTION 1.1 Given a $\delta > 0$, it is possible to find an ε such that

$$\inf_{|\boldsymbol{\theta} - \boldsymbol{\theta}^0| > \delta} \sum \pi_i(\boldsymbol{\theta}^0) \log \frac{\pi_i(\boldsymbol{\theta}^0)}{\pi_i(\boldsymbol{\theta})} \geqslant \varepsilon \tag{5e.2.1}$$

where $|\boldsymbol{\theta} - \boldsymbol{\theta}^0|$ is the distance between $\boldsymbol{\theta}$ and $\boldsymbol{\theta}^0$.

Assumption 1.1, which may be called *strong identifiability* condition, implies that outside the sphere $|\boldsymbol{\theta} - \boldsymbol{\theta}^0| \leqslant \delta$, there is no sequence of points $\boldsymbol{\theta}_r$ such that $\pi(\boldsymbol{\theta}_r) \to \pi(\boldsymbol{\theta}^0)$ as $r \to \infty$, that is, there are no values of $\boldsymbol{\theta}$ remote from $\boldsymbol{\theta}^0$ but yielding nearly the same set of probabilities $\pi_i(\boldsymbol{\theta}^0)$.

ASSUMPTION 1.2. $\pi_i(\boldsymbol{\theta}) \neq \pi_i(\boldsymbol{\beta})$ for at least one i when $\boldsymbol{\theta} \neq \boldsymbol{\beta}$, which is a weaker identifiability condition.

ASSUMPTION 2.1. The functions $\pi_i(\boldsymbol{\theta})$, $i = 1, \ldots, k$ are continuous in $\boldsymbol{\theta}$.

ASSUMPTION 2.2. The functions $\pi_i(\boldsymbol{\theta})$ admit first-order partial derivatives.

ASSUMPTION 2.3. The functions $\pi_i(\boldsymbol{\theta})$ admit first-order partial derivatives which are continuous at $\boldsymbol{\theta}^0$.

ASSUMPTION 2.4. The functions $\pi_i(\boldsymbol{\theta})$ are totally differentiable at $\boldsymbol{\theta}^0$.

ASSUMPTION 3. Let the information matrix (i_{rs}) be non-singular at $\boldsymbol{\theta}^0$. where

$$i_{rs} = \sum_j \frac{1}{\pi_j} \frac{\partial \pi_j}{\partial \theta_r} \frac{\partial \pi_j}{\partial \theta_s}.$$

As in **5e.1** we define $\boldsymbol{\theta}^*$ to be an approximate m.l. estimator if

$$L[\pi(\boldsymbol{\theta}^*)] \geqslant c \sup_{\theta \in \Theta} L[\pi(\boldsymbol{\theta})] \tag{5e.2.2}$$

(i) *Assumption* 1.1 $\Rightarrow \boldsymbol{\theta}^* \to \boldsymbol{\theta}^0$ *with probability* 1 *as* $n \to \infty$.

It is shown in [(i), **5e.1**], (equation 5c.1.6), that

$$\sum p_i \log \frac{\pi_i(\boldsymbol{\theta}^*)}{p_i} \xrightarrow{\text{a.s.}} 0,$$

from which it follows that, since $\log \pi_i(\boldsymbol{\theta}^*)$ are bounded as $\mathbf{p} \to \boldsymbol{\pi}^0$,

$$\sum \pi_i(\boldsymbol{\theta}^0) \log \frac{\pi_i(\boldsymbol{\theta}^*)}{\pi_i(\boldsymbol{\theta}^0)} \xrightarrow{\text{a.s.}} 0. \tag{5e.2.3}$$

Hence from Assumption 1.1, $|\boldsymbol{\theta}^* - \boldsymbol{\theta}^0| \leqslant \delta$ with probability 1. Since δ is arbitrary, $\boldsymbol{\theta}^* \to \boldsymbol{\theta}^0$ with probability 1.

(ii) *Assumptions* 1.1 *and* 2.1 $\Rightarrow$ *that an m.l. estimator of* $\boldsymbol{\theta}$ *exists and converges to* $\boldsymbol{\theta}^0$ *with probability* 1.

From (5e.1.8), we have as $\mathbf{p} \to \boldsymbol{\pi}^0$

$$\sup_{\pi \in A} L(\boldsymbol{\pi}) = \sup_{\pi \in A_\varepsilon} L(\boldsymbol{\pi}),$$

which in terms of the parameter reduces to

$$\sup_{\boldsymbol{\theta} \in \Theta} L[\pi(\boldsymbol{\theta})] = \sup_{|\boldsymbol{\theta} - \boldsymbol{\theta}^0| \leq \delta} L[\pi(\boldsymbol{\theta})].$$

Since $L[\pi(\boldsymbol{\theta})]$ is continuous in $\boldsymbol{\theta}$, the supremum is attained in the closed sphere $|\boldsymbol{\theta} - \boldsymbol{\theta}^0| \leqslant \delta$. We denote an m.l. estimator by $\hat{\boldsymbol{\theta}}$.

(iii) *Assumptions* 1.1 *and* 2.2 *imply that an m.l. estimator can be obtained as a root of the equations*

$$\frac{\partial L}{\partial \theta_i} = 0, \qquad i = 1, \ldots, k \tag{5e.2.4}$$

with probability 1.

We need only choose δ in such a way that the sphere $|\boldsymbol{\theta} - \boldsymbol{\theta}^0| \leqslant \delta$ is in the interior of Θ, so that the supremum of $L[\pi(\boldsymbol{\theta})]$ is attained in an open interval at, say, $\hat{\boldsymbol{\theta}}$. If then $L[\pi(\boldsymbol{\theta})]$ is differentiable, the partial derivatives must vanish at $\hat{\boldsymbol{\theta}}$.

(iv) *Let the Assumptions* 1.2, 2.3, *and* 3 *be true. Further let* (i^{rs}) *be the reciprocal of* (i_{rs}) *and*

$$z_r = \sum_{i=1}^{k} \frac{p_i}{\pi_i(\boldsymbol{\theta}^0)} \frac{\partial \pi_i}{\partial \theta_r^{\,0}}.$$

Then there exists a consistent root $\bar{\boldsymbol{\theta}}$ *of the likelihood equation* (5e.2.4), *which may not be an m.l. estimator and*

$$\sqrt{n}\,|(\bar{\theta}_r - \theta_r^{\,0}) - i^{r1} z_1 - \cdots - i^{rq} z_q\,| \xrightarrow{P} 0 \tag{5e.2.5}$$

$$r = 1, \ldots, q,$$

that is, the m.l. equation estimators are f.o.e. in the sense of (5c.2.6) *and their asymptotic distribution is q-variate normal.*

Consider the sphere $|\boldsymbol{\theta} - \boldsymbol{\theta}^0| \leqslant \delta$ of radius δ with its center at $\boldsymbol{\theta}^0$ and the function

$$\sum \pi_i(\boldsymbol{\theta}^0) \log \frac{\pi_i(\boldsymbol{\theta}^0)}{\pi_i(\boldsymbol{\theta})} \tag{5e.2.6}$$

over the sphere. Since $\pi_i(\boldsymbol{\theta})$, $i = 1, \ldots, k$ are continuous, the infimum of (5e.2.6) is attained on the sphere and Assumption 1.2 ensures that the infimum is not less than, say, $\varepsilon > 0$. Thus

$$\inf_{|\boldsymbol{\theta} - \boldsymbol{\theta}^0| = \delta} \sum \pi_i(\boldsymbol{\theta}^0) \log \frac{\pi_i(\boldsymbol{\theta}^0)}{\pi_i(\boldsymbol{\theta})} > \varepsilon > 0. \tag{5e.2.7}$$

Further, $\log \pi_i(\boldsymbol{\theta})$ are bounded on the sphere and when $\mathbf{p}$ is close to $\boldsymbol{\pi}(\boldsymbol{\theta}^0)$ we may write (5e.2.7) in the form

$$\inf_{|\boldsymbol{\theta} - \boldsymbol{\theta}^0| = \delta} \sum p_i \log \frac{\pi_i(\boldsymbol{\theta}^0)}{\pi_i(\boldsymbol{\theta})} > 0,$$

which implies that

$$\sum p_i \log \pi_i(\boldsymbol{\theta}^0) > \sum p_i \log \pi_i(\boldsymbol{\theta})$$

for every point on the sphere $|\boldsymbol{\theta} - \boldsymbol{\theta}^0| = \delta$. Hence there is a point $\bar{\boldsymbol{\theta}}$ in the open sphere $|\boldsymbol{\theta} - \boldsymbol{\theta}^0| < \delta$ at which the function $\sum p_i \log \pi_i(\boldsymbol{\theta})$ has a local maximum. Since $\pi_i(\boldsymbol{\theta})$ are differentiable, the derivative of $\sum p_i \log \pi_i(\boldsymbol{\theta})$ vanishes at $\bar{\boldsymbol{\theta}}$. Since δ is arbitrary, $\bar{\boldsymbol{\theta}}$ is a consistent solution of the likelihood equation.

The m.l. equations are

$$n \sum \frac{p_i}{\bar{\pi}_i} \frac{\partial \pi_i}{\partial \bar{\theta}_r} = 0, \qquad r = 1, \ldots, q. \tag{5e.2.8}$$

Dividing by $\sqrt{n}$, the rth equation can be written

$$\sum \frac{\sqrt{n}(p_i - \pi_i^0)}{\bar{\pi}_i} \frac{\partial \pi_i}{\partial \bar{\theta}_r} = \sum \frac{\sqrt{n}(\bar{\pi}_i - \pi_i^0)}{\bar{\pi}_i} \frac{\partial \pi_i}{\partial \bar{\theta}_r}. \tag{5e.2.9}$$

Expanding $(\bar{\pi}_i - \pi_i^0)$ at $\boldsymbol{\theta}^0$

$$\bar{\pi}_i - \pi_i^0 = \sum (\bar{\theta}_s - \theta_s^0) \frac{\partial \pi_i}{\partial \theta_s'}, \qquad \theta_s' \in (\bar{\theta}_s, \theta_s^0),$$

and substituting in the right-hand side of (5e.2.9), we see that

$$\sum_{i=1}^{k} \frac{\sqrt{n}(p_i - \pi_i^0)}{\bar{\pi}_i} \frac{\partial \pi_i}{\partial \bar{\theta}_r} = \sum_{s=1}^{q} \sqrt{n}(\bar{\theta}_s - \theta_s^0) j_{rs}, \tag{5e.2.10}$$

where

$$j_{rs} = \sum \frac{1}{\bar{\pi}_i} \frac{\partial \pi_i}{\partial \bar{\theta}_r} \frac{\partial \pi_i}{\partial \theta_s'} \to i_{rs} \quad \text{as} \quad \bar{\theta}_r \to \theta_r^0.$$

Since $\sqrt{n}(p_i - \pi_i^0)$ has a limiting distribution and $\bar{\pi}_i \to \pi_i^0$ and $(\partial \pi_i / \partial \bar{\theta}_r) \to (\partial \pi_i / \partial \theta_r^0)$, an application of the result [(x), **2c.4**] shows that the left-hand side of (5e.2.10) is asymptotically equivalent to

$$\sum \frac{\sqrt{n}\,(p_i - \pi_i^0)}{\pi_i^0} \frac{\partial \pi_i}{\partial \theta_r^0} = \sqrt{n}\, z_r,$$

which leads to the relation (using the symbol $\stackrel{a}{=}$ for asymptotic equivalence in probability)

$$\sum_{s=1}^{q} \sqrt{n}(\bar{\theta}_s - \theta_s^0) j^{rs} \stackrel{a}{=} \sqrt{n}\, z_r. \tag{5e.2.11}$$

By inverting the relation (5e.2.11), we have

$$\sqrt{n}(\bar{\theta}_s - \theta_s^0) \stackrel{a}{=} \sum \sqrt{n}\, z_r j^{rs} \stackrel{a}{=} \sum \sqrt{n}\, z_r i^{rs}, \tag{5e.2.12}$$

since $j^{rs} \to i^{rs}$, which is the desired result.

Note 1. If in the statement of (iv), Assumption 1.2 is replaced by Assumption 1.1, then the m.l. estimator is a consistent solution of the likelihood equation so that the property (5e.2.5) and the derived result on the asymptotic distribution can be claimed for the m.l. estimator.

Note 2. It has been shown by Birch (1964) that Assumption 2.3 of continuity of the first derivatives of $\pi_i(\boldsymbol{\theta})$ can be replaced by the weaker Assumption 2.4 of total differentiality of $\pi_i(\boldsymbol{\theta})$ for each i to assert the asymptotic normality of the m.l. estimator. More precisely, Birch's result is as follows: *Assumptions* 1.1, 2.4, *and* 3 *imply that the m.l. estimator is efficient in the sense* (5e.2.5), *and its asymptotic distribution is as stated in* (iv). The proof is a bit involved and the details can be found in Birch (1964). The result of (iv) is more general in the sense that it refers to a root of the likelihood equation whether it is an m.l. estimator or not, but it uses the stronger Assumption 2.3.

Note 3. The assumptions under which the results of Sections **5e.1** and **5e.2** are proved are much weaker than those found in the literature on estimation. A surprising result is the deduction of the asymptotic normality of the m.l. estimators of parameters assuming only the continuity of the first derivatives or the total differentiability of the cell probabilities as functions of $\boldsymbol{\theta}$.

It is possible to extend the results established for a finite multinomial to a general discrete distribution with a countable number of alternatives. For instance if $\boldsymbol{\pi} = (\pi_1, \pi_2, \ldots,)$ denotes a general discrete distribution belonging to an admissible set A, the condition $\sum \pi_i{}^0 \log \pi_i{}^0$ is finite, where $\boldsymbol{\pi}^0$ is the true distribution, is sufficient to ensure the consistency of an approximate m.l. estimator (Rao, 1957a, 1958c; Kiefer and Wolfowitz, 1956). For parametric estimation we need the assumption that it is a continuous functional of $\boldsymbol{\pi}$. It can be shown that, under this condition and continuity of π_i as functions of $\boldsymbol{\theta}$, an m.l. estimator of $\boldsymbol{\theta}$ exists with probability 1. If π_i are differentiable functions of $\boldsymbol{\theta}$, then an m.l. estimator is also an m.l. equation estimator. The asymptotic normality of the estimator may, however, require further assumptions (Rao, 1958c).

5f ESTIMATION OF PARAMETERS IN THE GENERAL CASE

5f.1 Assumptions and Notations

Very little is known about the estimation of a distribution function in the continuous case except through the estimation of parameters. Even then, the problems are complicated. We shall consider only the case of a single parameter and prove some results whose validity does not require undue assumptions on the probability density.

Let x be a random variable with probability density $p(\cdot, \theta)$ with respect to a σ-finite measure μ, where θ belongs to a nondegenerate interval of the real line. For a discrete random variable, $p(y, \theta)$ denotes the probability of $x = y$. The following assumptions are made.

ASSUMPTION 1. The derivatives $d \log p/d\theta$, $d^2 \log p/d\theta^2$, and $d^3 \log p/d\theta^3$ esist, for almost all x in an interval A of θ including the true value.

ASSUMPTION 2. At the true value of θ,

$$E\left[\frac{p'(x, \theta)}{p(x, \theta)}\bigg|\theta\right] = 0, \qquad E\left[\frac{p''(x, \theta)}{p(x, \theta)}\bigg|\theta\right] = 0$$

$$E\left[\frac{p'(x, \theta)^2}{p(x, \theta)}\bigg|\theta\right] > 0,$$

where the primes denote differentiation with respect to θ.

ASSUMPTION 3. For every θ in A,

$$\left|\frac{d^3 \log p}{d\theta^3}\right| < M(x), \qquad E[M(x)|\theta] < K,$$

where K is independent of θ.

Let $x_1, \ldots, x_n$ be n independent observations (r.v.'s) and put

$$l(\theta | x) = \sum \log p(x_i, \theta)$$

$$\frac{dl}{d\theta} = \sum \frac{p'(x_i, \theta)}{p(x_i, \theta)}.$$

5f.2 Properties of m.l. Equation Estimators

(i) *If* $\log p(x, \theta)$ *is differentiable in an interval including the true value, then the m.l. equation has a root with probability* 1 *as* $n \to \infty$, *which is consistent for* θ.

Let θ_0 be the true value and consider two values $\theta_0 \pm \delta$. Using the inequality (1e.6.6) we have

$$E\left\{\log \frac{p(\theta_0 - \delta)}{p(\theta_0)}\bigg|\theta_0\right\} < 0, \qquad E\left\{\log \frac{p(\theta_0 + \delta)}{p(\theta_0)}\bigg|\theta_0\right\} < 0. \qquad (5f.2.1)$$

Strict inequalities in (5f.2.1) are possible unless $p(x, \theta)$ is independent of θ in an interval enclosing θ_0. In such a case the result we are trying to prove is automatically true.

The inequalities (5f.2.1) imply, by an application of the strong law of large numbers, that, with probability 1, as $n \to \infty$

$$\frac{1}{n}[l(\theta_0 \pm \delta | x) - l(\theta_0 | x)] < 0 \tag{5f.2.2}$$

that is, for almost all sample sequences $l(\theta|x)$ will eventually be greater at θ_0 than at $\theta_0 + \delta$ and $\theta_0 - \delta$. Since $l(\theta|x)$ is continuous in $(\theta_0 \pm \delta)$, there is a local maximum of $l(\theta|x)$ within $(\theta_0 \pm \delta)$. If $l(\theta|x)$ is differentiable, its derivative must vanish at that point. Since δ is arbitrary a root $\bar{\theta}$ so located is consistent for θ_0.

(ii) *Let $\bar{\theta}$ be a consistent root of the likelihood equation and Assumptions 1–3 be true. Then*

$$\sqrt{n}\left[(\bar{\theta} - \theta_0)i(\theta_0) - \frac{1}{n}\frac{dl}{d\theta_0}\right] \to 0 \quad \text{in probability.} \tag{5f.2.3}$$

The result (5f.2.3) shows that $\bar{\theta}$ is f.o.e. in the sense of (5c.2.5) and the asymptotic distribution of $\sqrt{n}(\bar{\theta} - \theta_0)$ is $N(0, i^{-1})$.

Since $\bar{\theta}$ satisfies the likelihood equation $(dl/d\theta) = 0$, we have, by Taylor expansion of $(dl/d\bar{\theta})$ at θ_0,

$$0 = \frac{dl}{d\theta_0} + (\bar{\theta} - \theta_0)\frac{d^2 l}{d\theta_0^2} + \frac{(\bar{\theta} - \theta_0)^2}{2}\frac{d^3 l}{d\theta_1^3}, \qquad \theta_1 \in (\theta_0, \bar{\theta}), \tag{5f.2.4}$$

from which we derive the equation

$$\sqrt{n}(\bar{\theta} - \theta_0) = \frac{-\dfrac{1}{\sqrt{n}}\dfrac{dl}{d\theta_0}}{\dfrac{1}{n}\left\{\dfrac{d^2 l}{d\theta_0^2} + \dfrac{\bar{\theta} - \theta_0}{2}\dfrac{d^3 l}{d\theta_1^3}\right\}}. \tag{5f.2.5}$$

Consider

$$\sqrt{n}(\bar{\theta} - \theta_0)i(\theta_0) - \frac{1}{\sqrt{n}}\frac{dl}{d\theta_0} = -(b_n + 1)\frac{1}{\sqrt{n}}\frac{dl}{d\theta_0}, \tag{5f.2.6}$$

where

$$b_n = \frac{i(\theta_0)}{\dfrac{1}{n}\left\{\dfrac{d^2 l}{d\theta_0^2} + \dfrac{\bar{\theta} - \theta_0}{2}\dfrac{d^3 l}{d\theta_1^3}\right\}}.$$

We shall show that $b_n \to -1$ in probability, in which case the convergence of $(1/\sqrt{n})(dl/d\theta_0)$ to an asymptotic distribution implies that the right-hand side of (5f.2.6) $\to 0$, establishing the desired result.

Now, under Assumption 2, we see that

$$E\left\{\frac{d^2 \log p}{d\theta_0^{\,2}}\,\middle|\,\theta_0\right\} = -i(\theta_0),$$

and by applying the strong law of large numbers, we have

$$\frac{1}{n}\frac{d^2 l}{d\theta_0^{\,2}} \to -i(\theta_0). \tag{5f.2.7}$$

Again under Assumption 3,

$$\frac{1}{n}\left|\frac{d^3 l}{d\theta_1^{\,3}}\right| < K,$$

with probability 1 as $n \to \infty$ and since $(\bar\theta - \theta_0) \to 0$, the product

$$(\bar\theta - \theta_0)\frac{1}{n}\left|\frac{d^3 l}{d\theta_1^{\,3}}\right| \to 0 \quad \text{in probability}, \tag{5f.2.8}$$

using the convergence result [(x), **2c.4**]. By combining (5f.2.7) and (5f.2.8), $b_n \to -1$ in probability.

(iii) *If $\bar\theta_1$ and $\bar\theta_2$ are two consistent roots of the likelihood equation, they are equivalent in the sense that*

$$\sqrt{n}(\bar\theta_1 - \bar\theta_2) \to 0 \qquad \text{in probability}. \tag{5f.2.9}$$

The result follows since $\bar\theta_1$ and $\bar\theta_2$ are both efficient if we apply the result [(ii), **5c.2**] observing that $\gamma(\theta) = 1$ [see Huzurbazar (1948) for an equivalent result].

Results (i), (ii), and (iii) of this section do not enable us to identify a consistent root of the likelihood equation. It would be more satisfying to show that the consistent root corresponds to the supremum of the likelihood with probability 1. For this, the density function has to satisfy some further conditions such as those given by Wald (1949). For a rigorous treatment of the asymptotic properties of m.l. estimators, see Le Cam (1953, 1956). Other papers of interest in large sample estimation are by Bahadur (1958), Daniels (1961), Dugue (1937), Haldane (1951), and Neyman (1949).

5g. THE METHOD OF SCORING FOR THE ESTIMATION OF PARAMETERS

The maximum likelihood equations are usually complicated so that the solutions cannot be obtained directly. A general method would be to assume a trial solution and derive linear equations for small additive corrections. The process can be repeated until the corrections become negligible. A great mechanization is introduced by adopting the method known as the *scoring system* for obtaining the linear equations for the additive corrections.

The quantity $d \log L/d\theta$, where L is the likelihood of the parameter θ, is defined as the *efficient score for* θ. The maximum likelihood estimate is the value of θ for which the efficient score vanishes. If θ_0 is the *trial value* of the estimate, then expanding $d \log L/d\theta$ and retaining only the first power of $\delta\theta = \theta - \theta_0$ leads to

$$\frac{d \log L}{d\theta} \doteq \frac{d \log L}{d\theta_0} + \delta\theta \frac{d^2 \log L}{d{\theta_0}^2}$$

$$\doteq \frac{d \log L}{d\theta_0} - \delta\theta \mathscr{I}(\theta_0),$$

where $\mathscr{I}(\theta_0)$, the information at the value $\theta = \theta_0$, is the expected value of $-d^2 \log L/d{\theta_0}^2$. In large samples the difference between $-\mathscr{I}(\theta_0)$ and $d^2 \log L/d{\theta_0}^2$ will be $0(1/n)$, where n is the number of observations, so that the above approximation holds to the first order of small quantities. The correction $\delta\theta$ is obtained from the equation

$$\delta\theta \mathscr{I}(\theta_0) = \frac{d \log L}{d\theta_0}, \qquad \delta\theta = \frac{d \log L}{d\theta_0} \div \mathscr{I}(\theta_0).$$

The first approximation is $(\theta_0 + \delta\theta)$, and the foregoing process can be repeated with this as the new trial value.

Example 1. Consider a sample of size n from the Cauchy distribution;

$$\frac{1}{\pi} \frac{dx}{1 + (x - \theta)^2}.$$

The likelihood equation is

$$\frac{d \log L}{d\theta} = \sum \frac{2(x_i - \theta)}{1 + (x_i - \theta)^2} = 0.$$

The efficient score for any θ is

$$S(\theta) = \sum \frac{2(x_i - \theta)}{1 + (x_i - \theta)^2}.$$

Information in a single observation is $\frac{1}{2}$ so that $\mathscr{I}(\theta) = n/2$ and the additive correction to a trial value θ_0 is simply $2S(\theta_0)/n$. This process can be continued until a stable value is attained. Fortunately in the foregoing example $\mathscr{I}(\theta)$ happens to be independent of θ so that $\mathscr{I}(\theta)$ need not be calculated at each trial value.

Example 2. Score and Information for Grouped Data. Let $\pi_1, \ldots, \pi_k$, $(\sum \pi_i = 1)$, be the probabilities in k mutually exclusive classes defined as functions of a single parameter θ. If $f_1, \ldots, f_k$, $(\sum f_i = f)$ are the observed frequencies, then

$$\log L = f_1 \log \pi_1 + \cdots + f_k \log \pi_k.$$

The score at θ is

$$\frac{d \log L}{d\theta} = \frac{f_1}{\pi_1}\frac{d\pi_1}{d\theta} + \cdots + \frac{f_k}{\pi_k}\frac{d\pi_k}{d\theta}$$

$$\mathscr{I}(\theta) = V\left(\frac{d \log L}{d\theta}\right) = f \sum \frac{1}{\pi_i}\left(\frac{d\pi_i}{d\theta}\right)^2.$$

The quantities

$$\frac{1}{\pi_i}\frac{d\pi_i}{d\theta} \quad \text{and} \quad \frac{1}{\pi_i}\left(\frac{d\pi_i}{d\theta}\right)^2,$$

which may be called the score and information supplied by the ith class, will be computed in any particular problem before proceeding with the process of estimation.

Estimation of Linkage in Genetics. If two factors are linked with a recombination fraction θ, the intercrosses $AB/ab \times AB/ab$ (coupling) and $Ab/aB \times Ab/aB$ (repulsion) give rise to the expected proportions and information shown in Table 5g.1α. The amount of information can be used to judge

TABLE 5g.1α

	Coupling		Repulsion	
Class	Probability	Score	Probability	Score
	4π	$\frac{1}{\pi}\frac{d\pi}{d\theta}$	4π	$\frac{1}{\pi}\frac{d\pi}{d\theta}$
AB	$3 - 2\theta + \theta^2$	$\frac{-2(1-\theta)}{3 - 2\theta + \theta^2}$	$2 + \theta^2$	$\frac{2\theta}{2 + \theta^2}$
Ab	$2\theta - \theta^2$	$\frac{2(1-\theta)}{2\theta - \theta^2}$	$1 - \theta^2$	$\frac{-2\theta}{1 - \theta^2}$
aB	$2\theta - \theta^2$	$\frac{2(1-\theta)}{2\theta - \theta^2}$	$1 - \theta^2$	$\frac{-2\theta}{1 - \theta^2}$
ab	$1 - 2\theta + \theta^2$	$\frac{-2(1-\theta)}{1 - 2\theta + \theta^2}$	θ^2	$\frac{2}{\theta}$
Information.	$\frac{2(3 - 4\theta + 2\theta^2)}{\theta(2-\theta)(3 - 2\theta + \theta^2)}$		$\frac{2(1 + 2\theta^2)}{(2+\theta^2)(1-\theta^2)}$	

the relative efficiency of one type of cross with respect to the other for the estimation of the recombination fraction. For instance, if $\theta = \frac{1}{4}$, the amounts of information for coupling and repulsion are 3.7909 and 1.1636, respectively. This means that, using intercrosses with repulsion, the number of offsprings needed will be three times as great as that for coupling to estimate the recombination fraction with the same precision.

Consider coupling data with values for AB, Ab, aB, and ab as shown below. With the trial value $\theta = 0.21$, the score and information are obtained.

	4π	$4\dfrac{d\pi}{d\theta}$	$\dfrac{1}{\pi}\dfrac{d\pi}{d\theta}$	Observed Frequency
AB	2.6241	−1.58	−0.60211	125
Ab	0.3759	1.58	4.20325	18
aB	0.3759	1.58	4.20325	20
ab	0.6241	−1.58	−2.53164	34
	Absolute sum		11.54025	197

Information per observation:

$$\sum \frac{d\pi_i}{d\theta}\left(\frac{1}{\pi_i}\frac{d\pi_i}{d\theta}\right) = \frac{1.58}{4}(11.54025) = 4.55840$$

$$\begin{aligned}\text{Efficient score} &= -0.60211(125) + 4.20325(18 + 20) - 2.53164(34)\\ &= -1.61601\end{aligned}$$

$$\text{Correction term} = -\frac{1.61601}{197(4.55840)} = -0.0018$$

$$\text{First approximation} = 0.21 - 0.0018 = 0.2082.$$

The correction is small so that the process may not be repeated. The variance for determining the precision of the estimate is given by

$$\frac{1}{197(4.55840)} = 0.00111357.$$

A better estimate of the variance is the reciprocal of the information at the value $p = 0.2082$.

The scores for trial values from 1 to 50% are given in F.Y. Tables. They can be directly used by retaining two decimal places at each stage of approximation, and finally when two places are stabilized a complete round of calculations with more places may be carried out as indicated in the numerical illustration.

Example 3. Scoring for Several Parameters. The method of scoring developed in Example 2 can be extended to the case of the simultaneous estimation of several parameters. If $\theta_1, \theta_2, \ldots, \theta_q$ are the parameters, the ith efficient score is defined by

$$S_i = \frac{\partial \log L}{\partial \theta_i}, \qquad i = 1, 2, \ldots, q,$$

where L is the likelihood of the parameters, and the information matrix is defined by $(\mathscr{I}_{ij})$ where

$$\mathscr{I}_{ij} = E(S_i S_j).$$

If the values of the efficient scores and information at the trial values $\theta_1^0, \ldots, \theta_q^0$ are indicated with index zero, then small additive corrections $\delta\theta_1, \ldots, \delta\theta_q$ to the trial values are given by the simultaneous equations

$$\begin{array}{c} \mathscr{I}_{11}^0\, \delta\theta_1 + \mathscr{I}_{12}^0\, \delta\theta_2 + \cdots + \mathscr{I}_{1q}^0\, \delta\theta_q = S_1^0 \\ \cdot \quad\quad \cdot \quad\quad \cdots \quad\quad \cdot \quad\quad \cdot \\ \mathscr{I}_{q1}^0\, \delta\theta_1 + \mathscr{I}_{q2}^0\, \delta\theta_2 + \cdots + \mathscr{I}_{qq}^0\, \delta\theta_q = S_q^0. \end{array}$$

This operation is repeated with corrected values each time until stable values of $\theta_1, \ldots, \theta_q$ are obtained. The variance of the final estimate $\bar{\theta}_i$ of θ_i is given by $\mathscr{I}^{ii}$, the ith diagonal element in the reciprocal of $(\mathscr{I}_{ij})$.

In this method the main difficulty is the computation and inversion of the information matrix at each stage of approximation. In practice this is found to be unnecessary. The information matrix may be kept fixed after some stage and only the scores recalculated. This procedure should reduce the computations considerably. At the final stage when stable values are reached, the information matrix may be computed at the estimated values for obtaining the variances and covariances of estimates.

In grouped distributions with π_i and f_i as probability and frequency in the ith class,

$$S_t = \sum_i \frac{f_i}{\pi_i} \frac{\partial \pi_i}{\partial \theta_t}$$

and

$$\mathscr{I}_{ut} = f \sum_i \frac{1}{\pi_i} \frac{\partial \pi_i}{\partial \theta_u} \frac{\partial \pi_i}{\partial \theta_t}$$

so that the calculations become simple as illustrated in the next section.

Estimation of Gene Frequencies of Blood Groups, A B O Systems. Every human being can be classified into one of four blood groups O, A, B, AB. The inheritance of these blood groups is controlled by three allelomorphic genes O, A, B of which O is recessive to A and B. If r, p, and q are gene frequencies of O, A, and B, then the expected probabilities of the six genotypes (four phenotypes) in random mating are as follows:

Phenotype	Genotype	Probabilities	
O	OO	r^2	
A	AA	p^2	$p^2 + 2pr$
	AO	$2pr$	
B	BB	q^2	$q^2 + 2qr$
	BO	$2qr$	
AB	AB	$2pq$	

Given $\bar{O}$, $\bar{A}$, $\bar{B}$, and $\bar{A}\bar{B}$, are observed frequencies adding to N, the problem is to estimate the gene frequencies p, q, and r. A rough estimate is supplied by

$$r' = \sqrt{\frac{\bar{O}}{N}}$$

$$p' = 1 - \sqrt{\frac{\bar{O} + \bar{B}}{N}}$$

$$q' = 1 - \sqrt{\frac{\bar{O} + \bar{A}}{N}}.$$

These may not necessarily add to unity, whereas the true values should. Let D denote the deviation

$$-D = p' + q' + r' - 1.$$

Better estimates due to Bernstein are obtained as follows:

$$r = (1 + \tfrac{1}{2}D)(r' + \tfrac{1}{2}D)$$
$$p = (1 + \tfrac{1}{2}D)p'$$
$$q = (1 + \tfrac{1}{2}D)q'.$$

There is still some deviation, $(1 - p - q - r) = \frac{1}{4}D^2$. If this is small, then Bernstein's method supplies fairly good estimates. We shall now show how these estimates can be improved by the method of maximum likelihood, using the frequencies $\bar{O} = 176$, $\bar{A} = 182$, $\bar{B} = 60$, and $\bar{A}\bar{B} = 17$. Approximate solutions obtained by Bernstein's method are

$$p_0 = 0.26449, \qquad q_0 = 0.09317, \qquad r_0 = 0.644234.$$

In general, the probabilities and derivatives, with respect to the independent parameters p and q, are as follows:

	Probabilities	Derivatives	
	π	$\dfrac{\partial \pi}{\partial p}$	$\dfrac{\partial \pi}{\partial q}$
O	r^2	$-2r$	$-2r$
A	$p(p+2r)$	$2r$	$-2p$
B	$q(q+2r)$	$-2q$	$2r$
AB	$2pq$	$2q$	$2p$

The probabilities and coefficients for the calculation of efficient scores at the approximate values obtained above are set out in Table 5g.1β.

TABLE 5g.1β. Coefficients for Scores

	Probability π	$\dfrac{1}{\pi}\dfrac{\partial \pi}{\partial q}$	$\dfrac{1}{\pi}\dfrac{\partial \pi}{\partial q}$	Observed Frequency
O	0.41260	−3.11362	−3.11362	176
A	0.40974	3.13543	−1.27104	182
B	0.12838	−1.45217	10.00685	60
AB	0.04928	3.75086	10.73307	17
				435 = N

The scores are

$$\phi_p = (-3.11362)176 + (3.13543)182 + (-1.45217)60 + (3.75086)17$$
$$= -0.20444$$

$$\phi_q = (-3.11362)176 + (-1.27104)182 + (10.00685)60 + (10.73307)17$$
$$= -0.09324$$

The information matrix for a single observation is

$$i_{pp} = \sum \left(\frac{\partial \pi_i}{\partial p}\right)\left(\frac{1}{\pi_i}\frac{\partial \pi_i}{\partial p}\right) = 9.00315, \qquad i_{pq} = \sum \left(\frac{\partial \pi_i}{\partial p}\right)\left(\frac{1}{\pi}\frac{\partial \pi_i}{\partial q}\right) = 2.47676$$

$$i_{pq} = 2.47676 \qquad i_{qq} = \sum \left(\frac{\partial \pi_i}{\partial q}\right)\left(\frac{1}{\pi_i}\frac{\partial \pi_i}{\partial q}\right) = 23.2162$$

The inverse of the information matrix per single observation is

$$i^{pp} = 0.114430 \qquad i^{pq} = -0.012208$$
$$i^{pq} = -0.012208 \qquad i^{qq} = 0.044376$$

The solutions for corrections are

$$\delta p = \frac{i^{pp}\phi_p + i^{pq}\phi_q}{N} = -0.00005116$$

$$\delta q = \frac{i^{pq}\phi_p + i^{qq}\phi_q}{N} = -0.00000377$$

The corrections are hardly necessary in this particular case. If the corrections are not small, the whole process has to be repeated with the second approximations. It is important to note that after some stage the information matrix need not be recalculated for each approximation. Only the new scores have to be calculated at each stage and used in conjuction with the same inverse information matrix (kept constant from some stage) to obtain closer approximations. When convergence is achieved, the information matrix and scores may be calculated for obtaining the last approximate values.

The maximum likelihood estimates and the variances are

$$\hat{p} = 0.26444, \qquad V(\hat{p}) = \frac{i^{pp}}{N} = 0.00026305$$

$$\hat{q} = 0.09317, \qquad V(\hat{q}) = \frac{i^{qq}}{N} = 0.00010202$$

$$\hat{r} = 0.64239, \qquad V(\hat{r}) = \frac{i^{pp} + 2i^{pq} + i^{qq}}{N} = 0.00030893$$

For other applications of the scoring method in complicated cases reference may be made to Finney (1952), Fisher (1950), and Rao (1950a).

In the case of O, A, B blood group data there is another algorithm which yields the m.l. estimates, which appears to be more suitable for desk and electronic computers, and which avoids the repeated computation of the information matrix and its inverse.

Let p_0, q_0, r_0 be provisional estimates, and compute $P_0 = p_0/r_0$ and $Q_0 = q_0/r_0$. Let us represent by P_k and Q_k the kth approximations of p/r and q/r. The $(k + 1)$th approximations are found from the formulas

$$\frac{\bar{A} + \overline{AB}}{P_{k+1}} = 2(\bar{O}) + \frac{\bar{A}}{2 + P_k} + \frac{2\bar{B}}{2 + Q_k}$$

$$\frac{\bar{B} + \overline{AB}}{Q_{k+1}} = 2(\bar{O}) + \frac{2\bar{A}}{2 + P_{k+1}} + \frac{\bar{B}}{2 + Q_k}$$

The iteration may be repeated until stable values of P and Q are obtained, from which the estimates of p, q, r are computed as

$$\hat{r} = \frac{1}{P + Q + 1}, \qquad \hat{p} = \hat{r}P, \qquad \hat{q} = \hat{r}Q.$$

By adopting the provisional estimates

$$p_0 = 0.26449, \qquad q_0 = 0.09317, \qquad r_0 = 0.64234$$

with $P_0 = 0.41176$, $Q_0 = 0.14505$, the first round of computations yields $P_1 = 0.41167$, $Q_1 = 0.14503$ indicating stability. The m.l. estimates of p, q, r are

$$\hat{r} = 0.64238, \qquad \hat{p} = 0.26445, \qquad \hat{q} = 0.09316,$$

which agree closely with the estimates obtained by the iterative procedure using the information matrix. With the estimates so obtained, the expected probabilities of the phenotypic classes and the information matrix may be obtained as before to compute the standard errors of the estimates.

Combination of Data. The advantage of the scoring system can best be seen in the mechanization it introduces when various sets of data giving information on the same parameters have to be combined for estimation. If L is the joint likelihood based on all the data and L_i for the ith part, then

$$L = L_1 L_2 \ldots$$

$$\frac{\partial \log L}{\partial \theta_r} = \frac{\partial \log L_1}{\partial \theta_r} + \frac{\partial \log L_2}{\partial \theta_r} + \cdots,$$

which shows that the efficient scores are additive. Also, if $\mathscr{I}_{rs}$ is an element of the information matrix for the whole body of data and $\mathscr{I}_{rs}^{(j)}$ for the jth part, then

$$\mathscr{I}_{rs} = \mathscr{I}_{rs}^{(1)} + \mathscr{I}_{rs}^{(2)} + \cdots.$$

Thus, to obtain the best estimates it is necessary to replace each part of the data by the scores and information matrix at a trial value and obtain the total scores and information matrix by simple addition. The correction to trial values can be obtained by solving simultaneous equations as shown in the numerical examples.

COMPLEMENTS AND PROBLEMS

1 Consider a rectangular distribution in the range 0 to θ. Let $x_{\max}$ be the maximum of n independent observations from such a distribution. Show the following.

1.1 x_{max} is a complete sufficient statistic for θ.

1.2 $(n+1)x_{max}/n$ is the m.v.u. (minimum variance unbiased) estimator of θ. Find the minimum variance.

1.3 x_{max} is the m.l. estimator of θ and x_{max} is biased but consistent for θ.

1.4 Any function $t(x_{max})$ of x_{max} unbiased for θ has the terminal value $t(\theta) = (n+1)\theta/n$.

2 Let x_{max} and x_{min} be the maximum and minimum of n observations from a rectangular distribution in the range θ to ϕ. Show the following:

2.1 x_{max} and x_{min} are complete sufficient statistics for θ and ϕ.

2.2 The m.v.u. estimators of the midpoint $(\theta + \phi)/2$ and the range $(\phi - \theta)$ are

$$\frac{x_{max} + x_{min}}{2} \quad \text{and} \quad \frac{n+1}{n-1}(x_{max} - x_{min}).$$

2.3 Let $\phi = 2\theta$, so that there is only one parameter. Still, x_{max} and x_{min} are jointly sufficient for θ but they are not complete.

2.4 The m.l. estimator of θ, when $\phi = 2\theta$, is $x_{max}/2$, which is biased for θ. An unbiased estimator based on the m.l. estimator, that is, correcting for bias is $T_1 = (n+1)x_{max}/(2n+1)$.

2.5 Show that $T_2 = (n+1)(2x_{max} + x_{min})/(5n+4)$ is another estimator of θ, whose variance is uniformly smaller than that of T_1. Thus, we have an example of an m.l. estimator being inferior as judged by the criterion of minimum variance.

3 *Minimum variance unbiased estimators of probabilities of a discrete distribution.* Consider a random variable X taking integer values, 0, 1, ... and let $P(X = r) = \pi_r(\theta)$ where θ is an unknown parameter. Let $x_1, \ldots, x_n$ be n independent observations on X and let T be a complete sufficient estimator of θ. To estimate $\pi_r(\theta)$, consider the statistic $p =$ (number of x_i equal to r)/n.

3.1 Show that the m.v.u. estimator of $\pi_r(\theta)$ is $E(p \mid T)$, using the result [(iii), **5a.2**].

3.2 Apply the result to find the m.v.u. estimator of the Poisson probability, $P(X = 0) = e^{-\mu}$, on the basis of n independent observations $x_1, \ldots, x_n$.

3.3 Find the m.v.u. estimator of $P(X = 0)$ when X has the negative binomial distribution, on the basis of n independent observations.

4 *More stringent inequalities for the variance of an unbiased estimator.* Let T be an unbiased estimator of $g(\theta)$ and $P(X, \theta)$ the probability density at X. Differentiating the relation

$$\int TP(X, \theta)\, dv = g(\theta)$$

k times, we obtain

$$\int T \frac{d^k P}{d\theta^k} dv = \frac{d^k g}{d\theta^k}.$$

4.1 Show that $V(t) \geqslant [d^k g/d\theta^k]^2/J_{kk}$ where

$$J_{kk} = \int \left(\frac{d^k P}{d\theta^k}\right)^2 \frac{1}{P} dv$$

4.2 More generally, if $J_{kr} = \text{cov}\left(\frac{1}{P}\frac{d^k P}{d\theta^k}, \frac{1}{P}\frac{d^r P}{d\theta^r}\right)$, show that the multiple correlation of T on $[d^i P/P \, d\theta^i]$, $i = 1, 2, \ldots$, is

$$\left[\sum\sum J^{kr} \frac{d^k g}{d\theta^k} \frac{d^r g}{d\theta^r}\right] \div V(T),$$

where (J^{kr}) is the reciprocal of (J_{kr}). Hence deduce the inequality

$$V(T) \geqslant \sum\sum J^{kr} \frac{d^k g}{d\theta^k} \frac{d^r g}{d\theta^r}.$$

4.3 Show that the lower bound obtained in (4.2) is not smaller than $[g'(\theta)]^2/\mathscr{I}(\theta)$. The above inequality is due to Bhattacharya (1947).

5 Human twins are of two types, identical and nonidentical. Let π be the probability of a child being male and α the probability of twins being identical. If identical twins arise from the splitting of a fertilized egg, show that the probabilities of ♂♂, ♂♀, ♀♀ types of twins are $\pi(\pi + \alpha\phi)$, $2\pi\phi(1 - \alpha)$, $\phi(\alpha\pi + \phi)$, where $\phi = 1 - \pi$. Let n_1, n_2, n_3 be observed frequencies of these types in a sample of $n = n_1 + n_2 + n_3$ twins.

5.1 Obtain the estimates of α, π.

5.2 If $\pi = 0.516$, find the estimate of α by m.l. and min. chi-square methods.

5.3 Compare the standard errors of estimates in Problem 5.2 with that of Weinberg's estimate, $\alpha^* = (1 - n_2)/2n\pi\phi$.

6 *M.v.u. estimation of parameters of a power series distribution.* Consider a power series distribution with probability

$$P(X = r) = \frac{a_r}{f(\theta)} \theta^r, \qquad r = c, c + 1, \ldots \infty.$$

Let $x_1, \ldots, x_n$ be a sample of size n with a total, $T = x_1 + \cdots + x_n$.

6.1 Show that T is a complete sufficient statistic for θ and the distribution of T is of the same form

$$P(T = t) = b_t \theta^t \div [f(\theta)]^n, \qquad t = nc, nc + 1, \ldots \infty.$$

6.2 Show that the m.v.u. estimator of θ^r is

$$u_r(t) = 0, \qquad t < r$$
$$= \frac{b_{t-r}}{b_t}, \qquad t \geqslant r,$$

and a m.v.u. estimator of its variance is

$$[u_r(t)]^2 - u_{2r}(t)$$

[Roy and Mitra, 1957]

[Note: Examples 6.1 and 6.2 cover all the well-known discrete distributions with infinite range including those truncated at the left.]

6.3 Show that the results of (6.1) and (6.2) are not true if the range of the variable is not infinite. [Patil, 1963.]

7 *Maximum likelihood estimators of parameters in discrete distributions.* Consider a power series distribution

$$P(X = r) = a_r \theta^r / f(\theta), \qquad r = 0, 1, \ldots,$$

where a_r may be zero for some r. Let $x_1, \ldots, x_n$ be n independent observations and $\bar{x} = (x_1 + \cdots + x_n)/n$.

7.1 Show that the m.l. estimator of θ is a root of the equation $\bar{x} = \theta f'(\theta)/f(\theta) = \mu(\theta)$, the theoretical mean. The estimating equation is the same for the method of moments.

7.2 Show that $i(\theta)$, the information in a single observation, is $(d\mu/\theta\, d\theta) = \mu_2(\theta)/\theta^2$ where $\mu_2(\theta)$ is the second moment of the distribution.

7.3 Obtain the explicit equations for estimating the parameters for the complete Poisson, binomial, and logarithmic series and the truncated Poisson and binomial series omitting zero.

8 Let $x_1, \ldots, x_n$ be n independent observations on a random variable with a probability density $p(x, \theta)$. Further let $f(x_1, \ldots, x_n, \theta) = 0$ be an estimating equation for θ. Show that if $E(f) \equiv 0$, then (by differentiating under the integral sign and applying the C-S inequality to one of the terms)

$$V(f) \geqslant [E(f')]^2/\mathscr{I} \qquad \text{or} \qquad [E(f')]^2/V(f) \leqslant \mathscr{I},$$

where $f' = df(x_1, \ldots, x_n, \theta)/d\theta$ and $\mathscr{I}$ is the total information in the sample. Satisfy yourself that the upper bound of $[E(f')]^2/V(f)$ is attained for the choice $f = d \log L/d\theta$ where $L = p(x_1, \theta) \cdots p(x_n, \theta)$.

[Note: This example shows that if we define the efficiency of an estimating equation $f = 0$ by $E(f')^2/V(f)$, the maximum likelihood estimating equation has the maximum efficiency; an attempt to justify the method of maximum likelihood in small samples is due to Godambe, 1960.]

9 Let $x_1, \ldots, x_n$ be a sample of n independent observations on a random variable with a d.f. $F(x, \theta)$, $\theta \in A$, a nondegenerate interval of the real line. Denote by S_n the empirical d.f. Consider an estimating equation of the type

$$\int g(x, \theta)\, dS_n = 0 = [g(x_1, \theta) + \cdots + g(x_n, \theta)] \div n,$$

where $g(x, \theta)$ is such that

(a) $\int g(x, \theta)\, dF(x, \theta) \equiv 0,$

(b) $\int \frac{dg(x, \theta)}{d\theta}\, dF(x, \theta) = m(\theta),$

(c) $\int [g(x, \theta)]^2\, dF(x, \theta) = h(\theta),\ 0 < h(\theta) < \infty,$

(d) $\left| \frac{d^2 g(x, \theta)}{d\theta^2} \right| < M(x),\ E[M(x) \mid \theta] < K,$

where K is independent of θ.

9.1 Show that there exists a consistent root θ^* of the equation $\sum g(x_i, \theta) = 0$.

9.2 Prove, following the arguments of **5f.2**, that the asymptotic distribution of $\sqrt{n}(\theta^* - \theta)$ is $N(0, \sigma^2)$ where $\sigma^2 = h(\theta)/[m(\theta)]^2$.

9.3 Let $F(x, \theta)$ admit a density function $p(x, \theta)$. Show that under suitable conditions

$$\frac{h(\theta)}{[m(\theta)]^2} \geqslant \frac{1}{i(\theta)}$$

where $i(\theta) = E[p'(x, \theta)/p(x, \theta)]^2$.

9.4 Verify that the lower bound to the asymptotic variance is attained when $g(x, \theta) = p'(x, \theta)/p(x, \theta)$, that is, when the estimating equation is the m.l. equation.

[Thus we have a characterization of the m.l. method as providing estimates with minimum asymptotic variance, in a certain class of estimating equations.]

10 *Maximum likelihood characterization of distributions.*

10.1 Let $\{F(x - \theta), \theta \in R^1\}$ be a translation family of absolutely continuous distribution functions on the real line and let the version of the probability density function $f(x)$ be lower semicontinuous at $x = 0$. If for all samples of sizes 2 and 3, an m.l. estimate of θ is the sample arithmetic mean, then $F(x)$ is $N(0, \sigma^2)$.

10.2 Let $F(x/\sigma)$, $\sigma > 0$ constitute a scale parameter family of absolutely continuous distributions with the version of the probability density satisfying the following conditions:

(a) $f(x)$ is continuous in $(0, \infty)$,

(b) $\lim \dfrac{f(\lambda y)}{f(y)} = 1$ as $y \downarrow 0$.

If an m.l. estimate of σ is the arithmetic mean for all samples, the d.f. belongs to the exponential family, that is

$$f(x) = e^{-x}, \qquad x > 0$$
$$f(x) = 0, \qquad x \leqslant 0$$

10.3 If in (10.2), $f(x)$ satisfies the conditions

(a) $f(x)$ is continuous in $(-\infty, \infty)$

(b) $\lim f(\lambda y)/f(y) = 1$ as $y \to 0$

and an m.l. estimate of σ^2 is $\sum x_i^2/n$, then the distribution is $N(0, 1)$.
[Note: These propositions have been vaguely known in literature but have been stated in a rigorous form by Teicher, 1961.]

10.4 Let $\{F(x - \theta, \theta \in R^1)\}$ be the same family as defined in example 10.1. If the sample median is the m.l.e. of θ for $n = 4$, then $F'(x) = f(x) = (a/2) \exp(-a|x|)$, i.e., Laplace distribution. (See Kagan, Linnik and Rao, 1972B4 and Rao and Ghosh, 1971g).

11 Let X be a discrete random variable such that $P(X = -1) = \alpha$, $P(X = n) = (1 - \alpha)^2\alpha^n$, $n = 0, 1, \ldots$, where $0 < \alpha < 1$.

11.1 Show that X is minimal sufficient for α although it is not complete $[E(X) = 0]$.

11.2 A function of X unbiased for zero is necessarily of the form cX. Hence show that $T(X) = 1$ for $X = 0$ and zero otherwise is the m.v.u. estimator of $(1 - \alpha)^2$ for all α.

11.3 Show that there is no uniformly m.v.u. estimator of α itself, although unbiased estimators of α exist.

12 Let there be only two alternative values 0, 1 which θ can take, and define the probability densities

$$p(x|\theta = 0) = \begin{cases} 1, & 0 < x < 1 \\ 0 & \text{otherwise} \end{cases}$$

$$p^*(x|\theta = 1) = \begin{cases} \frac{1}{2}x^{-1/2}, & 0 < x < 1 \\ 0 & \text{otherwise} \end{cases}$$

Show that unbiased estimators of θ exist and the variance at $\theta = 0$ can be made arbitrarily small but the infimum zero is not attained for any unbiased estimator.

REFERENCES

Arnold, J. C. and S. K. Katti, (1972), An application of Rao-Blackwell theorem in preliminary test estimators, *J. Multivariate Analysis* **2**, 236–238.

Bahadur, R. R. (1958), Examples of inconsistency of maximum likelihood estimates, *Sankhyā* **20**, 207–210.

Bhattacharya, A. (1946), On some analogues of the amount of information and their uses in statistical estimation, *Sankhyā* **8**, 1–14, 201–218, 315–328.

Birch, M. W. (1964), A new proof of the Fisher-Pearson theorem, *Ann. Math. Statist.* **35**, 817–824.

Blackwell, D. (1947), Conditional expectation and unbiased sequential estimation, *Ann. Math. Statist.* **18**, 105–110.

Chapman Douglas, G. and Herbert Robbins (1951), Minimum variance estimation without regularity assumptions, *Ann. Math. Statist.* **22**, 581–586.

Cramer, H. (1946), *Mathematical Methods of Statistics*, Princeton University Press, Princeton, N.J.

Daniels, H. E. (1961), The asymptotic efficiency of a maximum likelihood estimator, *Proc. (Fourth) Berkeley Symp. on Math. Statist. Prob.* **1**, 151–163.

Dempster, A. P. (1963), On the difficulties inherent in Fisher's fiducial argument, *Proc. Inter. Stat. Conference*, Ottawa.

Doob, J. (1934), Probability and statistics, *Trans. Am. Math. Soc.* **36**, 759–775.

Doob, J. (1936), Statistical estimation, *Trans. Am. Math. Soc.* **39**, 410–421.

Dugue, D. (1937), Application des proprietes de la limite au sens du calcul des probabilities a l'etude des diverses questions d'estimation, *Ecol. Poly.* **3**, 305–372.

Finney, D. J. (1952), *Probit. Analysis* (second edition), Cambridge University Press, Cambridge, Mass.

Fisher, R. A. (1922), On the mathematical foundations of theoretical statistics, *Philos. Trans. Roy. Soc.* **A222**, 309–368.

Fisher, R. A. (1925), Theory of Statistical estimation, *Proc. Camb. Phil. Soc.* **22**, 700–725.

Fisher, R. A. (1950), *Statistical Methods for Research Workers* (first edition 1925, eleventh edition 1950), Oliver and Boyd, Edinburgh.

Fisher, R. A. (1956), *Statistical Methods and Scientific Inference*, Oliver and Boyd, Edinburgh and London.

Fraser, D. A. S. (1961), The fiducial method and invariance, *Biometrika* **48**, 261–280.

Gart, John J. (1959), An extension of the Cramer-Rao inequality, *Ann. Math. Statist.* **30**, 367–380.

Godambe, V. P. (1960), An optimum property of regular maximum likelihood estimation, *Ann. Math. Statist.* **31**, 1208–1212.

Godambe, V. P. and D. A. Sprott (1971), *Foundations of Statistical Inference*, Holt, Reinhart, and Winston, Canada.

Haldane, J. B. S. (1951), A class of efficient estimates of a parameter, *Bull. Int. Statist. Inst.* **33**, 231–248.

Hodges, J. L. (Jr.) and E. L. Lehmann (1951), Some applications of the Cramer-Rao inequality, *Proc. (Second) Berkeley Symp. on Math. Statist. Prob.*, 13–22.

Huzurbazar, V. S. (1948), The likelihood equation, consistency, and maxima of the likelihood function, *Ann. Eugen. (London)* **14**, 185–200.

Jeffreys, H. (1939), *Theory of Probability*, Clarendon Press, Oxford.

Kiefer, J. and J. Wolfowitz (1956), Consistency of the maximum likelihood estimator in the presence of infinitely many incidental parameters, *Ann. Math. Statist.* **27**, 887–906.

Kolmogorov, A. N. (1950), Unbiased estimates (in Russian), *Izvestia Acad. Nauk USSR* **14**, 303.

Le Cam, L. (1953), On some asymptotic properties of maximum likelihood estimates and related Bayes's estimates, *Uni. California Publ. Statist.* **1**, 277–330.

Le Cam, L. (1956), On the asymptotic theory of estimation and testing hypotheses, *Proc. (Third) Berkeley Symp. on Math. Statist. Prob.* **1**, 129–156.

Lehmann, E. L. and H. Scheffe (1950), Completeness, similar regions and unbiased estimation, *Sankhyā* **10**, 305–340.

Neyman, J. (1949), Contribution to the theory of the χ^2 test, *Proc. (First) Berkeley Symp. on Math. Statist. Prob.*, 239–273.

Patil, G. P. (1963), Minimum variance unbiased estimation and certain problems of additive number theory, *Ann. Math. Statist.* **34**, 1050–1056.

Perlman, M. D. (1972), Reduced mean square error estimation for several parameters, *Sankhyā*.

Robbins, H. (1955), An empirical Bayes approach to statistics, *Proc. (Third) Berkeley Symp. on Math. Statist. Prob.* **1**, 157–163.

Roy, Jogabratha and Sujitkumar Mitra (1957), Unbiased minimum variance estimation in a class of discrete distributions, *Sankhyā* **18**, 371–378.

Sprott, D. A. (1961), Similarities between likelihoods and associated distributions *a posteriori*, *J. Roy. Statist. Soc.* **24(B)**, 460–468.

Stein, C. (1950), Unbiased estimates with minimum variance, *Ann. Math. Statist.* **21**, 406–415.

Stein, C. M. (1962), Confidence sets for the mean of a multivariate normal distribution, *J. Roy. Statist. Soc.* **B24**, 265–296.

Teicher, H. (1961), Maximum likelihood characterization of distributions, *Ann. Math. Statist.* **32**, 1214–1222.

Wald, A. (1949), Note on the consistency of maximum likelihood estimate, *Ann. Math. Statist.* **20**, 595–601.

Chapter 6

LARGE SAMPLE THEORY AND METHODS

Introduction. This chapter deals with statistical inference when large samples are available. It covers a wide variety of statistical methods useful in practical applications. Special emphasis is given to the analysis of categorical data with reference to problems of specification, homogeneity of samples, independence of attributes, etc. Some general classes of large sample criteria for testing simple and composite hypotheses have also been discussed. The asymptotic efficiency of tests is discussed in **7a.7** of Chapter 7.

The theoretical developments in this chapter rest heavily on the basic convergence theorems discussed in Chapter 2 and the asymptotic properties of the maximum likelihood estimators given in Chapter 5. The reader is advised to study the propositions of the subsections **2c.3** to **2c.5** in Chapter 2 which deal with the laws of large numbers, univariate and multivariate central limit theorems, and convergence theorems for functions of random variables and Sections **5c** to **5f** in Chapter 5 which deal with asymptotic efficiency of estimators, large sample methods of estimation, and the method of maximum likelihood.

6a SOME BASIC RESULTS

6a.1 Asymptotic Distribution of Quadratic Functions of Frequencies

Let there be k mutually exclusive events and let the probability of the ith event be π_i, $i = 1, \ldots, k$. Suppose n independent events have been observed resulting in n_1 events of the first kind, n_2 of the second kind, and so on $\sum n_i = n$. Define the column vectors $\mathbf{V}$, $\boldsymbol{\phi}$

$$\mathbf{V}' = \left(\frac{n_1 - n\pi_1}{\sqrt{n\pi_1}}, \ldots, \frac{n_k - n\pi_k}{\sqrt{n\pi_k}}\right) \tag{6a.1.1}$$

$$\boldsymbol{\phi}' = (\sqrt{\pi_1}, \ldots, \sqrt{\pi_k}). \tag{6a.1.2}$$

Our object is to study the asymptotic distributions (a.d.) of linear and quadratic functions of $\mathbf{V}$ (i.e., of the frequencies $n_1, \ldots, n_k$ considered as r.v.'s). We establish the following results.

(i) *The a.d. of the linear function* $\mathbf{b}'\mathbf{V}$, *where* $\mathbf{b}$ *is a fixed vector, is normal with mean zero and variance* $\mathbf{b}'\mathbf{b} - (\mathbf{b}'\boldsymbol{\phi})^2 = \mathbf{b}'(\mathbf{I} - \boldsymbol{\phi}\boldsymbol{\phi}')\mathbf{b}$.

Consider a discrete random variable X which takes the value $b_i/\sqrt{\pi_i}$ when the ith kind of event occurs. Hence $P(X = b_i/\sqrt{\pi_i}) = \pi_i$, giving

$$E(X) = \sum \frac{b_i}{\sqrt{\pi_i}} \pi_i = \sum b_i \sqrt{\pi_i} = \mathbf{b}'\boldsymbol{\phi},$$

$$V(X) = \sum \frac{b_i^2}{\pi_i} \pi_i - (\mathbf{b}'\boldsymbol{\phi})^2 = \mathbf{b}'\mathbf{b} - (\mathbf{b}'\boldsymbol{\phi})^2.$$

We have n independent observations with $X = b_i/\sqrt{\pi_i}$ occurring n_i times, $i = 1, \ldots, k$. The average of the n observations is

$$\bar{X}_n = \left(\frac{b_1}{\sqrt{\pi_1}} n_1 + \cdots + \frac{b_k}{\sqrt{\pi_k}} n_k\right) \div n.$$

Hence by the central limit theorem [(i), **2c.5**], we see that

$$\sqrt{n}(\bar{X}_n - \mathbf{b}'\boldsymbol{\phi}) \xrightarrow{L} Y \sim N[0, \mathbf{b}'\mathbf{b} - (\mathbf{b}'\boldsymbol{\phi})^2]. \tag{6a.1.3}$$

But $\sqrt{n}(\bar{X}_n - \mathbf{b}'\boldsymbol{\phi}) = \mathbf{b}'\mathbf{V}$. Hence the result.

(ii) *The a.d. of p linear functions* $\mathbf{B}'\mathbf{V}$, *where* $\mathbf{B}$ *is $k \times p$ matrix of rank p is multivariate normal with mean zero and dispersion matrix* $\mathbf{B}'(\mathbf{I} - \boldsymbol{\phi}\boldsymbol{\phi}')\mathbf{B}$.

To establish this we need only examine the a.d. of a linear function of the p variables $\mathbf{B}'\mathbf{V}$. Consider $\boldsymbol{\lambda}'\mathbf{B}'\mathbf{V}$ which is linear in $\mathbf{V}$, however, and hence by result (i) we see that

$$\boldsymbol{\lambda}'\mathbf{B}'\mathbf{V} \xrightarrow{L} Y \sim N[0, \boldsymbol{\lambda}'\mathbf{B}'(\mathbf{I} - \boldsymbol{\phi}\boldsymbol{\phi}')\mathbf{B}\boldsymbol{\lambda}]. \tag{6a.1.4}$$

Let $\mathbf{Z}$ be a p-variate normal variable with mean zero and dispersion matrix $\mathbf{B}'(\mathbf{I} - \boldsymbol{\phi}\boldsymbol{\phi}')\mathbf{B}$. The distribution of a linear function $\boldsymbol{\lambda}'\mathbf{Z}$ is the same as that on the right-hand side of (6a.1.4). Hence by the multivariate normal convergence theorem [(iv), **2c.5**], we see that

$$\mathbf{B}'\mathbf{V} \xrightarrow{L} \mathbf{Z} \sim N_p[0, \mathbf{B}'(\mathbf{I} - \boldsymbol{\phi}\boldsymbol{\phi}')\mathbf{B}], \tag{6a.1.5}$$

where N_p denotes a p-variate normal distribution. (See **3b.3** for definition of N_p and Chapter 8 for a detailed study of N_p.)

(iii) *Let* $\mathbf{A}$ *be* $k \times k-1$ *matrix such that the partitioned matrix* $(\boldsymbol{\phi} \,\vdots\, \mathbf{A})$ *of order* $k \times k$ *is orthogonal. Then the a.d. of the* $(k-1)$ *linear functions* $\mathbf{G} = \mathbf{A}'\mathbf{V}$ *is that of* $(k-1)$ *independent normal variables each with zero mean and unit variance.*

Using result (ii) and the condition $\mathbf{A}'\boldsymbol{\phi} = 0$, we see that the dispersion matrix of the a.d. of $\mathbf{A}'\mathbf{V}$ is

$$\mathbf{A}'(\mathbf{I} - \boldsymbol{\phi}\boldsymbol{\phi}')\mathbf{A} = \mathbf{A}'\mathbf{A} = \mathbf{I}, \qquad \text{which is a} \quad (k-1) \times (k-1) \text{ matrix.}$$

Hence the result. Thus $\mathbf{G} \xrightarrow{L} \mathbf{Y} \sim N_{k-1}(0, \mathbf{I})$.

(iv) *A sufficient condition for a.d. of the quadratic form* $\mathbf{V}'\mathbf{C}\mathbf{V}$ *to be* χ^2 *is*

$$\mathbf{C}^2 = \mathbf{C} \qquad \text{and} \qquad \mathbf{C}\boldsymbol{\phi} = \alpha\boldsymbol{\phi} \tag{6a.1.6}$$

where α *is a constant, in which case the degrees of freedom (D.F.) of* χ^2 *is* $R(\mathbf{C})$ *if* $\alpha = 0$ *and* $[R(\mathbf{C}) - 1]$ *if* $\alpha \neq 0$ *where* $R(\mathbf{C})$ *denotes the rank of* $\mathbf{C}$.

Consider the equations

$$0 = \boldsymbol{\phi}'\mathbf{V}$$
$$\mathbf{G} = \mathbf{A}'\mathbf{V}$$

where $\mathbf{A}$ is as defined in (iii). Solving for $\mathbf{V}$, we have

$$\mathbf{V} = (\boldsymbol{\phi} \,\vdots\, \mathbf{A}) \begin{pmatrix} 0 \\ \hdashline \mathbf{G} \end{pmatrix} = \mathbf{A}\mathbf{G},$$

since $(\boldsymbol{\phi} \,\vdots\, \mathbf{A})$ is an orthogonal matrix. Hence

$$\mathbf{V}'\mathbf{C}\mathbf{V} = \mathbf{G}'\mathbf{A}'\mathbf{C}\mathbf{A}\mathbf{G}.$$

The a.d. of $\mathbf{V}'\mathbf{C}\mathbf{V}$ is the same as that of $\mathbf{G}'\mathbf{A}'\mathbf{C}\mathbf{A}\mathbf{G}$ which is a continuous function of $\mathbf{G}$. Hence the a.d. of $\mathbf{G}'\mathbf{A}'\mathbf{C}\mathbf{A}\mathbf{G}$ can be computed from the a.d. of $\mathbf{G}$ [see (xii), **2c.4**].

But the a.d. of $\mathbf{G}$ is that of $(k-1)$ independent normal variables each with zero mean and unit variance. A necessary and sufficient condition that $\mathbf{G}'\mathbf{A}'\mathbf{C}\mathbf{A}\mathbf{G}$ has a χ^2 distribution is then $\mathbf{A}'\mathbf{C}\mathbf{A}$ is idempotent, [(ii), **3b.4**], that is,

$$\mathbf{A}'\mathbf{C}\mathbf{A}\mathbf{A}'\mathbf{C}\mathbf{A} = \mathbf{A}'\mathbf{C}\mathbf{A}. \tag{6a.1.7}$$

Since the matrix $(\boldsymbol{\phi} \,\vdots\, \mathbf{A})$ is orthogonal, then

$$(\boldsymbol{\phi} \,\vdots\, \mathbf{A}) \begin{pmatrix} \boldsymbol{\phi}' \\ \hdashline \mathbf{A}' \end{pmatrix} = \boldsymbol{\phi}\boldsymbol{\phi}' + \mathbf{A}\mathbf{A}' = \mathbf{I}.$$

Substituting for $\mathbf{A}\mathbf{A}'$ in (6a.1.7), we have

$$\mathbf{A}'\mathbf{C}(\mathbf{I} - \boldsymbol{\phi}\boldsymbol{\phi}')\mathbf{C}\mathbf{A} = \mathbf{A}'\mathbf{C}\mathbf{A} \tag{6a.1.8}$$

for all $\mathbf{A}$ such that $(\boldsymbol{\phi} \vdots \mathbf{A})$ is an orthogonal matrix. A sufficient condition for (6a.1.8) to hold is

$$\mathbf{C}^2 = \mathbf{C} \quad \text{and} \quad \mathbf{C}\boldsymbol{\phi} = \alpha\boldsymbol{\phi}, \tag{6a.1.9}$$

that is, $\mathbf{C}$ *is idempotent and* $\boldsymbol{\phi}$ *is an eigenvector of* $\mathbf{C}$. The D.F. of χ^2 is $R(\mathbf{A}'\mathbf{C}\mathbf{A})$ the rank of $\mathbf{A}'\mathbf{C}\mathbf{A}$, if we use [(ii), **3b.4**]. Now

$$R(\mathbf{C}) = R\left\{\begin{pmatrix}\boldsymbol{\phi}' \\ \hdashline \mathbf{A}'\end{pmatrix}\mathbf{C}(\boldsymbol{\phi} \vdots \mathbf{A})\right\} = R\begin{pmatrix}\boldsymbol{\phi}'\mathbf{C}\boldsymbol{\phi} & \boldsymbol{\phi}'\mathbf{C}\mathbf{A} \\ \mathbf{A}'\mathbf{C}\boldsymbol{\phi} & \mathbf{A}'\mathbf{C}\mathbf{A}\end{pmatrix}$$

$$\begin{aligned} &= R(\mathbf{A}'\mathbf{C}\mathbf{A}), \quad &&\text{if } \mathbf{C}\boldsymbol{\phi} = 0 \\ &= R(\mathbf{A}'\mathbf{C}\mathbf{A}) + 1, \quad &&\text{if } \mathbf{C}\boldsymbol{\phi} = \alpha\boldsymbol{\phi}, \quad \alpha \neq 0. \end{aligned}$$

Hence the desired result.

(v) *A necessary and sufficient condition for the a.d. of* $\mathbf{V}'\mathbf{C}\mathbf{V}$ *to be* χ^2 *is*

$$(\mathbf{I} - \boldsymbol{\phi}\boldsymbol{\phi}')\mathbf{C}(\mathbf{I} - \boldsymbol{\phi}\boldsymbol{\phi}')\mathbf{C}(\mathbf{I} - \boldsymbol{\phi}\boldsymbol{\phi}') = (\mathbf{I} - \boldsymbol{\phi}\boldsymbol{\phi}')\mathbf{C}(\mathbf{I} - \boldsymbol{\phi}\boldsymbol{\phi}'). \tag{6a.1.10}$$

In (iv), it is shown that a necessary and sufficient condition is (6a.1.7):

$$\mathbf{A}'\mathbf{C}(\mathbf{I} - \boldsymbol{\phi}\boldsymbol{\phi}')\mathbf{C}\mathbf{A} = \mathbf{A}'\mathbf{C}\mathbf{A}. \tag{6a.1.11}$$

But the linear manifolds $\mathscr{M}(\mathbf{A})$ and $\mathscr{M}(\mathbf{I} - \boldsymbol{\phi}\boldsymbol{\phi}')$ are the same. Hence in the relation (6a.1.11), $\mathbf{A}$ can be replaced by $(\mathbf{I} - \boldsymbol{\phi}\boldsymbol{\phi}')$, which gives (6a.1.10).

(vi) *If* $\mathbf{V}'\mathbf{C}\mathbf{V} \xrightarrow{L} X \sim \chi^2(c)$, $\mathbf{V}'\mathbf{B}\mathbf{V} \xrightarrow{L} X \sim \chi^2(b)$, *and* $\mathbf{V}'\mathbf{D}\mathbf{V}$ *is non-negative, then* $\mathbf{B} = \mathbf{C} + \mathbf{D} \Rightarrow \mathbf{V}'\mathbf{D}\mathbf{V} \xrightarrow{L} X \sim \chi^2(b - c)$.

The result follows from [(iv), **3b.4**].

6a.2 Some Convergence Theorems

We shall be considering in this chapter the a.d. of a linear or a quadratic function h of the variables $(x_n, y_n, \ldots)$ or of the variables $(\alpha_n x_n, \beta_n y_n, \ldots)$ where $\alpha_n \xrightarrow{P} \alpha$, $\beta_n \xrightarrow{P} \beta$, etc., α, β, ... being constants. In such a case, it follows from [(x) and (xiv), **2c.4**] that

$$h(\alpha_n x_n, \beta_n y_n, \ldots) - h(\alpha x_n, \beta y_n, \ldots) \xrightarrow{P} 0,$$

provided $(x_n, y_n, \ldots)$ have a limiting distribution. Hence the a.d. of $h(\alpha_n x_n, \beta_n y_n \ldots)$ is the same as that of $h(\alpha x_n, \beta y_n, \ldots)$. Let us consider a few applications of this result.

(i) *Let* (T_n), $n = 1, 2, \ldots$, *be a sequence of statistics such that*

$$\sqrt{n}(T_n - \theta) \xrightarrow{L} X \sim N[0, \sigma^2(\theta)].$$

Let g *be a function of a single variable admitting the first derivative* g'. *Then*

$$\sqrt{n}[g(T_n) - g(\theta)] \xrightarrow{L} X \sim N(0, [g'(\theta)\sigma(\theta)]^2) \tag{6a.2.1}$$

if $g'(\theta) \neq 0$. *Further let* g' *be continuous, then*

$$\frac{\sqrt{n}[g(T_n) - g(\theta)]}{g'(T_n)} \xrightarrow{L} X \sim N[0, \sigma^2(\theta)], \tag{6a.2.2}$$

and if $\sigma(\theta)$ *is also continuous, then*

$$\frac{\sqrt{n}[g(T_n) - g(\theta)]}{g'(T_n)\sigma(T_n)} \xrightarrow{L} X \sim N(0, 1). \tag{6a.2.3}$$

By Taylor expansion, under the assumptions on g, we see that

$$g(T_n) - g(\theta) = (T_n - \theta)(g'(\theta) + \varepsilon_n),$$

where $\varepsilon_n \to 0$ as $T_n \to \theta$. This implies that for any given small quantity e, $|\varepsilon_n| < e$ whenever $|T_n - \theta| < \delta$. Hence $P(|\varepsilon_n| < e) \geqslant P(|T_n - \theta| < \delta) \to 1$ as $n \to \infty$ by assumption. Since e is arbitrary, $\varepsilon_n \xrightarrow{P} 0$. Now

$$\sqrt{n}[g(T_n) - g(\theta)] - \sqrt{n}(T_n - \theta)g'(\theta) = \sqrt{n}(T_n - \theta)\varepsilon_n \xrightarrow{P} 0,$$

since $\sqrt{n}(T_n - \theta)$ has an a.d. and $\varepsilon_n \xrightarrow{P} 0$, by using [(x) a, **2c.4**]. Then the a.d. of $\sqrt{n}[g(T_n) - g(\theta)]$ is the same as that of $\sqrt{n}(T_n - \theta)g'(\theta)$ which is $N(0, [g'(\theta)\sigma]^2$. Equation (6a.2.2) follows immediately because $g'(T_n) \xrightarrow{P} g'(\theta)$ as $T_n \xrightarrow{P} \theta$. Similarly, we have (6a.2.3) by using [(x), **2c.4**].

Standard Error. When the a.d. of $\sqrt{n}(T_n - \theta)$ is $N[0, \sigma^2(\theta)]$, it is customary to refer to $\sigma(\theta)/\sqrt{n}$ as the standard error (s.e.) of T_n. The estimated s.e. is $\sigma(T_n)/\sqrt{n}$. We may say that T_n is distributed approximately the same as $N(\theta, \sigma^2(\theta)/n)$. The results (6a.2.1 to 6a.2.3) show that the formula for the s.e. of $g(T_n)$ is

$$g'(\theta) \text{ (s.e. of } T_n). \tag{6a.2.4}$$

Note. The results (6a.2.1) to (6a.2.3) are not, ingeneral, applicable if, instead of g, we have a function g_n depending on n explicitly and we are required to find the a.d. of $g(T_n, n)$ with a suitable normalizing factor. For this purpose, consider g as a function of two variables (x, y) and suppose that $\partial g/\partial x$ exists and $\to G(\theta) \neq 0$ as $y \to \infty$, $x \to \theta$. *Under the conditions assumed on* g, *expanding* $g(T_n, n)$ *at* $T_n = \theta$ *for fixed* n *and following the proof of* (i), *we find that*

$$\frac{\sqrt{n}[g(T_n, n) - g(\theta, n)]}{h} \xrightarrow{L} X \sim N(0, 1), \tag{6a.2.5}$$

where h may be any one of the following forms:

(a) $G(\theta)\sigma(\theta)$,
(b) $G(T_n)\sigma(T_n)$, if G and σ are continuous,
(c) $(\partial g/\partial T_n)\sigma(T_n)$ if σ is continuous.

Furthermore, $g(\theta, n)$ can be replaced by $f(\theta)$, if $\sqrt{n}[g(\theta, n) - f(\theta)] \to 0$ as $n \to \infty$.

(ii) *Let $\mathbf{T}_n$ be a k-dimensional statistic $(T_{1n}, \ldots, T_{kn})$ such that the a.d. of $\sqrt{n}(T_{1n} - \theta_1), \ldots, \sqrt{n}(T_{kn} - \theta_k)$ is k-variate normal with mean zero and dispersion matrix $\mathbf{\Sigma} = (\sigma_{ij})$. Further let g be a function of k variables which is totally differentiable. Then the a.d. of*

$$\sqrt{n}u_n = \sqrt{n}[g(T_{1n}, \ldots, T_{kn}) - g(\theta_1, \ldots, \theta_k)]$$

is normal with mean zero and variance

$$v(\boldsymbol{\theta}) = \sum \sum \sigma_{ij} \frac{\partial g}{\partial \theta_i} \frac{\partial g}{\partial \theta_j}. \tag{6a.2.6}$$

provided $v(\boldsymbol{\theta}) \neq 0$.

Since g is a totally differentiable function, then

$$g(T_{1n}, \ldots, T_{kn}) - g(\theta_1, \ldots, \theta_k) = \sum (T_{in} - \theta_i)\left(\frac{\partial g}{\partial \theta_i}\right) + \varepsilon_n \|\mathbf{T}_n - \boldsymbol{\theta}\|,$$

where $\varepsilon_n \to 0$ as $T_{in} \to \theta_i$. By argument similar to that used in (i) $\varepsilon_n \xrightarrow{P} 0$ as $n \to \infty$ and since $\sqrt{n}\|\mathbf{T} - \boldsymbol{\theta}_n\| = [\sum n(T_{in} - \theta_i)^2]^{1/2}$ has an a.d.

$$\left| \sqrt{n}\, u_n - \sqrt{n} \sum (\mathbf{T}_{in} - \theta_i) \frac{\partial g}{\partial \theta_i} \right| \xrightarrow{P} 0.$$

But the a.d. of $\sqrt{n} \sum (T_{in} - \theta_i)\, \partial g/\partial \theta_i$, being a linear function of limiting normal variables is normal with zero mean and variance as given in (6a.2.6). By [(ix), **2c.4**], the a.d. of $\sqrt{n}\, u_n$ is the same.

As in (6a.2.3), if σ_{ij} and the partial derivatives of g are also continuous functions of $\boldsymbol{\theta}$,

$$[\sqrt{n}\, u_n/\sqrt{v(\mathbf{T}_n)}] \xrightarrow{L} X \sim N(0, 1), \tag{6a.2.7}$$

where $v(\mathbf{T}_n)$ is the value of $v(\boldsymbol{\theta})$ at $\boldsymbol{\theta} = \mathbf{T}_n$. For an extension of this result to cases where g is also a function of n, see Example 9 at the end of the chapter.

In practical applications we shall refer to the distribution of u_n, itself, as asymptotically normal with mean zero and asymptotic variance v/n, where v is as defined in (6a.2.6). It is also convenient to drop the suffix n when there is no chance of confusion.

δ **Method.** The method of determining the s.e. of $g(T_{1n}, \ldots, T_{kn})$ can be formally exhibited as follows. Taking the total differential of g, we have

$$dg = \frac{\partial g}{\partial \theta_1} dT_{1n} + \cdots + \frac{\partial g}{\partial \theta_k} dT_{kn}, \tag{6a.2.8}$$

where dT_{in} stands for $(T_{in} - \theta_i)$ etc. Now compute the variance, assuming that the variances and covariances of $(T_{in} - \theta_i)$ exist

$$V(dg) = \sum\sum \frac{\partial g}{\partial \theta_s} \frac{\partial g}{\partial \theta_s} \operatorname{cov}(T_{rn}, T_{sn})$$

$$= \frac{1}{n} \sum\sum \frac{\partial g}{\partial \theta_r} \frac{\partial g}{\theta \partial_s} \sigma_{rs}. \tag{6a.2.9}$$

(iii) *Let* $\mathbf{T}_n$ *be a k-dimensional statistic* $(T_{1n}, \ldots, T_{kn})$ *with the a.d. as in* (ii). *Let* $g_1, \ldots, g_q$ *be q functions of k variables and each* g_i *be totally differentiable. Then the a.d. of*

$$\sqrt{n}u_{in} = \sqrt{n}[g_i(T_{1n}, \ldots, T_{kn}) - g_i(\theta_1, \ldots, \theta_k)]$$
$$i = 1, \ldots, q$$

is q-variate normal with zero means and dispersion matrix $\mathbf{G\Sigma G}'$, *where* $\mathbf{G} = (\partial g_i/\partial \theta_j)$. *The rank of the distribution is equal to* $R(\mathbf{G\Sigma G}')$ (see Chapter 8).

A practically equivalent statement of the result of (iii) is as follows. If $(T_{1n}, \ldots, T_{kn})$ is distributed approximately as k-variate normal with mean $(\theta_1, \ldots, \theta_k)$ and dispersion matrix $n^{-1}\mathbf{\Sigma}$, then $(g_1, \ldots, g_q)$ is distributed approximately as q-variate normal with mean $[g_1(\boldsymbol{\theta}), \ldots, g_q(\boldsymbol{\theta})]$ and dispersion matrix $n^{-1}\mathbf{G\Sigma G}'$.

To prove the result we need only consider a linear function $g = b_1 g_1 + \cdots + b_q g_q$ and apply the result of (ii). Let $\delta = b_1 g_1(\boldsymbol{\theta}) + \cdots + b_q g_q(\boldsymbol{\theta})$. Then the a.d. of $\sqrt{n}(g - \delta)$ is normal with zero mean and variance

$$\sum\sum \frac{\partial g}{\partial \theta_r} \frac{\partial g}{\partial \theta_s} \sigma_{rs} = \sum\sum \left(\sum b_i \frac{\partial g_i}{\partial \theta_r}\right)\left(\sum b_i \frac{\partial g_i}{\partial \theta_s}\right)\sigma_{rs} = \mathbf{b}'\mathbf{G}\,\mathbf{\Sigma}\,\mathbf{G}'\mathbf{b} \tag{6a.2.10}$$

where $\mathbf{b}' = (b_1, \ldots, b_q)$. Since $\mathbf{b}$ is arbitrary we obtain the desired result.

The result of (iii) as stated is not useful in practical applications since the dispersion matrix $\mathbf{G\Sigma G}'$ depends on the parameters which may be unknown. We can state the stronger result as in (iv) below provided σ_{ij}, as functions of the parameters $\theta_1, \ldots, \theta_k$, are continuous.

(iv) *Let* $\mathbf{G}_n$, $\mathbf{\Sigma}_n$ *be the values of* $\mathbf{G}$ *and* $\mathbf{\Sigma}$ *at* $\theta_i = T_{in}$, $i = 1, \ldots, k$. *Further, let* F_n *be the d.f. of* $\sqrt{n}u_{in}$, $i = 1, \ldots, q$ *and* H_n *be the d.f. of a q-variate normal distribution with mean zero and dispersion matrix* $\mathbf{G}_n \mathbf{\Sigma}_n \mathbf{G}_n$. *Then if* $\mathbf{\Sigma}$ *and* $\mathbf{G}$ *are continuous in* $\mathbf{\theta}$,

$$\lim_{n \to \infty} \sup_{x_1, \ldots, x_q} (F_n - H_n) = 0. \tag{6a.2.11}$$

The result (6a.2.11) follows from the fact that both F_n and H_n have the same limit and the limit is continuous in the variables. The result shows that F_n can be approximated by a q-variate normal distribution with the dispersion matrix $\mathbf{G}_n \mathbf{\Sigma}_n \mathbf{G}_n'$, which involves only the sample statistics.

Pooling of Estimates. In practice, we are often faced with the problem of pooling of independent estimates of a parameter obtained from different sources. Generally, each estimate is reported as a number with an estimated standard error. In such cases two problems are to be considered. First, can all the estimates be considered as homogeneous, that is, are they estimating the same quantity? Second, if the estimates are homogeneous, what is the best way of combining them to obtain a single estimate? With these objects in view, we shall consider the following propositions.

Let $T_1, \ldots, T_k$ be independent consistent estimators of parameters $\theta_1, \ldots, \theta_k$, based on sample sizes $n_1, \ldots, n_k$ respectively. Let the a.d. of $\sqrt{n_i}(T_i - \theta_i)$ be $N[0, s_i^2(\theta_i)]$, as $n_i \to \infty$. Consider the statistic H (for homogeneity)

$$\mathrm{H} = \sum_{i=1}^{k} \frac{n_i(T_i - \hat{\theta})^2}{s_i^2(T_i)} = \sum \frac{n_i}{s_i^2} T_i^2 - \left(\sum \frac{n_i}{s_i^2} T_i\right)^2 \Big/ \sum \frac{n_i}{s_i^2} \tag{6a.2.12}$$

$$\hat{\theta} = \sum \frac{n_i T_i}{s_i^2(T_i)} \div \sum \frac{n_i}{s_i^2(T_i)}. \tag{6a.2.13}$$

If each T_i does not estimate the same quantity, the differences are reflected in the statistic (6a.2.12), and therefore it may be used to test the hypothesis $\theta_1 = \cdots = \theta_k$.

(v) *Let* $n = n_1 + \cdots + n_k$ *and denote by* π_i *the ratio* n_i/n, $i = 1, \ldots, k$. *The a.d. of*

$$\mathrm{H} = \sum \frac{n_i(T_i - \hat{\theta})^2}{s_i^2(T_i)} = n \sum \frac{\pi_i(T_i - \hat{\theta})^2}{s_i^2(T_i)} \tag{6a.2.14}$$

is χ^2 *on* $(k-1)$ *D.F. as* $n \to \infty$, *if we keep* $\pi_1, \ldots, \pi_k$ *fixed, under the hypothesis* $\theta_1 = \cdots = \theta_k$ *and the assumption that* s_i, $i = 1, \ldots, k$ *are continuous functions.*

Let us observe that, when $\theta_1 = \cdots = \theta_k = \theta$ (say),

$$n \sum \frac{\pi_i(T_i - \hat{\theta})^2}{s_i^2(T_i)} = n \sum \frac{\pi_i(T_i - \theta)^2}{s_i^2(T_i)} - n(\hat{\theta} - \theta)^2 \sum \frac{\pi_i}{s_i^2(T_i)} \tag{6a.2.15}$$

where

$$\hat{\theta} = \sum \frac{\pi_i T_i}{s_i^2(T_i)} \div \sum \frac{\pi_i}{s_i^2(T_i)}. \tag{6a.2.16}$$

It is then easy to establish, because of continuity of s_i, that

$$n \sum \frac{\pi_i(T_i - \theta)^2}{s_i^2(T_i)} - n \sum \frac{\pi_i(T_i - \theta)^2}{s_i^2(\theta)} \xrightarrow{P} 0$$

$$n(\hat{\theta} - \theta)^2 \sum \frac{\pi_i}{s_i^2(T_i)} - n(\theta^* - \theta)^2 \sum \frac{\pi_i}{s_i^2(\theta)} \xrightarrow{P} 0,$$

where θ^* is the same as $\hat{\theta}$ of (6a.2.16) with $s_i^2(T_i)$ replaced by $s_i^2(\theta)$. Hence the a.d. of H is the same as that of

$$\begin{aligned} \mathrm{H}' &= n \sum \frac{\pi_i(T_i - \theta)^2}{s_i^2(\theta)} - n(\theta^* - \theta)^2 \sum \frac{\pi_i}{s_i^2(\theta)} \\ &= n \sum \frac{(T_i - \theta)^2}{c_i^2} - n(\theta^* - \theta)^2 \sum \frac{1}{c_i^2}, \end{aligned} \tag{6a.2.17}$$

where $c_i^2 = s_i^2(\theta)/\pi_i$ and $\theta^* = (\sum T_i/c_i^2) \div \sum(1/c_i^2)$. Writing

$$y_i = \sqrt{n}(T_i - \theta)/c_i,$$

we see that

$$\mathrm{H}' = \sum y_i^2 - \left(\sum \frac{y_i}{c_i}\right)^2 \Big/ \left(\sum \frac{1}{c_i^2}\right), \tag{6a.2.18}$$

where the a.d. of y_i is $N(0, 1)$. Since H$'$ is quadratic (and therefore continuous in) in $y_1, \ldots, y_k$, the a.d. of H$'$ is the same as that of the quadratic form (6a.2.18) under the assumption that y_i is $N(0, 1)$ and y_i are independent. If we apply the criteria given in [(ii) **3b.4**], it is easily seen that the quadratic form (6a.2.18) is distributed as χ^2 on $(k - 1)$ D.F.

When H is not large, or not significant at a chosen probability level, the (natural) pooled estimate is

$$\hat{\theta} = \sum \frac{n_i T_i}{s_i^2(T_i)} \div \sum \frac{n_i}{s_i^2(T_i)}, \tag{6a.2.19}$$

with the estimated asymptotic variance

$$\left\{\sum \frac{n_i}{s_i^2(\hat{\theta})}\right\}^{-1} \tag{6a.2.20}$$

In practice, the formula (6a.2.12) in terms of T_i is useful in computing the statistic H.

Note. It is shown in (ii) that when the a.d. of $\sqrt{n_i}(T_i - \theta_i)$ is $N[0, s_i^2(\theta_i)]$, the a.d. of $\sqrt{n}[g(T_i) - g(\theta_i)]$ is $N[0, u_i^2(\theta_i)]$ where u_i depends on s_i and the derivative of g. If g is a one-to-one function of its argument, the hypothesis $\theta_1 = \cdots = \theta_k$ is equivalent to the hypothesis $g(\theta_1) = \cdots = g(\theta_k)$. We are then faced with the problem of choosing a suitable function g and applying the H test on $g(T_1), \ldots, g(T_k)$ instead of on $T_1, \ldots, T_k$. Similarly we may obtain a pooled estimate of $g(\theta)$ and then by inversion obtain a pooled estimate of θ. There is no satisfactory answer to this problem, but when $s_1^2 = \cdots = s_k^2$, the corresponding u_i^2 are also equal and g can be determined in such a way that the u_i are independent of θ_i (see **6g**). Such a choice of g seems to have some advantage. We consider applications of the H test on transformed statistics in **6g**.

6b CHI-SQUARE TESTS FOR THE MULTINOMIAL DISTRIBUTION

6b.1 Test of Departure from a Simple Hypothesis

Let $n_1, \ldots, n_k$ be the observed frequencies and $\pi_1, \ldots, \pi_k$ be the hypothetical probabilities of a multinomial distribution. To test the hypothesis $\pi_i = \pi_i^0$, $i = 1, \ldots, k$ (assigned values), K. Pearson (1900) suggested the criterion

$$\chi^2 = \sum \frac{(n_i - n\pi_i)^2}{n\pi_i} = \sum \frac{(\text{observed} - \text{expected})^2}{\text{expected}}. \tag{6b.1.1}$$

(i) *The a.d. of* χ^2 *as defined in* (6b.1.1) *is* $\chi^2(k-1)$.

To prove this we apply the sufficient condition of (6a.1.6). The statistic (6b.1.1) is the quadratic form $\mathbf{V'CV}$ with $\mathbf{C} = \mathbf{I}$. The conditions (6a.1.6)

$$\mathbf{C}^2 = \mathbf{I}^2 = \mathbf{I} = \mathbf{C}, \qquad \mathbf{C}\boldsymbol{\phi} = \mathbf{I}\boldsymbol{\phi} = \boldsymbol{\phi}$$

are satisfied with $\alpha = 1 \neq 0$. Hence a.d. of (6b.1.1) is χ^2 on D.F. equal to (rank $\mathbf{I} - 1) = (k-1)$. Hence the result.

6b.2 Chi-Square Test for Goodness of Fit

The general problem in judging goodness of fit is to test whether the cell probabilities are specified functions of a fewer number of parameters whose values may be unknown.

(i) *Let the cell probabilities be the specified functions* $\pi_1(\boldsymbol{\theta}), \ldots, \pi_k(\boldsymbol{\theta})$ *involving q unknown parameters* $(\theta_1, \ldots, \theta_q) = \boldsymbol{\theta}'$. *Further let (a)* $\hat{\boldsymbol{\theta}}$ *be an efficient estimator of* $\boldsymbol{\theta}$ *in the sense of* (5c.2.6), *(b) each* $\pi_i(\boldsymbol{\theta})$ *admit continuous partial derivatives of the first order (only) with respect to* θ_j, $j = 1, \ldots, q$ *or* *each $\pi_i(\boldsymbol{\theta})$ be a totally differentiable function of $\theta_1, \ldots, \theta_q$, and (c) the matrix $M = (\pi_r^{-1/2}\, \partial\pi_r/\partial\theta_s)$ of order $(k \times q)$ computed at the true values of $\boldsymbol{\theta}$ is of rank q. Then the a..d of*

$$\chi^2 = \sum \frac{(n_i - n\hat{\pi}_i)^2}{n\hat{\pi}_i} = \sum \frac{(0 - E)^2}{E} \tag{6b.2.1}$$

is $\chi^2(k - 1 - q)$, where $\hat{\pi}_i = \pi_i(\hat{\theta})$.

We note that efficient estimates in the sense of (5c.2.6) exist under the condition (b) as shown in [(iv), **5e.2**]. Hence the a.d. of (6b.2.1) is established under the sole assumption of the existence and continuity of first-order partial derivatives of $\pi_i(\boldsymbol{\theta})$ or simply the total differentiability of $\pi_i(\boldsymbol{\theta})$. For simplicity in the proof we write $X_n \stackrel{a}{=} Y_n$ (asymptotically equivalent) to represent the statement $X_n - Y_n \xrightarrow{P} 0$. Let

$$\mathbf{U}' = \left(\frac{n(\hat{\pi}_1 - \pi_1)}{\sqrt{n\pi_1}}, \ldots, \frac{n(\hat{\pi}_k - \pi_k)}{\sqrt{n\pi_k}}\right),$$

$$\mathbf{M} = \left(\frac{1}{\sqrt{\pi_r}} \frac{\partial\pi_r}{\partial\theta_s}\right), \text{ matrix of order } (k \times q) \text{ and rank } q.$$

$$\mathbf{D}' = [\sqrt{n}(\hat{\theta}_1 - \theta_1), \ldots, \sqrt{n}(\hat{\theta}_q - \theta_q)],$$

$$\mathbf{Z} = \mathbf{M}'\mathbf{V}, \text{ where } \mathbf{V} \text{ is as defined in (6a.1.1)}$$

The information matrix $(i_{rs}) = \mathbf{M}'\mathbf{M} = \mathscr{I}$ (say). By the assumption that the rank of $\mathbf{M}$ is q, $\mathscr{I}$ is nonsingular at the true value of $\boldsymbol{\theta}$ and since $\hat{\boldsymbol{\theta}}$ is efficient in the sense of (5c.2.6),

$$\mathbf{D} \stackrel{a}{=} \mathscr{I}^{-1}\mathbf{M}'\mathbf{V}. \tag{6b.2.2}$$

Expanding $\pi_i(\hat{\boldsymbol{\theta}})$ at $\boldsymbol{\theta}$ under the condition (b), we see that

$$\frac{\sqrt{n}[\pi_i(\hat{\boldsymbol{\theta}}) - \pi_i(\boldsymbol{\theta})]}{\sqrt{\pi_i(\boldsymbol{\theta})}} \stackrel{a}{=} \frac{1}{\sqrt{\pi_i(\boldsymbol{\theta})}} \sum_{s=1}^{q} \frac{\partial\pi_i}{\partial\theta_s} \sqrt{n}(\hat{\theta}_s - \theta_s). \tag{6b.2.3}$$

The relationship (6b.2.3) for $i = 1, \ldots, k$ can be written in matrix notation $\mathbf{U} \stackrel{a}{=} \mathbf{MD}$. By substitution for $\mathbf{D}$ from the relation (6b.2.2), $\mathbf{U} \stackrel{a}{=} \mathbf{M}\mathscr{I}^{-1}\mathbf{M}'\mathbf{V}$. Finally, subtracting $\mathbf{V}$ from either side, we have

$$\mathbf{V} - \mathbf{U} \stackrel{a}{=} (\mathbf{I} - \mathbf{M}\mathscr{I}^{-1}\mathbf{M}')\mathbf{V}. \tag{6b.2.4}$$

The χ^2 statistic of (6b.2.1) is, replacing $\hat{\pi}_i$ by π_i in the denominator,

$$\begin{aligned}(\mathbf{V} - \mathbf{U})'(\mathbf{V} - \mathbf{U}) &\stackrel{a}{=} \mathbf{V}'(\mathbf{I} - \mathbf{M}\mathscr{I}^{-1}\mathbf{M}')(\mathbf{I} - \mathbf{M}\mathscr{I}^{-1}\mathbf{M}')\mathbf{V} \\ &= \mathbf{V}'(\mathbf{I} - \mathbf{M}\mathscr{I}^{-1}\mathbf{M}')\mathbf{V}, \qquad \text{since} \quad \mathbf{M}'\mathbf{M} = \mathscr{I}.\end{aligned}$$

The sufficient condition (6a.1.6) for a χ^2 distribution

$$(\mathbf{I} - \mathbf{M}\mathscr{I}^{-1}\mathbf{M}')^2 = \mathbf{I} - \mathbf{M}\mathscr{I}^{-1}\mathbf{M}', \qquad \text{since} \quad \mathbf{M}'\mathbf{M} = \mathscr{I}$$
$$(\mathbf{I} - \mathbf{M}\mathscr{I}^{-1}\mathbf{M}')\boldsymbol{\phi} = \boldsymbol{\phi}, \qquad \text{since} \quad \mathbf{M}'\boldsymbol{\phi} = 0$$

is satisfied with $\alpha = 1 \neq 0$. Furthermore, because of idempotency

$$\begin{aligned} \text{rank } (\mathbf{I} - \mathbf{M}\mathscr{I}^{-1}\mathbf{M}') &= \text{trace } (\mathbf{I} - \mathbf{M}\mathscr{I}^{-1}\mathbf{M}') \\ &= k - \text{trace } \mathbf{M}\mathscr{I}^{-1}\mathbf{M}' = k - \text{trace } \mathscr{I}^{-1}\mathbf{M}'\mathbf{M} \\ &= k - \text{trace } \mathbf{I}_q = k - q \end{aligned}$$

where $\mathbf{I}_q$ is a unit matrix of order q. Hence by using the result [(iv), **6a.1**], the a.d. of (6b.2 1) is $\chi^2(k - q - 1)$, so that the rule for D.F. is ($k - 1$ − the number of parameters estimated) provided the rank of $(\pi_i^{-1/2}\partial\pi_i/\partial\theta_j)$ is q.

Note. One of the conditions under which the distribution of (6b.2.1) is established is that the estimator $\hat{\boldsymbol{\theta}}$ used in obtaining the expected frequencies is efficient. In Chapter 5 we discussed the various conditions under which efficient estimators exist and the methods leading to the computation of such estimators. There is indeed a large class of efficient estimators, the choice of any one of which would make the a.d. of (6b.2.1) valid. There is, however, some advantage in using the maximum likelihood estimators provided the conditions under which they are efficient are satisfied (see Rao, 1961a, 1962d).

6b.3 Test for Deviation in a Single Cell

The χ^2 goodness of fit test provides an overall test of differences between observed and expected frequencies specified by a model. Sometimes it is of interest to examine whether the departure from expected is due to deficiency or otherwise of the observed frequency in a particular cell. For instance, in studying a frequency distribution of accidents, we may suspect that there is underrecording of individuals with no accidents. Cochran (1954) proposed as the test criterion the residual

$$r_j = \frac{[n_j - n\pi_j(\hat{\boldsymbol{\theta}})]}{\sqrt{n\pi_j(\hat{\boldsymbol{\theta}})}} \tag{6b.3.1}$$

to examine the deviation in the jth cell. To determine the a.d. of the statistic (6b.1.3), it is enough to compute its asymptotic variance (a.v.), that is, the variance of its a.d.

From (6b.2.4), we see that

$$\mathbf{V} - \mathbf{U} \stackrel{a}{=} (\mathbf{I} - \mathbf{M}\mathscr{I}^{-1}\mathbf{M}')\mathbf{V}, \qquad \mathbf{U} \stackrel{a}{=} \mathbf{M}\mathscr{I}^{-1}\mathbf{M}'\mathbf{V}.$$

The covariance between the asymptotic equivalents of $\mathbf{V}-\mathbf{U}$ and $\mathbf{U}$ is

$$(\mathbf{I}-\mathbf{M}\mathscr{I}^{-1}\mathbf{M}')[E(\mathbf{V}\mathbf{V}')]\mathbf{M}\mathscr{I}^{-1}\mathbf{M}'$$
$$=(\mathbf{I}-\mathbf{M}\mathscr{I}^{-1}\mathbf{M}')(\mathbf{I}-\boldsymbol{\phi}\boldsymbol{\phi}')\mathbf{M}\mathscr{I}^{-1}\mathbf{M}'=\mathbf{0}, \qquad \text{since} \quad \boldsymbol{\phi}'\mathbf{M}=0.$$

Therefore each component of $\mathbf{V}-\mathbf{U}$ *is uncorrelated, asymptotically, with all the components of* $\mathbf{U}$, which is an important result, similar to the result in least squares theory that the residuals are uncorrelated with best estimates [(ii), **4a.4**]

Hence the a. cov. of the jth components of $\mathbf{U}$ and $\mathbf{V}$ is zero, which is the same as

$$\text{a. cov}\left\{\frac{n_j-n\pi_j(\hat{\boldsymbol{\theta}})}{\sqrt{n\pi_j(\hat{\boldsymbol{\theta}})}}, \frac{n\pi_j(\hat{\boldsymbol{\theta}})-n\pi_j(\boldsymbol{\theta})}{\sqrt{n\pi_j(\hat{\boldsymbol{\theta}})}}\right\}=0$$

that is,

$$\text{a.v.}\left\{\frac{n_j-n\pi_j(\boldsymbol{\theta})}{\sqrt{n\pi_j(\hat{\boldsymbol{\theta}})}}\right\}=\text{a.v.}\left\{\frac{n_j-n\pi_j(\hat{\boldsymbol{\theta}})}{\sqrt{n\pi_j(\hat{\boldsymbol{\theta}})}}\right\}+\text{a.v.}\left\{\frac{n\pi_j(\hat{\boldsymbol{\theta}})-n\pi_j(\boldsymbol{\theta})}{\sqrt{n\pi_j(\hat{\boldsymbol{\theta}})}}\right\}.$$

But

$$\text{a.v.}\left\{\frac{n_j-n\pi_j(\boldsymbol{\theta})}{\sqrt{n\pi_j(\hat{\boldsymbol{\theta}})}}\right\}=\text{v.}\left\{\frac{n_j-n\pi_j(\boldsymbol{\theta})}{\sqrt{n\pi_j(\boldsymbol{\theta})}}\right\}=1-\pi_j(\boldsymbol{\theta}),$$

and

$$\text{a.v.}\left\{\frac{n\pi_j(\hat{\boldsymbol{\theta}})-n\pi_j(\boldsymbol{\theta})}{\sqrt{n\pi_j(\hat{\boldsymbol{\theta}})}}\right\}=\text{a.v.}\left\{\frac{n\pi_j(\hat{\boldsymbol{\theta}})-n\pi_j(\boldsymbol{\theta})}{\sqrt{n\pi_j(\boldsymbol{\theta})}}\right\}=\sum_s\sum_t\frac{1}{\pi_j}\frac{\partial\pi_j}{\partial\theta_t}\frac{\partial\pi_j}{\partial\theta_s}i^{st}$$

by using the formula (6a.2.6). Hence

$$\text{a.v.}\left\{\frac{n_j-n\pi_j(\hat{\boldsymbol{\theta}})}{\sqrt{n\pi_j(\hat{\boldsymbol{\theta}})}}\right\}=v(\boldsymbol{\theta})=1-\pi_j(\boldsymbol{\theta})-\sum_s\sum_t\frac{1}{\pi_j}\frac{\partial\pi_j}{\partial\theta_s}\frac{\partial\pi_j}{\partial\theta_t}i^{st}. \qquad (6b.3.2)$$

The test based on a normal deviate $N(0, 1)$, for deviation in the jth cell, is

$$\frac{n_j-n\pi_j(\hat{\boldsymbol{\theta}})}{[nv(\hat{\boldsymbol{\theta}})\pi_j(\hat{\boldsymbol{\theta}})]^{1/2}}. \qquad (6b.3.3)$$

For a single parameter, formula (6b.3.2) becomes

$$v(\theta)=(1-\pi_j)-\frac{1}{\pi_j}\left(\frac{d\pi_j}{d\theta}\right)^2\frac{1}{i(\theta)}. \qquad (6b.3.4)$$

where $i(\theta)$ is the information per observation.

As an example let us consider the deviation in the zero cell of an observed distribution of a Poisson variable. Here,

$$\pi_0 = e^{-\mu}, \qquad \frac{d\pi_0}{d\mu} = -e^{-\mu}, \qquad \text{and} \quad i = \frac{1}{\mu}.$$

The value of (6b.3.4) is

$$1 - e^{-\mu} - \mu e^{-\mu}.$$

Suppose in an observed distribution we have

$$\begin{aligned} n = 200, \qquad n_0 &= 15, \qquad \text{and } \hat{\mu} \text{ (estimate of } \mu) = 2.1055. \\ n_0 - n\hat{\pi}_0 &= 15 - 200e^{-\hat{\mu}} \\ &= 15 - 24.356 = -9.356 \\ n\hat{\pi}_0 v(\hat{\mu}) &= ne^{-\hat{\mu}}(1 - e^{-\hat{\mu}} - \hat{\mu}e^{-\hat{\mu}}) \\ &= 24.356(0.62181) = 15.1448. \end{aligned}$$

The statistic (6b.3.3) has the absolute value

$$\frac{9.356}{\sqrt{15.1448}} = 2.40,$$

which is high for a normal deviate indicating a deficiency in the zero cell.

6b.4 Test Whether the Parameters Lie in a Subset

Assuming the specification $\pi_i(\boldsymbol{\theta})$ to be true, let us examine whether $\boldsymbol{\theta}$ belongs to a subset of the admissible values, specified by the locus

$$\theta_i = g_i(\tau_1, \ldots, \tau_r), \qquad i = 1, \ldots, q; \qquad r < q, \tag{6b.4.1}$$

where g_i admit continuous first derivatives and $\tau_1, \ldots, \tau_r$ are new parameters. To test the hypothesis (6b.4.1), the author proposed the test criterion (Rao, 1961a)

$$\chi^2 = \sum \frac{(E_1 - E_2)^2}{E_2}, \tag{6b.4.2}$$

where E_1 and E_2 stand for the estimated frequencies when there is no restriction on $\boldsymbol{\theta}$ and when subjected to the restrictions (6b.4.1) respectively.

More specifically, let $\hat{\boldsymbol{\theta}}$ be an efficient estimate of $\boldsymbol{\theta}$ without any restriction. When the hypothesis (6b.4.1) is true, π_i is a function of $\boldsymbol{\tau}' = (\tau_1, \ldots, \tau_r)$ and let $\hat{\boldsymbol{\tau}}$ be an efficient estimate of $\boldsymbol{\tau}$. Then the a.d. of

$$\chi^2 = \sum \frac{[n\pi_i(\hat{\boldsymbol{\theta}}) - n\pi_i(\hat{\boldsymbol{\tau}})]^2}{n\pi_i(\hat{\boldsymbol{\tau}})} \tag{6b.4.3}$$

is that of $\chi^2(q - r)$, provided $\pi_i(\boldsymbol{\tau})$ satisfy conditions similar to those of $\pi_i(\boldsymbol{\theta})$ as functions of $\boldsymbol{\theta}$.

This proposition can be proved on the same lines as in **6b.2**. We note that the information matrices $\mathscr{I}_1$ and $\mathscr{I}$ for $\boldsymbol{\tau}$ and $\boldsymbol{\theta}$ are connected by the relation $\mathscr{I}_1 = \Delta' \mathscr{I} \Delta$ where $\Delta = (\partial\theta_i/\partial\tau_j)$ is the matrix of the derivatives. The difference $\sqrt{n}[\pi_i(\hat{\boldsymbol{\theta}}) - \pi_i(\hat{\boldsymbol{\tau}})]$ can be expanded in terms of observed frequencies as in the case of $\sqrt{n}[\pi_i(\hat{\boldsymbol{\theta}}) - \pi_i(\boldsymbol{\theta})]$. The statistic (6b.4.3) is then equivalent to a quadratic form in **V**. An application of (6a.1.6) yields the result.

6b.5 Some Examples

Example 1. Bateson gives the following data concerning the segregation of two genes for purple-red flower color and long-round pollen shape in sweet peas.

The results come from an intercross so that the expected frequencies, on the hypothesis of independent segregation, are in the ratio 9 : 3 : 3 : 1. Are the data in agreement with the expected frequencies? We can apply the χ^2 test of **6b.1**.

	Purple-long	Red-long	Purple-Round	Red-Round	Total
Observed	296	27	19	85	427
Expected	3843 ÷ 16	1281 ÷ 16	1281 ÷ 16	427 ÷ 16	427
$\dfrac{(\text{Observed})^2}{\text{Expected}}$	364.7817	9.1054	4.5090	270.7260	649.1221

$$\chi^2 = \sum \frac{O^2}{E} - \sum O = 649.1221 - 427 = 222.1221 \text{ on 3 D.F.}$$

The probability of χ^2 on 3 D.F. exceeding 222.1221 is very small showing that there is departure from the expected. For the large sample test to hold, *it is necessary that each cell expected frequency should be greater than* 5. An exact test may be made in some situations when the expected frequencies are small, as shown in Example 2. Other devices for dealing with small expectations are considered in **6d.2** and **6d.3**.

Example 2. Test whether the frequencies 8, 3, 1 could have arisen from a trinomial with equal probabilities.

The expected values 4, 4, 4 are all small so that the χ^2 approximation may not be valid. If we ignore this condition, the χ^2 on 2 D.F. is

$$\frac{8^2}{4} + \frac{3}{4} + \frac{1^2}{4} - 12 = 6.5,$$

which has a probability between 2% and 5%. In the present example as the expected frequencies are all small, we are in doubt about the validity of the

χ^2 approximation. This doubt necessitates the evaluation of the actual probabilities of χ^2 being equal to or exceeding the observed value 6.5. The probability for any observed set of frequencies x, y, z is

$$\frac{12!}{x!\,y!\,z!}\left(\frac{1}{3}\right)^{12}$$

There are 91 partitions of 12 and the probabilities for partitions yielding a $\chi^2 \geqslant 6.5$ are computed as follows:

Partitional Type	Number	Probability	Chi-square
12 0 0	3	$0.0^5 1882$	24
11 1 0	6	$0.0^4 2258$	18.5
10 2 0	6	$0.0^3 1242$	14
10 1 1	3	$0.0^3 2484$	13.5
9 3 0	6	$0.0^3 4140$	10.5
9 2 1	6	$0.0^2 1242$	9.5
8 4 0	6	$0.0^3 9314$	8
8 3 1	6	$0.0^2 3726$	6.5
7 5 0	6	$0.0^2 1490$	6.5

When we add these probabilities, the exact probability of χ^2 being greater than or equal to 6.5 is found to be 0.04844, which is slightly higher than the value found from the asymptotic distribution. But in general the agreement may not be so close. The agreement will be fairly good when the expected value in each cell is greater than 5.

6b.6 Test for Deviations in a Number of Cells

In addition to **U**, **M** defined in **6b.2** and **V** in **6b.1**, we denote

$$\mathbf{W}' = \frac{n_1 - n\pi_1(\hat{\boldsymbol{\theta}})}{\sqrt{n\pi_1(\hat{\boldsymbol{\theta}})}}, \ldots, \frac{n_k - n\pi_k(\hat{\boldsymbol{\theta}})}{\sqrt{n\pi_k(\hat{\boldsymbol{\theta}})}} \tag{6b.6.1}$$

the vector denoting deviations in the k cells. It may seem that the χ^2 test of goodness of fit described in **6b.1** is $\mathbf{W}'\mathbf{W}$.

In **6b.3** we developed a test for examining deviation in a single cell. We shall now develop a test for deviations in r specified cells. For this we find the asymptotic dispersion (a.D) matrix of **W** which is equivalent to $\mathbf{V} - \mathbf{U}$. Then

$$\text{a.D}(\mathbf{W}) = \text{a.D}(\mathbf{V} - \mathbf{U}) = \text{a.D}(\mathbf{V}) - \text{a.D}(\mathbf{U}) \tag{6b.6.2}$$

since the asymptotic covariance matrix between $\mathbf{V}-\mathbf{U}$ and $\mathbf{U}$ is null as shown in **6b.3**. Hence

$$\text{a.D}(\mathbf{V}) - \text{a.D}(\mathbf{U}) = (\mathbf{I} - \boldsymbol{\varphi}\boldsymbol{\varphi}') - \mathbf{M}\mathscr{I}^{-1}\mathbf{M}'. \tag{6b.6.3}$$

Let $\mathbf{d}$ denote the sub-vector of $\mathbf{W}$ containing only the terms corresponding to the subset of cells in which the deviations have to be tested and let $\mathbf{B}$ denote the dispersions matrix of $\mathbf{d}$ obtained from (6b.4.3) by retaining only the relevant elements. The elements of $\mathbf{B}$ involve the unknown $\boldsymbol{\theta}$. Substituting $\hat{\boldsymbol{\theta}}$ for $\boldsymbol{\theta}$ we obtain the matrix $\hat{\mathbf{B}}$. The χ^2 test for examining the deviations in r specified cells is

$$\chi^2 = \mathbf{d}'\hat{\mathbf{B}}^{-1}\mathbf{d} \tag{6b.6.4}$$

on r D.F provided $R(\mathbf{B}) = r$. Otherwise a g-inverse of $\hat{\mathbf{B}}$ has to be used in which case the D.F. of χ^2 is $R(\mathbf{B})$.

6c TESTS RELATING TO INDEPENDENT SAMPLES FROM MULTINOMIAL DISTRIBUTIONS

6c.1 General Results

Suppose that we have samples of sizes $n_{1.}, \ldots, n_{m.}$ from m finite multinomial populations, all not necessarily having an equal number of cells. As for a single multinomial, two types of problems arise. First, to test the specifications (i.e., of cell probabilities as given functions of a certain number of parameters) simultaneously for all the multinomial distributions. Second, to test whether the parameters belong to a subset of the admissible values.

For example, we may have samples from two different populations providing the frequencies of O, A, B, AB blood groups. First we may examine whether the frequencies in each group are in accordance with Bernstein's hypothesis, as discussed in **5g** of Chapter 5. A second question that naturally arises is whether the gene frequencies are the same in both the populations. Let p, q represent the A, B gene frequencies in the first population and p', q' represent the corresponding gene frequencies in the second population. The hypothesis states that $p = p'$ and $q = q'$.

Let us consider a general problem where the cell probabilities of the ith population are specified functions

$$\pi_{i1}(\boldsymbol{\theta}), \ldots, \pi_{ik_i}(\boldsymbol{\theta}) \tag{6c.1.1}$$

of q unknown parameters $\theta_{1i}, \ldots, \theta_{qi}$. Let $n_{i1}, \ldots, n_{ik_i}$, $(\sum n_{ij} = n_{i.})$ be the observed frequencies in the ith sample, and $\hat{\theta}_{1i}, \ldots, \hat{\theta}_{qi}$ be the m.l. estimators based on the ith sample. Let the probabilities (6c.1.1) satisfy the conditions for an application of the χ^2 goodness of fit test of **6b.2**. Then the goodness of fit χ^2 for the ith sample is

$$\chi_i^{\,2} = \sum_{j=1}^{k_i} \frac{(n_{ij} - \pi_{i.}\hat{\pi}_{ij})^2}{n_{i.}\hat{\pi}_{ij}} \tag{6c.1.2}$$

on D.F. equal to $(k_i - q - 1)$, where $\hat{\pi}_{ij}$ are obtained by substituting $\hat{\theta}_{1i}, \ldots, \hat{\theta}_{qi}$ for the unknown parameters. The total χ^2 arising from the m samples (i.e., m individual goodness of fit χ^2's),

$$\sum_{i=1}^{m} \chi_i^{\,2} = \sum_{i=1}^{m} \sum_{j=1}^{k_i} \frac{(n_{ij} - n_{i.}\hat{\pi}_{ij})^2}{n_{i.}\hat{\pi}_{ij}}, \tag{6c.1.3}$$

has then $\sum k_i - mq - m$ D.F., which provides an overall test of the specifications of the probabilities.

Now let $\theta_{ri} = \theta_r$ for all r and i so that the same parameter values are applicable to all the m populations, and denote by $\theta_1^*, \ldots, \theta_q^*$ the m.l. estimators of $\theta_1, \ldots, \theta_q$ obtained from the combined sample. Let π_{ij}^* be the value of $\pi_{ij}(\boldsymbol{\theta})$ when $\theta_1^*, \ldots, \theta_q^*$ are substituted for $\theta_1, \ldots, \theta_q$. We thus have two estimated expected values for the observed frequency n_{ij},

$$\begin{aligned} E_1 &= (\text{expected})_1 = n_{i.}\hat{\pi}_{ij}, \\ E_2 &= (\text{expected})_2 = n_{i.}\pi_{ij}^*, \end{aligned}$$

where it may be noted that $\hat{\pi}_{ij}$ is obtained by substituting the m.l. estimators obtained from the ith sample alone as in (6c.1.2). The χ^2 test for the hypothesis that the same parameter values are applicable to all the samples is

$$\sum\sum \frac{(E_1 - E_2)^2}{E_2} = \sum_{i=1}^{m} \sum_{j=1}^{k_i} \frac{(n_{i.}\hat{\pi}_{ij} - n_{i.}\pi_{ij}^*)^2}{n_{i.}\pi_{ij}^*}, \tag{6c.1.4}$$

which is asymptotically $\chi^2[q(m-1)]$. The proof is similar to that of (6b.4.2).

6c.2 Tests of Homogeneity of Parallel Samples

Let the frequencies in k classes for two samples be as in Table 6c.2α.

TABLE 6c.2α. Observed Frequencies in Two Parallel Samples

	Classes				
Sample	(1)	(2)	$\cdots$	(k)	Total
First	n_{11}	n_{12}	$\cdots$	n_{1k}	$n_{1.}$
Second	n_{21}	n_{22}	$\cdots$	n_{2k}	$n_{2.}$
Total	$n_{.1}$	$n_{.2}$	$\cdots$	$n_{.k}$	$n_{..}$
First/total	p_1	p_2	$\cdots$	p_k	p

The classes may refer to a discrete classification or to intervals of a continuous variable. If nothing is specified about the hypothetical cell probabilities (i.e., as functions of a smaller number of parameters), how can it be tested that the two samples have come from the same population?

Let $\pi_{11}, \ldots, \pi_{1k}$ and $\pi_{21}, \ldots, \pi_{2k}$ be the probabilities of the cells in the two populations. The hypothesis to be tested is $\pi_{1i} = \pi_{2i}, i = 1, \ldots, k$. The estimate of π_{1i} from the first sample is

$$\hat{\pi}_{1i} = \frac{n_{1i}}{n_{1.}}$$

and that of π_{2i} from the second sample is

$$\hat{\pi}_{2i} = \frac{n_{2i}}{n_{2.}}.$$

If $\pi_{1i} = \pi_{2i} = \pi_i$ (say), the estimate of the common value (from the combined sample) is

$$\hat{\pi}_i = \frac{n_{1i} + n_{2i}}{n_{..}} = \frac{n_{.i}}{n_{..}}.$$

The χ^2 of (6c.1.4) for the hypothesis under test is

$$\sum\sum \frac{(E_1 - E_2)^2}{E_2} = \sum_{i=1}^{k} \frac{\left(n_{1i} - \dfrac{n_{1.}n_{.i}}{n_{..}}\right)^2}{\dfrac{n_{1.}n_{.i}}{n_{..}}} + \sum_{i=1}^{k} \frac{\left(n_{2i} - \dfrac{n_{2.}n_{.i}}{n_{..}}\right)^2}{\dfrac{n_{2.}n_{.i}}{n_{..}}}$$

$$= \frac{1}{p(1-p)} \sum n_{.i}(p_i - p)^2 = \frac{1}{p(1-p)} \left(\sum n_{.i} p_i^2 - n_{..} p^2\right)$$

$$= \frac{1}{p(1-p)} \left(\sum n_{1i} p_i - n_{1.} p\right), \tag{6c.2.1}$$

where p_i and p are as defined in Table 6c.2α. The χ^2 defined in (6c.2.1) has $(k - 1)$ D.F. as the hypothesis specifies the equality of $(k - 1)$ independent cell probabilities.

In general, when there are m samples to be compared, the formula is

$$\chi^2 = \sum\sum \frac{\left(n_{ij} - \dfrac{n_{i.}n_{.j}}{n_{..}}\right)^2}{\dfrac{n_{i.}n_{.j}}{n_{..}}}, \qquad D.F. = (m-1)(k-1). \tag{6c.2.2}$$

The formula for χ^2 cannot be simplified further in the general case.

6c.3 An Example

The distribution in four blood group classes O, A, B, AB of 353 individuals belonging to community C and 364 individuals of community D are given in Table 6c.3α. The test (6c.2.1) is applied to examine the equality of the hypothetical proportions of the blood group classes in the two communities. The $\chi^2 = 11.77$ on 3 D.F. computed in Table 6c.3α is significant at the 5% level indicating differences between communities.

TABLE 6c3α. Classification of Individuals by Blood Groups

	O	A	B	AB	Total
Community C	121	120	79	33	353
Community D	118	95	121	30	364
Total	239	215	200	63	717
$p = \dfrac{\text{Community C}}{\text{Total}}$	0.5063	0.5581	0.3950	0.5238	0.4923

$$\sum n_{1i} p_i - n_{1\cdot} p = 121(0.5063) + 120(0.5581) + \cdots - 353(0.4923) = 2.9428$$

$$p(1-p) = 0.24994, \qquad \chi^2 = 2.9428/0.24994 = 11.77 \text{ on 3 D.F.}$$

The χ^2 test on 3 D.F. is not, however, the best for the available data if the object is to establish differences between communities. The intrinsic (or genetic) difference, if any, between the communities, is in the relative frequencies of the O, A, B genes. A test of the hypothesis of equality of gene frequencies is more fundamental than that of the equality of frequencies of the phenotypes without recognizing that the latter are known functions of the former. Only when such functional dependence on more intrinsic characters like the gene frequencies is unknown or is in doubt, is a test of the type (6c.2.1) involving the phenotypical frequencies valid.

TABLE 6c.3β. Maximum Likelihood Estimates

	$\hat{p}$	$\hat{q}$	$\hat{r} = (1 - \hat{p} - \hat{q})$
Community C	0.24649	0.17317	0.58034
Community D	0.19025	0.23573	0.57401
Combined sample	0.21762	0.20435	0.57803

The specification of the probabilities of the phenotypes O, A, B, AB in terms of gene frequencies p, q, r is given in the illustrative example of Section **5g**. By applying the method of m.l. as indicated in that section, the estimates of p, q, r for the first sample, second sample, and the combined sample are obtained. Using these estimates, the expected values (E_1 and E_2) on the basis of individual estimates and combined estimates are obtained for each community and shown in Table 6c.3γ.

TABLE 6c.3γ. Expected Values under the Two Hypotheses

		Community C			Community D		
Classes	Probabilities	Observed	E_1	E_2	Observed	E_1	E_2
O	r^2	121	118.89	117.94	118	119.93	121.62
A	$p^2 + 2pr$	120	122.44	105.52	95	92.68	108.81
B	$q^2 + 2qr$	79	81.54	98.14	121	118.74	101.19
AB	$2pq$	33	30.13	31.40	30	32.65	32.37
Total	1	353	353	353	364	364	364

χ^2 (goodness of fit)

$$\text{for community C} = \sum \frac{(O - E_1)^2}{E_1} = 0.44, \quad 1 \text{ D.F.}$$

$$\text{for community D} = \sum \frac{(O - E_1)^2}{E_1} = 0.35, \quad 1 \text{ D.F.}$$

$$\left.\begin{array}{l}\chi^2 \text{ for homogeneity} \\ \text{of communities}\end{array}\right\} = \sum\sum \frac{(E_1 - E_2)^2}{E_2} = 11.04, \quad 2 \text{ D.F.}$$

The values of χ^2 for the individual communities are small, which indicates that the specification of cell probabilities in terms of gene frequencies are valid. The last χ^2 establishes the difference between communities in a better way than the χ^2 of 11.77 on 3 D.F. obtained from a direct comparison of phenotypic frequencies. The two χ^2's are nearly of the same magnitude, but the χ^2 based on a comparison of gene frequencies has 1 D.F. less, giving a lower probability in favor of the null hypothesis.

The test can be extended to examine differences between several communities simultaneously. We need to obtain m.l. estimates for individual samples and for the combined sample. We then compute two sets of expectations for comparison by the χ^2-test.

6d CONTINGENCY TABLES

6d.1 The Probability of an Observed Configuration and Tests in Large Samples

If the individuals of a population can be described as belonging to one of r categories, $A_1, \ldots, A_r$ with respect to an attribute A, and to one of s categories, $B_1, \ldots, B_s$ with respect to an attribute B, and so on, then we have a frequency distribution of individuals in $r \times s \times \cdots$ classes, a typical class being represented by $A_i B_j \ldots$. If there are k attributes the arrangement described earlier is called a k-fold contingency table. No new distribution problems arise since a contingency table as considered in the present section is only a special case of a multinomial with $r \times s \times \cdots$ classes and the general results of the section **6b** apply.

In this section the various problems connected with two attributes are discussed, the treatment being similar in the general case. Let the observed frequency in the class $A_i B_j$ be denoted by n_{ij} and the probability by π_{ij}. Also let

$$n_{i1} + n_{i2} + \cdots + n_{is} = n_{i.} \qquad \pi_{i1} + \pi_{i2} + \cdots + \pi_{is} = \pi_{i.}$$
$$n_{1j} + n_{2j} + \cdots + n_{rj} = n_{.j} \qquad \pi_{1j} + \pi_{2j} + \cdots + \pi_{rj} = \pi_{.j}$$
$$n_{1.} + n_{2.} + \cdots = n_{.1} + n_{.2} + \cdots = n_{..}$$
$$\pi_{1.} + \pi_{2.} + \cdots = \pi_{.1} + \pi_{.2} + \cdots = 1.$$

The probability of the observed frequencies is that of multinomial in $r \times s$ classes

$$n_{..}! \prod \prod \frac{(\pi_{ij})^{n_{ij}}}{n_{ij}!} = n_{..}! \prod \frac{(\pi_{i.})^{n_{i.}}}{n_{i.}!} \times n_{..}! \prod \frac{(\pi_{.j})^{n_{.j}}}{n_{.j}!}$$
$$\times \frac{\prod n_{i.}! \prod n_{.j}!}{n_{..}!} \prod \prod \frac{1}{n_{ij}!} \left(\frac{\pi_{ij}}{\pi_{i.} \pi_{.j}} \right)^{n_{ij}} \tag{6d.1.1}$$

If $\pi_{ij} = \pi_{i.} \pi_{.j}$, a situation in which the attributes (events) A_i and B_j are independent, then (6d.1.1) becomes

$$n_{..}! \prod \frac{(\pi_{i.})^{n_{..}}}{n_{i.}!} \times n_{..}! \prod \frac{(\pi_{.j})^{n_{.j}}}{n_{.j}!} \times \frac{\prod n_{i.}! \prod n_{.j}!}{n_{..}!} \prod \prod \frac{1}{n_{ij}!}. \tag{6d.1.2}$$

The first two expressions give the probability of the marginal totals, and the third gives the probability of the class frequencies for fixed values of the marginal totals. The conditional probability of all n_{ij} given the marginal totals $n_{1.}, \ldots, n_{r.}$ and $n_{.1}, \ldots, n_{.s}$, is

$$\frac{\prod n_{i.}! \prod n_{.j}!}{n_{..}!} \prod \prod \frac{1}{n_{ij}!} \tag{6d.1.3}$$

It is interesting to observe that the expression (6d.1.3) is independent of the hypothetical values of the proportions $\pi_{i.}$, $\pi_{.j}$, provided, of course, that the attributes are independent.

In some situations, especially in designed experiments, one set of marginals is determined in advance. [Thus, for instance, we might choose a number of individuals and inoculate them against an infection. Another chosen number of individuals could be kept as controls. In each group (inoculated or not inoculated) we obtain the actual numbers of individuals infected and not infected from which a 2 × 2 contingency table can be set up.] In general, if the row totals are fixed in advance, then, assuming the same set of probabilities $\pi_1, \ldots, \pi_s$ for different categories in each row, the joint probability of the observations is

$$P(n_{11}, \ldots n_{rs}, | n_{1.}, \ldots, n_{r.}) = \prod_{i=1}^{r} \frac{n_{i.}!}{n_{i1}! \ldots n_{is}!} \pi_1^{n_{i1}} \ldots \pi_s^{n_{is}}. \tag{6d.1.4}$$

The probability of the marginal totals $n_{.j}$ in this case is

$$P(n_{.1}, \ldots, n_{.s} | n_{1.}, \ldots, n_{r.}) = \frac{n_{..}!}{n_{.1}! \ldots n_{.s}!} \pi_1^{n_{.1}} \ldots \pi_s^{n_{.s}}, \tag{6d.1.5}$$

which is obtained by summing (6d.1.4) over $n_{.j} = \sum_i n_{ij}$, for $j = 1, \ldots, s$. Hence the conditional probability of n_{ij} given the marginal totals is (6d.1.4)/(6d.1.5),

$$\frac{\prod n_{.i}! \prod n_{.j}!}{n_{..}!} \prod \prod \frac{1}{n_{ij}!} \tag{6d.1.6}$$

which is the same as the expression (6d.1.3) obtained in the general case.

6d.2 Test of Independence in a Contingency Table

If the probabilities π_{ij} of the cells in a contingency table are assigned, then to test the hypothesis that the data are in agreement with these hypothetical values, the statistic

$$\chi^2 = \sum_i \sum_j \frac{(n_{ij} - n_{..}\pi_{ij})^2}{n_{..}\pi_{ij}} = \sum \frac{(O - E)^2}{E} \quad \text{(over all classes)} \tag{6d.2.1}$$

can be used as χ^2 on $(rs - 1)$ D.F., the only restriction being $\sum\sum n_{ij} = n_{.}$ If the *attributes are independent*, the cell probabilities satisfy the relations

$$\pi_{ij} = \pi_{i.}\pi_{.j} \qquad \text{for all} \quad i \text{ and } j. \tag{6d.2.2}$$

How can this hypothesis be tested on the basis of the observed data? Two situations may arise:

1. The hypothetical probabilities $\pi_{i.}$ and $\pi_{.j}$ specifying the marginal distributions may be known, in which case we are required to examine whether the cell probabilities could be constructed by the law (6d.2.2).
2. The hypothetical proportions of the marginal frequencies not being known, we are required to test whether the attributes are independent.

In the first problem we have

$$\chi^2 = \sum\sum \frac{(n_{ij} - n_{..}\pi_{i.}\pi_{.j})^2}{n_{..}\pi_{i.}\pi_{.j}}, \qquad \text{D.F.} = rs - 1, \tag{6d.2.3}$$

which measures the overall discrepancy between the observed and the expected frequencies. From this we can single out two components

$$\chi_1^2 = \sum_i \frac{(n_{i.} - n_{..}\pi_{i.})^2}{n_{..}\pi_{i.}}, \qquad \text{D.F.} = r - 1, \tag{6d.2.4}$$

$$\chi_2^2 = \sum_j \frac{(n_{.j} - n_{..}\pi_{.j})^2}{n_{..}\pi_{.j}}, \qquad \text{D.F.} = s - 1, \tag{6d.2.5}$$

which measure the discrepancy between the observed marginal frequencies and the expected. With these statistics we can test whether the observed marginal totals are as expected. On subtracting χ_1^2 and χ_2^2 from the total, we obtain

$$\begin{aligned} \chi_3^2 &= \chi^2 - \chi_1^2 - \chi_2^2 \\ &= \sum\sum \frac{[n_{ij} - n_{i.}\pi_{.j} - n_{.j}\pi_{i.} + n_{..}\pi_{i.}\pi_{.j}]^2}{n_{..}\pi_{i.}\pi_{.j}}. \end{aligned} \tag{6d.2.6}$$

Since the total χ^2 of (6d.2.3) has $(rs - 1)$ D.F., χ_1^2 and χ_2^2 have $(r - 1)$ and $(s - 1)$ D.F.; and χ_3^2 is positive, it follows from an extension of the result [(iv), **3b.4**] that χ_3^2 has $(rs - 1) - (r - 1) - (s - 1) = (r - 1)(s - 1)$ D.F. The component (6d.2.6) can be used to test the departure from independence.

As an example, consider the fourfold contingency table with the marginal hypothetical proportions (p_1, q_1) for A and (p_2, q_2) for B.

	B_1	B_2	Total
A_1	a	b	$a + b$
A_2	c	d	$c + d$
Total	$a + c$	$b + d$	n

The total χ^2 is

$$\chi^2 = \frac{(a - np_1p_2)^2}{np_1p_2} + \frac{(b - np_1q_2)^2}{np_1q_2} + \frac{(c - nq_1p_2)^2}{nq_1p_2} + \frac{(d - nq_1q_2)^2}{nq_1q_2}, \ 3 \text{ D.F.}$$

The components are

$$\chi_1^2 = \frac{(a+b-np_1)^2}{np_1} + \frac{(c+d-nq_1)^2}{nq_1} = \frac{(a+b-np_1)^2}{np_1q_1}, \text{ 1 D.F.}$$

$$\chi_2^2 = \frac{(a+c-np_2)^2}{np_2} + \frac{(b+d-nq_2)^2}{nq_2} = \frac{(a+c-np_2)^2}{np_2q_2}, \text{ 1 D.F.}$$

$$\chi_3^2 = \frac{(aq_1q_2 - bp_2q_1 - cp_1q_2 + dp_1p_2)^2}{np_1p_2q_1q_2}, \text{ 1 D.F.}$$

Example 1. Table 6d.2α gives Bateson's distribution of sweet pea plants obtained from an intercross. The marginal frequencies are expected to be in the ratio 3 : 1, and if the two characters, flower color and pollen shape, are independently inherited, then the cell frequencies are expected to be in the ratio 9 : 3 : 3 : 1.

TABLE 6d.2α. Distribution of Plants by Pollen Shape and Flower Color

Pollen Shape	Flower Color: Purple	Flower Color: Red	Total
Long	296	27	323
Round	19	85	104
Total	315	112	427

It was seen (Example 1, **6b.5**) that the total χ^2 of discrepancy on 3 D.F. is 222.1221, which is very high. The first component is

$$\chi_1^2 = \frac{(a+b-np_1)^2}{np_1q_1} = \frac{(323 - \frac{3}{4} \times 427)^2}{427 \times \frac{1}{4} \times \frac{3}{4}} = 0.0945,$$

which is quite small for 1 D.F., showing that the single factor segregation with respect to the pollen shape is as expected. Similarly, $\chi_2^2 = 0.3443$ which again is quite small, showing that the single factor segregation with respect to flower color is as expected. The third component is

$$\chi_3^2 = \frac{(aq_1q_2 - bp_2q_1 - cp_1q_2 + dp_1p_2)^2}{np_1p_2q_1q_2}$$

$$= \frac{(296 - 27 \times 3 - 19 \times 3 + 85 \times 9)^2}{427 \times 3 \times 3 \times 1 \times 1} = 221.6833,$$

which is very high for 1 D.F. The total of the three χ^2 values is

$$0.0945 + 0.3443 + 221.6833 = 222.1221,$$

and thus agrees with the total calculated earlier. It is seen that the whole discrepancy is concentrated in one component with a single degree of freedom. This shows that the departure of the observed from the expected cell frequencies is due to the dependence of the characters inherited but not to single factor segregations. The success of all statistical tests lies in isolating the more efficient components for judging the points at issue.

Suppose the hypothetical values of the marginal probabilities are not known. Then we estimate their values on the hypothesis

$$\pi_{ij} = \pi_{i.}\pi_{.j}.$$

The m.l. estimates are

$$\hat{\pi}_{i.} = \frac{n_{i.}}{n_{..}} \qquad \text{and} \qquad \hat{\pi}_{.j} = \frac{n_{.j}}{n_{..}}.$$

These values may be substituted in the total χ^2,

$$\chi^2 = \sum\sum \frac{\left(n_{ij} - \dfrac{n_{i.}n_{.j}}{n_{..}}\right)^2}{\dfrac{n_{i.}n_{.j}}{n_{..}}}. \tag{6d.2.7}$$

Since $(r-1)+(s-1)$ parameters have been estimated, the χ^2 of (6d.2.7) has $(rs-1)-(r-1)-(s-1)=(r-1)(s-1)$ degrees of freedom, using the result on the goodness of fit χ^2 distribution derived in **6b.2**. At the estimated values of $\pi_{i.}$ and $\pi_{.j}$, the components χ_1^2 and χ_2^2 have zero values so that $\chi_3^2 = \chi^2$. Thus χ_3^2 measures the departure from independence.

This test is useful in two situations:

1. Suppose that in Bateson's problem of the segregation of factors it is found that the marginal frequencies deviate significantly from the expected. This indicates that the assigned marginal probabilities may not be correct. In fact, if the single factor segregations are disturbed owing to unequal viability of the two types of plants, the marginal frequencies will not be in the ratio 3 : 1. In such a case the third component χ_3^2 of (6d.2.6) loses its validity, or, in other words, the significance of χ_3^2 may result from the use of wrong proportions. The best course is then to substitute the estimated proportions and use the test (6d.2.7) obtained earlier.
2. The second situation occurs when nothing is specified about the marginal proportions. In this case (6d.2.7) supplies the test for independence.

In a 2 × 2 table when the hypothetical marginal proportions are not known, the χ^2 test (6d.2.7) of independence is

$$\frac{[a-(a+b)(a+c)/n]^2}{(a+b)(a+c)/n}+\frac{[b-(a+b)(b+d)/n]^2}{(a+b)(b+d)/n}+\cdots+\cdots,$$

which reduces to

$$n(ad-bc)^2 \div (a+c)(b+d)(a+b)(c+d)$$

with the numerical value

$$\frac{427(296\times 85-27\times 19)^2}{315\times 112\times 323\times 104}=218.8722$$

on 1 D.F. in Bateson's problem of the segregation of factors. The χ^2 value of 218.8722 is lower than that obtained by using the hypothetical values of the marginal proportions. Such discrepancies will not in general lead to contradictory conclusions. The earlier test makes use of the information supplied by a total of 427 plants whereas the latter makes use of the information supplied by the set of configurations having the same marginal totals. Some marginal totals are more informative than the average, whereas others are less.

Example 2. The data of Table 6d.2β give the number of skulls excavated from a locality in three different seasons and the sex distribution as sexed by investigator A working in the first two seasons and by investigator B working in the third season. What information do the data supply on the sex ratio of the population buried in the locality?

TABLE 6d.2β. The Distribution of Skulls by Sex and Seasons

	Season			Total
	First	Second	Third	
♂	162	180	210	552
♀	110	125	200	435

How do we approach this problem? First, it is natural to examine whether the sex ratio is 1 : 1 in each season. The expected values under such a hypothesis are given in Table 6d.2γ. The χ^2 for the first season is

$$\frac{(162-136)^2}{136}+\frac{(110-136)^2}{136}=9.9412,$$

and similarly the χ^2's for the second and third seasons are 9.9180 and 0.2440 giving a total χ^2 of 20.1032 on 3 D.F. The χ^2 is too high, indicating departure from the 1 : 1 sex ratio. Individually the deviations in the first two seasons ($\chi^2 = 9.9412$ and 9.9180 on 1 D.F. each) are significant. The χ^2 from the marginal total measuring the deviation from the 1 : 1 sex ratio is

$$\frac{(552)^2}{493.5} + \frac{(435)^2}{493.5} - 987 = 13.8692,$$

which leaves a $\chi^2 = 20.1032 - 13.8692 = 6.2340$ on 2 D.F. for testing the differences in the ratio in the three seasons. This is no doubt significant at the 5% level, showing differences in sex ratio, but the test is not strictly valid owing to the fact that the marginal totals are not compatible with sex ratio 1 : 1, the $\chi^2 = 13.8692$ on 1 D.F. being large. Since we have observed

TABLE 6d.2γ. The Expected Values under the Hypothesis of Equal Sex Ratios

	Season		
	First	Second	Third
♂	136	152.5	205
♀	136	152.5	205

that there is an overall discrepancy from the 1 : 1 sex ratio, we might ask whether the sex ratio is the same for the three seasons although it may not be 1 : 1. A straight test of independence with fixed marginals (6d.2.7) or of homogeneity of parallel samples (6c.2.2) can now be calculated This gives a χ^2 equal to 6.3222 on 2 D.F., showing significant differences in sex ratios in the three seasons. The agreement of this χ^2 with the earlier value of 6.2340 obtained by using the hypothetical ratios is, perhaps, accidental.

We must be careful in drawing conclusions from data of this nature. It must be observed that the skulls are sexed by a subjective method of anatomical appreciation, and different investigators may have different methods of sexing, leading to different ratios. The observed discrepancy in the sex ratios for different seasons may be due to the result of there being a different investigator in the third season. The proportions of males as observed are:

	Investigator A		Investigator B	
Season	First	Second	Third	Overall
Proportion	0.5956	0.5902	0.5122	0.5593

The discrepancy between the two investigators is then tested by the χ^2 test of independence from the 2×2 table:

	First and Second Seasons	Third Seasons
♂	342	210
♀	235	200

The χ^2 on 1 D.F. is 6.3055, which is significant. Thus, out of a total χ^2 of 6.3222 on 2 D.F., measuring the differences in sex ratios, 1 degree alone accounts for 6.3055, which shows that the whole discrepancy between seasons is due only to the differences between investigators.

What then are our conclusions? There is difference between the investigators in the technique of sexing the skulls, assuming that the investigators had worked in the same regions of the locality, so that the observations of the two investigators are not comparable.

The data also suggest a preponderence of males but it is not known to what extent the observed deviation from 1 : 1 is due to a wrong technique of sexing of the skulls. There is also the question of differential preservation of male and female skulls, the latter being more delicate and fragile. *It appears that "we can never solve a problem without creating ten more."*

Example 3. Consider the following data, collected from a number of schools, regarding speech defects (S_1, S_2, S_3) and physical defects (P_1, P_2, P_3) of school children.

	S_1	S_2	S_3	Total
P_1	45	26	12	83
P_2	32	50	21	103
P_3	4	10	17	31
Total	81	86	50	217

The expected values on the hypothesis of independence are:

30.982	32.894	19.124
38.447	40.820	23.733
11.571	12.286	7.143

The χ^2 on $2 \times 2 = 4$ D.F. is 34.8828, which is significant at the 1 % level. It is seen that, although the frequency in one cell is as small as 4, the expected frequency is large enough for the χ^2 approximation to hold. The frequency also should be large enough, however, for the approximation to be good. In such cases it is reasonable to combine two cells by adding their frequencies and treating them as one cell for purposes of tests of significance. In the foregoing example, 4 and 10 may be added to yield an observed frequency of 14 with the corresponding expected value $11.571 + 12.286 = 23.857$. The new χ^2 computed by the formula $\sum (0 - E)^2/E$ over the 8 cells is 33.5763. But the distribution to which this value can be referred is no longer a χ^2, although it has been the practice to refer it to a χ^2 distribution with 1 D.F. less, that is, on $4 - 1 = 3$ D.F. Although the summation is now taken over one cell less, theoretically 1 D.F. is not lost. So to use the new χ^2 on $(4 - 1) = 3$ D.F. is to overestimate significance whereas to consider it as on 4 D.F. is to underestimate significance. Although definite conclusions can be drawn either when the new χ^2 is not significant on 3 D.F. or when it is significant on 4 D.F., as in the present case, it is not possible to infer about the actual level of significance when the new χ^2 is significant on 3 D.F. and not on 4 D.F.

The appropriate distribution of the χ^2 in such cases has been investigated by Chernoff and Lehmann (1954). They show it to be the distribution of a linear combination of χ^2 variables. Since it is difficult to use such a distribution the following approach is suggested. We first obtain a new set of expected values by considering the cells S_1P_3 and S_2P_3 as constituting a single cell. If p_1, q_1, r_1 and p_2, q_2, r_2 are the marginal proportions for physical defects and speech defects, the probability of the observed frequencies on the hypothesis of independence is proportional to

$$(p_1p_2)^{45}(p_1q_2)^{26}(p_1r_2)^{12}\cdots[(p_2+q_2)r_1]^{14}(r_1r_2)^{17}.$$

The maximum likelihood equations for p_1, q_1, p_2, q_2 are

$$\begin{aligned} 217p_1 &= 83, & 217q_1 &= 103 \\ 76p_2 &= 77q_2, & 217(1 - p_2 - q_2) &= 50 \end{aligned}$$

which give the estimates

$$\hat{p}_2 = 0.387308, \qquad \hat{q}_2 = 0.382278, \qquad \hat{r}_2 = 0.230414.$$

The estimates of p_1, q_1, r_1 are the same as before. The new expectations are

32.147	31.729	19.124
39.893	39.375	23.733
(12.006 + 11.851)		7.143

and the χ^2 on $(7 - 4)$ D.F. is 31.2472. This test is valid in the sense that when significance is noted the hypothetis of independence is rejected.

Fisher recommended a test based on likelihood (see **6e**), which is more appropriate when the cell frequencies are small. This is defined by

$$L = 2 \sum\sum n_{ij} \log_e n_{ij} - \sum n_{i.} \log_e n_{i.} - \sum n_{.j} \log_e n_{.j} + n_{..} \log_e n_{..}.$$

Here, L is approximately distributed as χ^2 on $(r-1)(s-1)$ D.F. The value of L in the foregoing problem is 30.4448, which is significant on 4 D.F. Even the use of L requires a large sample.

It must be emphasized that the object of a test is first to establish departure from independence in a general way. For this it is enough to use a valid test which is simple to compute. If necessary more refined tests may be used to examine different portions of the contingency table. For instance, we may inquire whether the two physical defects P_2 and P_3 and the speech defects S_1 and S_2 are associated. This needs a refined technique as discussed in the next section.

6d.3 Tests of Independence in Small Samples

In testing for any hypothesis specifying some relations satisfied by the parameters, we see that the unknown parameters do not enter into the large sample distribution of the χ^2 statistic. But in small samples it might happen that the χ^2 approximation breaks down and/or the unknown parameters appear in the exact distribution of χ^2. In the latter, no exact test of significance is possible, owing to the presence of the unknown parameters, called nuisance parameters, in the probability distribution.

One way to remove the nuisance parameters is to compare the particular observed sample, not with the whole population of samples with which a comparison might be made if the exact values of the nuisance parameters were known, but with a subpopulation selected with reference to the sample in such a way that the distribution of a statistic in this subpopulation does not involve any unknown parameter. For instance, it is shown that on the hypothesis of independence of two attributes the probability of cell frequencies given the marginal totals, is

$$\frac{\prod n_{i.}! \prod n_{.j}!}{n_{..}!} \prod\prod \frac{1}{n_{ij}!},$$

which does not contain the hypothetical values $\pi_{i.}$ and $\pi_{.j}$, thus admitting the possibility of determining the exact probabilities in tests of independence.

Example 1. Do the following data on sociability (S) and nonsociability (NS) of soldiers recruited in cities (C) and villages (V) suggest that the city soldiers are more sociable than the village soldiers?

	S	NS	Total
C	13	4	17
V	6	14	20
Total	19	18	37

The smaller frequencies in one diagonal suggest that city soldiers are more sociable. But it must be ascertained whether such a configuration as the observed and those indicating a higher degree of sociability of the city recruits can occur by chance if in fact there was no difference in the sociabilities of the city and the village recruits. Since for fixed marginals the probability of a given configuration a, b, c, d in the four cells is

$$\frac{17!\,20!\,19!\,18!}{37}\,\frac{1}{a!\,b!\,c!\,d!},$$

we find that the probabilities for configurations with 4, 3, 2, 1 and 0 in the northeast corner cell (these being less favorable to the hypothesis of independence and more to the alternative suggested) are, respectively, 0.0^25218, 0.0^35966, 0.0^43607, 0.0^51097, and 0.0^71075, adding up to 0.0059. The chances are very small, thus indicating that city soldiers are more sociable than village soldiers.

If the cell frequencies are not small, this result could be established by calculating a χ^2 for testing independence and determining the probability of a normal deviate with zero mean and unit variance exceeding χ. In the foregoing example,

$$\chi^2 = \frac{37(13 \times 14 - 6 \times 4)^2}{17 \times 20 \times 19 \times 18} = 7.9435,$$

$$\chi = \sqrt{7.9435} = 2.8181,$$

so that the normal probability is 0.0025, which is smaller than the actual value 0.0059, the discrepancy being due to the smallness of the sample.

Yates (1934) suggested that by calculating χ^2 from a table obtained by increasing the smaller frequency* by 1/2 without altering the marginal totals a closer approximation to the actual probability is realized. In the present example, the new χ^2, said to be corrected for continuity, is

$$\chi^2 = \frac{37(12.5 \times 13.5 - 6.5 \times 4.5)^2}{17 \times 20 \times 19 \times 18} = 6.1922.$$

* The reference is to the smaller frequency in the diagonal (6, 4) under consideration, indicating nonsociability of the village recruits. In the general test of independence, $\frac{1}{2}$ is to be added to a frequency to obtain a smaller χ^2.

The value of χ is 2.4884, so that the normal probability is 0.0064, which is closer to the actual value than in the case of the uncorrected χ^2.

A slightly different method suggested by V. M. Dandekar involves the calculation of χ_0^2, χ_{-1}^2, and χ_1^2 for the observed configuration and those obtained by increasing and decreasing the smallest frequency by unity. From these a corrected χ^2 can be obtained by the formula

$$\chi_c^2 = \chi_0^2 - \frac{\chi_0^2 - \chi_{-1}^2}{\chi_1^2 - \chi_{-1}^2}(\chi_1^2 - \chi_0^2)$$

In the present example, $\chi_0^2 = 7.9435$, $\chi_1^2 = 12.0995$, $\chi_{-1}^2 = 4.6587$, and

$$\chi_c^2 = 7.9435 - \frac{7.9435 - 4.6587}{12.0995 - 4.6587}(12.0995 - 7.9435) = 6.1086$$

$$\chi_c = 2.4715.$$

The normal probability is 0.0068, which is also close to the actual value.

The likelihood test

$$L = 2 \sum O \log_e \frac{E}{O}$$

in this case gives the value 8.2811. The value of χ is $\sqrt{8.2811} = 2.8778$, so that the probability is much smaller than the actual value. Thus the likelihood test does not improve the situation.

Example 2. In Example 1, the object of investigation was to study whether city soldiers are more sociable than village soldiers. For this purpose we considered the deviations from the expected in one way only. But, in general, if the object is to discover association between two attributes without specifying the nature of the association, it is necessary to consider deviations from the expected in either direction. The configurations giving a χ^2 value higher than the observed are not only those with 3, 2, 1, 0 frequency in the northeast cell but also 4, 3, 2, 1, 0 in the northwest cell. The probabilities for the latter are, respectively $0.0^2 2088$, $0.0^3 1864$, $0.0^5 8733$, $0.0^6 1828$, and $0.0^8 1132$ adding up to 0.0023. The probability of small frequencies in the northeast cell as calculated earlier is 0.0059 giving a total probability of $0.0059 + 0.0023 = 0.0082$, which is small, thus indicating departure from independence.

If the sample is large, we can find the probability by directly entering the uncorrected χ^2 in a χ^2 table on 1 D.F. In the present example χ^2 is 7.9435 with the associated probability of about 0.005, which is smaller than the actual value 0.0082.

Using Yate's correction for continuity, the χ^2 is 6.1922, with the corresponding probability 0.0128, which is higher than the actual value.

To extend Dandekar's correction to this example we first note that the χ^2 values below and above the observed $\chi^2 = 7.9435$ are 6.0598 and 9.7448 corresponding to the partitions

$$\begin{matrix} 5 & 12 \\ 14 & 6 \end{matrix} \quad \text{and} \quad \begin{matrix} 4 & 13 \\ 15 & 5 \end{matrix}$$

The corrected χ^2 is

$$\begin{aligned} \chi_c^2 &= 7.9435 - \frac{7.9435 - 6.0598}{9.7448 - 6.0598}(9.7448 - 7.9435) \\ &= 7.0228, \end{aligned}$$

which gives a probability 0.0082, almost exactly equal to the actual value. In general, Dandekar's correction is slightly better than that of Yates although the correction is simpler in Yates' method.

In testing for linkage on the basis of data classified according to two factors, it is enough to test for association one way if it is known that the recombination fraction is less than 1/2. It is now known that the recombination fraction can exceed 1/2, as demonstrated by Fisher. So it is better first to disprove the hypothesis of independence without inquiring as to the nature of association. Furthermore, it must be noted that departure from independence may occur in experimental data owing to various causes, and it is better to use a test which gives a direct appraisal of the data as to its compatibility with the hypothesis of independence.

For further examples illustrating the use of χ^2 tests on categorical data see Fisher (1925).

6e SOME GENERAL CLASSES OF LARGE SAMPLE TESTS

6e.1 Notations and Basic Results

Let $x_1, \ldots, x_n$ be i.i.d. random variables, which may be multidimensional and let $p(x, \boldsymbol{\theta})$ be the common probability density depending on a q dimensional parameter $\boldsymbol{\theta}' = (\theta_1, \ldots, \theta_q)$. The admissible set of $\boldsymbol{\theta}$ is taken to be a nondegenerate interval of the q-dimensional Euclidean space. The log likelihood of n independent observations and the ith efficient score as defined in **5f** and **5g** are

$$l(\boldsymbol{\theta}) = \log p(x_1, \boldsymbol{\theta}) + \cdots + \log p(x_n, \boldsymbol{\theta})$$

$$\begin{aligned} \phi_i(\boldsymbol{\theta}) &= \frac{1}{\sqrt{n}} \frac{\partial l}{\partial \theta_i} \\ &= \frac{1}{\sqrt{n}} \left\{ \frac{1}{p(x_1, \boldsymbol{\theta})} \frac{\partial p(x_1, \boldsymbol{\theta})}{\partial \theta_i} + \cdots + \frac{1}{p(x_n, \boldsymbol{\theta})} \frac{\partial p(x_n, \boldsymbol{\theta})}{\partial \theta_i} \right\}. \end{aligned} \tag{6e.1.1}$$

Let $\mathcal{I}$ be the information matrix on $\boldsymbol{\theta}$ in a single observation x, that is, if $\mathcal{I} = (\mathcal{I}_{rs})$, then

$$\mathcal{I}_{rs} = E\left\{\frac{1}{p(x, \boldsymbol{\theta})}\frac{\partial p}{\partial \theta_r} \cdot \frac{1}{p(x, \boldsymbol{\theta})}\frac{\partial p}{\partial \theta_s}\right\}. \tag{6e.1.2}$$

(i) *Let* $\mathbf{V}' = (\phi_1(\boldsymbol{\theta}), \ldots, \phi_q(\boldsymbol{\theta}))$. *Then the a.d. of* $\mathbf{V}$ *is* q *variate normal with mean zero and dispersion matrix* $\mathcal{I}$.

To prove the result, we need only show that the a.d. of a linear function $\mathbf{b}'\mathbf{V} = b_1\phi_1(\boldsymbol{\theta}) + \cdots + b_q\phi_q(\boldsymbol{\theta})$ is univariate normal with mean zero and variance $\mathbf{b}'\mathcal{I}\mathbf{b}$. Consider

$$y_i = \frac{b_1}{p(x_i, \boldsymbol{\theta})}\frac{\partial p(x_i, \boldsymbol{\theta})}{\partial \theta_1} + \cdots + \frac{b_q}{p(x_i, \boldsymbol{\theta})}\frac{\partial p(x_i, \boldsymbol{\theta})}{\partial \theta_q}. \tag{6e.1.3}$$

Then $E(y_i) = 0$ and

$$\begin{aligned} V(y_i) &= \sum\sum b_r b_s E\left\{\frac{1}{p(x_i, \boldsymbol{\theta})}\frac{\partial p}{\partial \theta_r} \cdot \frac{1}{p(x_i, \boldsymbol{\theta})}\frac{\partial p}{\partial \theta_s}\right\} \\ &= \sum\sum b_r b_s \mathcal{I}_{rs} = \mathbf{b}'\mathcal{I}\mathbf{b}. \end{aligned}$$

Thus $y_1, \ldots, y_n$ are i.i.d. variables each with mean zero and variance $\mathbf{b}'\mathcal{I}\mathbf{b}$. Hence the a.d. of $(y_1 + \cdots + y_n)/\sqrt{n}$ is $N(0, \mathbf{b}'\mathcal{I}\mathbf{b})$ as $n \to \infty$. But

$$(y_1 + \cdots + y_n)/\sqrt{n} = \mathbf{b}'\mathbf{V}. \tag{6e.1.4}$$

Hence the result of (i) is established.

(ii) *Let* $p(x, \boldsymbol{\theta})$ *satisfy the regularity conditions of* **5f.1**. *Denote by* $\bar{\boldsymbol{\theta}}$ *the consistent root of the likelihood equation based on* n *observations and by* $\mathbf{D}'$ *the vector of deviations* $[\sqrt{n}(\bar{\theta}_1 - \theta_1), \ldots, \sqrt{n}(\bar{\theta}_q - \theta_q)]$. *Furthermore, let* $\mathcal{I}$, *the information matrix, be nonsingular. Then we have the following asymptotic equivalences.*

$$\mathbf{V} \stackrel{a}{=} \mathcal{I}\mathbf{D} \quad \text{or} \quad \mathbf{D} \stackrel{a}{=} \mathcal{I}^{-1}\mathbf{V}, \tag{6e.1.5}$$

$$2[l(\bar{\boldsymbol{\theta}}) - l(\boldsymbol{\theta})] \stackrel{a}{=} \mathbf{D}'\mathcal{I}\mathbf{D}. \tag{6e.1.6}$$

The results (6e.1.5) and (6e.1.6) are easily established following the arguments of **5f.2**. In most of the examples the maximum likelihood estimator is found to be the consistent root of the likelihood equation, and we shall assume that they are the same in the rest of the discussion of the present section.

(iii) *Suppose that the* q *parameters* $\theta_1, \ldots, \theta_q$ *are subject to* $k < q$ *restrictions* $R_1(\boldsymbol{\theta}) = 0, \ldots, R_k(\boldsymbol{\theta}) = 0$, *where* $R_1, \ldots, R_k$ *admit continuous partial derivatives of the first order. Let these restrictions be equivalent to specifying each* θ_i *as a function of* $s = q - k$ *new parameters,* $\beta_1, \ldots, \beta_s$,

$$\theta_i = g_i(\beta_1, \ldots, \beta_s)$$

where g_i *admit continuous partial derivatives of the first order. Let* $\hat{\boldsymbol{\beta}}$ *be the m.l. estimator of* $\boldsymbol{\beta}$ *and denote by*

$$\Psi_i(\boldsymbol{\beta}) = \partial l/\sqrt{n}\, \partial\beta_i, \qquad \mathbf{U}' = [\Psi_1(\boldsymbol{\beta}), \ldots, \Psi_s(\boldsymbol{\beta})],$$
$$\mathbf{F}' = [\sqrt{n}(\hat{\beta}_1 - \beta_1)], \ldots, \sqrt{n}(\hat{\beta}_s - \beta_s)],$$

and by $\mathbf{H}$ *the information matrix on* $\boldsymbol{\beta}$ *in a single observation x. Then we have the following:*

(a) $\mathbf{U} \stackrel{u}{=} \mathbf{HF}$ or $\mathbf{H}^{-1}\mathbf{U} \stackrel{u}{=} \mathbf{F}$, applying (6e.1.5), (6e.1.7)
(b) $2[l(\hat{\boldsymbol{\beta}}) - l(\boldsymbol{\beta})] \stackrel{u}{=} \mathbf{F}'\mathbf{HF}$, (6e.1.8)
(c) $\mathbf{U} = \mathbf{M}'\mathbf{V}$ and $\mathbf{H} = \mathbf{M}'\boldsymbol{\mathscr{I}}\mathbf{M}$, (6e.1.9)

where $\mathbf{M} = (\partial g_i/\partial\beta_j)$ *is a matrix of order* $q \times s$.

Result (b) is similar to (6e.1.6) and it is easy to verify (c).

6e.2 Test of a Simple Hypothesis

Under the setup of **6e.1**, to test the hypothesis $\boldsymbol{\theta} = \boldsymbol{\theta}_0$ (an assigned value), Neyman and Pearson (1928) proposed the likelihood ratio criterion

$$\Lambda = \frac{L(\boldsymbol{\theta}_0)}{\sup\limits_{\theta} L(\boldsymbol{\theta})} \tag{6e.2.1}$$

where the supremum is taken over the admissible set of $\boldsymbol{\theta}$ values. Taking logarithms and using the m.l. estimator $\hat{\boldsymbol{\theta}}$ of $\boldsymbol{\theta}$, the test criterion (6e.2.1) may be written

$$-2 \log_e \Lambda = 2[l(\hat{\boldsymbol{\theta}}) - l(\boldsymbol{\theta}_0)]. \tag{6e.2.2}$$

Let $\mathbf{D}_0$, $\mathbf{V}_0$, $\boldsymbol{\mathscr{I}}_0$ denote the values of $\mathbf{D}$, $\mathbf{V}$, $\boldsymbol{\mathscr{I}}$ when $\boldsymbol{\theta}_0$ is substituted for $\boldsymbol{\theta}$ and $\boldsymbol{\mathscr{I}}(\hat{\boldsymbol{\theta}})$, the value of $\boldsymbol{\mathscr{I}}$ at $\boldsymbol{\theta} = \hat{\boldsymbol{\theta}}$. To test the hypothesis $\boldsymbol{\theta} = \boldsymbol{\theta}_0$, Wald (1943) proposed the criterion

$$W_0 = \mathbf{D}_0'[\boldsymbol{\mathscr{I}}(\hat{\boldsymbol{\theta}})]\mathbf{D}_0, \tag{6e.2.3}$$

and the author (Rao, 1947e) proposed the criterion based on efficient scores

$$S_0 = \mathbf{V}_0'\boldsymbol{\mathscr{I}}_0^{-1}\mathbf{V}_0, \tag{6e.2.4}$$

which does not require the explicit computation of the m.l. estimates. From (6e.1.6), we see that

$$2[l(\hat{\boldsymbol{\theta}}) - l(\boldsymbol{\theta}_0)] \overset{a}{=} \mathbf{D}_0' \mathscr{I}_0 \mathbf{D}_0 \overset{a}{=} \mathbf{D}_0'[\mathscr{I}(\hat{\boldsymbol{\theta}})]\mathbf{D}_0 .$$

and from (6e.1.5)

$$\mathbf{D}_0' \mathscr{I}_0 \mathbf{D}_0 \overset{a}{=} \mathbf{V}_0' \mathscr{I}_0^{-1} \mathbf{V}_0$$

so that the three statistics $-2 \log \Lambda_0$, W_0, and S_0 are equivalent in large samples, when the null hypothesis is true. From [(i), **6e.1**], the a.d. of $\mathbf{V}$ is q-variate normal with mean zero and dispersion matrix $\mathscr{I}$. Hence the a.d. of $\mathbf{V}_0'\mathscr{I}_0^{-1}\mathbf{V}_0$ is that of $\chi^2(q)$, and since all three statistics are equivalent to $\mathbf{V}_0'\mathscr{I}_0^{-1}\mathbf{V}_0$ they have the same a.d.

There is no adequate discussion as yet in statistical literature on the relative merits of these tests in detecting departures from the null hypothesis.

6e.3 Test of a Composite Hypothesis

Let the composite hypothesis specify θ_i, $i = 1, \ldots, q$ as functions of $\beta_1, \ldots, \beta_s$,

$$\theta_i = g_i(\beta_1, \ldots, \beta_s), \tag{6e.3.1}$$

or, alternatively let θ_i be subject to $k = q - s$ restrictions

$$R_i(\boldsymbol{\theta}) = 0, \qquad i = 1, \ldots, k, \tag{6e.3.2}$$

where g_i and R_i are functions admitting continuous partial derivatives of the first order. Let $\hat{\boldsymbol{\theta}}$ be the unrestricted and $\boldsymbol{\theta}^*$ the restricted m.l. estimators of $\boldsymbol{\theta}$. In terms of the m.l. estimator $\hat{\boldsymbol{\beta}}$ of $\boldsymbol{\beta}$,

$$\theta_i^* = g_i(\hat{\beta}_1, \ldots, \hat{\beta}_s). \tag{6e.3.3}$$

The Neyman-Pearson likelihood ratio test for a composite hypothesis is

$$-2 \log_e \Lambda_c = 2[l(\hat{\boldsymbol{\theta}}) - l(\boldsymbol{\theta}^*)] = 2[l(\hat{\boldsymbol{\theta}}) - l(\hat{\boldsymbol{\beta}})], \tag{6e.3.4}$$

where $l(\hat{\boldsymbol{\beta}})$ is computed with the log likelihood considered as a function of $\boldsymbol{\beta}$. It may be seen that

$$\Lambda_c = \frac{\sup_{R_i(\boldsymbol{\theta})=0} L(\boldsymbol{\theta})}{\sup_{\boldsymbol{\theta}} L(\boldsymbol{\theta})}, \tag{6e.3.5}$$

where in the numerator the supremum is when $\boldsymbol{\theta}$ is subject to the restrictions $R_i(\boldsymbol{\theta}) = 0$, $i = 1, \ldots, k$ and in the denominator, the supremum is over all admissible values of $\boldsymbol{\theta}$.

The efficient score test of Rao for the composite hypothesis is

$$S_c = \mathbf{V}^{*\prime}\mathscr{I}^{*-1}\mathbf{V}^*, \qquad \text{where} \quad \mathbf{V}^* = \mathbf{V}(\boldsymbol{\theta}^*) \text{ etc.}, \tag{6e.3.6}$$

and Wald's test is

$$W_c = \sum\sum \lambda^{ij}(\hat{\boldsymbol{\theta}})R_j(\hat{\boldsymbol{\theta}})R_i(\hat{\boldsymbol{\theta}}), \tag{6e.3.7}$$

which uses only the unrestricted m.l. estimators of $\boldsymbol{\theta}$. In the expression (6e.3.7), (λ^{ij}) is the reciprocal of (λ_{ij}), the asymptotic dispersion matrix of $R_i(\hat{\boldsymbol{\theta}})$, defined by

$$(\lambda_{ij}) = \mathbf{T}'\boldsymbol{\mathscr{I}}^{-1}\mathbf{T}, \qquad \mathbf{T} = \left(\frac{\partial R_i(\boldsymbol{\theta})}{\partial \theta_j}\right), \qquad q \times k \text{ matrix.} \tag{6e.3.8}$$

All three statistics (6e.3.5) to (6e.3.7), have the same a.d. $\chi^2(q - s)$. From (6e.1.6), we have

$$2[l(\hat{\boldsymbol{\theta}}) - l(\boldsymbol{\theta})] \stackrel{a}{=} \mathbf{D}'\boldsymbol{\mathscr{I}}\mathbf{D} \stackrel{a}{=} \mathbf{V}'\boldsymbol{\mathscr{I}}^{-1}\mathbf{V}$$

and from (6e.1.7) to (6e.1.9), we see that

$$2[l(\hat{\boldsymbol{\beta}}) - l(\boldsymbol{\beta})] \stackrel{a}{=} \mathbf{F}'\mathbf{H}\mathbf{F} \stackrel{a}{=} \mathbf{U}'\mathbf{H}^{-1}\mathbf{U} \stackrel{a}{=} \mathbf{V}'\mathbf{M}\mathbf{H}^{-1}\mathbf{M}'\mathbf{V}$$

$$2[l(\hat{\boldsymbol{\theta}}) - l(\hat{\boldsymbol{\beta}})] \stackrel{a}{=} \mathbf{V}'(\boldsymbol{\mathscr{I}}^{-1} - \mathbf{M}\mathbf{H}^{-1}\mathbf{M}')\mathbf{V}.$$

Therefore,

$$2[l(\hat{\boldsymbol{\theta}}) - l(\hat{\boldsymbol{\beta}})] \stackrel{a}{=} \mathbf{V}'(\boldsymbol{\mathscr{I}}^{-1} - \mathbf{M}\mathbf{H}^{-1}\mathbf{M}')\mathbf{V}. \tag{6e.3.9}$$

The necessary and sufficient condition that the right-hand side of (6e.3.9) has a χ^2 distribution is

$$(\boldsymbol{\mathscr{I}}^{-1} - \mathbf{M}\mathbf{H}^{-1}\mathbf{M}')\boldsymbol{\mathscr{I}}(\boldsymbol{\mathscr{I}}^{-1} - \mathbf{M}\mathbf{H}^{-1}\mathbf{M}') = (\boldsymbol{\mathscr{I}}^{-1} - \mathbf{M}\mathbf{H}^{-1}\mathbf{M}'). \tag{6e.3.10}$$

The left-hand side of (6e.3.10) is

$$\begin{aligned}\boldsymbol{\mathscr{I}}^{-1}\boldsymbol{\mathscr{I}}\boldsymbol{\mathscr{I}}^{-1} - \boldsymbol{\mathscr{I}}^{-1}\boldsymbol{\mathscr{I}}\mathbf{M}\mathbf{H}^{-}\mathbf{M}^{1\prime} - \mathbf{M}\mathbf{H}^{-1}\mathbf{M}'\boldsymbol{\mathscr{I}}\boldsymbol{\mathscr{I}}^{-1} + \mathbf{M}\mathbf{H}^{-1}(\mathbf{M}'\boldsymbol{\mathscr{I}}\mathbf{M})\mathbf{H}^{-1}\mathbf{M}\\ = \boldsymbol{\mathscr{I}}^{-1} - 2\mathbf{M}\mathbf{H}^{-1}\mathbf{M}' + \mathbf{M}\mathbf{H}^{-1}\mathbf{H}\mathbf{H}^{-1}\mathbf{M}' = \boldsymbol{\mathscr{I}}^{-1} - \mathbf{M}\mathbf{H}^{-1}\mathbf{M}',\end{aligned}$$

using (6e.1.9), so that the condition (6e.3.10) is satisfied. The degrees of freedom of χ^2 is

$$\begin{aligned}\text{trace}\,(\boldsymbol{\mathscr{I}}^{-1} - \mathbf{M}\mathbf{H}^{-1}\mathbf{M}')\boldsymbol{\mathscr{I}} &= \text{trace}\,(\mathbf{I} - \mathbf{M}\mathbf{H}^{-1}\mathbf{M}'\boldsymbol{\mathscr{I}})\\ &= q - \text{trace}\,\mathbf{M}\mathbf{H}^{-1}\mathbf{M}'\boldsymbol{\mathscr{I}}\\ &= q - \text{trace}\,\mathbf{H}^{-1}\mathbf{M}'\boldsymbol{\mathscr{I}}\mathbf{M} = q - \text{trace}\,\mathbf{H}^{-1}\mathbf{H}\\ &= q - s.\end{aligned}$$

We have thus proved the desired result for the Neyman-Pearson statistic (6e.3.4). The a.d. of the statistics S_c and W_c can be established in a similar manner.

In practice we need more general results than those considered in **6e**. For instance, the observations $x_1, \ldots, x_n$ may not have identical distributions.

In such a case we can define the test criteria Λ, W, S in exactly the same way by considering the appropriate likelihood function and the efficient scores. The derivation of the asymptotic distributions need some further conditions on the probability densities to ensure the applicability of the multivariate central limit theorem to nonidentically distributed random variables and to ensure the desired properties (6e.1.5) of the maximum likelihood estimators.

6f ORDER STATISTICS

6f.1 The Empirical Distribution Function

Let $x_1, \ldots, x_n$ be a sample of n independent observations on a random variable X with d.f. $F(x)$. Let $x_{(i)}$ be an observation such that the number of observations $<x_{(i)}$ is $\leqslant(i-1)$ and the number of observations $>x_{(i)}$ is $\leqslant(n-i)$. Then the observations can be arranged in increasing order, allowing repetitions and indexed by distinct numbers so that

$$x_{(1)} \leqslant x_{(2)} \leqslant \cdots \leqslant x_{(n)}, \tag{6f.1.1}$$

where $x_{(1)}$ is the minimum and $x_{(n)}$ the maximum of the observations.

The function $X_{(i)}$ of sample values such that $X_{(i)}(x_1, \ldots, x_n) = x_{(i)}$ is called the ith order statistic.

Let us define a function $S_n(\cdot)$ or more precisely $S_n(\cdot \mid x_1, \ldots, x_n)$ such that

$$S_n(x) = \begin{cases} 0, & x \leqslant x_{(1)} \\ \dfrac{k}{n}, & x_{(r)} < x \leqslant x_{(k+1)}, \text{ if } x_{(r-1)} < x_{(r)} = x_{(r+1)} = \cdots = x_{(k)} < x_{(k+1)} \\ 1, & x_{(n)} < x \end{cases}$$

Function $S_n(\cdot)$ is a random function since it depends on the observations, and it is called an *empirical distribution function*. It is easy to see that $S_n(x)$ is nondecreasing, continuous from the left and $S_n(-\infty) = 0$, $S_n(\infty) = 1$ so that it is in the strict sense a distribution function. It is a step function with discontinuities at n points. For any fixed point x, $S_n(x)$ *simply gives the proportion of observations less than* x *in the sample*, so that it is a random variable taking values $0/n, 1/n, \ldots, 1$. But the hypothetical probability of an observation being less than x is $F(x)$, the value of the d.f. at x. Hence the probability that $S_n(x) = m/n$ is the same as the binomial probability for m successes out of n with $F(x)$ as the probability of success, that is,

$$P\left[S_n(x) = \frac{m}{n}\right] = \binom{n}{m}[F(x)]^m[1 - F(x)]^{n-m}. \tag{6f.1.2}$$

As a consequence of (6f.1.2), using the strong law of large numbers, we have

$$P\left\{\lim_{n\to\infty} S_n(x) = F(x)\right\} = 1, \qquad \text{for fixed } x, \tag{6f.1.3}$$

or using the weak law of large numbers, we have

$$S_n(x) \xrightarrow{P} F(x), \qquad \text{for fixed } x. \tag{6f.1.4}$$

We shall state without proof some results of the limiting behavior of some functions of $S(_n x)$.

(i) GLIVENKO (1933) THEOREM. *The probability that the sequence* $S_n(x) \to F(x)$, *as* $n \to \infty$, *uniformly in* $x(-\infty < x < \infty)$ *is equal to unity. Or in symbols, if* $D_n(x) = \sup_x |S_n(x) - F(x)|$, *then*

$$P\left(\lim_{n\to\infty} D_n = 0\right) = 1. \tag{6f.1.5}$$

The result (6f.1.5) is much stronger than the results (6f.1.3, 6f.1.4).

(ii) KOLOMOGOROV (1933) THEOREM. *Let* D_n *be as defined in* (i) *and let*

$$Q_n(\lambda) = \begin{cases} P(\sqrt{n}\, D_n < \lambda), & \lambda > 0 \\ 0, & \lambda \leqslant 0. \end{cases}$$

If $F(x)$ *is continuous, then*

$$\lim_{n\to\infty} Q_n(\lambda) = Q(\lambda) = \begin{cases} \sum_{k=-\infty}^{\infty} (-1)^k \exp(-2k^2\lambda^2), & \lambda > 0 \\ 0, & \lambda \leqslant 0. \end{cases}$$

Observe that the limit distribution $Q(\lambda)$ is independent of F, except that it is continuous. This is a characteristic of all the large sample criteria discussed in this chapter. For a proof of this proposition see Doob (1949).

It is suggested that the statistic $\sqrt{n}\, D_n$ may be used for testing the hypothesis that a given sample arose from a specified distribution. Some of the percentage points of $\sqrt{n}\, D_n$ as obtained from the limit distribution are as follows (Smirnov, 1939):

	0.5%	1%	2.5%	5%	10%
Lower	0.42	0.44	0.48	0.52	0.57
Upper	1.73	1.63	1.48	1.36	1.23

As an alternative to Kolmogorov-Smirnov test based on $\sqrt{n}\, D_n$, we have the χ^2 test developed in **6b.1**. This test, however, needs the frequency distribution of observations in some chosen class intervals of x. The observed frequencies (O) are then compared with the hypothetical or expected frequen- ies (E) (computed from the specified distribution), by using the criterion

$$\chi^2 = \sum \frac{(O - E)^2}{E}$$

as defined in **6b.1**. Although the Kolmogorov-Smirnov test may be sensitive in some respects it does not afford a comparison of the observed distribution with the hypothetical over the entire range of x. The χ^2 test followed by an examination of the deviations in the individual class intervals would give a better insight into departures from specification.

(iii) SMIRNOV (1944) THEOREM. *Let $S_{1n_1}(x)$ and $S_{2n_2}(x)$ be two empirical distribution functions based on two independent samples of sizes n_1 and n_2 drawn from a population with a continuous d.f. Let $n = n_1 n_2/(n_1 + n_2)$ and*

$$D^+_{n_1 n_2} = \max_{-\infty < x < \infty} (S_{1n_1}(x) - S_{2n_2}(x))$$

$$D_{n_1 n_2} = \max_{-\infty < x < \infty} |S_{1n_1}(x) - S_{2n_2}(x)|.$$

Then

$$\lim_{n_1, n_2 \to \infty} P(\sqrt{n}\, D^+_{n_1 n_2} < \lambda) = \begin{cases} 1 - \exp(-2\lambda^2), & \lambda > 0 \\ 0, & \lambda \leqslant 0 \end{cases}$$

$$\lim_{n_1, n_2 < \infty} P(\sqrt{n}\, D_{n_1 n_2} < \lambda) = \begin{cases} \sum_{k=-\infty}^{\infty} (-1)^k \exp(-2k^2\lambda^2), & \lambda > 0 \\ 0, & \lambda \leqslant 0 \end{cases}$$

For a proof of this theorem see Gihman (1952). The statistic $\sqrt{n}\, D_{n_1 n_2}$ can be used to test the hypothesis that two samples arise from the same population. The percentage points are same as those of $\sqrt{n}\, D_n$ in (ii) as the limiting distributions are identical. An alternative to this is the χ^2 test for testing homogeneity of parallel samples, discussed in **6c.2**. The apparent difference between the two tests is again in the type of departures which we wish to detcct.

6f.2 Asymptotic Distribution of Sample Fractiles

The pth fractile of a d.f. F is defined as any value ξ_p such that

$$P(x \leqslant \xi_p) \geqslant p, \qquad P(x \geqslant \xi_p) \geqslant q, \tag{6f.2.1}$$

where $q = 1 - p$, which imply

$$F(\xi_p) \leqslant p, \qquad F(\xi_p) + P(x = \xi_p) \geqslant p. \tag{6f.2.2}$$

There may be a multiplicity of values of ξ_p satisfying (6f.2.1) and (6f.2.2) unless the d.f. $F(x)$ is strictly monotonic.

The pth fractile statistic of n observations is defined in an analogous manner. It is a value $\hat{\xi}_p$ such that the number of observations $\leqslant \hat{\xi}_p$ is $\geqslant [np]$ and the number of observations $\geqslant \hat{\xi}_p$ is $\geqslant [nq]$. It is easy to see then that in terms of quantities defined in (6f.1.1) the function $\hat{\xi}_p$ of the observations can be defined as

$$\begin{aligned}\hat{\xi}_p(x_1, \ldots, x_n)\\ &= x_{([np]+1)}, \quad \text{if} \quad np \neq \text{an integer}\\ &= \text{any value in the closed interval } [x_{(np)}, x_{(np+1)}] \text{ if } np \text{ is an integer}\end{aligned} \tag{6f.2.3}$$

where $[\cdot]$ denotes the greatest integer.

We shall consider two propositions concerning the asymptotic distribution of $\hat{\xi}_p$ when $p \neq 1$.

(i) *If there is a unique value ξ_p such that* (6f.2.1) *is true, then $\hat{\xi}_p \to \xi_p$ as $n \to \infty$ with probability* 1.

By definition for any $\epsilon > 0$,

$$F(\xi_p + \epsilon) \geqslant F(\xi_p) + P(x = \xi_p) \geqslant p, \qquad \text{using (6f.2.2).} \tag{6f.2.4}$$

Suppose $F(\xi_p + \varepsilon) = p$, then $\xi_p + \varepsilon$ also satisfies the condition (6f.2.1) contrary to assumption. Therefore $F(\xi_p + \varepsilon) > p$, and thus by Glivenko's theorem, $\lim S_n(\xi_p + \varepsilon) > p$ with probability 1. But $S_n(\xi_p + \varepsilon) > p$ implies $\hat{\xi}_p \leqslant \xi_p + \varepsilon$. Hence $\hat{\xi}_p \leqslant \xi_p + \varepsilon$ for sufficiently large n with probability 1, and a similar result is true of the inequality $\hat{\xi}_p \geqslant \xi_p - \varepsilon$. Since ε is arbitrary the desired result follows.

(ii) *Let $F(x)$ admit a p.d. $f(x)$ continuous in x. Further let ξ_p be unique and $f(\xi_p) > 0$. Then*

$$\sqrt{n}(\hat{\xi}_p - \xi_p) \xrightarrow{L} X \sim N\left(0, \frac{p(1-p)}{[f(\xi_p)]^2}\right). \tag{6f.2.5}$$

The probability that $\hat{\xi}_p < x$ is the same as the probability that the number of observations less than x is greater than or equal to np, which is equal to the sum of binomial probabilities

$$\begin{aligned}&\sum_m^n \frac{n!}{r!(n-r)!}[F(x)]^r[1-F(x)]^{n-r}\\ &\qquad = \frac{n!}{(m-1)!(n-m)!}\int_0^{F(x)} t^{m-1}(1-t)^{n-m}\,dt,\end{aligned} \tag{6f.2.6}$$

where $m = np$ if np is an integer and $m = [np] + 1$ if np is not an integer. The equivalence (6f.2.6) of the incomplete summation of binomial terms and an incomplete beta integral is easily established. The p.d. function of $\hat{\xi}_p$ is obtained by differentiating (6f.2.6) with respect to x,

$$\frac{n!}{(m-1!)(n-m)!}[F(x)]^{m-1}[1-F(x)]^{n-m}f(x). \tag{6f.2.7}$$

Let $y = F(x)$, so that $dy = f(x)\,dx$. The probability density of the variable y is

$$\frac{n!}{(m-1)!(n-m)!}y^{m-1}(1-y)^{n-m}. \tag{6f.2.8}$$

Let us introduce a new variable

$$z = \frac{\sqrt{n}(y-p)}{\sqrt{pq}}, \quad y = p\left(1 + z\sqrt{\frac{q}{np}}\right).$$

The p.d. of z is obtained as the product of two factors

$$\frac{1}{\sqrt{n}}\frac{n!}{(m-1)!(n-m)!}p^{m-1/2}q^{n-m+1/2}$$
$$\times\left(1 + z\sqrt{q}/\sqrt{np}\right)^{m-1}\left(1 - z\sqrt{p}/\sqrt{nq}\right)^{n-m}. \tag{6f.2.9}$$

Taking the logarithm of (6f.2.9) and using Stirling's approximation to the factorials, we find that the logarithm of the first term not involving z is

$$\log\left(\frac{1}{\sqrt{n}}\frac{n!}{(m-1)!(n-m)!}p^{m-1/2}q^{n-m+1/2}\right) \to \text{constant}$$

as $n \to \infty$. The logarithm of the terms involving z is

$$(m-1)\log\left(1 + \frac{z\sqrt{q}}{\sqrt{pn}}\right) + (n-m)\log\left(1 - \frac{z\sqrt{p}}{\sqrt{qn}}\right) \to -\frac{z^2}{2}$$

as $n \to \infty$. Hence the limit of (6f.2.9) as $n \to \infty$ is const. $e^{-z^2/2}$, which is the density of the normal distribution. Hence by Scheffe's theorem [(xv), **2c.4**], the a.d. of z is $N(0, 1)$. Hence the a.d. of $\sqrt{n}(y-p)$ is $N(0, pq)$.

But $y = F(\hat{\xi}_p)$ or $\hat{\xi}_p = F^{-1}(y)$ and $d\hat{\xi}_p/dy = f(\hat{\xi}_p)$. The value of the derivative at $y = p$ is $[f(\xi_p)]^{-1}$. Applying [(i), **6a.2**], the a.d. of $\sqrt{n}(\hat{\xi}_p - \xi_p)$ is $N(0, pq/[f(\xi_p)]^2)$.

In praticular, the a.d. of the median is normal with the population median μ as the mean and asymptotic variance

$$\frac{[f(\mu)]^{-2}}{4n} \tag{6f.2.10}$$

Thus if the original distribution is normal, the asymptotic variance of the sample median as an estimate of the mean of the normal distribution is $\pi\sigma^2/2n$, since the ordinate of the normal distribution at the median value is $1/\sigma\sqrt{2\pi}$. It may be observed that the variance of $\bar{x}$, the average of n observations, is σ^2/n giving the ratio of asymptotic variance of the average to that of the median as $2/\pi$. (Thus according to the earlier definition of efficiency of an estimator, the efficiency of the median as an estimator of the population mean is $2/\pi$ or of the order of 63%.)

In deriving (6f.2.6), which is the exact d.f. of $\hat{\xi}_p$, we used the equality

$$P(\hat{\xi}_p < x) = P(nS_n(x) \geq np) \tag{6f.2.11}$$

where $S_n(x)$ is the binomial variable defined in (6f. 1.2). Weiss (1970) used the relation (6f.2.11) to deduce the a.d. of $\hat{\xi}_p$ directly without going through the exact p.d. (6f.2.7). Substituting $x = \xi_p + t/\sqrt{n}$ in (6f.2.11),

$$\begin{aligned}
&P(\sqrt{n}(\hat{\xi}_p - \xi_p) < t) \\
&= P\left\{\frac{\sqrt{n}}{\sqrt{pq}}[S_n(\xi_p + t/\sqrt{n}) - F(\xi_p + t/\sqrt{n})] > \frac{\sqrt{n}}{\sqrt{pq}}[p - F(\xi_p + t/\sqrt{n})]\right\} \\
&\stackrel{a}{=} P\left\{\frac{\sqrt{n}}{\sqrt{pq}}[S_n(\xi_p) - F(\xi_p)] \geqslant -\frac{tf(\xi_p)}{\sqrt{pq}}\right\} \qquad (6f.2.12) \\
&\to P\left\{X \geq -\frac{tf(\xi_p)}{\sqrt{pq}}\right\} = P\left\{X < \frac{tf(\xi_p)}{\sqrt{pq}}\right\}
\end{aligned}$$

as $n \to \infty$ where $X \sim N(0, 1)$ by using the central limit theorem [(*i*), **2c.5**] on the binomial variable $S_n(\xi_p)$, which is exactly the result (6f.2.6).

Note that in deriving (6f.2.12), we use the result

$$\sqrt{n}\{[S_n(\xi_p + t/\sqrt{n}) - S_n(\xi_p)] - [F(\xi_p + t/\sqrt{n}) - F(\xi_p)]\} \xrightarrow{P} 0$$

since the expectation of the left hand side is zero and its (binomial) variance $\to 0$ as $n \to \infty$. Further

$$\sqrt{n}[p - F(\xi_p + t/\sqrt{n})] \to -tf(\xi_p)$$

as $n \to \infty$ because of the assumption on F, noting that $p = F(\xi_p)$.

Other papers of interest on the asymptotic behaviour of $\hat{\xi}_p$ are due to Bahadur (1966) and Kiefer (1967).

6g TRANSFORMATION OF STATISTICS

6g.1 A General Formula

In [(i), **6a.2**], it was shown that if $\{T_n\}$, $n = 1, 2, \ldots$, is a sequence of statistics, then

$$\sqrt{n}(T_n - \theta) \xrightarrow{L} X \sim N[0, \sigma^2(\theta)]^2$$

$$\Rightarrow \sqrt{n}[g(T_n) - g(\theta)] \xrightarrow{L} X \sim N(0, [g'(\theta)\sigma(\theta)]^2)$$

where g is a function admitting the first derivative and $g'(\theta) \neq 0$.

If we now choose the function g such that

$$g'(\theta)\sigma(\theta) = c \text{ (independent of } \theta), \tag{6g.1.1}$$

the asymptotic variance of the transformed statistic $g(T_n)$ will be independent of θ. The differential equation (6g.1.1) is solved to obtain the function g,

$$\frac{dg}{d\theta} = \frac{c}{\sigma(\theta)} \quad \text{or} \quad g = c\int \frac{d\theta}{\sigma(\theta)}. \tag{6g.1.2}$$

The result (6g.1.2) is applied to a number of important statistics in Sections **6g.2** to **6g.4**, where the use of transformed statistics is explained in greater detail.

6g.2 Square Root Transformation of the Poisson Variate

If x is a Poisson variate with

$$E(x) = \mu, \qquad V(x) = \mu,$$

then $(x - \mu)/\sqrt{\mu} \xrightarrow{L} X \sim N(0, 1)$ as $\mu \to \infty$. We wish to determine a function g such that the a.d. of $g(x) - g(\mu)$ is $N(0, c)$ where c is a constant independent of μ. Using the formula (6g.1.2), we see that

$$g(\mu) = \int \frac{c\, d\mu}{\sqrt{\mu}} = \sqrt{\mu} \tag{6g.2.1}$$

by choosing c suitably. The transformed variable $\sqrt{x}$ has the asymptotic mean and variance

$$\sqrt{\mu} \quad \text{and} \quad \tfrac{1}{4}$$

when μ is large. It was found by Anscombe (1948) that the transformation $\sqrt{x + b}$ where b is a suitably determined constant has some theoretical advantages. Let $(x - \mu) = t$ and $(\mu + b) = \mu'$; then by using Taylor's expansion we find

$$\sqrt{x+b} = \sqrt{\mu'}\left\{1 + a_1 \frac{t}{\mu'} - a_2\left(\frac{t}{\mu'}\right)^2 + \cdots + (-1)^s a_{s-1}\left(\frac{t}{\mu'}\right)^{s-1} + \cdots\right\}, \tag{6g.2.2}$$

where

$$a_s = (-1)^{s+1}\frac{(-1)(-3)\cdots(-2s+3)}{s!\,2^s}.$$

Observing that the Poisson moments are

$$E(t) = 0, \qquad E(t^2) = \mu, \qquad E(t^3) = \mu, \qquad E(t^4) = 3\mu^2 + \mu, \ldots,$$

we find, by taking expectations of both sides of the expansion (6g.2.2), that

$$E(\sqrt{x+b}) = \sqrt{\mu+b} - \frac{1}{8\sqrt{\mu}} + \frac{24b-7}{128\mu\sqrt{\mu}} + \cdots$$

$$V(\sqrt{x+b}) = \frac{1}{4}\left\{1 + \frac{3-8b}{8\mu} + \frac{32b^2 - 52b + 17}{32\mu^2} + \cdots\right\},$$

which, when we choose the value $b = \frac{3}{8}$, reduce to

$$E(\sqrt{x+\tfrac{3}{8}}) = \sqrt{\mu+\tfrac{3}{8}} - \frac{1}{8\sqrt{\mu}} + \frac{1}{64\mu\sqrt{\mu}} + \cdots$$

$$V(\sqrt{x+\tfrac{3}{8}}) = \frac{1}{4}\left(1 + \frac{1}{16\mu^2} + \cdots\right).$$

The variance of $\sqrt{x+\frac{3}{8}}$ is more stable than that of $\sqrt{x}$ because the second term in the expansion of the variance of $\sqrt{x+\frac{3}{8}}$ is $O(1/\mu^2)$.

6g.3 Sin^{-1} Transformation of the Square Root of the Binomial Proportion

The binomial proportion r/n has the mean value π and variance $\pi(1-\pi)/n$. The transformation is obtained by solving the equation

$$g(\pi) = \int \frac{c}{\sqrt{\pi(1-\pi)}}\,d\pi = \sin^{-1}\sqrt{\pi}, \text{ choosing } c \text{ suitably} \tag{6g.3.1}$$

$$\text{a.v.}\left(\sin^{-1}\sqrt{\frac{r}{n}}\right) = \frac{1}{4n},$$

Anscombe (1948) has shown that a slightly better transformation achieving more stability in variance is

$$\sin^{-1}\sqrt{(r + \tfrac{3}{8}) \div (n + \tfrac{3}{4})}, \qquad (6g.3.2)$$

which has the asymptotic variance $1/(4n + 2)$. If n is large, the simpler transformation $\sin^{-1}\sqrt{r/n}$ can be used; for moderately large sample sizes the transformation $\sin^{-1}\sqrt{(r + \frac{3}{8})/(n + \frac{3}{4})}$ may be used.

Example. R. A. Fisher (1949) found the following recombination fractions between undulated and agouti loci in house mice. The data relate to back-crosses so that the estimate of the fraction is the ratio of recombinants to the total offsprings.

TABLE 6g.3α. Recombination Fractions Observed for Twenty Classes of Heterozygous Parents at the Agouti Locus

	A^YA^L	A^YA	A^Ya^t	A^Ya	A^LA	
♀	$\frac{12}{194}$	$\frac{5}{118}$	$\frac{12}{235}$	$\frac{10}{146}$	$\frac{2}{78}$	
♂	$\frac{9}{128}$	$\frac{9}{126}$	$\frac{6}{160}$	$\frac{16}{243}$	$\frac{7}{214}$	
	A^La^t	A^La	Aa^t	Aa	a^ta	Total
♀	$\frac{9}{182}$	$\frac{4}{210}$	$\frac{10}{231}$	$\frac{8}{178}$	$\frac{11}{159}$	$\frac{83}{1731}$
♂	$\frac{7}{213}$	$\frac{4}{144}$	$\frac{13}{218}$	$\frac{3}{159}$	$\frac{13}{238}$	$\frac{87}{1843}$

In Table 6g.3α, the number in the denominator gives the number of animals contributing to the ratio.

Sex Difference. Is there sex difference in the recombination fraction for each heterozygote? Considering the heterozygote A^YA^L, we have the fourfold table:

	Recombinations	Old combinations	Total
♀	12	182	194
♂	9	119	128

The χ^2 test of independence has the value 0.0905 on 1 D.F. The sum of the χ^2 values for the ten types of heterozygotes is 4.827 on 10 D.F. The probability of χ^2 on 10 D.F. exceeding the value 4.827 is high, indicating no sex difference.

Differences between Heterozygotes. Since the sex differences can be ignored the data may be pooled over sex to obtain a 2×10 contingency table with the type of heterozygote as one attribute and nature of combinations (old and new) as the other.

The χ^2 test of homogeneity has the value 16.315 on 9 D.F. which is significant at the 5 % level indicating differences in the recombination values for the various heterozygotes.

Interaction between Sex and Heterozygotes. This example is not suitable for further analysis on sex differences. Suppose we find that sex differences exist in some or all the ten types of mating. Then the further problem arises as to whether the sex difference is the same in all the cases. That is, we need to test whether there is interaction between sex and the nature of the heterozygote. This can be done by using the angular transformation and then applying analysis of variance. Corresponding to each observed proportion p, an angle ϕ is determined such that $p = \sin^2 \phi$. If ϕ is given in degrees, as in R.M.M. or F.Y. Tables, then ϕ has the variance $8100/n\pi^2$ or approximately $820.7/n$. The 20 angles and the necessary computational steps are as given in Table 6g.3β.

The χ^2 on 10 D.F. for testing sex differences in each of the ten types is

$$\sum d^2 w \div 820.7 = 4434.10 \div 820.7 = 5.40.$$

This is slightly above the value 4.827 obtained earlier by a direct χ^2 analysis. From the 10 degrees of freedom χ^2 we subtract the χ^2 on 1 D.F.

$$\frac{(\sum dw)^2}{\sum w} \div 820.7 = \frac{(177.92)^2}{856.62} \div 820.7 = 0.05,$$

owing to overall sex differences. The residual

$$5.40 - 0.05 = 5.35$$

is χ^2 on 9 D.F. for testing the interaction between sex and type of heterozygote. The interaction χ^2 is not significant, nor is that because of sex difference. In such a case, the differences in the various types of heterozygotes can be studied by summing over sex. We shall illustrate, however, the appropriate test assuming that sex difference exists but not the interaction.

TABLE 6g.3β. Computation of the Interaction

	A^YA^L	A^YA	A^Ya^t	A^Ya	A^La	
♀	14.4	11.8	13.0	15.1	9.2	
♂	15.3	15.4	11.2	14.9	10.4	
$d =$ difference	−0.9	−3.6	1.8	0.2	−1.2	
$w = n_1n_2/(n_1 + n_2)$	77.12	60.93	95.19	91.20	57.16	
dw	−69.41	−219.35	171.34	18.24	−68.59	
$d^2w = d(dw)$	62.47	789.65	308.41	3.65	82.31	
	A^La^t	A^La	Aa^t	Aa	a^ta	
♀	12.7	7.9	11.9	12.2	15.2	
♂	10.4	9.6	14.2	7.9	13.5	
d	2.3	−1.7	−2.3	4.3	1.7	Total
w	98.14	85.42	112.16	83.98	95.32	856.92
dw	225.72	−145.21	−257.97	361.11	162.04	177.92
d^2w	519.16	246.86	593.33	1552.79	275.47	4434.10

TABLE 6g.3γ. Differences in Heterozygotes

	$\sum n$	$\sum n\phi$	$\sum n\phi^2$	$\sum n\phi^2 - (\sum n\phi)^2/\sum n$ $= 820.7\chi^2$	χ^2
♀	1731	21,470.9	274,666.21	8,346.43	10.17
♂	1843	22,699.4	290,576.08	10,997.81	13.40
Overall	3574	44,170.3	565,242.29	19.351.02	23.58

Differences in Heterozygotes Eliminating Sex Difference. The total χ^2 on $(20 - 1) = 19$ D.F. is

$$\left\{\sum n\phi^2 - \frac{(\sum n\phi)^2}{\sum n}\right\} \div 820.7,$$

the summation extending over the 20 angles. The further computations are shown in Table 6g.3γ where the summations extend over all heterozygotes. The total of χ^2 for ♀ and ♂

$$10.17 + 13.40 = 23.57$$

has 18 D.F. Subtracting the interaction component of 5.35 on 9 D.F. the residual χ^2 on 9 D.F. for testing differences in heterozygotes eliminating sex is

$$23.57 - 5.35 = 18.22$$

which is significant. This can also be calculated in a slightly different way. The sex χ^2, ignoring differences in heterozygotes, is obtained as follows

$$820.7\chi^2 = \frac{21{,}470.9^2}{1731} + \frac{22{,}699.4^2}{1843} - \frac{44{,}170.3^2}{3574}$$

$$\chi^2 = 0.01,$$

the values being obtained from the columns $\sum n\phi$ and $\sum n$ of Table 6g.3γ. The valid χ^2 for heterozygotes is obtained by subtracting from the total the value 0.01 and the interaction sum of squares. This leads to the value, $23.58 - 0.01 - 5.35 = 18.22$, the same as before.

TABLE 6g.3δ. Analysis of χ^2

	D.F.	χ^2
Sex	1	0.05
Heterozygotes	9	18.22
Interaction	9	5.35
Total	19	24.58

The total χ^2 on 19 D.F. is significant, showing overall differences. The various components of χ^2 do not add up to the total because the proportion are based on different numbers.

6g.4 Tanh^{-1} Transformation of the Correlation Coefficient

The asymptotic variance of the product moment correlation coefficient r is

$$\frac{(1-\rho^2)^2}{n}. \tag{6g.4.1}$$

Applying the formula (6g.1.2) for determining the transformation to remove the unknown parameter ρ in the expression for asymptotic variance,

$$g(\rho) = \int \frac{c}{1-\rho^2}\,d\rho = \tanh^{-1}\rho, \text{ choosing } c \text{ suitably}$$

$$\text{a.v.}\,(\tanh^{-1} r) = \frac{1}{n}.$$

We shall study the distribution of $\tanh^{-1} r$, through its moments. Define

$$\zeta = g(\rho) = \frac{1}{2}\log\frac{(1+\rho)}{(1-\rho)} = \tanh^{-1}\rho,$$

$$z = g(r) = \frac{1}{2}\log\frac{(1+r)}{(1-r)} = \tanh^{-1} r.$$

By putting $z - \zeta = x$, the distribution of x may be derived from the distribution of r. The first four moments of z were found by Fisher and later revised by Gayen (1951).

$$E(z) = \tanh^{-1}\rho + \frac{\rho}{2(n-1)}\left\{1 + \frac{5+\rho^2}{4(n-1)} + \cdots\right\},$$

$$\mu_2 = \frac{1}{n-1}\left\{1 + \frac{4-\rho^2}{2(n-1)} + \frac{22-6\rho^2-3\rho^4}{6(n-1)^2} + \cdots\right\},$$

$$\mu_3 = \frac{\rho^3}{(n-1)^3} + \cdots,$$

$$\mu_4 = \frac{1}{(n-1)^2}\left\{3 + \frac{14-3\rho^2}{n-1} + \frac{184-48\rho^2-21\rho^4}{4(n-1)^2} + \cdots\right\}.$$

Using the expansions for the first four moments, we find the β_1 and β_2 coefficients to be

$$\beta_1 = \frac{\rho^6}{(n-1)^3} + \cdots$$

$$\beta_2 = 3 + \frac{2}{n-1} + \frac{4 + 2\rho^2 - 3\rho^4}{(n-1)^2} + \cdots.$$

Since β_1 and $(\beta_2 - 3)$ are small, even for moderate n, it follows that $(z - \zeta)$ can be considered to be *approximately* a normal variate with

$$\text{Mean} = \frac{\rho}{2(n-1)}$$

$$\text{Variance} = \frac{1}{n-1} + \frac{4-\rho^2}{2(n-1)^2} \simeq \frac{1}{n-3} \tag{6g.4.2}$$

Test for a given ρ. In a sample of 28 independent pairs of observations, the correlation coefficient is found to be 0.6521. Can such a value have arisen from a population in which the coefficient ρ has the value 0.7211? Using appropriate tables (F.Y. or R.M.M.) of transformation of the correlation coefficient, we have

$$z = \frac{1}{2}\log_e \frac{1+r}{1-r} = 0.7790$$

$$\text{Mean } z = \frac{1}{2}\log_e \frac{1+\rho}{1-\rho} + \frac{\rho}{2(n-1)}$$

$$= 0.9100 + \frac{0.7211}{54} = 0.9233.$$

The normal deviate is

$$\sqrt{n-3}(z - \text{mean } z) = \sqrt{28-3}(0.7790 - 0.9233)$$

$$= 5(-0.1443) = -0.7215. \tag{6g.4.3}$$

The chance of exceeding the value 0.7215 in either direction is about 45% so that the hypothesis cannot be rejected.

The correction term $\rho/2(n-1)$ for mean z is unimportant if n is large. The probability will be more precisely obtained by its inclusion.

Test for the Equality of Two Correlation Coefficients. Two samples consisting of n_1 and n_2 observations give the correlation coefficients r_1 and r_2. Are these values compatible with the hypothesis that the samples arose from two populations having the same correlation coefficient? Let

$$z_1 = \frac{1}{2}\log\frac{1+r_1}{1-r_1} \quad \text{and} \quad z_2 = \frac{1}{2}\log\frac{1+r_2}{1-r_2}.$$

The statistic $z_1 - z_2$ is distributed about the mean

$$\frac{\rho}{2(n_1-1)} - \frac{\rho}{2(n_2-1)}, \tag{6g.4.4}$$

where ρ is the common correlation coefficient, with asymptotic variance

$$\frac{1}{n_1-3} + \frac{1}{n_2-3}. \tag{6g.4.5}$$

If the samples are not small, the statistic

$$\frac{z_1 - z_2}{\sqrt{1/(n_1-3) + 1/(n_2-3)}} \tag{6g.4.6}$$

can be used as a normal deviate.

Test for Homogeneity of a Set of Correlation Coefficients. Let $r_1, \ldots, r_k$ be k correlation coefficients based on samples of sizes $n_1, \ldots, n_k$. By means of the $\tanh^{-1}$ transformation, the quantities $z_1, \ldots, z_k$ corresponding to $r_1, \ldots, r_k$ can be obtained. If the bias in mean z can be neglected, the test for homogeneity of the correlation coefficients is equivalent to the test of equality of the mean values of z. The scheme of computation is as follows.

TABLE 6g.4α. Test of Homogeneity of Parallel Estimates of Correlation Coefficients

Sample No. t	Sample Size n	Correlation Coefficient r	$\tanh^{-1} r$ $= z$	Reciprocal of Variance $n - 3$	$(n-3)z$	$(n-3)z^2$
1	n_1	r_1	z_1	$n_1 - 3$	$(n_1-3)z_1$	$(n_1-3)z_1^2$
.	.	.	.	.	.	.
.	.	.	.	.	.	.
.	.	.	.	.	.	.
k	n_k	r_k	z_k	$n_k - 3$	$(n_k-3)z_k$	$(n_k-3)z_k^2$
Total				N	T_1	T_2

The best estimate of $\tanh^{-1} \rho$, when a common ρ is applicable, is T_1/N. The statistic for testing homogeneity is (see 6a.2.12 and (v), **6a.2**)

$$\text{H} = T_2 - \frac{T_1^2}{N} \tag{6g.4.7}$$

which can be used as χ^2 on $(k - 1)$ D.F.

As an example, consider the correlations obtained from 6 samples of sizes 10, 14, 16, 20, 25, 28 to be 0.318, 0.106, 0.253, 0.340, 0.116, 0.112. Can these be considered homogeneous?

Correlation Coefficient r	Sample size Minus 3 $n - 3$	z	$(n - 3)z$	$(n - 3)z^2$
0.318	7	0.3294	2.3058	0.7595
0.106	11	0.1064	1.1704	0.1245
0.253	13	0.2586	3.3618	0.8694
0.340	17	0.3541	6.0197	2.1316
0.116	22	0.1164	2.5608	0.2981
0.112	25	0.1125	2.8125	0.3164
Total	95		18.2310	4.4995

$$\frac{T_1}{95} = \frac{18.2310}{95} = 0.191905$$

$$T_2 - T_1 \frac{T_1}{95} = 4.9995 - 3.4986 = 1.0009$$

The value 1.0009 as χ^2 on 5 D.F. is not significant, so the estimates of correlations may be considered homogeneous.

Correction for Bias in Tests and the Best Estimate of ρ. When the sample sizes are not large and not nearly equal, there is a certain amount of bias (extraneous to the hypothesis tested) introduced in the H statistic used in (6g.4.7). This bias is due to the term $\rho/2(n - 1)$ in the mean value of z being neglected. Even if the bias introduced in the H statistic is small, the bias introduced in the best estimate of ρ when H is not significant will not be small when compared to the standard error of the estimate. This can be corrected by a slightly different procedure.

Since z can be considered as a normal deviate with mean

$$\tfrac{1}{2} \log_e [(1 + \rho)/(1 - \rho)] + \rho/2(n - 1)$$

and variance $1/(n-3)$, the score for ρ obtained from k samples is

$$S = \sum (n_i - 3)\left[\frac{1}{1-\rho^2} + \frac{1}{2(n_i-1)}\right]\left[z_i - \frac{1}{2}\log\frac{1+\rho}{1-\rho} - \frac{\rho}{2(n_i-1)}\right], \tag{6g.4.8}$$

and the information

$$\mathscr{I} = E(S^2) = \sum (n_i - 3)\left[\frac{1}{1-\rho^2} + \frac{1}{2(n_i-1)}\right]^2. \tag{6g.4.9}$$

If the value of ρ obtained in the last section is taken as a first approximation, then the additive correction $\delta\rho$ to this value is given by

$$\delta\rho = \frac{S_0}{\mathscr{I}_0}$$

where S_0 and $\mathscr{I}_0$ are the values of (6g.4.8) and (6g.4.9) calculated at the approximate value chosen. This process may be repeated until the correction becomes negligible. Having obtained the best estimate $\hat{\rho}$ of ρ, the H statistic with $(k-1)$ degrees of freedom for testing homogeneity is

$$\mathrm{H} = \sum (n_i - 3)\left\{z_i - \frac{1}{2}\log_e\frac{1+\hat{\rho}}{1-\hat{\rho}} - \frac{\hat{\rho}}{2(n_i-1)}\right\}^2.$$

6h STANDARD ERRORS OF MOMENTS AND RELATED STATISTICS

6h.1 Variances and Covariances of Raw Moments

Let $(x_1, \ldots, x_n)$ be independent observations from any population (on a random variable x). The rth raw and central moments of the r.v. x are denoted by

$$v_r = E(x^r) \qquad \text{and} \qquad \mu_r = E(x - v_1)^r.$$

The raw and corrected moment functions (of observations) are

$$0_r = \sum x_i^r/n \qquad \text{and} \qquad m_r = \sum (x_i - 0_1)^r \div n.$$

It is easy to see that

$$E(0_r) = \sum E(x_i^r)/n = v_r$$

$$E(0_r^2) = E\frac{\sum x_i^{2r} + 2\sum\sum x_i^r x_j^r}{n}$$

$$= \frac{nv_{2r} + n(n-1)v_r^2}{n^2}$$

$$V(0_r) = E(0_r^2) - E^2(0_r) = \frac{v_{2r} - v_r^2}{n}$$

Similarly,

$$\text{cov}(0_r, 0_s) = \frac{v_{r+s} - v_r v_s}{n}.$$

If the origin is the population mean, in the foregoing expressions v is replaced by μ.

6h.2 Asymptotic Variances and Covariances of Central Moments

The rth central corrected moment (about the observed mean) is

$$m_r = 0_r - \binom{r}{1} 0_{r-1} 0_1 + \binom{r}{2} 0_{r-2} 0_1{}^2 - \cdots (-1)^r 0_1{}^r.$$

Since this is invariant for origin, we may consider 0_r to be the raw moment about the population mean. We now use the δ-method (6a.2.9) for determining the asymptotic variance of m_r. We have to compute the derivatives of m_r with respect to $0_1, \ldots, 0_r$ at the expected values $E(0_1) = 0$, $E(0_i) = \mu_i$, $i = 2, \ldots, r$.

$$\frac{\partial m_r}{\partial 0_r} = 1, \qquad \frac{\partial m_r}{\partial 0_{r-s}} = (-1)^s \binom{r}{s} 0_1{}^s = 0 \quad \text{(at the expected values)}$$

$$\begin{aligned}\frac{\partial m_r}{\partial 0_1} &= -r0_{r-1} + \binom{r}{2} 0_{r-2}(20_1) - \cdots \\ &= -r\mu_{r-1} \quad \text{(at the expected values).}\end{aligned}$$

Hence we have the equivalence

$$\begin{aligned} m_r - \mu_r &\overset{a}{=} (0_r - \mu_r) - r\mu_{r-1}(0_1 - \mu_1) \\ \text{a.v.}\ (m_r - \mu_r) &= V(0_r) + r^2\mu_{r-1}^2 V(0_1) - 2r\mu_{r-1}\,\text{cov}(0_r, 0_1) \\ &= \frac{1}{n}[\mu_{2r} - \mu_r{}^2 - 2r\mu_{r-1}\mu_{r+1} + r^2\mu_{r-1}^2\mu_2]. \end{aligned}$$

Similarly

$$\text{a. cov}\ (m_r m_s) = \frac{1}{n}[\mu_{r+s} - \mu_r\mu_s + rs\mu_2\mu_{r-1}\mu_{s-1} - r\mu_{r-1}\mu_{s+1} - s\mu_{r+1}\mu_{s-1}].$$

As special cases we obtain the following:

$$\text{a.v.}\ (m_2) = \frac{1}{n}(\mu_4 - \mu_2{}^2)$$

$$\text{a.v.}\ (m_3) = \frac{1}{n}(\mu_6 - \mu_3{}^2 - 6\mu_2\mu_4 + 9\mu_2{}^3)$$

$$\text{a.v.}\ (m_4) = \frac{1}{n}(\mu_8 - \mu_4{}^2 - 8\mu_5\mu_3 + 16\mu_2\mu_3{}^2).$$

6h.3 Exact Expressions for Variances and Covariances of Central Moments

Since central moments involve only powers and products of powers of the observations, it is possible to find exact expressions for their variances and covariances. The computations are complicated, however, systematic methods were developed by Fisher (1928). We shall give the straightforward computation in one or two cases, which we need for statistical applications.

$$m_2 = (0_2 - 0_1{}^2)$$
$$V(m_2) = E(0_2{}^2) - 2E(0_2 0_1{}^2) + E(0_1{}^4) - [E(m_2)]^2.$$

Assuming, without loss of generality, that the origin is the population mean, we compute

$$E(0_2{}^2) = \frac{\mu_4 + (n-1)\mu_2{}^2}{n}, \qquad E(0_1{}^4) = \frac{\mu_4 + 3(n-1)\mu_2{}^2}{n^3}$$

$$E(0_2 0_1{}^2) = \frac{\mu_4 + (n-1)\mu_2{}^2}{n^2}, \qquad E^2(m_2) = \mu_2{}^2\left(\frac{n-1}{n}\right)^2,$$

which leads to

$$V(m_2) = \frac{(n-1)^2}{n^3}\left(\mu_4 - \frac{n-3}{n-1}\mu_2{}^2\right)$$
$$= \frac{(n-1)^2}{n^3}\left(\kappa_4 + \frac{2n\kappa_2{}^2}{n-1}\right) \quad \text{in terms of cumulants.} \tag{6h.3.1}$$

If k_2 is defined by $(0_2 - 0_1{}^2)/(n-1)$, then

$$V(k_2) = \frac{\kappa_4}{n} + \frac{2\kappa_2{}^2}{n-1} \tag{6h.3.2}$$

is obtained by dividing (6h.3.1) by $(n-1)^2/n^2$. Similarly, we find

$$\text{Cov}(m_2, 0_1) = \mu_3 \frac{(n-1)}{n^2} \tag{6h.3.3}$$

$$\text{Cov}(k_2, 0_1) = \frac{\kappa_3}{n}. \tag{6h.3.4}$$

Using the formulas (6h.3.2) and (6h.3.4) we can develop a test for examining the significance of the difference between the estimated mean $\bar{x}$ and the estimated variance k_2 of a sample of size n from a Poisson population (for which the theoretical mean and variance are the same). This test will be useful in detecting overdispersion or heterogeneity of the population sampled.

$$V(\bar{x} - k_2) = V(\bar{x}) + V(k_2) - 2\ \mathrm{Cov}\ (\bar{x}, k_2)$$

$$= \frac{\kappa_2}{n} + \frac{\kappa_4}{n} + \frac{2\kappa_2{}^2}{n-1} - \frac{2\kappa_3}{n}$$

$$= \frac{2\mu^2}{n-1}, \qquad \text{since} \quad \kappa_2 = \kappa_3 = \kappa_4 = \mu,$$

where μ is the Poisson parameter. If n is large, the statistic

$$(\bar{x} - k_2)/\sqrt{V(\bar{x} - k_2)},$$

which may be written, if we substitute $\bar{x}$ for μ in the expression for

$$V(\bar{x} - k_2),$$

as

$$\sqrt{n-1}\,\frac{(\bar{x} - k_2)}{\sqrt{2\bar{x}}}. \tag{6h.3.5}$$

can be used as a standard normal deviate. Suppose that in a sample of 29 observations the sample mean and variance k_2 were found to be 1.5172 and 1.3300 respectively. The value of (6h.3.5) is

$$\frac{\sqrt{28}(0.1872)}{\sqrt{2(1.5172)}} = 0.462,$$

which is small, indicating no significant difference.

COMPLEMENTS AND PROBLEMS

1 Four samples of sizes 120, 100, 100, and 125 from four Poisson populations gave the mean values, 251/120, 323/100, 180/100, and 426/125. Do the populations have the same mean value?

[Let T_1, T_2, T_3, T_4 be totals based on samples of sizes n_1, n_2, n_3, n_4. If the hypothesis is true the relative distribution of $T_1, \ldots, T_4$ given $T = T_1 + T_2 + T_3 + T_4$ is multinomial with cell probabilities proportional to n_1, n_2, n_3, n_4, which can be tested by χ^2. Thus the χ^2 on 3 D.F. is

$$\frac{(T_1 - n_1\bar{T})^2}{n_1\bar{T}} + \cdots + \frac{(T_4 - n_4\bar{T})^2}{n_4\bar{T}}$$

where $\bar{T} = T/(n_1 + n_2 + n_3 + n_4)$. Verify that the numerical value is 81.12 on 3 D.F.]

2 The following estimates of the correlation coefficients between intelligence test scores were found in an investigation of the relative influences of environmental and heredity factors. Comment on the figures using tests of significance if necessary.

	Two brothers		Twins	
	Reared Apart	Lving Together	Reared Apart	Living Together
Correlation coefficient	0.235	0.342	0.451	0.513
Sample size	50	40	45	55

3 In an investigation to estimate the number of births in a city during a particular month the following data were obtained. Out of 450 families surveyed, 350 reported no birth and 100 reported a single birth, of which 25 took place in a hospital. If the total number of births in the hospital is 1000, estimate the total number of births in the city. Find the standard error of the estimate.

4 A total number of 130 heads was observed when 100 rupee coins and 100 half rupee coins were thrown together. Assuming that the rupee coin is unbiased, what can you say about the half-rupee coin?

5 Consider a χ^2 variable on n D.F. and the transformation $\alpha\sqrt{n}\{(\chi^2/n)^\beta + (\gamma/n) - 1\}$. By comparing the first three moments of this statistic with those of a normal distribution with mean zero and unit variance, obtain the values of α, β, and γ. Show that

$$\alpha = \sqrt{\tfrac{9}{2}}, \qquad \beta = \tfrac{1}{3}, \quad \text{and} \quad \gamma = \tfrac{2}{9}$$

is a good choice for rapid approach of the transformed statistic to normality as $n \to \infty$.

6 The data in the following table relate to the distribution of animals bred for linkage between two factors A and B.

Sex of Heterozygotes	Phase	Sex of Animals Bred	Phenotype AB	Ab	aB	ab
	Coupling	♀	12	13	11	8
♂♂		♂	13	15	16	16
	Repulsion	♀	11	13	13	19
		♂	15	10	10	16
	Coupling	♀	30	17	20	13
♀♀		♂	18	18	20	24
	Repulsion	♀	17	12	13	17
		♂	15	12	11	14

Examine for differences in linkage between mating types (sex of heterozygotes and phase) and between sexes within a mating type.

7 Determine the likelihood ratio tests in the following situations.

(a) To test the hypothesis $\mu = 0$, when σ^2 is unknown on the basis of a sample of size n from $N(\mu, \sigma^2)$. Show that the test criterion can be reduced to Student's t-test.

(b) To test the hypothesis that σ^2 has a given value when μ is unknown on the basis of a sample of size n from $N(\mu, \sigma^2)$.

(c) To test the hypothesis ${\sigma_1}^2 = {\sigma_2}^2$ on the basis of samples of sizes n_1, n_2 from $N(\mu_1, {\sigma_1}^2)$, $N(\mu_2, {\sigma_2}^2)$ when the parameters μ_1, μ_2 are unknown.

(d) To test the hypothesis $\mu_1 = \mu_2 = \cdots = \mu_k$ on the basis of samples of sizes $n_1, n_2, \ldots, n_k$ from $N(\mu_1, \sigma^2), N(\mu_2, \sigma^2), \ldots, N(\mu_k, \sigma^2)$ when the common σ^2 is unknown. Show that this test reduces to the anslysis of variance test for one way classification **(4d.1)**.

(e) Determine the likelihood ratio test for the linear hypotheses under the general setup of the least squares theory **(4b.2)**. Show that the test is the same as the analysis of variance set.

8 Show that the distribution function of Students' t under the normality assumption for the observations, tends to normal as the degrees of freedom tends to infinity. [Hint: Consider the probability density function of the t distribution on n D.F. Find the limit as $n \to \infty$. Identify the limit as a density function and apply Scheffe's theorem [xv, **2c.4**].

More generally let $x_1, \ldots, x_n$ be independent observations on a random variable x with $E(x) = \mu$ and $V(x) = \sigma^2 < \infty$. Define $t = \sqrt{n}(\bar{x} - \mu)/s$. Show that $t \xrightarrow{L} X \sim N(0, 1)$.

[Hint: By the central limit theorem $\sqrt{n}(\bar{x} - \mu)/\sigma \xrightarrow{L} X \sim N(0, 1)$, and $s \xrightarrow{P} \sigma$. Then apply (x), **2c.4**.].

9 Let $(T_{1n}, \ldots, T_{kn})$ be a sequence of k-dimensional statistics such that the a.d. of $[\sqrt{n}(T_{1n} - \theta_1), \ldots, \sqrt{n}(T_{kn} - \theta_k)]$ is multivariate normal with mean zero and covariance matrix σ_{ij}. Consider a function $g(T_{1n}, \ldots, T_{kn}, n)$ involving n explicitly such that the first-order derivatives of the function $g(x_1, \ldots, x_k, n)$ with respect to $x_1, \ldots, x_k$ exist, and that $\partial g/\partial x_i \to G_i(\theta_1, \ldots, \theta_k)$ as $n \to \infty$ and $x_i \to \theta_i$, $i = 1, \ldots, k$. Then the a.d. of

$$v_n^{-1} \sqrt{n}[g(T_{1n}, \ldots, T_{kn}, n) - g(\theta_1, \ldots, \theta_k, n)]$$

is normal with mean zero and variance unity, where

$$v_n = \sum \sum \frac{\partial g}{\partial T_{in}} \frac{\partial g}{\partial T_{jn}} \sigma_{ij}$$

provided $v_n \neq 0$ when the true values of the parameters are substituted for T_{in}, $i = 1, \ldots, k$.

10 Determine the S_0 criterion of (6e.2.4) to test the hypothesis that the observed frequencies in a multinomial distribution are in agreement with assigned expected values. Show that it reduces to the χ^2 test of **6b.1**. Observe that the tests W_0 of (6e.2.3) and $(-2 \log_e \Lambda_0)$ of (6e.2.2) are different from the χ^2 test. Similarly the S_c criterion of (6e.3.6) for testing the goodness of fit is the same as the χ^2 test of goodness of fit of **6b.2**, whereas W_c and $(-2 \log_e \Lambda_c)$ are different from the χ^2 test.

11 Let $X_n \xrightarrow{L} X$. Then $g_n(X_n) \xrightarrow{L} g(X)$, provided $g_n(x) \to g(x)$ as a function of x uniformly in compacts of x and $g(x)$ is continuous.

REFERENCES

Anscombe, F. J. (1948), The transformation of Poisson, Binomial and negative-binomial data, *Biometrika* **35**, 246–254.

Bahadur, R. R. (1966), A note on quantiles in large samples, *Ann. Math. Statist.* **37**, 577–580.

Chernoff, H. and E. L. Lehmann (1954), The use of the maximum likelihood estimates in χ^2 test for goodness of fit, *Ann. Math. Statist.* **25**, 579–586.

Cochran, W. G. (1954), Some methods for strengthening the common χ^2 tests, *Biometrics* **10**, 417–451.

Doob, J. L. (1949), Heuristic approach to the Kolmogorov Smirnov theorems, *Ann. Math. Statist.* **20**, 393–403.

Fisher, R. A. (1935), *Statistical Methods for Research Workers* (first edition, twelfth edition, 1954), Oliver and Boyd, Edinburgh.

Fisher, R. A. (1928), Moments and product moments of sampling distributions, *Proc. London Math. Soc.* **30**, 199–238.

Fisher, R. A. (1949), A preliminary linkage test with agouti and undulated mice, *Heredity* **3**, 229–241.

Gayen, A. K. (1951), The frequency distribution of the product moment correlation coefficient in random samples of any size drawn from non-normal universes, *Biometrika* **38**, 219–247.

Gihman, I. I. (1952), On the empirical distribution function in the case of grouping of observations (in Russian), *Doklady Akademii Nauk, USSR* **82**, 837–840.

Glivenko, V. I. (1933), Sulla determinazione empirica delle leggi di probabilita, *Giorn. Ist. Ital. Attauri* **4**, 92.

Kiefer, J. (1967), On Bahadur representation of sample quantiles, *Ann. Math. Statist.* **38**, 1323–1342.

Neyman, J. and E. S. Pearson (1928), On the use and interpretation of certain test criteria for purposes of statistical inference, *Biometrika* **20A**, 175–240 and 263–294.

Pearson, Karl (1900), On the criterion that a given system of deviations from the probable in the case of a correlated system of variables is such that it can be reasonably supposed to have arisen from random sampling, *Phil. Mag. Ser.* (5) **50**, 157–172.

Smirnov, N. (1939a), On the estimation of the discrepancy between empirical curves of distribution for two independent samples, *Bull. Math. Univ. Moscow* **2**, 3–14.

Smirnov, N. (1939b), Sur les ecarts de la courbe de distribution empirique (in Russian, French summary), *Rec. Math.* N. S. (*Mat. Sborn.*) **6**, 3–26.

Wald, A. (1943), Tests of statistical hypotheses concerning several parameters when the number of observations is large, *Trans. Am. Math. Soc.* **54**, 426–482.

Weiss, Lionel (1970), Asymptotic distribution of quantiles in some nonstandard cases, *Nonparametric Techniques in Statistical Inference*, 343–348.

Yates, F. (1934), Contingency tables involving small numbers and the χ^2 test, *J. Roy. Statist. Soc.* (*suppl.*) **1**, 217–235.

Chapter 7

THEORY OF STATISTICAL INFERENCE

Introduction. In the previous chapters we have already encountered problems involving some aspects of inference such as testing of hypotheses, interval estimation, Bayesian procedures, principle of likelihood, and fiducial inference. Besides these there are other aspects such as the problem of identification (or discrimination), decision functions and acceptance sampling, sequential tests and estimation, and nonparametric methods. Each one of these aspects has a logical basis of its own and has an important part to play in drawing inferences from data. In recent literature on statistics, there has been a tendency on the part of each individual author to select a particular aspect of inference and advocate it in preference to the others. This is clearly not justifiable for the simple fact that the same form of inference is not applicable or not relevant in all situations.

In the present chapter we provide a theoretical discussion of all the important aspects of inference with the purpose of conveying to the reader the essential features and the different forms in which inferences may be drawn from given data.

Crisis in Statistics? Statistical inference is in the nature of inductive logic involving generalizations from the particular, and naturally any tool (statistical method) employed for this purpose will be subject to some controversy. A research worker using a suggested statistical procedure in the analysis of live data should be aware of these controversies and also of the modern trends (not necessarily improved techniques) in statistical inference. Some of these controversies have been referred to briefly in the discussion of the statistical methods presented in this book. The following literature is suggested for further study: Barnard (1949), Barnard, Jenkins, and Winston (1962), Birnbaum (1962), Fisher (1956), Hogben (1957), Jeffreys (1948), Kyburg (1961), Lindley (1953, 1957), Neyman (1961), and Savage (1954, 1963). See also the proceedings of a symposium edited by Godambe and Sprott (1971).

7a TESTING OF STATISTICAL HYPOTHESES

7a.1 Statement of the Problem

This section devoted to the theory of testing statistical hypotheses may appear to be a post mortem examination, since so many test procedures have been set forth and illustrated in the earlier chapters. Historically it is so. Furthermore, no consistent theory of testing statistical hypotheses exists from which all tests of significance can be deduced as acceptable solutions. In many situations test criteria may have to be obtained from intuitive considerations. Formal theories leading to a clear understanding of the problems of hypothesis testing are, nontheless, important. One such theory, contributed by Neyman and Pearson (1933), marked an important development because it unfolded the various complex problems in testing statistical hypotheses and led to the construction of general theories in problems of discrimination (identification), sequential analysis, decision functions, etc., which are discussed in this chapter.

Let S denote the sample space of outcomes of an experiment and x denote an arbitrary element of S. Further let H_0 be a hypothesis (to be called a null hypothesis) which specifies partly or completely the probability measure over a Borel field $\mathscr{B}$ of sets in S. The problem is to decide, on the basis of an observed x, whether H_0 is true or not.

For instance, H_0 may be the hypothesis that a coin is unbiased and we have to test this hypothesis on the basis of an observed number r of successes out of n independent trials. The null hypothesis in this situation *completely specifies* the probability distribution of r.

Or we may have an observed frequency distribution of heights, in suitable class intervals, of a certain number of individuals, and the null hypothesis H_0 specifies the theoretical distribution of heights as normal with unknown mean and variance. The null hypothesis in the latter instance does not provide the exact numerical values of the probabilities for the class intervals; but it does specify the probabilities as *particular functions* of two unknown parameters. The former type of hypothesis is called *simple* and the latter, *composite*.

Whatever procedure may be employed for testing a null hypothesis H_0, that is, *deciding on the basis of observed data whether to reject* H_0 *or not*, there are two types of errors involved, *viz.*, that of (a) rejecting H_0 when it is true, called the *first-kind*, and (b) not rejecting H_0 when, in fact, an alternative hypothesis is true, called the *second-kind.*

A (nonrandomized) test procedure consists in dividing the sample space into two regions w and $S - w$ and deciding to reject H_0 if the observed $x \in w$ and not reject H_0 otherwise. The region w is called the *critical region.*

In later sections we consider more general test procedures known as randomized decision rules, but for the purposes of the present discussion we shall confine our attention to nonrandomized procedures.

Let H_0 be a simple hypothesis in which case the probability of the first kind of error is [denoting the probability measure of a set $A \subset S$, under a given hypothesis g by $P(A \mid g)$]

$$P(w \mid H_0) = \alpha, \tag{7a.1.1}$$

which is called the *level of significance*. The probability of the second kind of error for a particular alternative $h \in H$, the class of alternative hypotheses, is

$$P(S - w \mid h) = \beta(h). \tag{7a.1.2}$$

The function $\gamma(h) = [1 - \beta(h)]$ defined over H is called the *power function*. When H_0 is a composite hypothesis, that is, a class of simple null hypotheses, we may define the level of significance as

$$\alpha = \sup_{h \in H_0} P(w \mid h). \tag{7a.1.3}$$

The problem posed by Neyman and Pearson is that of determining a critical region such that for a given level of significance the second kind of error is as *small* or the power function as *high* as possible. To solve this problem, and some related problems, we shall consider a few mathematical lemmas.

In practice, the points of S will be regarded as the realizations of a r.v. X such that $P(A \mid H) = P(X \in A \mid H)$ for $A \subset S$. We may then write a function T defined over S as a function $T(X)$ of X with the value $T(x)$ when $X = x$. Then $T(X) > \lambda \equiv \{x: T(x) > \lambda\}$ and $P(\{x: T(x) > \lambda\}) = P(T(X) > \lambda)$.

7a.2 Neyman-Pearson Fundamental Lemma and Generalizations

LEMMA 1. *Let $f_0, f_1, \ldots$ be integrable functions over space S with respect to a measure v and let w be any region such that*

$$\int_w f_i \, dv = c_i \text{ (given)}, \qquad i = 1, 2, \ldots. \tag{7a.2.1}$$

Further, let there exist constants $k_1, k_2, \ldots$ such that for the region w_0 within which $f_0 \geqslant k_1 f_1 + k_2 f_2 + \cdots$ and outside which $f_0 \leqslant k_1 f_1 + k_2 f_2 + \cdots$, the conditions (7a.2.1) *are satisfied. Then*

$$\int_{w_0} f_0 \, dv \geqslant \int_w f_0 \, dv. \tag{7a.2.2}$$

Let the common part of the regions w and w_0 be denoted by ww_0. From (7a.2.1), subtracting the common part, we have

$$\int_{w-ww_0} f_i\,dv = \int_{w_0-ww_0} f_i\,dv, \qquad i = 1, 2, \ldots. \tag{7a.2.3}$$

Consider the difference

$$\int_{w_0} f_0\,dv - \int_{w} f_0\,dv = \int_{w_0-ww_0} f_0\,dv - \int_{w-ww_0} f_0\,dv$$

$$\geqslant \int_{w_0-ww_0} \sum k_i f_i\,dv - \int_{w-ww_0} \sum k_i f_i\,dv. \tag{7a.2.4}$$

The expression (7a.2.4) is zero, using (7a.2.3). Hence the result (7a.2.2).

LEMMA 2. (A Generalization of Lemma 1.) *Let $f_0, f_1, f_2, \ldots$ be as in Lemma 1, and ϕ be a point function over S such that $0 \leqslant \phi \leqslant 1$ and*

$$\int f_i \phi\,dv = c_i \text{ (given)}, \qquad i = 1, 2, \ldots \tag{7a.2.5}$$

Further, let there exist a ϕ^* such that

(a) *the conditions* (7a.2.5) *are satisfied,*
(b) $\phi^* = 0$, *if* $f_0 < k_1 f_1 + k_2 f_2 + \cdots$
$= 1$, *if* $f_0 > k_1 f_1 + k_2 f_2 + \cdots$
$=$ *arbitrary, if* $f_0 = k_1 f_1 + k_2 f_2 + \cdots$

Then

$$\int f_0 \phi^*\,dv \geqslant \int f_0 \phi\,dv. \tag{7a.2.6}$$

Let S_1, S_2, and S_3 be the regions $f_0 < \sum k_i f_i$, $f_0 > \sum k_i f_i$ and $f_0 = \sum k_i f_i$. In S_2, $\phi^* - \phi \geqslant 0$ and in S_1, $(\phi^* - \phi) \leqslant 0$ since $\phi^* = 0$. Therefore,

$$f_0(\phi^* - \phi) \geqslant (\sum k_i f_i)(\phi^* - \phi)$$

in each of S_1, S_2, and trivially in S_3 and therefore at all points in S. Thus

$$\int_S f_0(\phi^* - \phi)\,dv \geqslant \int_S (\sum k_i f_i)(\phi^* - \phi)\,dv = 0,$$

using the condition (7a.2.5), and hence the result (7a.2.6).

LEMMA 3. *Let $f_1, \ldots, f_k$ be integrable functions over S with respect to a measure v and $\phi_i(x) \geqslant 0$ be such that $(\phi_1 + \cdots + \phi_k = 1)$. Consider the choice $\phi^*_{i_{r+1}} = \cdots = \phi^*_{i_k} = 0$ and $\phi^*_{i_1}, \ldots, \phi^*_{i_r}$ are non-negative but arbitrary subject to* *the condition* $\phi_{i_1}^* + \cdots + \phi_{i_r}^* = 1$, *when* $f_{i_1} = \cdots = f_{i_r} > f_{i_{r+1}} \geqslant f_{i_{r+2}} \geqslant \cdots \geqslant f_{i_k}$. *Then*

$$\sum_1^k \int f_i \phi_i^* \, dv \geqslant \sum_1^k \int f_i \phi_i \, dv. \tag{7a.2.7}$$

If $f_{i_1} = \cdots = f_{i_r} > f_{i_{r+1}} \geqslant \cdots \geqslant f_{i_k}$, then

$$\sum_1^k f_i \phi_i^* = \frac{(f_{i_1} + \cdots + f_{i_r})}{r} \geqslant \sum f_i \phi_i. \tag{7a.2.8}$$

By taking integrals of both sides of (7a.2.8), we obtain the result (7a.2.7). A slight variation of Lemma 3 is Lemma 4, which is easily proved.

LEMMA 4. *Let* $S_1, \ldots, S_k$ *be a division of the space* S *into* k *mutually exclusive regions. Let* $S_1^*, \ldots, S_k^*$ *be a division into* k *mutually exclusive regions such that*

$$x \in S_i^* \Rightarrow f_i(x) \geqslant f_j(x), \qquad j = 1, \ldots, k.$$

Then

$$\int_{S^*_1} f_1 \, dv + \cdots + \int_{S^*_k} f_k \, dv \geqslant \int_{S_1} f_1 \, dv + \cdots + \int_{S_k} f_k \, dv. \tag{7a.2.9}$$

7a.3 Simple H_0 against Simple H

Let $P(x \mid H_0)$ and be$P(x \mid H)$ the densities at x under H_0 and H respectively with respect to a σ-finite measure v. The problem is that of determining a critical region w such that

$$\int_w P(x \mid H_0) \, dv = \alpha \text{ (assigned value)}, \tag{7a.3.1}$$

$$\int_w P(x \mid H) \, dv \text{ is a maximum.} \tag{7a.3.2}$$

The solution is provided by Lemma 1. The optimum region w is defined by [choosing $f_0 = P(x \mid H)$ and $f_1 = P(x \mid H_0)$]

$$\{x : P(x \mid H) \geqslant kP(x \mid H_0)\}, \tag{7a.3.3}$$

provided there exists a k such that (7a.3.1) is satisfied.

We observe that using the r.v. X the test (7a.3.3) can be written as

$$T = \frac{P(X \mid H)}{P(X \mid H_0)} \geqslant k. \tag{7a.3.4}$$

We determine the distribution of T with respect to H_0. If the distribution is continuous, then there exists a k such that

$$P(T \geqslant k \mid H_0) = \alpha \tag{7a.3.5}$$

for any assigned α. The test, $T \geqslant k$, depends on the simple alternative H. But in some situations the critical region $T \geqslant k$ is independent of the particular alternative used in its derivation among a class of alternative hypothesis. Then we have a test which is uniformly most powerful (U.M.P.) with respect to all such alternatives against a simple hypothesis H_0. Let us consider some examples.

In most of the illustrations, the space S is R^n, the Euclidean real space of n dimensions in which case the r.v. X is the vector variable $\mathbf{x} = (x_1, \ldots, x_n)$. We refer to the r.v.'s $x_1, \ldots, x_n$ as observations from a population or independent observations on a r.v. Occasionally we use the same symbols $x_1, \ldots, x_n$ as dummy variables in integrating with respect to the density at $x_1, \ldots, x_n$.

Consider the class of normal distributions $N(\mu, \sigma)$ with σ fixed and known and μ lying in some range of the real line. Let $x_1, \ldots, x_n$ be n independent observations on a random variable and let H_0 be the null hypothesis specifying its distribution as $N(\mu_0, \sigma)$, where μ_0 is a specified value of the mean μ. In the present example the space S is R^n and X is an n dimensional vector. For an alternative value of μ, the optimum test (7a.3.4) is, with the vector $(x_1, \ldots, x_n)$ denoted by $\mathbf{x}$,

$$P(\mathbf{x} \mid \mu) \geqslant kP(\mathbf{x} \mid \mu_0)$$

that is,

$$\exp\left[-\sum \frac{(x_i - \mu)^2}{2\sigma^2}\right] \geqslant k \exp\left[-\sum \frac{(x_i - \mu_0)^2}{2\sigma^2}\right]. \tag{7a.3.6}$$

With logarithms, we can reduce (7a.3.6) to

$$(\mu - \mu_0) \sum x_i \geqslant c \text{ (constant)}. \tag{7a.3.7}$$

If the class of alternatives is defined by $\mu > \mu_0$, then the region (7a.3.7) is the same as $\bar{x} \geqslant c_1$, where c_1 is chosen such that $P(\bar{x} \geqslant c_1 \mid \mu_0) = \alpha$. Hence c_1 is independent of μ so that a U.M.P. test exists for the class of alternatives $\mu > \mu_0$. Similarly the test $\bar{x} \leqslant c_2$ is U.M.P. for the class of alternatives $\mu < \mu_0$. But if the entire class of alternatives, $\mu < \mu_0$, $\mu > \mu_0$, is considered there is no U.M.P. test.

Since $\bar{x} \sim N(\mu_0, \sigma^2/n)$ under the null hypothesis $\mu = \mu_0$, we can find c_1 such that $P(\bar{x} \geqslant c_1 \mid \mu_0) = \alpha$ (for any given value of α). Thus an optimum test exists for any assigned level of significance for the set of alternatives $\mu > \mu_0$ and similarly for $\mu < \mu_0$.

Consider the problem of testing that the probability of success is π_0 on the basis of an observed number of successes r in n *Bernoullian trials*. If π is an alternative value of the probability of success, an application of Lemma 1 or the result (7a.3.4) gives the critical region

$$\pi^r(1 - \pi)^{n-r} \geqslant k\pi_0{}^r(1 - \pi_0)^{n-r}. \tag{7a.3.8}$$

With logarithms, the region (7a.3.8) is seen to be $r \geqslant r_0$ when the class of alternatives is $\pi > \pi_0$. We have to determine r_0 such that

$$P(r \geqslant r_0 \mid \pi_0) = \alpha. \tag{7a.3.9}$$

Since the distribution of r is discrete there may not exist an r_0 such that (7a.3.9) is true. There exist, however, r_1 and r_2 $(=r_1 + 1)$ such that

$$P(r \geqslant r_1 \mid \pi_0) = \alpha_1 > \alpha > \alpha_2 = P(r \geqslant r_2 \mid \pi_0). \tag{7a.3.10}$$

Of course, when $r < r_1$ the hypothesis cannot be rejected at the level of significance α. Suppose we decide to reject H_0 when $r \geqslant r_2$, and when $r = r_1$ we indulge in a *random experiment* by tossing a coin with a chosen probability θ for heads and decide to reject the hypothesis when a head appears. It is easy to see that when $\theta = (\alpha - \alpha_2)/(\alpha_1 - \alpha_2)$, the ultimate probability of rejection of H_0 is $\alpha_2 + (\alpha_1 - \alpha_2)\theta = \alpha$, which satisfies the requirement of an assigned level of significance.

The *randomized rule* employed in the case of the discrete distribution (due to discontinuity in the d.f.) may be generalized to *any problem*. Given an observed $x \in S$, we toss a coin with a probability $\phi(x)$ for heads and decide to reject the hypothesis H_0 when a head appears. The function ϕ is called a *critical function* where as the critical region w is a special case, obtained by choosing $\phi = 1$ inside w and $\phi = 0$ outside w. The problem of testing H_0 against H reduces to the determination of ϕ such that

$$\int_S P(x \mid H_0)\phi(x)\, dv = \alpha$$

$$\int_S P(x \mid H)\phi(x)\, dv \text{ is a maximum,}$$

where S is the whole space. Lemma 2 provides the answer. The optimum critical function is

$$\begin{aligned} \phi^*(x) &= 1, \text{ if } P(x \mid H) > kP(x \mid H_0) \\ &= 0, \text{ if } P(x \mid H) < kP(x \mid H_0) \\ &= \text{arbitrary if } P(x \mid H) = kP(x \mid H). \end{aligned} \tag{7a.3.11}$$

It may be noted the k in (7a.3.11) *always exists to satisfy the condition* $E[\phi(x) \mid H_0] = \alpha$ [*any given value in* (0, 1)].

To prove this, let us consider the random variable $Y = P(X \mid H)/P(X \mid H_0)$ and its distribution under H_0. If F, the d.f. of Y, is continuous, then for any $0 < \alpha < 1$, there exists y_α such that $1 - F(y_\alpha) = \alpha$ so that $k = y_\alpha$. In other cases, F is at least continuous from the left [see (iii) of **2a**] and for given α there exists a y_α such that

$$1 - F(y_\alpha) \geqslant \alpha \geqslant 1 - F(y_\alpha + 0).$$

Then $k = y_\alpha$, and when the observed Y has the value y_α, we reject H_0 with probability θ, where θ is determined from the equation

$$1 - F(y_\alpha + 0) + \theta[F(y_\alpha + 0) - F(y_\alpha)] = \alpha$$

The solution is practically the same as that obtained in (7a.3.3), except that it covers the situations where a k does not exist—as in the binomial example—such that the region (7a.3.3) has a given size α. We have three regions corresponding to the solution (7a.3.11), w_1, w_2, and $S - w_1 - w_2$. If $x \in w_1$, H_0 is rejected, and if $x \in w_2$, H_0 is not rejected. If $x \in S - w_1 - w_2$, we decide by tossing a coin with a chosen probability θ for heads as in (7a.3.10). It may be seen that θ is determined from the equation $P(w_1) + \theta P(S - w_1 - w_2) = \alpha$.

Thus the randomized rule suggested for binomial distribution is optimum and leads to the maximum power for the class of alternatives $\pi > \pi_0$. Such a procedure involving a random experiment after an event is observed however, may appear arbitrary to a scientific worker who wants to interpret and draw inferences from the data he has collected. We must concede that his mental reservation is justified, as the arbitrariness is introduced just to satisfy the requirement that the so-called level of significance should be exactly equal to a value specified in advance. In examining the hypothesis $\pi = \pi_0$, he may, judging from the circumstances of the case, decide to adopt the lower level attained α_2, and reject the hypothesis whenever $r \geqslant r_2$. Or if α_1 is not much greater than α, he may decide to include r_1 also in the critical region.

We shall consider some general properties of the test (7a.3.11), which may be called the probability ratio test (P.R.T.).

(i) *The P.R.T. is a function of the minimal sufficient statistic when the factorization theorem for the density holds.*

The result is true since the likelihood ratio is a function of the sufficient statistic using the factorization (2d.3.2).

(ii) *The P.R.T. is unbiased in the sense that the power of the test is greater than the level of significance α if $\alpha < 1$ and the distributions under H and H_0 are different.*

Consider the test $\phi(x) = \alpha$ for all x. Then the level of significance $E(\phi \mid H_0) = \alpha$ and so also the power $E(\phi \mid H) = \alpha$. Then obviously $E(\phi^* \mid H) \geqslant \alpha$. If $E(\phi^* \mid H) = \alpha < 1$, then it would imply that $P(X \mid H) = P(X \mid H_0)$ a.e. v contrary to assumption. Hence $E(\phi^* \mid H) > \alpha$.

(iii) *Consider independent observations $x_1, \ldots, x_n$ on a random variable with density $p(\cdot \mid H_0)$ under H_0 and $p(\cdot \mid H)$ under H. Let*

$$\int p(x \mid H_0) \log \frac{p(x \mid H)}{p(x \mid H_0)}\, dx = -\delta, \tag{7a.3.14}$$

where $\delta > 0$ (*by the inequality* 1e.6.6), *and*

$$\frac{P(\mathbf{x}\,|\,H)}{P(\mathbf{x}\,|\,H_0)} = \frac{p(x_1\,|\,H)\ldots p(x_n\,|\,H)}{p(x_1\,|\,H_0)\ldots p(x_n\,|\,H_0)} \geqslant \lambda_n \tag{7a.3.15}$$

be a sequence of likelihood ratio tests such that the first kind of error $\alpha_n \to \alpha$, $0 < \alpha < 1$. *Then*

$$\overline{\lim_{n\to\infty}} \frac{\log \beta_n}{n} \leqslant -\delta, \tag{7a.3.16}$$

where β_n *is the second kind of error.*

By definition

$$\beta_n = \int_{w_n^c} p(x_1\,|\,H) \cdots p(x_n\,|\,H)\, dv \leqslant \lambda_n(1-\alpha_n) \tag{7a.3.17}$$

where $w_n^c = S - w_n$, the complement of the critical region w_n. But under H_0, by the law of large numbers,

$$n^{-1} \sum_{i=1}^{n} \log \frac{p(x_i\,|\,H)}{p(x_i\,|\,H_0)} \xrightarrow{\text{a.s.}} -\delta, \tag{7a.3.18}$$

which implies that

$$\overline{\lim_{n\to\infty}}\, n^{-1} \log \lambda_n \leqslant -\delta \tag{7a.3.19}$$

for α_n to approach $\alpha > 0$. Taking logarithms of both sides of (7a.3.17) and dividing by n, we have

$$n^{-1} \log \beta_n \leqslant n^{-1} \log(1-\alpha_n) + n^{-1} \log \lambda_n .$$

Since $\alpha_n \to \alpha < 1$, $n^{-1}\log(1-\alpha_n) \to 0$ as $n \to \infty$, and thus we have the required result

$$\lim_{n\to\infty} n^{-1} \log \beta_n \leqslant \overline{\lim_{n\to\infty}}\, n^{-1} \log \lambda_n \leqslant -\delta \tag{7a.3.20}$$

by using (7a.3.19).

It is shown further (see Rao, 1962d, for details) that for any sequence of critical regions such that α_n is bounded away from unity

$$\underline{\lim_{n\to\infty}}\, n^{-1} \log \beta_n \geqslant -\delta. \tag{7a.3.21}$$

Combining (7a.3.20) and (7a.3.21), we find that for a L.R.T.

$$\lim_{n\to\infty} n^{-1} \log \beta_n = -\delta. \tag{7a.3.22}$$

The result (7a.3.22) implies, however, that $\beta_n < 0(e^{-n\delta'})$ for $\delta' < \delta$ so that $\beta_n \to 0$ at an exponential rate.

(iv) *Suppose that it is decided to accept H_0 if the ratio in* (7a.3.15) *is* <1 *and reject H_0 otherwise. Let α_n and β_n be the first and second kind of errors for such a procedure. Then $\alpha_n \to 0$ and $\beta_n \to 0$ as $n \to \infty$.*

The result implies that complete discrimination between the hypotheses H_0 and H_1 is possible as the sample size becomes indefinitely large. By the law of large numbers,

$$\frac{1}{n}\sum \log \frac{p(x_i|H)}{p(x_i|H_0)} \xrightarrow{P} -\delta$$

when H_0 is true. Therefore

$$\alpha_n = P\left[\frac{1}{n}\sum \log \frac{p(x_i|H)}{p(x_i|H_0)} > 0 \,|\, H_0\right] \to 0 \qquad \text{as} \quad n \to \infty.$$

By symmetry β_n also $\to 0$.

Chernoff (1952) has obtained more precise estimates of α_n and β_n and of the minimum value of $(\beta_n + \lambda\alpha_n)$ for a test of the type $P(\mathbf{x}|H)/P(\mathbf{x}|H_0) \geqslant k$, when minimized with respect to k.

7a.4 Locally Most Powerful Tests

When a U.M.P. test does not exist, there is not a single region which is best for all alternatives. We may, however, find regions which are best for alternatives close (in some sense) to the null hypothesis and hope that such regions will also do well for distant alternatives. We consider the special case where the density $P(x|H)$ is a function $P(x, \theta)$ of θ, a (single) parameter assuming values in an interval of the real line. The null hypothesis H_0 specifies a value $\theta = \theta_0$ of the parameter. If w is any critical region such that

$$\int_w P(x, \theta_0)\, dv = \alpha,$$

then the power of the test as a function of θ is

$$\gamma(\theta) = \int_w P(x, \theta)\, dv.$$

Let $\gamma(\theta)$ admit Taylor expansion [writing $\gamma(\theta_0) = \alpha$]

$$\gamma(\theta) = \alpha + (\theta - \theta_0)\gamma'(\theta_0) + \frac{(\theta - \theta_0)^2}{2!}\gamma''(\theta_0) + \cdots. \tag{7a.4.1}$$

If H is the class of one-sided alternatives $\theta > \theta_0$, we need to maximize $\gamma'(\theta_0)$ to obtain a locally most powerful one-sided test. We shall assume differentiation under the integral sign so that the quantity to be maximized is

$$\gamma'(\theta_0) = \int_w P'(x, \theta_0)\, dv.$$

(i) *Let w be any region such that*

$$\gamma(\theta_0) = \int_w P(x, \theta_\theta)\, dv = \alpha \tag{7a.4.2}$$

and w_0 be the region $\{x : P'(x, \theta_0) \geqslant kP(x, \theta_0)\}$, where k exists such that the condition (7a.4.2) *is satisfied for w_0. Then*

$$\int_{w_0} P'(x, \theta_0)\, dv \geqslant \int_w P'(x, \theta_0)\, dv. \tag{7a.4.3}$$

The result (7a.4.3) is obtained by an application of Lemma 1. Note that x stands for the entire sample and $P(x, \theta)$ the density at x.

Similarly if the alternatives are $\theta < \theta_0$, the locally most powerful (one-sided) critical region is

$$\{x : P'(x, \theta_0) \leqslant kP(x, \theta_0)\}. \tag{7a.4.4}$$

If the alternatives are both sided we impose the local unbiased restriction $\gamma'(\theta_0) = 0$, in which case the locally most powerful test is obtained by maximizing $\gamma''(\theta_0)$, the coefficient of $(\theta - \theta_0)^2/2$ in the expansion (7a.4.1). Such a test is called a locally most powerful unbiased test. The problem is that of determining a region w such that

$$\gamma(\theta_0) = \int_w P(x, \theta_0)\, dv = \alpha, \tag{7a.4.5}$$

$$\gamma'(\theta_0) = \int_w P'(x, \theta_0)\, dv = 0, \tag{7a.4.6}$$

$$\gamma''(\theta_0) = \int_w P''(x, \theta_0)\, dv \text{ is a maximum.} \tag{7a.4.7}$$

The optimum region is obtained again by the use of Lemma 1.

(ii) *The optimum region w_0 for which* (7a.4.7) *is a maximum is defined by*

$$\{x : P''(x, \theta_0) \geqslant k_1 P'(x, \theta_0) + k_2 P(x, \theta_0)\}, \tag{7a.4.8}$$

provided there exist k_1 and k_2 such that the conditions (7a.4.5) *and* (7a.4.6) *are satisfied.*

Note. (i) and (ii) only establish the sufficiency of the solutions (7a.4.3), (7a.4.4), and (7a.4.8) provided k, k_1, k_2 exist to satisfy the side conditions. Under some conditions on the probability densities, Dantzig and Wald (1951) established the necessity of these types of solutions and also the existence of k_1 and k_2. We shall consider some applications of the results of (i) and (ii).

Let $x_1, \ldots, x_n$ be n independent observations from $N(\mu, \sigma^2)$ and let it be required to test the hypothesis $\mu = \mu_0$ when the alternatives are both sided assuming that σ is known.

It is easy to verify that (representing the n observations by $\mathbf{x}$)

$$\frac{P'(\mathbf{x}, \mu_0)}{P(\mathbf{x}, \mu_0)} = \frac{n(\bar{x} - \mu_0)}{\sigma^2}, \qquad \frac{P''(\mathbf{x}, \mu_0)}{P(\mathbf{x}, \mu_0)} = \frac{-n}{\sigma^2} + \frac{n^2(\bar{x} - \mu_0)^2}{\sigma^4}$$

so that the optimum test (region w_0) defined by (7a.4.8) is

$$\frac{n(\bar{x} - \mu_0)^2}{\sigma^2} \geqslant k_1(\bar{x} - \mu_0) + k_2. \tag{7a.4.9}$$

By symmetry it may be inferred that the unbiased condition (7a.4.6) is satisfied if $k_1 = 0$, so that the test (7a.4.9) can be written

$$\frac{n(\bar{x} - \mu_0)^2}{\sigma^2} \geqslant k_2. \tag{7a.4.10}$$

The constant k_2 can be determined such that the probability of the relation (7a.4.10) is a given value α, since $[n(\bar{x} - \mu_0)^2/\sigma^2] \sim \chi^2(1)$. We thus see that the test (7a.4.10) for both-sided alternatives is different from the test (7a.3.7) for one-sided alternatives.

Let $x_1, \ldots, x_n$ be n independent observations on a random variable with probability density $p(\cdot, \theta)$. The locally powerful one-sided test of $H_0: \theta = \theta_0$, against $\theta > \theta_0$ is, applying (7a.4.3),

$$\sum_{r=1}^{n} \frac{p'(x_r, \theta_0)}{p(x_r, \theta_0)} \geqslant k. \tag{7a.4.11}$$

Let the density function $p(\cdot, \theta)$ satisfy regularity conditions such that

$$E\left[\frac{p'(x, \theta_0)}{p(x, \theta_0)} \middle| \theta_0\right] = 0, \qquad V\left[\frac{p'(x, \theta_0)}{p(x, \theta_0)} \middle| \theta_0\right] = i(\theta_0) \tag{7a.4.12}$$

exist, where it may be noted that $i(\theta_0)$ is Fisher's information at θ_0. If n is large, by an application of the central limit theorem,

$$\frac{1}{\sqrt{ni(\theta_0)}} \sum_{1}^{n} \frac{p'(x_r, \theta_0)}{p(x_r, \theta_0)} \sim N(0, 1) \tag{7a.4.13}$$

asymptotically. Hence if d_α is the upper α-point of $N(0, 1)$ an approximate value of k (in 7a.4.11) is

$$d_\alpha \sqrt{ni(\theta_0)}. \tag{7a.4.14}$$

We thus have a general solution for the problem of a locally powerful one-sided test when n is large.

On the basis of the statistic (7a.4.13) we may set up a two-sided test

$$\left| n^{-1} \sum \frac{p'(x_r, \theta_0)}{p(x_r, \theta_0)} \right| > \frac{d_{\alpha/2}\sqrt{i(\theta_0)}}{\sqrt{n}}, \tag{7a.4.15}$$

which may not be locally most powerful unbiased but is shown to have some optimum properties in large samples (Rao, 1947e; Rao and Poti, 1946i; Wald, 1941).

7a.5 Testing a Composite Hypothesis

As mentioned earlier, a null hypothesis H_0 is said to be composite if it does not specify completely the probability measure over the sample space S, but only specifies the true probability measure as a member of a subset of a wider set of possible probability measures. Some examples of composite hypotheses involving the parameters μ and σ of a normal distribution are:

(a) $H_0: \mu \leqslant 0, \sigma^2 = 1$,
(b) $H_0: \mu = 0, \sigma^2$ arbitrary
(c) $H_0: \sigma^2 = \sigma_0{}^2$ (a given value), μ arbitrary,
(d) $\mu \leqslant 0, \sigma^2$ arbitrary.

In a composite hypothesis the size of the critical region w is defined by

$$\sup_{h \in H_0} \alpha(h) = \sup_{h \in H_0} P(w \mid h). \tag{7a.5.1}$$

If $\alpha(h) = \alpha$ for every member $h \in H_0$, the critical region w has size α and is said to be *similar* to the sample space. It is of interest to determine the entire class of similar regions in any given problem where we may restrict the choice of an optimum critical region to similar regions only. The following results are concerned with the existence and construction of similar region tests.

(i) *Let a similar region w of size α exist and T be a boundedly complete sufficient statistic for the family of measures admissible under H_0. Then the conditional size of w given $T = t$ is α.*

Define the indicator function of the region w

$$\begin{aligned} \chi(x) &= 1, \quad \text{if } x \in w \\ &= 0, \quad \text{if } x \notin w \end{aligned}$$

and choose a simple hypothesis $h \in H_0$. Then the sufficiency of T implies

$$E[\chi \mid T = t, h] = \eta(t) \text{ independent of } h \in H_0. \tag{7a.5.2}$$

Since w is a similar region, we have for every $h \in H_0$.

$$\alpha = E[\chi \mid h] = E_T(E[\chi \mid T, h]) = E[\eta(T) \mid h]. \tag{7a.5.3}$$

Since T is boundedly complete, $\eta(t) = \alpha$ for almost all t, that is, the conditional size of w given $T = t$ is α. [Note that T is said to be boundedly complete if $f(.)$ is bounded and $E[f(T)] = 0$ for all $h \in H_0 \Rightarrow f(t) = 0$ for almost all t.]

(ii) *Let $h \in H_0$ and $g \in H$, the class of alternative hypotheses. The problem of determining w such that*

$$\int_w P(x \mid h)\, dv = \alpha \text{ for every } h \in H_0$$

$$\int_w P(x \mid g)\, dv \text{ is a maximum for a given } g \in H$$

is the same as that of determining an indicator function χ such that

$$E[\chi \mid T = t, h] = \alpha \tag{7a.5.4}$$

$$E[\chi \mid T = t, g] \text{ is a maximum} \tag{7a.5.5}$$

when a statistic T exists such that it is sufficient and boundedly complete for the family of probability measures specified by H_0 and not necessarily for those of H.

The proof is simple and is based on the result obtained in (i). The problem of determining an optimum similar region is thus reduced to that of determining an optimum critical region of a given size considering the conditional probability measures given T over S with respect to the hypotheses h and g.

(iii) *Let $P(\cdot \mid h)$ and $P(\cdot \mid g)$ be the probability densities corresponding to $h \in H_0$ and $g \in H$ and T, a boundedly complete sufficient statistic for the family of measures specified by H_0. Consider a region w_0, inside which*

$$P(x \mid g) \geqslant \lambda(T) P(x \mid h) \tag{7a.5.6}$$

and outside which

$$P(x \mid g) \leqslant \lambda(T) P(x \mid h). \tag{7a.5.7}$$

If $\lambda(T)$ exists such that the conditional size of w_0 given $T = t$ is α, then w_0 is the best similar region for testing H_0. Further, the test is unbiased.

The proof is similar to that of the problem of a simple null hypothesis against a simple alternative considered in **7a.3** except that the discussion is restricted to similar regions which satisfy the property deduced in (i). If in (iii) T is known to be only sufficient and not necessarily boundedly complete, the construction of w_0 as in (7a.5.6) and (7a.5.7) does provide a similar region test satisfying the property of (i). But we would not be able to say that w_0 so obtained is the best for there may exist a similar region for which the property of (i) is not true when T is not boundedly complete and no comparison with such regions has been made. We shall consider some examples to explain the method described in (iii).

Let H_0 be the hypothesis—$\mu = \mu_0$ and σ^2 is arbitrary—to be tested on the basis of n independent observations $x_1, \ldots, x_n$ from $N(\mu, \sigma^2)$. The probability density under H_0, regarding σ^2 as a parameter, is

$$(2\pi\sigma^2)^{-n/2}e^{-\Sigma(x_i-\mu_0)^2/2\sigma^2},$$

which shows that $T = \sum (x_i - \mu_0)^2$ is sufficient for σ^2. Choose a particular hypothesis $h \in H_0: \mu = \mu_0,\ \sigma^2 = \sigma_0^2$, and $g \in H: \mu = \mu_1,\ \sigma^2 = \sigma_1^2$. Then using (7a.5.6, 7a.5.7), the critical region we have to try is

$$\sigma_1^{-n}e^{-\Sigma(x_i-\mu_1)^2/2\sigma_1^2} \geqslant \lambda(T)\sigma_0^{-n}e^{-\Sigma(x_i-\mu_0)^2/2\sigma_0^2}. \tag{7a.5.8}$$

If we take logarithms and write $T = \sum (x_i - \mu_0)^2$, (7a.5.8) reduces to

$$(\mu_1 - \mu_0)(\bar{x} - \mu_0) \geqslant \lambda_1(T) \Rightarrow (\bar{x} - \mu_0) > \lambda_2(T)$$

if $\mu_1 > \mu_0$. The test is equivalent to

$$\frac{\sqrt{n}(\bar{x} - \mu_0)}{\sqrt{T}} \geqslant \lambda_3(T). \tag{7a.5.9}$$

We have to determine $\lambda_3(T)$ such that the conditional size of the region (7a.5.9) given T is α. It is easy to see that

$$\frac{\sqrt{n}(\bar{x} - \mu_0)}{\sqrt{T}} \quad \text{and} \quad T$$

are independently distributed and hence the conditional distribution of $\sqrt{n}(\bar{x} - \mu_0)/\sqrt{T}$ given T is the same as the marginal distribution of

$$\frac{\sqrt{n}(\bar{x} - \mu_0)}{\sqrt{T}}.$$

The distribution of $\sqrt{n}(\bar{x} - \mu_0)/\sqrt{T}$ under the hypothesis H_0 is obtained in (3b.1.9) of Chapter 3. The distribution is seen to be independent of the unknown σ. If b_α is the upper α-point of the distribution, then the region

$$\frac{\sqrt{n}(\bar{x} - \mu_0)}{\sqrt{T}} \geqslant b_\alpha \tag{7a.5.10}$$

has the required properties. Thus with the choice $\lambda_3(T) = b_\alpha$ in (7a.5.9), the condition of result (iii) is satisfied and, therefore, (7a.5.9) is the best similar region test for any alternative hypothesis with $\mu > \mu_0$. We note that

$$\frac{\sqrt{n}(\bar{x} - \mu_0)}{\sqrt{T}} = \frac{\sqrt{n}(\bar{x} - \mu_0)}{\sqrt{n(\bar{x} - \mu_0)^2 + \sum (x_i - \bar{x})^2}} = \frac{t}{\sqrt{t^2 + n - 1}}$$

where

$$t = \frac{\sqrt{n}(\bar{x} - \mu_0)}{\sqrt{\sum (x_i - \bar{x})^2/(n - 1)}}$$

is the Student's t-statistic introduced in Chapters 3 and 4. The region (7a.5.9) may be written in terms of t as

$$\frac{t}{\sqrt{t^2 + (n - 1)}} \geqslant b_\alpha \qquad \text{or equivalently} \quad t \geqslant t_\alpha$$

where t_α is the upper α-point of the t distribution on $(n - 1)$ degrees of freedom.

Similarly, when $\mu < \mu_0$, the test is $t \leqslant -t_\alpha$. When the nature of the alternatives is unknown we may apply the concept of the locally most powerful unbiased region. It may be shown that such a test is provided by $|t| \geqslant t_{\alpha/2}$, the upper $(\alpha/2)$-point of the t-distribution.

The reader may apply the same method to deduce the similar region test for the hypothesis—$\sigma^2 = \sigma_0{}^2$ and μ is arbitrary. If $S = \sum (x_i - \bar{x})^2$, the critical region is

$$\frac{S}{\sigma_0{}^2} \geqslant a_1 \qquad \text{or} \qquad \frac{S}{\sigma_0{}^2} \leqslant a_2,$$

where a_1, a_2 are constants, according as the alternatives are $\sigma^2 > \sigma_0{}^2$ or $\sigma^2 < \sigma_0{}^2$. To obtain the locally most powerful unbiased region we first find the distribution of $u = S/\sigma_0{}^2$ under an alternative value σ^2. The density of u for an alternative σ^2 is

$$P(u, \sigma^2) = \text{c.}(\sigma_0/\sigma)^{n-1} e^{-u\sigma_0{}^2/2\sigma^2} u^{(n-3)/2}.$$

We must determine two values (u_1, u_2) such that

$$\int_{u_1}^{u_2} P(u, \sigma_0{}^2)\, du = 1 - \alpha \qquad (\alpha \text{ being the level of significance}) \qquad (7a.5.11)$$

$$\int_{u_1}^{u_2} \frac{dP(u, \sigma_0{}^2)}{d\sigma_0{}^2}\, du = 0 \qquad \text{(unbiased).} \qquad (7a.5.12)$$

It is easy to verify by straightforward computation that

$$\frac{dP(u, \sigma_0^2)}{d\sigma_0^2} = \frac{(-1)}{\sigma_0^2} \frac{d}{du} [u^{(n-1)/2} e^{-u/2}].$$

Hence the equation (7a.5.12) reduces to

$$u_1^{(n-1)/2} e^{-u_1/2} = u_2^{(n-1)/2} e^{-u_2/2}. \tag{7a.5.13}$$

Equation (7a.5.11) is

$$\int_{u_1}^{u_2} \frac{1}{2^{(n-1)/2} \Gamma\left(\frac{n-1}{2}\right)} e^{-u/2} u^{(n-3)/2} \, du = \alpha. \tag{7a.5.14}$$

The values of u_1, u_2 can be found by solving equations (7a.5.13) and (7a.5.14). The values of unbiased partitions (u_1, u_2) for various values of n and $\alpha = 0.05$ and 0.01 are given in R.M.M. Tables.

Let $x_1, \ldots, x_n$ be n independent observations from $N(\theta, 1)$ and H_0 be $\theta \leqslant 0$. Let us construct a similar region test for H_0. The statistic $\bar{x}$ is sufficient for θ, for its entire range, $-\infty < \theta < \infty$. If χ is the indicator function of a similar region w, then

$$E[\chi \,|\, \bar{x}, \theta] = \alpha, \qquad \theta \leqslant 0. \tag{7a.5.15}$$

Since $\bar{x}$ is sufficient for θ, for the whole range $-\infty < \theta < \infty$, the result (7a.5.15) is true for $\theta > 0$ also. The power of w for any $\theta > 0$ is then

$$\underset{\bar{x}}{E} \{E[\chi \,|\, \bar{x}, \theta]\} = E(\alpha) = \alpha,$$

which is independent of θ and is the same as the first kind of error. The condition that the test should be based on a similar region thus gives a useless test. On the other hand, let us choose the simple hypothesis $\theta = 0$ in H_0 and construct a test against an alternative $\theta > 0$. As shown earlier the test criterion is $\bar{x} \geqslant \lambda$ where λ is determined such that $P(\bar{x} \geqslant \lambda \,|\, \theta = 0) = \alpha$. In such a case, it may be seen that

$$P(\bar{x} \geqslant \lambda \,|\, \theta \leqslant 0) \leqslant \alpha \qquad \text{and} \qquad P(\bar{x} \geqslant \lambda \,|\, \theta > 0) > \alpha$$

so that the condition of the size as defined in (7a.5.1) is satisfied and the power is always greater than α and in fact increases with θ, which is a desirable property. Thus a reasonable test of the composite hypothesis $\theta \leqslant 0$ exists and is superior to the similar region test. This example shows that a certain amount of caution is necessary in accepting test criteria derived *by any of the principles stated* without examination of their performance (nature of the power curve, etc.).

Let us now consider composite hypotheses in one parameter family of distributions, where similar regions may not exist. If θ is the parameter, we may have the following null (H_0) and alternative (H) hypotheses:

	H_0	H
(a)	$\theta \leqslant \theta_0$	$\theta > \theta_0$
(b)	$\theta \leqslant \theta_0$ or $\theta \geqslant \theta_1$	$\theta_0 < \theta < \theta_1$
(c)	$\theta_0 \leqslant \theta \leqslant \theta_1$	$\theta < \theta_0$, or $\theta > \theta_1$

Let $\alpha(\theta) = E(\phi \mid \theta)$, where ϕ is a critical function. In general, the problem is one of determining ϕ such that $\alpha(\theta')$ is a maximum when $\theta' \in H$ subject to the condition

$$\alpha(\theta) \leqslant \alpha \text{ when } \theta \in H_0. \tag{7a.5.16}$$

The problem may not have a reasonable solution in general. Proposition (iv) relates to a satisfactory situation when the families of densities $P(x, \theta)$ have a *monotone likelihood ratio*, i.e., there exists a real-valued function $T(x)$ such that for any $\theta < \theta'$, the distributions of X are distinct and the ratio $P(x, \theta')/P(x, \theta)$ is a nondecreasing function of $T(x)$.

(iv) *Let θ be a real parameter, and let the random variable X have probability density $P(x, \theta)$ with a monotone likelihood ratio in $T(x)$. For testing $H_0\colon \theta \leqslant \theta_0$ against $\theta > \theta_0$, there exists a U.M.P. test, which is given by*

$$\phi^*(x) = \begin{cases} 1 & \text{if } T(x) > c \\ \delta & \text{if } T(x) = c \\ 0 & \text{if } T(x) < c \end{cases}$$

where c and δ are determined such that $E(\phi^* \mid \theta_0) = \alpha$. Further $\alpha^*(\theta) = E(\phi^* \mid \theta)$ is a strictly increasing function for points θ at which $\alpha^*(\theta) < 1$.

Consider $H_0\colon \theta_0$ against $H\colon \theta_1 > \theta_0$. The most powerful test (7a.3.11) is based on the regions $P(x \mid \theta_1) >, =, < kP(x \mid \theta_0)$ which are equivalent to $T(x) >, =, < c$ so that the test ϕ^* is U.M.P. for $\theta = \theta_0$ against $\theta > \theta_0$. We show that ϕ^* is also U.M.P. for $\theta \leqslant \theta_0$ against $\theta > \theta_0$.

Of course, the same test is most powerful for any θ' against $\theta'' > \theta'$ at a level $\alpha^*(\theta')$, in which case $\alpha^*(\theta'') > \alpha^*(\theta')$ using the proposition (ii), **7a.3**. Then $\alpha^*(\theta) \leqslant \alpha$ for $\theta \leqslant \theta_0$, so that ϕ^* satisfies (7a.5.16). Suppose ϕ is another test satisfying (7a.5.16), i.e., $E(\phi \mid \theta_0) \leqslant \alpha$. Then $E(\phi \mid \theta) \leqslant E(\phi^* \mid \theta)$ for $\theta > \theta_0$ for otherwise there is a contradiction, since ϕ^* is the most powerful test of $H_0 : \theta_0$ at level α against $H : \theta > \theta_0$.

A typical example of monotone likelihood ratio family is the one-parameter exponential family of distributions (See **3b.7**)

$$P(x, \theta) = a(\theta)b(x)\exp[d(\theta)T(x)] \tag{7a.5.17}$$

where $d(\theta)$ is strictly monotone in the range of θ. The family has a monotone likelihood ratio in $T(x)$ or in $-T(x)$ according as $d(\theta)$ is increasing or decreasing in θ. It may be seen that a number of well-known, discrete (binomial, Poisson, negative binomial, etc.), and continuous distributions (normal, gamma, etc.) belong to this family. Proposition (v) relates to the family (7a.5.17).

(v) *For testing the hypothesis $H_0 : \theta \leqslant \theta_1$ or $\theta \geqslant \theta_2$ against the alternative $H : \theta_1 < \theta < \theta_2$ in the family* (7a.5.17) *there exists a U.M.P. test given by*

$$\phi^*(x) = \begin{cases} 1 & \text{when } c_1 < T(x) < c_2, \\ \delta_i & \text{when } T(x) = c_i,\ i = 1, 2, \\ 0 & \text{when } T(x) < c_1 \text{ or } T(x) > c_2, \end{cases} \tag{7a.5.18}$$

where the c's and δ's are determined such that

$$E(\phi^* \mid \theta_1) = E(\phi^* \mid \theta_2) = \alpha. \tag{7a.5.19}$$

Let us consider $H_0 : \theta = \theta_1$ or $\theta = \theta_2$ and determine ϕ such that $E(\phi \mid \theta)$ for given $\theta_1 < \theta < \theta_2$ is a maximum subject to the conditions $E(\phi \mid \theta_1) = E(\phi \mid \theta_2) = \alpha$. In such a case, Lemma 2 of **7a.2** is applicable giving

$$\phi^*(x) = \begin{cases} 1 & \text{if } P(x \mid \theta) > k_1 P(x \mid \theta_1) + k_2 P(x \mid \theta_2) \\ \delta & \text{if } P(x \mid \theta) = k_1 P(x \mid \theta_1) + k_2 P(x \mid \theta_2) \\ 0 & \text{if } P(x \mid \theta) < k_1 P(x \mid \theta_1) + k_2 P(x \mid \theta_2). \end{cases} \tag{7a.5.20}$$

Substituting the special form (7a.5.17) for $P(x \mid \theta)$ in (7a.5.20), we find that the test (7a.5.20) is equivalent to (7a.5.18). [The result is proved by showing that the three regions in (7a.5.20) correspond to $y^a \gtreqless b + cy$, where $y = e^{dT}$ and hence to intervals of T specified in (7a.5.18)]. Then the test (7a.5.20) depending only on T is U.M.P. for $H_0 : \theta = \theta_1$ or $\theta = \theta_2$. Now we apply an argument similar to that used in proposition (iv) to establish that the test is U.M.P. for $H_0 : \theta \leqslant \theta_1$ or $\theta \geqslant \theta_2$.

Unbiased Test. A test is said to be unbiased if $E(\phi \mid \theta) \leqslant \alpha$ when $\theta \in H_0$ and $\geqslant \alpha$ when $\theta \in H$. We have already seen in **7a.3** that U.M.P. does not exist for testing $H_0 : \theta = \theta_0$ against $\theta \neq \theta_0$ where θ is the mean of the normal distribution. In such a situation, we may impose the restriction of unbiasedness and look for an U.M.P. test. Proposition (vi) provides an answer in particular cases.

(vi) *Let $P(x \mid \theta)$ belong to the one-parameter exponential family of distributions. Then U.M.P.U.* (*U.M.P. unbiased*) *tests* (*a*) *of $H_0 : \theta = \theta_0$ against*

$\theta \neq \theta_0$ and (b) of $H_0 : \theta_1 \leqslant \theta \leqslant \theta_2$ against $H_1 : \theta < \theta_1$ or $\theta > \theta_2$ at level α exist and each test is defined by the critical function.

$$\phi(x) = \begin{cases} 1 & \text{if } T(x) < c_1 \text{ or } > c_2 \\ \delta_i & \text{if } T(x) = c_i,\ i = 1, 2, \\ 0 & \text{if } c_1 < T(x) < c_2 \end{cases}$$

where $c_1, c_2, \delta_1, \delta_2$ are determined by the equations $E(\phi \mid \theta_0) = \alpha, E(T\phi \mid \theta_0) = \alpha E(T \mid \theta_0)$ for problem (a) and $E(\phi \mid \theta_1) = E(\phi \mid \theta_2) = \alpha$ for problem (b).

For a detailed proof, the reader is referred to Lehmann (1959), Chapter 4.

7a.6 Fisher-Behrens Problem

Let n_1, $\bar{x}_1$, s_1^2 and n_2, $\bar{x}_2$, s_2^2 be the size, sample mean, and variance of independent observations from $N(\mu_1, \sigma_1^2)$ and $N(\mu_2, \sigma_2^2)$ respectively. What is an appropriate test based on these statistics of the hypothesis, $\mu_1 - \mu_2 = \xi$ (a given value) when σ_1^2 and σ_2^2 are unknown and possibly different?

It may be noted that when $\sigma_1 = \sigma_2$, the statistic

$$t = \sqrt{\frac{n_1 n_2}{n_1 + n_2}} \frac{\bar{x}_1 - \bar{x}_2 - (\mu_1 - \mu_2)}{s}. \tag{7a.6.1}$$

where $s^2 = [(n_1 - 1)s_1^2 + (n_2 - 1)s_2^2]/(n_1 + n_2 - 2)$, has a t distribution on $(n_1 + n_2 - 2)$ D.F., and provides a similar region test for any hypothesis concerning the difference $\mu_1 - \mu_2$. The test (7a.6.1) may be derived in the same way as the t-test in **7a.5**, first, by observing that when $\mu_1 - \mu_2 = \xi$ (given) and $\sigma_1 = \sigma_2$, complete sufficient statistics exist for the unknown parameters and, second, by applying the method of [(iii), **7a.5**].

When $\sigma_1 \neq \sigma_2$ and $\mu_1 - \mu_2 = \xi$ (given), it is not known whether complete sufficient statistics exist for the unknown parameters and hence the method of [(iii), **7a.5**] cannot be applied for the determination of a similar region test. But does any similar region test exist? The question has been recently answered by Linnik, who showed that similar regions based on incomplete sufficient statistics exist; the test procedure is, however, complicated.

Fisher and Behrens (see Fisher, 1935, 1956) proposed a test based on the fiducial distribution of $\mu_1 - \mu_2$, which does not satisfy the criteria of testing of hypotheses as laid down in **7a.1**. For instance, the first kind of error is not below a specified value for all alternatives of the null hypothesis.

Various approximate tests have been suggested; one, due to Banerji (1960), strictly satisfies the inequality concerning the first kind of error. The proposed test is

$$|\bar{x}_1 - \bar{x}_2 - (\mu_1 - \mu_2)| \geqslant \left(\frac{t_1^2 s_1^2}{n_1} + \frac{t_2^2 s_2^2}{n_2}\right)^{1/2},$$

where t_i is the upper $(\alpha/2)$-point of the t distribution on $(n_i - 1)$ D.F. The first kind of error for the test is $\leqslant \alpha$. For other approximate tests see, Chernoff (1949), Cochran and Cox (1957), Linnik (1963) and Welch (1947).

7a.7 Asymptotic Efficency of Tests

Consider the problem of testing a simple hypothesis that the value of a parameter θ is θ_0 against the alternatives such as $\theta \neq \theta_0$, $\theta > \theta_0$, or $\theta < \theta_0$. Let the data consist of n observations and denote by $\alpha_n(\theta_0)$ the first kind of error and $\beta_n(\theta)$ the second kind of error for the alternative θ. The power function is $\gamma_n(\theta) = 1 - \beta_n(\theta)$.

We have already seen that there is usually no test for which $\gamma_n(\theta)$ has the largest value for each θ compared to any other test with the same value of $\alpha_n(\theta_0)$. Nonetheless, we will have occasion to compare two given tests and choose one in preference to the other. But given two tests with the same value of the first kind of error $\alpha_n(\theta_0)$, we may find that the power functions $\gamma_n^{(1)}(\theta)$ and $\gamma_n^{(2)}(\theta)$ satisfy one kind of inequality for some values of θ and the reverse inequality for the other values of θ in the admissible range. Such a knowledge is extremely useful, but in practice the computation of the power functions is extremely difficult and we may have to depend on some criterion, preferably a single numerical measure which is easily computable, for making a choice between tests. It is clear that such an approach is unsatisfactory for test criteria based on finite samples. We shall therefore consider the case of the large samples, where the distinctions between tests are expected to be clearcut.

For any reasonable sequence of tests $\gamma_n(\theta) \to 1$ as $n \to \infty$, for any fixed alternative θ ($\neq \theta_0$). Such a test is called a consistent test. A test which is not consistent has, indeed, a poor performance in the sense that it does not enable us to detect the alternative—with certainty if it is true—as $n \to \infty$. Since for any consistent test, $\gamma_n(\theta) \to 1$ as $n \to \infty$ for fixed θ, the limiting value of the power function $\gamma_n(\theta)$ cannot serve as a criterion for distinguishing between tests. But for any n, however large it may be, there will be alternative values of θ possibly close to θ_0 such that $\gamma_n(\theta)$ is less than unity. We shall therefore study the behavior of $\gamma_n(\theta)$ as $\theta \to \theta_0$ and $n \to \infty$ in some specified way and develop some measures of asymptotic efficiency.

Measures of Asymptotic Efficiency. In the context of sequences of tests based on i.i.d. observations, some of the alternative measures of efficiency proposed are as follows:

$$\text{(a)} \quad e_1 = \varliminf_{n \to \infty} n^{-1}[\gamma_n'(\theta_0)]^2 \tag{7a.7.1}$$

where $\gamma_n'(\theta_0)$ is the first derivative of $\gamma_n(\theta)$ at $\theta = \theta_0$.

(b) $e_2 = \varlimsup_{n\to\infty} n^{-1}\gamma_n''(\theta_0)$ (7a.7.2)

where $\gamma_n''(\theta_0)$ is the second derivative of $\gamma_n(\theta)$ at $\theta = \theta_0$.

(c) $$e_3 = -\varliminf_{\theta\to\theta_0}\varliminf_{n\to\infty}\frac{2n^{-1}\log\beta_n(\theta)}{(\theta-\theta_0)^2}.$$ (7a.7.3)

(d) $$e_4 = \varlimsup_{n\to\infty}\gamma_n\left(\theta_0 + \frac{\delta}{\sqrt{n}}\right),$$ for some chosen δ. (7a.7.4)

Each measure, e_1 to e_4, provides a description of the *local behavior* of the power curve at $\theta = \theta_0$ in large samples. Let us denote by $e_r(T_n)$, the value of the measure e_r for a sequence of tests $\{T_n\}$. If $e_r(T_{1n}) > e_r(T_{2n})$, it would imply that in large samples, the power curve for $\{T_{1n}\}$ is higher than that for $\{T_{2n}\}$ in the neighborhood of θ_0.

Under mild regularity conditions on the common p.d. (probability density) of each observation, all the measures e_1 to e_4 are bounded above by quantities independent of any test procedure and which are easily computable. If for any sequence of tests $\{T_n\}$, the upper bound is attained for $e_r(T_n)$, then $\{T_n\}$ is asymptotically best according to the measure e_r. The upper bounds are quoted without proof and with reference to the original papers where the proofs can be found. The upper bounds are in fact for lim sup in (7a.7.1–7a.7.4).

(a′) $e_1 \leqslant (i/2\pi)^{1/2}\exp(-a^2/2)$, where i is Fisher's information on θ contained in a single observation and a is the upper α-point of $N(0, 1)$, (Rao, 1962d).

(b′) $e_2 \leqslant i$.

(c′) $e_3 \leqslant i$, (Rao, 1962d).

(d′) $e_4 \leqslant 1 - \Phi(a - \delta\sqrt{i})$, where Φ is the distribution function of $N(0, 1)$, (Rao, 1963a).

The computations of these measures are not easy for any general test criterion. The most important cases are, however, tests based on asymptotically normally distributed statistics. We shall illustrate the computation of some of the measures in such cases.

Computation of e_1. From the definition it is seen that e_1 provides an estimate of the slope of the power curve at $\theta = \theta_0$ and the measure is useful for comparing one-sided tests (i.e., when the alternatives are $\theta > \theta_0$ or $\theta < \theta_0$). For unbiased tests, the value of e_1 would be zero, in which case e_2, which measures the curvature of the power curve at $\theta = \theta_0$, is relevant.

Let us consider i.i.d. observations each with p.d. $p(x, \theta)$ such that $i = E(d \log p/d\theta)^2 < \infty$, at $\theta = \theta_0$. Let $x_1, \ldots, x_n$ denote the first n observations and $P(X_n, \theta) = p(x_1, \theta) \ldots p(x_n, \theta)$. Then

$$Z_n = n^{-1/2} \frac{P'(X_n, \theta_0)}{P(X_n, \theta)} \tag{7a.7.5}$$

has asymptotic normal distribution with mean zero and variance i. Let T_n be a consistent estimator of θ such that the a.d. of $\sqrt{n}(T_n - \theta_0)$ is $N[0, \sigma^2(\theta_0)]$. We have the following proposition leading to the computation of e_1 for the sequence of tests based on $\sqrt{n}(T_n - \theta_0)$.

(i) *Let the sequence of test criteria be $U_n \geqslant a$, where $U_n = \sqrt{n}(T_n - \theta_0)/\sigma(\theta_0)$ and a is the α probability point of $N(0, 1)$. Further, let the asymptotic joint distribution of U_n and Z_n be bivariate normal with the correlation coffiecient equal to ρ. If differentiation under the integral sign is valid for the function $P(X_n, \theta)$, then $e_1 = (i\rho^2/2\pi)\exp(-a^2/2) \propto \rho^2$.*

By definition

$$n^{-1/2}\gamma_n'(\theta_0) = \int_{U_n \geqslant a} n^{-1/2}P'(X_n, \theta_0)\, dv_n$$

$$= \int_{U_n \geqslant a} Z_n P(X_n, \theta_0)\, dv_n,$$

$$\to \int_{U \geqslant a} ZP(Z, U)\, dZ\, dU, \quad \text{as} \quad n \to \infty, \tag{7a.7.6}$$

where $P(Z, U)$ is the bivariate normal density with $V(Z) = i$, $V(U) = 1$, and $\operatorname{cov}(Z, U) = \rho\sqrt{i}$. The integral (7a.7.6) is easily evaluated to be

$$\left(\frac{i\rho^2}{2\pi}\right)^{1/2} e^{-a^2/2}, \tag{7a.7.7}$$

which proves the result of (i).

If the statistic T_n is efficient in the sense of (5c.2.5 and 5c.2.6) then $\rho = 1$ and e_1 attains the upper bound. Thus efficient estimators provide asymptotically best test criteria according to the e_1 measure. If T_{1n} and T_{2n} are two sequences of statistics satisfying the conditions of (i) with ρ_1 and ρ_2 as the asymptotic correlations with Z_n, the relative efficiency of T_{2n} compared with T_{1n} is $(\rho_2/\rho_1)^2$, so that we may define the efficiency of a test by ρ^2 itself instead of by (7a.7.7) as in the case of estimating efficiency. We will denote ρ^2 as the measure e_1'.

The computations of e_2 and e_3 are difficult and require far more assumptions on the sequence of statistics T_n. We shall now consider e_4, which is called *Pitman efficiency* and which has been frequently used to compare large sample tests.

(ii) *Let T_n be a test statistic based on the first n observations and let the critical region be $T_n \geqslant \lambda_n$. Suppose that*

(a) $\lim_{n \to \infty} P(T_n \geqslant \lambda_n) = \alpha$ *(fixed value)* > 0,

(b) *there exist a positive quantity r and functions $\mu(\theta)$, $\sigma(\theta)$ such that*

$$\lim_{n \to \infty} P\left\{n^r \frac{T_n - \mu(\theta_0 + \delta n^{-r})}{\sigma(\theta_0 + \delta n^{-r})} < y \,|\, \theta_0 + \delta n^{-r}\right\} = \Phi(y) \quad (7a.7.8)$$

for every real y, where $\Phi(y)$ is the d.f. of $N(0, 1)$,

(c) *$\mu(\theta)$ has a derivative $\mu'(\theta_0)$ at $\theta = \theta_0$, which is positive and $\sigma(\theta)$ is continuous at θ_0.*

Then

$$\lim_{n \to \infty} \gamma(\theta_0 + \delta n^{-r}) = \Phi\left(\frac{\delta \mu'(\theta_0)}{\sigma(\theta_0)} - a\right) \quad (7a.7.9)$$

where a is the upper α probability point of $N(0, 1)$.

In many practical applications $r = \frac{1}{2}$, which is chosen in the definition of e_4. We could have defined e_4 as the limit of $\gamma(\theta_0 + \delta n^{-r})$ by choosing the value of r to satisfy assumption (b). Now, by definition

$$\gamma_n(\theta_0 + \delta n^{-r}) = P(T_n \geqslant \lambda_n \,|\, \theta_0 + \delta n^{-r})$$

$$= P\left\{n^r \frac{T_n - \mu(\theta_0 + \delta n^{-r})}{\sigma(\theta_0 + \delta n^{-r})} \geqslant n^r \frac{\lambda_n - \mu(\theta_0 + \delta n^{-r})}{\sigma(\theta_0 + \delta n^{-r})} \,\bigg|\, \theta_0 + \delta n^{-r}\right\}$$

$$= \Phi\left(-n^r \frac{\lambda_n - \mu(\theta_0 + \delta n^{-r})}{\sigma(\theta_0 + \delta n^{-r})}\right) + \varepsilon_n(\delta), \quad (7a.7.10)$$

where $\varepsilon_n(\delta) \to 0$ as $n \to \infty$, observing that the convergence of (7a.7.8) is uniform in y, although not in δ. Substituting $\delta = 0$, we have

$$\gamma_n(\theta_0) = \Phi\left(-n^r \frac{\lambda_n - \mu(\theta_0)}{\sigma(\theta_0)}\right) + \varepsilon_n(0). \quad (7a.7.11)$$

Taking limits of both sides of (7a.7.11)

$$\alpha = \lim_{n \to \infty} \Phi\left(-n^r \frac{\lambda_n - \mu(\theta_0)}{\sigma(\theta_0)}\right)$$

which shows that

$$n^{-r} \frac{\lambda_n - \mu(\theta_0)}{\sigma(\theta_0)} = a + \eta_n, \qquad \eta_n \to 0 \quad \text{as} \quad n \to \infty.$$

Therefore $\lambda_n = n^r(a + \eta_n)\sigma(\theta_0) + \mu(\theta_0)$. Hence the argument of Φ in (7a.7.10),

$$-n^r \frac{\lambda_n - \mu(\theta_0 + \delta n^{-r})}{\sigma(\theta_0 + \delta n^{-r})} = \frac{\delta \dfrac{\mu(\theta_0 + \delta n^{-r}) - \mu(\theta_0)}{\delta n^{-r}} - (a + \eta_n)\sigma(\theta_0)}{\sigma(\theta_0 + \delta n^{-r})}$$

$$\rightarrow \frac{\delta\mu'(\theta_0) - a\sigma(\theta_0)}{\sigma(\theta_0)} = \frac{\delta\mu'(\theta_0)}{\sigma(\theta_0)} - a,$$

which proves the required result.

(iii) *A sufficient condition for the assumption (b) of (ii) to hold is*

$$\lim_{n\to\infty} P\left\{n^r \frac{T_n - \mu(\theta)}{\sigma(\theta)} < y \,|\, \theta\right\} = \Phi(y) \tag{7a.7.12}$$

uniformly in θ for $\theta_0 \leqslant \theta \leqslant \theta_0 + \eta$, where η is any positive number.

The result of (iii) is easy to prove. In practice it is easier to verify the condition (7a.7.12) instead of (b) of (ii). Furthermore, if we consider only test criteria satisfying the conditions of (ii) or (iii), the efficiency may be defined as $[\mu'(\theta_0)/\sigma(\theta_0)]^{1/r}$, which is an increasing function of (7a.7.9) for given δ and a. We call the measure $[\mu'(\theta_0)/\sigma(\theta_0)]^{1/r}$ as e_4'.

The Concept of Relative Sample Sizes. Let us consider the measure e_1 and two sequences of tests T_{1n} and T_{2n}. Denote by $\gamma_{in}'(\theta_0)$, the slope of the power curve for given n for the ith test, $i = 1, 2$. Let us suppose that for a given n, there exists a number N_n such that

$$\gamma_{1n}'(\theta_0) = \gamma_{2N_n}'(\theta_0), \tag{7a.7.13}$$

that is, the slopes are equal and that $N_n \to \infty$ as $n \to \infty$. The efficiencies $e_1(T_1)$ and $e_1(T_2)$ are, by definition,

$$e_1(T_1) = \lim_{n\to\infty} [n^{-1/2}\gamma_{1n}'(\theta_0)]^2,$$

$$e_1(T_2) = \lim_{N_n\to\infty} [N_n^{-1/2}\gamma_{2N_n}'(\theta_0)]^2.$$

But $e_1(T_1)$ is equal to, from the relation (7a.7.13),

$$\lim_{n\to\infty}[n^{-1/2}\gamma_{1n}'(\theta_0)]^2 = \lim_{n\to\infty}[n^{-1/2}\gamma_{2N_n}'(\theta_0)]^2$$

$$= \lim_{n\to\infty}\left[\frac{n^{-1/2}}{N_n^{-1/2}} \cdot N_n^{-1/2}\gamma_{2N_n}'(\theta_0)\right]^2$$

$$= e_1(T_2)\lim_{n\to\infty}\left(\frac{n^{-1}}{N_n^{-1}}\right). \tag{7a.7.14}$$

Hence $\lim (n/N_n) = e_1(T_2)/e_1(T_1)$ as $n \to \infty$, that is, in large samples the ratio of the asymptotic efficiencies is equal to the limit of the inverse ratio of sample sizes needed for the two tests to have the same slope for the power curves. Thus, we have a practically useful interpretation of the measure e_1.

Let us consider the measure e_4'. It is seen that under the assumptions of the proposition (ii) or (iii)

$$\lim_{n\to\infty} \gamma_n(\theta_0 + \delta n^{-r}) = \Phi\left(\frac{\delta\mu'(\theta_0)}{\sigma(\theta_0)} - a\right).$$

Let us consider two sequences of test statistics $\{T_{1n}\}$ and $\{T_{2n}\}$ and denote by μ_1, σ_1, and μ_2, σ_2 the relevant functions for T_{1n} and T_{2n}. As before, let n and N_n be such that

$$\gamma_{1n}(\theta_0 + \delta n^{-r}) = \gamma_{2N_n}(\theta_0 + \delta n^{-r}) \tag{7a.7.15}$$

and suppose that $N_n \to \infty$ as $n \to \infty$. Now

$$\begin{aligned} \Phi\left(\frac{\delta\mu_1'(\theta_0)}{\sigma_1(\theta_0)} - a\right) &= \lim_{n\to\infty} \gamma_{1n}(\theta_0 + \delta n^{-r}) \\ &= \lim_{n\to\infty} \gamma_{2N_n}[\theta_0 + \delta N_n^{-r}(n^{-r}/N_n^{-r})] \\ &= \lim_{N_n\to\infty} \gamma_{2N_n}(\theta_0 + \delta\lambda^{-r}N_n^{-r}) && (7a.7.16) \\ &= \Phi\left(\frac{\lambda^{-r}\delta\mu_2'(\theta_0)}{\sigma_2(\theta_0)} - a\right), && (7a.7.17) \end{aligned}$$

where $\lambda = \lim(n/N_n)$ as $n \to \infty$. The transition from (7a.7.16) to (7a.7.17) needs careful demonstration following the arguments used in the proof of (ii). The equation (7a.7.17) gives

$$\frac{\delta\mu_1'(\theta_0)}{\sigma_1(\theta_0)} = \frac{\delta\mu_2'(\theta_0)}{\lambda^r\sigma_2(\theta_0)}$$

or

$$\begin{aligned} \lambda = \lim_{n\to\infty} \frac{n}{N_n} &= \left\{\frac{\mu_2'(\theta_0)}{\sigma_2(\theta_0)} \div \frac{\mu_1'(\theta_0)}{\sigma_1(\theta_0)}\right\}^{1/r} \\ &= \frac{e_4'(T_2)}{e_4'(T_1)}. \end{aligned}$$

Thus, the relative efficiency of T_2 with respect to T_1 is equal to the inverse ratio of sample sizes needed to have the same power at a sequence of values $\theta_n \to \theta_0$ such that $\gamma_n(\theta_n) \to \gamma$ (any assigned value) < 1, as $n \to \infty$.

There are a few other measures of interest of asymptotic efficiency of tests due to Bahadur (1960), Chernoff (1952), Noether (1955), and others, but they are somewhat more specialized than those considered in this section. For further details on asymptotic efficiency of tests and applications to certain parametric and nonparametric tests, reference may be made to Chernoff and Savage (1958), Pitman (1949), Hodges and Lehman (1956), Hoeffding and Rosenblatt (1955).

7b CONFIDENCE INTERVALS

7b.1 The General Problem

Numerous examples have been given in the previous chapters of estimating parameters by intervals. We shall formulate the problem in a general way due to Neyman (1935, 1937) and examine the extent to which satisfactory solutions can be obtained.

Let x denote a sample point and $\boldsymbol{\theta}$ a parameter (which may be multi-dimensional) specifying the probability distribution over the sample space. Let Θ denote the set of admissible values of $\boldsymbol{\theta}$. For given x let $I(x)$ denote a set of $\boldsymbol{\theta}$ values. I is said to be a confidence set estimator of $\boldsymbol{\theta}$ with a confidence coefficient $(1-\alpha)$, or in short $(1-\alpha)$ *confidence set estimator*, if

$$P[x\colon \boldsymbol{\theta} \in I(x) \,|\, \boldsymbol{\theta}] = 1 - \alpha, \qquad \text{for every} \quad \boldsymbol{\theta} \in \Theta. \tag{7b.1.1}$$

A confidence set estimator I is said to be unbiased if

$$P[\boldsymbol{\theta}_1 \in I \,|\, \boldsymbol{\theta}_2] \leqslant 1 - \alpha, \boldsymbol{\theta}_1, \boldsymbol{\theta}_2 \in \Theta. \tag{7b.1.2}$$

Let I and J be two set estimators of $\boldsymbol{\theta}$ with the same confidence coefficient. I is said to be shorter than J if

$$P[\boldsymbol{\theta}_1 \in I \,|\, \boldsymbol{\theta}_2] \leqslant P[\boldsymbol{\theta}_1 \in J \,|\, \boldsymbol{\theta}_2], \boldsymbol{\theta}_1, \boldsymbol{\theta}_2 \in \Theta. \tag{7b.1.3}$$

The conditions (7b.1.2) and (7b.1.3) correspond to unbias and power of a test in the theory of testing of hypothesis.

In practice we are generally interested in interval estimators of individual parameters and not of any general set estimators. In theory, however, it is simpler to consider the more general problem of set estimation.

7b.2 A General Method of Constructing a Confidence Set

The problem of set estimation is closely linked with that of testing a simple hypothesis concerning the unknown parameters. We shall prove some results which establish the relationship between the two problems.

(i) *Let $w_{\mathbf{a}}$ be a critical region of size α for testing the simple hypothesis $\boldsymbol{\theta} = \mathbf{a}$ (given value) and let $w_{\mathbf{a}}^c$ be the region complementary to $w_{\mathbf{a}}$. Then for given x, the set of* $\mathbf{a}$

$$I(x) = \{\mathbf{a} : x \in w_{\mathbf{a}}^c\} \tag{7b.2.1}$$

is a $(1 - \alpha)$ confidence set for $\boldsymbol{\theta}$ for given x.

By construction

$$x \in w_{\mathbf{a}}^c \Leftrightarrow \mathbf{a} \in I(x). \tag{7b.2.2}$$

Hence

$$P(w_{\mathbf{a}}^c \mid \boldsymbol{\theta} = \mathbf{a}) = P(x : \mathbf{a} \in I(x) \mid \boldsymbol{\theta} = \mathbf{a}). \tag{7b.2.3}$$

The left-hand expression of (7b.2.3) has the value $(1 - \alpha)$, which proves the required result.

(ii) *Let $I(x)$ be a $(1-\alpha)$ confidence set of $\boldsymbol{\theta}$. Then the set of x values*

$$w_{\mathbf{a}}^c = \{x : \mathbf{a} \in I(x)\} \tag{7b.2.4}$$

constitutes an acceptance region for the hypothesis $\boldsymbol{\theta} = \mathbf{a}$.

(iii) *If $w_{\mathbf{a}}$ is unbiased as a critical region for testing the hypothesis $\boldsymbol{\theta} = \mathbf{a}$, the set estimator I based on $w_{\mathbf{a}}$ is unbiased in the sense of* (7b.1.2). *Conversely if I is unbiased in the sense of* (7b.1.2), *the critical region $w_{\mathbf{a}}$ based on I provides an unbiased test of the hypothesis $\boldsymbol{\theta} = \mathbf{a}$.*

The result follows from the equivalence relation (7b.2.2) by taking probabilities for $\boldsymbol{\theta} = \mathbf{b} \neq \mathbf{a}$.

(iv) *Let I be shorter than J, another set estimator, and $w_{\mathbf{a}}(I)$ and $w_{\mathbf{a}}(J)$ be the associated critical regions. Then the test based on $w_{\mathbf{a}}(I)$ is more powerful than that based on $w_{\mathbf{a}}(J)$.*

From the equivalence relation (7b.2.2), we see that

$$\begin{aligned} P(w_{\mathbf{a}}^c(I) \mid \theta) &= P(\mathbf{a} \in I \mid \boldsymbol{\theta}), \qquad \boldsymbol{\theta} \neq \mathbf{a} \\ &\leqslant P(\mathbf{a} \in J \mid \boldsymbol{\theta}), \quad \text{since } I \text{ is shorter} \\ &= P(w_{\mathbf{a}}^c(J) \mid \boldsymbol{\theta}). \end{aligned}$$

(v) *The converse of* (iv) *is also true.*

Results (iv) and (v) show that if a U.M.P. test exists for the hypothesis $\boldsymbol{\theta} = \mathbf{a}$ (any given value), then the shortest confidence set exists and vice-versa. In general we have a satisfactory set estimator if there is a reasonable test criterion for simple hypotheses concerning $\boldsymbol{\theta}$. For example, on the basis of a sample $(x_1, \ldots, x_n)$ from $N(\mu, \sigma^2)$, a reasonable test for $\mu = a$ when σ^2 is unknown is Student's t-test with the critical region

$$w_a = \left\{x : \frac{\sqrt{n}|\bar{x} - a|}{s} \geqslant k\right\},$$

where k is the upper $(\alpha/2)$ point of the t-distribution on $(n - 1)$ D.F. Then by (7b.2.1),

$$\begin{aligned} I(x) &= \{a : x \in w_a{}^c\} \\ &= \left\{a : \frac{\sqrt{n}|\bar{x} - a|}{s} \leqslant k\right\} \\ &= \left\{a : \bar{x} - \frac{ks}{\sqrt{n}} \leqslant a \leqslant \bar{x} + \frac{ks}{\sqrt{n}}\right\}, \end{aligned}$$

which gives

$$\left(\bar{x} - \frac{ks}{\sqrt{n}}, \bar{x} + \frac{ks}{\sqrt{n}}\right)$$

as a $(1 - \alpha)$ confidence set (interval) estimator of μ.

It is also seen that, corresponding to the one-sided test,

$$w_a = \left\{x : \frac{\sqrt{n}(\bar{x} - a)}{s} \geqslant k\right\},$$

where k is the upper α-point of the t-distribution on $(n - 1)$ D.F., we obtain the one-sided $(1 - \alpha)$ confidence interval

$$\begin{aligned} I(x) &= \left\{a : \frac{\sqrt{n}(\bar{x} - a)}{s} \leqslant k\right\} \\ &= \left\{a : a \geqslant \bar{x} - \frac{ks}{\sqrt{n}}\right\}. \end{aligned} \tag{7b.2.5}$$

In the two examples considered, the set estimator turned out to be an interval. We shall now consider an example to show that the method of inversion (7b.2.1) does not always lead to an interval estimator even for a one-dimensional parameter.

Let $x \sim N(\mu, 1)$ and $y \sim N(\lambda\mu, 1)$ and be independent. To test the hypothesis $\lambda = a$ we may use the test criterion

$$\frac{(ax - y)^2}{a^2 + 1} \sim \chi^2(1). \tag{7b.2.6}$$

The critical region of size α is $(ax - y)^2 \geqslant (a^2 + 1)k$, where k is the upper α probability point of $\chi^2(1)$. The set estimator of λ is the set of values of a such that $(ax - y)^2 \leqslant k(a^2 + 1)$, that is,

$$a^2(x^2 - k) - 2axy + y^2 - k \leqslant 0. \tag{7b.2.7}$$

The set derived from the inequality (7b.2.7) will not be an interval if, for instance, $(x^2 - k)$ is negative and the roots of the quadratic in a are real. This can happen since x and y can have any arbitrary values. The set of values of (x, y) for which the estimate is not an interval, however, may have a low probability, in which case the suggested solution may be acceptable. A different procedure is needed if the set estimator is restricted to intervals only.

7b.3 Set Estimators for Functions of $\boldsymbol{\theta}$

Let θ be a one-dimensional parameter, I be a set estimator of θ, and $g(\theta)$ be any function of θ. Consider the set $g[I(x)]$ of $g(\theta)$ values corresponding to the θ values of the set $I(x)$. If $g(\theta)$ is a one-to-one function of θ, then

$$g(\theta) \in g[I(x)] \Leftrightarrow \theta \in I(x),$$

in which case $g[I]$ is a set estimator of $g(\theta)$ with the same confidence as I is of θ. On the other hand, if $g(\theta)$ is not a one-to-one function, then

$$\theta \in I(x) \Rightarrow g(\theta) \in g[I(x)],$$

and therefore the confidence coefficient of the estimator $g[I]$ is higher than that of I. This may not be the desired solution, however. To obtain a set estimator of $g(\theta)$ with a given confidence coefficient for such instances, it may be necessary to find a test criterion for testing simple hypotheses concerning $g(\theta)$ itself.

Suppose $\boldsymbol{\theta}$ is a multidimensional parameter and I is a $(1 - \alpha)$-confidence set estimator. As before, let $g[I(x)]$ be the set of $g(\boldsymbol{\theta})$ values corresponding to the $\boldsymbol{\theta}$ values of the set $I(x)$. In general, we can only assert that

$$\boldsymbol{\theta} \in I(x) \Rightarrow g(\boldsymbol{\theta}) \in g[I(x)],$$

and therefore $g[I]$ is a set estimator of $g(\boldsymbol{\theta})$ with a confidence coefficient greater than $(1 - \alpha)$. The reader is referred to the examples and the discussion on simultaneous confidence intervals given in **4b.2**.

7c SEQUENTIAL ANALYSIS

7c.1 Wald's Sequential Probability Ratio Test

In practice all investigations are sequential. Experiments are continued until sufficient evidence accumulates to come to a decision or are discontinued after some stage owing to lack of resources or because further continuance is judged to be unprofitable. It is also commonplace that evidence collected up to any stage is utilized in planning further investigations. The concept of a pilot survey, introduced by Mahalanobis (1940) to collect preliminary information for efficient planning of a larger survey of jute acreage in Bengal, is a typical application of sequential analysis. In a slightly different context of acceptance sampling where the quality of a lot or a batch of items is judged by examining a small sample of items, Dodge and Roming (1929) introduced a two-stage sampling plan. Let r_1 be the number of defective items in a first sample of size n_1. If $r_1 \leqslant c_1$, accept the lot, and if $r_1 \geqslant c_2$, reject the lot. If $c_1 < r_1 < c_2$ take a further sample of size n_2 and if the total number of defects r in both the samples is not greater than c, accept the lot and reject otherwise. A general theory of sequential analysis was developed by Wald in connection with acceptance sampling during the second world war. We shall consider in this section some basic concepts and methods of sequential analysis as developed by Wald (1947).

Sequential Probability Ratio Test. Let H_0 and H_1 be two alternative hypotheses concerning a sequence of random variables $(x_1, x_2, \ldots)$ not necessarily i.i.d. Denote by $P(\cdot, \ldots, \cdot \,|\, H_i)$ $i = 0, 1$ the probability densities under H_0 and H_1 based on the first m observations. The likelihood ratio based on the first m observations is

$$R_m = \frac{P(x_1, \ldots, x_m \,|\, H_1)}{P(x_1, \ldots, x_m \,|\, H_0)}. \tag{7c.1.1}$$

Wald's *sequential probability ratio test* (S.P.R.T) for deciding between the two alternatives H_0 and H_1 (or of testing H_0 against an alternative H_1) is defined as follows.

Choose and fix two constants A and B such that $0 < B < 1 < A < \infty$. Observe the random variables $(x_1, x_2, \ldots)$ one after the other. At each stage compute the likelihood ratio. At the mth stage, if (a) $R_m \leqslant B$, stop sampling and accept H_0, (b) $R_m \geqslant A$, stop sampling and accept H_1, and (c) $B < R_m < A$, continue sampling by taking an additional observation. The constants A and B are called the boundary points of the S.P.R.T.

As in the theory of testing of hypotheses based on a fixed sample size, two kinds of errors arise. If α is the probability of rejecting H_0 when it is true and β

of accepting H_0 when H_1 is true, the pair (α, β) is called the strength of the sequential test.

7c.2 Some Properties of the S.P.R.T.

We shall prove some results concerning the performance of an S.P.R.T.

(i) *If the S.P.R.T. of strength* (α, β) *and the boundary points* (A, B) *terminates with probability* 1, *then*

$$A \leqslant \frac{1-\beta}{\alpha}, \qquad B \geqslant \frac{\beta}{1-\alpha}. \tag{7c.2.1}$$

Let w_m be the region $B < R_i < A$, $i = 1, \ldots, m-1$ and $R_m \geqslant A$. The probability of rejecting H_0 when it is true is

$$\begin{aligned}\alpha &= \sum_{m=1}^{\infty} \int_{w_m} P(x_1, \ldots, x_m \mid H_0) dv^{(m)} \\ &\leqslant \sum_{m=1}^{\infty} \int_{w_m} A^{-1} P(x_1, \ldots, x_m \mid H_1) dv^{(m)} = A^{-1}(1-\beta),\end{aligned}$$

where $dv^{(m)} = dx_1 \cdots dx_m$, which establishes the first inequality in (7c.2.1). Similarly, the second inequality follows.

(ii) *If for the choice*

$$A = \frac{1-\beta}{\alpha}, \qquad B = \frac{\beta}{1-\alpha}, \tag{7c.2.2}$$

the S.P.R.T. terminates with probability 1 *and is of strength* (α', β'), *then*

$$\alpha' \leqslant \frac{\alpha}{1-\beta}, \qquad \beta' \leqslant \frac{\beta}{1-\alpha}, \qquad \text{and} \qquad (\alpha' + \beta') \leqslant \alpha + \beta. \tag{7c.2.3}$$

Applying (7c.2.1) and substituting the expressions for A and B as in (7c.2.2), we have

$$\frac{1-\beta}{\alpha} \leqslant \frac{1-\beta'}{\alpha'}, \qquad \frac{\beta}{1-\alpha} \geqslant \frac{\beta'}{1-\alpha'}. \tag{7c2.4}$$

From the first inequality in (7c.2.4), we see that

$$\alpha' \leqslant (1-\beta') \frac{\alpha}{1-\beta} \leqslant \frac{\alpha}{1-\beta}.$$

Similarly, $\beta' \leqslant \beta/(1-\alpha)$. Further from (7c.2.4), we have

$$\frac{1-\alpha'}{\beta'} \geqslant \frac{1-\alpha}{\beta} \Rightarrow \frac{1-\alpha'-\beta'}{\beta'} \geqslant \frac{1-\alpha-\beta}{\beta}$$

$$\frac{1-\beta'}{\alpha'} \geqslant \frac{1-\beta}{\alpha} \Rightarrow \frac{1-\alpha'-\beta'}{\alpha'} \geqslant \frac{1-\alpha-\beta}{\alpha}.$$

Hence if $(1-\alpha-\beta) > 0$, which is true when α and β are small, then

$$\frac{\alpha'+\beta'}{1-\alpha'-\beta'} \leqslant \frac{\alpha+\beta}{1-\alpha-\beta} \Rightarrow \alpha'+\beta' \leqslant \alpha+\beta.$$

The inequalities (7c.2.3) show that when α and β are small, α' and β' are *close to α and β and at least one of the inequalities $\alpha' \leqslant \alpha$, $\beta' \leqslant \beta$ is true* (since $\alpha'+\beta' \leqslant \alpha+\beta$). Generally, both the inequalities are likely to be true and the special choice of A, B as in (7c.2.2) leads to a more stringent test than the one with the correct choice of A and B.

(iii) *Suppose the successive observations $(x_1, x_2, \ldots)$ are independent and identically distributed. Let $z(x) = \log[p(x \mid H_1)/p(x \mid H_0)]$ where $p(\cdot \mid H_0)$ and $p(\cdot \mid H_1)$ are the probability densities of a single observation x under H_0 and H_1 respectively. Further, denote by n the number of observations (which is a random variable) needed for reaching a decision by using the S.P.R.T. With respect to any hypothesis H (not necessarily H_0 or H_1) for which $P(|z(x)| > 0 \mid H) > 0$, the following results are true.*

(a) *$P(n < \infty) = 1$, that is, the S.P.R.T. eventually terminates.*
(b) *$E(e^{tn}) < \infty$, for $-\infty < t < t_0$, where $t_0 > 0$.*

Result (a) shows that the S.P.R.T. terminates with probability 1, whatever may be the distribution from which the observations are drawn provided only $P(|z(x)| > 0) > 0$. Result (b) shows that all moments of n are finite.

Let $z_i = \log[p(x_i \mid H_1)/p(x_i \mid H_0)]$. Then $(z_1, z_2, \ldots)$ is a sequence of i.i.d. random variables. The sequential process continues as long as

$$b = \log B < (z_1 + \cdots + z_i) = S_i < \log A = a, \qquad i = 1, 2, \ldots,$$

that is, $S_i \in (b, a)$. Let r be the greatest integer in (m/k) where m and k are fixed and consider $S_k, S_{2k} - S_k, S_{3k} - S_{2k}, \ldots, S_{rk} - S_{(r-1)k}$. If n, the sample size needed, exceeds m, then $S_i \in (b, a)$ for $i = 1, 2, \ldots, m$ and in particular for $i = k, 2k, \ldots, rk$. Hence $|T_i| = |S_{ik} - S_{(i-1)k}| < (|b| + |a|) = c$, $i = 1, \ldots, r$, which implies that

$$\begin{aligned} P(n > m) &\leqslant P(|T_i| < c, \quad i = 1, \ldots, r) \\ &= [P(|T_1| < c)]^r, \qquad \text{since } T_i \text{ are independent.} \qquad (7c.2.5) \\ \lim_{m\to\infty} P(n > m) &= \lim_{r\to\infty} [P(|T_1| < c]^r, \quad \text{keeping } k \text{ fixed} \\ &= 0, \qquad \text{if} \quad P(|T_1| < c) < 1. \end{aligned}$$

Since $P(|z| > 0) > 0$, there exists a constant h such that $P(z > h)$ and/or $P(z < -h)$ is >0. Now

$$\begin{aligned} P(|T_1| > c) &= P(|z_1 + \cdots + z_k| > c) \\ &\geqslant P\left(z_i > \frac{c}{k}, i = 1, \ldots, k\right) + P\left(z_i < -\frac{c}{k}, i = 1, \ldots, k\right) \\ &\geqslant [P(z > h)]^k + [P(z < -h)]^k > 0 \end{aligned}$$

choosing k to be a fixed integer greater than c/h. Hence $P(|T_1| < c) = \delta < 1$ and the proposition (a) of (iii) is proved.

To prove (b), consider

$$\begin{aligned} E(e^{tn}) &= \sum_{m=1}^{\infty} e^{tm} P(n = m) \\ &\leqslant \sum e^{tm} P(n > m - 1) \\ &\leqslant \sum e^{tm}\, \delta^r,\ r = \left[\frac{m-1}{k}\right],\ \delta = P(|T_1| < c) \qquad \text{using (7c.2.5)} \\ &= \sum e^{tm}\, \delta^{m/k}\, \delta^{(r - m/k)} \\ &\leqslant \delta^{-1/k} \sum e^{tm}\, \delta^{m/k} = \delta^{-1/k} \sum (e^t\, \delta^{1/k})^m. \end{aligned}$$

The series $\sum (e^t\, \delta^{1/k})^m$ is convergent if $e^t\, \delta^{1/k} < 1$ or $t < (-\log \delta)/k = t_0$, where it may be noted that $t_0 > 0$ and that its value can actually be determined.

(iv) *For any random variable n which takes the values* 0, 1, 2, ...

$$E(n) = \sum_{m=1}^{\infty} P(n \geqslant m). \tag{7c.2.6}$$

By definition

$$\begin{aligned} E(n) &= P(n = 1) + 2P(n = 2) + 3P(n = 3) + \cdots \\ &= P(n = 1) + P(n = 2) + P(n = 3) + \cdots \\ &\qquad\qquad\quad + P(n = 2) + P(n = 3) + \cdots \\ &\qquad\qquad\qquad\qquad\quad + \quad \cdots \quad + \\ &= P(n \geqslant 1) + P(n \geqslant 2) + P(n \geqslant 3) + \cdots. \end{aligned}$$

(v) A GENERAL LEMMA. *Consider a sequence of i.i.d. observations,* $x_1, x_2,$ *... on a random variable x and a sequential decision procedure with a given stopping rule. Let n be the number of observations needed to come to a decision, z(x) be any measurable function of x, and H be some hypothesis specifying the probability distribution of x. Then*

$$E(|z(x)| \mid H) < \infty \text{ and } E(n|H) < \infty \Rightarrow E(S_n|H) = E(z|H)E(n|H),$$

where $S_n = z(x_1) + \cdots + z(x_n)$, that is, the cumulative sum of z values up to the stage of taking a decision.

Note that the problem need not be one of testing of hypothesis but can be of a more general nature with specified rules for continuing sampling and for stopping to make a decision.

Define

$$y_i = \begin{cases} 1 & \text{if no decision is taken up to the } (i-1)\text{th stage} \\ 0 & \text{if a decision is taken at an earlier stage.} \end{cases}$$

Then y_i is clearly a function of $x_1, \ldots, x_{i-1}$ only and is, therefore, independent of x_i and hence independent of $z_i = z(x_i)$. Consider the sum

$$z_1 y_1 + z_2 y_2 + \cdots + z_n y_n + z_{n+1} y_{n+1} + \cdots,$$

which is easily seen to be S_n. Taking expectations

$$\begin{aligned} E(S_n) &= E(\sum z_i y_i) = \sum E(z_i y_i) \\ &= \sum E(z_i)E(y_i) = E(z) \sum E(y_i) \\ &= E(z) \sum P(n \geqslant i) = E(z)E(n), \qquad \text{using (7c.2.6).} \end{aligned}$$

The interchanging of E and $\sum$ is justified since

$$\sum_1^\infty E(|z_i y_i|) = E(|z|) \sum P(n \geqslant i) = E(|z|)E(n) < \infty$$

by assumption.

7c.3 Efficiency of the S.P.R.T.

Let the sequence of observations $(x_1, x_2, \ldots)$ be independent and identically distributed and let the probability densities of a single observation x under H_0 and H_1 be $p(\cdot | H_0)$ and $p(\cdot | H_1)$. As in [(iii), **7c.2**], let

$$z(x) = \log \left[\frac{p(x | H_1)}{p(x | H_0)} \right], \qquad a = \log A \qquad \text{and} \qquad b = \log B.$$

Under these conditions we shall show that the S.P.R.T. is superior to a fixed sample size test in the sense that the average sample number (A.S.N.) for the former is smaller than the fixed size of the latter, provided the two types of tests are of the same strength (α, β). Furthermore, compared to any other sequential test procedure the S.P.R.T. has the least A.S.N. We prove the following results.

(i) *The S.P.R.T. terminates with probability* 1 *both under H_0 and H_1.*

It is shown in (1e.6.6) that unless $z(x) = 1$ with probability 1

$$E[z(x) | H_0] < 0$$

which implies

$$P(z < 0 | H_0) > 0 \qquad \text{or} \qquad P(|z| > 0 | H_0) > 0.$$

Hence by applying the result (a) of [(iii), **7c.2**], the S.P.R.T. terminates with probability 1 when H_0 is true. The proof is similar when H_1 is true.

(ii) *Let* $E(|z| \mid H_0) < \infty$ *and* $E(|z| \mid H_1) < \infty$. *Then approximate expressions for the A.S.N. under* H_0, H_1 *are*

$$E(n | H_0) \doteq \frac{b(1-\alpha) + a\alpha}{E(z | H_0)} \tag{7c.3.1}$$

$$E(n | H_1) \doteq \frac{b\beta + a(1-\beta)}{E(z | H_1)} \tag{7c.3.2}$$

where (α, β) *is the strength of the S.P.R.T.*

From (b) of [(iii), **7c.2**], $E(n) < \infty$ and from [(v), **7c.2**],

$$E(n | H_0) E(z | H_0) = E(S_n | H_0). \tag{7c.3.3}$$

$$\begin{aligned} E(S_n | H_0) &= P(S_n \leqslant b) E(S_n | S_n \leqslant b) + P(S_n \geqslant a) E(S_n | S_n \geqslant a) \\ &\doteq (1-\alpha) b + \alpha a \end{aligned} \tag{7c.3.4}$$

if we use the approximations $E(S_n | S_n \leqslant b) \doteq b$, $E(S_n | S_n \geqslant a) \doteq a$, that is, by neglecting the excess of S_n over the boundary when the S.P.R.T. terminates. Combining (7c.3.3) and (7c.3.4), we obtain (7c.3.1). Similarly (7c.3.2) is established.

(iii) *For any sequential test procedure which terminates with probability* 1,

$$E(n | H_0) \geqslant \frac{(1-\alpha) \log \dfrac{\beta}{1-\alpha} + \alpha \log \dfrac{1-\beta}{\alpha}}{E(z | H_0)}. \tag{7c.3.5}$$

For the S.P.R.T., as shown in (7c.3.1),

$$E(n | H_0) \doteq \frac{(1-\alpha) b + \alpha a}{E(z | H_0)} \tag{7c.3.6}$$

where a and b have the approximate values

$$a = \log \frac{1-\beta}{\alpha}, \qquad b = \log \frac{\beta}{1-\alpha}.$$

Suppose the result (7c.3.5) is true. Then (7c.3.6) is approximately the lower bound in (7c.3.5) thus showing that the A.S.N. for the S.P.R.T. is nearly the least.

To prove the inequality (7c.3.5) we use the result [(v), **7c.2**],

$$E(n|H_0)E(z|H_0) \equiv E(S_n|H_0) \tag{7c.3.7}$$

and compute $E(S_n|H_0)$ for any sequential test.

$$\begin{aligned} E(S_n|H_0) &= P(H_0 \text{ is accepted})E(S_n|H_0,\ H_0 \text{ is accepted}) \\ &\quad + P(H_0 \text{ is rejected})E(S_n|H_0,\ H_0 \text{ is rejected}) \\ &= (1-\alpha)E(S_n|H_0,\ H_0 \text{ is accepted}) \\ &\quad + \alpha E(S_n|H_0,\ H_0 \text{ is rejected}). \end{aligned} \tag{7c.3.8}$$

By Jensen's inequality (1e.5.6),

$$\begin{aligned} &E(S_n|H_0,\ H_0 \text{ is accepted}) \\ &\quad \leqslant \log E(e^{S_n}|H_0,\ H_0 \text{ is accepted}) \\ &\quad = \log \frac{1}{1-\alpha} \sum_1^\infty \int_{w_m} e^{S_m} p(x_1|H_0) \cdots p(x_m|H_0)\, dv^{(m)} \end{aligned} \tag{7c.3.9}$$

where w_m is the region in R^m leading to the acceptance of H_0. Observing that

$$e^{Sm} = \frac{p(x_1|H_1) \cdots p(x_m|H_1)}{p(x_1|H_0) \cdots p(x_m|H_0)},$$

we can reduce the expression (7c.3.9) to

$$\log \frac{1}{1-\alpha} \sum_1^\infty \int_{w_m} p(x_1|H_1) \cdots p(x_m|H_1)\, dv^{(m)} = \log \frac{\beta}{1-\alpha}.$$

Similarly, $E(S_n|H_0,\ H_0 \text{ is rejected}) < \log[(1-\beta)/\alpha]$. Hence from (7c.3.8), we see that

$$E(S_n|H_0) \leqslant (1-\alpha) \log \frac{\beta}{1-\alpha} + \alpha \log \frac{1-\beta}{\alpha}.$$

Furthermore, $E(z|H_0) < 0$, and hence the equation (7c.3.7) gives the inequality (7c.3.5).

7c.4 An Example of Economy of Sequential Testing

Let $x \sim N(\theta_0, \sigma^2)$ under H_0 and $\sim N(\theta_1, \sigma^2)$ under H_1, where $\theta_1 > \theta_0$ and σ^2 is known. The logarithm of the likelihood ratio based on the first m independent observations is

$$\begin{aligned} \log R_m &= -\frac{1}{2\sigma^2}[\sum (x_i - \theta_1)^2 - \sum (x_i - \theta_0)^2] \\ &= \frac{\theta_1 - \theta_0}{2\sigma^2}\left[2 \sum_1^m x_i - m(\theta_1 + \theta_\theta)\right]. \end{aligned}$$

At the mth stage the procedure is as follows:

(a) Accept H_0, if

$$\sum_1^m x_i - \frac{m(\theta_1 + \theta_0)}{2} \leqslant \frac{b\sigma^2}{\theta_1 - \theta_0}.$$

(b) Reject H_0, if

$$\sum_1^m x - \frac{m(\theta_1 + \theta_0)}{2} \geqslant \frac{a\sigma^2}{\theta_1 - \theta_0}.$$

(c) Continue sampling if

$$\frac{b\sigma^2}{\theta_1 - \theta_0} < \sum_1^m x_i - \frac{m(\theta_1 + \theta_0)}{2} < \frac{a\sigma^2}{\theta_1 - \theta_0},$$

where $a = \log[(1 - \beta)/\alpha]$ and $b = \log[\beta/(1 - \alpha)]$.

To determine the A.S.N. we use the approximate expression

$$E(n \mid H_0) \doteq \frac{b(1 - \alpha) + a\alpha}{E(z \mid H_0)}. \tag{7c.4.1}$$

Now

$$z = -\frac{(x - \theta_1)^2 - (x - \theta_0)^2}{2\sigma^2}$$

$$E(z \mid H_0) = -\frac{(\theta_1 - \theta_0)^2}{2\sigma^2}. \tag{7c.4.2}$$

For given values of α, β, a, b, the value of (7c.4.1) can be computed substituting the value (7c.4.2) for $E(z \mid H_0)$. Similarly,

$$E(n \mid H_1) = \frac{b\beta + a(1 - \beta)}{(\theta_1 - \theta_0)^2/2\sigma^2} \tag{7c.4.3}$$

can be computed.

Let us consider a fixed sample size test of the same strength (α, β). The fixed sample size test at α level of significance is

$$\sqrt{N}(\bar{x} - \theta_0) \geqslant d_\alpha \sigma,$$

where N is the sample size and d_α is the upper α-point of $N(0, 1)$. Now

$$\begin{aligned} \beta &= P(\sqrt{N}(\bar{x} - \theta_0) < d_\alpha \sigma \mid \theta_1) \\ &= P(\sqrt{N}(\bar{x} - \theta_1) < d_\alpha \sigma - \sqrt{N}(\theta_1 - \theta_0) \mid \theta_1). \end{aligned}$$

Hence, if d_β is the upper β-point of $N(0, 1)$, then

$$-d_\alpha \sigma + \sqrt{N}(\theta_1 - \theta_0) = d_\beta \sigma,$$

which gives

$$N = \frac{\sigma^2(d_\alpha + d_\beta)^2}{(\theta_1 - \theta_0)^2}. \tag{7c.4.4}$$

The ratio of A.S.N. (7c.4.1) when H_0 is true to the fixed sample size N, (7c.4.4) is

$$-\frac{2[b(1-\alpha) + a\alpha]}{(d_\alpha + d_\beta)^2}, \tag{7c.4.5}$$

which is independent of $(\theta_1 - \theta_0)$. Similarly, the ratio when H_1 is true is

$$\frac{2[b\beta + a(1-\beta)]}{(d_\alpha + d_\beta)^2}. \tag{7c.4.6}$$

When $\alpha = 0.05$, $\beta = 0.05$, we find that $d_\alpha = d_\beta = 1.6449$ and the values of a and b are

$$a = \log\frac{(1-\beta)}{\alpha} = 2.9444, \qquad b = \log\frac{\beta}{1-\alpha} = -2.9444.$$

The values of the ratios (7c.4.5) and (7c.4.6) are each equal to 0.4897, which shows that the saving in the observations by the sequential method is of the order of 50%.

Chernoff (1959) investigated the saving in observations in the general case, when α and β are small and the alternative hypothesis is close to the null hypothesis. It appears that the ratio of fixed sample size to the expected size of the S.P.R.T. is of the order of 4 : 1 in large samples.

7c.5 The Fundamental Identity of Sequential Analysis

LEMMA 1. *Let z be a random variable such that*

(a) $P(z > 0) > 0$ *and* $P(z < 0) > 0$,
(b) $\phi(t) = E(e^{tz})$ *exists for any real value* t, *and*
(c) $E(z) \neq 0$.

Then there exists a $\tau \neq 0$ *such that* $\phi(\tau) = 1$. *If* $E(z) < 0$, *then* $\tau > 0$ *and if* $E(z) > 0$, *then* $\tau < 0$.

Since $P(z > 0) > 0$, there exists a c such that $P(z > c) = \delta > 0$. Hence

$$\phi(t) = E(e^{tz}) = \int e^{tz}\, dF$$

$$\geqslant \int_{z>c} e^{tz}\, dF > e^{tc}P(z > c), \quad \text{if} \quad t > 0. \tag{7c.5.1}$$

The result (7c.5.1) $\Rightarrow \phi(t) \to \infty$ as $t \to \infty$. Similarly, $\phi(t) \to +\infty$ as $t \to -\infty$. Furthermore, $\phi(0) = 1$ and the slope of $\phi(t)$,

$$\phi'(t) = E(ze^{tz})$$

at $t = 0$ is $E(z)$. If $E(z) < 0$, the curve of $\phi(t)$ has value unity at $t = 0$, slopes downwards, and then tends to ∞ as $t \to \infty$. Since $\phi(t)$ is analytic, it must assume the value unity again for some other value $\tau > 0$ of t. Similarly, when $E(z) > 0$, there exists a $\tau < 0$ such that $\phi(\tau) = 1$.

Since $\phi''(t) = E(z^2 e^{tz})$, the condition (a) of lemma 1 implies that $\phi''(t) > 0$ for all t, which in turn implies that it can have one minimum at the utmost. Hence there is only one value τ of t other than 0 at which $\phi(t) = 1$, for if $\phi(t) = 1$ at τ_1, τ_2, besides zero, then at least two minima are implied.

LEMMA 2. *Let $x_1, x_2, \ldots$ be a sequence of i.i.d. observations on a random variable x with probability density $p(\cdot \mid H)$ under a hypothesis H. Consider any sequential procedure with a given stopping rule and let n be the number of observations needed for coming to a decision. If $z_i = z(x_i)$ is any measurable function, then $P(n < \infty \mid H) = 1 \Rightarrow$*

$$E\{e^{tS_n}[\phi(t)]^{-n} \mid H\} = P(n < \infty \mid H_t), \tag{7c.5.2}$$

where under H_t the probability density of x is

$$p(\cdot \mid H_t) = \frac{e^{tz(\cdot)}}{\phi(t)} p(\cdot \mid H), \tag{7c.5.3}$$

$S_n = z_1 + \cdots + z_n$, and $\phi(t) = E(e^{tz} \mid H)$.

Let w_m denote the region in R^m leading to the termination of the sequential procedure at the mth stage. Then

$$\begin{aligned} E\{e^{tS_n}[\phi(t)]^{-n} \mid H\} &= \sum_1^\infty \int_{w_m} [\phi(t)]^{-m} e^{tS_m} p(x_1 \mid H) \cdots p(x_m \mid H)\, dv^{(m)} \\ &= \sum_1^\infty \int_{w_m} p(x_1 \mid H_t) \cdots p(x_m \mid H_t)\, dv^{(m)} \\ &= P(n < \infty \mid H_t). \end{aligned}$$

The Fundamental Identity of Sequential Analysis. Consider the S.P.R.T. based on a sequence of i.i.d. observations and let z be as defined in [(iii), **7c.2**], the logarithm of the ratio of probability densities under the hypotheses H_0 and H_1. If H is any hypothesis such that $P(|z| > 0 \mid H) > 0$, then

$$E\{e^{tS_n}[\phi(t)]^{-n} \mid H\} = 1, \tag{7c.5.4}$$

which is called the fundamental identity of sequential analysis.

It is easy to see that $P(|z| > 0|H) > 0 \Rightarrow P(|z| > 0|H_t) > 0$, where H_t is as defined in (7c.5.3). Hence by [(iii) (a), **7c.2**], $P(n < \infty|H_t) = 1$. The result (7c.5.4) follows from (7c.5.2). The proofs of propositions in **7c.5** are due to Bahadur.

Operating Characteristic (O.C.) Function of a Sequential Plan. Let us consider a sequential plan for coming to a decision like accepting or rejecting a batch of goods offered for inspection or taking some action or not. A sequential plan consists in drawing goods one by one, making measurements, and applying certain rules of continuing sampling and stopping to take a decision, based on the measurements available at each stage. Let us denote by $\pi(H)$ the probability of accepting a batch under a given procedure when the probability distribution of the measurements is as specified by hypothesis H. Then π, considered as a function of H belonging to a set of possible hypotheses, is called a *O.C. function* of the sequential plan. We shall compute the O.C. function for an important class of sequential plans.

We impose the restriction that the probability of accepting a batch is $(1 - \alpha)$ if the probability distribution of the measurements is as specified by a chosen hypothesis H_0 and β if it is according to another chosen hypothesis H_1. Suppose that this specification is realized by using the S.P.R.T. based on the likelihood ratio of the hypotheses H_0 and H_1 with the convention of accepting a batch when H_0 is accepted under S.P.R.T. and of rejecting a batch when H_1 is accepted. We shall obtain an approximate value of $\pi(H)$ for such a plan under the conditions of lemmas 1 and 2 on the distribution of z—the logarithm of the likelihood ratio of the hypotheses H_0 and H_1 for a single observation.

Let $\phi(t)$ be the c.f. of z under H. Then by lemma 1, τ exists such that $\phi(\tau) = 1$. Substituting this value in the fundamental identity (7c.5.4), we have

$$\begin{aligned} 1 = E(e^{\tau S_n}|H) &= P\{S_n \leqslant b\}E(e^{\tau S_n}|S_n \leqslant b) + P\{S_n \geqslant a\}E(e^{\tau S_n}|S_n \geqslant a) \\ &\doteq \pi(H)e^{\tau b} + [1 - \pi(H)]e^{\tau a}, \end{aligned}$$

which gives

$$\pi(H) \doteq \frac{1 - e^{\tau a}}{e^{\tau b} - e^{\tau a}}. \tag{7c.5.5}$$

Average Sample Number (A.S.N.). Using an argument similar to that of (7c.3.4), we have the approximate expression

$$E(S_n|H) \doteq b\pi(H) + a[(1 - \pi(H)].$$

But from [(v), **7c.2**]

$$E(n|H)E(z|H) = E(S_n|H);$$

and since $E(z|H) \neq 0$, then

$$E(n|H) \doteq \frac{b\pi(H) + a[1 - \pi(H)]}{E(z|H)}, \tag{7c.5.6}$$

which is an approximate evaluation of the average sample number for coming to a decision under the hypothesis H by using the sequential plan based on the S.P.R.T. for testing H_0 against H_1.

Let us compute the approximate O.C. and A.S.N. functions for the normal distribution for the S.P.R.T. constructed in **7c.4**. It may be recalled that H_0 is $N(\theta_0, \sigma^2)$ and H_1 is $N(\theta_1, \sigma^2)$. Let H be $N(\theta, \sigma^2)$ where θ is arbitrary. We shall study the behavior of the test for θ_0 against the alternative θ_1 when the true value is θ different from θ_0 and θ_1. Now

$$z = -\frac{1}{2\sigma^2}[(x - \theta_1)^2 - (x - \theta_0)^2]$$

$$\phi(t) = E(e^{tz}|H) = e^{q/2\sigma^2}$$

$$q = t(\theta_1 - \theta_0)[2\theta - \theta_1 - \theta_0 + t(\theta_1 - \theta_0)].$$

The value of $\tau \neq 0$, such that $\phi(\tau) = 1$ is

$$\tau(\theta) = \frac{\theta_1 + \theta_0 - 2\theta}{\theta_1 - \theta_0},$$

in which case using (7c.5.5), the O.C. function is

$$\pi(\theta) = \frac{1 - e^{a\tau(\theta)}}{e^{b\tau(\theta)} - e^{a\tau(\theta)}},$$

and by using (7c.5.6), the A.S.N. function is

$$E(n|\theta) = \frac{b(1 - e^{a\tau(\theta)}) + a(e^{b\tau(\theta)} - 1)}{(e^{b\tau(\theta)} - e^{a\tau(\theta)})E(z|\theta)}$$

where $E(z|\theta) = [(\theta - \theta_0)^2 - (\theta - \theta_1)^2]/2\sigma^2$.

7c.6 Sequential Estimation

The basic idea of sequential estimation of a parameter is to do just enough sampling to be able to obtain an estimate with a *predetermined precision* or to maximize the precision (i.e., minimizing the risk) for a given cost of sampling. In the context of interval estimation, the precision may be expressed by the length of the interval with a given confidence coefficient. For a point estimate it may be the mean square error or some suitable loss function. The subject of sequential estimation is not, however, fully developed. Some problems have been considered by Anscombe (1952, 1953), Wolfowitz (1947), and others. In this section we shall consider an interesting method of Stein (1945) for determining an interval estimate of a given length for the mean of a normal distribution when the variance is unknown, by sampling in two stages.

It may be recalled [v, **7b.2**] that on the basis of a sample of fixed size n providing an average $\bar{x}$ and an unbiased estimator of variance s^2, we found the $(1-\alpha)$ confidence interval estimator for the mean μ of a normal distribution as

$$\bar{x} \pm \frac{s t_{\alpha/2}}{\sqrt{n}} \tag{7c.6.1}$$

where $t_{\alpha/2}$ is the upper $(\alpha/2)$-point of the t-distribution on $(n-1)$ D.F. But the length of the confidence interval $(2st_{\alpha/2})/\sqrt{n}$ is itself a random variable and in any particular case it may be large. It is known, however, that it is not possible to find a confidence interval of an assigned length, when σ^2 is unknown on the basis of a single sample (Dantzig, 1940). The problem can be solved, however, by using sequential sampling in two stages.

Draw a sample $x_1, \ldots, x_m$ of size m, fixed in advance and compute $\bar{x}$, s^2. Let n be the smallest integer $\geqslant m$ for which

$$\frac{s^2}{n} \leqslant \frac{c^2}{b^2}, \tag{7c.6.2}$$

where b^2 is the variance of the t-distribution on $(m-1)$ D.F. and c is an assigned constant. Then draw a further sample of $(n-m)$ observations. Let $\bar{x} = (x + \cdots + x_n)/n$ be the average based on all the observations. Then we have the following results.

(i) *$\bar{x}$ is an unbiased estimate of μ with a standard deviation $\leqslant c$.*

It is easy to verify that

$$\frac{\sqrt{n}(\bar{x}-\mu)}{\sigma} \quad \text{and} \quad \frac{(m-1)s^2}{\sigma^2} \tag{7c.6.3}$$

are independently distributed as $N(0, 1)$ and χ^2 on $(m-1)$ D.F. respectively, and hence

$$\frac{\sqrt{n}(\bar{x}-\mu)}{s} \sim t \quad \text{on} \quad (m-1) \text{ D.F.} \tag{7c.6.4}$$

$$\frac{E\sqrt{n}(\bar{x}-\mu)}{\sigma} = 0. \tag{7c.6.5}$$

Observe that (7c.6.5) does not automatically imply that $E(\bar{x} - \mu) = 0$, since n itself is a random variable. Now

$$E(\bar{x} - \mu) = E\left[\frac{\sqrt{n}(\bar{x} - \mu)}{\sigma} \cdot \frac{\sigma}{\sqrt{n}}\right]$$

$$= E\left[\frac{\sqrt{n}(\bar{x} - \mu)}{\sigma} \cdot f(s^2, \sigma)\right], \qquad \text{since } n \text{ depends on } s^2$$

$$= E\left[\frac{\sqrt{n}(\bar{x} - \mu)}{\sigma}\right] E[f(s^2, \sigma)] \tag{7c.6.6}$$

$$= 0, \qquad \text{using (7c.6.5).}$$

The step in (7c.6.6) follows because of the independence of $\sqrt{n}(\bar{x} - \mu)/\sigma$ and s^2 as observed in (7c.6.3).

Furthermore, we see that

$$E(\bar{x} - \mu)^2 = E\left[\frac{n(\bar{x} - \mu)^2}{s^2} \cdot \frac{s^2}{n}\right]$$

$$\leqslant E\left[\frac{n(\bar{x} - \mu)^2}{s^2} \cdot \frac{c^2}{b^2}\right]$$

$$= \frac{c^2}{b^2} E\left[\frac{n(\bar{x} - \mu)^2}{s^2}\right] = \frac{c^2}{b^2} \cdot b^2 = c^2. \tag{7c.6.7}$$

(ii) *Let $2l$ be the preassigned length of the confidence interval estimator of μ. Then the confidence coefficient of the estimator $\bar{x} \pm l$ is not less than $1 - \alpha$ by choosing $c = bl \div t_{\alpha/2}$.*

$$P(|\bar{x} - \mu| < l) = P\left(\frac{|\sqrt{n}(\bar{x} - \mu)|}{s} < \frac{\sqrt{n}\, l}{s}\right)$$

$$\geqslant P\left(\frac{|\sqrt{n}(\bar{x} - \mu)|}{s} < \frac{bl}{c}\right). \tag{7c.6.8}$$

The last probability (7c.6.8) has the value $1 - \alpha$ if

$$\frac{bl}{c} = t_{\alpha/2} \qquad \text{or} \qquad c = \frac{bl}{t_{\alpha/2}}.$$

The two sample procedure for obtaining an interval estimator of assigned length is thus completely specified for any fixed choice of m, the size of the first stage sample. The consequences of different choices of m may be examined. It may be observed that the information supplied by the second sample on σ^2 is not utilized in the solution of the problem, which throws open the possibility of improving on Stein's procedure.

7c.7 Sequential Tests with Power One

In **7c.1–7c.5** we have considered sequential tests which terminate with probability one in favor of one or the other of the two alternative hypotheses. On the other hand, there are situations, as in continuous production, where no action need be taken so long as goods are being produced according to specification (null hypothesis) but necessarily take action otherwise (when the alternative hypothesis is true). Or, in other words, we need a sequential plan,

(a) which terminates rarely when the null hypothesis is true, and
(b) which terminates with probability one, as quickly as possible, when the alternative is true.

Such sequential plans have been recently suggested by Barnard (1969), Darling and Robbins (see Robbins and Siegmund, 1969) using the celebrated law of iterated logarithm mentioned in **2c.6** which may be restated as follows.

Law of Iterated Logarithm. Let $X_1, X_2, \ldots$ be i.i.d., r.v.'s such that $E(X_1) = \mu$ and $V(X_1) = \sigma^2 < \infty$. Let $Z_n = (X_1 + \cdots + X_n - n\mu)/\sigma\sqrt{n}$. Then

$$P\{Z_n > (1 + \varepsilon)(2 \log \log n)^{1/2} \text{ infinitely often}\} = 0, \tag{7c.7.1}$$

$$P\{Z_n > (1 - \varepsilon)(2 \log \log n)^{1/2} \text{ infinitely often}\} = 1. \tag{7c.7.2}$$

As a consequence of (7c.7.1) and (7c.7.2) we have

$$P\{Z_n > (1 - \varepsilon)(2 \log \log n)^{1/2} \text{ for some } n \geqslant 1\} = 1, \tag{7c.7.3}$$

$$\lim_{m \to \infty} P\{Z_n < (1 + \varepsilon)(2 \log \log n)^{1/2} \text{ for all } n \geqslant m\} = 1. \tag{7c.7.4}$$

We shall consider the sequential plan with power one first given by Barnard in 1964 (see Barnard, 1969) for testing $H_0 : p < 0.01$ against $H : p \geqslant 0.01$, where p is the probability of an item produced being defective. The plan consists of drawing items one after the other and rejecting H_0 as and when

$$R_n > 0.01\, n + a\sqrt{n} \qquad \text{for some} \quad n \geqslant 1 \tag{7c.7.5}$$

where R_n is the number of defectives in n successive items and a is a suitably chosen constant. If $H_0 : p < \delta$ and $H : p \geqslant \delta$, we replace .01 by δ in (7c.7.5).

Let $Z_n = (R_n - np)/\sqrt{npq}$, $q = (1 - p)$ in which case (7c.7.5) can be written as

$$Z_n > \frac{(0.01 - p)\sqrt{n}}{\sqrt{pq}} + \frac{a}{\sqrt{pq}} = b_n \tag{7c.7.6}$$

In (7c.7.6), $b_n < (1-\varepsilon)(2\log\log n)^{1/2}$ if $p \geqslant 0.01$ so that with probability one (7c.7.5) holds by using (7c.7.3). Thus the sequential plan terminates with probability one in favor of the alternative hypothesis, if true. On the other hand, if $p < 0.01$ then b_n in (7c.7.6) exceeds $(1+\varepsilon)(2\log\log n)^{1/2}$, and the probability of (7c.7.5) can be made small by choosing a suitably. The larger the a the smaller is the Probability. In fact we can choose a such that the Probability of (7c.7.5) has an assigned value α for given $p = p_0 < .01$.

A second example due to Robbins and Darling (see Robbins and Siegmund, 1969) is that of testing $\theta \leqslant 0$ against $\theta > 0$, where θ is the mean of a normal population with variance one. Independent observations $X_1, X_2, \ldots$ are drawn one by one, and sampling is terminated as and when

$$s_n = X_1 + \cdots + X_n \geqslant \sqrt{n}\, f_n \qquad \text{for some } n \geqslant 1 \tag{7c.7.7}$$

in favor of the hypothesis $\theta > 0$, where

$$f_n = \left[\frac{(n+m)}{n}\left(a^2 + \log\frac{n+m}{m}\right)\right]^{1/2} \tag{7c.7.8}$$

for any chosen values of a and m.

Writing $Z_n = (s_n - n\theta)/\sqrt{n}$, (7c.7.7) can be written as

$$Z_n \geqslant f_n - \sqrt{n}\,\theta = b_n. \tag{7c.7.9}$$

If $\theta > 0$, then b_n in (7c.7.9) $< (1-\varepsilon)(2\log\log n)^{1/2}$ so that (7c.7.7) holds with probability one by using (7c.7.3). Thus the power of detecting the alternative is one. If $\theta < 0$, then it is shown by Robbins and Siegmund (1969) that

$$P\{Z_n < f_n - \sqrt{n}\,\theta \text{ for all } n \mid \theta < 0\} \tag{7c.7.10}$$

$$> P\{Z_n < f_n \text{ for all } n \mid \theta = 0\} > 1 - \frac{e^{-a^2/2}}{2\Phi(a)}, \tag{7c.7.11}$$

where Φ (a) is $P(X < a)$ and $X \sim N(0, 1)$. For example, if $m = 1$ and $a = 3$,

$$f_n = (1 + n^{-1})^{1/2}[(9 + \log(n+1)]^{1/2} \tag{7c.7.12}$$

and the probability (7c.7.10) that the sampling continues indefinitely is not less than

$$1 - \frac{e^{-9/2}}{2\Phi(3)} = 1 - 0.0056 = 0.9944 \tag{7c.7.13}$$

so that the test will rarely terminate.

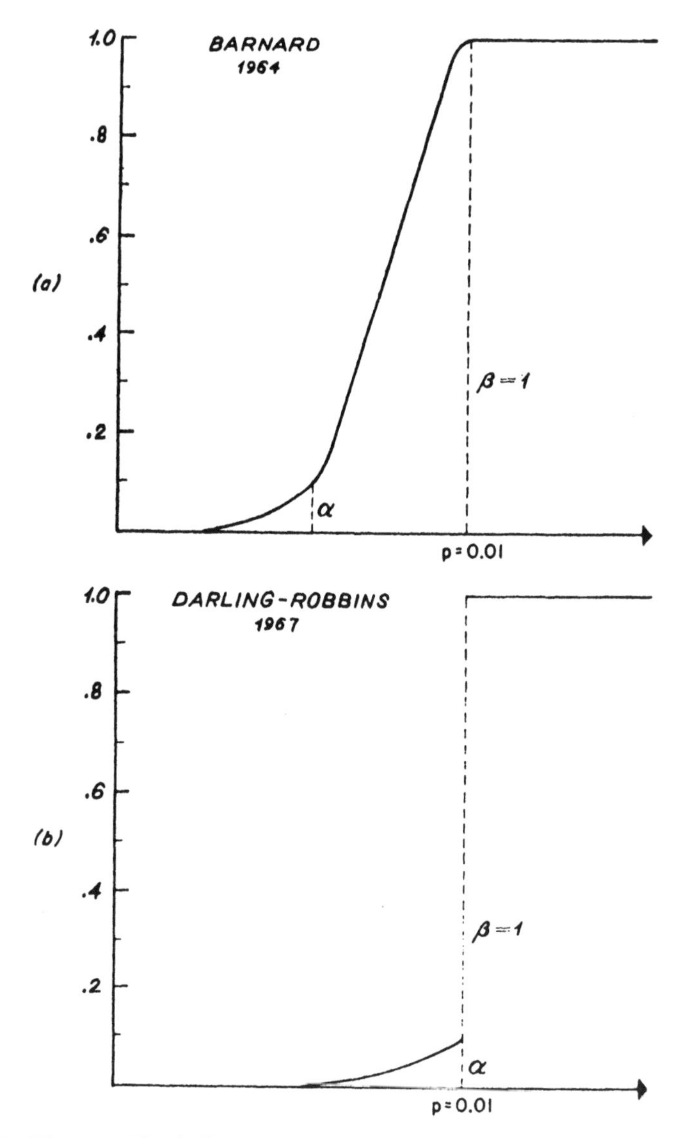

FIGURE 1. The O.C. curve (probability of termination as a function of the parameter) is continuous in the case of Barnard's plan as shown in *(a)* and discontinuous in the case of Darling–Robbins as shown in *(b)*. The first kind of error is always less than an assigned value α for H_0 in the latter while it is so only for values of the parameter less than a chosen value in H_0 in the former.

7d PROBLEM OF IDENTIFICATION—DECISION THEORY

7d.1 Statement of the Problem

Consider the problem of assigning an individual to one of a finite number of groups to which he may belong on the basis of a set of characteristics observed. Such a problem arises when an anthropologist tries to identify a jaw bone excavated from a burial ground as having belonged to a male or a female, or when a doctor decides on the basis of some diagnostic tests whether a patient suffering from jaundice requires surgery or not. Or a biologist wants to identify an observed specimen as a member of one out of k possible known species. In all these situations the problem is one of deciding among a number of alternative hypotheses and not of testing any particular hypothesis against a set of alternatives as considered in Sections **7a** and **7c**. In statistical literature such a problem is referred to as one of *classification* or *discrimination*, but *identification* seems to be more appropriate terminology.

In any practical situation we must consider the possibility that an observed individual does not belong to any of the specified groups as when a biologist discovers a member of a new species. There appears to be no general theory including such a possibility. We shall, however, consider an example in **8e.2** of Chapter 8 and show how the problem can be approached in particular cases. In the discussion of the present section the knowledge regarding the alternative hypotheses is taken to be complete.

7d.2 Randomized and Nonrandomized Decision Rules

Let x denote the measurements on an individual and S the (sample) space of possible values of x. On the basis of observed x, a decision has to be reached about the membership of an individual to one of k specified populations. The problem is the same as that of choosing one among a given set of alternative hypotheses as appropriate to an observed event.

A *nonrandomized* decision rule consists in dividing the space S into k mutually exclusive regions $w_1, \ldots, w_k$ with the rule of assigning an individual with measurements x to the ith population if $x \in w_i$, that is, choosing the ith alternative hypothesis if $x \in w_i$.

A *randomized* decision rule consists in determining a vector function of x, $[\lambda_1(x), \ldots, \lambda_k(x)]$, such that $\lambda_i(x) \geqslant 0$, and $\sum \lambda_i(x) = 1$, with the rule of assigning an individual with measurements x to the ith population with probability $\lambda_i(x)$, $i = 1, \ldots, k$. That is, after x is observed, a random experiment is performed to generate an observation on a random variable which takes the values $1, \ldots, k$ with probabilities $\lambda_1(x), \ldots, \lambda_k(x)$ respectively. If the result is i, the individual is assigned to the ith population.

Loss Vector. Let $P_1(x), \ldots, P_k(x)$ be the probability densities at x with respect to a σ-finite measure v in the k populations and r_{ij} the loss in assigning a member of ith population to the jth. The expected loss in applying a given rule, when in fact the individuals come from the ith population, is

$$L_i = \int_{w_1} r_{i1} P_i(x)\, dv + \cdots + \int_{w_k} r_{ik} P_i(x)\, dv \tag{7d.2.1}$$

for a nonrandomized rule and

$$L_i = \int [r_{i1}\lambda_1(x) + \cdots + r_{ik}\lambda_k(x)] P_i(x)\, dv \tag{7d.2.2}$$

for a randomized rule. In either case we have *a loss vector*

$$(L_1, \ldots, L_k) \tag{7d.2.3}$$

corresponding to the k alternative hypotheses, as the *operating characteristic* of a decision rule, which plays a key role in our investigation.

If there are two decision rules δ_1, δ_2 with associated loss vectors $(L_{11}, \ldots, L_{k1})$, $(L_{12}, \ldots, L_{k2})$, then δ_1 is obviously better than δ_2 if

$$L_{i1} \leqslant L_{i2}, \qquad i = 1, \ldots, k, \tag{7d.2.4}$$

and for at least one i, $L_{i1} < L_{i2}$. If the equality holds for all i, the two rules are equivalent. On the other hand, any two decision rules may not be comparable, that is, for some values of i, $L_{i1} > L_{i2}$, and for the rest the reverse relationship may hold. There is no method of choosing between such rules without bringing in additional criteria. This leads us to the concept of *admissible decision rules.*

Admissible Rules. A decision rule is admissible if there is no other decision rule which is better in the sense of (7d.2.4). If the class of admissible decision rules consists of a single member, we have an optimum solution. But generally the class is very wide and no two decision rules in the admissible set are comparable. We need additional criteria for selecting one among the admissible set of decision rules. The problem, however, admits a simple solution when the prior probabilities of the k populations are known as shown in **7d.3.**

7d.3 Bayes Solution

Let $\pi_1, \ldots, \pi_k$ be *a priori* probabilities of the k populations, that is, an observed individual is regarded as chosen from a mixture of populations with individuals of the k populations in the ratios $\pi_1, \ldots, \pi_k$. The expected loss in such a situation reduces to a single quantity

$$L = \pi_1 L_1 + \cdots + \pi_k L_k, \tag{7d.3.1}$$

where $L_1, \ldots, L_k$ are as defined in (7d.2.3). In a nonrandomized decision rule which uses the expression (7d.2.1) for L_i, (7d.3.1) reduces to

$$L = \sum_{i=1}^{k} \int_{w_i} (\pi_1 r_{1i} P_1 + \cdots + \pi_k r_{ki} P_k)\, dv$$

$$= \int_{w_1} - S_1\, dv + \cdots + \int_{w_k} - S_k\, dv, \tag{7d.3.2}$$

where $S_i = -(\pi_1 r_{1i} P_1 + \cdots + \pi_k r_{ki} P_k)$ is called the ith *discriminant score* of an individual (for the ith population). In a randomized decision rule,

$$L = \int -[\sum \lambda_i(x) S_i(x)]\, dv. \tag{7d.3.3}$$

(i) *Let $w_1^*, \ldots, w_k^*$ be mutually exclusive regions covering the whole sample space such that*

$$x \in w_i^* \Rightarrow S_i(x) \geqslant S_j(x) \quad \text{for all } j, \quad i = 1, \ldots, k. \tag{7d.3.4}$$

Then for such a choice of w_i, the expected loss L of (7d.3.2) *is a minimum.*

The result is an immediate consequence of lemma 4 of **7a.2**.

It may be observed that the condition (7d.3.4) does not specify w_i uniquely because of the equality signs. For instance, if for a particular x

$$S_{i_1}(x) = \cdots = S_{i_r}(x) > S_{i_{r+1}}(x) \geqslant \cdots \geqslant S_{i_k}(x), \tag{7d.3.5}$$

then x may be placed in any one of the regions $w_{i_1}^*, \ldots, w_{i_r}^*$ arbitrarily. The minimum loss remains the same, however, so long as the condition (7d.3.4) is satisfied. Indeed in a situation like (7d.3.5) we may choose to place x in $w_{i_j}^*$ with probability $\lambda_{i_j}(x)$ where $\lambda_{i_1}(x), \ldots, \lambda_{i_r}(x)$ are chosen arbitrarily with the condition that the sum is unity.

(ii) *Let $\lambda_i(x) = 1$, $\lambda_j(x) = 0$, $j \neq i$, if $S_i(x)$ is greater than every other $S_j(x)$ and let $\lambda_{i_1}(x), \ldots, \lambda_{i_r}(x)$ be arbitrary subject to the condition $\lambda_{i_1}(x) + \cdots + \lambda_{i_r}(x) = 1$ and $\lambda_{i_{r+1}}(x) = \cdots = \lambda_{i_k}(x) = 0$ if*

$$S_{i_1}(x) = \cdots = S_{i_r}(x) > S_{i_{r+1}}(x) \geqslant S_{i_{r+2}}(x) \ldots, \geqslant S_{i_k}(x).$$

For such a choice of $\lambda_i(x)$, the loss L of (7d.3.3) *is a minimum.*

The result is an immediate consequence of lemma 3 of **7a.2**.

Let us observe that the optimum solutions for the randomized and nonrandomized rules are essentially the same. We shall call the decision rule specified by (ii) as the Bayes rule with respect to the prior distribution or prior probability vector $\boldsymbol{\pi}' = (\pi_1, \ldots, \pi_k)$. The corresponding loss vector is denoted by $\mathbf{L}'_\pi = (L_{1\pi}, \ldots, L_{k\pi})$. The Bayes solution is then characterized by the property

$$\boldsymbol{\pi}'\mathbf{L}_\pi \leqslant \boldsymbol{\pi}'\mathbf{L}, \tag{7d.3.6}$$

where $\mathbf{L}$ is the loss vector for any other decision rule (randomized or non-randomized).

The Bayes rule can also be deduced by a different kind of argument. The conditional probability of an individual with measurements x belonging to the ith population or the posterior probability of the ith population when x is observed is, by the Bayes theorem,

$$\frac{\pi_i P_i}{\sum \pi_i P_i}, \qquad i = 1, \ldots, k. \tag{7d.3.7}$$

If it is *decided* to assign an individual with measurements x to the jth population, the conditional loss is

$$\frac{r_{1j}\pi_1 P_1 + \cdots + r_{kj}\pi_k P_k}{\sum \pi_i P_i}, \tag{7d.3.8}$$

which is the same as $-S_j(x)/\sum \pi_i P_i$ where S_i is as defined in (7d.3.2). For given x the conditional loss is a minimum when the individual is *assigned to a population for which* (7d.3.8) *is the least or the discriminant score is the highest.* Since the conditional loss is a minimum for such a decision rule, the overall expected loss is obviously at a minimum. This decision rule is, indeed, the same as Bayes with respect to $\boldsymbol{\pi}$.

When $r_{ij} = 1, i \neq j$, and $r_{ii} = 0$, L_i represents the expected proportion of wrong identifications for individuals of the ith population and L for all individuals using the prior probability vector $\boldsymbol{\pi}$. The discriminant score for the ith population when $r_{ij} = 1$, $i \neq j$, and $r_{ii} = 0$ is

$$S_i = -\sum_1^k \pi_j P_j + \pi_i P_i = \pi_i P_i + c, \tag{7d.3.9}$$

where c is a constant independent of i. Then the individual with measurement x is assigned to any one of the populations i such that

$$\pi_i P_i(x) + c \geqslant \pi_j P_j(x) + c \qquad \text{for all} \quad j,$$

or

$$\pi_i P_i(x) \geqslant \pi_j P_j(x) \qquad \text{for all} \quad j. \tag{7d.3.10}$$

The rule (7d.3.10) minimizes the expected number of wrong identifications. For this purpose we may define S_i, the ith discriminant score as simply $\pi_i P_i$.

7d.4 Complete Class of Decision Rules

In many practical problems the prior probabilities may not be known or may not be relevant, in which cases we need some criterion for choosing one among all possible decision rules. With this end in view, let us first examine the method by which we can arrive at the minimal class of rules from which a choice has to be made.

A class of decision rules C is said to be *complete* if corresponding to any decision rule δ outside C, there exists a member of C better than δ.

A complete class which does not contain a complete subset is said to be a *minimal complete class*. We now prove some propositions characterizing complete and minimal complete classes of decision rules.

(i) *A necessary and sufficient condition for the existence of a minimal complete class is that the class of admissible decision rules be complete.*

The proof is easy.

(ii) *The class $\mathscr{L}$ of all loss vectors corresponding to all decision functions is a closed convex set in E_k (k-dimensional Euclidean space).*

The proof is omitted. Reference may be made to Wald (1950).

(iii) *The class A of all admissible decision rules is a minimal complete class.*

We need only prove that A is complete. Consider a rule $\delta_1 \notin A$ and therefore inadmissible. Let Δ be the class of decision rules better than δ_1 and denote by $f(\delta)$, the sum of the components of the loss vector corresponding to δ. Choose δ_j such that

$$f(\delta_j) \leqslant \inf_{\delta \in \Delta} f(\delta) + \frac{1}{j}.$$

Since $\mathscr{L}$ is closed by (ii) then there exists a subsequence of j values such that $\lim \delta_j = \delta^*$ exists and is better than δ_1. Further for any rule $\delta_0 \in \Delta$

$$f(\delta_0) \geqslant \inf_{\delta \in \Delta} f(\delta) \geqslant f(\delta_j) - \frac{1}{j} \to f(\delta^*)$$

so that δ^* is admissible. A is thus complete since $\delta^* \in A$.

(iv) *Every admissible rule is a Bayes rule with respect to some prior distribution.*

Let $\mathbf{L}_0$ be the loss vector corresponding to a rule $a_0 \in A$. Then the totality of vectors $\mathbf{L} - \mathbf{L}_0$, for $\mathbf{L}$ varying and $\mathbf{L}_0$ fixed in $\mathscr{L}$, is a convex set and no member of this has all its components negative since a_0 is admissible. Hence by an application of the supporting plane theorem [(iv), **1d.2**], there exists a vector $\boldsymbol{\pi}$ with non-negative components such that

$$\boldsymbol{\pi}'(\mathbf{L} - \mathbf{L}_0) \geqslant 0 \qquad \text{for all} \quad \mathbf{L} \in \mathscr{L}. \tag{7d.4.1}$$

The result (7d.4.1) shows that $\mathbf{L}_0$ is the loss vector for the Bayes rule with respect to $\boldsymbol{\pi}$.

(v) *The class* $\mathscr{B}$ *of all Bayes rules is a complete class.*

This result follows from (iii).

(vi) *The Bayes rule with respect to any* $\boldsymbol{\pi}$ *is admissible if all the components of* $\boldsymbol{\pi}$ *are positive.*

Let $\mathbf{L}_\pi$ be the loss vector corresponding to the Bayes rule with respect to $\boldsymbol{\pi}$. Then

$$\boldsymbol{\pi}'\mathbf{L}_\pi \leqslant \boldsymbol{\pi}'\mathbf{L}, \qquad \mathbf{L} \in \mathscr{L}$$

If there exists $\mathbf{L}_0 \in \mathscr{L}$ such that $\mathbf{L}_0$ is better than $\mathbf{L}_\pi$, we have

$$\boldsymbol{\pi}'\mathbf{L}_0 \leqslant \boldsymbol{\pi}'\mathbf{L}_\pi \leqslant \boldsymbol{\pi}'\mathbf{L}_0.$$

Since all the components of $\boldsymbol{\pi}$ are positive, $\mathbf{L}_\pi = \mathbf{L}_0$.

7d.5 Minimax Rule

Section **7d.4** shows that the class of admissible rules is very wide, and therefore it is necessary to have further criteria for choosing a decision rule. One such principle is known as minimax, which leads to the choice of a decision rule whose maximum component in the loss vector has the minimum value. It is that rule δ^* with the loss vector $(L_1^*, \ldots, L_k^*)$, if it exists, such that

$$\max_i (L_1^* \ \ldots, L_k^*) = \min_\delta \max_i (L_1, \ldots, L_k), \tag{7d.5.1}$$

where i runs over the indices $1, \ldots, k$ and δ over the set of admissible decision rules. The determination of such a rule, even if it exists, may be difficult. But there exist situations where a decision rule may be identified as a minimax rule.

(*i*) *Let there exist a vector of positive probabilities* $\bar{\boldsymbol{\pi}}' = (\bar{\pi}_1, \ldots, \bar{\pi}_k)$ *such that for the corresponding Bayes solution the components of the loss vector are all equal. Then* $\bar{\boldsymbol{\pi}}$ *is called the least favourable prior distribution, and the Bayes decision rule corresponding to* $\bar{\boldsymbol{\pi}}$ *is a minimax rule.*

By definition, the loss vector for such a rule is

$$(\bar{L}, \ldots, \bar{L}), \tag{7d.5.2}$$

and since every Bayes decision rule with positive prior probabilities is admissible, there is no loss vector whose maximum component is not greater than $\bar{L}$, for otherwise (7d.5.2) would not be admissible.

For a proof of the existence of the minimax solution and an extension of the results of **7d.4** and **7d.5** to countable or uncountably many alternatives, the reader is referred to books by Blackwell and Girshick (1954) and Wald (1950). The problem of identification discussed in **7d** is a particular case of the general decision theory or the more general game theory. For a good and elementary introduction to the subject the reader is referred to the book on decision theory by Chernoff and Moses (1959).

7.e NONPARAMETRIC INFERENCE

7e.1 Concept of Robustness

The aim of statistical methods discussed in the various chapters has been the following. In any problem we have a set of observations and some partial information regarding the probability distribution of the observations. Assuming the given information to be true, optimum procedures have been found for analyzing the data and drawing inferences on the unknown aspects (or characteristics) of the probability distribution. Strictly speaking, the validity of any particular method depends on the information supplied being correct. Let us review the situation with reference to Student's t-statistic used extensively in practice for drawing inferences on the unknown mean of a population from which observations are obtained. Now the distribution of the t-statistic defined by

$$t = \frac{\sqrt{n}\,|\bar{x} - \mu_0|}{s}, \tag{7e.1.1}$$

where $\bar{x}$ and s are computed from n observations $x_1, \ldots, x_n$ on a random variable X, is obtained on the following assumptions:

(a) The distribution of the random variable X is normal.

(b) The observations drawn are mutually independent.

(c) There are no errors in recording the observations.

(d) The mean in the population is exactly μ_0.

Suppose we wish to use the t-distribution to test the hypothesis that the average height of individuals in a given population is $\mu_0 = 70$ inches on the basis of n observations. What inference can we draw when a large value of t, significant at some (small) chosen level, occurs? One or more of the assumptions (a) to (d) may be wrong, thus setting aside the possibility that a rare event has occurred while all the assumptions are true. Let us examine each.

The number of individuals in a population, although large, is finite. Consequently a characteristic, such as the height of an individual, has only a finite number of alternatives. The assumption of a continuous distribution such as the normal for the heights of individuals, cannot be strictly true. But it is known from partly theoretical and partly empirical studies that the t-distribution is not sensitive to moderate departures from normality so that its application is not strictly governed by the normality assumption (a). Such a property is described as robustness (Box and Anderson, 1955). A significant t may not, therefore, be interpreted as indicating departure from normality of the observations.

What is the effect of dependence of observations on the t distribution? Suppose that all the observations are mutually correlated with a common correlation ρ for any two. Then

$$V(\bar{x}) = \frac{\sigma^2}{n}(1 + \overline{n-1}\rho) = E(\bar{x} - \mu_0)^2 \tag{7e.1.2}$$

$$E(s^2) = \sigma^2(1 - \rho). \tag{7e.1.3}$$

Instead of the t-statistic (7e.1.1), let us consider

$$t^2 = n(\bar{x} - \mu_0)^2 \div s^2, \tag{7e.1.4}$$

which is distributed as F on 1 and $(n - 1)$ D.F. From (7e.1.2) and (7e.1.3) the expected values of the numerator and denominator of t^2 are

$$\sigma^2(1 + \overline{n-1}\rho) \quad \text{and} \quad \sigma^2(1 - \rho). \tag{7e.1.5}$$

The ratio of the expectations is unity when $\rho = 0$, but is greater than unity when $\rho > 0$ and $\rightarrow$ infinity as $\rho \rightarrow 1$. Thus a large value of t is expected to occur when ρ is positively large, even when μ_0 is the true value of the mean. A significant t may therefore be due to departure in the assumption (b).

There is no way to study the effect of recording errors on the distribution of t, but with a little care departure from assumption (c) can be avoided.

Finally, when the assumptions (a) to (c) are true and the true value of the mean is $\mu \neq \mu_0$, the ratio of the expected values of the numerator and denominator of t^2 of (7e.1.4) is

$$\frac{n(\mu - \mu_0)^2}{\sigma^2} + 1, \tag{7e.1.6}$$

compared to 1 when $\mu = \mu_0$, so that large values of $|t|$ do occur when assumption (d) is wrong. This is exactly the reason why the t-test is used to test the null hypothesis concerning the mean of a distribution.

In computing expressions (7e.1.5) we have considered the extreme case of mutual dependence with a common correlation ρ. But in general we may expect that any dependence giving positive correlations to pairs of variables to overestimate the significance of t. Once we are satisfied that the random mechanism employed, such as the use of random numbers in selecting individuals, is not likely to introduce correlations among the observations, the significance of t may be interpreted as indicating departure from the specified mean value.

For a similar examination of other test criteria, see Examples 1, 2, and 3 at the end of the chapter, and for examination of test criteria based on the analysis of variance technique, see books by Plackett (1960) and Scheffe (1959).

7e.2 Distribution-Free Methods

In **7e.1** we have seen the importance of studying the performance or the operating characteristic of a statistical procedure under widely different conditions. This is necessary in examining the *nature of the new knowledge any statistical procedure is expected to provide and the prerequisites to be met to validate particular inferences.* It would not be a tool for creating new knowledge if a statistical procedure were to depend on assumptions that have to be taken for granted, or that cannot be verified by data themselves, or that cannot be justified on some other logical grounds. Thus a statistical tool is not to be considered as appropriate for a particular model but more appropriately as an aid in interpreting experimental data through different models.

Having observed a significant t, it would not be a useful statement to make that, "if the assumptions (a) to (c) of **7e.1** are satisfied, then possibly the specified mean value is wrong." Motivated by this philosophy, statisticians tried to develop statistical procedures which depend on the least possible assumptions. Thus were evolved procedures known as nonparametric methods—more appropriately called distribution-free methods since their application is independent of the distribution of the random variable on which observations have been obtained.

Such tests may not be very sensitive. On the other hand, in many practical situations it may be possible to make a transformation of the observations in such a way that the assumption of approximate normality is met, in which case many of the optimum test procedures based on the normal theory will be applicable. We shall briefly consider some non-parametric tests to indicate their variety.

7e.3 Some Nonparametric Tests

Sign Test. Let $x_1, \ldots, x_n$ be independent and identically distributed (i.i.d.) observations on a r.v. X with a continuous d.f. Consider the hypothesis that the median of X is μ_0 (a given value).

Let r of the quantitites $(x_1 - \mu_0), \ldots, (x_n - \mu_0)$ be negative and s, positive, where $r + s \leqslant n$. The reader may verify that the conditional distribution of r given $r + s$ is binomial with probability of success π equal to 1/2. The hypothesis under examination is equivalent to testing the hypothesis that $\pi = 1/2$, given that r successes have occurred out of $r + s$ Bernoulli trials. We may use an exact test based on the binomial probability tables or a chi-square test on 1 D.F. when $r + s$ is large.

Two Sample Tests. Let $x_1, \ldots, x_m$ be i.i.d. observations on X with a continuous d.f. F_1 and similarly $y_1, \ldots, y_n$ on Y with a continuous d.f. F_2. To test the hypothesis $F_1 = F_2$ we have a number of nonparametric tests. The chi-square and Kolmogorov-Smirnov tests for this purpose have already been described in Sections **6b**, **6c**, and **6f**. They are useful in detecting departures of a wide variety from the null hypothesis $F_1 = F_2$. But if the main interest is in detecting differences in the mean values alone, some of the following tests will be useful.

Wald-Wolfowitz (1940) Run Test. The $(m + n)$ observations of the two samples put together are first arranged in increasing order of magnitude. Each observation of the first sample is replaced by 0 and of the second sample by 1 to obtain a sequence of 0's and 1's. If $F_1 = F_2$, then all possible $[(n + m)!/(n!\, m!)]$ arrangements of m zeros and n unities are equally likely and hence the distribution of any statistic based on a sequence of 0's and 1's can be exactly determined. One statistic is the number of runs of 0's and 1's. If $F_1 \neq F_2$ and specially when F_1 differs from F_2 by a shift in the distribution, the expected number of runs will be small and hence the observed number of runs can be declared as significant if it is smaller than a number specified by a chosen level of significance.

Wilcoxon (1945) Test. The $(m + n)$ observations are arranged in increasing order of magnitude as in the Wald-Wolfowitz test, and the ranks of the observations of the second sample are obtained. Let the ranks be $s_1, \ldots, s_n$. The test criterion is the sum of the ranks. If the null hypothesis is true, any selection of n ranks (numbers) from $1, 2, \ldots, (m + n)$ has the same probability $n!\, m!/(n + m)!$ and hence the distribution of the sum of ranks can be obtained. If one-sided alternatives are considered, a large value of the statistic (sum of ranks) would indicate departure from null hypothesis. If the alternatives are two sided, both small and large values of the statistic have to be considered in the specification of the region of rejection.

Fisher-Yates Test. Let $s_1, \ldots, s_n$ be the rank values for the second sample. The test criterion is $\sum c(s_i)$ where $c(s_i)$ is the expected value of the s_ith order statistic in a sample of size $(m + n)$ from $N(0, 1)$. Tables of expected values of order statistics up to sample size 20 from $N(0, 1)$ are given in R.M.M. Tables and up to 50 in Biometrika Tables, Vol. 1. One- and two-sided tests can be built up on the basis of the statistic $\sum c(s_i)$ as in the case of Wilcoxon test.

Van der Waerden's Test. The test criterion is $\sum \Phi^{-1}[s_i/(m + n + 1)]$ where Φ is the cumulative distribution of the standard normal variable. Tables facilitating this test are given by van der Waerden and Nievergelt (1956).

These are only a few nonparametric tests now available. Applications of nonparametric tests and significance values of some of the test statistics are contained in a book by Siegel (1956). Some theory of nonparametric tests and inference is found in books by Fraser (1957), Puri and Sen (1970) and Lehmann (1959). A bibliography of nonparametric statistics and related topics is compiled by Savage (1953) and Walsh (1962).

7e.4 Principle of Randomization

Suppose we wish to test the difference between the effects of two treatments A and B on a population of individuals. We obtain a random sample of m individuals and measure the effects of treatment A,

$$x_1, \ldots, x_m, \tag{7e.4.1}$$

which constitute a set of independent observations on a random variable X which represents the effect of A on an individual. Similarly, we obtain another independent sample of n individuals and measure the effects of B,

$$y_1, \ldots, y_n \tag{7e.4.2}$$

which are then observations on a random variable Y representing the effect of B. If F_1 and F_2 denote the d.f.'s of X and Y, the null hypothesis may be interpreted as the equality of F_1 and F_2. In such a case some of the nonparametric tests of **7e.3** can be applied on the observations (7e.4.1) and (7e.4.2) to draw valid inferences on the difference between A and B *with respect to the population of individuals under consideration.*

But, generally, at least in the preliminary stages of an enquiry, it is not desirable to conduct an experiment involving samples of individuals from the population for which an ultimate recommendation has to be made. Experiments may be conducted on a laboratory scale by using volunteers or individuals specially chosen on the basis of some characteristics or even experimental animals which respond to the treatments. The results of such experiments may not be applicable to the whole population but they would provide the basis for a decision to plan and try out a suitable experiment on the individuals of a population.

We then have the problem of experimenting with the available subjects and generating observations which yield information on the difference between treatments. Since the basis of statistical inference is the available knowledge on the probability model governing the observations, it is necessary to generate the observations in such a way that the probability measure related to the various outcomes can be specified at least to the extent necessary for carrying out a test of significance of the null hypothesis. This is exactly what is achieved by the principle of *randomization*, a remarkable contribution by Fisher, which formed the basis of the now widely explored field of *Design of Experiments.*

According to the principle of randomization, the assignment of a test subject to a treatment is based on the result of a random experiment. If there are two treatments A and B, we may in each individual case decide on A or B by tossing a coin (giving A if head appears and B otherwise). Or we may divide the available subjects at random into two groups (of chosen sizes), by using random sampling numbers, and administer A to members of one group and B to another. If we decide to have m subjects for A and n for B, there are $(m+n)!/m!\,n!$ ways of dividing the total number $N = (m+n)$ of individuals into two groups of sizes m and n. By the random mechanism used, we give equal chance to each such division being chosen for the experiment. How does such a procedure specify the probability measure on the space of outcomes of the experiment?

Let the null hypothesis be that the *effect of A or of B is the same on any given individual, although this effect may be different from individual to individual* (*natural variation*). Under the null hypothesis, let $z_1, \ldots, z_N$ represent the effects of A or B on the N individuals. If the individuals are divided into two groups of sizes m and n, the first group receiving A and the second B, the *observed effects of A* would be a subset of $z_1, \ldots, z_N$,

$$x_1, \ldots, x_m, \tag{7e.4.3}$$

in which case the complementary subset

$$y_1, \ldots, y_n \tag{7e.4.4}$$

would be the *observed effects of B*. Thus a single experiment would provide us with the observations (7e.4.3), (7e.4.4), which taken together are the same as the set $z_1, \ldots, z_N$. Since every subset of m individuals has the same chance of being assigned to A, every subset of m values of $z_1, \ldots, z_N$ has the same chance, $n!\,m!/N!$, of being observed as effects of A. If we define the statistic

$$d = |\bar{x} - \bar{y}|, \tag{7e.4.5}$$

which is the difference between the averages of (7e.4.3) and (7e.4.4), its distribution can be computed by considering all divisions into two sets of given sizes of $z_1, \ldots, z_N$, which values are known from any particular experiment. Such a distribution is known as a *permutation distribution*, and its validity depends only on the *perfection of the mechanism* used in assigning individuals to the treatments. The permutation distribution of d provides us with a value d_α such that $P(d \geqslant d_\alpha) \leqslant \alpha$, a chosen level of significance. The null hypothesis is rejected when an observed d is $\geqslant d_\alpha$.

In practice, it may be difficult to determine the exact distribution of d as the number of divisions to be considered is large unless N is small. But if m and n are large, the statistic

$$t = \sqrt{\frac{mn}{m+n}}\,\frac{\bar{x}-\bar{y}}{s} \tag{7e.4.6}$$

$$s^2 = \frac{\sum (x_i - \bar{x})^2 + \sum (y_i - \bar{y})^2}{m+n-2}$$

can be used as Student's t on $(m + n - 2)$ D.F.

How should a significant t or d be interpreted? The null hypothesis that the treatments A and B are identical in their effects on any individual stands rejected. But there may be the possibility that in *each* individual case there is difference between A and B but the *overall* (or average) difference is zero. It is therefore necessary to examine what specific departures from the null hypothesis the test is capable of detecting. This question can be answered by studying the O.C. (operating characteristic) of the test procedure when the null hypothesis is not true.

We therefore consider the most general case by letting arbitrary numbers

$$\begin{gathered} z_{1A}, \ldots, z_{NA} \\ z_{1B}, \ldots, z_{NB} \end{gathered} \tag{7e.4.7}$$

denote the effects of A and B on the N individuals. Define

$$\sum_{i=1}^{N} z_{ir} = \bar{z}_r, \qquad r = A, B,$$

$$\bar{z}_A - \bar{z}_B = \Delta$$

$$\sum (z_{ir} - \bar{z}_r)^2 = (N-1)V_r, \qquad r = A, B$$

$$\sum (z_{iA} - \bar{z}_A)(z_{iB} - \bar{z}_B) = (N-1)C_{AB}.$$

Let $x_1, \ldots, x_m$ and $y_1, \ldots, y_n$ be the observed effects of A and B. Under the general set up (7e.4.7) the expectations, variances, and covariances of x_i, y_i of the permutation distribution are

$$E(x_i) = \bar{z}_A, \qquad E(y_i) = \bar{z}_B$$

$$V(x_i) = \frac{(N-1)V_A}{N}, \qquad \text{cov}(x_i, x_j) = -\frac{V_A}{N} \tag{7e.4.8}$$

$$V(y_i) = \frac{(N-1)V_B}{N}, \qquad \text{cov}(y_i, y_j) = -\frac{V_B}{N}$$

$$\text{cov}(x_i, y_j) = -\frac{C_{AB}}{N}.$$

Let us consider the statistic t^2, instead of t defined in (7e.4.6); then

$$t^2 = \frac{mn}{m+n}(\bar{x} - \bar{y})^2 \div s^2 = a^2 \div b^2 \quad \text{(say)} \tag{7e.4.9}$$

By using the formulas (7e.4.8) it is a straightforward computation to show that

$$E(a^2) = \frac{nm}{n+m}\left[\Delta^2 - \frac{1}{N}(V_A + V_B - 2C_{AB})\right] + \frac{nV_A + mV_B}{m+n} \tag{7e.4.10}$$

$$E(b^2) = \frac{(m-1)V_A + (n-1)V_B}{m+n-2}. \tag{7e.4.11}$$

The second expression in (7e.4.10) is of the same order of magnitude as (7e.4.11), especially when $m = n$. If $\Delta = 0$, the expected value of the numerator of t^2 actually tends to be smaller than that of the denominator since $V_A + V_B - 2C_{AB} \geqslant 0$; so a significantly large t could not arise if in fact the treatment effects were different for each individual but the average difference $\Delta = 0$. If $\Delta \neq 0$, the numerator of t^2 is expected to have a larger value and, therefore, it appears to be reasonable to assert that a significant t is mainly due to $\Delta \neq 0$, that is, a real difference in the averages rather than in some other aspects of the effects (7e.4.7).

For the variety of ways that the principle of randomization can be exploited to enable valid inferences to be drawn from experimental data, the reader is referred to books on Design of Experiments by Cochran and Cox (1957), Fisher (1935), and Kempthorne (1952). Some papers of interest in connection with permutation distribution and nonparametric inference are by Chernoff and Savage (1958), Heoffding (1951), Mitra (1955), Neyman (1935), Pitman (1937, 1938), Rao (1959b), and Wolfowitz (1949).

7f ANCILLARY INFORMATION

In **6e.3**, we introduced a conditional test of independence of attributes in a 2×2 contingency table, where the significance of the test statistic was judged using its conditional distribution given the marginal totals as observed in the sample. Barnard (1945) has shown that an alternative test exists, which is more powerful, although the suggestion did not find favor with Fisher (1956) who advocated conditional tests on the basis of what is called *ancillary information*. The exact ground rules of such conditional tests have not been fully worked out. However, we shall explain the principle enumerated by Fisher. The interested reader is referred to papers by Barnard (1945), Barnard, Jenkins, and Winsten (1962), Basu (1964), Birnbaum (1962), Cox (1958) and Fisher (1956).

Definition. Consider a family of probability measures, P_θ defined on $(X, \mathscr{B})$, and P_θ^T, the family of distributions induced by a statistic T defined on X. T is said to be an ancillary statistic iff P_θ^T is independent of θ.

Fisher suggested that for purposes of inference one should consider the family of conditional probability measures on X given $T = t$, the observed value of the ancillary statistic in the sample, as a starting point. It would have been nice if there existed what may be called a maximal ancillary T_* in the sense that every other ancillary T is a function of T_*, in which case we could condition on T_*. Unfortunately, a maximal ancillary may not exist as pointed out by Basu (1964). However, we shall consider some examples to underline the importance of ancillary information. Suppose there is an urn containing five balls out of which an unknown number α are red and others are white. An experiment consists in first tossing an unbiased coin and drawing five balls, one after the other, with replacement if the coin falls head upwards and observing all the five balls in the urn otherwise.

In such a situation the m.l. estimate of α is r, the observed number of red balls, and the standard error of r is $\sqrt{\alpha(5-\alpha)}/5\sqrt{10}$. Such an inference, or reduction of data to an m.l. estimate and its standard error, does not recognize ancillary information.

Let us observe that the event head or tail provides ancillary information in which case the estimator of α is as follows, where the standard error is based on the conditional distribution depending on the ancillary statistic.

Estimator	Standard Error	Ancillary Information
r	$\sqrt{\alpha(5-\alpha)}/5\sqrt{5}$	Head
r	0	Tail

It appears that nonrecognition of ancillary information has led to a peculiar situation where we are forced to attach a standard error although there is no error in the estimate! Thus the reduction of data to an m.l. estimator without supplementing it with ancillary information may result in an anomalous situation as in the present example.

More generally, let us consider a mixture of experiments $E_1, \ldots, E_k$ where E_i is chosen with a known probability π_i but each experiment contains information on the same unknown parameter θ. If the problem is one of testing the null hypothesis $H_0 : \theta = \theta_0$, against the alternative $H : \theta = \theta_1$, the optimum procedure according to Neyman Pearson lemma is to reject H_0 if

$$P_1(x \mid E_i)/P_0(x \mid E_i) \geqslant k \tag{7f.1}$$

where $P_j(x \mid E_i)$ is the probability density at the observed value x when E_i is performed and H_j is true. In (7f.1), k is determined such that

$$\sum_{i=1}^{k} \pi_i \text{ Prob. } \{P_1(x \mid E_i)/P_0(x \mid E_i) \geqslant k \mid E_i\} = \alpha \text{ (given).} \tag{7f.2}$$

However, the recognition of ancillary information tells us that when E_i is chosen, the inference should be based solely on the conditional probability given E_i, ignoring other possibilities of experimentation. Thus the test statistic when E_i is chosen would be

$$P_1(x \mid E_i)/P_0(x \mid E_i) \geqslant k_i \tag{7f.3}$$

where k_i is determined such that

$$\text{Prob. } \{P_1(X \mid E_i)/P_0(X \mid E_i) \geqslant k_i \mid E_i\} = \alpha \text{ (given).} \tag{7f.4}$$

The test procedure (7f.3) has lower power than that of (7f.1) in the long run. Which procedure should one prefer? It is left to the reader to reflect and decide although the author is inclined to choose the latter.

COMPLEMENTS AND PROBLEMS

1 *Departure from Normality Assumption on the t Test.* The t-statistic based on a sample of size n is $\sqrt{n}(\bar{x} - \mu)/s$ where $\bar{x}$ is the average and s^2 is the corrected sum of squares divided by $(n - 1)$.

1.1 Determine the distribution of t when the parent population is exponential, $\theta^{-1} \exp(-x/\theta)$, $x \geqslant 0$ for the sample size $n = 2$. Compute the actual probability of a type I error by using the 5% point of $|t|$ distribution based on the normal theory and show that it is divided unequally between the two tails of the distribution of t. [Note that for the exponential distribution the measures of skewness and kurtosis $\sqrt{\beta_1}$ and β_2 have the values 2 and 9 widely different from the values 0 and 3 for a normal distribution].

1.2 Draw size-5 samples from a rectangular population and compute the t-statistic in each case. Obtain the frequency distribution of the observed t values and compare with the t distribution on 4 D.F. derived from the normal theory. Estimate from the observed frequency distribution, the actual probabilities of type I error corresponding to the 5% and 1% values of t on normal theory. [It is necessary to have from 200 to 500 samples to make proper comparisons. This is an example which can be given as a joint project to the students in a class.]

2 *Departure from Independence Assumption on the t Test.* Let $x_1, \ldots, x_n$ be observations such that $E(x_i) = 0$, $V(x_i) = \sigma^2$, and $\text{cov}(x_i, x_j) = \rho\sigma^2$ if $|i - j| = 1$ and 0 otherwise.

2.1 Show that the asymptotic distribution of the t-statistic $\sqrt{n}\bar{x}/s$, is $N[0, (1 + 2\rho)]$, and hence that the significance of t is overestimated when $\rho > 0$ and underestimated when $\rho < 0$ in large samples. For finite n, compute the expected values of the numerator and denominator of $t^2 = n\bar{x}^2/s^2$ and study the effect of $\rho \neq 0$ in interpreting significant values of t as contradicting the hypothesis $E(x_i) = 0$.

2.2 Show that a valid inference can be drawn on the expected value of x_i by considering the observations

$$y_1 = \frac{x_1 + x_2}{2}, \ldots, y_{n/2} = \frac{x_{n-1} + x_n}{2},$$

when n is even and by applying a t test on the y sample (with $(n/2) - 1$ D.F.).

3 *Departure from normality assumption on the χ^2 test for an assigned value of σ^2.* Consider the statistic

$$\chi^2 = \frac{(n-1)s^2}{\sigma_0^2}$$

where σ_0 is the assigned value of σ, which is distributed as $\chi^2(n-1)$ when the observations $x_1, \ldots, x_n$ are on a r.v. $X \sim N(\mu, \sigma^2)$.

3.1 Let X be distributed with variance σ^2 and $\beta_2 - 3 = \gamma_2 \neq 0$ (non-normal). Show that

$$E(s^2) = \sigma^2 \qquad \text{and} \qquad V(s^2) = \sigma^4\left(\frac{2}{n-1} + \frac{\gamma_2}{n}\right).$$

3.2 Show that the asymptotic distribution of

$$z = \frac{\chi^2 - (n-1)}{\sqrt{2(n-1)}} = \sqrt{\frac{n-1}{2}}\left(\frac{s^2}{\sigma_0^2} - 1\right),$$

where ${\sigma_0}^2$ is the true value of σ^2, is $N(0, 1)$ when $\gamma_2 = 0$ and $N(0, 1 + \gamma_2/2)$ in general.

3.3 Deduce that the χ^2 test based on $(n-1)s^2/{\sigma_0}^2$ for testing the null hypothesis, $\sigma^2 = {\sigma_0}^2$, is not robust for departures from normality.

4 Let $x_i \sim G(\alpha_i, p_i)$, $i = 1, \ldots, k$ and independent. Derive appropriate tests for the hypotheses:

(a) $\alpha_1 = \cdots = \alpha_k$ given $p_1 = \cdots = p_k$
(b) $p_1 = \cdots = p_k$ given $\alpha_1 = \cdots = \alpha_k$
(c) $\alpha_1 = \cdots = \alpha_k$ and $p_1 = \cdots = p_k$

[Hint: Try likelihood ratio criteria.]

5 Let $x_1, \ldots, x_m$ be m independent observations on a r.v. X with density ${\theta_1}^{-1} \exp(-x/\theta_1)$ and $y_1, \ldots, y_n$ on Y with density ${\theta_2}^{-1} \exp(-y/\theta_2)$.

5.1 Show that the likelihood ratio test for $H_0\colon \theta_1 = \theta_2$ is a function of $\bar{x}/\bar{y}$. Find the distribution of $\bar{x}/\bar{y}$.

5.2 Characterize the class of similar region tests and derive optimum similar region tests when (a) the alternatives are $\theta_1 > \theta_2$, and (b) the alternatives are both sided (locally unbiased and most powerful test). Show that these tests also depend on the statistic $\bar{x}/\bar{y}$. [Observe that when $\theta_1 = \theta_2 = \theta$, $T = \sum x_i + \sum y_j$ is sufficient for θ, and then apply the method of **7a.5**.]

5.3 Derive the power functions of these tests.

6 Let $x_1, x_2, \ldots, x_m$ be independent observations from a Poisson distribution $p(\mu)$ and $f_0, f_1, \ldots$ the frequencies of 0, 1, ...

6.1 Find the conditional probability of $f_0, f_1, \ldots$ given $T = x_1 + x_2 + \cdots + x_m$, which is sufficient for μ.

6.2 Derive in particular the exact conditional probability distribution of f_0, the frequency of zero observation, given T. Show that

$$E(f_0 \mid T) = m\left(1 - \frac{1}{m}\right)^T$$

$$V(f_0 \mid T) = m(m-1)\left(1 - \frac{2}{m}\right)^T + m\left(1 - \frac{1}{m}\right)^T - m^2\left(1 - \frac{1}{m}\right)^{2T}.$$

6.3 Show that the asymptotic distribution of

$$[f_0 - E(f_0 \mid T)] \div \sqrt{V(f_0 \mid T)}$$

is $N(0, 1)$ as m and $T \to \infty$.

6.4 Show that the similar region test for the hypothesis that the frequency of zero observation is as expected in sampling from a Poisson distribution is based on the conditional distribution of f_0 given T (Rao and Chakravarthy 1956d).

6.5 Let $z_1, \ldots, z_{m-1}$ be theoretical fractiles of a continuous d.f. $F(x)$ such that $F(z_i) = i/m$, $i = 1, \ldots, m-1$. Further let $x_1, \ldots, x_T$ be T independent samples from $F(x)$ and f_r the number of intervals $(-\infty, z_1)$, (z_1, z_2), $\ldots$, (z_{m-1}, ∞) containing r of the observations $x_1, \ldots, x_T$. Show that the distribution of $f_0, f_1, \ldots$ is the same as that derived in Example (6.1) and that of f_0 is the same as that derived in Example (6.2).

7 *Nonparametric Tests.*

7.1 Let $x_1, \ldots, x_m$ and $y_1, \ldots, y_n$ be independent observations from populations with continuous d.f.'s F_1 and F_2. Define by m_1 and n_1 the number of x's and y's exceeding the kth order statistic of the combined sample. If $F_1 = F_2$ show that the probability of given values of m_1, n_1 is

$$P(m_1, n_1) = \binom{m}{m_1}\binom{n}{n_1} \div \binom{m+n}{k}, \qquad m_1 + n_1 = m + n - k.$$

Based on this probability distribution of (m_1, n_1), derive a nonparametric test of the hypothesis, $F_1 = F_2$.

7.2 In an experiment to test the difference between two skin ointments A and B, the treatments were assigned at random to the right and left hand of each patient and the improvement is scored as bad, fair, and good. The data on 50 patients are summarized as follows.

		Treatment A		
		Bad	Fair	Good
Treatment B	Bad	2	7	10
	Fair	4	4	16
	Good	3	4	0

Test for treatment differences.

7.3 Let $x_1, \ldots, x_m$ and $y_1, \ldots, y_n$ be samples from continuous d.f.'s F_1 and F_2. Wilcoxon statistic for testing $F_1 = F_2$ is $\sum_1^n s_i$ where $s_1, \ldots, s_n$ are the ranks of $y_1, \ldots, y_n$ when the two samples are combined. If $U = \sum s_i - n(n+1)/2$, show that

$$E\left(\frac{U}{mn}\right) = \int F_1 \, dF_2$$

$$mnV\left(\frac{U}{mn}\right) = \int F_1 \, dF_2 + (n-1)\int (1-F_2)^2 \, dF_1 + (m-1)\int F_1^2 \, dF_2 - (m+n-1)\left(\int F_1 \, dF_2\right)^2.$$

In particular, when $F_1 = F_2$

$$\mathrm{E}\left(\frac{U}{mn}\right) = \frac{1}{2}, \qquad \mathrm{V}\left(\frac{U}{mn}\right) = \frac{m+n+1}{12mn}.$$

8 Show that the sequential test for testing a binomial probability π_0 against an alternative $\pi_1 > \pi_0$, can be written in the following form. At the mth stage let m_1 be the number of successes. Then:

(a) accept π_0 if $m_1 \leqslant cm + d_1$,
(b) reject π_0 if $m_1 \geqslant cm + d_2$,
(c) continue sampling otherwise,

where c, d_1, d_2 are constants depending on (α, β), the strength of the test, and π_0, π_1.

Obtain the O.C. function and A.S.N. of the foregoing S.P.R.T.

9 *Truncation of S.P.R.T.* If the S.P.R.T. does not terminate for $n = 1, \ldots, N-1$, suppose the following rule is adopted for terminating the test at the Nth stage; accept H_0 if $\log B < z_1 + \cdots + z_N \leqslant 0$ and accept H_1 if $0 < z_1 + \cdots + z_N < \log A$.

Let $y_0 = -\sqrt{N}E(z\,|\,\theta_0)/\sigma(z\,|\,\theta_0)$ and $y_0' = \sqrt{N}\,[(1/N)\log A - E(z\,|\,\theta_0)]/\sigma(z\,|\,\theta_0)$, where $\sigma^2(z\,|\,\theta_0)$ is the variance of z. Show that an upper bound to the first kind of error is

$$\alpha + \Phi(y_0') - \Phi(y_0) + 0\left(\frac{1}{\sqrt{N}}\right)$$

where Φ is the d.f. of $N(0, 1)$. Find the upper bound to β_N.

10 *O.C. of a Control Chart.* In operating a control chart with a central line L_0, two lines L_1, L_2 above L_0 and two lines L_2', L_1' below L_0, some of the suggested rules for taking action are one or a combination of the following depending on the configuration of the successive plotted points:

(a) If a point falls above L_1 or below L_1'.
(b) Two successive points between L_1, L_2 or between L_1', L_2'.
(c) A configuration of three points such that the first and third are between L_1, L_2 and the second between L_0, L_2, and a similar situation with respect to L_0, L_2', L_1'.

Let α_1, α_2, and α_3 be the probabilities of a plotted point being above L_1, between L_2, L_1, and between L_0, L_2. Similarly, α_1', α_2', and α_3' are defined with respect to L_1', L_2', L_0. Let P_n be the probability of taking action at the nth stage and $Q_n(L_i, L_j)$ be the probability of a point falling between L_i, L_j and no action being taken.

10.1 Show that when rules (a) and (b) are jointly used, P_n, Q_n satisfy the recurrence relations

$$P_n = \alpha_1 + \alpha_1' + \alpha_2 + \alpha_2' - Q_n(L_1, L_2) - Q_n(L_1', L_2')$$
$$Q_n(L_1, L_2) = \alpha_2 - \alpha_2 Q_{n-1}(L_1, L_2)$$
$$Q_n(L_1', L_2') = \alpha_2' - \alpha_2' Q_{n-1}(L_1', L_2').$$

10.2 If $\lambda = \lim \sum_{r=1}^{n} P_r/n$, as $n \to \infty$ and $p_{ij} = \lim \sum_{r=1}^{n} Q_r(L_i, L_j)/n$, $p_{ij}' = \lim \sum_{r=1}^{n} Q_r(L_i', L_j')/n$ as $n \to \infty$, show that

$$\lambda = \alpha_1 + \alpha_1' + \alpha_2 + \alpha_2' - p_{12} - p_{12}'$$
$$p_{12} = \alpha_2 - \alpha_2 p_{12}, \qquad p_{12}' = \alpha_2' - \alpha_2' p_{12}'.$$

Observe that λ is the proportion of occasions when action is taken in the long run on any hypothesis specifying the values of α_i, α_i'.

10.3 Deduce that the O.C. function for the rules (a) and (b) jointly is

$$\lambda = \alpha_1 + \alpha_1' + \alpha_2 + \alpha_2' - \frac{\alpha_2}{1 + \alpha_2} - \frac{\alpha_2'}{1 + \alpha_2'}.$$

10.4 Similarly obtain the O.C. function for the rules (a) and (c) jointly

$$\lambda = \alpha_1 + \alpha_1' + \alpha_2 + \alpha_2' - \frac{\alpha_2}{1 + \alpha_2 + \alpha_2 \alpha_3} - \frac{\alpha_2'}{1 + \alpha_2' + \alpha_2' \alpha_3'}.$$

10.5 Suppose that each observation is drawn from $N(\theta, \sigma^2)$ and a control chart is made up of a horizontal line L_0 at $\theta = 50$ with L_2, L_2' placed at distances of $\pm 2\sigma$ from L_0 and L_1, L_1' placed at distances of $\pm 3\sigma$. Tabulate the O.C. function λ of Examples (10.3) and (10.4) for the following values of θ

$$50, \quad 50 + \frac{\sigma}{2}, \quad 50 - \frac{\sigma}{2}, \quad 50 + \sigma, \quad 50 - \sigma, \quad 50 + 2\sigma, \quad 50 - 2\sigma.$$

10.6 In the situation of Example (10.5) and for the rules of Examples (10.3) and (10.4), obtain an optimum placing of the lines L_1, L_2, L_1', L_2' when the value of λ at $\theta = 50$ is fixed at a given value (first kind of error). You may have to impose some restrictions such as symmetrical placing of L_1, L_1' and L_2, L_2' and so on and then maximize the O.C. function in some sense (locally or at specified values of θ).

11 Consider the sequence of children *NNNNDDDND* scored as normal (N) and defective (D) in the sequence of their birth orders. Satisfy yourself that the sum of the parities of defective children is a good criterion with which to test the association between birth order (parity) and defectiveness. In this example the sum is $5 + 6 + 7 + 9 = 27$. Find the probability of the sum being equal to or greater than 27 under the hypothesis that all sequences of 5 N's and 4 D's are equally likely. [Haldane and Smith, 1948.]

12 Suggest a sequential test of the hypothesis H_0 against H_1, setting an upper limit to the number of observations to be drawn. [See Rao (1950e) for a test of the null hypothesis placing an upper limit on the number of observations.]

13 *A Statistical Paradox* (Lindley, 1957). Let $\bar{x}$ be the average of n observations from $N(\theta, 1)$, and the prior probability of θ be c at the single point $\theta = \theta_0$ and uniform in an interval around θ_0 but excluding θ_0. Suppose we want to test the hypothesis $\theta = \theta_0$, and $\bar{x} = \theta_0 + \lambda_\alpha/\sqrt{n}$ is observed, where λ_α is the α percent significant value of $N(0, 1)$. Show that the posterior probability that $\theta = \theta_0$ at $\bar{x} = \theta_0 + \lambda_\alpha/\sqrt{n}$ is

$$\bar{c} = ce^{-(1/2)\lambda_\alpha^2} \div [ce^{-(1/2)\lambda_\alpha^2} + (1-c)\sqrt{2\pi/n}]$$

and $\bar{c} \to 1$ as $n \to \infty$. Thus there exists an n such that $\bar{c} = (1 - \alpha)$ where α is the level of significance. If $\alpha = 0.05$, we have then situations where the null hypothesis is rejected at 5% since $(\bar{x} - \theta_0)\sqrt{n} = \lambda_\alpha$, but the posterior probability of the null hypothesis being true is 95%.

REFERENCES

Anscombe, F. J. (1952), Large sample theory of sequential estimation, *Proc. Camb. Phil. Soc.* **48**, 600–607.

Anscombe, F. J. (1953), Sequential estimation, *J. Roy. Statist. Soc.* **B 15**, 1–29.

Bahadur, R. R. (1960), Stochastic comparison of tests, *Ann. Math. Statist.* **31**. 276–295.

Banerji Saibal Kumar (1960), Approximate confidence interval for linear functions of means of k populations when the population variances are not equal, *Sankhyā* **22**, 357–358.

Barnard, G. A. (1942), A new test for 2 × 2 tables, *Nature* **156**, 177.

Barnard, G. A. (1949), Statistical inference, *J. Roy. Statist. Soc.* **B 11**, 115–149.

Barnard, G. A. (1969), Practical applications of tests with power one, *Bull. Inst. Inter. Statist.* **XLIII**(i), 389–393.

Barnard, G. A., G. M. Jenkins and C. B. Winsten (1962), Likelihood inference and time series (with discussion), *J. Roy. Statist. Soc.* **A 125**, 321–372.

Basu, D. (1964), Recovery of ancillary information, *Sankhyā* **26**, 3–16.

Birnbaum, A. (1962), On the foundations of statistical inference (with discussion), *J. Am. Stat. Assn.* **57**, 269–326.

Blackwell, D. and M. A. Girshick (1954), *Theory of Games and Statistical Decisions*, Wiley, New York.

Box, G. E. P. and S. L. Anderson (1955), Permutation theory in the derivation of robust criteria and study of departures from assumptions, *J. Roy. Statist. Soc.* **B 17**, 1–34.

Chernoff, H. (1949), Asymptotic studentization in testing of hypothesis, *Ann. Math. Statist.* **20**, 268–278.

Chernoff, H. (1952), A measure of asymptotic efficiency of tests of a hypothesis based on the sum of observations, *Ann. Math. Statist.* **23**, 493–507.

Chernoff, H. (1959), Sequential design of experiments, *Ann. Math. Statist.* **30**, 755–770.

Chernoff, H. and L. E. Moses (1959), *Elementary Decision Theory*, Chapman and Hall, London.

Chernoff, H. and R. I. Savage (1958), Asymptotic normality and efficiency of certain nonparametric test statistics, *Ann. Math. Statist.* **29**, 972–994.

Cochran, W. G. and G. M. Cox (1957), *Experimental Designs* (second edition), Wiley, New York.

Cox, D. R. (1958), Some problems connected with statistical inference, *Ann. Math. Statist.* **29**, 357–372.

Dantzig, G. B. (1940), On the non-existence of tests of 'Student's' hypothesis having power functions independent of σ, *Ann. Math. Statist.* **11**, 186–192.

Dantzig, G. B. and A. Wald (1951), On the fundamental lemma of Neyman and Pearson, *Ann. Math. Statist.* **22**, 87–93.

Dodge, H. F. and H. G. Romig (1929), A method of sampling inspection, *Bell System Tech. Jour.* **8**, 613–631.

Fisher, R. A. (1935a), *The Design of Experiments*, Oliver and Boyd, Edinburgh.

Fisher, R. A. (1935b), The fiducial argument in statistical inference, *Ann. Eugen.* **6**, 391–398.

Fisher, R. A. (1936), Uncertain inference, *Proc. Am. Acad. of Arts and Sciences* **71**, 245–258.

Fisher, R. A. (1956), *Statistical Methods and Scientific Inference*, Oliver and Boyd, London.

Fraser, D. A. S. (1957), *Nonparametric Methods in Statistics*, Wiley, New York.

Godambe, V. P. and D. A. Sprott (1971), *Foundations of Statistical Inference*, Holt, Reinhart and Winston, Canada.

Haldane, J. B. S. and C. A. B. Smith (1948), A simple exact test for birth-order effect, *Ann. Eugen.* **14**, 117–124.

Hodges, J. L. Jr. and E. L. Lehmann (1956), The efficiency of some nonparametric competitors of the t-test, *Ann. Math. Statist.* **27**, 324–335.

Hoeffding, Wassily (1951), "Optimum" nonparametric tests, *Proc. (Second) Berkeley Symp. on Math. Statist. Prob.* Berkeley Univ. California Press, 83–92.

Hoeffding, Wassily and J. R. Rosenblatt (1955), The efficiency of tests, *Ann. Math. Statist.* **26**, 52–63.

Hogben, L. (1957), *Statistical Theory: the relationship of probability, credibility, and error; an examination of the contemporary crisis in statistical theory from a behaviouristic viewpoint*. George Allen and Unwin, London.

Jeffreys, H. (1948), *Theory of Probability* (second edition), Clarendon Press, Oxford.

Kempthorne, O. (1952), *Design and Analysis of Experiments*, Wiley, New York.

Kyburg, H. E., Jr (1961), *Probability and the Logic of Rational Belief*, Wesleyan Univ. Press, Middletown, Connecticut.

Lehmann, E. L. (1959), *Testing Statistical Hypotheses*, Wiley, New York.

Lindley, D. V. (1953), Statistical inference, *J. Roy. Stat. Soc.* **B 15**, 30–76.

Lindley, D. V. (1957), A statistical paradox, *Biometrika* **44**, 187–192.

Linnik, Yu. V. (1963), On the Behrens-Fisher problem, *Proc. Int. Stat. Conference*, Ottawa.

Mahalanobis, P. C. (1940), A sample survey of acreage under jute in Bengal with discussion on planning of experiments, *Proc. (Second) Indian Stat. Conference, Calcutta.*

Mitra, Sujit Kumar (1960), On the F-test in the intrablock analysis of a balanced incomplete block design, *Sankhyā* **17**, 279–284.

Neyman, J. (1935), On the problem of confidence intervals, *Ann. Math. Statist.* **6**, 111–116.

Neyman, J. (1937), Outline of a theory of statistical estimation based on the classical theory of probability, *Philos. Trans. Roy. Soc.* **A 236**, 333–380.

Neyman, J. (1961), Silver jubilee of my dispute with Fisher, *J. Operations Research Soc. (Japan)* **3**, 145–154.

Neyman, J. with the cooperation of K. Iwaszkiewicz and St. Kolodziesezyk (1935), Statistical problems in agricultural experimentation, *J. Roy. Statist. Soc.* (suppl.) **2**, 107–154.

Neyman, J. and E. S. Pearson (1933), On the problem of the most efficient tests of statistical hypotheses, *Phil. Trans. Roy. Soc.* **A 231**, 289–337.

Noether, G. E. (1955), On a theorem of Pitman, *Ann. Math. Statist.* **26**, 64–68.

Pitman, E. J. G. (1937), Significance tests which may be applied to samples from any population, *J. Roy. Statist. Soc. (suppl.)* **4**, 119–130, 225–232, and *Biometrika* **29**, 322–335.

Pitman, E. J. G. (1949), Lecture notes on nonparametric statistical inference, Columbia University.

Plackett, R. L. (1960), *Principles of Regression Analysis*, Clarendon Press, Oxford.

Puri, M. L. and P. K. Sen (1970), *Nonparametric Methods in Multivariate Analysis*, Wiley, New York.

Robbins, H. and D. Seigmund (1969), Confidence sequences and interminable tests, *Bull. Inst. Int. Statist.* **XLIII**(1), 379–387.

Savage, I. R. (1953), Bibliography of nonparametric statistics and related topics, *J. Am. Stat. Assoc.* **48**, 844–906.

Savage, L. J. (1954), *Foundation of Statistics*, Wiley, New York.

Savage, L. J. (1962), *Foundations of Statistical Inference*, A discussion, Methuen, London.

Scheffe, H. (1959), *The Analysis of Variance*, Wiley, New York.

Siegel, Sidney (1956), *Nonparametric Statistics for the Behavioural Sciences*, McGraw-Hill, New York, Toronto, London.

Stein, Charles (1945), A two-samples test for a linear hypothesis whose power is independent of the variance, *Ann. Math. Statist.* **16**, 243–258.

Van der Waerden, B. L. and E. Nievergelt (1956), *Tables for Comparing Two Samples by X-Test and Sign Test*, Springer-Verlag, Berlin.

Wald, A. (1941), Asymptotically most powerful tests of statistical hypotheses, *Ann. Math. Statist.* **12**, 1–19.

Wald, A. (1947), *Sequential Analysis*, Wiley, New York.

Wald, A. (1950), *Statistical Decision Functions*, Wiley, New York.

Wald, A. and J. Wolfowitz (1940), On a test whether two samples are from the same population, *Ann. Math. Statist.* **11**, 147–162.

Walsh, J. E. (1962), *Handbook of Nonparametric Statistics. Investigation of Randomness, Moments, Percentiles and Distributions*, Nostrand, London.

Welch, W. L. (1947), The generalization of Student's problem when several population variances are involved, *Biometrika* **34**, 28–35.

Wilcoxon, Frank (1945), Individual comparisons by ranking methods, *Biometrics* **1**, 80–83.

Wolfowitz, J. (1947), The efficiency of sequential estimates and Wald's equation for sequential processes, *Ann. Math. Statist.* **18**, 215–230.

Wolfowitz, J. (1949), Nonparametric statistical inference, *Proc.* (*First*) *Berkeley Symp. on Math. Statist. Prob.* Berkeley Univ. California, 93–113.

Chapter 8

MULTIVARIATE ANALYSIS

Introduction. The general concepts of a multivariate distribution and a conditional distribution are introduced in Chapter 2. The properties of regression, which is the mean of a conditional distribution of one random variable given the others, have been studied in **4g.1** to **4g.2**. We shall now consider a particular class of distributions known as the multivariate normal, which plays an important role in statistical inference involving multiple measurements.

The multivariate normal distribution has already been encountered in **3b.3** as the *distribution of linear functions of independent univariate normal variables*. The special cases of the p-variate normal distribution when all the variances are equal and all the covariances are equal and the general bivariate normal distribution have been considered in great detail in **3c.1** and **3d.1**, respectively. In this chapter we shall investigate the properties of a general p-variate normal distribution, derive the distributions of some sample statistics, and illustrate their use in practical problems.

The theoretical approach to the p-variate normal distribution in this chapter is not along the usual lines and therefore needs some attention. First, the distribution is not defined by a probability density function. It is characterized by the property *that every linear function of the p-variables has a univariate normal distribution*. Second, such a characterization is exploited in deriving the distributions of sample statistics. It is shown that corresponding to any known result in the univariate theory, the generalization to the multivariate theory can be written down with a little or no further analysis. The method of doing this is stated in **8b.1** to **8b.2**. For instance, knowing the joint distribution of the sample mean and sample variance in the univariate theory we can write down the joint distribution of the sample means of multiple measurements and the sample variances and covariances. The entire theory of multivariate tests of significance by analysis of dispersion is obtained as a generalization of the univariate analysis of variance. The reader should familiarize himself with this general approach.

A number of other characterizations of the multivariate normal distribution have been given, which the reader may omit on first reading since they are not used in the derivation of the later results. But these characterizations will be useful for those who want to study the theory of normal distributions in Hilbert and more general spaces.

An important aspect of the multivariate analysis which is not emphasized in statistical literature is the examination of additional information provided by some measurements when some others are already available. This is important. Since in practice there is no limit to the number of multiple measurements which can be obtained in any problem, it is relevant to examine whether some are superfluous in the presence of the others. This problem is investigated in **8c.4**. A number of practical problems are considered in **8d.1** to **8d.5**.

8a MULTIVARIATE NORMAL DISTRIBUTION

8a.1 Definition

We have already encountered the multivariate density in **3b.3**, while studying the sampling distribution of linear functions of independent univariate normal variables. Denoting by $\mathbf{U}$, a p-dimensional random variable $(U_1, \ldots, U_p)$ and (for convenience) by the same symbol the values it can take, the density function of a p-variate normal distribution was defined to be

$$(2\pi)^{-p/2}|\mathbf{A}|^{1/2} \exp[-\tfrac{1}{2}(\mathbf{U}-\boldsymbol{\mu})'\mathbf{A}(\mathbf{U}-\boldsymbol{\mu})], \tag{8a.1.1}$$

where $\mathbf{A}$ is a positive definite matrix. It was also observed that $E(\mathbf{U}) = \boldsymbol{\mu}$ and $D(\mathbf{U}) = \mathbf{A}^{-1}$, where $D(\mathbf{U})$ is the matrix of variances and covariances of the components of $\mathbf{U}$. We shall refer to the variance covariance matrix as the *dispersion* matrix and use the symbol D for the operator leading to the variances and covariances.

We shall not, however, choose the density function (8a.1.1) as basic to our study of the multivariate normal distribution. Instead, we shall characterize the distribution in such a way that the concepts involved can be extended to more complex random variables with countable or uncountable dimensions. Further, the density function such as (8a.1.1) does not exist if $D(\mathbf{U})$ is singular; a more general approach is necessary to include such cases.

For this purpose we shall exploit a result due to Cramer and Wold which states that the distribution of a p-dimensional random variable is completely determined by the one-dimensional distributions of linear functions $\mathbf{T}'\mathbf{U}$, for every fixed real vector $\mathbf{T}$. To prove the result, let $\phi(t, \mathbf{T})$ be the c.f. of $\mathbf{T}'\mathbf{U}$. Then

$$\phi(t, T) = E[\exp(it\mathbf{T}'\mathbf{U})], \tag{8a.1.2}$$

giving $\phi(1, \mathbf{T}) = E[\exp(i\mathbf{T}'\mathbf{U})]$, which as a function of $\mathbf{T}$ is the c.f. of $\mathbf{U}$. The distribution of $\mathbf{U}$ is then uniquely defined by the inversion theorem of the c.f. Motivated by this result we define the p-variate normal distribution as follows.

Definition 1. A p-dimensional random variable $\mathbf{U}$, that is, a random variable $\mathbf{U}$ taking values in E_p (Euclidean space of p-dimensions) is said to have a p-variate normal distribution N_p if and only if every linear function of $\mathbf{U}$ has a univariate normal distribution.

The result (8a.1.2) only shows that if a random variable $\mathbf{U}$ exists satisfying definition 1, then its distribution is uniquely determined. We shall, however, establish the existence of N_p as defined, in Section **8a.2**.

It may be noted that the definition 1 of N_p can be extended to the definition of a normal probability measure on more general spaces such as Hilbert or Banach spaces by demanding that the induced distribution of every linear functional is univariate normal (Frechet, 1951). In this chapter we shall confine our attention to the study of only finite dimensional random variables. But the treatment and the nature of problems discussed should serve as a good preparation for those interested in studying the distributions of more general random variables which are considered in many areas of applied research. (See Grenander and Rosenblatt, 1957; Prohorov and Fisz, 1957; Rao and Varadarajan, 1963b; Wold, 1938; etc.)

Notations and Operations. The following notations and operations with vector random variables will be used in the proofs of the various propositions. Let $\mathbf{X}' = (X_1, \ldots, X_p)$ and $\mathbf{Y}' = (Y_1, \ldots, Y_q)$ be two random variables. Then denoting by E, D, C, V, expectation, dispersion matrix, covariance, and variance respectively, we have

$$E(\mathbf{X}') = [E(X_1), \ldots, E(X_p)]$$

$$D(\mathbf{X}) = \begin{pmatrix} V(X_1) & \cdots & C(X_1, X_p) \\ \cdot & \cdots & \cdot \\ C(X_p, X_1) & \cdots & V(X_p) \end{pmatrix} \tag{8a.1.3}$$

$$C(\mathbf{X}, \mathbf{Y}) = \operatorname{cov}(\mathbf{X}, \mathbf{Y}) = \begin{pmatrix} C(X_1, Y_1) & \cdots & C(X_1, Y_q) \\ \cdot & \cdots & \cdot \\ C(X_p, Y_1) & \cdots & C(X_p, Y_q) \end{pmatrix} \tag{8a.1.4}$$

$$C(\mathbf{Y}, \mathbf{X}) = [C(\mathbf{X}, \mathbf{Y})]'$$

$$V(\mathbf{L}'\mathbf{X}) = \mathbf{L}'D(\mathbf{X})\mathbf{L}, \qquad C(\mathbf{L}'\mathbf{X}, \mathbf{M}'\mathbf{Y}) = \mathbf{L}'C(\mathbf{X}, \mathbf{Y})\mathbf{M}, \tag{8a.1.5}$$

where $\mathbf{L}$ and $\mathbf{M}$ are column vectors.

If $\mathbf{X}_1, \ldots, \mathbf{X}_k$ are k, p-dimensional r.v.'s and $A_1, \ldots, A_k$ are fixed constants then

$$\begin{aligned} D(A_1\mathbf{X}_1 + \cdots + A_k\mathbf{X}_k) &= \sum A_i^2 D(\mathbf{X}_i) + 2\sum\sum A_i A_j\, C(\mathbf{X}_i, \mathbf{X}_j) \\ &= \sum A_i^2 D(\mathbf{X}_i), \quad \text{when } \mathbf{X}_i \text{ are uncorrelated} \\ &= \left(\sum A_i^2\right) D(\mathbf{X}), \quad \text{when } \mathbf{X}_i \text{ are uncorrelated and have the same dispersion.} \end{aligned} \tag{8a.1.6}$$

If $\mathbf{X}$ is a p-dimensional r.v. and $\mathbf{B}$ is a $(q \times p)$ matrix, then $\mathbf{BX}$ is a q-dimensional r.v. and

$$\left.\begin{aligned} E(\mathbf{BX}) &= \mathbf{B}E(\mathbf{X}) \\ D(\mathbf{BX}) &= \mathbf{B}D(\mathbf{X})\mathbf{B}' \end{aligned}\right\}. \tag{8a.1.7}$$

We denote the rank of a matrix $\mathbf{A}$ by rank $\mathbf{A}$ or $R(\mathbf{A})$.

8a.2 Properties of the Distribution

Let $\mathbf{U} \sim N_p$ according to definition 1. Then the following results hold.

(i) *$E(\mathbf{U})$ and $D(\mathbf{U})$ exist which we denote by $\boldsymbol{\mu}$ and $\boldsymbol{\Sigma}$ respectively. Further, for a fixed vector $\mathbf{T}$, $\mathbf{T}'\mathbf{U} \sim N_1(\mathbf{T}'\boldsymbol{\mu}, \mathbf{T}'\boldsymbol{\Sigma}\mathbf{T})$, that is univariate normal with mean $\mathbf{T}'\boldsymbol{\mu}$ and variance $\mathbf{T}'\boldsymbol{\Sigma}\mathbf{T}$.*

Let the components of $\mathbf{U}$ be $(U_1, \ldots, U_p)$. By definition U_i is univariate normal so that $E(U_i) < \infty$ and $V(U_i) < \infty$. $\text{Cov}(U_i, U_j)$ exists since $V(U_i)$ and $V(U_j)$ exist. Let $E(U_i) = \mu_i$, $V(U_i) = \sigma_{ii}$, and $\text{cov}(U_i, U_j) = \sigma_{ij}$. Then

$$\begin{aligned} E(\mathbf{T}'\mathbf{U}) &= \mathbf{T}'\boldsymbol{\mu}, \quad \text{where } \boldsymbol{\mu}' = (\mu_1, \ldots, \mu_p) \\ V(\mathbf{T}'\mathbf{U}) &= \mathbf{T}'\boldsymbol{\Sigma}\mathbf{T}, \quad \text{where } \boldsymbol{\Sigma} = (\sigma_{ij}), \end{aligned} \tag{8a.2.1}$$

and therefore $\mathbf{T}'\mathbf{U} \sim N_1(\mathbf{T}'\boldsymbol{\mu}, \mathbf{T}'\boldsymbol{\Sigma}\mathbf{T})$.

(ii) *The characteristic function of $\mathbf{U}$ is*

$$\exp(i\mathbf{T}'\boldsymbol{\mu} - \tfrac{1}{2}\mathbf{T}'\boldsymbol{\Sigma}\mathbf{T}). \tag{8a.2.2}$$

The c.f. of a univariate normal variable $u \sim N_1(\mu, \sigma^2)$ is (p. 103)

$$\exp(it\mu - \tfrac{1}{2}t^2\sigma^2), \tag{8a.2.3}$$

By applying (8a.2.3) the c.f. of the r.v. $\mathbf{T}'\mathbf{U} \sim N_1(\mathbf{T}'\boldsymbol{\mu}, \mathbf{T}'\boldsymbol{\Sigma}\mathbf{T})$ is

$$\exp(it\mathbf{T}'\boldsymbol{\mu} - \tfrac{1}{2}t^2\mathbf{T}'\boldsymbol{\Sigma}\mathbf{T}). \tag{8a.2.4}$$

Substituting $t = 1$ in (8a.2.4), we see that the c.f. of $\mathbf{U}$ is as in (8a.2.2). The following are immediate consequences of (8a.2.2).

The p-variate normal distribution is completely specified by the mean vector $\boldsymbol{\mu}$ and the dispersion matrix $\boldsymbol{\Sigma}$ of the random variable, since the c.f. (8a.2.2) involves only $\boldsymbol{\mu}$ and $\boldsymbol{\Sigma}$. We may therefore, denote a p-variate normal distribution by $N_p(\boldsymbol{\mu}, \boldsymbol{\Sigma})$, involving $\boldsymbol{\mu}$ and $\boldsymbol{\Sigma}$ as parameters.

(a) *If there exist a vector* $\boldsymbol{\mu}$ *and matrix* $\boldsymbol{\Sigma}$ *such that for every* $\mathbf{T}$, $\mathbf{T}'\mathbf{U} \sim N_1(\mathbf{T}'\boldsymbol{\mu}, \mathbf{T}'\boldsymbol{\Sigma}\mathbf{T})$, *then* $\mathbf{U} \sim N_p(\boldsymbol{\mu}, \boldsymbol{\Sigma})$.

(b) *If* $\boldsymbol{\Sigma} = \Delta$ *(a diagonal matrix), the components* $U_1, \ldots, U_p$ *are independent and each is univariate normal.*

Note that the c.f. (8a.2.2) is then the product of p factors showing independence of variables. *Thus, zero values of all the product moment correlations* $\Leftrightarrow$ *independence of the components of* $\mathbf{U}$.

(c) *Let* $\mathbf{U}_1$ *and* $\mathbf{U}_2$ *be two subsets of the variable* $\mathbf{U}$. *We can write*

$$\boldsymbol{\Sigma} = \begin{pmatrix} \boldsymbol{\Sigma}_{11} & \boldsymbol{\Sigma}_{12} \\ \boldsymbol{\Sigma}_{21} & \boldsymbol{\Sigma}_{22} \end{pmatrix} = \begin{pmatrix} D(\mathbf{U}_1) & C(\mathbf{U}_1, \mathbf{U}_2) \\ C(\mathbf{U}_2, \mathbf{U}_1) & D(\mathbf{U}_2) \end{pmatrix} \tag{8a.2.5}$$

where $\boldsymbol{\Sigma}_{11}$ *and* $\boldsymbol{\Sigma}_{22}$ *are the dispersion matrices of* $\mathbf{U}_1, \mathbf{U}_2$, *and* $\boldsymbol{\Sigma}_{21}$ *is covariance matrix of* $\mathbf{U}_2, \mathbf{U}_1$ *(i.e., the matrix of covariances of the components of* $\mathbf{U}_2$ *with those of* $\mathbf{U}_1$*). The random variables* $\mathbf{U}_1, \mathbf{U}_2$ *are independently distributed if and only if* $\boldsymbol{\Sigma}_{12} = 0$.

The result follows since the quadratic form in (8a.2.2) can be written in subsets $\mathbf{T}_1, \mathbf{T}_2$ of components of $\mathbf{T}$ as

$$\begin{aligned} \mathbf{T}'\boldsymbol{\Sigma}\mathbf{T} &= \mathbf{T}_1'\boldsymbol{\Sigma}_{11}\mathbf{T}_1 + 2\mathbf{T}_1'\boldsymbol{\Sigma}_{12}\mathbf{T}_2 + \mathbf{T}_2'\boldsymbol{\Sigma}_{22}\mathbf{T}_2 \\ &= \mathbf{T}_1'\boldsymbol{\Sigma}_{11}\mathbf{T}_1 + \mathbf{T}_2'\boldsymbol{\Sigma}_{22}\mathbf{T}_2, \qquad \text{when} \quad \boldsymbol{\Sigma}_{12} = 0 \end{aligned}$$

in which case, the c.f. of $\mathbf{U}$ is seen to be the product of two factors. The result is important since *the independence of two subsets of a multivariate normal variable can be deduced by examining the covariance matrix between the subsets.*

(d) *If the subsets* $\mathbf{U}_1, \mathbf{U}_2, \ldots, \mathbf{U}_k$ *of* $\mathbf{U}$ *are independent pairwise, they are mutually independent.*

(e) *The function* (8a.2.2) *is indeed a c.f. so that* N_p *of definition 1 exists.*

Since $\boldsymbol{\Sigma}$ is non-negative definite (being a dispersion matrix), the quadratic form $\mathbf{T}'\boldsymbol{\Sigma}\mathbf{T}$ can be written (p. 36)

$$\mathbf{T}'\boldsymbol{\Sigma}\mathbf{T} = (\mathbf{B}_1'\mathbf{T})^2 + \cdots + (\mathbf{B}_m'\mathbf{T})^2$$

where $\mathbf{B}_1, \ldots, \mathbf{B}_m$ are linearly independent vectors and $m = R(\boldsymbol{\Sigma})$. The function (8a.2.2) can be written

$$e^{-i\mathbf{T}'\boldsymbol{\mu}} \cdot e^{-(\mathbf{B}_1'\mathbf{T})^2/2} \cdots e^{-(\mathbf{B}_m'\mathbf{T})^2/2}, \tag{8a.2.6}$$

and the desired result is proved if each factor of (8a.2.6) is a c.f. since the product of c.f.'s is a c.f.

The first factor of (8a.2.6) is the c.f. of the p-dimensional r.v.,

$$\mathbf{Y}_0' = (\mu_1, \ldots, \mu_p) \quad \text{with probability 1.} \tag{8a.2.7}$$

The factor $\exp[-\frac{1}{2}(\mathbf{B}_i'\mathbf{T})^2]$ is easily shown to be the c.f. of the p-dimensional r.v.

$$\mathbf{Y}_i' = (B_{1i}G_i, \ldots, B_{pi}G_i), \qquad G_i \sim N_1(0, 1), \tag{8a.2.8}$$

where $\mathbf{B}_i' = (B_{1i}, \ldots, B_{pi})$. The number m is called the rank of the distribution $N_p(\boldsymbol{\mu}, \boldsymbol{\Sigma})$.

(f) $\mathbf{U} \sim N_p(\boldsymbol{\mu}, \boldsymbol{\Sigma})$ *with rank* m *if, and only if,*

$$\mathbf{U} = \boldsymbol{\mu} + \mathbf{BG}, \qquad \mathbf{BB}' = \boldsymbol{\Sigma} \tag{8a.2.9}$$

where $\mathbf{B}$ *is* $(p \times m)$ *matrix of rank* m *and* $\mathbf{G} \sim N_m(\mathbf{0}, \mathbf{I})$, *that is, the components* $G_1, \ldots, G_m$ *are independent and each is distributed as* $N_1(0, 1)$. [See **8g** for an alternative derivation of (8a.2.9).]

To prove the "if" part, consider a linear function of $\mathbf{U}$

$$\mathbf{T}'\mathbf{U} = \mathbf{T}'\boldsymbol{\mu} + \mathbf{T}'\mathbf{BG}.$$

But $\mathbf{G} \sim N_m(\mathbf{0}, \mathbf{I})$, and from the univariate theory $(\mathbf{T}'\mathbf{B})\mathbf{G}$, which is a linear function of independent $N_1(0, 1)$ variables, is distributed as $N_1(0, \mathbf{T}'\mathbf{BB}'\mathbf{T})$. Hence

$$\begin{aligned} \mathbf{T}'\mathbf{U} = \mathbf{T}'\boldsymbol{\mu} + \mathbf{T}'\mathbf{BG} &\sim N_1(\mathbf{T}'\boldsymbol{\mu}, \mathbf{T}'\mathbf{BB}'\mathbf{T}) \\ \Rightarrow \mathbf{U} \sim N_p(\boldsymbol{\mu}, \mathbf{BB}') &= N_p(\boldsymbol{\mu}, \boldsymbol{\Sigma}) \end{aligned}$$

since $\mathbf{T}$ is arbitrary.

To prove the "only if" part we observe from the factorization (8a.2.6) that the distribution of $\mathbf{U}$ is a convolution of the distributions of independent p-dimensional random variables

$$\mathbf{Y}_0, \mathbf{Y}_1, \ldots, \mathbf{Y}_m$$

defined in (8a.2.7) and (8a.2.8). Thus $\mathbf{U}$ can be written

$$\begin{aligned} \mathbf{U} &= \mathbf{Y}_0 + \mathbf{Y}_1 + \cdots + \mathbf{Y}_m \\ &= \boldsymbol{\mu} + G_1\mathbf{B}_1 + \cdots + G_m\mathbf{B}_m = \boldsymbol{\mu} + \mathbf{BG}, \end{aligned}$$

choosing $\mathbf{B}$ as the partitioned matrix $(\mathbf{B}_1 \vdots \cdots \vdots \mathbf{B}_m)$, where $\mathbf{B}_i$ and G_j are as defined in (8a.2.8). Thus we have an alternative definition of the multivariate normal distribution.

Definition 2. A p-dimensional random variable $\mathbf{U}$ is said to have a normal distribution N_p if it can be expressed in the form $\mathbf{U} = \boldsymbol{\mu} + \mathbf{BG}$, where $\mathbf{B}$ is $p \times m$ matrix of rank m and $\mathbf{G}$ is a $m \times 1$ vector of independent N_1 (univariate normal) variables, each with mean zero and unit variance.

Note 1. The relationship $\mathbf{U} = \boldsymbol{\mu} + \mathbf{BG}$ shows that the random vector $\mathbf{U} \in \mathcal{M}(\boldsymbol{\mu} : \boldsymbol{\Sigma})$, the linear manifold generated by the columns of $\boldsymbol{\mu}$ and $\boldsymbol{\Sigma}$, with probability 1.

Note 2. Let $\mathbf{U}_i \sim N_{p_i}(\boldsymbol{\mu}_i, \boldsymbol{\Sigma}_i)$, $i = 1, \ldots, k$, be independent and T be a function (statistic) of $\mathbf{U}_1, \ldots, \mathbf{U}_k$. Using the representation $\mathbf{U}_i = \boldsymbol{\mu}_i + \mathbf{B}_i \mathbf{G}_i$, we have

$$\begin{aligned} T(\mathbf{U}_1, \ldots, \mathbf{U}_k) &= T(\boldsymbol{\mu}_1 + \mathbf{B}_1\mathbf{G}_1, \ldots, \boldsymbol{\mu}_k + \mathbf{B}_k\mathbf{G}_k) \\ &= t(\mathbf{G}_1, \ldots, \mathbf{G}_k). \end{aligned}$$

Then the study of T, a statistic based on $\mathbf{U}_1, \ldots, \mathbf{U}_k$, reduces to the study of a function t of independent univariate normal variables $\mathbf{G}_1, \ldots, \mathbf{G}_k$. Such a reduction helps, as the known results in the univariate theory can be immediately applied to deduce the results in the multivariate theory.

Note 3. It is easy to see that $\mathbf{U} = \boldsymbol{\mu} + \mathbf{BG}$, with $\boldsymbol{\mu}$, $\mathbf{B}$, $\mathbf{G}$ as in definition 2 and $\mathbf{U} = \boldsymbol{\mu} + \mathbf{CF}$, where $\mathbf{C}$ is $p \times q$ matrix and $\mathbf{F}$ is a q-vector of independent $N(0, 1)$ variables, have the same distribution if $\mathbf{BB}' = \mathbf{CC}'$, so that no restriction on q or $R(\mathbf{C})$ need be imposed. But a representation with restriction on $\mathbf{B}$ as in definition 2 is useful in practical applications.

(iii) *If* $\mathbf{U} \sim N_p$, *the marginal distribution of any subset of q components of* $\mathbf{U}$ *is* N_q. [Follows from definition 1 since every linear function of the subset is also univariate normal.]

(iv) *The joint distribution of q linear functions of* $\mathbf{U}$ *is* N_q, *which follows from definition 1. If* $\mathbf{Y} = \mathbf{CU}$, *where* $\mathbf{C}$ *is* $(q \times p)$, *represents the q linear functions, then*

$$\mathbf{Y} \sim N_q(\mathbf{C}\boldsymbol{\mu}, \mathbf{C}\boldsymbol{\Sigma}\mathbf{C}'). \tag{8a.2.10}$$

Consider a linear function $\mathbf{L}'\mathbf{Y} = \mathbf{L}'\mathbf{CU} = (\mathbf{L}'\mathbf{C})\mathbf{U}$. From (i),

$$(\mathbf{L}'\mathbf{C})\mathbf{U} \sim N_1[(\mathbf{L}'\mathbf{C})\boldsymbol{\mu}, (\mathbf{L}'\mathbf{C})\boldsymbol{\Sigma}(\mathbf{C}'\mathbf{L})]$$

or

$$\mathbf{L}'\mathbf{Y} \sim N_1(\mathbf{L}'\mathbf{C}\boldsymbol{\mu}, \mathbf{L}'\mathbf{C}\boldsymbol{\Sigma}\mathbf{C}'\mathbf{L}) \Rightarrow \mathbf{Y} = N_q(\mathbf{C}\boldsymbol{\mu}, \mathbf{C}\boldsymbol{\Sigma}\mathbf{C}').$$

(v) *Let* $\mathbf{U}_1' = (U_1, \ldots, U_r)$, $\mathbf{U}_2' = (U_{r+1}, \ldots, U_p)$ *be two subsets of* $\mathbf{U}$ *and* $\boldsymbol{\Sigma}_{11}, \boldsymbol{\Sigma}_{12}, \boldsymbol{\Sigma}_{22}$ *be the partitions of* $\boldsymbol{\Sigma}$ *as defined in* (8a.2.5). *Then the conditional distribution of* $\mathbf{U}_2$ *given* $\mathbf{U}_1$ *is*

$$N_{p-r}(\boldsymbol{\mu}_2 + \boldsymbol{\Sigma}_{21}\boldsymbol{\Sigma}_{11}^{-}(\mathbf{U}_1 - \boldsymbol{\mu}_1), \boldsymbol{\Sigma}_{22} - \boldsymbol{\Sigma}_{21}\boldsymbol{\Sigma}_{11}^{-}\boldsymbol{\Sigma}_{12}), \tag{8a.2.11}$$

where $E(\mathbf{U}_i) = \boldsymbol{\mu}_i$, $i = 1, 2$, *and* $\boldsymbol{\Sigma}_{11}^{-}$ *is a generalized inverse of* $\boldsymbol{\Sigma}_{11}$.

The reader may verify that $\mathcal{M}(\mathbf{\Sigma}_{12}) \subset \mathcal{M}(\mathbf{\Sigma}_{11})$, where $\mathcal{M}$ represents the linear space spanned by the columns of a matrix (by showing that any vector orthogonal to columns of $\mathbf{\Sigma}_{11}$ is also orthogonal to columns of $\mathbf{\Sigma}_{12}$). Hence there exists a matrix $\mathbf{B}$ such that $\mathbf{\Sigma}_{21} = \mathbf{B\Sigma}_{11}$. Now by using the property $\mathbf{\Sigma}_{11}\mathbf{\Sigma}_{11}^{-}\mathbf{\Sigma}_{11} = \mathbf{\Sigma}_{11}$ of a generalized inverse, we see that

$$\mathbf{\Sigma}_{21}\mathbf{\Sigma}_{11}^{-}\mathbf{\Sigma}_{11} = \mathbf{B\Sigma}_{11}\mathbf{\Sigma}_{11}^{-}\mathbf{\Sigma}_{11} = \mathbf{B\Sigma}_{11} = \mathbf{\Sigma}_{21} \tag{8a.2.12}$$

which is obvious if $\mathbf{\Sigma}_{11}^{-}$ is a true inverse. Readers not familiar with a generalized inverse may assume that $\mathbf{\Sigma}_{11}^{-1}$ exists and thus avoid the argument leading to (8a.2.12).

Consider the covariance matrix:

$$\begin{aligned} &C[\mathbf{U}_2 - \boldsymbol{\mu}_2 - \mathbf{\Sigma}_{21}\mathbf{\Sigma}_{11}^{-}(\mathbf{U}_1 - \boldsymbol{\mu}_1), \mathbf{U}_1 - \boldsymbol{\mu}_1] \\ &\quad = C(\mathbf{U}_2 - \boldsymbol{\mu}_2, \mathbf{U}_1 - \boldsymbol{\mu}_1) - \mathbf{\Sigma}_{21}\mathbf{\Sigma}_{11}^{-}C(\mathbf{U}_1 - \boldsymbol{\mu}_1, \mathbf{U}_1 - \boldsymbol{\mu}_1) \\ &\quad = \mathbf{\Sigma}_{21} - \mathbf{\Sigma}_{21}\mathbf{\Sigma}_{11}^{-}\mathbf{\Sigma}_{11} = \mathbf{0}, \qquad \text{using (8a.2.12).} \end{aligned} \tag{8a.2.13}$$

Similarly,

$$\begin{aligned} &D[\mathbf{U}_2 - \boldsymbol{\mu}_2 - \mathbf{\Sigma}_{21}\mathbf{\Sigma}_{11}^{-}(\mathbf{U}_1 - \boldsymbol{\mu}_1)] \\ &\quad = C(\mathbf{U}_2 - \mathbf{\Sigma}_{21}\mathbf{\Sigma}_{11}^{-}\mathbf{U}_1, \mathbf{U}_2 - \mathbf{\Sigma}_{21}\mathbf{\Sigma}_{11}^{-}\mathbf{U}_1) \\ &\quad = C(\mathbf{U}_2, \mathbf{U}_2) - C(\mathbf{U}_2, \mathbf{U}_1)\mathbf{\Sigma}_{11}^{-}\mathbf{\Sigma}_{12} \\ &\qquad - \mathbf{\Sigma}_{21}\mathbf{\Sigma}_{11}^{-}C(\mathbf{U}_1, \mathbf{U}_2) + \mathbf{\Sigma}_{21}\mathbf{\Sigma}_{11}^{-}C(\mathbf{U}_1, \mathbf{U}_1)\mathbf{\Sigma}_{11}^{-}\mathbf{\Sigma}_{12} \\ &\quad = \mathbf{\Sigma}_{22} - 2\mathbf{\Sigma}_{21}\mathbf{\Sigma}_{11}^{-}\mathbf{\Sigma}_{12} + \mathbf{\Sigma}_{21}\mathbf{\Sigma}_{11}^{-}\mathbf{\Sigma}_{11}\mathbf{\Sigma}_{11}^{-}\mathbf{\Sigma}_{12} \\ &\quad = \mathbf{\Sigma}_{22} - \mathbf{\Sigma}_{21}\mathbf{\Sigma}_{11}^{-}\mathbf{\Sigma}_{12}, \quad \text{using } \mathbf{\Sigma}_{21}\mathbf{\Sigma}_{11}^{-}\mathbf{\Sigma}_{11} = \mathbf{\Sigma}_{21}. \end{aligned} \tag{8a.2.14}$$

Since $E[\mathbf{U}_2 - \boldsymbol{\mu}_2 - \mathbf{\Sigma}_{21}\mathbf{\Sigma}_{11}^{-}(\mathbf{U}_1 - \boldsymbol{\mu}_1)] = \mathbf{0}$, the result (8a.2.14) shows that

$$\mathbf{U}_2 - \boldsymbol{\mu}_2 - \mathbf{\Sigma}_{21}\mathbf{\Sigma}_{11}^{-}(\mathbf{U}_1 - \boldsymbol{\mu}_1) \sim N_{p-r}(\mathbf{0}, \mathbf{\Sigma}_{22} - \mathbf{\Sigma}_{21}\mathbf{\Sigma}_{11}^{-}\mathbf{\Sigma}_{12}). \tag{8a.2.15}$$

The result (8a.2.13) shows that $\mathbf{U}_2 - \boldsymbol{\mu}_2 - \mathbf{\Sigma}_{21}\mathbf{\Sigma}_{11}^{-}(\mathbf{U}_1 - \boldsymbol{\mu}_1)$ and $\mathbf{U}_1 - \boldsymbol{\mu}_1$ are independently distributed. Hence (8a.2.15) can be interpreted as the conditional distribution of $\mathbf{U}_2 - \boldsymbol{\mu}_2 - \mathbf{\Sigma}_{12}\mathbf{\Sigma}_{11}^{-}(\mathbf{U}_1 - \boldsymbol{\mu}_1)$ given $\mathbf{U}_1$, which is equivalent to the result (8a.2.11) on the conditional distribution of $\mathbf{U}_2$ given $\mathbf{U}_1$. The computations (8a.2.13) and (8a.2.14) are simpler when the true inverse of $\mathbf{\Sigma}_{11}$ exists.

An important special case of (8a.2.15) is when $\mathbf{U}_2$ consists of a single component, say U_p and $\mathbf{U}_1' = (U_1, \ldots, U_{p-1})$. The conditional distribution of U_p given $\mathbf{U}_1$ is univariate normal with mean (regression of U_p on $U_1, \ldots, U_{p-1}$) as $\mu_p + \mathbf{\Sigma}_{21}\mathbf{\Sigma}_{11}^{-}(\mathbf{U}_1 - \boldsymbol{\mu}_1)$, which is of the form

$$\alpha + \beta_1 U_1 + \cdots + \beta_{p-1}U_{p-1} \tag{8a.2.16}$$

linear in $U_1, \ldots, U_{p-1}$ and variance (residual variance) as $\sigma_{pp} - \mathbf{\Sigma}_{21}\mathbf{\Sigma}_{11}^{-}\mathbf{\Sigma}_{12}$. When $\mathbf{\Sigma}_{11}$ is nonsingular (using Example **2.4** at the end of **1b.8**), then

$$|\mathbf{\Sigma}_{11}|(\sigma_{pp} - \mathbf{\Sigma}_{21}\mathbf{\Sigma}_{11}^{-1}\mathbf{\Sigma}_{12}) = |\mathbf{\Sigma}|,$$

which gives the explicit expression to the variance

$$\sigma_{pp} - \Sigma_{21}\Sigma_{11}^{-1}\Sigma_{12} = \frac{|\Sigma|}{|\Sigma_{11}|} = \frac{1}{\sigma^{pp}}. \tag{8a.2.17}$$

where σ^{pp} is the last element in the reciprocal of Σ.

(vi) REPRODUCTIVE PROPERTY OF N_p. *Let* $\mathbf{U}_i \sim N_p(\boldsymbol{\mu}_i, \Sigma_i)$, $i = 1, \ldots, n$ *be all independent. Then for fixed constants* $A_1, \ldots, A_n$,

$$\mathbf{Y} = A_1\mathbf{U}_1 + \cdots + A_n\mathbf{U}_n \sim N_p(\textstyle\sum A_i\boldsymbol{\mu}_i, \sum A_i^2\Sigma_i). \tag{8a.2.18}$$

To establish (8a.2.18) we need consider a linear function $\mathbf{L}'\mathbf{Y}$,

$$\mathbf{L}'\mathbf{Y} = A_1(\mathbf{L}'\mathbf{U}_1) + \cdots + A_n(\mathbf{L}'\mathbf{U}_n),$$

which is univariate normal since $\mathbf{L}'\mathbf{U}_1, \ldots, \mathbf{L}'\mathbf{U}$ are all independent and univariate normal. Hence $\mathbf{Y} \sim N_p$. Furthermore,

$$E(\mathbf{L}'\mathbf{Y}) = A_1\mathbf{L}'\boldsymbol{\mu}_1 + \cdots + A_n\mathbf{L}'\boldsymbol{\mu}_n = \mathbf{L}'(A_1\boldsymbol{\mu}_1 + \cdots + A_n\boldsymbol{\mu}_n)$$

$$V(\mathbf{L}'\mathbf{Y}) = A_1^2V(\mathbf{L}'\mathbf{U}_1) + \cdots + A_n^2V(\mathbf{L}'\mathbf{U}_n)$$

$$= \textstyle\sum A_i^2\mathbf{L}'\Sigma_i\mathbf{L} = \mathbf{L}'(\sum A_i^2\Sigma_i)\mathbf{L},$$

which shows that $\sum A_i\boldsymbol{\mu}_i$ and $\sum A_i^2\Sigma_i$ are the parameters of the distribution of $\mathbf{Y}$.

(vii) *Let* $\mathbf{U}_i$, $i = 1, \ldots, n$, *be independent and identically distributed as* $N_p(\boldsymbol{\mu}, \Sigma)$. *Then*

$$\frac{1}{n}\sum \mathbf{U}_i = \bar{\mathbf{U}} \sim N_p\left(\boldsymbol{\mu}, \frac{1}{n}\Sigma\right), \tag{8a.2.19}$$

The result (8a.2.19), which follows from (8a.2.18), provides the distribution of the mean of n independent observations. [Choose $A_i = 1/n$ in 8a.2.18].

(viii) *Let* $\mathbf{U} \sim N_p(\boldsymbol{\mu}, \Sigma)$. *Then an n.s. condition that*

$$Q = (\mathbf{U} - \boldsymbol{\mu})'\mathbf{A}(\mathbf{U} - \boldsymbol{\mu}) \sim \chi^2(k)$$

is $\Sigma(\mathbf{A}\Sigma\mathbf{A} - \mathbf{A})\Sigma = 0$ *in which case* $k = \text{trace}\,(\mathbf{A}\Sigma)$.

From (8a.2.9), $\mathbf{U} - \boldsymbol{\mu} = \mathbf{BG}$ where $\mathbf{G} \sim N_m(\mathbf{0}, \mathbf{I})$ and $\Sigma = \mathbf{BB}'$.

$$Q = (\mathbf{U} - \boldsymbol{\mu})'\mathbf{A}(\mathbf{U} - \boldsymbol{\mu}) = \mathbf{G}'\mathbf{B}'\mathbf{ABG}, \tag{8a.2.20}$$

so that Q is expressed as a quadratic form involving independent normal variables G_i. In independent variables, an n.s. condition for a quadratic form Q to have a χ^2 distribution is that the matrix of Q is idempotent [(ii), **3b.4**]. Hence,

$$(\mathbf{B}'\mathbf{AB})(\mathbf{B}'\mathbf{AB}) = \mathbf{B}'\mathbf{AB}, \qquad \text{that is,} \quad \mathbf{B}'(\mathbf{A}\Sigma\mathbf{A} - \mathbf{A})\mathbf{B} = \mathbf{0}.$$

But $\mathbf{B}'(\mathbf{A\Sigma A} - \mathbf{A})\mathbf{B} = \mathbf{0} \Leftrightarrow \mathbf{BB}'(\mathbf{A\Sigma A} - \mathbf{A})\mathbf{BB}' = \mathbf{0} = \mathbf{\Sigma}(\mathbf{A\Sigma A} - \mathbf{A})\mathbf{\Sigma}$. The D.F. of χ^2 is (from the univariate theory)

$$\text{rank}\,(\mathbf{B}'\mathbf{AB}) = \text{trace}\,(\mathbf{B}'\mathbf{AB}) = \text{trace}\,(\mathbf{ABB}') = \text{trace}\,(\mathbf{A\Sigma}).$$

If $\mathbf{\Sigma}$ is of full rank, an n.s. condition is $\mathbf{A\Sigma A} = \mathbf{A}$.

8a.3 Some Characterizations of N_p

(i) CHARACTERIZATION 1 OF N_p. *Let* $\mathbf{U}_1$ *and* $\mathbf{U}_2$ *be two independent p-dimensional variables such that*

$$\mathbf{U} = \mathbf{U}_1 + \mathbf{U}_2 \sim N_p.$$

Then both $\mathbf{U}_1$ *and* $\mathbf{U}_2$ *are* N_p.

Consider $\mathbf{L}'\mathbf{U} = \mathbf{L}'\mathbf{U}_1 + \mathbf{L}'\mathbf{U}_2 \sim N_1$ by hypothesis. But $\mathbf{L}'\mathbf{U}_1$ and $\mathbf{L}'\mathbf{U}_2$ are independent. Hence by Cramer's (1937) theorem, $\mathbf{L}'\mathbf{U}_1$ and $\mathbf{L}'\mathbf{U}_2$ are each N_1. Since L is arbitrary, $\mathbf{U}_1$ and $\mathbf{U}_2$ are each N_p.

(ii) CHARACTERIZATION 2 OF N_p. *Consider two linear functions of n independent p-dimensional random variables* $\mathbf{U}_1, \ldots, \mathbf{U}_n$,

$$\mathbf{V}_1 = B_1\mathbf{U}_1 + \cdots + B_n\mathbf{U}_n, \mathbf{V}_2 = C_1\mathbf{U}_1 + \cdots + C_n\mathbf{U}_n.$$

Let $\mathbf{B}' = (B_1, \ldots, B_n)$ *and* $\mathbf{C}' = (C_1, \ldots, C_n)$. *Then:*

(a) $\mathbf{U}_i$ *are i.i.d. as* N_p *and* $\mathbf{B}'\mathbf{C} = 0 \Rightarrow \mathbf{V}_1$ *and* $\mathbf{V}_2$ *are independently distributed.*

(b) $\mathbf{V}_1$, $\mathbf{V}_2$ *are independently distributed* $\Rightarrow \mathbf{U}_i \sim N_p$ *for any i such that* $B_i C_i \neq 0$ *and* $\mathbf{U}_i$ *need not be identically distributed.*

To prove (a), it follows from definition 1 that the joint distribution of $\mathbf{V}_1$, $\mathbf{V}_2$ is N_{2p}. Hence to establish independence we need only show that $\text{cov}(\mathbf{V}_1, \mathbf{V}_2) = 0$.

$$\begin{aligned}\text{cov}(\mathbf{V}_1, \mathbf{V}_2) &= \text{cov}(B_1\mathbf{U}_1 + \cdots + B_n\mathbf{U}_n, C_1\mathbf{U}_1 + \cdots + C_n\mathbf{U}_n)\\ &= B_1C_1\,\text{cov}(\mathbf{U}_1, \mathbf{U}_1) + \cdots + B_nC_n\,\text{cov}(\mathbf{U}_n, \mathbf{U}_n)\\ &= (B_1C_1 + \cdots + B_nC_n)\mathbf{\Sigma} = 0.\end{aligned}$$

The result (b) is a generalization of the univariate Darmois-Skitovic theorem to a p-dimensional random variable (Ex.**21**, Chapter 3). Darmois-Skitovic theorem states that if $u_1, \ldots, u_n$ are independent one-dimensional variables the independence of

$$B_1u_1 + \cdots + B_nu_n \qquad \text{and} \qquad C_1u_1 + \cdots + C_nu_n$$

for fixed values of B_i, C_i implies that $u_i \sim N_1$ provided $B_i C_i \neq 0$ and can be arbitrarily distributed otherwise, $i = 1, \ldots, n$. To apply the theorem we need only consider linear functions of $\mathbf{V}_1$, $\mathbf{V}_2$:

$$\mathbf{L}'\mathbf{V}_1 = \mathbf{L}'(B_1\mathbf{U}_1 + \cdots + B_n\mathbf{U}_n) = B_1(\mathbf{L}'\mathbf{U}_1) + \cdots + B_n(\mathbf{L}'\mathbf{U}_n),$$
$$\mathbf{L}'\mathbf{V}_2 = \mathbf{L}'(C_1\mathbf{U}_1 + \cdots + C_n\mathbf{U}_n) = C_1(\mathbf{L}'\mathbf{U}_1) + \cdots + C_n(\mathbf{L}'\mathbf{U}_n),$$

where now $\mathbf{L}'\mathbf{U}_i$ are all one-dimensional independent variables. Hence $\mathbf{L}'\mathbf{U}_i$ is N_1 and therefore $\mathbf{U}_i$ is N_p.

Characterization 3 of N_p. According to definition 2 of N_p, $\mathbf{Y} = \boldsymbol{\mu} + \mathbf{BG}$ where $\mathbf{G} \sim N_m(\mathbf{0}, \mathbf{I})$, that is, the components of $\mathbf{G}$ are independent and each component is distributed as $N_1(0, 1)$. The representation $\mathbf{Y} = \boldsymbol{\mu} + \mathbf{BG}$ is not, however, unique. If $\mathbf{Y} = \boldsymbol{\mu} + \mathbf{B}_1\mathbf{G}_1$ and $\mathbf{Y} = \boldsymbol{\mu} + \mathbf{B}_2\mathbf{G}_2$ are two different representations, then $D(\mathbf{Y}) = \mathbf{B}_1\mathbf{B}_1' = \mathbf{B}_2\mathbf{B}_2' = \boldsymbol{\Sigma}$. We shall utilize the non-uniqueness of the representation of Y to characterize N_p.

(iii) *Let* $\mathbf{Y} = \boldsymbol{\mu}_1 + \mathbf{B}_1\mathbf{G}_1$ *and* $\mathbf{Y} = \boldsymbol{\mu}_2 + \mathbf{B}_2\mathbf{G}_2$ *be two representations of a p-dimensional random variable* $\mathbf{Y}$ *in terms of vectors* $\mathbf{G}_1$, $\mathbf{G}_2$ *of nondegenerate independent random variables (not necessarily univariate normal), where* $\mathbf{B}_1$ *and* $\mathbf{B}_2$ *are* $p \times m$ *matrices of rank m such that no column of* $\mathbf{B}_1$ *is a multiple of some column of* $\mathbf{B}_2$. *Then* $\mathbf{Y}$ *is a p-variate normal variable.*

Let $m < p$ in which case consider any p-vector $\mathbf{C}$ orthogonal to the columns of $\mathbf{B}_1$. Then

$$\mathbf{C}'\mathbf{Y} = \mathbf{C}'\boldsymbol{\mu}_1 = \mathbf{C}'\boldsymbol{\mu}_2 + \mathbf{C}'\mathbf{B}_2\mathbf{G}_2$$

which implies that $\mathbf{C}'\mathbf{B}_2\mathbf{G}_2$ is a degenerate random variable contrary to assumption unless $\mathbf{C}'\mathbf{B}_2 = 0$. Hence, the columns of $\mathbf{B}_1$ and $\mathbf{B}_2$ belong to the same vector space implying the existence of a square matrix $\mathbf{H}$ of order and rank m such that $\mathbf{B}_1 = \mathbf{B}_2\mathbf{H}$. If $m = p$, the relationship $\mathbf{B}_1 = \mathbf{B}_2\mathbf{H}$ is automatically true. Now

$$\mathbf{Y} - \boldsymbol{\mu}_2 = \mathbf{B}_2\mathbf{G}_2 \Rightarrow (\mathbf{B}_2'\mathbf{B}_2)^{-1}\mathbf{B}_2'(\mathbf{Y} - \boldsymbol{\mu}_2) = \mathbf{G}_2$$
$$\mathbf{Y} - \boldsymbol{\mu}_1 = \mathbf{B}_1\mathbf{G}_1 = \mathbf{B}_2\mathbf{H}\mathbf{G}_1 \Rightarrow (\mathbf{B}_2'\mathbf{B}_2)^{-1}\mathbf{B}_2'(\mathbf{Y} - \boldsymbol{\mu}_1) = \mathbf{H}\mathbf{G}_1$$

Thus $\mathbf{G}_2$ and $\mathbf{HG}_1$ have the same distribution apart from translation. Hence the components of $\mathbf{HG}_1$ are independently distributed. Thus the transformation $\mathbf{G}_1 \Rightarrow \mathbf{HG}_1$ preserves independence of random variables.

The condition that no column vector of $\mathbf{B}_1$ depends on a column vector of $\mathbf{B}_2$ implies that no column of $\mathbf{H}$ contains $(m - 1)$ zeros. This means that we can find two linear combinations of exclusive sets of the rows of $\mathbf{H}$ such that the coefficients in the ith and jth positions in both are different from zero. Thus we have two linear combinations of the components of $\mathbf{G}_1$ which are independently distributed, and by Darmois-Skitovic theorem mentioned in (ii), the ith and jth components are normally distributed. Since i and j are arbitrary, the desired result is established.

For other characterizations of N_p, the reader is referred to the book by Kagan, Linnik, and Rao (1972B4), and papers by Khatri and Rao (1972f) and Rao (1966a, 1967d, 1969a, b).

8a.4 Density Function of the Multivariate Normal Distribution

We have not made use of the density function of the multivariate normal variable (or established the existence of the density function) in the discussion of **8a.1** and **8a.2.** In what form can we determine the density function? The answer depends on $R(\mathbf{\Sigma})$, the rank of the dispersion matrix $\mathbf{\Sigma}$.

CASE 1. $R(\mathbf{\Sigma}) = p$. Let $\mathbf{U} \sim N_p(\boldsymbol{\mu}, \mathbf{\Sigma})$. Then as shown in (8a.2.9),

$$\mathbf{U} - \boldsymbol{\mu} = \mathbf{BG}, \qquad \text{where} \quad \mathbf{G} \sim N_m(\mathbf{0}, \mathbf{I}), \qquad \mathbf{BB}' = \mathbf{\Sigma} \tag{8a.4.1}$$

and $m = \text{rank}\, \mathbf{\Sigma} = p$. Then $|\mathbf{\Sigma}| = |\mathbf{BB}'| \neq 0$ or $|\mathbf{B}| \neq 0$, in which case the inverse relationship exists as

$$\mathbf{G} = \mathbf{B}^{-1}(\mathbf{U} - \boldsymbol{\mu}). \tag{8a.4.2}$$

Since $\mathbf{G} \sim N_p(\mathbf{0}, \mathbf{I})$, the density of $\mathbf{G}$ is

$$(2\pi)^{-p/2} \exp[-\tfrac{1}{2}\mathbf{G}'\mathbf{G}]. \tag{8a.4.3}$$

By changing over to $(\mathbf{U} - \boldsymbol{\mu})$, using the relation (8a.4.2), the Jacobian is $|\mathbf{B}|^{-1} = |\mathbf{\Sigma}|^{-1/2}$ (since $\mathbf{BB}' = \mathbf{\Sigma}$) and the density (8a.4.3) transforms to

$$(2\pi)^{-p/2}|\mathbf{\Sigma}|^{-1/2} \exp[-\tfrac{1}{2}(\mathbf{U} - \boldsymbol{\mu})'\mathbf{\Sigma}^{-1}(\mathbf{U} - \boldsymbol{\mu})], \tag{8a.4.4}$$

which is the density of $\mathbf{U}$ with respect to the Lebesgue measure in R^p.

CASE 2. $R(\mathbf{\Sigma}) = k < p$. When $R(\mathbf{\Sigma}) < p$, the inversion (8a.4.2) is not possible and therefore no explicit determination of the density function with respect to Lebesgue measure in R^p is possible. However, the density function exists on a subspace.

Let $\mathbf{B}$ be a $p \times k$ matrix of orthonormal column vectors belonging to $\mathcal{M}(\mathbf{\Sigma})$ and $\mathbf{N}$ be $p \times (p - k)$ matrix of rank $(p - k)$ such that $\mathbf{N}'\mathbf{\Sigma} = \mathbf{0}$. Consider the transformation $\mathbf{U}' \to (\mathbf{X}' : \mathbf{Z}')$, $\mathbf{X} = \mathbf{B}'\mathbf{U}$, $\mathbf{Z} = \mathbf{N}'\mathbf{U}$. Then

$$E(\mathbf{Z}) = \mathbf{N}'\boldsymbol{\mu}, \qquad D(\mathbf{Z}) = \mathbf{N}'\mathbf{\Sigma}\mathbf{N} = \mathbf{0} \tag{8a.4.5}$$

so that

$$\mathbf{Z} = \mathbf{N}'\boldsymbol{\mu} \text{ with probability 1.} \tag{8a.4.6}$$

Further,

$$E(\mathbf{X}) = \mathbf{B}'\boldsymbol{\mu}, \quad D(\mathbf{X}) = \mathbf{B}'\mathbf{\Sigma}\mathbf{B} \tag{8a.4.7}$$

so that

$$\mathbf{X} \sim N_k(\mathbf{B}'\boldsymbol{\mu}, \mathbf{B}'\boldsymbol{\Sigma}\mathbf{B}). \tag{8a.4.8}$$

It may be seen that $|\mathbf{B}'\boldsymbol{\Sigma}\mathbf{B}| = \lambda_1, \ldots, \lambda_k$, the product of the nonzero eigenvalues of $\boldsymbol{\Sigma}$, and hence $\mathbf{B}'\boldsymbol{\Sigma}\mathbf{B}$ nonsingular. In such a case $\mathbf{X}$ has the density

$$\frac{(2\pi)^{-k/2}}{|\mathbf{B}'\boldsymbol{\Sigma}\mathbf{B}|^{1/2}} e^{-(\mathbf{X}-\mathbf{B}'\boldsymbol{\mu})'(\mathbf{B}'\boldsymbol{\Sigma}\mathbf{B})^{-1}(\mathbf{X}-\mathbf{B}'\boldsymbol{\mu})/2}. \tag{8a.4.9}$$

Thus the distribution of $\mathbf{U}$ or equivalently that of $\mathbf{X}$, $\mathbf{Z}$ is specified by (8a.4.6) and (8a.4.9).

Let us observe that

$$\begin{aligned}(\mathbf{X} - \mathbf{B}'\boldsymbol{\mu})'&(\mathbf{B}'\boldsymbol{\Sigma}\mathbf{B})^{-1}(\mathbf{X} - \mathbf{B}'\boldsymbol{\mu}) \\ &= (\mathbf{U} - \boldsymbol{\mu})'\mathbf{B}(\mathbf{B}'\boldsymbol{\Sigma}\mathbf{B})^{-1}\mathbf{B}'(\mathbf{U} - \boldsymbol{\mu}) \\ &= (\mathbf{U} - \boldsymbol{\mu})'\boldsymbol{\Sigma}^{-}(\mathbf{U} - \boldsymbol{\mu})\end{aligned} \tag{8a.4.10}$$

for any choice of the g-inverse $\boldsymbol{\Sigma}^{-}$. Then the density (8a.4.9) can be written as

$$\frac{(2\pi)^{-k/2}}{\sqrt{\lambda_1 \cdots \lambda_k}} e^{-(\mathbf{U}-\boldsymbol{\mu})'\boldsymbol{\Sigma}^{-}(\mathbf{U}-\boldsymbol{\mu})/2} \tag{8a.4.11}$$

while (8a.4.6) can be written as

$$\mathbf{N}'\mathbf{U} = \mathbf{N}'\boldsymbol{\mu} \text{ with probability 1.} \tag{8a.4.12}$$

Thus the distribution of $\mathbf{U}$ is specified by (8a.4.11) and (8a.4.12) where the former is interpreted as the density on the hyperplane $\mathbf{N}'\mathbf{U} = \mathbf{N}'\boldsymbol{\mu}$. (The above representation is due to Khatri, 1968.)

8a.5 Estimation of Parameters

We consider n, p-variate populations and an independent observation from each. Let $\mathbf{U}_i$ be the observation from the ith p-variate population. The observations can be written in the form of a matrix

$$\begin{array}{c|cccc|} & \mathbf{U}_1 & \mathbf{U}_2 & & \mathbf{U}_n \\ \hline \mathbf{Y}_1' & U_{11} & U_{12} & \cdots & U_{1n} \\ & \cdot & \cdot & \cdots & \cdot \\ \mathbf{Y}_p' & U_{p1} & U_{p2} & \cdots & U_{pn} \\ \hline \end{array} = \mathscr{U}' \tag{8a.5.1}$$

where the ith row vector $\mathbf{Y}_i'$ represents n independent observations on the *ith component* of the p-dimensional variable. $\mathscr{U}$ represents the overall matrix of the order $n \times p$. We shall use the notations (8a.5.1) throughout this chapter.

Estimation of $\boldsymbol{\mu}$ and $\boldsymbol{\Sigma}$ from i.i.d. Variables. Let $\mathbf{U}_1, \ldots, \mathbf{U}_n$ be independent and have the same distribution with mean $\boldsymbol{\mu}$ and dispersion matrix $\boldsymbol{\Sigma}$. Then

$$U_{i1}, \ldots, U_{in}$$

are i.i.d. observations from a univariate distribution with mean μ_i and variance σ_{ii}. Hence from the univariate computations, we see that

$$\bar{Y}_i = \bar{U}_i = \frac{1}{n}\sum_j U_{ij} \qquad \text{is unbiased for} \quad \mu_i$$

$$S_{ii} = \sum_j (U_{ij} - \bar{U}_i)^2 \qquad \text{is unbiased for} \quad (n-1)\sigma_{ii}$$

$$s_{ii} = \frac{S_{ii}}{n-1} \qquad \text{is unbiased for} \quad \sigma_{ii}.$$

Consider

$$U_{i1} + U_{j1}, \ldots, U_{in} + U_{jn},$$

which are i.i.d. observations from a univariate distribution with mean $\mu_i + \mu_j$ and variance $\sigma_{ii} + 2\sigma_{ij} + \sigma_{jj}$. Hence

$$\sum (U_{ir} + U_{jr} - \bar{U}_i - \bar{U}_j)^2 = S_{ii} + S_{jj} + 2S_{ij}$$

is unbiased for $(n-1)(\sigma_{ii} + 2\sigma_{ij} + \sigma_{jj})$, where

$$S_{ij} = \sum (U_{ir} - \bar{U}_i)(U_{jr} - \bar{U}_j).$$

Hence $S_{ij}/(n-1)$ is unbiased for σ_{ij}, since $S_{ii}/(n-1)$ and $S_{jj}/(n-1)$ are unbiased for σ_{ii} and σ_{jj} respectively.

Thus we have unbiased estimators of $\boldsymbol{\mu}'$ and $\boldsymbol{\Sigma}$ as

$$\bar{\mathbf{U}}' = (\bar{U}_1, \ldots, \bar{U}_p), \qquad \frac{1}{n-1}(S_{ij}) = \frac{1}{n-1}\mathbf{S}$$

in the general case without assuming normality of the distributions of $\mathbf{U}_i$. It is of interest to give the different matrix representations of (S_{ij}) using the notation (8a.5.1), which the reader may verify.

$$S_{ij} = \mathbf{Y}_i'\mathbf{Y}_j - n\bar{U}_i\bar{U}_j$$

$$\mathbf{S} = (S_{ij}) = \sum_1^n (\mathbf{U}_r - \bar{\mathbf{U}})(\mathbf{U}_r - \bar{\mathbf{U}})' = \sum_1^n \mathbf{U}_r\mathbf{U}_r' - n\bar{\mathbf{U}}\bar{\mathbf{U}}' \qquad (8a.5.2)$$

$$= \mathcal{U}'\mathcal{U} - n\bar{\mathbf{U}}\bar{\mathbf{U}}'$$

Estimation by the Method of Maximum Likelihood. When it is known that $R(\boldsymbol{\Sigma}) = p$, the density function for a single observation $\mathbf{U}$ is, as shown in (8a.4.4),

$$(2\pi)^{-p/2}|(\sigma^{ij})|^{1/2} \exp[-\tfrac{1}{2}\operatorname{trace}\{\boldsymbol{\Sigma}^{-1}(\mathbf{U} - \boldsymbol{\mu})(\mathbf{U} - \boldsymbol{\mu})'\}]$$

where $(\sigma^{ij}) = (\sigma_{ij})^{-1} = \mathbf{\Sigma}^{-1}$. The joint density of the observations $(\mathbf{U}_1, \ldots, \mathbf{U}_n)$ is apart from a constant

$$|(\sigma^{ij})|^{n/2} \exp\left[-\tfrac{1}{2}\operatorname{trace}\left\{\mathbf{\Sigma}^{-1}\sum_1^n(\mathbf{U}_r - \boldsymbol{\mu})(\mathbf{U}_r - \boldsymbol{\mu})'\right\}\right]. \tag{8a.5.3}$$

Observing that

$$\sum_1^n(\mathbf{U}_r - \boldsymbol{\mu})(\mathbf{U}_r - \boldsymbol{\mu})' = \sum_1^n(\mathbf{U}_r - \overline{\mathbf{U}})(\mathbf{U}_r - \overline{\mathbf{U}})' + n(\overline{\mathbf{U}} - \boldsymbol{\mu})(\overline{\mathbf{U}} - \boldsymbol{\mu})'$$
$$= \mathbf{S} + n(\overline{\mathbf{U}} - \boldsymbol{\mu})(\overline{\mathbf{U}} - \boldsymbol{\mu})',$$

we see that

$$\operatorname{trace}\left\{\mathbf{\Sigma}^{-1}\sum_1^n(\mathbf{U}_r - \boldsymbol{\mu})(\mathbf{U}_r - \boldsymbol{\mu})'\right\}$$
$$= \operatorname{trace}\{\mathbf{\Sigma}^{-1}\mathbf{S}\} + n\operatorname{trace}\{\mathbf{\Sigma}^{-1}(\overline{\mathbf{U}} - \boldsymbol{\mu})(\overline{\mathbf{U}} - \boldsymbol{\mu})'\}$$
$$= \sum\sum \sigma^{ij}S_{ij} + n(\overline{\mathbf{U}} - \boldsymbol{\mu})'\mathbf{\Sigma}^{-1}(\overline{\mathbf{U}} - \boldsymbol{\mu}).$$

Hence the logarithm of (8a.5.3) can be written

$$\frac{n}{2}\log|(\sigma^{ij})| - \frac{1}{2}\sum\sum \sigma^{ij}S_{ij} - \frac{n}{2}(\overline{\mathbf{U}} - \boldsymbol{\mu})'\mathbf{\Sigma}^{-1}(\overline{\mathbf{U}} - \boldsymbol{\mu}), \tag{8a.5.4}$$

or in the alternative form

$$\frac{n}{2}\log|(\sigma^{ij})| - \frac{1}{2}\sum\sum \sigma^{ij}[S_{ij} + n(\overline{U}_i - \mu_i)(\overline{U}_j - \mu_j)]. \tag{8a.5.5}$$

We use the form (8a.5.4) to differentiate with respect to $\boldsymbol{\mu}$ and obtain

$$\mathbf{\Sigma}^{-1}(\overline{\mathbf{U}} - \boldsymbol{\mu}) = \mathbf{0} \qquad \text{or} \qquad \overline{\mathbf{U}} - \boldsymbol{\mu} = \mathbf{0},$$

which leads to the estimator $\hat{\boldsymbol{\mu}} = \overline{\mathbf{U}}$ or $\hat{\mu}_i = \overline{U}_i$, $i = 1, \ldots, p$.

The second form (8a.5.5) is convenient for differentiating with respect to σ^{ij}, thus giving the equation

$$\frac{n}{|(\sigma^{ij})|}\frac{\partial|(\sigma^{ij})|}{\partial\sigma^{ij}} = [S_{ij} + n(\overline{U}_i - \mu_i)(\overline{U}_j - \mu_j)]. \tag{8a.5.6}$$

It is easy to see that

$$\frac{\partial|(\sigma^{ij})|}{\partial\sigma^{ij}} = \text{cofactor of } \sigma^{ij} \text{ in } (\sigma^{ij})$$

so that

$$\frac{n}{|(\sigma^{ij})|}\frac{\partial|(\sigma^{ij})|}{\partial\sigma^{ij}} = n\sigma_{ij}.$$

Furthermore, by substituting $(\bar{U}_i - \mu_i) = 0$, the equation (8a.5.6) reduces to

$$n\sigma_{ij} - S_{ij} = 0,$$

thus giving the estimator of σ_{ij} as

$$\hat{\sigma}_{ij} = \frac{S_{ij}}{n}, \qquad E(\hat{\sigma}_{ij}) = \frac{E(S_{ij})}{n} = \frac{n-1}{n}\sigma_{ij},$$

which is slightly biased. The estimator of σ^{ij} is then nS^{ij}, where $(S^{ij}) = (S_{ij})^{-1}$.

The value of log likelihood (8a.5.4) at the estimated values of the parameters is

$$\frac{n}{2}\log n^p + \frac{n}{2}\log|(S^{ij})| - \frac{n}{2}\sum\sum S^{ij}S_{ij} = \frac{n}{2}\log n^p - \frac{n}{2}\log|(S_{ij})| - \frac{np}{2}. \tag{8a.5.7}$$

We now establish that *the estimators* $\hat{\boldsymbol{\mu}} = \bar{\mathbf{U}}$ *and* $(\hat{\sigma}_{ij}) = n^{-1}(S_{ij}) = n^{-1}\mathbf{S}$ *obtained by equating the derivatives of the log likelihood to zero are, in fact, the maximum likelihood estimators.*

It is enough to prove that (8a.5.5) $\leqslant$ (8a.5.7) for all $\boldsymbol{\mu}$ and $\boldsymbol{\Sigma}$. The difference (8a.5.7) – (8a.5.5) is equal to

$$\begin{aligned} -\frac{n}{2}\log\frac{|n^{-1}\mathbf{S}|}{|\boldsymbol{\Sigma}|} &- \frac{np}{2} + \frac{1}{2}\operatorname{trace}\{\boldsymbol{\Sigma}^{-1}[\mathbf{S} + n(\bar{\mathbf{U}} - \boldsymbol{\mu})(\bar{\mathbf{U}} - \boldsymbol{\mu})']\} \\ &\geqslant -\frac{n}{2}\log\frac{|n^{-1}\mathbf{S}|}{|\boldsymbol{\Sigma}|} - \frac{np}{2} + \frac{1}{2}\operatorname{trace}(\boldsymbol{\Sigma}^{-1}\mathbf{S}) \\ &= \frac{n}{2}[-\log(\lambda_1 \cdots \lambda_p) - p + (\lambda_1 + \cdots + \lambda_p)], \end{aligned} \tag{8a.5.8}$$

where $\lambda_1, \ldots, \lambda_p$ are the roots of the determinantal equation $|n^{-1}\mathbf{S} - \lambda\boldsymbol{\Sigma}| = 0$. But for any non-negative x, $x \leqslant e^{x-1}$, or by taking logarithms we have

$$-\log x - 1 + x \geqslant 0. \tag{8a.5.9}$$

The quantity inside the bracket of (8a.5.8) is

$$\sum_{i=1}^{p}(-\log\lambda_i - 1 + \lambda_i), \tag{8a.5.10}$$

which is $\geqslant 0$, if we apply the inequality (8a.5.9) to each term in (8a.5.10). This elegant proof is by Watson (1964).

It may be verified that the m.l. estimators $\bar{\mathbf{U}}$ and $n^{-1}\mathbf{S}$ are also sufficient for the parameters $\boldsymbol{\mu}$ and $\boldsymbol{\Sigma}$, since the log likelihood is an explicit function of the estimators and the parameters only.

If $R(\boldsymbol{\Sigma})$ is unknown, we have to admit the possibility of $R(\boldsymbol{\Sigma})$ being less than p, in which case we have to use the density function (8a.4.11) or (8a.4.9), determining the appropriate hyperplane (8a.4.12) or (8a.4.6).

Adopting the density (8a.4.9) and utilizing the results already established, the m.l. estimators of $\mathbf{B}'\boldsymbol{\mu}$ and $\mathbf{B}'\boldsymbol{\Sigma}\mathbf{B}$ satisfy the equations

$$\mathbf{B}'\bar{\mathbf{U}} = \mathbf{B}'\boldsymbol{\mu}, \qquad \mathbf{B}'\mathbf{S}\mathbf{B} = n\mathbf{B}'\boldsymbol{\Sigma}\mathbf{B},$$

while (8a.4.6) gives

$$\mathbf{B}'\bar{\mathbf{U}} = \mathbf{N}'\boldsymbol{\mu}, \qquad \mathbf{N}'\mathbf{S}\mathbf{N} = n\,\mathbf{N}'\boldsymbol{\Sigma}\mathbf{N} = \mathbf{0},$$

which yield the same solution as in the case of full rank for $\boldsymbol{\Sigma}$,

$$\hat{\boldsymbol{\mu}} = \bar{\mathbf{U}} \quad \text{and} \quad \hat{\boldsymbol{\Sigma}} = n^{-1}\mathbf{S}.$$

8a.6 N_p as a Distribution with Maximum Entropy

In **3a.1**, we have shown that the univariate normal density has the maximum entropy subject to the condition that mean and variance are fixed. A similar result is true for N_p.

Let $P(\mathbf{U})$ be the density of a p-dimensional random variable $\mathbf{U}$. The problem is to maximize

$$-\int P(\mathbf{U})\log P(\mathbf{U})\,dv \tag{8a.6.1}$$

where dv stands for the volume element, subject to the conditions that the mean and dispersion matrix have given values

$$\left.\begin{aligned} &\int P(\mathbf{U})\,dv = 1, \qquad \int \mathbf{U}P(\mathbf{U})\,dv = \boldsymbol{\mu} \\ &\int [(\mathbf{U}-\boldsymbol{\mu})(\mathbf{U}-\boldsymbol{\mu})']P(\mathbf{U})\,dv = \boldsymbol{\Sigma} \end{aligned}\right\}. \tag{8a.6.2}$$

Choose the normal density

$$Q(\mathbf{U}) = (2\pi)^{-p/2}|\boldsymbol{\Sigma}|^{-1/2}\exp[-\tfrac{1}{2}(\mathbf{U}-\boldsymbol{\mu})'\boldsymbol{\Sigma}^{-1}(\mathbf{U}-\boldsymbol{\mu})], \tag{8a.6.3}$$

which satisfies the conditions (8a.6.2). From the information theory inequality (1e.6.6) we have for any two alternative densities $P(\mathbf{U})$, $Q(\mathbf{U})$,

$$-\int P\log P\,dv \leqslant -\int P\log Q\,dv,$$

where the equality is attained when $P = Q$ almost everywhere (dv). Substituting the choice (8a.6.3) for Q we have

$$-\int P \log P\,dv \leqslant -\int P(\mathbf{U})\left[-\frac{p}{2}\log(2\pi) - \frac{1}{2}\log|\boldsymbol{\Sigma}| - \frac{1}{2}(\mathbf{U}-\boldsymbol{\mu})'\boldsymbol{\Sigma}^{-1}(\mathbf{U}-\boldsymbol{\mu})\right]dv,$$

$$= \frac{p}{2}\log(2\pi) + \frac{1}{2}\log|\boldsymbol{\Sigma}| + \frac{1}{2}\operatorname{trace}\left\{\boldsymbol{\Sigma}^{-1}\int[(U-\boldsymbol{\mu})(\mathbf{U}-\boldsymbol{\mu})']P(\mathbf{U})\,dv\right\}$$

$$= \frac{p}{2}\log(2\pi) + \frac{1}{2}\log|\boldsymbol{\Sigma}| + \frac{1}{2}\operatorname{trace}\,\boldsymbol{\Sigma}^{-1}\boldsymbol{\Sigma} \tag{8a.6.4}$$

by using (8a.6.2). The expression (8a.6.4) is thus a fixed upper bound to entropy which is attained for the normal density (8a.6.3). The entropy of the distribution so determined is

$$\frac{p}{2}\log(2\pi) + \frac{p}{2} + \frac{1}{2}\log|\boldsymbol{\Sigma}|.$$

8b WISHART DISTRIBUTION

8b.1 Definition and Notation

Let $\mathbf{U}_i \sim N_p(\boldsymbol{\mu}_i, \Sigma)$, $i = 1, \ldots, k$ be all independent and $\mathbf{Y}_j$, $\mathscr{U}$ be as defined in (8a.5.1) with k in the place of n. Further, let $\mathbf{M}'$ be a $p \times k$ matrix with $\boldsymbol{\mu}_1, \ldots, \boldsymbol{\mu}_k$ as its columns. Thus

$$\begin{array}{c|ccc|l} & \mathbf{U}_1 & \ldots & \mathbf{U}_k & \\ \hline \mathbf{Y}_1' & U_{11} & \ldots & U_{1k} & \\ \cdot & \cdot & \ldots & \cdot & = \mathscr{U}'. \\ \mathbf{Y}_p' & U_{p1} & \ldots & U_{pk} & \\ \hline \mathbf{Y}' & \mathbf{L}'\mathbf{U}_1 & \ldots & \mathbf{L}'\mathbf{U}_k & = \mathbf{L}'\mathscr{U}' \end{array} \tag{8b.1.1}$$

where $\mathbf{L}$ is a fixed vector with elements $L_1, \ldots, L_p$.

In our discussion, there will be frequent references to a *linear function* $\mathbf{L}'\mathbf{U}_i$ of $\mathbf{U}_i$ for each i. Now $E(\mathbf{L}'\mathbf{U}_i) = \mathbf{L}'\boldsymbol{\mu}_i$ and $V(\mathbf{L}'\mathbf{U}_i) = \mathbf{L}'\boldsymbol{\Sigma}\mathbf{L}$. Then

$$\mathbf{L}'\mathbf{U}_i \sim N_1(\mathbf{L}'\boldsymbol{\mu}_i, \sigma_L{}^2), \qquad \sigma_L{}^2 = \mathbf{L}'\boldsymbol{\Sigma}\mathbf{L}, \qquad i = 1, \ldots, k, \tag{8b.1.2}$$

and are all independent since $\mathbf{U}_i$ are independent. The vector $\mathbf{Y}' = (\mathbf{L}'\mathbf{U}_1, \ldots, \mathbf{L}'\mathbf{U}_k)$ of k independent N_1 variables has the representation

$$\mathbf{Y} = L_1\mathbf{Y}_1 + \cdots + L_p\mathbf{Y}_p = \mathscr{U}\mathbf{L}. \tag{8b.1.3}$$

The distribution of $\mathbf{Y} = \mathscr{U}\mathbf{L}$ is that of k independent N_1 variables with the same variance $\sigma_L{}^2$ but different means $\mathbf{L}'\boldsymbol{\mu}_1, \ldots, \mathbf{L}'\boldsymbol{\mu}_k$, that is

$$\mathbf{Y} \sim N_k(\mathbf{ML}, \sigma_L{}^2\mathbf{I}). \tag{8b.1.4}$$

We also consider *linear combinations* of the vector variables $\mathbf{U}_i$, which are also distributed as N_p by the reproductive property [(vi), **8a.2**]. Let $\mathbf{B}' = (B_1, \ldots, B_k)$ be a fixed vector. Then

$$B_1\mathbf{U}_1 + \cdots + B_k\mathbf{U}_k = \mathscr{U}'\mathbf{B} \sim N_p(\mathbf{M}'\mathbf{B}, (\textstyle\sum B_i{}^2)\boldsymbol{\Sigma}), \tag{8b.1.5}$$

from the result (8a.2.18). Furthermore, it is shown in [(ii), **8a.3**], that $\mathscr{U}'\mathbf{B}_1$ and $\mathscr{U}'\mathbf{B}_2$ are independently distributed if $\mathbf{B}_1'\mathbf{B}_2 = 0$. Hence it follows that the linear combinations

$$\mathbf{V}_1 = \mathscr{U}'\mathbf{B}_1, \ldots, \mathbf{V}_r = \mathscr{U}'\mathbf{B}_r, \qquad r \leqslant k \tag{8b.1.6}$$

are all independently distributed as N_p with the same dispersion matrix $\boldsymbol{\Sigma}$, when the vectors $\mathbf{B}_1, \ldots, \mathbf{B}_r$ are orthonormal.

Let $\mathbf{H}$ be an orthogonal matrix of order k. Then the transformation

$$\mathscr{V} = \mathscr{U}'\mathbf{H} \tag{8b.1.7}$$

takes the column vectors $\mathbf{U}_1, \ldots, \mathbf{U}_k$ into column vectors $\mathbf{V}_1, \ldots, \mathbf{V}_k$ where each $\mathbf{V}_i$ is a linear combination of $\mathbf{U}_1, \ldots, \mathbf{U}_k$. The result (8b.1.6) shows that $\mathbf{V}_i$ are independently distributed (as in the univariate case, *an orthogonal transformation preserves independence*).

Definition of Wishart (1928) Distribution. The joint distribution of the elements of the matrix

$$(S_{ij}) = \mathbf{S} = \sum_1^k \mathbf{U}_r\mathbf{U}_r' = (\mathbf{Y}_i'\mathbf{Y}_j) = \mathscr{U}'\mathscr{U}, \tag{8b.1.8}$$

where $\mathbf{U}_r$, $\mathbf{Y}_i$, and $\mathscr{U}$ and $\mathbf{M}$ are as defined in (8b.1.1), is said to be Wishart on k D.F. and is denoted by $W_p(k, \boldsymbol{\Sigma}, \mathbf{M})$. The distribution is said to be central when $\mathbf{M} = \mathbf{0}$ and written $W_p(k, \boldsymbol{\Sigma})$. We use the symbol $W_p(k, \boldsymbol{\Sigma}, \cdot)$ to indicate a central or a noncentral distribution. When $p = 1$, $W_1(k, \sigma^2)$ is the same as $\sigma^2\chi^2(k)$, the χ^2 distribution on k D.F. Thus the Wishart distribution is a multivariate generalization of the χ^2 distribution.

As in the case of N_p, the density function of W_p does not always exist. It exists when $k \geqslant p$ as derived in Example 11 at the end of the Chapter. We shall not need the density of W_p in our study. We derive very general results from the definition of W_p.

The matrix $\mathbf{S} = (S_{ij})$ in a more explicit form is

$$(S_{ij}) = \mathcal{U}'\mathcal{U} = \begin{pmatrix} \mathbf{Y}_1'\mathbf{Y}_1 & \mathbf{Y}_1'\mathbf{Y}_2 & \cdots & \mathbf{Y}_1'\mathbf{Y}_p \\ \cdot & \cdot & \cdots & \cdot \\ \mathbf{Y}_p'\mathbf{Y}_1 & \mathbf{Y}_p'\mathbf{Y}_2 & \cdots & \mathbf{Y}_p'\mathbf{Y}_p \end{pmatrix},$$

where the ith diagonal entry is the sum of squares of the ith components of the vector variables $\mathbf{U}_r$ and the (i, j) entry is the sum of products of the ith and jth components. The representation $\mathcal{U}'\mathcal{U}$ resembles that of the sum of squares in the univariate theory. A generalization of the quadratic form is the matrix $\mathcal{U}'\mathbf{A}\mathcal{U}$,

$$\mathcal{U}'\mathbf{A}\mathcal{U} = \begin{pmatrix} \mathbf{Y}_1'\mathbf{A}\mathbf{Y}_1 & \cdots & \mathbf{Y}_1'\mathbf{A}\mathbf{Y}_p \\ \cdot & \cdots & \cdot \\ \mathbf{Y}_p'\mathbf{A}\mathbf{Y}_1 & \cdots & \mathbf{Y}_p'\mathbf{A}\mathbf{Y}_p \end{pmatrix}, \tag{8b.1.9}$$

which consists of quadratic and bilinear forms involving $\mathbf{Y}_1, \ldots, \mathbf{Y}_p$. Then we raise the following question. Under what conditions does $\mathcal{U}'\mathbf{A}\mathcal{U}$ have a Wishart distribution? The answer is

$$\mathcal{U}'\mathbf{A}\mathcal{U} \sim W_p \Leftrightarrow \mathbf{L}'\mathcal{U}'\mathbf{A}\mathcal{U}\mathbf{L} \sim \chi^2 \tag{8b.1.10}$$

for any fixed $\mathbf{L}$. But the R.H.S. of (8b.1.10) is $(\mathcal{U}\mathbf{L})'\mathbf{A}(\mathcal{U}\mathbf{L})$ which is the quadratic form $\mathbf{Y}'\mathbf{A}\mathbf{Y}$ of the k independent N_1 variables $(\mathbf{L}'\mathbf{U}_1, \ldots, \mathbf{L}'\mathbf{U}_k)$. Since the distribution of quadratic forms of independent N_1 variables is extensively studied in **3b.4**, the result (8b.1.10) enables us to obtain suitable generalization to the distribution of matrices. A proof of the result (8b.1.10) is given in **8b.2**.

8b.2 Some Results on Wishart Distribution

(i) *Let $S \sim W_p(k, \Sigma, \cdot)$ and* $\mathbf{L}$ *be any fixed vector. Then* $\mathbf{L}'\mathbf{S}\mathbf{L} \sim \sigma_L{}^2\chi^2(k, \cdot)$, *and if W_p is central then χ^2 is so.*

By definition $\mathbf{S} = \sum \mathbf{U}_r\mathbf{U}_r'$. Pre- and postmultiplication by $\mathbf{L}'$, $\mathbf{L}$ gives

$$\begin{aligned} \mathbf{L}'\mathbf{S}\mathbf{L} &= \sum (\mathbf{L}'\mathbf{U}_r)(\mathbf{U}_r'\mathbf{L}) = \sum (\mathbf{L}'\mathbf{U}_r)^2 \\ &= \mathbf{Y}'\mathbf{Y} \sim \sigma_L{}^2\chi^2(k, \cdot), \end{aligned}$$

since $\mathbf{Y}$ is a vector of independent N_1 variables with a common variance $\sigma_L{}^2$ (see 8b.1.4). If W_p is central, $\mathbf{M} = \mathbf{0}$, in which case $E(\mathbf{Y}) = \mathbf{ML} = \mathbf{0}$ and hence the χ^2 is central.

Note that the strict converse of (i) is not true as shown by Mitra (1970). However, the weaker version of the converse given in (ii) is useful.

(ii) *Let* $\mathbf{U}_i \sim N_p(\boldsymbol{\mu}_i, \Sigma)$, $i = 1, \ldots, k$ *be independent and* $\mathcal{U}$, $\mathbf{Y}_i$, $\mathbf{Y}$ *be as defined in* (8b.1.1). *An n.s. condition that* $\mathcal{U}'\mathbf{A}\mathcal{U} \sim W_p(r, \Sigma, \cdot)$ *is that* $\mathbf{Y}'\mathbf{A}\mathbf{Y} \sim \sigma_L{}^2\chi^2(r, \cdot)$ *for any fixed* $\mathbf{L}$, *in which case r = rank* $\mathbf{A}$ = *trace* $\mathbf{A}$. *And* $\mathcal{U}'\mathbf{A}\mathcal{U} \sim W_p(r, \Sigma)$, *if and only if* $\mathbf{Y}'\mathbf{A}\mathbf{Y} \sim \sigma_L{}^2\chi^2(r)$ *for every* $\mathbf{L}$.

The necessity follows from the more general result (i). To prove sufficiency let us observe (from the univariate theory, (ii), **3b.4**)

$$\mathbf{Y'AY} \sim \sigma_L^2\chi^2(r, \cdot) \Rightarrow \mathbf{A} \text{ is idempotent of rank } r.$$

Then there exist r orthonormal vectors $\mathbf{B}_1, \ldots, \mathbf{B}_r$ such that

$$\mathbf{A} = \mathbf{B}_1\mathbf{B}_1' + \cdots + \mathbf{B}_r\mathbf{B}_r' \tag{8b.2.1}$$

$$\mathscr{U}'\mathbf{A}\mathscr{U} = \mathscr{U}'\mathbf{B}_1\mathbf{B}_1'\mathscr{U} + \cdots + \mathscr{U}'\mathbf{B}_r\mathbf{B}_r'\mathscr{U}$$

$$= \mathbf{V}_1\mathbf{V}_1' + \cdots + \mathbf{V}_r\mathbf{V}_r', \tag{8b.2.2}$$

where $\mathbf{V}_i = \mathscr{U}'\mathbf{B}_i$. Since $\mathbf{B}_i$ are orthonormal, $\mathbf{V}_i$ are independent N_p variables (8b.1.6). The result, then, follows from the definition of Wishart distribution.

Since $\mathbf{Y} \sim N_k(\mathbf{ML}, \sigma_L^2\mathbf{I})$, the noncentral parameter of the distribution of $\mathbf{Y'AY}$ is the value of $\mathbf{Y'AY}/\sigma_L^2$ at $\mathbf{Y} = \mathbf{ML}$, which is $\mathbf{L'M'AML}/\sigma_L^2$ (see 3b.2.2). If $\mathbf{Y'AY} \sim \chi^2(r)$, a central χ^2 for every $\mathbf{L}$, then

$$\frac{\mathbf{L'M'AML}}{\mathbf{L'\Sigma L}} = 0 \qquad \text{for every} \quad \mathbf{L} \Rightarrow \mathbf{M'AM} = \mathbf{0}.$$

But from (8b.2.1), we have

$$\mathbf{A} = \mathbf{B}_1\mathbf{B}_1' + \cdots + \mathbf{B}_r\mathbf{B}_r'$$

$$\mathbf{M'AM} = \sum (\mathbf{M'B}_i)(\mathbf{M'B}_i)' = 0$$

$$\Rightarrow \mathbf{M'B}_i = \mathbf{0}, \qquad i = 1, \ldots, r.$$

Hence from (8b.2.2), we see that

$$\mathscr{U}'\mathbf{A}\mathscr{U} = \sum \mathbf{V}_i\mathbf{V}_i', \qquad \text{where} \quad E(\mathbf{V}_i) = \mathbf{M'B}_i = \mathbf{0},$$

and by definition the distribution is Wishart and central.

We could have stated the result of (ii) in a more direct form that an n.s. condition that $\mathscr{U}'\mathbf{A}\mathscr{U} \sim W_p$ is that $\mathbf{A}$ is idempotent and the distribution is central when $\mathbf{AM} = 0$. The reason for stating the result in an indirect form is that it enables us to generalize known results of the univariate theory in the following way.

Suppose it is known that the quadratic form $\mathbf{Y'AY} \sim \chi^2$ *where* $\mathbf{Y}$ *is a vector of independent* N_1 *variables. Then we substitute* $\mathscr{U}\mathbf{L}$ *for* $\mathbf{Y}$ *and obtain the quadratic form* $\mathbf{L}'\mathscr{U}'\mathbf{A}\mathscr{U}\mathbf{L}$ *in* $\mathbf{L}$ *and by dropping* $\mathbf{L}$ *assert that the associated matrix* $\mathscr{U}'\mathbf{A}\mathscr{U} \sim W_p$.

It is easy to establish the result stated in (iii) following the arguments of (ii).

(iii) *Under the conditions and notations of (ii), the matrices* $\mathscr{U}'\mathbf{A}_1\mathscr{U}$ *and* $\mathscr{U}'\mathbf{A}_2\mathscr{U}$ *have independent Wishart distributions if and only if* $\mathbf{Y'A}_1\mathbf{Y}$ *and* $\mathbf{Y'A}_2\mathbf{Y}$ *have independent* χ^2 *distributions for any* $\mathbf{L}$. *Furthermore,* $\mathscr{U}'\mathbf{B}$ *and*

$\mathscr{U}'\mathbf{A}\mathscr{U}$ are independently distributed as N_p and W_p if $\mathbf{Y}'\mathbf{B}$ and $\mathbf{Y}'\mathbf{AY}$ are independently distributed as N_1 and χ^2 for any $\mathbf{L}$.

We could have stated the result (iii) in the form that the n.s. condition for $\mathscr{U}'\mathbf{A}_1\mathscr{U}$, $\mathscr{U}'\mathbf{A}_2\mathscr{U}$ to be independently distributed (whether in Wishart's form or not) is that $\mathbf{A}_1\mathbf{A}_2=0$ and for $\mathscr{U}'\mathbf{B}$ and $\mathscr{U}'\mathbf{A}\mathscr{U}$ to be independently distributed (whether the latter as Wishart or not) is that $\mathbf{B}'\mathbf{A}=0$.

But what we wish to do is to consider quadratic forms in independent N_1 variables known to have independent χ^2 distributions and write the corresponding generalizations to the multivariate case without explicitly determining the matrices involved. As an immediate example we have the result (iv). We shall continue to explore this technique throughout this chapter and avoid any direct attempt to derive distributions in the multivariate case as is done in books on this subject using heavy transformations.

(iv) THE JOINT DISTRIBUTION OF SAMPLE MEAN AND DISPERSION MATRIX. Let $\mathbf{U}_1, \ldots, \mathbf{U}_n$ be i.i.d. each as $N_p(\boldsymbol{\mu}, \boldsymbol{\Sigma})$. Consider the r.v.'s

$$\mathbf{L}'\mathbf{U}_1, \ldots, \mathbf{L}'\mathbf{U}_n$$

which are i.i.d. each as $N_1(\mathbf{L}'\boldsymbol{\mu}, \mathbf{L}'\boldsymbol{\Sigma}\mathbf{L})$. From the univariate theory we know that the sample mean

$$\frac{1}{n}\sum \mathbf{L}'\mathbf{U}_r = \mathbf{L}'\bar{\mathbf{U}} \sim N_1\left(\mathbf{L}'\boldsymbol{\mu}, \frac{1}{n}\mathbf{L}'\boldsymbol{\Sigma}\mathbf{L}\right), \tag{8b.2.3}$$

and the corrected sum of squares

$$\begin{aligned}\sum_1^n (\mathbf{L}'\mathbf{U}_r)^2 - n(\mathbf{L}'\bar{\mathbf{U}})^2 &= \mathbf{L}'(\sum \mathbf{U}_r\mathbf{U}_r' - n\bar{\mathbf{U}}\bar{\mathbf{U}}')\mathbf{L} \\ &= \mathbf{L}'(S_{ij})\mathbf{L} \sim (\mathbf{L}'\boldsymbol{\Sigma}\mathbf{L})\chi^2(n-1), \text{ a central } \chi^2, \quad (8b.2.4)\end{aligned}$$

for any $\mathbf{L}$, and that they are independent. Thus the independence of

$$\mathbf{L}'\bar{\mathbf{U}} \quad \text{and} \quad \mathbf{L}'(S_{ij})\mathbf{L}$$

where $(S_{ij}) = \sum \mathbf{U}_r\mathbf{U}_r' - n\bar{\mathbf{U}}\bar{\mathbf{U}}'$ is the corrected sum of squares and products defined in (8a.5.2), for any $\mathbf{L}$ implies (by dropping $\mathbf{L}$) *the independence of the distributions of* $\bar{\mathbf{U}}$ and (S_{ij}). From (8b.2.3) $\bar{\mathbf{U}} \sim N_p(\mu, (1/n)\boldsymbol{\Sigma})$ and from (8b.2.4), $(S_{ij}) \sim W_p(n-1, \boldsymbol{\Sigma})$, a central Wishart [using (ii)].

(v) *Let $\mathbf{S}_1 \sim W_p(k_1, \boldsymbol{\Sigma})$ and $\mathbf{S}_2 \sim W_p(k_2, \boldsymbol{\Sigma})$ be independent. Then $\mathbf{S} = \mathbf{S}_1 + \mathbf{S}_2 \sim W_p(k_1 + k_2, \boldsymbol{\Sigma})$, a similar result being true for noncentral distributions.*

We can write

$$\mathbf{S}_1 = \sum_1^{k_1} \mathbf{U}_i\mathbf{U}_i' \quad \text{and} \quad S_2 = \sum_{k_1+1}^{k_1+k_2} \mathbf{U}_i\mathbf{U}_i'$$

where $\mathbf{U}_1, \ldots, \mathbf{U}_{k_1+k_2}$ are i.i.d. each as $N_p(\mathbf{0}, \mathbf{\Sigma})$. Hence by definition

$$\mathbf{S} = \sum_1^{k_1+k_2} \mathbf{U}_i \mathbf{U}_i' \sim W_p(k_1 + k_2, \mathbf{\Sigma}).$$

(vi) *Let* $\mathbf{S} \sim W_p(k, \mathbf{\Sigma})$ *and* $\mathbf{B}$ *be* $(q \times p)$ *matrix. Then* $\mathbf{BSB}' \sim W_q(k, \mathbf{B\Sigma B}')$.

$$\mathbf{S} = \sum_{i=1}^{k} \mathbf{U}_i \mathbf{U}_i' \Rightarrow \mathbf{BSB}' = \mathbf{B}(\sum \mathbf{U}_i \mathbf{U}_i')\mathbf{B}'$$

$$\mathbf{BSB}' = \sum_{i=1}^{k} (\mathbf{BU}_i)(\mathbf{BU}_i)' \sim W_q(k, \mathbf{B\Sigma B}')$$

since $\mathbf{BU}_i \sim N_q(\mathbf{0}, \mathbf{B\Sigma B}')$, $i = 1, \ldots, k$ and are independent.

(vii) *Let* $W_p(\mathbf{S}|k, \mathbf{\Sigma})$ *denote the density function of* $\mathbf{S}$ *and let* $\mathbf{C}$ *be a non-singular square matrix of order p. Then the density function of* $\mathbf{S}_* = \mathbf{CSC}'$ *is*

$$|\mathbf{C}|^{-p-1} W_p(\mathbf{C}^{-1}\mathbf{S}_*\mathbf{C}^{-1'}|k, \mathbf{C}^{-1}\mathbf{\Sigma}\mathbf{C}^{-1'}).$$

The result is obtained by change of variable and a computation of the Jacobian

$$\frac{D(\mathbf{S})}{D(\mathbf{S}_*)} = |\mathbf{C}|^{-p-1}.$$

(viii) *Let* (S^{ij}) *be the reciprocal of* $\mathbf{S} = (S_{ij})$ *and* (σ^{ij}) *the reciprocal of* $(\sigma_{ij}) = \mathbf{\Sigma}$ *exist. If* $\mathbf{S} \sim W_p(k, \mathbf{\Sigma})$, *then*

(a) $\dfrac{\sigma^{pp}}{S^{pp}} \sim \chi^2(\overline{k - p - 1})$ *and is independent of* (S_{ij}), $i, j = 1, \ldots, (p-1)$.

(b) $\dfrac{\mathbf{L}'\mathbf{\Sigma}^{-1}\mathbf{L}}{\mathbf{L}'\mathbf{S}^{-1}\mathbf{L}} \sim \chi^2(\overline{k - p - 1})$ *for any fixed vector* $\mathbf{L}$.

Consider k independent variables $\mathbf{U}_1, \ldots, \mathbf{U}_k$ each distributed as $N_p(\mathbf{0}, \mathbf{\Sigma})$. The conditional distribution of U_{pj} given $U_{1j}, \ldots, U_{p-1j}$ is shown to be (see 8a.2.16, 8a.2.17)

$$N_1(\beta_1 U_{1j} + \cdots + \beta_{p-1} U_{p-1,j}, 1/\sigma^{pp}) \qquad j = 1, \ldots, k. \tag{8b.2.4}$$

The setup (8b.2.4) is the same as in the theory of least squares with Gauss-Markoff model. Hence by using the first fundamental theorem [(i), **3b.5**] of the theory of least squares

$$R_0^2 = \min_\beta \sum_1^k (U_{pj} - \beta_1 U_{1j} - \cdots - \beta_{p-1} U_{p-1,j})^2 \tag{8b.2.5}$$

is distributed as $(\sigma^{pp})^{-1}\chi^2(k - r)$ where r is the rank of (U_{ij}), $i = 1, \ldots, p - 1$; $j = 1, \ldots, k$. The distribution of R_0^2 is conditional but since it does not involve the conditioning variables (U_{ij}), $i = 1, \ldots, p-1$; $j = 1, \ldots, k$, it is also unconditional and is independent of (U_{ij}), $i = 1, \ldots, p-1$; $j = 1, \ldots, k$.

Since $\mathbf{S} \sim W_p(k, \mathbf{\Sigma})$ and $|\mathbf{\Sigma}| \neq 0$, we can represent $\mathbf{S}$ in the form

$$\mathbf{S} = \sum_{i=1}^{k} \mathbf{U}_j \mathbf{U}_j', \qquad \mathbf{U}_j \sim N_p(\mathbf{0}, \mathbf{\Sigma}), \qquad j = 1, \ldots, k, \tag{8b.2.6}$$

and the rank of (U_{ij}), $i = 1, \ldots, p-1$; $j = 1, \ldots, k$ is therefore $(p-1)$ with probability 1, if $k > (p-1)$. The result (a) is proved if $R_0^2 = 1/S^{pp}$ which is already established in the univariate theory (see 4a.5.2).

Instead of U_{pi}, we could have chosen any component of $\mathbf{U}_i$ and considered its conditional distribution. Hence the distribution of σ^{ii}/S^{ii} is the same for all i.

The result (b) follows from (a) if we consider $\mathbf{BSB}'$ where $\mathbf{B}$ is an orthogonal matrix with its first row proportional to $\mathbf{L}$. We have by the result (vi), $\mathbf{BSB}' \sim W_p(k, \mathbf{B\Sigma B}')$. Furthermore,

$$(\mathbf{BSB}')^{-1} = \mathbf{BS}^{-1}\mathbf{B}', \qquad (\mathbf{B\Sigma B}')^{-1} = \mathbf{B\Sigma}^{-1}\mathbf{B}'.$$

The first diagonal element in $\mathbf{BS}^{-1}\mathbf{B}' \propto \mathbf{L}'\mathbf{S}^{-1}\mathbf{L}$ and the first element in $\mathbf{B\Sigma}^{-1}\mathbf{B}' \propto \mathbf{L}'\mathbf{\Sigma}^{-1}\mathbf{L}$ with the same constant of proportionality. Hence applying result (a), $(\mathbf{L}'\mathbf{\Sigma}^{-1}\mathbf{L}/\mathbf{L}'\mathbf{S}^{-1}\mathbf{L}) \sim \chi^2(\overline{k - p - 1})$.

(ix) *Let* $\mathbf{S} \sim W_p(k, \mathbf{\Sigma})$ *and consider the partition of* $\mathbf{S}$

$$\mathbf{S} = \begin{pmatrix} \mathbf{S}_{11} & \mathbf{S}_{12} \\ \mathbf{S}_{21} & \mathbf{S}_{22} \end{pmatrix} \tag{8b.2.7}$$

where $\mathbf{S}_{11}, \mathbf{S}_{12}, \mathbf{S}_{22}$ *are* $(r \times r)$, $(r \times s)$, *and* $(s \times s)$ *matrices, with* $(r + s) = p$. *Then*

$$\mathbf{S}_{22} - \mathbf{S}_{21}\mathbf{S}_{11}^{-1}\mathbf{S}_{12} \sim W_s(k - r, \mathbf{\Sigma}_{22} - \mathbf{\Sigma}_{21}\mathbf{\Sigma}_{11}^{-1}\mathbf{\Sigma}_{12})$$

where $\mathbf{\Sigma}_{ij}$ *constitute a partition of* $\mathbf{\Sigma}$ *similar to* (8b.2.7).

Since $\mathbf{S} \sim W_p(k, \mathbf{\Sigma})$

$$\mathbf{S} = \sum_{1}^{k} \mathbf{U}_i \mathbf{U}_i', \qquad \mathbf{U}_i \sim N_p(\mathbf{0}, \mathbf{\Sigma}).$$

Consider a partition of $\mathbf{U}_i' = (\mathbf{U}_{1i}' \mid \mathbf{U}_{2i}')$ and the $(r+1)$ dimensional variable $(\mathbf{U}_{1i}' \mid \mathbf{L}'\mathbf{U}_{2i})$ with the Wishart matrix

$$\begin{pmatrix} \mathbf{S}_{11} & \mathbf{S}_{12}\mathbf{L} \\ \mathbf{L}'\mathbf{S}_{21} & \mathbf{L}'\mathbf{S}_{22}\mathbf{L} \end{pmatrix}$$

where $\mathbf{L}$ is a fixed column vector of s elements.

Using the result (a) of (viii), we see that

$$\begin{vmatrix} \mathbf{S}_{11} & \mathbf{S}_{12}\mathbf{L} \\ \mathbf{L}'\mathbf{S}_{21} & \mathbf{L}'\mathbf{S}_{22}\mathbf{L} \end{vmatrix} \div |\mathbf{S}_{11}| \sim c\chi^2(k-r).$$

The left-hand side expression is

$$\mathbf{L}'\mathbf{S}_{22}\mathbf{L} - \mathbf{L}'\mathbf{S}_{21}\mathbf{S}_{11}{}^{-1}\mathbf{S}_{12}\mathbf{L} = \mathbf{L}'(\mathbf{S}_{22} - \mathbf{S}_{21}\mathbf{S}_{11}{}^{-1}\mathbf{S}_{12})\mathbf{L}, \qquad (8b.2.8)$$

and similarly $c = \mathbf{L}'(\mathbf{\Sigma}_{22} - \mathbf{\Sigma}_{21}\mathbf{\Sigma}_{11}{}^{-1}\mathbf{\Sigma}_{12})\mathbf{L}$. By dropping $\mathbf{L}$ from (8b.2.8), the distribution of

$$\mathbf{S}_{22} - \mathbf{S}_{12}\mathbf{S}_{11}{}^{-1}\mathbf{S}_{12} \sim W_s(k-r, \mathbf{\Sigma}_{22} - \mathbf{\Sigma}_{21}\mathbf{\Sigma}_{11}{}^{-1}\mathbf{\Sigma}_{12}).$$

(x) *Let* $\mathbf{S} \sim W_p(k, \mathbf{\Sigma})$ *and* $|\mathbf{\Sigma}| \neq 0$. *Then* $|\mathbf{S}|/|\mathbf{\Sigma}|$ *is distributed as the product of* p *independent central* χ^2 *variables with D.F.,* $k-p+1, \ldots, k-1, k$.

Observe the decomposition (writing $|\mathbf{S}|_r = |(S_{ij})|$, $i, j = 1, \ldots, r$)

$$\frac{|\mathbf{S}|}{|\mathbf{\Sigma}|} = \frac{|\mathbf{S}|_p}{|\mathbf{S}|_{p-1}} \cdot \frac{|\mathbf{\Sigma}|_{p-1}}{|\mathbf{\Sigma}|_p} \times \frac{|\mathbf{S}|_{p-1}}{|\mathbf{S}|_{p-2}} \cdot \frac{|\mathbf{\Sigma}|_{p-2}}{|\mathbf{\Sigma}|_{p-1}} \times \cdots \times \frac{|\mathbf{S}|_1}{|\mathbf{\Sigma}|_1}$$

where each factor, by an application of (a) of (viii) is an independent χ^2 with appropriate D.F.

(xi) *Let* $\mathbf{S}_1 \sim W_p(k_1, \mathbf{\Sigma})$ *and* $\mathbf{S}_2 \sim W_p(k_2, \mathbf{\Sigma})$ *be independent. Then if* $k_1 \geqslant p$, $\Lambda = |\mathbf{S}_1|/|\mathbf{S}_1 + \mathbf{S}_2|$ *is distributed as the product of* p *independent beta variables, with parameters*

$$\left(\frac{k_1-p+1}{2}, \frac{k_2}{2}\right), \left(\frac{k_1-p+2}{2}, \frac{k_2}{2}\right), \ldots, \left(\frac{k_1}{2}, \frac{k_2}{2}\right),$$

In the special case $k_2 = 1$, *the product of beta variables has the same distribution (applying* iii, **3a.3**) *as a single beta variable with the parameters*

$$\left(\frac{k_1-p+1}{2}, \frac{p}{2}\right).$$

Consider independent vector variables $\mathbf{U}_1, \ldots, \mathbf{U}_{k_1}, \mathbf{U}_{k_1+1}, \ldots, \mathbf{U}_{k_1+k_2}$ such that

$$\mathbf{S}_1 = \sum_1^{k_1} \mathbf{U}_i \mathbf{U}_i' \qquad \text{and} \qquad \mathbf{S}_2 = \sum_{k_1+1}^{k_1+k_2} \mathbf{U}_i \mathbf{U}_i'.$$

The conditional distribution of U_{pi} given $U_{1i}, \ldots, U_{p-1,i}$ is

$$N_1\left(\beta_1 U_{1i} + \cdots + \beta_{p-1} U_{p-1,i}, \frac{1}{\sigma^{pp}}\right)$$

$$i = 1, \ldots, k_1 + k_2.$$

We now use the third fundamental theorem of the least square theory (see **3b.5**) which states that

$$B_p = \frac{\min_{\beta} \sum_{1}^{k_1} (U_{pi} - \beta_1 U_{1i} - \cdots - \beta_{p-1} U_{p-1,i})^2}{\min_{\beta} \sum_{1}^{k_1+k_2} (U_{pi} - \beta_1 U_{1i} - \cdots - \beta_{p-1} U_{p-1,i})^2} \tag{8b.2.9}$$

is distributed as

$$B\left(\frac{k_1 - p + 1}{2}, \frac{k_2}{2}\right).$$

The numerator of (8b.2.9) is $|\mathbf{S}_1|_p \div |\mathbf{S}_1|_{p-1}$ and the denominator $|\mathbf{S}_1 + \mathbf{S}_2|_p \div |\mathbf{S}_1 + \mathbf{S}_2|_{p-1}$ if we use formula (8b.2.5). We, then use the decomposition of Λ

$$\frac{|\mathbf{S}_1|}{|\mathbf{S}_1 + \mathbf{S}_2|} = \left[\frac{|\mathbf{S}_1|_p}{|\mathbf{S}_1|_{p-1}} \div \frac{|\mathbf{S}_1 + \mathbf{S}_2|_p}{|\mathbf{S}_1 + \mathbf{S}_2|_{p-1}}\right] \times \left[\frac{|\mathbf{S}_1|_{p-1}}{|\mathbf{S}_1|_{p-2}} \div \frac{|\mathbf{S}_1 + \mathbf{S}_2|_{p-1}}{|\mathbf{S}_1 + \mathbf{S}_2|_{p-2}}\right]$$
$$\cdots \times [|\mathbf{S}_1|_1 \div |\mathbf{S}_1 + \mathbf{S}_2|_1]$$
$$= B_p \times B_{p-1} \times \cdots \times B_1, \tag{8b.2.10}$$

where $B_p, B_{p-1}, \ldots, B_1$ are all independent. Since the distribution of (8b.2.9) is independent of (U_{ij}), $i \neq p$, so that B_p is independent of the rest of B_i, $i < p$. Similarly, B_{p-1} is independent of B_i, $i < p - 1$, and so on. The parameters of the distribution of B_j are substituting j for p in (8b.2.9)

$$\left(\frac{k_1 - j + 1}{2}, \frac{k_2}{2}\right).$$

We denote the distribution of the Λ statistic by $\Lambda(p, k_1, k_2)$.

(xii) HOTELLING'S DISTRIBUTION. Let $S \sim W_p(k, \Sigma)$ and $d \sim N_p(\delta, c^{-1}\Sigma)$ be independent. Hotelling's generalized T^2 statistic is defined by

$$T^2 = ck\mathbf{d}'\mathbf{S}^{-1}\mathbf{d}$$
$$= \frac{k\mathbf{d}'\mathbf{S}^{-1}\mathbf{d}}{\mathbf{d}'\boldsymbol{\Sigma}^{-1}\mathbf{d}} c\mathbf{d}'\boldsymbol{\Sigma}^{-1}\mathbf{d}.$$

For given **d**,

$$\frac{\mathbf{d}'\boldsymbol{\Sigma}^{-1}\mathbf{d}}{\mathbf{d}'\mathbf{S}^{-1}\mathbf{d}} \sim \chi^2(k - p + 1) \tag{8b.2.11}$$

if we use result (b) of (viii). The distribution (8b.2.11) does not involve **d**. Therefore the ratio $\mathbf{d}'\boldsymbol{\Sigma}^{-1}\mathbf{d}/\mathbf{d}'\mathbf{S}^{-1}\mathbf{d}$ is distributed independently of **d**. Since $\mathbf{d} \sim N_p(\boldsymbol{\delta}, c^{-1}\boldsymbol{\Sigma})$,

$$c\mathbf{d}'\boldsymbol{\Sigma}^{-1}\mathbf{d} \sim \chi^2(p, c\tau^2) \tag{8b.2.12}$$

where $\tau^2 = \boldsymbol{\delta}'\boldsymbol{\Sigma}^{-1}\boldsymbol{\delta}$. The distributions (8b.2.11) and (8b.2.12) are independent. Thus the statistic T^2/k is the ratio of a noncentral $\chi^2(p, c\tau^2)$ to an independent central $\chi^2(k-p+1)$. Hence

$$\frac{k-p+1}{p}\frac{T^2}{k} \sim F(p, k-p+1, c\tau^2) \tag{8b.2.13}$$

which is a noncentral F distribution defined in Example **17.1** of Chapter 3.

If $\boldsymbol{\delta} = \mathbf{0}$, both the χ^2's are central and hence

$$\frac{k-p+1}{p}\frac{T^2}{k} \sim F(p, k-p+1) \tag{8b.2.14}$$

which is a central F distribution (3a.2.13).

The representation of T^2 as the ratio of independent χ^2's leading to an elegant derivation of its distribution is due to Wijsman (1957). See also Bowker (1960). For earlier literature see Bose and Roy (1938), Hotelling (1931), Hsu (1938, 1947), K. S. Rao (1951) and Rao (1946f, 1949b).

It may be seen that

$$\left(1 + \frac{T^2}{k}\right)^{-1} = \frac{|\mathbf{S}|}{|\mathbf{S} + c\mathbf{dd}'|} \tag{8b.2.15}$$

where $c\mathbf{dd}' \sim W_p(1, \boldsymbol{\Sigma})$ when $\boldsymbol{\delta} = \mathbf{0}$. Thus Hotelling's T^2, after a monotonic transformation is a special case of the statistic $\Lambda = |\mathbf{S}_1|/|\mathbf{S}_1 + \mathbf{S}_2|$ considered in (xi) with $k_2 = 1$. From the relationship (8b.2.15), and by using the distribution (8b.2.14) of T^2 we find that

$$\Lambda = \frac{|\mathbf{S}|}{|\mathbf{S} + c\mathbf{dd}'|} \sim B\left(\frac{k-p+1}{2}, \frac{p}{2}\right)$$

which has been independently derived in (xi). Then using the relation (8b.2.15) we could have obtained the distribution (8b.2.14) of T^2. Thus we have two simple proofs for the Hotelling's distribution.

(xiii) ROOTS OF A DETERMINANTAL EQUATION. Fisher introduced some tests of hypotheses depending on the roots of the determinantal equation

$$|\mathbf{S}_1 - \lambda(\mathbf{S}_1 + \mathbf{S}_2)| = 0 \tag{8b.2.16}$$

where $\mathbf{S}_1 \sim W_p(k_1, \boldsymbol{\Sigma})$ and $\mathbf{S}_2 \sim W_p(k_2, \boldsymbol{\Sigma})$ are independent. We shall not derive the exact distribution of the roots but shall observe that the density function involves only p, k_1, and k_2 and not the unknown $\boldsymbol{\Sigma}$. For a derivation of the density function the reader is referred to the original papers by Fisher (1939), Girshick (1939), Hsu (1939), Roy (1939), and an interesting proof by James (1954). Detailed proofs are given in books by Anderson (1958), Roy

(1957), and Wilks (1962). We do not need the density function in our study. The reader is referred to Rao (1972a) for a review of the work on the roots of (8b.2.16) by Bagai, Constantine, Khatri, Krishnaih, Mathai, Pillai and others.

8c ANALYSIS OF DISPERSION

8c.1 The Gauss-Markoff Setup for Multiple Measurements

The least squares theory under the Gauss-Markoff setup, as developed in Chapter 4, is applicable in situations where a *single measurement* is taken (i.e., the value of a particular characteristic is ascertained) on an experimental unit or individual. The expectation of the measurement (i.e., of a one-dimensional r.v.), is considered to be a linear function of unknown parameters, the compounding coefficients being specified by a particular design of experiment adopted. If a second measurement is taken (or another characteristic is examined) on the same chosen individual, it is natural to consider its expectation as the same linear function as in the case of the first measurement, but of a different set of unknown parameters, and so on. Thus the *parameters are specific to the characteristic* examined whereas the *coefficients of the linear function are specific to the design of experiment* adopted.

Thus, if the expectation of yield of a plant receiving a particular treatment j in an experimental block i is written $\beta_{i1} + \tau_{j1}$ representing the effect of the ith block and jth treatment, the expectation of another character such as the height of a plant may be written as the sum of two other parameters $\beta_{i2} + \tau_{j2}$ representing similar effects.

Let p characters be observed on each of n independent experimental units or individuals. The observations can be written

Individual	Character (1)	(2)	(p)

$$\begin{array}{c} (1) \\ \cdot \\ (n) \end{array} \quad \begin{pmatrix} U_{11} & U_{21} & U_{p1} \\ \cdot & \cdot & \cdot \\ U_{1n} & U_{2n} & U_{pn} \end{pmatrix} = \mathscr{U} \tag{8c.1.1}$$

Let $\mathbf{Y}_i$ denote the ith column vector in $\mathscr{U}$. Then the model we consider is

$$\begin{aligned} E(\mathbf{Y}_i) &= \mathbf{X}\boldsymbol{\beta}_i, \qquad i = 1, \ldots, p, \\ \operatorname{Cov}(\mathbf{Y}_i, \mathbf{Y}_j) &= \sigma_{ij}\mathbf{I}, \qquad i, j = 1, \ldots, p. \end{aligned} \tag{8c.1.2}$$

Or, in other words, the model is Gauss-Markoff $(\mathbf{Y}_i, \mathbf{X}\boldsymbol{\beta}_i, \sigma_{ii}\mathbf{I})$ for each $\mathbf{Y}_i$ with a specific unknown vector parameter $\boldsymbol{\beta}_i$ for each i but the same $n \times m$

design matrix $\mathbf{X}$ for all i. The variance of the ith character is σ_{ii}. The dependence of the characters examined is expressed by the relation

$$\text{cov}(\mathbf{Y}_i, \mathbf{Y}_j) = \sigma_{ij}\mathbf{I}, \tag{8c.1.3}$$

where σ_{ij} is the covariance between the ith and jth characters of an individual.

Let $\mathbf{B}$ be an $m \times p$ matrix with $\boldsymbol{\beta}_1, \ldots, \boldsymbol{\beta}_p$ as its columns and $\boldsymbol{\Sigma} = (\sigma_{ij})$. Then the multivariate model (8c.1.2) with (8c.1.3) can be simply expressed as

$$(\mathscr{U}, \mathbf{XB}, \boldsymbol{\Sigma} \otimes \mathbf{I}) \tag{8c.1.4}$$

where the Kronecker product matrix $\boldsymbol{\Sigma} \otimes \mathbf{I}$ is to be interpreted as the covariance matrix of the np-vector variable $\mathbf{Y}$ obtained by writing the columns of $\mathscr{U}$ one below the other, i.e., $\mathbf{Y}' = (\mathbf{Y}_1' \mid \cdots \mid \mathbf{Y}_p')$.

We can also write the model (8c.1.4) as a univariate Gauss-Markoff model involving $\mathbf{Y}$, its expectation and dispersion matrix. Let $\boldsymbol{\beta}$ be the mp-vector defined by $\boldsymbol{\beta}' = (\boldsymbol{\beta}_1' \mid \boldsymbol{\beta}_2' \mid \cdots \mid \boldsymbol{\beta}_p')$. Then it is readily seen that

$$\begin{aligned} E(\mathbf{Y}) &= (\mathbf{I} \otimes \mathbf{X})\boldsymbol{\beta} \\ D(\mathbf{Y}) &= \boldsymbol{\Sigma} \otimes \mathbf{I} \end{aligned} \tag{8c.1.5}$$

so that the multivariate linear model may be written as the triplet

$$(\mathbf{Y}, (\mathbf{I} \otimes \mathbf{X})\boldsymbol{\beta}, \boldsymbol{\Sigma} \otimes \mathbf{I}) \tag{8c.1.6}$$

of the Gauss-Markoff model of Chapter 4. One may then use the results on estimation in linear models developed in Chapter 4. However, because of the special covariance structure, it is better to exhibit the model as (8c.1.4) or as (8c.1.2), i.e., as Gauss-Markoff for each individual character supplemented by (8c.1.3) to show the correlations between characters. This has great advantage in showing the intimate connection between the univariate and multivariate theories (see **8c.2–8c.5**).

8c.2 Estimation of Parameters

As in the univariate Gauss-Markoff setup we shall consider the estimation of unknown parameters, utilizing only the structure (8c.1.2), (8c.1.3) of the first- and second-order moments of the variables and not assuming any particular distribution.

Let us first consider the estimation of a parametric function of the type $\mathbf{P}'\boldsymbol{\beta}_i$ which involves only the parameters specific to the ith character, by a linear function of *all the observations* such that the estimator is (a) unbiased and (b) has minimum variance. A linear function of all the observations can be written as

$$\mathbf{A}_1'\mathbf{Y}_1 + \cdots + \mathbf{A}_p'\mathbf{Y}_p. \tag{8c.2.1}$$

Using criterion (a) of unbiasedness, we see that

$$E(\mathbf{A}_1'\mathbf{Y}_1) + \cdots + E(\mathbf{A}_p'\mathbf{Y}_p) = \mathbf{A}_1'\mathbf{X}\boldsymbol{\beta}_1 + \cdots + \mathbf{A}_p'\mathbf{X}\boldsymbol{\beta}_p = \mathbf{P}'\boldsymbol{\beta}_i$$
$$\Rightarrow \mathbf{A}_i'\mathbf{X} = \mathbf{P}', \mathbf{A}_j'\mathbf{X} = 0, j \neq i. \quad (8c.2.2)$$

Since $\mathbf{A}_i'\mathbf{X} = \mathbf{P}'$, there exists a vector $\boldsymbol{\lambda}$ such that $\mathbf{P} = \mathbf{X}'\mathbf{X}\boldsymbol{\lambda}$. Consider the linear function $\boldsymbol{\lambda}'\mathbf{X}'\mathbf{Y}_i$ [which we know (see 4a.2.3) is the least squares estimator of $\mathbf{P}'\boldsymbol{\beta}_i$ utilizing only $\mathbf{Y}_i$, with the setup $E(\mathbf{Y}_i) = \mathbf{X}\boldsymbol{\beta}_i$]. The estimator is unbiased, since

$$E(\boldsymbol{\lambda}'\mathbf{X}'\mathbf{Y}_i) = \boldsymbol{\lambda}'\mathbf{X}'\mathbf{X}\boldsymbol{\beta}_i = \mathbf{P}'\boldsymbol{\beta}_i. \quad (8c.2.3)$$

Furthermore,

$$V(\textstyle\sum \mathbf{A}_j'\mathbf{Y}_j) = V(\sum \mathbf{A}_j'\mathbf{Y}_j - \boldsymbol{\lambda}'\mathbf{X}'\mathbf{Y}_i) + V(\boldsymbol{\lambda}'\mathbf{X}'\mathbf{Y}_i)$$
$$+ 2\,\mathrm{cov}(\boldsymbol{\lambda}'\mathbf{X}'\mathbf{Y}_i, \textstyle\sum \mathbf{A}_j'\mathbf{Y}_j - \boldsymbol{\lambda}'\mathbf{X}'\mathbf{Y}_i). \quad (8c.2.4)$$

But

$$\mathrm{cov}\big(\boldsymbol{\lambda}'\mathbf{X}'\mathbf{Y}_i, \textstyle\sum \mathbf{A}_j'\mathbf{Y}_j - \boldsymbol{\lambda}'\mathbf{X}'\mathbf{Y}_i\big) = \sum_j \boldsymbol{\lambda}'\mathbf{X}'\,\mathrm{cov}(\mathbf{Y}_i, \mathbf{Y}_j)\mathbf{A}_j - \boldsymbol{\lambda}'\mathbf{X}'\,\mathrm{cov}(\mathbf{Y}_i, \mathbf{Y}_i)\mathbf{X}\boldsymbol{\lambda}$$
$$= \textstyle\sum \boldsymbol{\lambda}'\mathbf{X}'\mathbf{A}_j\sigma_{ij} - \boldsymbol{\lambda}'\mathbf{X}'\mathbf{X}\boldsymbol{\lambda}\sigma_{ii}$$
$$= \boldsymbol{\lambda}'\mathbf{X}'\mathbf{A}_i\sigma_{ii} - \boldsymbol{\lambda}'\mathbf{X}'\mathbf{X}\boldsymbol{\lambda}\sigma_{ii} = (\boldsymbol{\lambda}'\mathbf{P} - \boldsymbol{\lambda}'\mathbf{P})\sigma_{ii} = 0$$

if we use (8c.2.2). Hence from (8c.2.4), we see that

$$V(\textstyle\sum \mathbf{A}_j'\mathbf{Y}_j) \geqslant V(\boldsymbol{\lambda}'\mathbf{X}'\mathbf{Y}_i)$$

that is, $\boldsymbol{\lambda}'\mathbf{X}'\mathbf{Y}_i$ is the minimum variance unbiased estimator (m.v.u.e.) of $\mathbf{P}'\boldsymbol{\beta}_i$. Thus the desired estimator of $\mathbf{P}'\boldsymbol{\beta}_i$ is the same as the least squares estimator based on $\mathbf{Y}_i$—the observations on the ith character only. Let $\hat{\boldsymbol{\beta}}_i$ be a solution of the normal equations (considering only $\mathbf{Y}_i$, $\boldsymbol{\beta}_i$)

$$\mathbf{X}'\mathbf{X}\boldsymbol{\beta}_i = \mathbf{X}'\mathbf{Y}_i. \quad (8c.2.5)$$

Then the least squares estimator of $\mathbf{P}'\boldsymbol{\beta}_i$ can be written as $\mathbf{P}'\hat{\boldsymbol{\beta}}_i$. The variances and covariances of estimators of $\mathbf{P}'\boldsymbol{\beta}_i$ and $\mathbf{Q}'\boldsymbol{\beta}_i$ are same as in the least squares theory. The variances and covariances of $\mathbf{P}'\hat{\boldsymbol{\beta}}_i$ and $\mathbf{Q}'\hat{\boldsymbol{\beta}}_i$ are then of the form

$$V(\mathbf{P}'\hat{\boldsymbol{\beta}}_i) = a\sigma_{ii}, \qquad V(\mathbf{Q}'\hat{\boldsymbol{\beta}}_i) = b\sigma_{ii}$$
$$\mathrm{cov}(\mathbf{P}'\hat{\boldsymbol{\beta}}_i, \mathbf{Q}'\hat{\boldsymbol{\beta}}_i) = c\sigma_{ii}, \quad (8c.2.6)$$

where the coefficients, a, b, c are *independent of i* and depend only on $\mathbf{X}'\mathbf{X}$, $\mathbf{P}$, $\mathbf{Q}$ (see **4a.3**). For instance, if $(\mathbf{X}'\mathbf{X})^-$ is a g-inverse of $\mathbf{X}'\mathbf{X}$, then $a = \mathbf{P}'(\mathbf{X}'\mathbf{X})^-\mathbf{P}$, $b = \mathbf{Q}'(\mathbf{X}'\mathbf{X})^-\mathbf{Q}$, and $c = \mathbf{P}'(\mathbf{X}'\mathbf{X})^-\mathbf{Q}$.

Let $\hat{\boldsymbol\beta}_1, \ldots, \hat{\boldsymbol\beta}_p$ be the solutions of the corresponding normal equations (8c.2.5) for $i = 1, \ldots, p$, $\mathbf{P}_i \in \mathcal{M}(\mathbf{x}')$, $i = 1, \ldots, p$. Then it follows that

$$\mathbf{P}_1'\hat{\boldsymbol\beta}_1 + \cdots + \mathbf{P}_p'\hat{\boldsymbol\beta}_p \tag{8c.2.7}$$

is the m.v.u.e. of

$$\mathbf{P}_1'\boldsymbol\beta_1 + \cdots + \mathbf{P}_p'\boldsymbol\beta_p,$$

since the sum of m.v. estimators is also an m.v. estimator [see (i), **5a.2**]. From the structure of the expressions (8c.2.6), it easily follows that

$$\operatorname{cov}(\mathbf{P}'\hat{\boldsymbol\beta}_i, \mathbf{P}'\hat{\boldsymbol\beta}_j) = a\sigma_{ij}$$
$$\operatorname{cov}(\mathbf{Q}'\hat{\boldsymbol\beta}_i, \mathbf{Q}'\hat{\boldsymbol\beta}_j) = b\sigma_{ij}$$
$$\operatorname{cov}(\mathbf{P}'\hat{\boldsymbol\beta}_i, \mathbf{Q}'\hat{\boldsymbol\beta}_j) = c\sigma_{ij},$$

where a, b, c are the same as in (8c.2.6) depending only on $\mathbf{P}$ and $\mathbf{Q}$, which enable us to write the variance of (8c.2.7). We need only to know the expressions for the variances and covariances of $\mathbf{P}_1'\hat{\boldsymbol\beta}_i, \ldots, \mathbf{P}_p'\hat{\boldsymbol\beta}_i$ for any i, which can be determined from the univariate least squares theory. The variance of (8c.2.7) is seen to be tr $\mathbf{P}'(\mathbf{X}'\mathbf{X})^-\mathbf{P}\boldsymbol\Sigma$ where $\mathbf{P} = (\mathbf{P}_1, \vdots \cdots \vdots, \mathbf{P}_p)$.

The estimators of σ_{ii} and σ_{ij} do not present any difficulty. From the univariate theory of least squares, we know that an unbiased estimator of σ_{ii} is $R_0(i, i)/(n - r)$ where

$$\begin{aligned} R_0(i, i) &= \min_{\boldsymbol\beta_i} (\mathbf{Y}_i - \mathbf{X}\boldsymbol\beta_i)'(\mathbf{Y}_i - \mathbf{X}\boldsymbol\beta_i) \\ &= (\mathbf{Y}_i - \mathbf{X}\hat{\boldsymbol\beta}_i)'(\mathbf{Y}_i - \mathbf{X}\hat{\boldsymbol\beta}_i) = \mathbf{Y}_i'\mathbf{Y}_i - \mathbf{Y}_i'\mathbf{X}\hat{\boldsymbol\beta}_i. \end{aligned} \tag{8c.2.8}$$

Now consider the vector random variable $\mathbf{Y}_i + \mathbf{Y}_j = \mathbf{Y}$ with its expectation $\mathbf{X}(\boldsymbol\beta_i + \boldsymbol\beta_j) = \mathbf{X}\boldsymbol\beta$ and $D(\mathbf{Y}) = \sigma^2\mathbf{I}$, where $\sigma^2 = \sigma_{ii} + 2\sigma_{ij} + \sigma_{jj}$. The normal equation for $\boldsymbol\beta$ is

$$\mathbf{X}'\mathbf{X}\boldsymbol\beta = \mathbf{X}'\mathbf{Y}$$

with a solution $\hat{\boldsymbol\beta} = \hat{\boldsymbol\beta}_i + \hat{\boldsymbol\beta}_j$, where $\hat{\boldsymbol\beta}_i$ and $\hat{\boldsymbol\beta}_j$ are solutions of (8c.2.5) for i and j. An unbiased estimator of σ^2 is $R_0{}^2/(n - r)$, where

$$\begin{aligned} R_0{}^2 &= (\mathbf{Y} - \mathbf{X}\hat{\boldsymbol\beta})'(\mathbf{Y} - \mathbf{X}\hat{\boldsymbol\beta}) = (\mathbf{Y} - \mathbf{X}\hat{\boldsymbol\beta}_i - \mathbf{X}\hat{\boldsymbol\beta}_j)'(\mathbf{Y} - \mathbf{X}\hat{\boldsymbol\beta}_i - \mathbf{X}\hat{\boldsymbol\beta}_j) \\ &= (\mathbf{Y}_i - \mathbf{X}\hat{\boldsymbol\beta}_i)'(\mathbf{Y}_i - \mathbf{X}\hat{\boldsymbol\beta}_i) + (\mathbf{Y}_j - \mathbf{X}\hat{\boldsymbol\beta}_j)'(\mathbf{Y}_j - \mathbf{X}\hat{\boldsymbol\beta}_j) \\ &\qquad + 2(\mathbf{Y}_i - \mathbf{X}\hat{\boldsymbol\beta}_i)'(\mathbf{Y}_j - \mathbf{X}\hat{\boldsymbol\beta}_j) \\ &= R_0(i, i) \qquad + R_0(j, j) \\ &\qquad + 2R_0(i, j) \end{aligned}$$

where $R_0(i, j) = (\mathbf{Y}_i - \mathbf{X}\hat{\boldsymbol{\beta}}_i)'(\mathbf{Y}_j - \mathbf{X}\hat{\boldsymbol{\beta}}_j)$. Now

$$\begin{aligned} E(R_0{}^2) &= (n-r)\sigma^2 = (n-r)(\sigma_{ii} + 2\sigma_{ij} + \sigma_{jj}), \text{ (by the univariate theory)} \\ &= E[R_0(i, i)] + E[R_0(j, j)] + 2E[R_0(i, j)] \\ &= (\sigma_{ii} + \sigma_{jj})(n-r) + 2E[R_0(i, j)] \\ &\Rightarrow E[R_0(i, j) \div (n-r)] = \sigma_{ij}. \end{aligned} \tag{8c.2.9}$$

We thus have unbiased estimators of all the elements of $\boldsymbol{\Sigma}$. The expression

$$\begin{aligned} R_0(i, j) &= (\mathbf{Y}_i - \mathbf{X}\hat{\boldsymbol{\beta}}_i)'(\mathbf{Y}_j - \mathbf{X}\hat{\boldsymbol{\beta}}_j) \\ &= \mathbf{Y}_i'\mathbf{Y}_j - \mathbf{Y}_i'\mathbf{X}\hat{\boldsymbol{\beta}}_j = \mathbf{Y}_i'\mathbf{Y}_j - \mathbf{Y}_j'\mathbf{X}\hat{\boldsymbol{\beta}}_i \end{aligned} \tag{8c.2.10}$$

is called the *residual sum of products* just as $R_0(i, i)$ is the *residual sum of squares*. The entire matrix

$$\mathbf{R}_0 = \begin{pmatrix} R_0(1, 1) & \cdots & R_0(1, p) \\ \cdot & \cdots & \cdot \\ R_0(p, 1) & \cdots & R_0(p, p) \end{pmatrix} \tag{8c.2.11}$$

may be called the residual sum of squares and products matrix (residual S.S.P. or residual S.P. matrix) on $(n - r)$ D.F. as a generalization of the residual sum of squares in the univariate least square theory.

A solution of the normal equation $\mathbf{X}'\mathbf{X}\boldsymbol{\beta}_j = \mathbf{X}'\mathbf{Y}_j$ can be written $\boldsymbol{\beta}_j = \mathbf{C}\mathbf{X}'\mathbf{Y}_j$ where $\mathbf{C}$ is a generalized inverse of $\mathbf{X}'\mathbf{X}$. Then $R_0(i, j) = \mathbf{Y}_i'\mathbf{Y}_j - \mathbf{Y}_i'\mathbf{X}\mathbf{C}\mathbf{X}'\mathbf{Y}_j = \mathbf{Y}_i'(\mathbf{I} - \mathbf{X}\mathbf{C}\mathbf{X}')\mathbf{Y}_j$. Hence the S.P. matrix $\mathbf{R}_0$ can be written explicitly as

$$\mathbf{R}_0 = \mathscr{U}'(\mathbf{I} - \mathbf{X}\mathbf{C}\mathbf{X}')\mathscr{U} = \mathscr{U}'[\mathbf{I} - \mathbf{X}(\mathbf{X}'\mathbf{X})^{-}\mathbf{X}']\mathscr{U}$$

in terms of the matrix of observations $\mathscr{U}'$ as in (8c.1.1) and the design matrix $\mathbf{X}$. If $R(\mathbf{X}) = m$, then $(\mathbf{X}'\mathbf{X})^{-} = (\mathbf{X}'\mathbf{X})^{-1}$ the true inverse.

The estimation of parameters of the model (8c.1.2, 8c.1.3) is thus complete without assuming any particular distribution of the random variables involved. But for the purpose of testing linear hypotheses we shall assume a p-variate normal distribution for each $\mathbf{U}_i$.

8c.3 Tests of Linear Hypotheses, Analysis of Dispersion (A.D.)

Consider the problem of testing simultaneously the sets of linear hypotheses

$$\mathbf{H}'\boldsymbol{\beta}_i = \boldsymbol{\xi}_i, \qquad i = 1, \ldots, p, \tag{8c.3.1}$$

where the matrix $\mathbf{H}'$, of order $(q \times m)$ and rank q, and vectors $\boldsymbol{\xi}_i$ are given.

Let us reduce the problem to the univariate case by considering the hypothesis $\mathbf{H}'\boldsymbol{\beta} = \boldsymbol{\xi}$, where $\boldsymbol{\beta} = L_1\boldsymbol{\beta}_1 + \cdots + L_p\boldsymbol{\beta}_p$ and $\boldsymbol{\xi} = L_1\boldsymbol{\xi}_1 + \cdots + L_p\boldsymbol{\xi}_p$ and set of variables $\mathbf{Y} = L_1\mathbf{Y}_1 + \cdots + L_p\mathbf{Y}_p$. We then have a univariate

Gauss-Markoff setup, $E(\mathbf{Y}) = \mathbf{X}\boldsymbol{\beta}$, $D(\mathbf{Y}) = \sigma^2\mathbf{I}$, with the hypothesis $\mathbf{H}'\boldsymbol{\beta} = \boldsymbol{\xi}$ to be tested. The appropriate test criterion is based on the two minimum sums of squares

$$\begin{aligned} R_0^2 &= \min_{\boldsymbol{\beta}} (\mathbf{Y} - \mathbf{X}\boldsymbol{\beta})'(\mathbf{Y} - \mathbf{X}\boldsymbol{\beta}) \\ R_1^2 &= \min_{\mathbf{H}'\boldsymbol{\beta}=\boldsymbol{\xi}} (\mathbf{Y} - \mathbf{X}\boldsymbol{\beta})'(\mathbf{Y} - \mathbf{X}\boldsymbol{\beta}). \end{aligned} \tag{8c.3.2}$$

Under the null hypothesis $\mathbf{H}'\boldsymbol{\beta} = \boldsymbol{\xi}$, $R_1^2 - R_0^2$, and R_0^2 are independently distributed as central χ^2's on q and $(n - r)$ D.F. respectively where $r = R(\mathbf{x})$.

Let $\hat{\boldsymbol{\beta}}_i$ be the value of $\boldsymbol{\beta}_i$ for which $(\mathbf{Y}_i - \mathbf{X}\boldsymbol{\beta}_i)'(\mathbf{Y}_i - \mathbf{X}\boldsymbol{\beta}_i)$ is a minimum and

$$\begin{aligned} R_0(i, j) &= (\mathbf{Y}_i - \mathbf{X}\hat{\boldsymbol{\beta}}_i)'(\mathbf{Y}_j - \mathbf{X}\hat{\boldsymbol{\beta}}_j) \\ &= \mathbf{Y}_i'\mathbf{Y}_j - \mathbf{Y}_i'\mathbf{X}\hat{\boldsymbol{\beta}}_j = \mathbf{Y}_i'\mathbf{Y}_j - \mathbf{Y}_j'\mathbf{X}\hat{\boldsymbol{\beta}}_i, \end{aligned} \tag{8c.3.3}$$

as defined earlier. It is easy to see that $\hat{\boldsymbol{\beta}} = L_1\hat{\boldsymbol{\beta}}_1 + \cdots + L_p\hat{\boldsymbol{\beta}}_p$ minimizes $(\mathbf{Y} - \mathbf{X}\boldsymbol{\beta})'(\mathbf{Y} - \mathbf{X}\boldsymbol{\beta})$ and

$$\begin{aligned} (\mathbf{Y} - \mathbf{X}\hat{\boldsymbol{\beta}})'(\mathbf{Y} - \mathbf{X}\hat{\boldsymbol{\beta}}) &= [\sum L_i(\mathbf{Y}_i - \mathbf{X}\hat{\boldsymbol{\beta}}_i)]'[\sum L_i(\mathbf{Y}_i - \mathbf{X}\hat{\boldsymbol{\beta}}_i)] \\ &= \sum\sum L_i L_j(\mathbf{Y}_i - \mathbf{X}\hat{\boldsymbol{\beta}}_i)'(\mathbf{Y}_j - \mathbf{X}\hat{\boldsymbol{\beta}}_j) \\ &= \sum\sum L_i L_j R_0(i, j) = \mathbf{L}'\mathbf{R}_0\mathbf{L}, \end{aligned} \tag{8c.3.4}$$

where $\mathbf{R}_0 = (R_0(i, j))$ is the residual S.P. matrix defined in (8c.2.11).

Similarly let $\boldsymbol{\beta}_i^*$ be the value of $\boldsymbol{\beta}_i$ for which $(\mathbf{Y}_i - \mathbf{X}\boldsymbol{\beta}_i)'(\mathbf{Y}_i - \mathbf{X}\boldsymbol{\beta}_i)$ is a minimum subject to the hypothesis $\mathbf{H}'\boldsymbol{\beta}_i = \boldsymbol{\xi}_i$ and

$$R_1(i, j) = (\mathbf{Y}_i - \mathbf{X}\boldsymbol{\beta}_i^*)'(\mathbf{Y}_i - \mathbf{X}\boldsymbol{\beta}_i^*), \qquad \mathbf{R}_1 = (R_1(i, j)).$$

Then as in (8c.3.4)

$$R_1^2 = \min_{\mathbf{H}'\boldsymbol{\beta}=\boldsymbol{\xi}} (\mathbf{Y} - \mathbf{X}\boldsymbol{\beta})'(\mathbf{Y} - \mathbf{X}\boldsymbol{\beta}) = \mathbf{L}'\mathbf{R}_1\mathbf{L} \tag{8c.3.5}$$

giving

$$R_1^2 - R_0^2 = \mathbf{L}'(\mathbf{R}_1 - \mathbf{R}_0)\mathbf{L}.$$

The matrix $\mathbf{R}_1 - \mathbf{R}_0$ may be called the S.P. matrix due to deviation from the hypothesis. We thus have two matrices $\mathbf{R}_0$ and $\mathbf{R}_1 - \mathbf{R}_0$, the S.P. matrices due to "residual" and "deviation from hypothesis" just as the residual and deviation from hypothesis sum of squares in the univariate case. Such a decomposition of the matrix $\mathbf{R}_1 = \mathbf{R}_0 + (\mathbf{R}_1 - \mathbf{R}_0)$, obtained as a generalization of analysis of variance, $R_1^2 = R_0^2 + (R_1^2 - R_0^2)$ is called analysis of dispersion (A.D.).

Joint Distribution of $\mathbf{R}_0$ and $(\mathbf{R}_1 - \mathbf{R}_0)$. If the null hypothesis (8c.3.1) is true, then for every $\mathbf{L}$ [with $\mathbf{R}_0^2$ and $\mathbf{R}_1^2$ as defined in (8c.3.2)]

$$R_0^2 \sim \sigma_L^2\chi^2(n - r), \qquad R_1^2 - R_0^2 \sim \sigma_L^2\chi^2(q)$$

and are independent. But $R_0^2 = \mathbf{L}'\mathbf{R}_0\mathbf{L}$ and $R_1^2 = \mathbf{L}'\mathbf{R}_1\mathbf{L}$. Hence (dropping $\mathbf{L}$) by applying the result [(iii), **8b.2**], the matrices $\mathbf{R}_0$ and $\mathbf{R}_1 - \mathbf{R}_0$ have the independent Wishart distributions:

$$\mathbf{R}_0 \sim W_p(n-r, \boldsymbol{\Sigma}), \qquad \mathbf{R}_1 - \mathbf{R}_0 \sim W_p(q, \boldsymbol{\Sigma}). \tag{8c.3.6}$$

If the null hypothesis (8c.3.1) is not true, then for each L

$$R_0^2 \sim \sigma_L^2\chi^2(n-r), \qquad R_1^2 - R_0^2 \sim \sigma_L^2\chi^2(q, \cdot)\text{(central or noncentral)}$$

and are independent (again using the univariate theory). Hence the matrices $\mathbf{R}_0$ and $\mathbf{R}_1 - \mathbf{R}_0$ have the independent Wishart distributions:

$$\mathbf{R}_0 \sim W_p(n-r, \boldsymbol{\Sigma}), \qquad \mathbf{R}_1 - \mathbf{R}_0 \sim W_p(q, \boldsymbol{\Sigma}, \cdot)\text{(noncentral)}.$$

The latter has a central distribution if and only if $R_1^2 - R_0^2$ has a central distribution for every $\mathbf{L}$. Departure from the null hypothesis (8c.3.1) may then be detected by comparing the matrices $\mathbf{R}_0$ and $\mathbf{R}_1 - \mathbf{R}_0$.

Test Criteria. If the null hypothesis (8c.3.1) is true, then for each $\mathbf{L}$, the hypothesis $\mathbf{H}'\boldsymbol{\beta} = \boldsymbol{\xi}$ is true. An appropriate test is the analysis of variance F test based on the least squares theory,

$$F = \frac{R_1^2 - R_0^2}{q} \div \frac{R_0^2}{n-r}.$$

where R_0^2, R_1^2 are as defined in (8c.3.2).

Instead of F, we may use the statistic

$$B = \frac{R_0^2}{R_0^2 + (R_1^2 - R_0^2)} = \frac{R_0^2}{R_1^2} = \frac{1}{1 + \dfrac{qF}{n-r}},$$

which is a monotonically decreasing function of F and which has the beta distribution,

$$B\left(\frac{q}{2}, \frac{n-r}{2}\right).$$

Small values of B would indicate significance. In terms of $\mathbf{L}$

$$B = \frac{\mathbf{L}'\mathbf{R}_0\mathbf{L}}{\mathbf{L}'\mathbf{R}_1\mathbf{L}}. \tag{8c.3.7}$$

Up to now $\mathbf{L}$ is arbitrary but fixed.

We have already observed that the effect of deviation from the hypothesis (8c.3.1) is to decrease the value of B (stochastically). Let us then choose $\mathbf{L}$

for which B has the smallest value in an attempt to discredit the hypothesis. The test criterion then reduces to

$$\lambda = \min_{\mathbf{L}} B = \min_{\mathbf{L}} \frac{\mathbf{L}'\mathbf{R}_0\mathbf{L}}{\mathbf{L}'\mathbf{R}_1\mathbf{L}},$$

or λ is the smallest root of the determinantal equation

$$|\mathbf{R}_0 - \lambda\mathbf{R}_1| = 0. \tag{8c.3.8}$$

To determine the extent to which deviations from the null hypothesis (8c.3.1) are reflected in the roots of the equation (8c.3.8), let us replace $\mathbf{R}_0$ and $\mathbf{R}_1$ by their expected values, which can again be found using the univariate formulas. In (4b.2.3) the expectations of $R_0{}^2$ and $R_1{}^2 - R_0{}^2$ have been found to be

$$E(R_0{}^2) = (n-r)\sigma_L{}^2 = E(\mathbf{L}'\mathbf{R}_0\mathbf{L})$$
$$E(R_1{}^2 - R_0{}^2) = q\sigma_L{}^2 + (\mathbf{H}'\boldsymbol{\beta} - \boldsymbol{\xi})'\mathbf{D}^{-1}(\mathbf{H}'\boldsymbol{\beta} - \boldsymbol{\xi}) = E[\mathbf{L}'(\mathbf{R}_1 - \mathbf{R}_0)\mathbf{L}]$$

where $\mathbf{D}$ is the dispersion matrix of the estimate of $(\mathbf{H}'\boldsymbol{\beta} - \boldsymbol{\xi})$. Writing $\sigma_L{}^2$ and $\mathbf{H}\boldsymbol{\beta} - \boldsymbol{\xi}$ in terms of $\mathbf{L}$,

$$E(\mathbf{L}'\mathbf{R}_0\mathbf{L}) = (n-r)\mathbf{L}'\boldsymbol{\Sigma}\mathbf{L} \tag{8c.3.9}$$

$$E[\mathbf{L}'(\mathbf{R}_1 - \mathbf{R}_0)\mathbf{L}] = q\mathbf{L}'\boldsymbol{\Sigma}\mathbf{L} + \left[\sum_1^p \mathbf{L}_i(\mathbf{H}\boldsymbol{\beta}_i - \xi_i)\right]\mathbf{D}^{-1}\left[\sum_1^p \mathbf{L}_i(\mathbf{H}\boldsymbol{\beta}_i - \xi_i)\right]$$
$$= q\mathbf{L}'\boldsymbol{\Sigma}\mathbf{L} + \mathbf{L}'\boldsymbol{\Delta}\mathbf{L}, \tag{8c.3.10}$$

where $\boldsymbol{\Delta} = \mathbf{0}$ when the null hypothesis (8c.3.1) is true and otherwise non-negative definite. Since (8c.3.9) and (8c.3.10) are true for every $\mathbf{L}$,

$$E(\mathbf{R}_0) = (n-r)\boldsymbol{\Sigma}$$
$$E(\mathbf{R}_1 - \mathbf{R}_0) = q\boldsymbol{\Sigma} + \boldsymbol{\Delta}, \qquad E(\mathbf{R}_1) = \boldsymbol{\Delta} + (n-r+q)\boldsymbol{\Sigma}.$$

The equation (8c.3.8) with the expected values of $\mathbf{R}_0$ and $\mathbf{R}_1$ can be written

$$|(n-r)\boldsymbol{\Sigma} - \lambda^*(\boldsymbol{\Delta} + \overline{n-r+q\boldsymbol{\Sigma}})| = 0,$$

which may be written in the alternative form

$$|\boldsymbol{\Delta} - \theta^*\boldsymbol{\Sigma}| = 0, \qquad \theta^* = \frac{(n-r) - \lambda^*(n-r+q)}{\lambda^*}.$$

Small values of λ^* correspond to large values of θ^*. The number of nonzero roots θ_i^* depends on the rank of $\boldsymbol{\Delta}$, and the deviation from $\boldsymbol{\Delta} = \mathbf{0}$ is thus reflected in all the nonzero roots θ_i^* or the corresponding λ_i^*. If $\boldsymbol{\Delta} = \mathbf{0}$, all θ_i^* are zero and when the rank of $\boldsymbol{\Delta}$ is 1, the entire deviation is concentrated in the one nonzero root θ^* which corresponds to the smallest λ^*. But when nothing is known about the alternative $\boldsymbol{\Delta}$, more than one root may be useful in detecting departures from the null hypothesis. We have only estimates of the roots λ^* in the roots of the equation (8c.3.8). But a study of the true roots in relation to the deviation matrix has shown the relevance of the estimated roots in tests of hypotheses concerning $\boldsymbol{\Delta}$.

Although the minimum root of (8c.3.8) is the single root which records the maximum deviation from null hypothesis it may not provide an adequate comparison of the "residual" and "deviation from the hypothesis" matrices $\mathbf{R}_0$ and $\mathbf{R}_1 - \mathbf{R}_0$. We, therefore, consider some functions (symmetrical) of all the roots which seem to provide an overall comparison of the matrices involved. The individual roots, however, play an important part in investigating the rank of $\boldsymbol{\Delta}$ as already indicated. A detailed discussion of such tests is given in **8c.6**.

Some possible test criteria involving all the roots are

(a) $\Lambda = \lambda_1 \lambda_2 \cdots \lambda_p = |\mathbf{R}_0| \div |\mathbf{R}_1|$, product of the roots,

(b) $\dfrac{1-\lambda_1}{\lambda_1} + \cdots + \dfrac{1-\lambda_p}{\lambda_p}$,

(c) $\dfrac{1-\lambda_1}{\lambda_1} \times \dfrac{1-\lambda_2}{\lambda_2} \cdots \times \dfrac{1-\lambda_p}{\lambda_p}$,

and so on. The first criterion is known as the Λ criterion derived by Wilks (1932) applying the principle of likelihood ratio in some special cases and later extended by Bartlett (1947) for general use in multivariate analysis.

Observe that the computation of $\mathbf{R}_0$ and $\mathbf{R}_1$ involve no new formulas other than those used in the univariate analysis of variance (A.V.). Since formulas for obtaining the various entries of the A.V. table are available in many situations, the generalization to analysis of dispersion (A.D.) in such cases is immediate. We have only to *consider a linear function* $\mathbf{L}'\mathbf{U}$ *of the p measurements* $\mathbf{U}$ *and apply the known formulas of A.V. to obtain* R_0^2 *and* R_1^2, *which are, now, quadratic forms in the compounding coefficients* $\mathbf{L}$. *The matrices of these quadratic forms in* $\mathbf{L}$ *are precisely* $\mathbf{R}_0$ *and* $\mathbf{R}_1$. Under the null hypothesis, the distribution of $\Lambda = |\mathbf{R}_0| \div |\mathbf{R}_1|$ is $\Lambda(p, n-r, q)$ as denoted in [(xi), **8b.2**].

8c.4 Test for Additional Information

In **8c.3** we have considered simultaneous hypotheses concerning parameters associated with each character. Thus in an experiment with different diets on chickens, differences between diets as reflected in a number of characters (such as age at maturity, size of egg, weight of bird) are tested simultaneously for significance. Each character chosen for study may be of some special interest to the experimenter but the differences in some characters may be concomitant on the differences in others so that the additional information supplied by the former (independently of the latter) may be negligible. Then it may be uneconomical to obtain observations on such characters. Furthermore, it is of interest to determine, if possible, those characters directly influenced by diets and those in which the changes are only of a concomitant nature. In this section we shall develop some tests for this purpose.

Out of the p characters measured, let us ask whether the (additional) information supplied by the last $(p - s)$ characters independently of the first s characters is significant. This is different from asking whether there are significant differences in the last $(p - s)$ characters. The difference between these questions will be clear from the mathematical formulations in (8c.4.1) and (8c.4.2). To answer the former problem we consider the conditional distribution of the observations $\mathbf{Y}_{s+1}, \ldots, \mathbf{Y}_p$, for which case the Gauss-Markoff model may be written

$$\begin{aligned} E(\mathbf{Y}_i) &= \mathbf{X}\boldsymbol{\tau}_i + \gamma_{i1}\mathbf{Y}_1 + \cdots + \gamma_{is}\mathbf{Y}_s \\ &= \mathbf{X}\boldsymbol{\tau}_i + \mathscr{U}_1\boldsymbol{\gamma}_i \\ i &= s + 1, \ldots, p, \end{aligned} \tag{8c.4.1}$$

where $\mathbf{X}$ is the design matrix as before in the model (8c.1.2) and $\mathscr{U}_1 = (\mathbf{Y}_1 \cdots \mathbf{Y}_s)$ and $\boldsymbol{\gamma}_i' = (\gamma_{i1}, \ldots, \gamma_{is})$. In terms of the original parameters $\boldsymbol{\beta}_1, \ldots, \boldsymbol{\beta}_p$, we see that

$$\begin{aligned} \boldsymbol{\tau}_i &= \boldsymbol{\beta}_i - \gamma_{i1}\boldsymbol{\beta}_1 - \cdots - \gamma_{is}\boldsymbol{\beta}_s \\ i &= s + 1, \ldots, p. \end{aligned} \tag{8c.4.2}$$

In terms of unconditional expectations the Gauss-Markoff setup for $\mathbf{Y}_{s+1}, \ldots, \mathbf{Y}_p$ is

$$E(\mathbf{Y}_i) = \mathbf{X}\boldsymbol{\beta}_i, \qquad i = s + 1, \ldots, p. \tag{8c.4.3}$$

Let $\mathbf{H}'\boldsymbol{\beta}_i = 0$ be a hypothesis of interest concerning the ith character without reference to the other characters. The same hypothesis applied to the parameters of the conditional expectation (8c.4.1) is $\mathbf{H}'\boldsymbol{\tau}_i = \mathbf{0}$. In terms of $\boldsymbol{\beta}_1, \ldots, \boldsymbol{\beta}_p$

$$\mathbf{H}'\boldsymbol{\tau}_i = \mathbf{0} = \mathbf{H}'\boldsymbol{\beta}_i - \gamma_{i1}\mathbf{H}'\boldsymbol{\beta}_1 - \cdots - \gamma_{is}\mathbf{H}'\boldsymbol{\beta}_s,$$

so that the hypotheses $\mathbf{H}'\boldsymbol{\beta}_i = \mathbf{0}$ and $\mathbf{H}'\boldsymbol{\tau}_i = \mathbf{0}$ are different. It may so happen that $\mathbf{H}'\boldsymbol{\beta}_i \neq \mathbf{0}$, but $\mathbf{H}'\boldsymbol{\tau}_i = \mathbf{0}$, that is, when the deviation in $\mathbf{H}'\boldsymbol{\beta}_i$ is a particular linear function of (or explained by) the magnitudes $\mathbf{H}'\boldsymbol{\beta}_1, \ldots, \mathbf{H}'\boldsymbol{\beta}_s$.

The hypotheses $\mathbf{H}'\boldsymbol{\tau}_i = \mathbf{0}$, $i = s + 1, \ldots, p$ will be referred to as the null hypotheses concerning additional information provided by the last $p - s$ characters. We shall consider the problem of testing such hypotheses under the setup of (8c.4.1). The problem is, in fact, a generalization of the least squares theory with two sets of parameters or with covariance adjustment for concomitant variables as discussed in **4h.2**.

No new problem of testing arises since the setup (8c.4.1) is again Gauss-Markoff, that is, of the same form as (8c.1.2) except that more parameters are involved. To derive the explicit expressions for the S.P. matrices due to deviation from hypothesis and residual, we need only consider a linear function of $\mathbf{Y}_{s+1}, \ldots, \mathbf{Y}_p$ and determine the corresponding sum of squares using the formulas of **4h.2**. The S.P. matrices for testing the hypotheses $\mathbf{H}'\boldsymbol{\tau}_i = \mathbf{0}$ are then obtained as follows. First consider the setup (8c.1.2) and obtain in the usual way the S.P. matrices $\mathbf{R}_0$ and $\mathbf{R}_1$ for testing $\mathbf{H}'\boldsymbol{\beta}_i = 0$, $i = 1, \ldots, p$. Let us write these matrices in the partitioned form

$$\mathbf{R}_0 = \begin{pmatrix} \mathbf{R}_0(1,1) & \mathbf{R}_0(1,2) \\ \mathbf{R}_0(2,1) & \mathbf{R}_0(2,2) \end{pmatrix}, \qquad \mathbf{R}_1 = \begin{pmatrix} \mathbf{R}_1(1,1) & \mathbf{R}_1(1,2) \\ \mathbf{R}_1(2,1) & \mathbf{R}_1(2,2) \end{pmatrix}$$
$$\text{D.F.} = n - r \qquad\qquad \text{D.F.} = (n - r) + q$$

where the arguments 1 and 2 stand for the first and second sets of s and $p - s$ variables and $R(\mathbf{H}) = q$. The S.P. matrix due to residual for the setup (8c.4.1) is

$$\mathbf{R}_0(2.1) = \mathbf{R}_0(2,2) - \mathbf{R}_0(2,1)[\mathbf{R}_0(1,1)]^{-1}\mathbf{R}_0(1,2) \tag{8c.4.4}$$
$$\text{D.F.} = n - r - s,$$

and due to "residual + deviation" from hypothesis is

$$\mathbf{R}_1(2.1) = \mathbf{R}_1(2,2) - \mathbf{R}_1(2,1)[\mathbf{R}_1(1,1)]^{-1}\mathbf{R}_1(1,2) \tag{8c.4.5}$$
$$\text{D.F.} = (n - r - s) + q.$$

We then have the two matrices

$$\mathbf{R}_0(2.1) \sim W_{p-s}(n - r - s, \boldsymbol{\Sigma}_{2.1}),$$
$$\mathbf{R}_1(2.1) - \mathbf{R}_0(2.1) \sim W_{p-s}(q, \boldsymbol{\Sigma}_{2.1}),$$

which are independently distributed as central Wishart when the null hypothesis is true. Hence we can apply any test criterion based on the roots of

$$|\mathbf{R}_0(2.1) - \lambda\mathbf{R}_1(2.1)| = 0.$$

If the test criterion is the product of the roots

$$\Lambda_{(p-s)-s} = \frac{|\mathbf{R}_0(2.1)|}{|\mathbf{R}_1(2.1)|}, \tag{8c.4.6}$$

then its distribution is that of $\Lambda(p - s, n - r - s, q)$. (8c.4.6) can be written in the simpler form depending explicitly on the determinants of the matrices $\mathbf{R}_0$, $\mathbf{R}_1$, and their submatrices

$$\frac{|\mathbf{R}_0(2.1)|}{|\mathbf{R}_1(2.1)|} = \frac{|\mathbf{R}_0|_p}{|\mathbf{R}_1|_p} \div \frac{|\mathbf{R}_0|_s}{|\mathbf{R}_1|_s}, \tag{8c.4.7}$$

where

$$|\mathbf{R}_0|_k = \begin{vmatrix} R_0(1, 1) & \cdots & R_0(1, k) \\ \cdot & \cdots & \cdot \\ R_0(k, 1) & \cdots & R_0(k, k) \end{vmatrix},$$

and so on. The result (8c.4.7) follows from the identities

$$|\mathbf{R}_0|_p = |\mathbf{R}_0|_s |\mathbf{R}_0(2.1)|, \qquad |\mathbf{R}_1|_p = |\mathbf{R}_1|_s |\mathbf{R}_1(2.1)|.$$

We can write the relationship (8c.4.7) as a decomposition of the Λ criterion for testing the hypotheses $\mathbf{H}'\boldsymbol{\beta}_i = \mathbf{0}$ simultaneously for all the p characters as follows:

$$\frac{|\mathbf{R}_0|_p}{|\mathbf{R}_1|_p} = \frac{|\mathbf{R}_0|_s}{|\mathbf{R}_1|_s} \times \frac{|\mathbf{R}_0(2.1)|}{|\mathbf{R}_1(2.1)|}$$

$$\Lambda_p = \Lambda_s \times \Lambda_{(p-s).s}, \tag{8c.4.8}$$

where Λ_s is the Λ criterion for the first s characters. It is important to note that the derived distribution of $\Lambda_{(p-s)\cdot s}$ depends *only* on the hypotheses $\mathbf{H}'\boldsymbol{\tau}_i = \mathbf{0}$, $i = s + 1, \ldots, p$ being true irrespective of what the true values are of $\boldsymbol{\beta}_1, \ldots, \boldsymbol{\beta}_s$—the parameters associated with the first s characters (i.e., irrespective of whether $\mathbf{H}'\boldsymbol{\beta}_i = \mathbf{0}$ or not for $i = 1, \ldots, s$).

Simplification when $q = 1$. It is shown in [(xii), **8b.2**] that when the D.F. of the hypothesis $q = 1$, the ratio $|\mathbf{R}_0|_p/|\mathbf{R}_1|_p$ can be written in terms of Hotelling's T^2:

$$\frac{|\mathbf{R}_0|_p}{|\mathbf{R}_1|_p} = \frac{1}{1 + T_p^2/k}, \qquad \text{where} \quad k = \text{D.F. of } \mathbf{R}_0.$$

If the null hypothesis $\mathbf{H}'\boldsymbol{\beta}_i = \mathbf{0}$ is true for all i, then

$$\frac{k - p + 1}{p}\frac{T_p^2}{k} \sim F(p, k - p + 1). \tag{8c.4.9}$$

If T_s^2 is the Hotelling's statistic based on the first s characters (i.e., to test the single hypothesis $\mathbf{H}'\boldsymbol{\beta}_i = 0$, $i = 1, \ldots, s$), then

$$\frac{|\mathbf{R}_0|_s}{|\mathbf{R}_1|_s} = \frac{1}{1 + T_s^2/k}.$$

If we then write

$$\frac{|\mathbf{R}_0|_p}{|\mathbf{R}_1|_p} \div \frac{|\mathbf{R}_0|_s}{|\mathbf{R}_1|_s} = \frac{|\mathbf{R}_0(2.1)|}{|\mathbf{R}_1(2.1)|} = \frac{1}{1 + T^2_{(p-s).s}/(k - s)},$$

we have the relationship

$$U_{(p-s).s} = \frac{T^2_{(p-s).s}}{k - s} = \frac{T_p^2 - T_s^2}{k + T_s^2}, \tag{8c.4.10}$$

which leads to the variance ratio on $(p-s, k-p+1)$ D.F.

$$\frac{k-p+1}{p-s} \cdot \frac{T_p^2 - T_s^2}{k+T_s^2} \tag{8c.4.11}$$

when the null hypothesis $\mathbf{H}'\boldsymbol{\tau}_i = 0$, $i = s+1, \ldots, p$, of no additional information is true whatever $\mathbf{H}'\boldsymbol{\beta}_i$ may be for $i = 1, \ldots, s$.

8c.5 The Distribution of Λ

The following notation will generally be used in the application of Λ test.

TABLE 8c.5α. Analysis of Dispersion (p Characters)

Due to	D.F.	S.P. matrix
Deviation from hypothesis	q	$\mathbf{R}_1 - \mathbf{R}_0$
Residual (error)	$t-q$	$\mathbf{R}_0$
Total	t	$\mathbf{R}_1$

$$\Lambda = |\mathbf{R}_0| \div |\mathbf{R}_1|$$

It is shown in [(xi), **8b.2**] that Λ with parameters $(p, t-q, q)$ is distributed as the product of independent beta variables with parameters

$$\left(\frac{t-q-p+1}{2}, \frac{q}{2}\right), \ldots, \left(\frac{t-q}{2}, \frac{q}{2}\right). \tag{8c.5.1}$$

In some cases the distribution of the product reduces to an exact variance ratio after a transformation (Nair, 1939, Wilks, 1932). The results are summarized in Table 8c.5β.

TABLE 8c.5β. The Exact Variance Ratios in Special Cases

Values of p, q	Variance ratio	D.F.
$q = 1$, for any p	$\dfrac{1-\Lambda}{\Lambda}\,\dfrac{t-p}{p}$	$p, t-p$
$q = 2$, for any p	$\dfrac{1-\sqrt{\Lambda}}{\sqrt{\Lambda}}\,\dfrac{t-p-1}{p}$	$2p, 2(t-p-1)$
$p = 1$, for any q	$\dfrac{1-\Lambda}{\Lambda}\,\dfrac{t-q}{q}$	$q, t-q$
$p = 2$, for any q	$\dfrac{1-\sqrt{\Lambda}}{\sqrt{\Lambda}}\,\dfrac{t-q-1}{q}$	$2q, 2(t-q-1)$

For other values of p, q, good approximations are available for $\Lambda(p, t-q, q)$. One due to Bartlett (1947) is to use

$$-m \log_e \Lambda = -\left(t - \frac{p+q+1}{2}\right) \log_e \Lambda \tag{8c.5.2}$$

as central χ^2 on pq degrees of freedom. Another due to the author (Rao, 1951d) which provides a better approximation, is to use

$$R = \frac{1 - \Lambda^{1/s}}{\Lambda^{1/s}} \frac{ms - 2\lambda}{pq} \tag{8c.5.3}$$

as a variance ratio on pq and $(ms - 2\lambda)$ degrees of freedom, where

$$m = t - \frac{p+q+1}{2}, \qquad s = \sqrt{\frac{p^2q^2 - 4}{p^2 + q^2 - 5}}$$

$$\lambda = \frac{pq - 2}{4}. \tag{8c.5.4}$$

It may be noted that $(ms - 2\lambda)$ need not be integral. But this does not cause any difficulty in consulting a table of significant values of the variance ratio. The appropriate value lies between the significant values for $[ms - 2\lambda]$ and $[ms - 2\lambda] + 1$. For practical purposes, it is safer to use $[ms - 2\lambda]$, the greatest integer in $ms - 2\lambda$, as the degrees of freedom of the denominator. For other approximations and asymptotic expansions of the distribution function the reader is referred to Rao (1948b, 1951d) and J. Roy (1951). For references to work on exact distribution of Λ, see Rao (1972a).

8c.6 Test for Dimensionality (Structural Relationship)

Hypothesis of Dimensionality. Consider k, p-variate normal populations $N_p(\boldsymbol{\mu}_1, \boldsymbol{\Sigma}), \ldots, N_p(\boldsymbol{\mu}_k, \boldsymbol{\Sigma})$. It may be of interest to examine whether there is any structural relationship among the components of the mean values (Fisher, 1939). A linear structural relationship is of the form

$$\mathbf{H}\boldsymbol{\mu}_i = \boldsymbol{\xi}, \qquad i = 1, \ldots, k \tag{8c.6.1}$$

where $\mathbf{H}$ is $(s \times p)$ matrix and $\boldsymbol{\xi}$ is $(s \times 1)$ vector. $\mathbf{H}$ and $\boldsymbol{\xi}$ are unknown but they are fixed for all i in the equation (8c.6.1). The hypothesis (8c.6.1) implies that the points $\boldsymbol{\mu}_1, \ldots, \boldsymbol{\mu}_k$ in E_p actually lie on a $r = p - s$ dimensional plane. Indeed, the equation $\mathbf{H}\boldsymbol{\mu} = \boldsymbol{\xi}$ where $\boldsymbol{\mu}$ represents a point of E_p is that of a $(p - s)$-dimensional plane. It may be noted that the hypothesis (8c.6.1) is of interest only when $r < k - 1$.

Data. Let N_i observations be available from $N_p(\boldsymbol{\mu}_i, \boldsymbol{\Sigma})$ and

$$\overline{\mathbf{U}}_i \quad \text{and} \quad (S_{rs}^{(i)})$$

be the sample mean and corrected S.P. matrix based on the N_i observations, $i = 1, \ldots, k$.

The analysis of dispersion between and within populations is as follows:

	D.F.	S.P. matrix
Between	$k-1$	$\mathbf{B} = \sum N_i \overline{\mathbf{U}}_i \overline{\mathbf{U}}_i' - (\sum N_i)\overline{\mathbf{U}}\overline{\mathbf{U}}'$
Within	$\sum N_i - k$	$\mathbf{W} = \left(\sum_i S_{rs}^{(i)}\right)$
Total	$\sum N_i - 1$	$\mathbf{T} = (S_{rs})$

where $\overline{\mathbf{U}}$ is the grand average and (S_{rs}) the corrected S.P. matrix from the combined sample of $N_1 + \cdots + N_k$ observations.

Test Criterion. Let $\boldsymbol{\Sigma}$ be known and nonsingular. Then the density of the observed mean values, is, apart from a constant,

$$L(\boldsymbol{\mu}_1, \ldots, \boldsymbol{\mu}_k) = \exp\left[-\frac{1}{2}\sum_{i=1}^{k} N_i(\overline{\mathbf{U}}_i - \boldsymbol{\mu}_i)'\boldsymbol{\Sigma}^{-1}(\overline{\mathbf{U}}_i - \boldsymbol{\mu}_i)\right]. \tag{8c.6.2}$$

Considering (8c.6.2) as a function of $\boldsymbol{\mu}_1, \ldots, \boldsymbol{\mu}_k$, we derive the likelihood ratio criterion for testing the hypothesis H_0: $\boldsymbol{\mu}_i$ lie on an r-dimensional plane,

$$\chi^2 = -2 \log_e \frac{\sup_{H_0} L(\boldsymbol{\mu}_1, \ldots, \boldsymbol{\mu}_k)}{\sup L(\boldsymbol{\mu}_1, \ldots, \boldsymbol{\mu}_k)}. \tag{8c.6.3}$$

In the numerator of (8c.6.3), the supremum is sought with the restriction that the $\boldsymbol{\mu}_i$, $i = 1, \ldots, k$ are confined to an r-dimensional plane whereas in the denominator the $\boldsymbol{\mu}_i$ are unrestricted. From the general theory of the likelihood ratio criteria the asymptotic distribution of the statistic (8c.6.3), when the sample sizes $N_1, \ldots, N_k$ are large, is χ^2 on D.F. equal to the number of restrictions on the pk parameters imposed by the hypothesis H_0. This number can be easily computed by observing that an r-dimensional plane is specified by $(r + 1)$ points and any point on the plane can be expressed as a combination of $(r + 1)$ points with coefficients adding to unity. Each point is represented by p coordinates or parameters. The number of free (arbitrary) parameters is therefore $p(r + 1) + (k - r - 1)r$, giving the number of restrictions

$$pk - p(r + 1) - (k - r - 1)r = (p - r)(k - r - 1).$$

Thus the D.F. of the χ^2 is $(p - r)(k - r - 1)$.

Now sup log $L(\boldsymbol{\mu}_1, \ldots, \boldsymbol{\mu}_k) = 0$ when $\boldsymbol{\mu}_i$ are unrestricted, and it will be shown later in this section that

$$\sup_{H_0} \log L = -\tfrac{1}{2}(\lambda_{r+1} + \cdots + \lambda_p) \tag{8c.6.4}$$

where $\lambda_{r+1}, \ldots, \lambda_p$ are the smallest $p - r$ eigenroots of the determinantal equation $|\mathbf{B} - \lambda\boldsymbol{\Sigma}| = 0$. Hence the χ^2 of (8c.6.3) is

$$\chi^2 = \lambda_{r+1} + \cdots + \lambda_p, \qquad \text{D.F.} = (p - r)(k - r - 1). \tag{8c.6.6}$$

If $n = \sum N_i - k$ is large and $\boldsymbol{\Sigma}$ is unknown, the estimate $n^{-1}\mathbf{W}$ may be substituted for $\boldsymbol{\Sigma}$, in which case the χ^2 test of (8c.6.6) would remain valid, although approximate. If we consider the determinantal equation

$$|\mathbf{B} - \nu\mathbf{W}| = 0 \tag{8c.6.7}$$

without dividing $\mathbf{W}$ by n, then

$$\chi^2 = n(\nu_{r+1} + \cdots + \nu_p), \qquad \text{D.F.} = (p - r)(k - r - 1)$$

where $\nu_{r+1}, \ldots, \nu_p$ are the smallest $p - r$ roots of the equation (8c.6.7).

In many situations the problem will be one of determining the dimensionality of the configuration of mean values. This may be approached as follows by listing the roots in decreasing order and the corresponding D.F. and cumulating upwards.

Root	D.F.	Cumulated	
		Roots	D.F.
λ_1	$p + k - 2$	$\chi_0^2 = \lambda_1 + \cdots + \lambda_p$	$p(k - 1)$
λ_2	$p + k - 4$	$\chi_1^2 = \lambda_2 + \cdots + \lambda_p$	$(p - 1)(k - 2)$
.	.	.	.
λ_{r+1}	$p + k - 2r - 2$	$\chi_r^2 = \lambda_{r+1} + \cdots + \lambda_p$	$(p - r)(k - r - 1)$
.	.	.	.
.	.	.	.

If $\chi_r^2, \chi_{r+1}^2, \ldots$ are significantly small compared to their D.F. while χ_{r-1}^2, $\chi_{r-2}^2, \ldots$ are large, then the dimensionality may be inferred as r.

The procedure described suggests an alternative test criterion based on the product of the residual roots of

$$|\mathbf{W} - \theta(\mathbf{B} + \mathbf{W})| = 0 \tag{8c.6.8}$$

where λ and θ are connected by the relation

$$\left(1 + \frac{\lambda}{n}\right) = \frac{1}{\theta}.$$

The statistic corresponding to $\lambda_{r+1} + \cdots + \lambda_p$ is

$$-n \log(\theta_{r+1} \ldots \theta_p) = -n(\log \theta_{r+1} + \cdots + \log \theta_p).$$

Bartlett (1947b) suggests that

$$-\left(\sum n_i - 1 - \frac{p+k}{2}\right)(\log \theta_{r+1} + \cdots + \log \theta_p) \tag{8c.6.9}$$

may be used approximately as χ^2 on $(p-r)(k-r-1)$ D.F.

Proof of the Result (8c.6.4). The problem is one of showing

$$\min_{H_0} \sum N_i(\overline{\mathbf{U}}_i - \boldsymbol{\mu}_i)'\boldsymbol{\Sigma}^{-1}(\overline{\mathbf{U}}_i - \boldsymbol{\mu}_i) = \lambda_{r+1} + \cdots + \lambda_p, \tag{8c.6.10}$$

where $\lambda_1 \geqslant \cdots \geqslant \lambda_p$ are the roots of

$$|\mathbf{B} - \lambda\boldsymbol{\Sigma}| = 0, \tag{8c.6.11}$$

the matrix $\mathbf{B}$ being $\sum N_i \overline{\mathbf{U}}_i \overline{\mathbf{U}}_i' - (\sum N_i)\overline{\mathbf{U}}\overline{\mathbf{U}}'$. Let $\mathbf{Z}_i = \boldsymbol{\Sigma}^{-1/2}\overline{\mathbf{U}}_i$ and $\zeta_i = \boldsymbol{\Sigma}^{-1/2}\boldsymbol{\mu}_i$, in which case the expression (8c.6.10) is the same as

$$\min_{H_0} \sum_{i=1}^{k} N_i(\mathbf{Z}_i - \zeta_i)'(\mathbf{Z}_i - \zeta_i), \tag{8c.6.12}$$

where H_0 is the same hypothesis on ζ_i, that is, they are confined to an r-dimensional plane. An r-dimensional plane is determined by a point $\mathbf{X}_0$ and r orthonormal directions $\mathbf{X}_1, \ldots, \mathbf{X}_r$. And a point ζ_i on this plane has the representation

$$\zeta_i = \mathbf{X}_0 + C_{i1}\mathbf{X}_1 + \cdots + C_{ir}\mathbf{X}_r. \tag{8c.6.13}$$

The ith term in (8c.6.12) without the multiplier N_i can be written

$$(\mathbf{Z}_i - \mathbf{X}_0 - \sum C_{ij}\mathbf{X}_j)'(\mathbf{Z}_i - \mathbf{X}_0 - \sum C_{ij}\mathbf{X}_j). \tag{8c.6.14}$$

For given $\mathbf{X}_0, \mathbf{X}_1, \ldots, \mathbf{X}_r$, the expression (8c.6.14) has the minimum value (using the usual regression theory and the expression for residual sum of squares) of

$$(\mathbf{Z}_i - \mathbf{X}_0)'(\mathbf{Z}_i - \mathbf{X}_0) - [(\mathbf{Z}_i - \mathbf{X}_0)'\mathbf{X}_1]^2 - \cdots - [(\mathbf{Z}_i - \mathbf{X}_0)'\mathbf{X}_r]^2 \tag{8c.6.15}$$

by making use of the fact that $\mathbf{X}_1, \ldots, \mathbf{X}_r$ are orthonormal. Multiplying by N_i and taking the summation over i, we find the expression to be minimized with respect to $\mathbf{X}_0, \mathbf{X}_1, \ldots, \mathbf{X}_r$ to be

$$\sum_{i=1}^{k} N_i(\mathbf{Z}_i - \mathbf{X}_0)'(Z_i - \mathbf{X}_0) - \sum_{i=1}^{k}\sum_{j=1}^{r} N_i[(\mathbf{Z}_i - \mathbf{X}_0)'\mathbf{X}_j]^2. \tag{8c.6.16}$$

By introducing $\bar{\mathbf{Z}} = \sum N_i \mathbf{Z}_i / \sum N_i$, (8c.6.16) can be written

$$\sum_{i=1}^{k} N_i(\mathbf{Z}_i - \bar{\mathbf{Z}})'(\mathbf{Z}_i - \bar{\mathbf{Z}}) - \sum_{i=1}^{k} \sum_{j=1}^{r} N_i[(\mathbf{Z}_i - \bar{\mathbf{Z}})'\mathbf{X}_j]^2$$

$$+ N(\bar{\mathbf{Z}} - \mathbf{X}_0)'(\bar{\mathbf{Z}} - \mathbf{X}_0) - N \sum_{j=1}^{r} [(\bar{\mathbf{Z}} - \mathbf{X}_0)'\mathbf{X}_j]^2. \quad (8c.6.17)$$

The expression involving $(\bar{\mathbf{Z}} - \mathbf{X}_0)$ is positive and therefore for given $\mathbf{X}_1, \ldots, \mathbf{X}_r$, (8c.6.17) is a minimum when $\mathbf{X}_0 = \bar{\mathbf{Z}}$. Then the expression to be minimized with respect to $\mathbf{X}_1, \ldots, \mathbf{X}_r$ is

$$\sum_{i=1}^{k} N_i(\mathbf{Z}_i - \bar{\mathbf{Z}})'(\mathbf{Z}_i - \bar{\mathbf{Z}}) - \sum_{i=1}^{k} \sum_{j=1}^{r} N_i[(\mathbf{Z}_i - \bar{\mathbf{Z}})'\mathbf{X}_j]^2. \quad (8c.6.18)$$

Since the first term is independent of $\mathbf{X}_i$, we need to maximize the second term with respect to $\mathbf{X}_1, \ldots, \mathbf{X}_r$. Now

$$\sum_{i=1}^{k} \sum_{j=1}^{r} N_i[\mathbf{X}_j'(\mathbf{Z}_i - \bar{\mathbf{Z}})]^2 = \sum_{j=1}^{r} \mathbf{X}_j' \left[\sum_{i=1}^{k} N_i(\mathbf{Z}_i - \bar{\mathbf{Z}})(\mathbf{Z}_i - \bar{\mathbf{Z}})' \right] \mathbf{X}_j$$

$$= \sum_{j=1}^{r} \mathbf{X}_j' \mathbf{C} \mathbf{X}_j,$$

where $\mathbf{C} = \sum N_i(\mathbf{Z}_i - \bar{\mathbf{Z}})(\mathbf{Z}_i - \bar{\mathbf{Z}})'$. Let $\mathbf{P}_1, \mathbf{P}_2, \ldots, \mathbf{P}_p$ be the eigenvectors and $\lambda_1, \ldots, \lambda_p$ the eigenroots associated with the determinantal equation $|\mathbf{C} - \lambda \mathbf{I}| = 0$. Then, as shown in (1f.2.8), we have

$$\max_{\mathbf{X}_1, \ldots, \mathbf{X}_r} \sum_{1}^{r} \mathbf{X}_j' \mathbf{C} \mathbf{X}_j = \mathbf{P}_1' \mathbf{C} \mathbf{P}_1 + \cdots + \mathbf{P}_r' \mathbf{C} \mathbf{P}_r$$

$$= \lambda_1 + \cdots + \lambda_r.$$

But $\lambda_1 + \cdots + \lambda_p = \text{trace } \mathbf{C} = \sum N_i(\mathbf{Z}_i - \bar{\mathbf{Z}})'(\mathbf{Z}_i - \bar{\mathbf{Z}})$, and therefore the minimum of (8c.6.18) is $\lambda_{r+1} + \cdots + \lambda_p$. Since

$$\mathbf{Z}_i = \mathbf{\Sigma}^{-1/2}\mathbf{U}_i, \qquad \bar{\mathbf{Z}} = \mathbf{\Sigma}^{-1/2}\bar{\mathbf{U}}$$

$$\sum N_i(\mathbf{Z}_i - \bar{\mathbf{Z}})(\mathbf{Z}_i - \bar{\mathbf{Z}})' = \mathbf{\Sigma}^{-1/2}[\sum N_i(\mathbf{U}_i - \bar{\mathbf{U}})(\mathbf{U}_i - \bar{\mathbf{U}})']\mathbf{\Sigma}^{-1/2}$$

$$\mathbf{C} = \mathbf{\Sigma}^{-1/2}\mathbf{B}\mathbf{\Sigma}^{-1/2}.$$

Hence

$$|\mathbf{C} - \lambda \mathbf{I}| = 0 = |\mathbf{B} - \lambda \mathbf{\Sigma}|.$$

Thus $\lambda_1, \ldots, \lambda_p$ are the roots of the determinantal equation $|\mathbf{B} - \lambda \mathbf{\Sigma}| = 0$ which proves the desired result.

8c.7 Analysis of Dispersion with Structural Parameters (Growth Model)

Let us consider the general multivariate linear model

$$(\mathcal{U}, \mathbf{XB}, \mathbf{\Sigma} \otimes \mathbf{I}_n) \tag{8c.7.1}$$

introduced in **8c.1**, where $\mathcal{U}$ is of order $n \times p$, $\mathbf{X}$ of order $n \times m$, $\mathbf{B}$ of order $m \times p$ and $\mathbf{\Sigma}$ of order $p \times p$. In the analysis of **8c.2–8.c.6**, we considered the mp parameters in $\mathbf{B}$ as free. Let us now introduce linear restrictions on the parameters in each row of $\mathbf{B}$, which can be done by considering a relation of the form

$$\mathbf{B} = \mathbf{\Theta}_0 + \mathbf{\Theta H}$$

where $\mathbf{\Theta}$ is $m \times k$ matrix of free parameters, and $\mathbf{\Theta}_0$ of order $m \times p$ and $\mathbf{H}$ of order $k \times p$ are given matrices. In such a case, the model (8c.7.1) can be written as

$$(\mathcal{U} - \mathbf{X\Theta}_0, \mathbf{X\Theta H}, \mathbf{\Sigma} \otimes \mathbf{I}_n) \tag{8c.7.2}$$

which is used in a number of investigations (Rao, 1959a, 1961g, 1965b) on growth data. The general formulation of the model (8c.7.2) is due to Pothoff and Roy (1964).

Reduction of the Model. Let $R(\mathbf{H}) = k$ and $\mathbf{Z}$ be a $p \times (p-k)$ matrix of rank $(p-k)$ such that $\mathbf{HZ} = \mathbf{0}$. Further let $\mathbf{\Sigma}_0$ be an a priori value of $\mathbf{\Sigma}$, assumed to be nonsingular. [For the purpose of our theory $\mathbf{\Sigma}_0$ may be any p.d. matrix. It may be chosen as $\mathbf{I}$, if no a priori value is available.] Consider

$$\begin{aligned} \mathbf{W}_1 &= (\mathcal{U} - \mathbf{X}\mathbf{\Theta}_0)\mathbf{\Sigma}_0^{-1}\mathbf{H}'(\mathbf{H}\mathbf{\Sigma}_0^{-1}\mathbf{H}')^{-1}, \\ \mathbf{W}_2 &= (\mathcal{U} - \mathbf{X}\mathbf{\Theta}_0)\mathbf{Z}. \end{aligned} \tag{8c.7.3}$$

Then $E(\mathbf{W}_1) = \mathbf{X\Theta}$, $E(\mathbf{W}_2) = \mathbf{0}$, and in terms of the variables $\mathbf{W}_1$, $\mathbf{W}_2$, the model (8c.7.1) becomes

$$[(\mathbf{W}_1 \mid \mathbf{W}_2), (\mathbf{X\Theta} \mid \mathbf{0}), \mathbf{\Lambda} \otimes \mathbf{I}_n] \tag{8c.7.4}$$

where $\mathbf{\Lambda}$ is the matrix of variances and covariances of the variables in $(\mathbf{W}_1 \mid \mathbf{W}_2)$. Since $E(\mathbf{W}_2) = \mathbf{0}$, we can use $\mathbf{W}_2$ as observations on concomitant variables and write our model (8c.7.4) as

$$(\mathbf{W}_1, \mathbf{X\Theta} + \mathbf{W}_2\mathbf{\Gamma}, \mathbf{\Lambda}_1 \otimes \mathbf{I}_n) \tag{8c.7.5}$$

with additional regression parameters $\mathbf{\Gamma}$ as in (8c.4.1). Thus the model (8c.7.1) is reduced to a model with free parameters $\mathbf{\Theta}$, $\mathbf{\Gamma}$. Then the analysis using covariance adjustment applies, and no new problem is involved.

If $R(\mathbf{H}) = r < k$, we proceed as follows. Construct a nonsingular matrix $(\mathbf{H}_1 \,\vdots\, \mathbf{H}_2)$ such that $\mathbf{H}\mathbf{H}_2 = 0$ and the columns of $\mathbf{H}_1$ form a basis of the vector space generated by the rows of $\mathbf{H}$. Then make the transformation

$$\mathbf{W}_1 = (\mathscr{U} - \mathbf{X}\boldsymbol{\Theta}_0)\boldsymbol{\Sigma}_0^{-1}\mathbf{H}_1'$$
$$\mathbf{W}_2 = (\mathscr{U} - \mathbf{X}\boldsymbol{\Theta}_0)\mathbf{H}_2 \tag{8c.7.6}$$

so that $E(\mathbf{W}_1) = \mathbf{X}\boldsymbol{\Theta}\mathbf{H}\boldsymbol{\Sigma}_0^{-1}\mathbf{H}_1'$ and $E(\mathbf{W}_2) = \mathbf{0}$. Now $R(\mathbf{H}\boldsymbol{\Sigma}_0^{-1}\mathbf{H}_1') = r$, and hence $\boldsymbol{\Theta}\mathbf{H}\boldsymbol{\Sigma}_0^{-1}\mathbf{H}_1'$ can be replaced by a $m \times r$ matrix $\boldsymbol{\Phi}$ of free parameters. Then the model (8c.7.1) reduces to

$$(\mathbf{W}_1, \mathbf{X}\boldsymbol{\Phi} + \mathbf{W}_2\boldsymbol{\Gamma}, \boldsymbol{\Lambda}_1 \otimes \mathbf{I}_n) \tag{8c.7.7}$$

which is again of the same form as (8c.7.5).

In practice, we have the problem of examining the usefulness of the concomitant observations $\mathbf{W}_2$. If $\boldsymbol{\Sigma}_0$ is a good approximation to $\boldsymbol{\Sigma}$, then we may omit $\mathbf{W}_2$ and consider the model $(\mathbf{W}_1, \mathbf{X}\boldsymbol{\Theta}, \boldsymbol{\Lambda}_1 \otimes \mathbf{I}_n)$ or $(\mathbf{W}_1, \mathbf{X}\boldsymbol{\Phi}, \boldsymbol{\Lambda}_1 \otimes \mathbf{I}_n)$, for drawing inferences on $\boldsymbol{\Theta}$ or $\boldsymbol{\Phi}$. The usefulness of covariance adjustment is, however, a matter of empirical verification (Rao, 1965b).

8d SOME APPLICATIONS OF MULTIVARIATE TESTS

8d.1 Test for Assigned Mean Values

Observations were obtained on 28 trees for thickness of cork borings in the north (N), east (E), south (S), and west (W) directions. The problem is to examine whether the bark deposit is same in all the directions. We may consider the three characters (contrasts)

$$U_1 = (N + S) - (E + W); \qquad U_2 = N - S, \qquad U_3 = E - W, \tag{8d.1.1}$$

and test whether the expected values of U_1, U_2, U_3 are zero. Instead of the constrasts (8d.1.1), any others could have been chosen. But the choice (8d.1.1) is motivated by the suggestion that bark deposit is likely to be uniform in north and south directions and so also in east and west directions. The maximum deviation is then expected in $(N + S) - (E + W)$. The contrasts (8d.1.1) chosen enable a direct examination of these hypothetical suggestions.

The sample means and the sample dispersion matrix (i.e., corrected sum of squares and products after dividing by 27, D.F.) are

$$\bar{U}_1 = 8.8571, \qquad \bar{U}_2 = 0.8571, \qquad \bar{U}_3 = 1.0000$$

$$(s_{ij}) = \begin{pmatrix} 128.7200 & -21.0211 & -26.9259 \\ & 63.5344 & 21.3544 \\ & & 96.5761 \end{pmatrix}.$$

The reciprocal of the matrix (s_{ij}) is represented by (s^{ij}).

Hotelling's T^2 can be applied to test for assigned values of the means, the general formula being [for number of characters p, sample size N, estimated means $\bar{U}_1, \ldots, \bar{U}_p$ and estimated dispersion matrix s_{ij} on $(N-1)$ D.F. and expected values $\xi_1, \xi_2, \ldots$, under test.]

$$T_p{}^2 = N \sum \sum s^{ij}(\bar{U}_i - \xi_i)(\bar{U}_j - \xi_j). \tag{8d.1.2}$$

If N is large, $T_p{}^2$ has the χ^2 distribution on p D.F. For small N the appropriate procedure is to use

$$F = \frac{N-p}{(N-1)p} T_p{}^2 \tag{8d.1.3}$$

as the variance ratio statistic on p and $(N-p)$ D.F. In the present sample $p = 3$, $N = 28$, and $\xi_i = 0$, $i = 1, 2, 3$. A simple method of computing the quadratic form

$$\sum \sum s^{ij}(\bar{U}_i - \xi_i)(\bar{U}_j - \xi_j) \tag{8d.1.4}$$

is as follows. Start with the matrix

Dispersion matrix (s_{ij})			Deviation ($\bar{U}_i - \xi_i$)
128.7200	−21.0211	−26.9259	8.8571
	63.5344	21.3544	0.8571
		96.5761	1.0000
			0

and reduce it by pivotal condensation. The last value with the sign changed is the value of the quadratic form (8d.1.4). In the present example the value is found to be 0.7408, thus giving

$$T_p{}^2 = N(0.7408) = 20.7424$$

$$F = \frac{N-p}{(N-1)p} T_p{}^2 = \frac{25 \times 28}{27 \times 3}(0.7408) = 6.4019.$$

The value 6.4019 as a variance ratio on 3 and 25 D.F. is significant at the 1% level.

Let us make individual tests based on each $\bar{U}_i$ ignoring the rest. Each is $T_1{}^2$ with a variance ratio on 1 and 27 D.F.

$$F \text{ for } N+S-E-W = T_1{}^2 = \frac{N(\bar{U}_1 - \xi_1)^2}{s_{11}} = \frac{28(8.8571)^2}{128.7200} = 17.0645$$

$$F \text{ for } N-S = T_1{}^2 = \frac{N(\bar{U}_2 - \xi_2)^2}{s_{22}} = \frac{28(0.8571)^2}{63.5344} < 1$$

$$F \text{ for } E-W = T_1{}^2 = \frac{N(\bar{U}_3 - \xi_3)^2}{s_{33}} = \frac{28(1.000)^2}{76.5761} < 1$$

That the F for $N + S - E - W$ is large and the rest are small indicates the truth of the suggested hypotheses. Suppose the F values for $N - S$ and $E - W$ had been large. We could then ask the question whether the large differences in $(N - S)$ and $(E - W)$ were solely due to their correlations with $(N + S - E - W)$. This could be examined by the test for additional information developed in **8c.4**. Since the D.F. of the hypothesis is unity the formula (8c.4.11) is applicable. The criterion for testing the significance of additional information due to $p - s$ characters independently of s characters is the variance ratio (see 8c.4.11)

$$F = \frac{N - p}{p - s} \frac{T_p^2 - T_s^2}{(N - 1) + T_s^2}, \tag{8d.1.5}$$

by substituting $(N - 1)$ for k in the formula (8c.4.11). In the present example, $p = 3$, $s = 1$, and the value of T_1^2 for $(N + S - E - W)$ is 17.0645. Substituting in (8d.1.5), we have

$$F = \frac{25}{2} \frac{20.7424 - 17.0645}{44.0645} = 1.0431,$$

which is small as a variance ratio on 2 and 25 D.F.

Note. Since T_p^2 can be computed in a number of ways, we shall give some alternative formulas. Let $(S_{ij}) = \mathbf{S}$ be the matrix of the corrected sum of squares and products, $(s_{ij}) = \mathbf{s} = (N - 1)^{-1}\mathbf{S}$ and $\mathbf{d}' = (\bar{U}_1 - \xi_1, \ldots, \bar{U}_p - \xi_p)$. Then

$$\begin{aligned}
T_p^2 &= N \sum \sum s^{ij}(\bar{U}_i - \xi_i)(\bar{U}_j - \xi_j) \\
&= N(N - 1) \sum \sum S^{ij}(\bar{U}_i - \xi_i)(\bar{U}_j - \xi_j) \\
&= -N \begin{vmatrix} \mathbf{s} & \mathbf{d} \\ \mathbf{d}' & 0 \end{vmatrix} \div |\mathbf{s}| = N\left\{\frac{|\mathbf{s} + \mathbf{dd}'|}{|\mathbf{s}|} - 1\right\} = \frac{|\mathbf{s} + N\mathbf{dd}'|}{|\mathbf{s}|} - 1 \\
&= N(N - 1) \begin{vmatrix} \mathbf{S} & \mathbf{d} \\ \mathbf{d}' & 0 \end{vmatrix} \div |\mathbf{S}| = N(N - 1)\left\{\frac{|\mathbf{S} + \mathbf{dd}'|}{|\mathbf{S}|} - 1\right\} \\
&= (N - 1)\left\{\frac{|\mathbf{S} + N\mathbf{dd}'|}{|\mathbf{S}|} - 1\right\}.
\end{aligned}$$

8d.2 Test for a Given Structure of Mean Values

Suppose N independent observations are available from a p-variate population. Let $\bar{U}_1, \ldots, \bar{U}_p$ be sample means and (s_{ij}) the sample (estimated) dispersion matrix on $(N - 1)$ D.F. as in **8d.1**. Further, let the theoretical mean values be $\mu_1, \ldots, \mu_p$. Some hypotheses of interest are as follows:

1. $\mu_i = \mu$ for all i (i.e., all the characters have the same mean value).

2. $\mu_i = \mu_{r+i}$, $i = 1, \ldots, r$ with $p = 2r$. Such a hypothesis arises when we have multiple measurements on the right and left sides of an organism and we wish to test for symmetry.
3. $\mu_i = \alpha + \beta_1 t_i + \cdots + \beta_r t_i^r$, where α, β_1, $\ldots$, β_r are unknown parameters and t_i are known. For instance, t_i may be the time at which the ith measurement is taken and the hypothesis specifies a polynomial trend for the mean values.

All such hypotheses can be tested by the statistic

$$R_0^2 = \min_{H_0} \sum \sum s^{ij}(\bar{U}_i - \mu_i)(\bar{U}_j - \mu_j), \tag{8d.2.1}$$

where H_0 stands for the restiictions on μ_i, $i = 1, \ldots, p$. The variance ratio based on (8d.2.1) is

$$F = \frac{N-q}{q} \cdot \frac{N}{N-1} R_0^2 \tag{8d.2.2}$$

on q and $(N - q)$ D.F., where q is the number of *independent* restrictions on $\mu_1, \ldots, \mu_p$. Thus $q = p - 1$ in situation (1), $q = r$ in (2), and $q = p - (r + 1)$ in (3) and so on.

The situation in (1), involving a composite hypothesis on the mean values, is the same as in the illustrative example of **8d.1**. We have seen that the problem can be reduced to one of testing a simple hypothesis on the mean values by considering $(p - 1)$ contrasts of the p components of the random variable. This meant transformation of the original p variables to $(p - 1)$ variables and working with the new variables. We will now give an explicit formula for the Hotelling's T^2 in terms of the estimated means and variances and covariances of the original variables. Let $\bar{U}_1, \ldots, \bar{U}_p$ be the mean values and (s_{ij}) be the sample dispersion matrix on $(N - 1)$ D.F. The statistic R_0^2 of (8d.2.1) for testing the hypothesis $\mu_1 = \cdots = \mu_p$ is

$$R_0^2 = \min_{\mu} \sum \sum s^{ij}(\bar{U}_i - \mu)(\bar{U}_j - \mu),$$

which, as the reader may easily verify, has the explicit form

$$\sum \sum s^{ij}\bar{U}_i\bar{U}_j - \frac{[\sum \sum s^{ij}(\bar{U}_i + \bar{U}_j)]^2}{4 \sum \sum s^{ij}}.$$

For explicit forms in other situations, see the illustrative examples in Rao (1959a).

8d.3 Test for Differences between Mean Values of Two Populations

Let N_1, N_2 samples be available from two p-variate populations. The sample means in the two cases are represented by

$$\bar{U}_{11}, \ldots, \bar{U}_{p1} \qquad \text{and} \qquad \bar{U}_{12}, \ldots, \bar{U}_{p2}$$

and the corrected S.P. matrices within samples by

$$(S_{ij}^{(1)}), \quad \text{D.F.} = N_1 - 1, \qquad (S_{ij}^{(2)}) \quad \text{D.F.} = N_2 - 1$$

and the pooled matrix by

$$(S_{ij}^{(0)}) = (S_{ij}^{(1)} + S_{ij}^{(2)}) \qquad \text{on} \quad N_1 + N_2 - 2 \text{ D.F.}$$

To test the hypothesis that the differences in population means (simultaneously for all characters) are zero, Hotelling's test which is same as Mahalanobis D^2 test can be used. Mahalanobis D^2 is defined by

$$D_p^{\,2} = \sum \sum s^{ij} \, d_i \, d_j \tag{8d.3.1}$$

where $d_i = \overline{U}_{i2} - \overline{U}_{i1}$, $(s^{ij}) = (s_{ij})^{-1}$ and $s_{ij} = S_{ij}^{(0)}/(N_1 + N_2 - 2)$. The variance ratio based on $D_p^{\,2}$ is

$$F = \frac{N_1 + N_2 - p - 1}{p} \cdot \frac{N_1 N_2}{(N_1 + N_2)(N_1 + N_2 - 2)} \cdot D_p^{\,2} \tag{8d.3.2}$$

on p and $(N_1 + N_2 - p - 1)$ D.F. Let us consider the example shown in Table 8d.3α.

TABLE 8d.3α.

Sample Means Based on 50 Observations Each for Two Species (Source: Fisher, 1936)

Character	*Iris Versicolor*	*Iris Setosa*	Difference
Sepal length	5.936	5.006	0.930
Sepal width	2.770	3.428	−0.658
Petal length	4.260	1.462	2.789
Petal width	1.326	0.246	1.080

TABLE 8d.3β.

Pooled Matrix (s_{ij}) on 98 D.F.

0.195340	0.092200	0.099626	0.033055
	0.121079	0.047175	0.025251
		0.125488	0.039586
			0.025106

Fisher's Discriminant Function. Fisher defined the discriminant function between two populations as that linear function of the characters for which the ratio

$$(\text{mean difference})^2 \div \text{variance}$$

is a maximum. Let $l_1 U_1 + \cdots + l_p U_p$ be the linear function and δ_i the difference of the expected values of U_i in the two populations. Then the quantity to be maximized is

$$(\sum l_i \delta_i)^2 \div \sum \sum l_i l_j \sigma_{ij}. \qquad (8d.3.3)$$

Differentiating with respect to l_i and equating to zero, we obtain the equations

$$l_1 \sigma_{i1} + \cdots + l_p \sigma_{ip} = c\delta_i, \qquad i = 1, \ldots, p \qquad (8d.3.4)$$

where c is a constant, which may be chosen as unity since only the ratios of l_i can be uniquely determined.

The coefficients can be estimated by solving the equations

$$l_1 s_{i1} + \cdots + l_p s_{ip} = d_i, \qquad i = 1, \ldots, p. \qquad (8d.3.5)$$

In the problem of *Iris Setosa* and *Iris Versicolor*, using the mean differences of Table 8d.3α and (s_{ij}) of Table 8d.3β, we find the solution as

$$l_1 = -3.0692, \qquad l_2 = -18.0006, \qquad l_3 = 21.7641, \qquad l_4 = 30.7549.$$

The estimated discriminant function is therefore

$$-3.0692 U_1 - 18.0006 U_2 + 21.7641 U_3 + 30.7549 U_4.$$

Computation of D_p^2. It is easy to see that

$$D_p^2 = \sum \sum s^{ij} d_i d_j = \sum l_i d_i \qquad (8d.3.6)$$

so that D_p^2 can be easily computed once the discriminant function coefficients are obtained by solving the equations (8d.3.5). If the discriminant function is not required, the quadratic form (8d.3.6) may be computed as in (8d.1.4). In the present example, using the formula (8d.3.6), we have

$$D_4^2 = (-3.0692)(0.930) + \cdots + (30.7549)(1.080) = 103.2119.$$

To test the hypothesis $\delta_i = 0$, $i = 1, \ldots, 4$, the F of (8d.3.2) is

$$F = \frac{95}{4} \cdot \frac{50 \times 50}{100 \times 98} \cdot 103.2119 = 625.3256,$$

which as a variance ratio on 4 and 95 D.F. is significant at the 1% level.

Test for Additional Information. Denote by D_q^2, the D^2 based on a subset q of the characters. To test the hypothesis that the rest of $(p - q)$ characters do not provide additional discrimination, the variance ratio test is (8c.4.11)

$$F = \frac{N_1 + N_2 - p - 1}{p - q} \cdot \frac{N_1 N_2 (D_p^2 - D_q^2)}{(N_1 + N_2)(N_1 + N_2 - 2) + N_1 N_2 D_q^2} \tag{8d.3.7}$$

on $p - q$ and $(N_1 + N_2 - p - 1)$ D.F.

In the example of *Iris versicolor and Iris setosa,* we may ask whether sepal and petal lengths alone are sufficient for discrimination. The D_2^2 corresponding to lengths alone is found to be 76.7082, and D_4^2 for all the measurements is 103.2119 giving

$$F = \frac{95}{2} \cdot \frac{50 \times 50(103.2119 - 76.7082)}{100 \times 98 + 50 \times 50(76.7082)} = \frac{95}{2}(0.3287) = 15.6132,$$

which is significantly high on 2 and 95 D.F. indicating that widths provide additional discrimination and that the differences in widths are not solely dependent on differences in lengths.

Test for an Assigned Discriminant Function (Fisher, 1940). The discriminant function between the two species was estimated to be

$$-3.0692U_1 - 18.0006U_2 + 21.7641U_3 + 30.7549U_4$$

with the value of $D_4^2 = 103.2119$. Since the mean measurements for *versicolor* exceed those for *setosa* except in sepal width, a discriminant function of the type

$$y = U_1 - U_2 + U_3 + U_4 \tag{8d.3.8}$$

may be suggested. In such a case it might be of interest to know whether the linear function (8d.3.8) is sufficient to discriminate between the two species. For this purpose we can apply the test (8d.3.7) computing D_1^2 based on the linear function y,

$$D_1^2 = \frac{(\bar{y}_1 - \bar{y}_2)^2}{s_{yy}}.$$

where $\bar{y}_1, \bar{y}_2$ are the mean values of y in the two samples and s_{yy} is estimated variance of y. Now

$$\begin{aligned} \bar{y}_1 - \bar{y}_2 &= d_1 - d_2 + d_3 + d_4 \\ &= 0.930 - 0.658 + 2.798 + 1.080 = 5.466 \\ V(y) &= V(U_1) + V(U_2) + V(U_3) + V(U_4) - 2C(U_1 U_2) + 2C(U_1 U_3) \\ &\quad + 2C(U_1, U_4) - 2C(U_2, U_3) - 2C(U_2, U_4) + 2C(U_3 U_4), \end{aligned}$$

which gives the estimate of $V(y)$

$$\begin{aligned} s_{yy} &= s_{11} + s_{22} + s_{33} + s_{44} - 2s_{12} + 2s_{13} + 2s_{14} - 2s_{23} - 2s_{24} + 2s_{34} \\ &= 0.482295 \end{aligned}$$

if we use s_{ij} values of Table 8d.3β. The value of D_1^2 is

$$\frac{(5.466)^2}{0.482295} = 61.9479.$$

The variance ratio on $(p - 1)$ and $(N_1 + N_2 - p - 1)$ D.F. for testing the adequacy of the assigned discriminant function is

$$F = \frac{N_1 + N_2 - p - 1}{p - 1} \cdot \frac{N_1 N_2 (D_p^2 - D_1^2)}{(N_1 + N_2)(N_1 + N_2 - 2) + N_1 N_2 D_1^2}.$$

Substituting the values of D_4^2 and D_1^2, we obtain

$$F = \tfrac{95}{3}(0.6265) = 19.8392$$

which is significant at the 1% level, on 3 and 95 D.F., showing that the linear function (8d.3.8) is not the best discriminant.

Test for Discriminant Function Coefficients. Standard errors of discriminant function coefficients have been evaluated in an attempt to judge the significance of any single coefficient. There is some difficulty in this approach, however, because the discriminant function coefficients are not unique in the sense that they are estimates of definite population parameters. But what is unique is the ratio of any two coefficients, and an exact test is possible for any assigned ratio. For instance, if the hypothetical ratio for the ith and jth characters is ρ, we consider the characters

$$U_1, \ldots, U_{i-1}, U_{i+1}, \ldots, U_{j-1}, U_{j+1}, \ldots, U_p, U_i + \rho U_j \qquad (8d.3.9)$$

$(p - 1)$ in number, obtained by replacing U_i and U_j by $U_i + \rho U_j$. Let D_{p-1}^2 be the D^2 based on the set (8d.3.9) and let D_p^2 be that based on the entire set of p characters. The variance ratio for testing whether ρ is the true value is

$$F = \frac{N_1 + N_2 - p - 1}{1} \cdot \frac{N_1 N_2 (D_p^2 - D_{p-1}^2)}{(N_1 + N_2)(N_1 + N_2 - 2) + N_1 N_2 D_{p-1}^2} \qquad (8d.3.10)$$

on 1 and $N_1 + N_2 - p - 1$ D.F. [Also see example **15**, p. 600.]

8d.4 Test for Differences in Mean Values between Several Populations

Let $N_1, \ldots, N_k$, $(N = \sum N_i)$ samples be available from k populations and denote by

$$\bar{U}_{1r}, \ldots, \bar{U}_{pr}; \qquad (S_{ij}^{(r)}) \quad \text{on} \quad N_r - 1 \text{ D.F.}$$

the sample mean values and corrected S.P. matrix from the rth sample of size N_r. Further let

$$\bar{U}_1, \ldots, \bar{U}_p; \qquad (S_{ij}) \text{ with } \sum N_r - 1 \text{ D.F.}$$

be the overall mean values and corrected S.P. matrix by throwing *all the samples* together. The sum of products between populations is computed by the formula

$$B_{ij} = \sum_{r=1}^{k} N_r \bar{U}_{ir} \bar{U}_{jr} - N \bar{U}_i \bar{U}_j$$
$$= \sum_{r=1}^{k} \frac{T_{ir} T_{jr}}{N_r} - \frac{T_i T_j}{N},$$

by using the totals instead of the means. The sum of products within populations is

$$W_{ij} = \sum_{r=1}^{k} S_{ij}^{(r)} = S_{ij} - B_{ij}$$

so that W_{ij} can be obtained without actually computing the individual corrected S.P. matrices. We then obtain the analysis of dispersion as between and within populations as in the univariate analysis of variance.

Analysis of dispersion

	D.F.	S.P. matrix
Between	$k - 1$	(B_{ij})
Within	$N - k$	(W_{ij})
Total	$N - 1$	(S_{ij})

The Λ criterion is

$$\Lambda = |\mathbf{W}| \div |\mathbf{W} + \mathbf{B}|$$

and in the notation of **8c.5**, for an application of (8c.5.3)

$$m = t - \frac{p + q + 1}{2}, \qquad t = N - 1, \qquad q = k - 1$$

$$\lambda = \frac{pq - 2}{4}, \quad s = \sqrt{(p^2q^2 - 4)}/\sqrt{p^2 + q^2 - 5}, \qquad r = \frac{pq}{2}.$$

We may use $-m \log_e \Lambda$ as χ^2 on $p(k - 1)$ D.F. or more accurately

$$F = \frac{ms - 2\lambda}{2r} \frac{1 - \Lambda^{1/s}}{\Lambda^{1/s}} \tag{8d.4.1}$$

as variance ratio on $2r$ and $(ms - 2\lambda)$ D.F.

Table 8d.4α gives the analysis of dispersion for three characters, head

TABLE 8d.4α. Analysis of Dispersion

		S.P. matrix					
	D.F.	U_1^2	U_2^2	U_3^2	U_1U_2	U_1U_3	U_2U_3
Between schools	5	752.0	151.3	1,612.7	214.2	521.3	401.2
Within schools	134	12,809.3	1499.6	21,009.6	1003.7	2671.2	4123.6
Total	139	13,561.3	1650.9	22,622.3	1217.9	3192.5	4524.8

$$\Lambda = \begin{vmatrix} 12,809.3 & 1003.7 & 2,671.2 \\ 1,003.7 & 1499.6 & 4,123.6 \\ 2,671.2 & 4123.6 & 21,009.6 \end{vmatrix} \div \begin{vmatrix} 13,561.3 & 1217.9 & 3,192.5 \\ 1,217.9 & 1650.9 & 4,524.8 \\ 3,192.5 & 4524.8 & 22,622.3 \end{vmatrix}$$

$$-\log_e \Lambda = 0.193724, \qquad m = 139 - \tfrac{1}{2}(5 + 3 + 1) = 134.5$$

$$\chi^2 = -m \log_e \Lambda = 26.0559, \qquad \text{D.F.} = 15$$

length U_1, head breadth U_2, and weight U_3 measured on 140 school boys of almost the same age belonging to six different schools in an Indian city. The value 26.0559 as χ^2 on 15 D.F. is significant at the 5% level. To use the variance ratio approximation we find $2r = 15$, $s = 2.67$, $ms - 2\lambda = 364.79$. The variance ratio (8d.4.1) is found to be 1.77, which is significant at the 5% level on 15 and 364.79 D.F. Both the approximations agree closely in view of the large number of degrees of freedom for within S.P. matrix.

8d.5 Barnard's Problem of Secular Variations in Skull Characters

The measurement of secular variations in skull characters studied by Barnard (1935) is of some importance in biometric research. Two problems involved in her study are:

1. The selection of a smaller number, out of seven skull characters, which give significant information, so far as is possible, as to changes taking place with time in Egyptian skulls belonging to four different time periods.
2. The determination of an expression linear in the measurements, which characterizes most effectively an individual skull with respect to progressive secular changes.

Taking the four measurements, basialveolar length U_1, nasal height U_2, maximum breadth U_3, and basibregmatic height U_4, we have the relevant analysis of dispersion given in Table 8d.5α.

TABLE 8d.5. Analysis of Dispersion

	Between				
Characters	Regression (1 D.F.)	Deviation (2 D.F.)	Total (3 D.F.)	Within (394 D.F.)	Total (397 D.F.)
U_1^2	119.9303	3.2503	123.1806	9661.9975	9785.1781
U_2^2	459.7344	26.6126	486.3459	9073.1150	9559.4609
U_3^2	39.0429	61.3686	100.4115	3938.3203	4088.7319
U_4^2	124.8741	515.8598	640.7339	8741.5088	9382.2427
U_1U_2	−234.8108	3.4352	−231.3756	445.5733	214.1977
U_1U_3	68.4282	18.8771	87.3053	1130.6239	1217.9292
U_1U_4	−122.3772	−6.3868	−128.7640	2148.5842	2019.8202
U_2U_3	−133.9752	26.4696	−107.5056	1239.2220	1131.7164
U_2U_4	−149.6016	274.9149	125.3133	2255.8127	2381.1260
U_3U_4	−69.8243	−67.7565	−137.5808	1271.0547	1133.4739
Matrices	(R_{ij})	(D_{ij})	(B_{ij})	(W_{ij})	$(S_{ij}^{(0)})$

The S.P. matrices for between (3 D.F.) and within the four series (397 D.F.) are obtained as in **8d.4**. Then a regression analysis is carried out to analyze the between S.P. elements as due to linear regression (1 D.F.) with time and deviation from regression (2 D.F.). The values of t, the time variable for the four series, are taken as -5, -1, 1, and 5. The calculation of linear regression involves the quantities

$$\sum (t-\bar{t})^2 = 4307.6683$$

$$\sum U_1(t-\bar{t}) = 718.7628, \qquad \sum U_3(t-\bar{t}) = +410.1019$$

$$\sum U_2(t-\bar{t}) = -1407.2608, \qquad \sum U_4(t-\bar{t}) = -733.4276$$

The sum of products due to regression are obtained from the formulas

$$R_{11} = \left(\sum U_1(t-\bar{t})\right)^2 \div \sum (t-\bar{t})^2 = 119.9304$$

$$R_{12} = \left(\sum U_1(t-\bar{t})\right)\left(\sum U_2(t-\bar{t})\right) \div \sum (t-\bar{t})^2 = -234.8108,$$

and so on.

Problem 1. Do the characters U_3, U_4 show significant variation between the series independently of U_1, U_2?

The test criterion for examining additional information is the ratio of Λ_4 to Λ_2, the values of Wilks Λ for U_1, U_2, U_3, U_4, and U_1, U_2 respectively.

$$\Lambda_4 = \frac{|W_{ij}|}{|S_{ij}^{(0)}|}, \quad i,j = 1, \ldots, 4; \quad \Lambda_2 = \frac{|W_{ij}|}{|S_{ij}^{(0)}|}, \quad i,j = 1, 2.$$

Substituting the values of $S_{ij}^{(0)}$ and W_{ij} from Table 8d.5α, we see that $\Lambda_4/\Lambda_2 = 0.87806$, in the notation of **8c.5**, $p = 2$, $q = 3$, $t = 395$ giving $m = 392$ and

$$\chi^2 = -392 \log_e (0.8781) = 51.39 \qquad \text{on} \qquad 6 \text{ D.F.}$$

which is high thus showing the importance of all the four characters.

Problem 2. Can the regression of each of the characters on time be considered linear?

The appropriate test for this is

$$\Lambda = \frac{|W|}{|D+W|} = 0.9031$$

with $p = 4$, $q = 2$, $t = 396$. The χ^2 approximation 40.02 on 8 D.F. is high, indicating nonlinearlity of regression. It is easy to see that the test can be extended to examine whether parabolic regression with time can explain the differences in mean values, and so on.

For other applications refer to Bartlett (1947), Riffenburgh and Clunies-Ross (1960), Williams (1952, 1955, 1959), and the papers by the author, Rao (1948b, 1949b, 1954c, 1955a, 1958b, 1959a, 1961g, 1962c, 1965b, 1965d, 1966c, f, 1967c, 1972a). See also Roy, Gnanadesikan and Srivastava (1972) and Williams (1967) for a discussion of related problems and applications.

8e DISCRIMINATORY ANALYSIS (IDENTIFICATION)

8e.1 Discriminant Scores for Decision

In **7d.1** to **7d.6** of Chapter 7 on statistical inference, we have discussed the problem of identification, that is, of deciding on the membership of an *observed* individual to *one of a given set* of populations to which he can possibly belong. For a satisfactory solution of the problem we need to know the following:

1. The probability densities, $P_1(\mathbf{U}), \ldots, P_k(\mathbf{U})$, for a given set of measurements $\mathbf{U}$ on an individual in the k alternative populations.
2. Prior probabilities $\pi_1, \ldots, \pi_k$ for the populations, which are relative frequencies of individuals of the k populations in the composite population from which an individual to be identified has been observed.
3. The assignment of a loss function, that is, the specification of values r_{ij} representing the loss in identifying an individual of the ith population as a member of the jth population.

Then, given an individual with measurements $\mathbf{U}$, his discriminant score for the ith population is computed as (see 7d.3.8)

$$S_i = -[\pi_1 P_1(\mathbf{U}) r_{1i} + \cdots + \pi_k P_k(\mathbf{U}) r_{ki}]$$
$$i = 1, \ldots, k, \qquad (8\text{e}.1.1)$$

and the individual is assigned to that population for which his discriminant score is the highest. Such a rule is shown to minimize the expected loss (in the long run, see **7d.3**).

The decision rule (8e.1.1) is ideal when individuals have to be identified in a routine manner as in vocational guidance and selection problems. In such cases the true identification may become available at a later point of time, and the accumulating data would provide progressively satisfactory estimates of the densities $P_i(\mathbf{U})$ and prior probabilities π_i so that an application of the rule (8e.1.1) becomes feasible.

In many practical problems, the losses due to wrong identification may be difficult to assess, in which case the criterion of minimizing the frequency of wrong identification (in the long run) may serve the purpose. The optimum rule in such a case is to assign the individual with measurements $\mathbf{U}$ to that population for which the posterior probability has the highest value. Or in other words the discriminant score for the ith population is

$$S_i = \pi_i P_i(\mathbf{U}). \qquad (8\text{e}.1.2)$$

Let us consider the case where the distribution of **U** is p-variate normal in each of the populations. Choose $P_i(\mathbf{U})$ as the normal density

$$(2\pi)^{-p/2}|\mathbf{\Sigma}_i|^{-1/2}\exp[-\tfrac{1}{2}(\mathbf{U}-\boldsymbol{\mu}_i)'\mathbf{\Sigma}_i^{-1}(\mathbf{U}-\boldsymbol{\mu}_i)] \qquad (8e.1.3)$$
$$i=1,\ldots,k,$$

that is, with mean $\boldsymbol{\mu}_i$ and dispersion $\mathbf{\Sigma}_i$ for the ith population. Taking the logarithm of $\pi_i P_i(\mathbf{U})$ and omitting the factor $(2\pi)^{-p/2}$ common to all i, we see that the *equivalent discriminant score* (8e.1.2) for the ith population is

$$S_i = -\tfrac{1}{2}\log|\mathbf{\Sigma}_i| - \tfrac{1}{2}(\mathbf{U}-\boldsymbol{\mu}_i)'\mathbf{\Sigma}_i^{-1}(\mathbf{U}-\boldsymbol{\mu}_i) + \log \pi_i \qquad (8e.1.4)$$

involving the mean $\boldsymbol{\mu}_i$ and the dispersion matrix $\mathbf{\Sigma}_i$ of the ith population. The function (8e.1.4) is quadratic in **U** and may be called a quadratic discriminant score. An individual is assigned to that population for which the quadratic discriminant score has the highest value.

If the populations do not differ in the dispersion matrices, the terms

$$-\tfrac{1}{2}\log|\mathbf{\Sigma}| - \tfrac{1}{2}\mathbf{U}'\mathbf{\Sigma}^{-1}\mathbf{U} \qquad (8e.1.5)$$

are common to all S_i. Subtracting the terms (8e.1.5) from (8e.1.4), the *equivalent discriminant score* for the ith population may be written

$$S_i = (\boldsymbol{\mu}_i'\mathbf{\Sigma}^{-1})\mathbf{U} - \tfrac{1}{2}\boldsymbol{\mu}_i'\mathbf{\Sigma}^{-1}\boldsymbol{\mu}_i + \log \pi_i \qquad (8e.1.6)$$

which is linear in **U** and may be called a linear discriminant score.

It may be seen that when there are only two populations, only one comparison is involved and the decision can be taken by computing the difference $S_1 - S_2$, which is $L(\mathbf{U}) - c$ where

$$L(\mathbf{U}) = (\boldsymbol{\mu}_1' - \boldsymbol{\mu}_2')\mathbf{\Sigma}^{-1}\mathbf{U}$$
$$c = \tfrac{1}{2}(\boldsymbol{\mu}_1'\mathbf{\Sigma}^{-1}\boldsymbol{\mu}_1 - \boldsymbol{\mu}_2'\mathbf{\Sigma}^{-1}\boldsymbol{\mu}_2) + \log\pi_2 - \log\pi_1. \qquad (8e.1.7)$$

The function (8e.1.7) is the linear discriminant function of Fisher derived in **8d.3**. The decision rule may then be expressed in the form

$$\text{assign to the first population if} \quad L(\mathbf{U}) \geqslant c,$$
$$\text{assign to the second population if} \quad L(\mathbf{U}) \leqslant c. \qquad (8e.1.8)$$

But the approach in terms of discriminant scores (8e.1.6) brings the special case of two populations in the general frame work of identification rules. The computation of individual scores even for two populations is preferable in practice since they are more informative.

As an application, let us consider the problem of identifying the neurotic state of an individual referred for psychiatric examination. Table 8e.1α gives the mean values and the dispersion matrix estimated from previous data for three measurements A, B, C.

TABLE 8e.1α. Mean Scores for Neurotic Groups (Rao and Slater, 1949)

Group	Sample Size	Mean Score A	B	C
Anxiety state	114	2.9298	1.1667	0.7281
Hysteria	33	3.0303	1.2424	0.5455
Psychopathy	32	3.8125	1.8438	0.8125
Obsession	17	4.7059	1.5882	1.1176
Personality change	5	1.4000	0.2000	0.0000
Normal	55	0.6000	0.1455	0.2182

	Within Group Dispersion Matrix (σ_{ij}) A	B	C		Reciprocal (σ^{ij}) A	B	C
A	2.3008	0.2516	0.4742	A	0.5432	−0.2002	−0.4208
B	0.2516	0.6075	0.0358	B	−0.2002	1.7258	0.0558
C	0.4742	0.0358	0.5951	C	−0.4208	0.0558	2.0123

For any group, the linear discriminant score is

$$l_1 A + l_2 B + l_3 C - \tfrac{1}{2}(l_1 m_1 + l_2 m_2 + l_3 m_3) + \log \pi,$$

where A, B, C are the observed values of the measurements, m_1, m_2, m_3 are the mean values, and π is the prior probability. The coefficients l_1, l_2, l_3 are computed from the formulas

$$l_1 = \sigma^{11} m_1 + \sigma^{12} m_2 + \sigma^{13} m_3$$
$$l_2 = \sigma^{21} m_1 + \sigma^{22} m_2 + \sigma^{23} m_3$$
$$l_3 = \sigma^{31} m_1 + \sigma^{32} m_2 + \sigma^{33} m_3 .$$

The coefficients l_1, l_2, l_3 and the other terms needed for computing the discriminant score for each neurotic group are given in Table 8e.1β.

TABLE 8c.1β. The Coefficients and Other Terms for Discriminant Scores

Group	l_1	l_2	l_3	$\frac{1}{2}\sum l_j m_j$	$\log \pi - \frac{1}{2}\sum l_j m_j$
Anxiety state	1.0515	1.4676	0.2974	2.5047	−3.3137
Hysteria	1.1678	1.5679	−0.1081	2.7139	−4.7626
Psychopathy	1.3599	2.4641	0.1336	4.9182	−6.9977
Obsession	1.7680	1.8611	0.3573	5.8375	−8.5495
Personality change	0.7204	0.0649	−0.5780	0.5107	−4.4465
Normal	0.2050	0.1431	0.1947	0.0931	−1.6311

In computing the last column of Table 8e.1β we take the prior probabilities to be proportional to observed sample sizes, which may not be estimates of true prior probabilities. The appropriate estimates are the relative frequencies of individuals of different psychiatric conditions referred to the hospital in the past. Given the measurements A, B, C for any individual, the six scores

$$S_i = l_1 A + l_2 B + l_3 C + (\log \pi_i - \tfrac{1}{2} \sum l_j m_j)$$
$$i = 1, \ldots, 6,$$

are computed, and the individual is assigned to the group for which the score is the highest.

8e.2 Discriminant Analysis in Research Work

The procedure of **8e.1** is feasible and appropriate in a routine decision-making situation where the object is to minimize the loss (or frequency of wrong identification) in the long run. Now, consider a problem like that of examining whether *Australopethicus africanus*, a fossil discovered in Africa, was more hominid than anthropoid. It is clear that the measurements on *A. africanus* have to be compared with those of known groups of hominid and anthropoid fossils in order to identify his affinities.

It is not realistic in such situations to accept as a *datum* of the problem that the observed fossil is a member of one of the known groups of fossils. In statistical analysis we have to take into account the possibility that he might belong to an unknown group whose existence has yet to be established. Although it may be possible to have even crude estimates of the means and dispersion of measurements for each known group, there does not seem to be any method of obtaining the prior probabilities. The available numbers of fossils in different groups need not be good indicators of their relative abundance in the population of fossils. Furthermore, there is no possibility of attaching prior probabilities to the undiscovered groups. Finally, the concept of loss function is not meaningful in the present context. Consequently the method of **8e.1** is neither applicable nor feasible in answering the present problem.

We shall now discuss a general approach for examining the membership of an individual to one of *two* known groups or *possibly to a third* unknown group. With this end in view we establish the following.

(i) *Given two p-variate normal populations* $N_p(\boldsymbol{\mu}_1, \boldsymbol{\Sigma})$ *and* $N_p(\boldsymbol{\mu}_2, \boldsymbol{\Sigma})$ *with the same dispersion matrix, the linear discriminant function* (*which is same as the logarithm of the ratio of densities apart from a constant*) *as defined in* (8e.1.7) *is*

$$L(\mathbf{U}) = (\boldsymbol{\mu}_2 - \boldsymbol{\mu}_1)'\boldsymbol{\Sigma}^{-1}\mathbf{U}. \tag{8e.2.1}$$

Let $\boldsymbol{\mu}_1$, $\boldsymbol{\mu}_2$, *and* $\boldsymbol{\Sigma}$ *be all known. Then the statistic* $L(\mathbf{U})$ *is sufficient for the class of normal densities* $N_p(\alpha\boldsymbol{\mu}_1 + \beta\boldsymbol{\mu}_2, \boldsymbol{\Sigma})$ *where* α, β *vary such that* $\alpha + \beta = 1$, *that is, the conditional distribution of* $\mathbf{U}$ *given* $L(\mathbf{U})$ *is the same for all values of* $\alpha\boldsymbol{\mu}_1 + \beta\boldsymbol{\mu}_2$, $\alpha + \beta = 1$.

Consider the density of $N_p(\alpha\boldsymbol{\mu}_1 + \beta\boldsymbol{\mu}_2, \boldsymbol{\Sigma})$,

$$(2\pi)^{-p/2}|\boldsymbol{\Sigma}|^{-1/2}\exp[-\tfrac{1}{2}(\mathbf{U} - \boldsymbol{\mu}_1 - \beta\boldsymbol{\delta})'\boldsymbol{\Sigma}^{-1}(\mathbf{U} - \boldsymbol{\mu}_1 - \beta\boldsymbol{\delta})], \quad (8e.2.2)$$

where $\alpha = 1 - \beta$ and $\boldsymbol{\delta} = \boldsymbol{\mu}_2 - \boldsymbol{\mu}_1$. The expression in the exponential (omitting $-\frac{1}{2}$) is

$$\begin{aligned}(\mathbf{U} - \boldsymbol{\mu}_1)'\boldsymbol{\Sigma}^{-1}(\mathbf{U} - \boldsymbol{\mu}_1) &- 2\beta\boldsymbol{\delta}'\boldsymbol{\Sigma}^{-1}\mathbf{U} + 2\beta\boldsymbol{\delta}'\boldsymbol{\Sigma}^{-1}\boldsymbol{\mu}_1 + \beta^2\boldsymbol{\delta}'\boldsymbol{\Sigma}^{-1}\boldsymbol{\delta} \\ = H(\mathbf{U}) \qquad &- 2\beta L(\mathbf{U}) + G(\beta),\end{aligned} \quad (8e.2.3)$$

where $H(\mathbf{U})$ and $L(\mathbf{U})$ do not involve β and $G(\beta)$ does not involve $\mathbf{U}$. The density (8e.2.2) can then be written as the product (omitting the constant multiplier)

$$\exp[-\tfrac{1}{2}H(\mathbf{U})] \times \exp\{\tfrac{1}{2}[2\beta L(\mathbf{U}) - G(\beta)]\},$$

where the first factor is independent of β and the second factor involves only $L(\mathbf{U})$ and β. Hence $L(\mathbf{U})$ is sufficient for β, that is, the conditional distribution of $\mathbf{U}$ given $L(\mathbf{U})$ is the same for all values of $\alpha\boldsymbol{\mu}_1 + \beta\boldsymbol{\mu}_2$, $(\alpha + \beta) = 1$.

The result of (i) is a generalization of a theorem due to Smith (1947) who established the sufficiency of $L(\mathbf{U})$ only for the special cases $\beta = 0$ and $\beta = 1$.

Suppose we have the problem of examining whether an observed individual belongs to one of two given populations A, B, or to an unspecified population. The result of (i) shows that the discriminant function constructed from the densities for A and B is sufficient to examine an individual's membership not only with respect to A and B but to the wider set of populations with mean values lying on the line joining the mean values of A and B. Thus, to use the discriminant function constructed with reference to A and B to infer on the membership of an individual, it is necessary to test whether the individual belongs to the wider set of populations defined by the line joining the mean vectors of A and B. Such a test is considered in (ii) below.

(ii) *Let* H_0 *be the null hypothesis that an individual with measurements* $\mathbf{U}$ *comes from* $N_p(\alpha\boldsymbol{\mu}_1 + \beta\boldsymbol{\mu}_2, \boldsymbol{\Sigma})$, *where* β *is arbitrary such that* $\alpha + \beta = 1$ *and* $\boldsymbol{\mu}_1$, $\boldsymbol{\mu}_2$, $\boldsymbol{\Sigma}$ *are known or estimated on large samples. The test criterion for testing* H_0,

$$(\mathbf{U} - \boldsymbol{\mu}_1)'\boldsymbol{\Sigma}^{-1}(\mathbf{U} - \boldsymbol{\mu}_1) - \frac{[(\mathbf{U} - \boldsymbol{\mu}_1)'\boldsymbol{\Sigma}^{-1}(\boldsymbol{\mu}_2 - \boldsymbol{\mu}_1)]^2}{(\boldsymbol{\mu}_2 - \boldsymbol{\mu}_1)'\boldsymbol{\Sigma}^{-1}(\boldsymbol{\mu}_2 - \boldsymbol{\mu}_1)} \quad (8e.2.4)$$

is distributed as χ^2 *on* $(p - 1)$ *D.F.*

The reader should not find any difficulty in establishing the distribution of (8e.2.4) by using the results on quadratic forms. Observe that the first term is distributed as $\chi^2(p)$, the second as $\chi^2(1)$, and the difference is positive. Then apply [(iv), **3b.4**].

By applying the test (8e.2.4) to *A. africanus*, choosing chimpanzee and human fossils as alternative groups and length and breadth of tooth as characters, we find the value of χ^2 as nearly 5, which is high for $(p - 1 = 1)$ D.F. Thus the *A. africanus* possibly belongs to a group with mean dimensions not related in any simple way to those of chimpanzee and human fossils.

If the test (8e.2.4) had not been significant, some further analysis based on the discriminant function would have been necessary. The membership to one of the groups (chimpanzee) is examined by the statistic

$$\frac{[(\boldsymbol{\mu}_2 - \boldsymbol{\mu}_1)\boldsymbol{\Sigma}^{-1}(\mathbf{U} - \boldsymbol{\mu}_1)]^2}{(\boldsymbol{\mu}_2 - \boldsymbol{\mu}_1)'\boldsymbol{\Sigma}^{-1}(\boldsymbol{\mu}_2 - \boldsymbol{\mu}_1)} \sim \chi^2(1),$$

and to the other (human) by

$$\frac{[(\boldsymbol{\mu}_2 - \boldsymbol{\mu}_1)'\boldsymbol{\Sigma}^{-1}(\mathbf{U} - \boldsymbol{\mu}_2)]^2}{(\boldsymbol{\mu}_2 - \boldsymbol{\mu}_1)'\boldsymbol{\Sigma}^{-1}(\boldsymbol{\mu}_2 - \boldsymbol{\mu}_1)} \sim \chi^2(1).$$

If both are significantly large as χ^2's on 1 D.F., then again there is the possibility of the individual belonging to a third group but with the mean vector collinear with the mean vectors for chimpanzee and human fossils. If one χ^2 is small and the other is large, there is an indication that the individual belongs to one of the groups. There is also the possibility that both the χ^2's are small, in which case the affinity of the individual to any one of the groups is not clearcut.

In the example of *A. africanus*, the χ^2 for chimpanzee is 3.80, which is nearly at the 5% level, and that for human is 0.03, which is very small. But these tests are not valid (may be misleading) since the discriminant function is not appropriate to infer about the position of *A. africanus* in relation to the two groups considered. An indiscriminate use of the disciminant function without the prior test (8e.2.4) would have led to the wrong conclusion that *A. africanus* has strong affinities with the human group. On the other hand we have demonstrated that the fossil possibly belongs to a third unknown group (by the test of 8e.2.4).

The foregoing analysis can be extended to the case of several alternative groups (see Rao, 1962c). For further details on *A. africanus* and references to previous work the reader is referred to Ashton, Healy, and Lipton (1957).

The problem becomes complicated when the parameters of the alternative hypotheses are unknown, but can be estimated (see John, 1961; Rao, 1954a; and Wald, 1944).

8e.3 Discrimination between Composite Hypotheses

Let $\mathbf{U}$ be a p-variate random variable and $P(\cdot|\theta)$ its density depending on a parameter $\theta \in \Theta$. Let H_1 be the hypothesis that $\theta \in \Theta_1$ and H_2 that $\theta \in \Theta_2$ where Θ_1 and Θ_2 are two disjoint subsets of Θ. The problem is one of choosing between H_1 and H_2 on the basis of an observed value of $\mathbf{U}$.

The problem admits a nice solution if there exists an ancillary S such that its density

$$P(s|\theta) = \begin{cases} P_1(s) & \text{for any } \theta \in \Theta_1, \\ P_2(s) & \text{for any } \theta \in \Theta_2. \end{cases} \tag{8e.3.1}$$

When $P_1(\cdot)$ and $P_2(\cdot)$ are different and independent of θ, the discriminant function for choosing between H_1 and H_2 is provided by the likelihood ratio

$$\log \frac{P_1(s)}{P_2(s)} \tag{8e.3.2}$$

which is independent of the particular hypotheses in Θ_1 and Θ_2. Let us consider some examples.

Example 1. Let $\mathbf{U}$ be a p-vector normal variable such that

$$\begin{aligned} E(\mathbf{U}|\boldsymbol{\theta}_1, H_1) &= \boldsymbol{\alpha}_1 + \mathbf{B}'\boldsymbol{\theta}_1, \quad D(\mathbf{U}|\boldsymbol{\theta}_1, H_1) = \boldsymbol{\Sigma}, \\ E(\mathbf{U}|\boldsymbol{\theta}_2, H_2) &= \boldsymbol{\alpha}_2 + \mathbf{B}'\boldsymbol{\theta}_2, \quad D(\mathbf{U}|\boldsymbol{\theta}_2, H_2) = \boldsymbol{\Sigma}. \end{aligned} \tag{8e.3.3}$$

The parameters $\boldsymbol{\theta}_1$ and $\boldsymbol{\theta}_2$ are unknown but belong to given sets specified by the composite hypotheses H_1 and H_2. The problem is to decide on the basis of observed $\mathbf{U}$ whether H_1 or H_2 is true. Such a problem was faced by Burnaby (1964) when he had to discriminate between two fossil groups, each group being a mixture of subpopulations with different stages of growth of individuals. The problem is to identify a fossil as belonging to one of the two groups when the age of the fossil is not known. In Burnaby's case

$$E(\mathbf{U}|t, H_1) = \boldsymbol{\alpha}_1 + t\boldsymbol{\beta}, \; E(\mathbf{U}|t, H_2) = \boldsymbol{\alpha}_2 + t\boldsymbol{\beta} \tag{8e.3.4}$$

where t represents age and $\boldsymbol{\beta}$ is the vector of growth rates for different characters. Let us consider the general setup (8e.3.3) of which (8e.3.4) is a particular case. Consider a $(p-k) \times p$ matrix $\mathbf{C}$ of rank $p-k$ such that $\mathbf{BC}' = 0$ where $k = R(\mathbf{B})$. Then

$$\begin{aligned} E(\mathbf{CU}|H_1) &= \mathbf{C}\boldsymbol{\alpha}_1, \qquad D(\mathbf{CU}|H_1) = \mathbf{C\Sigma C}', \\ E(\mathbf{CU}|H_2) &= \mathbf{C}\boldsymbol{\alpha}_2, \qquad D(\mathbf{CU}|H_2) = \mathbf{C\Sigma C}', \end{aligned} \tag{8e.3.5}$$

independent of $\boldsymbol{\theta}_1, \boldsymbol{\theta}_2$, which shows that $\mathbf{CU}$ is an ancillary of the type (8e.3.1). Using the normal density, the discriminant function based on $\mathbf{CU}$ is

$$(\mathbf{C}\boldsymbol{\alpha}_1 - \mathbf{C}\boldsymbol{\alpha}_2)'(\mathbf{C\Sigma C}')^{-1}\mathbf{CU} = (\boldsymbol{\alpha}_1 - \boldsymbol{\alpha}_2)'\mathbf{C}'(\mathbf{C\Sigma C}')^{-1}\mathbf{CU}. \tag{8e.3.6}$$

Using the identity of Example 33 (p. 77) at the end of Chapter 1

$$\mathbf{C}'(\mathbf{C}\mathbf{\Sigma}\mathbf{C}')^{-1}\mathbf{C} = \mathbf{\Sigma}^{-1} - \mathbf{\Sigma}^{-1}\mathbf{B}'(\mathbf{B}\mathbf{\Sigma}^{-1}\mathbf{B}')^{-1}\mathbf{B}\mathbf{\Sigma}^{-1} \tag{8e.3.7}$$

(8e.3.6) reduces to

$$(\boldsymbol{\alpha}_1 - \boldsymbol{\alpha}_2)'(\mathbf{\Sigma}^{-1} - \mathbf{\Sigma}^{-1}\mathbf{B}'(\mathbf{B}\mathbf{\Sigma}^{-1}\mathbf{B}')^{-1}\mathbf{B}\mathbf{\Sigma}^{-1})\mathbf{U}. \tag{8e.3.8}$$

The result (8e.3.8) involves only the known matrices $\boldsymbol{\alpha}_1$, $\boldsymbol{\alpha}_2$, $\mathbf{B}$, and $\mathbf{\Sigma}$.

Example 2. Let us now consider the alternative composite hypotheses

$$\begin{aligned} E(\mathbf{U}\,|\,\boldsymbol{\theta}_1, H_1) &= \boldsymbol{\alpha}_1 + \mathbf{B}'\boldsymbol{\theta}_1, \qquad D(\mathbf{U}\,|\,\boldsymbol{\theta}_1, H_1) = \mathbf{\Sigma}_1, \\ E(\mathbf{U}\,|\,\boldsymbol{\theta}_2, H_2) &= \boldsymbol{\alpha}_2 + \mathbf{B}'\boldsymbol{\theta}_2, \qquad D(\mathbf{U}\,|\,\boldsymbol{\theta}_2, H_2) = \mathbf{\Sigma}_2, \end{aligned} \tag{8e.3.9}$$

where $\boldsymbol{\theta}_1$ and $\boldsymbol{\theta}_2$ are unknown.

The same statistic $\mathbf{CU}$ defined in Example 1 is ancilliary under both the hypotheses in (8e.3.9). The distributions under H_1 and H_2 are specified by

$$\begin{aligned} E(\mathbf{CU}\,|\,H_1) &= \mathbf{C}\boldsymbol{\alpha}_1, \qquad D(\mathbf{CU}\,|\,H_1) = \mathbf{C}\mathbf{\Sigma}_1\mathbf{C}', \\ E(\mathbf{CU}\,|\,H_2) &= \mathbf{C}\boldsymbol{\alpha}_2, \qquad D(\mathbf{CU}\,|\,H_2) = \mathbf{C}\mathbf{\Sigma}_2\mathbf{C}' \end{aligned} \tag{8e.3.10}$$

so that the log likelihood ratio is the quadratic function in $\mathbf{U}$

$$\begin{aligned} &\mathbf{U}'\mathbf{C}'[(\mathbf{C}\mathbf{\Sigma}_1\mathbf{C}')^{-1} - (\mathbf{C}\mathbf{\Sigma}_2\mathbf{C}')^{-1}]\mathbf{CU} \\ &\qquad - 2\,[\boldsymbol{\alpha}_1'\mathbf{C}'(\mathbf{C}\mathbf{\Sigma}_1\mathbf{C}')^{-1} - \boldsymbol{\alpha}_2'\mathbf{C}'(\mathbf{C}\mathbf{\Sigma}_2\mathbf{C}')^{-1}]\mathbf{CU}. \end{aligned} \tag{8e.3.11}$$

Using the identity (8e.3.7) for each $\mathbf{\Sigma}_i$, (8e.3.11) can be expressed in terms of $\boldsymbol{\alpha}_i$, $\mathbf{\Sigma}_i$, and $\mathbf{B}$ only. We shall study the properties of the discriminant functions (8e.3.8) and (8e.3.11) without making any assumption on the distribution of $\mathbf{U}$. While doing so we will also consider the situation where the past data on $\mathbf{U}$ is available only on mixtures of populations in H_1 and H_2. Thus we may not know or have estimates of $\boldsymbol{\alpha}_1$, $\boldsymbol{\alpha}_2$, $\mathbf{\Sigma}_1$, $\mathbf{\Sigma}_2$ but only of

$$\begin{aligned} E(\mathbf{U}\,|\,H_1) &= \boldsymbol{\alpha}_1 + \mathbf{B}'\bar{\boldsymbol{\theta}}_1, \qquad D(\mathbf{U}\,|\,H_1) = \mathbf{\Sigma}_1 + \mathbf{B}'\mathbf{D}_1\mathbf{B} \\ E(\mathbf{U}\,|\,H_2) &= \boldsymbol{\alpha}_2 + \mathbf{B}'\bar{\boldsymbol{\theta}}_2, \qquad D(\mathbf{U}\,|\,H_2) = \mathbf{\Sigma}_2 + \mathbf{B}'\mathbf{D}_2\mathbf{B} \end{aligned}$$

where $\bar{\boldsymbol{\theta}}_1$, $\bar{\boldsymbol{\theta}}_2$ are the unknown average values of $\boldsymbol{\theta}_1$ and $\boldsymbol{\theta}_2$ and $\mathbf{D}_1$, $\mathbf{D}_2$ are unknown matrices depending on mixtures in H_1 and in H_2.

Let $\boldsymbol{\delta}_1 = \boldsymbol{\alpha}_1 - \boldsymbol{\alpha}_2 + \mathbf{B}'(\bar{\boldsymbol{\theta}}_1 - \bar{\boldsymbol{\theta}}_2)$, $\mathbf{\Lambda}_1 = \mathbf{\Sigma}_1 + \mathbf{B}'\mathbf{D}_1\mathbf{B}$ and $\mathbf{\Lambda}_2 = \mathbf{\Sigma}_2 + \mathbf{B}'\mathbf{D}_2\mathbf{B}$. We show that the discriminant functions (8e.3.8) and (8e.3.11) remain unchanged if we use $\boldsymbol{\delta}_1$ for $\boldsymbol{\alpha}_1 - \boldsymbol{\alpha}_2$, $\mathbf{\Lambda} = \mathbf{\Sigma} + 2^{-1}\mathbf{B}'(\mathbf{D}_1 + \mathbf{D}_2)\mathbf{B}$ for $\mathbf{\Sigma}$ and $\mathbf{\Lambda}_1 = \mathbf{\Sigma}_1 + \mathbf{B}'\mathbf{D}_1\mathbf{B}$, $\mathbf{\Lambda}_2 = \mathbf{\Sigma}_2 + \mathbf{B}'\mathbf{D}_2\mathbf{B}$ for $\mathbf{\Sigma}_1$, $\mathbf{\Sigma}_2$ whatever $\bar{\boldsymbol{\theta}}_1$, $\bar{\boldsymbol{\theta}}_2$, $\mathbf{D}_1$, $\mathbf{D}_2$ may be, which is indeed a very fortunate situation.

First we characterize the linear discriminant function (8e.3.8).

(i) *Let* **U** *be a vector random variable such that*

$$E(\mathbf{U}|H_1) = \boldsymbol{\alpha}_1 + \mathbf{B}'\bar{\boldsymbol{\theta}}_1, \qquad D(\mathbf{U}|H_1) = \boldsymbol{\Sigma} + \mathbf{B}'\mathbf{D}_1\mathbf{B},$$
$$E(\mathbf{U}|H_2) = \boldsymbol{\alpha}_2 + \mathbf{B}'\bar{\boldsymbol{\theta}}_2, \qquad D(\mathbf{U}|H_2) = \boldsymbol{\Sigma} + \mathbf{B}'\mathbf{D}_2\mathbf{B}. \tag{8e.3.12}$$

Then

$$\sup_{\mathbf{BL}=\mathbf{0}} \frac{[E(\mathbf{L}'\mathbf{U}|H_1) - E(\mathbf{L}'\mathbf{U}|H_2)]^2}{2^{-1}[V(\mathbf{L}'\mathbf{U}|H_1) + V(\mathbf{L}'\mathbf{U}|H_2)]} \tag{8e.3.13}$$

is attained at

$$L_* = [\boldsymbol{\Sigma}^{-1} - \boldsymbol{\Sigma}^{-1}\mathbf{B}'(\mathbf{B}'\boldsymbol{\Sigma}^{-1}\mathbf{B})^{-1}\mathbf{B}\boldsymbol{\Sigma}^{-1}](\boldsymbol{\alpha}_1 - \boldsymbol{\alpha}_2). \tag{8e.3.14}$$

Under the condition $\mathbf{BL} = \mathbf{0}$, the expression (8e.3.13) reduces to

$$\sup_{\mathbf{BL}=\mathbf{0}} \frac{[\mathbf{L}'(\boldsymbol{\alpha}_1 - \boldsymbol{\alpha}_2)]^2}{\mathbf{L}\boldsymbol{\Sigma}\mathbf{L}}, \tag{8e.3.15}$$

and the result follows by an application of (1c.6.3), (p. 50).

Note that the discriminant function (8e.3.8) is $\mathbf{L}_*'\mathbf{U}$ where $\mathbf{L}_*$ is as defined in (8e.3.14). It may be recalled that the linear discriminant for two simple hypotheses is obtained by taking the supremum in (8e.3.13) without the restriction $\mathbf{BL} = \mathbf{0}$. $\mathbf{L}_x$ is same if we use $\boldsymbol{\delta}$ for $\boldsymbol{\alpha}_1 - \boldsymbol{\alpha}_2$ and $(\boldsymbol{\Lambda}_1 + \boldsymbol{\Lambda}_2)/2$ for $\boldsymbol{\Sigma}$.

(ii) *The quadratic discriminant function* (8e.3.11) *is invariant if* $\boldsymbol{\delta}$ *is used for* $\boldsymbol{\alpha}_1 - \boldsymbol{\alpha}_2$ *and* $\boldsymbol{\Lambda}_1, \boldsymbol{\Lambda}_2$ *are used for* $\boldsymbol{\Sigma}_1, \boldsymbol{\Sigma}_2$.

The result can be proved by direct verification.

8f RELATION BETWEEN SETS OF VARIATES

8f.1 Canonical Correlations

Multiple correlation as defined in (4g.1.11) measures association between one variable and a set of other variables. In fact, it is shown to be the maximum correlation between one variable and a linear function of the others, which is also the same as the correlation between a variable and its minimum mean square error predictor based on other variables [iv, **4g.1**]. This concept was generalized by Hotelling (1935, 1936) to study the association between two sets of variables.

Let $\mathbf{U}_1' = (U_1, \ldots, U_r)$, $\mathbf{U}_2' = (U_{r+1}, \ldots, U_p)$ be two sets of variables with

$$D(\mathbf{U}_1) = \boldsymbol{\Sigma}_{11}, \qquad \text{cov}(\mathbf{U}_1, \mathbf{U}_2) = \boldsymbol{\Sigma}_{12}, \qquad D(\mathbf{U}_2) = \boldsymbol{\Sigma}_{22} \tag{8f.1.1}$$

We consider two linear functions $\mathbf{L}'\mathbf{U}_1$ and $\mathbf{M}'\mathbf{U}_2$ of unit variance and choose $\mathbf{L}$, $\mathbf{M}$ such that the correlation between $\mathbf{L}'\mathbf{U}_1$ and $\mathbf{M}'\mathbf{U}_2$ is a maximum. The problem is then one of maximizing $\mathbf{L}'\mathbf{\Sigma}_{12}\mathbf{M}$ subject to the conditions $\mathbf{L}'\mathbf{\Sigma}_{11}\mathbf{L} = 1 = \mathbf{M}'\mathbf{\Sigma}_{22}\mathbf{M}$. Introducing Lagrangian multipliers, the expression to be differentiated is

$$\mathbf{L}'\mathbf{\Sigma}_{12}\mathbf{M} - \frac{\lambda_1}{2}\mathbf{L}'\mathbf{\Sigma}_{11}\mathbf{L} - \frac{\lambda_2}{2}\mathbf{M}'\mathbf{\Sigma}_{22}\mathbf{M}$$

giving the equations

$$\mathbf{\Sigma}_{12}\mathbf{M} - \lambda_1\mathbf{\Sigma}_{11}\mathbf{L} = 0, \qquad -\lambda_2\mathbf{\Sigma}_{22}\mathbf{M} + \mathbf{\Sigma}_{21}\mathbf{L} = \mathbf{0} \tag{8f.1.2}$$

From the first equation, $\mathbf{L}'\mathbf{\Sigma}_{12}\mathbf{M} = \lambda_1$, and from second, $\lambda_2 = \mathbf{M}'\mathbf{\Sigma}_{21}\mathbf{L}$, which shows that $\lambda_1 = \lambda_2 = \rho$ (say). Multiplying the first equation of (8f.1.2) by $\mathbf{\Sigma}_{21}\mathbf{\Sigma}_{11}^{-1}$ and by adding it to ρ times the second we obtain

$$(\mathbf{\Sigma}_{21}\mathbf{\Sigma}_{11}^{-1}\mathbf{\Sigma}_{12} - \rho^2\mathbf{\Sigma}_{22})\mathbf{M} = \mathbf{0}. \tag{8f.1.3}$$

Thus ρ^2 and $\mathbf{M}$ are an eigenroot and vector corresponding to the determinantal equation

$$|\mathbf{\Sigma}_{21}\mathbf{\Sigma}_{11}^{-1}\mathbf{\Sigma}_{12} - \rho^2\mathbf{\Sigma}_{22}| = 0. \tag{8f.1.4}$$

Let $\rho_1^2, \ldots, \rho_s^2 (s = p - r)$ be the roots and $\mathbf{M}_1, \ldots, \mathbf{M}_s$ the corresponding vectors. Further let $\mathcal{M} = (\mathbf{M}_1 \vdots \ldots \vdots \mathbf{M}_s)$. Then we have the following

$$\begin{aligned} \mathcal{M}'\mathbf{\Sigma}_{22}\mathcal{M} &= \mathbf{I}, & \mathbf{\Sigma}_{22} &= \mathcal{M}'^{-1}\mathcal{M}^{-1} \\ \mathcal{M}'\mathbf{\Sigma}_{21}\mathbf{\Sigma}_{11}^{-1}\mathbf{\Sigma}_{12}\mathcal{M} &= \mathbf{R}_2, & \mathbf{\Sigma}_{21}\mathbf{\Sigma}_{11}^{-1}\mathbf{\Sigma}_{12} &= \mathcal{M}'^{-1}\mathbf{R}_2\mathcal{M}^{-1}, \end{aligned} \tag{8f.1.5}$$

where $\mathbf{R}_2$ is the diagonal matrix of the eigenroots $\rho_1^2, \ldots, \rho_s^2$ (see ii, **1c.3** on canonical reduction of two quadratic forms).

Similarly we obtain the determinantal equation

$$|\mathbf{\Sigma}_{12}\mathbf{\Sigma}_{22}^{-1}\mathbf{\Sigma}_{21} - \rho^2\mathbf{\Sigma}_{11}| = 0, \tag{8f.1.6}$$

with roots $\rho_1^2, \ldots, \rho_r^2$ and eigenvectors $\mathbf{L}_1, \ldots, \mathbf{L}_r$. The non-zero roots of (8f.1.4) and (8f.1.6) are the same so that the same symbols can be used to represent the roots. The multiplicity of the zero root is however different in the two cases. If $\mathcal{L}$ and $\mathbf{R}_1$ correspond to $\mathcal{M}$ and $\mathbf{R}_2$ in (8f.1.5). then

$$\begin{aligned} \mathcal{L}'\mathbf{\Sigma}_{11}\mathcal{L} &= \mathbf{I}, & \mathbf{\Sigma}_{11} &= \mathcal{L}'^{-1}\mathcal{L}^{-1} \\ \mathcal{L}'\mathbf{\Sigma}_{12}\mathbf{\Sigma}_{22}^{-1}\mathbf{\Sigma}_{21}\mathcal{L} &= \mathbf{R}_1, & \mathbf{\Sigma}_{12}\mathbf{\Sigma}_{22}^{-1}\mathbf{\Sigma}_{21} &= \mathcal{L}'^{-1}\mathbf{R}_1\mathcal{L}^{-1}. \end{aligned} \tag{8f.1.7}$$

The non-zero roots $\rho_1 \geqslant \rho_2 \geqslant \cdots$ are called the canonical correlations and the linear functions

$$\mathbf{L}_1'\mathbf{U}_1, \ldots, \mathbf{L}_r'\mathbf{U}_1 \quad \text{and} \quad \mathbf{M}_1'\mathbf{U}_2, \ldots, \mathbf{M}_s'\mathbf{U}_2$$

the canonical variables.

8f.2 Properties of Canonical Variables

It may be seen that $\mathbf{L}_i$ and $\mathbf{M}_i$ as determined from (8f.1.4) and (8f.1.6) satisfy (8f.1.2) with $\pm\rho_i$ where ρ_i is the positive square root of ρ_i^2. But it is possible, by changing the sign of $\mathbf{L}_i$ or $\mathbf{M}_i$ if necessary, to have $\mathbf{L}_i$, $\mathbf{M}_i$ and ρ_i as a solution of (8f.1.2). By convention we take all the canonical correlations to be positive. We have the following results.

(i) *The number of non-zero roots of* (8f.1.4) *or* (8f.1.6) *is equal to the rank of* $\mathbf{\Sigma}_{12}$.

(ii) (a) $\text{cov}(\mathbf{L}_i'\mathbf{U}_1, \mathbf{L}_j'\mathbf{U}_1) = 1$ *for* $i = j$, *and* 0 *for* $i \neq j$

(b) $\text{cov}(\mathbf{M}_i'\mathbf{U}_2, \mathbf{M}_j'\mathbf{U}_2) = 1$ *for* $i = j$, *and* 0 *for* $i \neq j$

that is, the canonical variables arising out of $\mathbf{U}_1$ *are all uncorrelated and have unit standard deviation and the same is true of the canonical variables of* $\mathbf{U}_2$.

The results (a) and (b) follow from the equations $\mathscr{L}'\mathbf{\Sigma}_{11}\mathscr{L} = \mathbf{I}$ and $\mathscr{M}'\mathbf{\Sigma}_{22}\mathscr{M} = \mathbf{I}$ of (8f.1.7) and (8f.1.5) respectively.

(iii) (a) $\text{cov}(\mathbf{L}_i'\mathbf{U}_1, \mathbf{M}_i'\mathbf{U}_2) = \rho_i \neq 0, \ i = 1, \ldots, k$
$= 0, \ i > k$

(b) $\text{cov}(\mathbf{L}_i'\mathbf{U}_1, \mathbf{M}_j'\mathbf{U}_2) = 0, \qquad i \neq j$

where $k = R(\mathbf{\Sigma}_{12})$.

Consider one of the equations of (8f.1.2)

$$\mathbf{\Sigma}_{12}\mathbf{M}_i = \rho_i \mathbf{\Sigma}_{11}\mathbf{L}_i.$$

Multiplying by $\mathbf{L}_i'$, we have

$$\mathbf{L}_i'\mathbf{\Sigma}_{12}\mathbf{M}_i = \rho_i \mathbf{L}_i'\mathbf{\Sigma}_{11}\mathbf{L}_i = \rho_i.$$

Multiplying by $\mathbf{L}_j'$, we have

$$\mathbf{L}_j'\mathbf{\Sigma}_{12}\mathbf{M}_i = \rho_i \mathbf{L}_j'\mathbf{\Sigma}_{11}\mathbf{L}_i = 0.$$

But $\mathbf{L}_i'\mathbf{\Sigma}_{12}\mathbf{M}_i = \text{cov}(\mathbf{L}_i'\mathbf{U}_1, \mathbf{M}_i'\mathbf{U}_2)$ and $\mathbf{L}_j'\mathbf{\Sigma}_{12}\mathbf{M}_i = \text{cov}(\mathbf{L}_j'\mathbf{U}_1, \mathbf{M}_i'\mathbf{U}_2)$ which prove the required results.

We thus see that the result of the simultaneous transformation $\mathscr{L}'\mathbf{U}_1$, $\mathscr{M}'\mathbf{U}_2$ of $\mathbf{U}_1$, $\mathbf{U}_2$ is to reduce the dispersion matrix of the transformed variables $(\mathscr{L}'\mathbf{U}_1, \mathscr{M}'\mathbf{U}_2)$ to the simpler form

$$\begin{pmatrix} \mathbf{I}_r & \mathbf{R} \\ \mathbf{R}' & \mathbf{I}_s \end{pmatrix},$$

where $\mathbf{I}_r$ and $\mathbf{I}_s$ are unit matrices of order r and s and $\mathbf{R}$ is an $r \times s$ matrix with the first k diagonal elements as $\rho_1, \ldots, \rho_k$ and the rest of the elements zero.

(iv) *Let* $\mathbf{q}_i$ *be the* ith *column of* $\mathscr{L}'^{-1}$ *and* $\mathbf{m}_i$ *the* ith *column of* $\mathscr{M}'^{-1}$. Then we have the following representations:

(a) $\Sigma_{11} = (\mathscr{L}^{-1})'(\mathscr{L}^{-1}) = \mathbf{q}_1\mathbf{q}_1' + \cdots + \mathbf{q}_r\mathbf{q}_r'$
(b) $\Sigma_{22} = (\mathscr{M}^{-1})'(\mathscr{M}^{-1}) = \mathbf{m}_1\mathbf{m}_1' + \cdots + \mathbf{m}_s\mathbf{m}_s'$
(c) $\Sigma_{12} = (\mathscr{L}^{-1})'\mathbf{R}(\mathscr{M}^{-1}) = \rho_1\mathbf{q}_1\mathbf{m}_1' + \cdots + \rho_k\mathbf{q}_k\mathbf{m}_k'$

The results (a), (b), and (c) follow if we consider the dispersion matrix of the transformed variables $\mathscr{L}'\mathbf{U}_1$, $\mathscr{M}'\mathbf{U}_2$.

$$D\begin{pmatrix}\mathscr{L}'\mathbf{U}_1\\ \hdashline \mathscr{M}'\mathbf{U}_2\end{pmatrix} = \begin{pmatrix}\mathscr{L}'\\ \hdashline \mathscr{M}'\end{pmatrix} D\begin{pmatrix}\mathbf{U}_1\\ \hdashline \mathbf{U}_2\end{pmatrix}(\mathscr{L} \mid \mathscr{M}) = \begin{pmatrix}\mathscr{L}'\Sigma_{11}\mathscr{L} & \mathscr{L}'\Sigma_{12}\mathscr{M}\\ \mathscr{M}'\Sigma_{12}\mathscr{L} & \mathscr{M}'\Sigma_2\mathscr{M}\end{pmatrix}$$

$$= \begin{pmatrix}\mathbf{I}_r & \mathbf{R}\\ \mathbf{R}' & \mathbf{I}_s\end{pmatrix}.$$

8f.3 Effective Number of Common Factors

To understand the practical significance of canonical correlations let us consider the following situation. Suppose that the observed measurements are linearly dependent on a number of uncorrelated hypothetical (unobservable) factors (measurements). Certain factors may influence the components of both the variables $\mathbf{U}_1$ and $\mathbf{U}_2$, whereas others may influence only the components of $\mathbf{U}_1$ or $\mathbf{U}_2$. The former are called common factors of $\mathbf{U}_1$, $\mathbf{U}_2$ and the latter are specific factors. The structure of the variables $\mathbf{U}_1$, $\mathbf{U}_2$ can be expressed in the form

$$\begin{aligned}\mathbf{U}_1 &= \mathbf{A}_1\mathbf{F} + \mathbf{G}_1\\ \mathbf{U}_2 &= \mathbf{A}_2\mathbf{F} + \mathbf{G}_2,\end{aligned} \tag{8f.3.1}$$

where $\mathbf{F}' = (F_1, \ldots, F_m)$ is the vector of common factors, $\mathbf{A}_1$ and $\mathbf{A}_2$ are the matrices of compounding coefficients, and $\mathbf{G}_1$, $\mathbf{G}_2$ are the total effects of specific factors. By definition

$$\text{cov}(\mathbf{F}, \mathbf{G}_1) = 0, \quad \text{cov}(\mathbf{F}, \mathbf{G}_2) = 0, \quad \text{cov}(\mathbf{G}_1, \mathbf{G}_2) = 0. \tag{8f.3.2}$$

Let $D(\mathbf{F}) = \mathbf{I}$ without loss of generality. Then using (8f.3.2)

$$\begin{aligned}D(\mathbf{U}_1) &= \Sigma_{11} = \mathbf{A}_1\mathbf{A}_1' + D(\mathbf{G}_1)\\ D(\mathbf{U}_2) &= \Sigma_{22} = \mathbf{A}_2\mathbf{A}_2' + D(\mathbf{G}_2)\\ C(\mathbf{U}_1, \mathbf{U}_2) &= \Sigma_{12} = C(\mathbf{A}_1\mathbf{F}, \mathbf{A}_2\mathbf{F}) = \mathbf{A}_1\mathbf{A}_2'.\end{aligned} \tag{8f.3.3}$$

It is seen that the association between $\mathbf{U}_1$, $\mathbf{U}_2$ is solely due to influence of common factors.

The question naturally arises as to what is the smallest number (m) of common factors such that the representation (8f.3.1) with the restrictions

(8f.3.2, 8f.3.3) is possible. This number is called the *effective number of common factors.*

We shall show that the effective number of common factors is equal to $R(\mathbf{\Sigma}_{12})$, which is same as the number of non-zero roots of (8f.1.4) or (8f.1.6).

From (8f.3.3), $\mathbf{\Sigma}_{12} = \mathbf{A}_1\mathbf{A}_2' \Rightarrow R(\mathbf{\Sigma}_{12}) \leqslant \min\ (R(\mathbf{A}_1),\ R(\mathbf{A}_2))$. But $m \geqslant \max(R(\mathbf{A}_1), R(\mathbf{A}_2))$. Hence $m \geqslant R(\mathbf{\Sigma}_{12})$. In fact, m can be arbitrary subject to this restriction. However, $m = R(\mathbf{\Sigma}_{12})$ is an admissible value. The last result follows from the representation

$$\mathbf{\Sigma}_{12} = \rho_1\mathbf{q}_1\mathbf{m}_1' + \cdots + \rho_k\mathbf{q}_k\mathbf{m}_k' \tag{8f.3.4}$$

by choosing

$$\mathbf{A}_1 = (\sqrt{\rho_1}\mathbf{q}_1 \mid \cdots \mid \sqrt{\rho_k}\mathbf{q}_k)$$
$$\mathbf{A}_2 = (\sqrt{\rho_1}\mathbf{m}_1 \mid \cdots \mid \sqrt{\rho_k}\mathbf{m}_k),$$

where it may be noted that $k = \text{rank}\ \mathbf{\Sigma}_{12}$. To prove that the choice of $\mathbf{A}_1$, $\mathbf{A}_2$ is consistent, we have to show that

$$D(\mathbf{G}_1) = \mathbf{\Sigma}_{11} - \mathbf{A}_1\mathbf{A}_1' \quad \text{and} \quad D(\mathbf{G}_2) = \mathbf{\Sigma}_{22} - \mathbf{A}_2\mathbf{A}_2' \tag{8f.3.4}$$

are non-negative definite [c.f. (8f.3.3)]. But

$$\mathbf{\Sigma}_{11} - \mathbf{A}_1\mathbf{A}_1' = \mathbf{q}_1\mathbf{q}_1' + \cdots + \mathbf{q}_r\mathbf{q}_r' - (\rho_1\mathbf{q}_1\mathbf{q}_1' + \cdots + \rho_k\mathbf{q}_k\mathbf{q}_k')$$

where $\rho_i \leqslant 1$, $i = 1, \ldots, k$, which shows that $\mathbf{\Sigma}_{11} - \mathbf{A}_1\mathbf{A}_1'$ is non-negative definite. The same is true of $\mathbf{\Sigma}_{22} - \mathbf{A}_2\mathbf{A}_2'$.

The estimation of effective number of factors from an estimated dispersion matrix poses a serious problem. The rank of $\mathbf{S}_{12}$, the estimate of $\mathbf{\Sigma}_{12}$, will be $\min(r, s)$ with probability 1, even when rank $\mathbf{\Sigma}_{12} < \min(r, s)$. Hence no inference can be made on the effective number of factors from the rank of $\mathbf{S}_{12}$. The only method is then to determine the roots of the determinantal equation

$$|\mathbf{S}_{21}\mathbf{S}_{11}^{-1}\mathbf{S}_{12} - r^2\mathbf{S}_{22}| = 0, \tag{8f.3.5}$$

where estimates $\mathbf{S}_{11}, \mathbf{S}_{12}, \mathbf{S}_{22}$ are substituted for $\mathbf{\Sigma}_{11}, \mathbf{\Sigma}_{12}, \mathbf{\Sigma}_{22}$ in the equation (8f.1.4), and to examine the magnitude of the different roots. If only k of the roots $\rho_1^2, \rho_2^2, \ldots$ of (8f.1.4) are non-zero, then the roots $r_{k+1}^2, r_{k+2}^2, \ldots$ of (8f.3.5) corresponding to the zero values of the hypothetical roots are likely to be small. Hence it may be possible to infer on the effective number of common factors by testing the significance of the smaller roots.

But in any real problem the effective number of common factors is likely to be very large so that the problem we have posed and the method of estimation suggested are not meaningful. But what may be true is that a smaller number of common factors account for a large portion of the association between the two sets of variables. Such factors may be called *dominant common factors*, and the problem of estimation of their number can be formulated in a meaningful way and in rigorous mathematical terms if necessary.

The representation (8f.3.4)

$$\mathbf{\Sigma}_{12} = \rho \mathbf{q}_1 \mathbf{m}_1 + \cdots + \rho_k \mathbf{q}_k \mathbf{m}_k'$$

shows that the major contribution to $\mathbf{\Sigma}_{12}$ comes from terms with larger values of ρ_i, and consequently the number of dominant factors may be determined by the number of dominant roots ρ_i. We may then utilize the estimated values of the roots (of equation 8f.3.5) for drawing inferences on the number of dominant hypothetical roots and hence on the number of dominant factors.

8f.4 Factor Analysis

Let $\mathbf{U}$ be a p-dimensional r.v. with the structure

$$\mathbf{U} = \mathbf{AF} + \mathbf{G} \tag{8f.4.1}$$

where $\mathbf{F}' = (F_1, \ldots, F_m)$ and $\mathbf{G}' = (G_1, \ldots, G_p)$ are all uncorrelated variables and $\mathbf{A}$ is a $p \times m$ matrix of (unknown) constants. Using the terminology of **8f.3**, we can call $F_1, \ldots, F_m$ common factors and $G_1, \ldots, G_p$ specific factors. Let $V(G_i) = \delta_i$ and $V(F_i) = 1$ without loss of generality. Then

$$D(\mathbf{U}) = \mathbf{\Sigma} = \mathbf{AA}' + \mathbf{\Delta} \tag{8f.4.2}$$

where $\mathbf{\Delta}$ is the diagonal matrix with $\delta_1, \ldots, \delta_p$ in the diagonal. As in the problem of relation between two sets of variables, let us define as the effective number of common factors the minimum value of m for which the representation (8f.4.1) with the restriction (8f.4.2) is possible. The problem as stated is not realistic since the effective number of common factors in any given situation is likely to be large. But it may be reasonable to enquire how many of the common factors are important in explaining the mutual correlations of the variables. A suitable approach to such a problem is given by the author (Rao, 1955b). There exists a vast literature on the subject, and the interested reader is referred to a recent book by Lawley and Maxwell (1963). The problem of simultaneous factor analysis in several populations has been investigated by Rasch (1953) see also Rao (1966a, 1967d, 1969a).

8g ORTHONORMAL BASIS OF A RANDOM VARIABLE

8g.1 The Gram-Schmidt Basis

Let $U_1, \ldots, U_p$ be p (one-dimensional) r.v.'s such that $E(U_i) = 0$ and the dispersion matrix $\mathbf{\Sigma}$ is of rank $m \leqslant p$. Denote by $\mathscr{M}(U_1, \ldots, U_p)$, or simply by $\mathscr{M}(\mathbf{U})$, the collection of all r.v.'s $c_1 U_1 + \cdots + c_p U_p$ with real coefficients c_i.

By definition $\mathcal{M}(\mathbf{U})$ is a vector space. Let us define the inner product of any two elements Y_1, Y_2 of $\mathcal{M}(\mathbf{U})$ as $\text{cov}(Y_1, Y_2)$. The norm of Y_1 is then the standard deviation, $\|Y_1\| = [V(Y_1)]^{1/2}$. The vector space $\mathcal{M}(\mathbf{U})$ of one-dimensional r.v.'s is then a normed vector space.

From the theory of vector spaces it follows that $\mathcal{M}(\mathbf{U})$ has an o.n.b. (orthonormal basis) $G_1, \ldots, G_m$, each with unit norm, and inner products zero (i.e., G_i are r.v.'s with unit standard deviation and zero covariances). Every element (r.v.) of $\mathcal{M}(\mathbf{U})$ can then be expressed as a linear combination of the elements (r.v.'s) $G_1, \ldots, G_m$. In particular

$$U_i = a_{i1}G_1 + \cdots + a_{im}G_m, \qquad i = 1, \ldots, p. \tag{8g.1.1}$$

If $\mathbf{G}' = (G_1, \ldots, G_m)$, the relationship (8g.1.1) can be expressed in matrix notation

$$\mathbf{U} = \mathbf{AG}, \qquad \mathbf{\Sigma} = \mathbf{AA}'. \tag{8g.1.2}$$

Conversely, since $\mathcal{M}(\mathbf{U}) = \mathcal{M}(\mathbf{G})$, $\mathbf{G}$ can be expressed in terms of $\mathbf{U}$,

$$\mathbf{G} = \mathbf{BU}, \qquad \mathbf{I} = \mathbf{B\Sigma B}'. \tag{8g.1.3}$$

Furthermore, we know from the theory of vector spaces that the dimension of $\mathcal{M}(\mathbf{U})$ is the rank of the matrix of inner products of the elements $U_1, \ldots, U_p$ which in the present case is $\mathbf{\Sigma}$. Hence the number of variables in the o.n.b. is $m = R(\mathbf{\Sigma})$. An o.n.b. is not unique. Some special bases are, however, of statistical interest, which we shall consider.

Linear Predictor as a Projection. Let $P(U_i)$ be the projection of U_i on $\mathcal{M}(U_1, \ldots, U_{i-1})$. By definition it is a linear function of $U_1, \ldots, U_{i-1}$ such that

$$\left\| U_i - \sum_1^{i-1} b_r U_r \right\|^2 = E(U_i - \sum b_r U_r)^2 \tag{8g.1.4}$$

is a minimum. Hence $P(U_i)$ is the minimum mean square error linear predictor of U_i based on $U_1, \ldots, U_{i-1}$. Furthermore, $U_i - P(U_i)$ is $\perp$er to

$$\mathcal{M}(U_1, \ldots, U_{i-1}),$$

so that the coefficients of the best linear function are determined by expressing the condition that the inner products of $(U_i - \sum b_r U_r)$ with U_s, $s = 1, \ldots, i-1$ are all zero. Since inner product is a covariance, the equations are

$$\text{cov}\left(U_s, U_i - \sum_1^{i-1} b_r U_r\right) = 0, \quad s = 1, \ldots, i-1, \tag{8g.1.5}$$

which imply that

$$\begin{aligned} \text{cov}[U_s, U_i - P(U_i)] &= 0, \qquad s = 1, \ldots, i-1 \\ \text{cov}[U_i - P(U_i), U_j - P(U_j)] &= 0, \qquad i \neq j. \end{aligned} \tag{8g.1.6}$$

If we denote the residual $U_i - P(U_i)$ by $U_{i\cdot 12\cdots(i-1)}$, then (8g.1.6) is equivalent to the statement that the residuals

$$U_1, U_{2\cdot 1}, U_{3\cdot 12}, \ldots, U_{p\cdot 12\cdots(p-1)} \tag{8g.1.7}$$

are all uncorrelated.

Gram-Schmidt o.n.b. Let $t_{ii} = \|U_i - P(U_i)\|$, which is the standard deviation of the residual $U_{i\cdot 12\cdots(i-1)}$. Consider

$$\begin{aligned} t_{11}G_1 &= U_1 &&= U_1 \\ t_{22}G_2 &= U_2 - P(U_1) &&= U_2 - t_{21}G_1 \\ &\cdots \quad \cdots && \quad \cdots \\ t_{pp}G_p &= U_p - P(U_p) &&= U_p - t_{p1}G_1 - \cdots - t_{p,\,p-1}G_{p-1}. \end{aligned} \tag{8g.1.8}$$

Note that the projection of U_i on $\mathcal{M}(U_1, \ldots, U_{i-1})$ can be expressed in terms of $G_1, \ldots, G_{i-1}$ and the coefficients can be sequentially determined as in Gram-Schmidt orthogonalization process [(ii), **1a.4**]. If $t_{ii} = 0$, U_i is completely determined by the previous variables. Then put $G_i = 0$. Otherwise

$$G_i = [U_i - P(U_i)] \div t_{ii} = U_{i\cdot 12\cdots(i-1)} \div t_{ii}$$

so that $\|G_i\| = 1$, and by using (8g.1.7) the G_i corresponding to non-zero t_{ii} constitute an o.n.b. The inverse relationship of (8g.1.8) is

$$\begin{aligned} U_1 &= t_{11}G_1 \\ U_2 &= t_{21}G_1 + t_{22}G_{22} \\ &\cdot \qquad \cdot \qquad \cdot \\ U_p &= t_{p1}G_1 + t_{p2}G_2 + \cdots + t_{pp}G_p \end{aligned} \tag{8g.1.9}$$

which can be written as $\mathbf{U} = \mathbf{TG}$, where $\mathbf{T}$ is the lower triangular matrix as in (8g.1.9). Then

$$\mathbf{\Sigma} = E(\mathbf{UU'}) = \mathbf{T}E(\mathbf{GG'})\mathbf{T'} = \mathbf{TT'}. \tag{8g.1.10}$$

Equation (8g.1.10) shows that $\mathbf{T}$ is exactly the matrix obtained by the *square root* method of reduction of $\mathbf{\Sigma}$, which is simple and well known. When $R(\mathbf{\Sigma}) = p$, all the diagonal elements of $\mathbf{T}$ are non-zero. Otherwise $(p - m)$ diagonal elements and the corresponding columns are zero so that the relationship (8g.1.9) involves only m of the G_i variables.

In practice the two relationships expressing (correlated variables) $\mathbf{U}$ in terms of (uncorrelated variables) $\mathbf{G}$, and $\mathbf{G}$ in terms of $\mathbf{U}$ may be obtained simultaneously if necessary by applying the square root technique to $\mathbf{\Sigma}$ with a unit matrix $\mathbf{I}$ appended to the right:

$$\mathbf{\Sigma} \;\vdots\; \mathbf{I}. \tag{8g.1.11}$$

After reduction by the square root method, one obtains

$$\mathbf{T}' \ \vdots \ \mathbf{W}, \tag{8g.1.12}$$

where $\mathbf{T}'$ is an upper triangular matrix and $\mathbf{W}$ is a lower triangular matrix. Then the transformation from $\mathbf{G}$ to $\mathbf{U}$ and from $\mathbf{U}$ to $\mathbf{G}$ are given by

$$\mathbf{U} = \mathbf{TG} \quad \text{and} \quad \mathbf{G} = \mathbf{WU}. \tag{8g.1.13}$$

8g.2 Principal Component Analysis

Consider a p-dimensional r.v. $\mathbf{U}$ with the dispersion matrix $\mathbf{\Sigma}$. Let $\lambda_1 \geqslant \cdots \geqslant \lambda_p$ be the eigenvalues and $\mathbf{P}_1, \ldots, \mathbf{P}_p$ be the corresponding eigenvectors of $\mathbf{\Sigma}$. Then it is shown in (1c.3.5) that

$$\begin{aligned} \mathbf{\Sigma} &= \lambda_1 \mathbf{P}_1 \mathbf{P}_1' + \cdots + \lambda_p \mathbf{P}_p \mathbf{P}_p' \\ \mathbf{I} &= \mathbf{P}_1 \mathbf{P}_1' + \cdots + \mathbf{P}_p \mathbf{P}_p' \end{aligned} \tag{8g.2.1}$$

$$\mathbf{P}_i'\mathbf{\Sigma}\mathbf{P}_i = \lambda_i, \qquad \mathbf{P}_i'\mathbf{\Sigma}\mathbf{P}_j = 0, \qquad i \neq j. \tag{8g.2.2}$$

Consider the transformed r.v.'s

$$Y_i = \mathbf{P}_i'\mathbf{U}, \qquad i = 1, \ldots, p.$$

If $\mathbf{Y}$ denotes the vector of the new r.v.'s, and $\mathbf{P}$ denotes the orthogonal matrix with $\mathbf{P}_1, \ldots, \mathbf{P}_p$ as its columns, then $\mathbf{Y}$ is obtained from $\mathbf{U}$ by the orthogonal transformation $\mathbf{Y} = \mathbf{PU}$. The r.v. Y_i is called the ith principal component of $\mathbf{U}$. We shall examine some of the properties of the principal components and their interpretations.

(i) *The principal components are all uncorrelated. The variance of the* ith *principal components is* λ_i.

These results follow from (8g.2.2)

$$\begin{aligned} V(\mathbf{P}_i'\mathbf{U}) &= \mathbf{P}_i'\mathbf{\Sigma}\mathbf{P}_i = \lambda_i \\ \operatorname{cov}(\mathbf{P}_i'\mathbf{U}, \mathbf{P}_j'\mathbf{U}) &= \mathbf{P}_i'\mathbf{\Sigma}\mathbf{P}_j = 0, \qquad i \neq j. \end{aligned}$$

Thus the linear transformation $\mathbf{Y} = \mathbf{PU}$ reduces a correlated set of variables into an uncorrelated set by an orthogonal transformation.

(ii) *Let* $G_i = \lambda_i^{-1/2}\mathbf{P}_i'\mathbf{U}$ *for* $\lambda_i \neq 0$ *and further let* r *be the rank of* $\mathbf{\Sigma}$ *so that only the first* r *eigenvalues of* $\mathbf{\Sigma}$ *are non-zero. Then* $G_1, \ldots, G_r$ *is an o.n.b. of the r.v.* $\mathbf{U}$.

(iii) *Let* $\mathbf{B}$ *be any vector such that* $\|\mathbf{B}\| = 1$. *Then* $V(\mathbf{B}'\mathbf{U})$ *is a maximum when* $\mathbf{B} = \mathbf{P}_1$ *and the maximum variance is* λ_1.

Since $V(\mathbf{B}'\mathbf{U}) = \mathbf{B}'\mathbf{\Sigma}\mathbf{B}$, we need the max $\mathbf{B}'\mathbf{\Sigma}\mathbf{B}$ subject to the condition $\|\mathbf{B}\| = 1$. But it is shown in (1f.2.1) that the maximum is attained when $\mathbf{B} = \mathbf{P}_1$.

In fact we have the following as a consequence of the results established in **1f.2.**

(iv)

(a) $\min\limits_{\|\mathbf{B}\|=1} V(\mathbf{B}'\mathbf{U}) = \lambda_p = V(\mathbf{P}_p'\mathbf{U})$.

(b) $\max\limits_{\|\mathbf{B}\|=1, \mathbf{B}\perp \mathbf{P}_1,\ldots,\mathbf{P}_{i-1}} V(\mathbf{B}'\mathbf{U}) = \lambda_i = V(\mathbf{P}_i'\mathbf{U})$.

(c) $\min\limits_{S_{i-1}} \max\limits_{\|\mathbf{B}\|=1, \mathbf{B}\perp S_{i-1}} V(\mathbf{B}'\mathbf{U}) = \lambda_i = V(\mathbf{P}_i'\mathbf{U})$.

where S_{i-1} is a space of $(i-1)$ dimensions in E_p.

(d) *Let $\mathbf{B}_1, \ldots, \mathbf{B}_k$ be a set of orthonormal vectors in E_p. Then*

$$\begin{aligned}\lambda_1 + \cdots + \lambda_k &= \max_{\mathbf{B}_1,\ldots,\mathbf{B}_k} [V(\mathbf{B}_1'\mathbf{U}) + \cdots + V(\mathbf{B}_k'\mathbf{U})] \\ &= V(\mathbf{P}_1'\mathbf{U}) + \cdots + V(\mathbf{P}_k'\mathbf{U}).\end{aligned}$$

(v) *Let $\mathbf{B}_1'\mathbf{U}, \ldots, \mathbf{B}_k'\mathbf{U}$ be k linear functions of $\mathbf{U}$ and σ_i^2 be the residual variance in predicting U_i by the best linear predictor based on $\mathbf{B}_1'\mathbf{U}, \ldots, \mathbf{B}_k'\mathbf{U}$. Then*

$$\min_{\mathbf{B}_1,\ldots,\mathbf{B}_k} \sum_{i=1}^{p} \sigma_i^2$$

is attained when the set $\mathbf{B}_1'\mathbf{U}, \ldots, \mathbf{B}_k'\mathbf{U}$ is equivalent to $\mathbf{P}_1'\mathbf{U}, \ldots, \mathbf{P}_k'\mathbf{U}$, that is, each $\mathbf{B}_i'\mathbf{U}$ is a linear combination of the first k principal components.

We can, without loss of generality replace $\mathbf{B}_1'\mathbf{U}, \ldots, \mathbf{B}_k'\mathbf{U}$ by an equivalent set of uncorrelated functions and each with variance unity. It is also clear that the functions $\mathbf{B}_1'\mathbf{U}, \ldots, \mathbf{B}_k'\mathbf{U}$ should be linearly independent for the optimum solution. The residual variance in predicting U_i on the basis of $\mathbf{B}_1'\mathbf{U}, \ldots, \mathbf{B}_k'\mathbf{U}$ is

$$\begin{aligned}\sigma_i^2 &= \sigma_{ii} - [\mathrm{cov}(U_i, \mathbf{B}_1'\mathbf{U})]^2 - \cdots - [\mathrm{cov}(U_i, \mathbf{B}_k'\mathbf{U})]^2 \\ &= \sigma_{ii} - (\mathbf{B}_1'\mathbf{\Sigma}_i)^2 - \cdots - (\mathbf{B}_k'\mathbf{\Sigma}_i)^2 \\ &= \sigma_{ii} - \mathbf{B}_1'\mathbf{\Sigma}_i\mathbf{\Sigma}_i'\mathbf{B}_1 - \cdots - \mathbf{B}_k'\mathbf{\Sigma}_i\mathbf{\Sigma}_i'\mathbf{B}_k,\end{aligned}$$

where $\mathbf{\Sigma}_i$ is the ith column of $\mathbf{\Sigma}$ and σ_{ii} is the variance of U_i. Now

$$\begin{aligned}\sum_{i=1}^{p} \sigma_i^2 &= \sum_{i=1}^{p} \sigma_{ii} - \mathbf{B}_1'\left(\sum_{i=1}^{p} \mathbf{\Sigma}_i \mathbf{\Sigma}_i'\right)\mathbf{B}_1 - \cdots - \mathbf{B}_k'\left(\sum_{i=1}^{p} \mathbf{\Sigma}_i \mathbf{\Sigma}_i'\right)\mathbf{B}_k \\ &= \mathrm{trace}\ \mathbf{\Sigma} - \mathbf{B}_1'\mathbf{\Sigma}\mathbf{\Sigma}\mathbf{B}_1 - \cdots - \mathbf{B}_k'\mathbf{\Sigma}\mathbf{\Sigma}\mathbf{B}_k.\end{aligned}$$

To minimize $\sum \sigma_i^2$, we need maximize

$$\mathbf{B}_1'\boldsymbol{\Sigma\Sigma}\mathbf{B}_1 + \cdots + \mathbf{B}_k'\boldsymbol{\Sigma\Sigma}\mathbf{B}_k \tag{8g.2.3}$$

subject to the conditions

$$\mathbf{B}_i'\boldsymbol{\Sigma}\mathbf{B}_i = 1, \qquad \mathbf{B}_i'\boldsymbol{\Sigma}\mathbf{B}_j = 0, \qquad i \neq j \tag{8g.2.4}$$

(i.e., $\mathbf{B}_1'\mathbf{U}, \ldots, \mathbf{B}_k'\mathbf{U}$ are uncorrelated and each has variance unity). The optimum choice of $\mathbf{B}_i$ in such a case, as shown in Example **22** at the end of Chapter 1, is

$$\mathscr{M}(\mathbf{B}_1, \ldots, \mathbf{B}_k) = \mathscr{M}\mathbf{P}_1, \ldots, \mathbf{P}_k) \tag{8g.2.5}$$

where $\mathbf{P}_1, \ldots, \mathbf{P}_k$ are the first k eigenvectors of

$$|\boldsymbol{\Sigma} - \lambda\mathbf{I}| = 0. \tag{8g.2.6}$$

The result (v) provides an interesting interpretation of the principal components. Suppose we wish to replace the p-dimensional r.v. $\mathbf{U}$ by $k < p$ linear functions without much loss of information. How are the best k linear functions to be chosen? The efficiency of any choice of k linear functions depends on the extent to which the k linear functions enable us to reconstruct the p original variables. One method of reconstructing the variable U_i is by determining its best linear predictor on the basis of the k linear functions, in which case the efficiency of prediction may be measured by the residual variance σ_i^2. An overall measure of the predictive efficiency is $\sum \sigma_i^2$. The best choice of the linear functions, for which $\sum \sigma_i^2$ is a minimum, is the first k principal components of $\mathbf{U}$. We shall show in (vi) that a much stronger measure of overall predictive efficiency yields the same result.

Let us represent the k linear functions $\mathbf{B}_1'\mathbf{U}, \ldots, \mathbf{B}_k'\mathbf{U}$ by the transformation $\mathbf{Y} = \mathbf{B}'\mathbf{U}$ where $\mathbf{B}$ is a $p \times k$ matrix, with $\mathbf{B}_1, \ldots, \mathbf{B}_k$ as its columns. Now the joint dispersion matrix of $\mathbf{U}$ and $\mathbf{Y}$ is

$$\begin{pmatrix} \boldsymbol{\Sigma} & \boldsymbol{\Sigma}\mathbf{B} \\ \mathbf{B}'\boldsymbol{\Sigma} & \mathbf{B}'\boldsymbol{\Sigma}\mathbf{B} \end{pmatrix} \tag{8g.2.7}$$

and the residual dispersion matrix of $\mathbf{U}$ subtracting its best linear predictor in terms of $\mathbf{Y}$ is

$$\boldsymbol{\Sigma} - \boldsymbol{\Sigma}\mathbf{B}(\mathbf{B}'\boldsymbol{\Sigma}\mathbf{B})^{-1}\mathbf{B}'\boldsymbol{\Sigma}. \tag{8g.2.8}$$

The smaller the values of the elements of (8g.2.8), the greater is the predictive efficiency of $\mathbf{Y}$. In (v) we found $\mathbf{B}$ in such a way that

$$\text{trace}\,(\boldsymbol{\Sigma} - \boldsymbol{\Sigma}\mathbf{B}(\mathbf{B}'\boldsymbol{\Sigma}\mathbf{B})^{-1}\mathbf{B}'\boldsymbol{\Sigma}) \tag{8g.2.9}$$

is a minimum. We now prove the stronger result.

(vi) *The Euclidean norm*

$$\|\mathbf{\Sigma} - \mathbf{\Sigma B}(\mathbf{B}'\mathbf{\Sigma B})^{-1}\mathbf{B}'\mathbf{\Sigma}\| \tag{8g.2.10}$$

is a minimum when the set of vectors $\mathbf{B}_1, \ldots, \mathbf{B}_k$ *is equivalent to the eigenvectors* $\mathbf{P}_1, \ldots, \mathbf{P}_k$.

Thus the two measures of overall predictive efficiency, *viz.*,

(a) $\operatorname{trace}(\mathbf{\Sigma} - \mathbf{\Sigma B}(\mathbf{B}'\mathbf{\Sigma B})^{-1}\mathbf{B}'\mathbf{\Sigma})$,
(b) $\|\mathbf{\Sigma} - \mathbf{\Sigma B}(\mathbf{B}'\mathbf{\Sigma B})^{-1}\mathbf{B}'\mathbf{\Sigma}\|$,

lead to the same optimum choice of the vectors $\mathbf{B}_i$.

To prove the result of (vi) we apply the result (1f.2.10) which shows that the minimum of (8g.2.10) is attained when

$$\mathbf{\Sigma B}(\mathbf{B}'\mathbf{\Sigma B})^{-1}\mathbf{B}'\mathbf{\Sigma} = \lambda_1 \mathbf{P}_1\mathbf{P}_1' + \cdots + \lambda_k \mathbf{P}_k\mathbf{P}_k'. \tag{8g.2.11}$$

It is easy to verify that the choice $\mathbf{B}_i = \mathbf{P}_i$ satisfies the equality in (8g.2.11).

For a study of estimated principal components and tests of significance associated with them reference may be made to Kshirsagar (1961, 1972). The book by Kshirsagar (1972) contains a good account of multivariate distribution problems.

COMPLEMENTS AND PROBLEMS

1 Let $\mathbf{U}_1, \ldots, \mathbf{U}_n$ be n points in a p-dimensional Euclidean space E_p. The point $\overline{\mathbf{U}} = n^{-1} \sum \mathbf{U}_i$ is the center of gravity of the points. Denote by $\mathscr{U}$, the $p \times n$ matrix with its ith column as $\mathbf{U}_i - \overline{\mathbf{U}}$. Consider the matrices $\mathscr{U}\mathscr{U}'$ and $\mathscr{U}'\mathscr{U}$. It is known that both the matrices have the same non-zero eigenvalues. If $\mathbf{P}_i$ and $\mathbf{Q}_i$ are the eigenvectors of $\mathscr{U}\mathscr{U}'$ and $\mathscr{U}'\mathscr{U}$ corresponding to the eigenvalue $\lambda_i \neq 0$, then $\sqrt{\lambda_i}\mathbf{Q}_i = \mathscr{U}'\mathbf{P}_i$ and $\sqrt{\lambda_i}\mathbf{P}_i = \mathscr{U}\mathbf{Q}_i$.

1.1 Determine points $\mathbf{V}_1, \ldots, \mathbf{V}_n$ in E_p such that they lie on a k-dimensional plane and

$$\sum_{i=1}^{n} (\mathbf{U}_i - \mathbf{V}_i)'(\mathbf{U}_i - \mathbf{V}_i) \quad \text{is a minimum.}$$

[Show that the k-dimensional plane S_k on which the optimum points $\mathbf{V}_1^*, \ldots, \mathbf{V}_n^*$ lie is specified by the point $\overline{\mathbf{U}}$ and the k orthonormal vectors $\mathbf{P}_1, \ldots, \mathbf{P}_k$, the first k eigenvectors of $\mathscr{U}\mathscr{U}'$. Hence show that

$$\mathbf{V}_i^* = \overline{\mathbf{U}} + \mathbf{P}_1'(\mathbf{U}_i - \overline{\mathbf{U}})\mathbf{P}_1 + \cdots + \mathbf{P}_k'(\mathbf{U}_i - \overline{\mathbf{U}})\mathbf{P}_k \qquad i = 1, \ldots, n.]$$

Also show that $\min \sum (\mathbf{U}_i - \mathbf{V}_i)'(\mathbf{U}_i - \mathbf{V}_i) = \lambda_{k+1} + \cdots + \lambda_p$ where $\lambda_1, \ldots, \lambda_p$ are the eigenvalues of $\mathscr{U}\mathscr{U}'$.

1.2 Show that the k-dimensional plane S_k defined in (**1.1**) is the best fitting k-dimensional plane to the given points $\mathbf{U}_1, \ldots, \mathbf{U}_n$ in the sense that the sum of squares of the perpendiculars from the points to S_k is a minimum compared to any other k dimensional plane.

1.3 Let $\mathbf{Y} = \mathbf{B}\mathscr{U}$, where $\mathbf{B}$ is a $k \times p$ matrix, represent a transformation of the points in E_p into points in E_k, $k < p$. Further let $\mathbf{Y}_i = \mathbf{B}\mathbf{U}_i$, $i = 1, \ldots, n$ be the transformed points corresponding to the points $\mathbf{U}_1, \ldots, \mathbf{U}_n$ in E_p. Show that

$$\sum_{i,j=1}^{n}\sum (\mathbf{Y}_i - \mathbf{Y}_j)'(\mathbf{Y}_i - \mathbf{Y}_j) \quad \text{is a maximum}$$

when the columns of $\mathbf{B}'$ are $\mathbf{P}_1, \ldots, \mathbf{P}_k$, the first k eigenvectors of $\mathscr{U}\mathscr{U}'$. Show that the maximum value of the above sum is $\lambda_1 + \cdots + \lambda_k$, where λ_i are eigenvalues of $\mathscr{U}\mathscr{U}'$.

1.4 Consider the matrix $\mathscr{U}'\mathscr{U}$. The ith diagonal element of the matrix is $\|\mathbf{U}_i - \bar{\mathbf{U}}\|^2$, the square of the distance of the point $\mathbf{U}_i$ from the centre of gravity. The (i, j)th element is the product of the cosine of the angle between the vectors $\mathbf{U}_i - \bar{\mathbf{U}}$ and $\mathbf{U}_j - \bar{\mathbf{U}}$ and the distances $\|\mathbf{U}_i - \bar{\mathbf{U}}\|$, $\|\mathbf{U}_j - \bar{\mathbf{U}}\|$. Thus $\mathscr{U}'\mathscr{U}$ is a matrix characterizing the configuration of the points $\mathbf{U}_1, \ldots, \mathbf{U}_n$ as measured by the distances and mutual angles between the lines joining the points to the center of gravity. Let $\mathbf{Y} = \mathbf{B}\mathscr{U}$ be a transformation of the points of E_p into E_k as in Example 1.3. The configuration of the transformed points is represented by the matrix $\mathbf{Y}'\mathbf{Y} = \mathscr{U}'\mathbf{B}'\mathbf{B}\mathscr{U}$. Determine the matrix $\mathbf{B}$ such that there is a minimum distortion in the configuration, due to the transformation, as measured by the Euclidean norm $\|\mathscr{U}'\mathscr{U} - \mathscr{U}'\mathbf{B}'\mathbf{B}\mathscr{U}\|$.

[Hint: Use the result (1f.2.10) which shows that for the optimum choice of **B**

$$\mathscr{U}'\mathbf{B}'\mathbf{B}\mathscr{U} = \lambda_1\mathbf{Q}_1\mathbf{Q}_1' + \cdots + \lambda_k\mathbf{Q}_k\mathbf{Q}_k'$$

where $\mathbf{Q}_1, \ldots, Q_k$ are the eigenvectors of $\mathscr{U}'\mathscr{U}$. But $\mathbf{Q}_i = \lambda_i^{-1/2}\mathscr{U}'\mathbf{P}_i$. Hence

$$\mathscr{U}'\mathbf{B}'\mathbf{B}\mathscr{U} = \mathscr{U}'(\mathbf{P}_1\mathbf{P}_1' + \cdots + \mathbf{P}_k\mathbf{P}_k')\mathscr{U},$$

which gives $\mathbf{B}'\mathbf{B} = \mathbf{P}_1\mathbf{P}_1' + \cdots + \mathbf{P}_k\mathbf{P}_k'$ or the columns of $\mathbf{B}'$ are the eigenvectors $\mathbf{P}_1, \ldots, \mathbf{P}_k$].

2 Let $u_1, \ldots, u_k$ and $v_1, \ldots, v_m$ be two sets of k and m one-dimensional random variables. Their joint dispersion matrix may be written in the partitioned form

$$\begin{pmatrix} \boldsymbol{\Sigma}_{11} & \boldsymbol{\Sigma}_{12} \\ \boldsymbol{\Sigma}_{21} & \boldsymbol{\Sigma}_{22} \end{pmatrix}.$$

Some or all of u_i may be a subset of the variables v_i.

2.1 Consider a linear function $b_1 v_1 + \cdots + b_m v_m$. The residual variance in predicting u_i by its linear regression on $b_1 v_1 + \cdots + b_m v_m$ is

$$\sigma_i^2 = V(u_i) - \frac{[\text{cov}(u_i, b_1 v_1 + \cdots + b_m v_m)]^2}{V(b_1 v_1 + \cdots + b_m v_m)}.$$

Show that the vector $(b_1, \ldots, b_m)$ which minimizes $\sum_1^k \sigma_i^2$ is the eigenvector that corresponds to the dominant root of

$$|\boldsymbol{\Sigma}_{21}\boldsymbol{\Sigma}_{12} - \lambda\boldsymbol{\Sigma}_{22}| = 0.$$

2.2 Show that the best r linear functions of $v_1, \ldots, v_m$ for predicting $u_1, \ldots, u_k$ in the sense of minimizing the sum of residual variances correspond to the first r eigenvectors that arise out of the determinantal equation in (**2.1**).

2.3 Observe that when $u_1, \ldots, u_k$ is a subset of $v_1, \ldots, v_m$, the first k eigenvectors give rise to linear functions of $u_1, \ldots, u_k$ only, which are the principal components of $u_1, \ldots, u_k$.

3 Let $u_1, \ldots, u_k$ be k one-dimensional random variables. Determine a linear function $b_1 u_1 + \cdots + b_k u_k$ such that

$$\sum_1^k w_i^2 \left\{ V(u_i) - \frac{[\text{cov}(u_i, b_1 u_1 + \cdots + b_k u_k)]^2}{V(b_1 u_1 + \cdots + b_k u_k)} \right\}$$

is a minimum where $w_1^2, \ldots, w_k^2$ are given weights.

4 Let $\mathbf{U}_1, \ldots, \mathbf{U}_n$ be n points in E_p. Determine n points $\mathbf{V}_1, \ldots, \mathbf{V}_n$ such that they lie on a k-dimensional plane and

$$\sum_1^n (\mathbf{U}_i - \mathbf{V}_i)'\boldsymbol{\Lambda}(\mathbf{U}_i - \mathbf{V}_i) \quad \text{is a minimum}$$

where $\boldsymbol{\Lambda}$ is a given positive definite matrix.

5 Show that Hotelling's T^2 test for testing the hypothesis that the mean vector of a p-variate normal population has an assigned value, on the basis of a sample of size n, can be obtained by the likelihood ratio principle.

6 Consider two alternative distributions of a p-dimensional random variable $\mathbf{U}$, with means $\boldsymbol{\delta}_1, \boldsymbol{\delta}_2$ and common dispersion matrix $\boldsymbol{\Sigma}$. Mahalanobis (1936) distance D^2 between the populations is defined by

$$D^2 = (\boldsymbol{\delta}_1 - \boldsymbol{\delta}_2)'\boldsymbol{\Sigma}^{-1}(\boldsymbol{\delta}_1 - \boldsymbol{\delta}_2).$$

Show that D^2 is invariant under a linear transformation of the variable $\mathbf{U}$. Find the linear transformation which reduces D^2 to a sum of squares.

7 Let $\mathbf{U}_i \sim N_p(a_i \boldsymbol{\mu}, \boldsymbol{\Sigma})$, $i = 1, \ldots, n$ be independent. Show that

$$\mathbf{Q} = \sum \mathbf{U}_r \mathbf{U}_r' - \mathbf{Z}\mathbf{Z}' \quad \text{and} \quad \mathbf{Z}\mathbf{Z}'$$

are independently distributed, where $\mathbf{Z} = \sum b_i \mathbf{U}_i$, $b_i = a_i/\sqrt{\sum a_i^2}$.

[Hint: From the notation of (8b.1.1), $\mathbf{Z} = \mathcal{U}'\mathbf{B}, \mathbf{B}' = (b_1, \ldots, b_n)$, and $\mathbf{Q} = \mathcal{U}'\mathcal{U} - \mathcal{U}'\mathbf{B}\mathbf{B}'\mathcal{U} = \mathcal{U}'\mathbf{A}\mathcal{U}$. Verify that $\mathbf{AB} = 0$ and apply (iii), **8b.2.**]

8 Let $\mathbf{S} \sim W_p(k, \Sigma)$ and $|\Sigma| \neq 0$. Show that the c.f. of $\mathbf{S}$ is

$$E \exp(i \sum \sum t_{rs} S_{rs}) = \frac{|\sigma^{rs}|^{k/2}}{|\sigma^{rs} - 2it_{rs}|^{k/2}}.$$

The result follows from the definition of $W_p(k, \Sigma)$ by first obtaining

$$E \exp(i \sum \sum t_{rs} U_r U_s)$$

where $(U_1, \ldots, U_p) \sim N_p(\mathbf{0}, \Sigma)$.

9 Consider a partition $\mathbf{U}' = (\mathbf{U}_1' \vdots \mathbf{U}_2')$ of the random variable and the corresponding partition of Σ.

$$\Sigma = \begin{pmatrix} \Sigma_{11} & \Sigma_{12} \\ \Sigma_{21} & \Sigma_{22} \end{pmatrix}.$$

Consider independent observations on $\mathbf{U}$ and let (S_{ij}) be the corrected sum of squares and products. Choose a linear function $\mathbf{L}'\mathbf{U}_2$ of $\mathbf{U}_2$ and carry out a regression analysis of $\mathbf{L}'\mathbf{U}_2$ on the components of $\mathbf{U}_1$.

9.1 Show that the sum of squares due to regression and deviation from regression are (using the univariate formula)

$$\mathbf{L}'\mathbf{S}_{21}\mathbf{S}_{11}^{-1}\mathbf{S}_{12}\mathbf{L} \quad \text{and} \quad \mathbf{L}'(\mathbf{S}_{22} - \mathbf{S}_{21}\mathbf{S}_{11}^{-1}\mathbf{S}_{12})\mathbf{L},$$

where $\mathbf{S}_{11}, \mathbf{S}_{12}, \mathbf{S}_{22}$ are the partitions of $\mathbf{S}$ similar to that of Σ.

9.2 Let the number of components in $\mathbf{U}_1$ and $\mathbf{U}_2$ be r and s. Observe that when $\Sigma_{12} = 0$, from the univariate theory, then

$$\mathbf{L}'\mathbf{S}_{21}\mathbf{S}_{11}^{-1}\mathbf{S}_{12}\mathbf{L} \quad \text{and} \quad \mathbf{L}'(\mathbf{S}_{22} - \mathbf{S}_{21}\mathbf{S}_{11}^{-1}\mathbf{S}_{12})\mathbf{L}$$

are independently distributed as central χ^2's on r and $(n - 1 - r)$ D.F. Deduce that, by the technique of dropping $\mathbf{L}$ (ii, **8b.2**),

$$\mathbf{S}_{21}\mathbf{S}_{11}^{-1}\mathbf{S}_{12} \sim W_s(r, \Sigma), \qquad \mathbf{S}_{22} - \mathbf{S}_{21}\mathbf{S}_{11}^{-1}\mathbf{S}_{12} \sim W_s(n - 1 - r, \Sigma)$$

and are independent.

9.3 Show that when $\Sigma_{12} = 0$,

$$\frac{|\mathbf{S}|}{|\mathbf{S}_{11}|\,|\mathbf{S}_{22}|} = \frac{|\mathbf{S}_{22} - \mathbf{S}_{21}\mathbf{S}_{11}^{-1}\mathbf{S}_{12}|}{|\mathbf{S}_{22}|}$$

is distributed as a product of independent beta variables.

10 Let p measurements on each of two brothers be made from n families. Determine the best linear compound of the measurements which maximizes the sample intraclass correlation. Devise a test for examining the significance of correlation between measurements of brothers. [For details see Rao (1945a).]

11 *Rectangular Coordinates.* Let $U_i \sim N_p(\mathbf{0}, \mathbf{\Sigma})$, $i = 1, \ldots, k$ be independent. The Wishart matrix is defined by $\mathbf{S} = \sum \mathbf{U}_r \mathbf{U}_r'$. Furthermore, let $\mathbf{T}$ be a (lower) *triangular* square root of $\mathbf{S}$. The elements T_{ij} of $\mathbf{T}$ are called rectangular coordinates by Mahalanobis, Bose and Roy (1937).

11.1 Consider the transformation from (S_{ij}) to (T_{ij}). Show that the Jacobian of transformation is (using the relation $\mathbf{S} = \mathbf{TT}'$)

$$\frac{D(S_{11}, \ldots, S_{pp})}{D(T_{11}, \ldots, T_{pp})} = (2T_{11})(2T_{11}T_{22}) \cdots (2T_{11}T_{22} \cdots T_{pp})$$
$$= 2^p T_{11}{}^p T_{22}^{p-1} \cdots T_{pp}.$$

11.2 Let $(S^{ij}) = (S_{ij})^{-1}$. Then

(a) $|\mathbf{S}|^{1/2} = T_{11}T_{22} \cdots T_{pp}$,

(b) $T_{pp} = \dfrac{1}{S^{pp}}$,

(c) $T_{pp}{}^2 \sim \dfrac{1}{\sigma^{pp}} \chi^2(k - p + 1)$.

[Note that (c) is same as the result (a), (viii), **8b.2**.]

11.3 Let $\mathbf{\Sigma} = \mathbf{I}$. From the relation $\mathbf{S} = \mathbf{TT}'$, observe that $T_{p1}, \ldots, T_{p,p-1}$ are linear and $T_{pp}{}^2$ is quadratic in $U_{p1}, \ldots, U_{pk}$. Furthermore,

$$\sum_{r=1}^{k} U_{pr}{}^2 = S_{pp} = T_{p1}{}^2 + \cdots + T_{p,p-1}^2 + T_{pp}{}^2,$$

and $T_{pp}{}^2$ is a quadratic form of rank $(k - p + 1)$ since it is a χ^2 on $(k - p + 1)$ D.F. Since $\mathbf{\Sigma} = \mathbf{I}$, $U_{p1}, \ldots, U_{pk}$ are independent $N_1(0, 1)$ variables. Hence, by applying Cochran's theorem, we see that

$$T_{pi}{}^2 \sim \chi^2(1) \qquad \text{or} \qquad T_{pi} \sim N_1(0, 1), \qquad i = 1, \ldots, (p-1)$$

and $T_{pp}{}^2 \sim \chi^2(k - p + 1)$ and are all independent. Now establish that all T_{ij} are independently distributed and that

$$T_{ij} \sim N_1(0, 1), \quad i \neq j \qquad \text{and} \qquad T_{ii}{}^2 \sim \chi^2(k - i + 1).$$

11.4 The joint density of T_{ij} when $\mathbf{\Sigma} = \mathbf{I}$ is, therefore,

$$\left[2^{kp/2}\pi^{p(p-1)/4} \prod_{i=1}^{p} \Gamma\left(\frac{k-i+1}{2}\right)\right]^{-1} \tag{A}$$
$$\times\, e^{-[T_{11}{}^2 + (T_{21}{}^2 + T_{22}{}^2) + \cdots + (T_{p1}{}^2 + \cdots + T_{pp}{}^2)]/2} \tag{B}$$
$$\times\, 2^p T_{11}^{k-1} T_{22}^{k-2} \cdots T_{pp}^{k-p}. \tag{C}$$

11.5 The joint density of (S_{ij}) when $\boldsymbol{\Sigma} = \mathbf{I}$ is obtained by dividing the density in (**11.4**) by the Jacobian in (**11.1**) and writing the expressions in terms of (S_{ij}). Thus, we obtain the density of (S_{ij})

$$Ae^{-(S_{11}+S_{22}+\cdots+S_{pp})/2}|S|^{(k-p-1)/2}$$

where the constant A is as in (11.4).

11.6 The joint density of (S_{ij}) for general $\boldsymbol{\Sigma}$ is obtained by making the transformation

$$\mathbf{S} = \mathbf{CS^*C'}, \qquad \mathbf{CC'} = \boldsymbol{\Sigma}$$

where $\mathbf{S^*}$ has the density as in (**11.5**). Observe that

(a) $\dfrac{D(\mathbf{S})}{D(\mathbf{S^*})} = |\mathbf{C}|^{p+1}$ (Jacobian)

(b) $|\mathbf{S}| = |\mathbf{S^*}|\,|\boldsymbol{\Sigma}|$

(c) $\sum \mathbf{S}^*_{ii} = \text{trace } \mathbf{S^*} = \text{trace}(\mathbf{C}^{-1}\mathbf{SC}'^{-1}) = \text{trace}(\mathbf{C}'^{-1}\mathbf{C}^{-1}\mathbf{S})$
$= \text{trace } \boldsymbol{\Sigma}^{-1}\mathbf{S} = \sum\sum \sigma^{ij}S_{ij}.$

Hence establish the required density as

$$A|\boldsymbol{\Sigma}|^{-k/2}|\mathbf{S}|^{(k-p-1)/2}e^{-\text{trace}(\boldsymbol{\Sigma}^{-1}\mathbf{S})/2}$$

which is the density of Wishart distribution $W_p(k, \boldsymbol{\Sigma})$ when $k \geqslant p$. The density function does not exist when $k < p$.

11.7 Hence, or directly from Example **11.4**, show that the density of rectangular coordinates for general $\boldsymbol{\Sigma}$ is

$$2^p A|\boldsymbol{\Sigma}|^{-k/2}T_{11}^{k-1}T_{22}^{k-2}\cdots T_{pp}^{k-p}e^{-\text{trace}(\mathbf{T}'\boldsymbol{\Sigma}^{-1}\mathbf{S})/2}.$$

11.8 Observe that the density function $P(\mathbf{U}|\boldsymbol{\Sigma}) = p(\mathbf{U}_1|\boldsymbol{\Sigma})\ldots p(\mathbf{U}_n|\boldsymbol{\Sigma})$ of $\mathbf{U}_1, \ldots, \mathbf{U}_n$ involves only $\mathbf{S}$ and $\boldsymbol{\Sigma}$. Then show that (due to Malay Ghosh)

$$P(\mathbf{S}|\boldsymbol{\Sigma}) = \frac{P(\mathbf{U}|\boldsymbol{\Sigma})}{P(\mathbf{U}|\mathbf{I})}P(\mathbf{S}|\mathbf{I})$$

where $P(\mathbf{S}|\boldsymbol{\Sigma})$ is the density of $\mathbf{S}$ for general $\boldsymbol{\Sigma}$. (Since $P(\mathbf{S}|\mathbf{I})$ is derived in **11.4**, the expression for $P(\mathbf{S}|\boldsymbol{\Sigma})$ can be written down readily using the above formula without the heavy algebra involved in **11.6**).

Hint: Let us transform from $\mathbf{U}$ to $\mathbf{S}, \mathbf{Z}$ and let the Jacobian be $J(S, Z)$. Then

$$P(\mathbf{S}|\boldsymbol{\Sigma}) = \int \mathbf{P}(\mathbf{U}|\boldsymbol{\Sigma})J(\mathbf{S}, \mathbf{Z})\,d\mathbf{Z} = P(\mathbf{U}|\boldsymbol{\Sigma})\int J(\mathbf{S}, \mathbf{Z})\,d\mathbf{Z}.$$

Putting $\boldsymbol{\Sigma} = \mathbf{I}$

$$p(\mathbf{S}|\mathbf{I}) = P(\mathbf{U}|\mathbf{I})\int J(\mathbf{S}, \mathbf{Z})\,d\mathbf{Z}$$

Then $P(\mathbf{S}|\boldsymbol{\Sigma})/P(\mathbf{S}|\mathbf{I}) = P(\mathbf{U}|\boldsymbol{\Sigma})/P(\mathbf{U}|\mathbf{I})$ which also follows from sufficiency of $\mathbf{S}$ for $\boldsymbol{\Sigma}$.]

12 Let $\overline{\mathbf{U}}$ be the sample mean and $(\mathbf{S}_{ij})$ be the corrected S.P. matrix of n independent samples $\mathbf{U}_1, \ldots, \mathbf{U}_n$ from $N_p(\boldsymbol{\mu}, \boldsymbol{\Sigma})$. Then Hotelling's T^2 is defined by

$$T^2 = n(\overline{\mathbf{U}} - \boldsymbol{\mu})(s_{ij})^{-1}(\overline{\mathbf{U}} - \boldsymbol{\mu}), \qquad s_{ij} = S_{ij}/(n-1),$$

and it was shown in [(xii), **8b.2**] that

$$\frac{n-p}{p}\frac{T^2}{n-1} \sim F(p, n-p).$$

12.1 Show that

$$T^2 = n \max_{\mathbf{L}} \frac{[\mathbf{L}'(\overline{\mathbf{U}} - \boldsymbol{\mu})]^2}{\mathbf{L}'(s_{ij})\mathbf{L}}.$$

12.2 Hence, if F_α is the upper α point of the F distribution on p and $(n-p)$ D.F., deduce that

$$P\{n[\mathbf{L}'(\overline{\mathbf{U}} - \boldsymbol{\mu})]^2 \leqslant \mathbf{L}'(s_{ij})\mathbf{L}[(n-1)pF_\alpha/(n-p)] \text{ for all } \mathbf{L}\} = 1 - \alpha.$$

Observe that

$$P\{\mathbf{L}'\overline{\mathbf{U}} - c\sqrt{\mathbf{L}'(s_{ij})\mathbf{L}} \leqslant \mathbf{L}'\boldsymbol{\mu} \leqslant \mathbf{L}'\overline{\mathbf{U}} + c\sqrt{\mathbf{L}'(s_{ij})\mathbf{L}}, \text{ for any } \mathbf{L}\} \geqslant 1 - \alpha$$

where $c^2 = (n-1)pF_\alpha/n(n-p)$, leading to simultaneous confidence intervals for all linear functions of $\boldsymbol{\mu}$. [Roy and Bose, 1953.]

13 *Distribution of the Multiple Correlation* (Fisher, 1928). Let $\mathbf{U}_1, \ldots, \mathbf{U}_n$ be n independent observations from $N_p(\boldsymbol{\mu}, \boldsymbol{\Sigma})$ and $(S_{ij})_k$ be the corrected S.P. matrix for the first k components of $\mathbf{U}$. The multiple correlation of the pth component on the others is estimated by

$$R^2 = 1 - \frac{1}{S_{pp}}\frac{|S_{ij}|_p}{|S_{ij}|_{p-1}} = \frac{R_1{}^2 - R_0{}^2}{R_1{}^2},$$

where $R_0{}^2 = \min\limits_{\alpha, \beta_i} \sum (U_{pr} - \alpha - \beta_1 U_{1r} - \cdots - \beta_{p-1}U_{p-1,r})^2 = \dfrac{|S_{ij}|_p}{|S_{ij}|_{p-1}}$

$$R_1{}^2 = \min_{\beta_i = 0, \alpha} \sum (U_{pr} - \alpha - \beta_1 U_{1r} - \cdots - \beta_{p-1}U_{p-1,r})^2 = S_{pp}.$$

13.1 From the second fundamental theorem of least squares (p. 191) deduce that

$$R_0{}^2 \sim \sigma^2\chi^2(n-p), \qquad R_1{}^2 - R_0{}^2 \sim \sigma^2\chi^2(p-1, \delta^2),$$

where $\delta^2 = \sum\limits_{i,j=1}^{p-1}\sum S_{ij}\beta_i\beta_j \div \sigma^2$ in which σ^2 is the residual variance of U_p given

$U_1, \ldots, U_{p-1}$. Hence the conditional density of R^2 given U_{ir}, $i = 1, \ldots, p-1$; $r = 1, \ldots, n$ is that of $\chi^2(p-1, \delta^2)/[\chi^2(p-1, \delta^2) + \chi^2(n-p)]$ which is noncentral beta (Examples 17.1 and 17.2 of Chapter 3):

$$e^{-\delta^2/2} \sum \left(\frac{\delta^2}{2}\right)^r \frac{1}{r!} B\left(R^2 \,\middle|\, \frac{p-1}{2} + r, \frac{n-p}{2}\right)$$

$$= e^{-\delta^2/2} B\left(R^2 \,\middle|\, \frac{p-1}{2}, \frac{n-p}{2}\right) {}_1F_1\left(\frac{n-1}{2}, \frac{p-1}{2}, \frac{\delta^2 R^2}{2}\right),$$

where $B(x|\alpha, \beta) = \Gamma(\alpha + \beta)x^{\alpha-1}(1-x)^{\beta-1}/\Gamma(\alpha)\Gamma(\beta)$.

13.2 Observe that the conditional distribution of R^2 involves only δ^2. But $\delta^2\sigma^2 \sim (\sum\sum \sigma_{ij}\beta_i\beta_j)\chi^2(n-1)$. Multiplying the conditional density of R^2 by the density of δ^2 and integrating term by term with respect to δ^2, obtain the unconditional density of R^2 as

$$(1-\rho^2)^{(n-1)/2} B\left(R^2 \,\middle|\, \frac{p-1}{2}, \frac{n-p}{2}\right) {}_2F_1\left(\frac{n-1}{2}, \frac{n-1}{2}, \frac{p-1}{2}, \rho^2 R^2\right),$$

where ρ^2 is the population multiple correlation coefficient,

$$\rho^2 = \sum_{i,j=1}^{p-1}\sum \sigma_{ij}\beta_i\beta_j \div \sigma^2 = 1 - \frac{1}{\sigma_{pp}\sigma^{pp}}$$

$$= 1 - \frac{1}{\sigma_{pp}} \frac{|\sigma_{ij}|_p}{|\sigma_{ij}|_{p-1}}.$$

14 Let $\mathbf{S}_1 \sim \mathbf{W}_p(k_1, \Sigma)$ and $\mathbf{S}_2 \sim \mathbf{W}_p(k_2, \Sigma)$ be independently distributed. Use only the definition of Wishart distribution or the density function for the random matrix $\mathbf{S}_1$ to establish the following.

14.1 The roots of the equation $|\mathbf{S}_1 - \lambda(\mathbf{S}_1 + \mathbf{S}_2)| = 0$ and the elements of $(\mathbf{S}_1 + \mathbf{S}_2)$ are independently distributed.

14.2 In particular $|\mathbf{S}_1|/|\mathbf{S}_1 + \mathbf{S}_2|$ is distributed independently of $\mathbf{S}_1 + \mathbf{S}_2$.

14.3 Hence deduce that if $\mathbf{S}_i \sim \mathbf{W}_p(k_i, \Sigma)$, $i = 1, 2, \ldots$ and are all independent and $k_i \geqslant p$, then

$$\frac{|\mathbf{S}_1|}{|\mathbf{S}_1 + \mathbf{S}_2|}, \quad \frac{|\mathbf{S}_1 + \mathbf{S}_2|}{|\mathbf{S}_1 + \mathbf{S}_2 + \mathbf{S}_3|}, \quad \ldots$$

are all independently distributed.

15 In **8d.3** we have considered the tests for an assigned discriminant function and for an assigned ratio of the coefficients of any two specified variables. What is the test for assigned ratios $\rho_1, \ldots, \rho_s$ of s specified variables say $x_1, \ldots, x_s$ out of $x_1, \ldots, x_p$?

[Hint: The test is the same as that of examining the sufficiency of the $p - s + 1$ variables

$$\rho_1 x_1 + \cdots + \rho_s x_s, \qquad x_{s+1}, \ldots, x_p.$$

If D^2_{p-s+1} is the D^2 based on them and $D_p^{\,2}$ is the D^2 based on all the variables, the variance ratio is

$$F = \frac{N_1 + N_2 - p - 1}{s - 1} \frac{N_1 N_2 (D_p^{\,2} - D^2_{p-s+1})}{(N_1 + N_2)(N_1 + N_2 - 2) + N_1 N_2 D^2_{p-s+1}}$$

on $(s - 1)$ and $(N_1 + N_2 - p - 1)$ D.F. See **8d.3** for notation.]

16 Let $\Delta_p^{\,2}$ and $\Delta_q^{\,2}$ denote the Mahalanobis distances between two populations based on p characters and a subset of q characters respectively. Show that the test (8d.3.2) is for the hypothesis $\Delta_p^{\,2} = 0$ whereas the test (8d.3.7) is for the hypothesis $\Delta_p^{\,2} = \Delta_q^{\,2}$. Observe that $\Delta_p^{\,2} \geqslant \Delta_q^{\,2}$ in general.

17 Complex Normal Distribution. A complex random p-vector $\mathbf{Z} = \mathbf{X}_1 + i\mathbf{X}_2$ is said to be distributed as complex p-variate normal distribution if for every complex vector $\mathbf{L}$,

$$\text{R.P.}(\mathbf{L}^*\mathbf{Z}) \sim N_1(\text{R.P.}(\mathbf{L}^*\boldsymbol{\mu}), \mathbf{L}^*\mathbf{Q}\mathbf{L})$$

where R.P. denotes real part, $\boldsymbol{\mu}$ is a complex p-vector and $\mathbf{Q}$ is $p \times p$ hermitian, non-negative definite matrix. In such a case we write:

$$\mathbf{Z} \sim CN_p(\boldsymbol{\mu}, \mathbf{Q})$$

Denote $\boldsymbol{\mu} = \boldsymbol{\mu}_1 + i\boldsymbol{\mu}_2$, $\mathbf{Q} = \mathbf{Q}_1 + i\mathbf{Q}_2$, $\mathbf{v}' = (\boldsymbol{\mu}_1' \,\vdots\, \boldsymbol{\mu}_2')$ and

$$\boldsymbol{\Sigma} = \begin{pmatrix} \mathbf{Q}_1 & -\mathbf{Q}_2 \\ \mathbf{Q}_2 & \mathbf{Q}_1 \end{pmatrix}$$

Show that

(i) $\begin{pmatrix} \mathbf{X}_1 \\ \mathbf{X}_2 \end{pmatrix} \sim N_{2p}(\mathbf{v}, \boldsymbol{\Sigma})$

(ii) R.P.($\mathbf{L}^*\mathbf{Z}$) and I.P.($\mathbf{L}^*\mathbf{Z}$) are independently distributed for any $\mathbf{L}$.

(iii) I.P.($\mathbf{L}^*\mathbf{Z}$) $\sim N_1$(I.P.($\mathbf{L}^*\boldsymbol{\mu}$), $\mathbf{L}^*\mathbf{Q}\mathbf{L}$) where I.P. denotes imaginary part.

REFERENCES

Anderson, T. W. (1958), *An Introduction to Multivariate Statistical Analysis,* Wiley, New York.

Ashton, E. H., M. J. R. Healy and S. Lipton (1957), The descriptive use of discriminant functions in physical anthropology, *Proc. Roy. Soc.* (*series B*) **146**, 552–572.

Barnard, M. M. (1935), The secular variations of skull characters in four series of Egyptian skulls, *Ann. Eugen.* **6**, 352–371.

Bartlett, M. S. (1947a), Multivariate analysis, *J. Roy. Stat. Soc. Supple* **9(B)**, 176–197.

Bartlett, M. S. (1947b), The general canonical correlation distribution, *Ann. Math. Statist.* **18**, 1–17.

Bartlett, M. S. (1950), Tests of significance in factor analysis, *Brit. J. Psych.* (*Stat. Sec.*) **3**, 77–85.

Bose, R. C. and S. N. Roy (1938), The distribution of Studentised D^2-statistic, *Sankhyā* **4**, 19–38.

Bowker, A. H. (1960), A representation of Hotelling's T^2 and Anderson's classification statistic, *Contributions to Probability and Statistics* (*Essays in honour of Harold Hotelling*), 142–149, Stanford University Press.

Burnaby, T. P. (1966), Growth invariant discriminant functions and generalized distances, *Biometrics* **22**, 96–110.

Cramer, H. (1937), *Random Variables and Probability Distributions*, Cambridge Tracts in Mathematics and Math. Phy. **36**, Cambridge University Press.

Fisher, R. A. (1928), The general sampling distribution of the multiple correlation coefficient, *Proc. Roy. Soc.* **A 121**, 654–673.

Fisher, R. A. (1936), The use of multiple measurements in taxonomic problems, *Ann. Eugen.* **7**, 179–188.

Fisher, R. A. (1939), The sampling distribution of some statistics obtained from nonlinear equations, *Ann. Eugen.* **9**, 238–249.

Fisher, R. A. (1940), The precision of discriminant functions, *Ann. Eugen.* **10**, 422–429.

Frechet, M. (1951), Generalisation de la loi probabilite de Laplace, *Ann. de l'Institute Henri Poincare* **12**.

Girshick, M. A. (1939), On the sampling theory of roots of determinantal equations, *Ann. Math. Statist.* **10**, 203–224.

Grenander, U. and M. Rosenblatt (1957), *Statistical Analysis of Stationary Time Series*, Wiley, New York, Almqvist and Wicksell, Stockholm.

Hotelling, H. (1931), The generalization of Student's ratio, *Ann. Math. Statist.* **2**, 360–378.

Hotelling, H. (1935), The most predictable criterion, *J. Educ. Psych.* **26**, 139–142.

Hotelling, H. (1936), Relations between two sets of variates, *Biometrika* **28**, 321–377.

Hsu, P. L. (1938), Notes on Hotelling's generalised T^2, *Ann. Math. Statist.* **9**, 231–243.

Hsu, P. L. (1939), On the distribution of the roots of certain determinantal equations, *Ann. Eugen.* **9**, 250–258.

James, A. T. (1954), Normal multivariate analysis and the orthogonal group, *Ann. Math. Statist.* **25**, 40–75.

John, S. (1961), Errors in discrimination, *Ann. Math. Statist.* **32,** 1125–1144.

Khatri, C. G. (1968), Some results for the singular normal multivariate regression models, *Sankhyā* **A 30,** 267–280.

Kshirsagar, A. M. (1961), The goodness of fit of a single (non-isotropic) hypothetical principal component, *Biometrika* **48,** 397–407.

Kshirsagar, A. M. (1972), *Multivariate Analysis,* Marcell Dekker, Inc. New York.

Lawley, D. N. and A. E. Maxwell (1963), *Factor Analysis as a Statistical Method,* Butterworth's Mathematical Texts, London.

Mahalanobis, P. C. (1936), On the generalized distance in statistics, *Proc. Nat. Inst. Sci. India* **12,** 49–55.

Mahalanobis, P. C., R. C. Bose and S. N. Roy (1937), Normalization of variates and the use of rectangular coordinates in the theory of sampling distributions, *Sankhyā* **3,** 1–40.

Mitra, S. K. (1970), Some characteristic and non-characteristic properties of the Wishart distribution, *Sankhyā* **A 31,** 19–22.

Nair, U. S. (1938), The application of the moment function in the study of distribution laws in statistics, *Biometrika* **30,** 274–294.

Prohorov, Yu. V. and M. Fisz (1957), A characteristic property of the normal distribution in Hilbert space, *Teoria Veroiatnostei iee Primeneniia* **2,** 475–477.

Potthoff, R. F. and S. N. Roy (1964), A generalized multivariate analysis of variance model useful especially for growth curve problems, *Biometrika* **51,** 313–326.

Rao, K. S. (1951), On the mutual independence of a set of Hotelling's T^2 derivable from a sample of size n from a k-variate normal population, *Bull. Inst. Internat. Stat.* **33,** Part II, 171–176.

Rasch, G. (1953), On simultaneous factor analysis in several populations, *Uppsala Symposium on Psychological Factor Analysis, Uppsala,* Almqvist and Wicksell, 65–71.

Riffenburg, R. H. and C. W. Clunies-Ross (1960), Linear discriminant analysis, *Pacific Science* **14,** 251–256.

Roy, J. (1951), The distribution of certain likelihood criteria useful in multivariate analysis, *Bull. Inst. Internat. Stat.* **33,** Part II, 219–230.

Roy, S. N. (1939), p-Statistics or some generalizations in the analysis of variance appropriate to multivariate problems, *Sankhyā* **4,** 381–396.

Roy, S. N. (1957), *Some Aspects of Multivariate Analysis,* Wiley, New York.

Roy, S. N. and R. C. Bose (1953), Simultaneous confidence interval estimation, *Ann. Math. Statist.* **24,** 513–536.

Roy, S. N., R. Gnanadesikan and J. N. Srivastava (1972) *Analysis and Design of Certain Quantitative Multiresponse Experiments,* Pergamon Press.

Smith, C. A. B. (1947), Some examples of discrimination, *Ann. Eugen.* **13,** 272–282.

Wald, A. (1944), On a statistical problem arising in the classification of an individual into one of two groups, *Ann. Math. Statist.* **15,** 145–162.

Watson, G. W. (1964), A note on maximum likelihood, *Sankhyā* **26A,** 303–304.

Wijsman, R. A. (1957), Random orthogonal transformations and their use in some classical distribution problems in multivariate analysis, *Ann. Math. Statist.* **28**, 415–423.

Wilks, S. S. (1932), Certain generalizations in the analysis of variance, *Biometrika* **24**, 471–494.

Wilks, S. S. (1962), *Mathematical Statistics*, Wiley, New York.

Williams, E. J. (1952), Some exact tests in multivariate analysis, *Biometrika* **39**, 17–31.

Williams, E. J. (1955), Significance tests for discriminant functions and linear functional relationships, *Biometrika* **42**, 360–381.

Williams, E. J. (1959), *Regression Analysis*, Wiley, New York.

Williams, E. J. (1967), The analysis of association among many variates. *J. Roy Statist. Soc.* **B 29**, 199-228.

Wishart, John (1928), The generalized product moment distribution in samples from a normal multivariate population, *Biometrika* **20 A**, 32—52.

Wold, H. (1938), *A Study in the Analysis of Stationary Time Series*, University of Uppsala.

PUBLICATIONS OF THE AUTHOR

BOOKS

B1. *Linear Statistical Inference and Its Applications.* John Wiley, 1965, 1973 (second edition).

B2. *Advanced Statistical Methods in Biometric Research.* John Wiley, 1952 and Heffner, 1971.

B3. (with S. K. Mitra) *Generalized Inverse of Matrices and Its Applications.* John Wiley, 1971.

B4. (with A. Kagan and Yu. V. Linnik) *Characterization Problems of Mathematical Statistics* (in Russian). U.S.S.R. Academy of Sciences, 1972. (English translation, in press, John Wiley)

B5. *Computers and the Future of Human Society.* Andhra University and Statistical Publishing Society, 1970.

B6. (with R. K. Mukherjee and J. C. Trevor) *The Ancient Inhabitants of Jebel Moya.* Cambridge University Press, 1955.

B7. (with P. C. Mahalanobis and D. N. Majumdar) *Anthropometric Survey of the United Provinces, 1941.* A statistical study. *Sankhyā* **9**, 90–324, 1949.

B8. (with Majumdar) *Bengal Anthropometric Survey, 1945.* A statistical study. *Sankhyā* **19**, 201–408; also reprinted as a book under the title *Race Elements of Bengal, a Quantitative Study.* Asia Publishing House, 1958.

B9. (with A. Matthai and S. K. Mitra) *Formulae and Tables for Statistical Work.* Statistical Publishing Society, 1966.

RESEARCH PAPERS

1941

a. (with K. R. Nair) Confounded designs for asymmetrical factorial experiments. *Science and Culture* **7**, 361.

1942

a. On the volume of a prismoid in N-space and some problems in continuous probability. *Math. Student* **10**, 68–74.

b. On the sum of observations from different Gamma type populations. *Science and Culture* **7**, 614.

c. (with K. R. Nair) Confounded designs for $k \times p^m \times q^n \times \cdots$ type of factorial experiments. *Science and Culture* **7**, 361.

d. (with K. R. Nair) A general class of quasi-factorial designs leading to confounded designs for factorial experiments. *Science and Culture* **7**, 457.

e. (with K. R. Nair) A note on partially balanced incomplete block designs. *Science and Culture* 7, 568.

f. (with K. R. Nair) Incomplete block designs involving several groups of varieties. *Science and Culture* 7, 625.

1943

a. Certain experimental arrangements in quasi-latin squares. *Current Science* **12**, 322.

b. On bivariate correlation surfaces. *Science and Culture* **8**, 236.

1944

a. (with R. C. Bose and S. Chowla) On the integral order mod p of quadratics $x^2 + ax + b$ with applications to the construction of minimum functions for $GF(p^2)$ and to some number theory results. *Bull. Cal. Math. Soc.* **36**, 153–174.

1945

a. Familial correlations or the multivariate generalization of the intraclass correlation. *Current Science* **14**, 66.

b. Generalization of Markoff's theorem and tests of linear hypotheses. *Sankhyā* **7**, 9–16.

c. Markoff's theorem with linear restrictions on parameters. *Sankhya* **7**, 16–19.

d. Information and accuracy attainable in the estimation of statistical parameters. *Bull. Cal. Math. Soc.* **37**, 81–91.

e. Finite geometries and certain derived results in theory of numbers. *Proc. Nat. Inst. Sc.* **11**, 136–149.

f. (with R. C. Bose and S. Chowla) On the roots of a well-known congruence. *Proc. Nat. Acad. Sc.* **14**, 193.

g. (with R. C. Bose and S. Chowla) Minimum functions in Galois fields. *Proc. Nat. Acad. Sc.* **14**, 191.

1946

a. Difference sets and combinatorial arrangements derivable from finite geometries. *Proc. Nat. Inst. Sc.* **12**, 123–135.

b. On the linear combination of observations and the general theory of least squares. *Sankhyā* **7**, 237–256.

c. Confounded factorial designs in quasi-latin squares. *Sankhyā* **7**, 295–304.

d. Hypercubes of strength 'd' leading to confounded designs in factorial experiments. *Bull. Cal. Math. Soc.* **38**, 67–78.

e. On the mean conserving property. *Proc. Ind. Acad. Sc.* **23**, 165-173.

f. Tests with discriminant functions in multivariate analysis. *Sankhyā* **7**, 407–414.

g. On the most efficient designs in weighing. *Sankhyā* **7**, 440.

h. Minimum variance and the estimation of several parameters. *Proc. Camb. Phil. Soc.* **43**, 280–283.

i. (with S. J. Poti) On locally most powerful tests when alternatives are one sided. *Sankhyā* **7**, 441.

1947

a. The problem of classification and distance between two populations. *Nature* **159**, 30.

b. Note on a problem of Ragnar Frisch. *Econometrika* **15**, 245–249. A correction to Note on a problem of Ragnar Frisch. *Econometrika* **17**, 212.

c. General methods of analysis for incomplete block designs. *J. Am. Statist. Assoc.* **42**, 541–561.

d. Factorial experiments derivable from combinatorial arrangements of arrays. *J. Roy. Statist. Soc.* **B9**, 128–140.

e. Large sample tests of statistical hypotheses concerning several parameters with applications to problems of estimation. *Proc. Camb. Phil. Soc.* **44**, 50–57.

1948

a. A statistical criterion to determine the group to which an individual belongs. *Nature* **160**, 835.

b. Tests of significance in multivariate analysis. *Biometrika* **35**, 58–79.

c. The utilization of multiple measurements in problems of biological classification. *J. Roy. Statist. Soc.* **B10**, 159–203.

d. Sufficient statistics and minimum variance estimates. *Proc. Camb. Phil. Soc.* **45**, 215–218.

e. (with D. C. Shaw) On a formula for the prediction of cranial capacity. *Biometrics* **4**, 247–253.

f. (with K. R. Nair) Confounding in asymmetrical factorial experiments. *J. Roy. Statist. Soc.* **B10**, 109–131.

1949

a. On a class of arrangements. *Edin. Math. Proc.* **8**, 119–125.

b. On some problems arising out of discrimination with multiple characters. *Sankhyā* **9**, 343–364.

c. A note on unbiased and minimum variance estimates. *Cal. Statist. Bull.* **3**, 36.

d. (with P. Slater) Multivariate analysis applied to differences between neurotic groups. *British J. Psychology, Statist. Sec.* **2**, 17–29.

1950

a. Methods of scoring linkage data giving the simultaneous segregation of three factors. *Heredity* **4**, 37–59.

b. The theory of fractional replication in factorial experiments. *Sankhyā* **10**, 84–86.

c. Statistical inference applied to classificatory problems. *Sankhyā* **10**, 229–256.

d. A note on the distribution of $D^2_{p+q} - D_p{}^2$ and some computational aspects of D^2 statistic and discriminant function. *Sankhyā* **10**, 257–268.

e. Sequential tests of null hypotheses. *Sankhyā* **10**, 361–370.

1951

a. A theorem in least squares. *Sankhyā* **11**, 9–12.

b. Statistical inference applied to classificatory problems. Part II. Problems of selecting individuals for various duties in a specified ratio. *Sankhyā* **11**, 107–116.

c. A simplified approach to factorial experiments and the punched card technique in the construction and analysis of designs. *Bull. Inst. Inter. Statist.* **XXXIII(2)**, 1–28.

d. An asymptotic expansion of the distribution of Wilks' criterion. *Bull. Inst. Inter. Statist.* **XXXIII(2)**, 177–180.

e. The applicability of large sample tests for moving average and auto regressive schemes to series of short length—an experimental study. Part 3: The discriminant function approach in the classification of time series. *Sankhyā* **11**, 257–272.

f. Progress of statistics in India. *Progress of Science in India* (1939–1950), 68–94, National Institute of Sceinces, India.

1952

a. Some theorems on minimum variance estimation. *Sankhyā* **12**, 27–42.

b. Minimum variance estimation in distributions admitting ancillary statistics. *Sankhyā* **12**, 53–56.

1953

a. Discriminant function for genetic differentiation and selection. *Sankhyā* **12**, 229–246.

b. On transformations useful in the distribution problems of least squares. *Sankhyā* **12**, 339–346.

1954

a. A general theory of discrimination when the information about alternative population distributions is based on samples. *Ann. Math. Statist.* **25**, 651–670.

b. On the use and interpretation of distance functions in statistics. *Bull. Inst. Inter. Statist.* **34**, 90–100.

c. Estimation of relative potency from multiple response data. *Biometrics* **10,** 208–220.

1955

a. Analysis of dispersion for multiply classified data with unequal numbers in cells. *Sankhyā* **15,** 253–280.
b. Estimation and tests of significance in factor analysis. *Psychometrika* **20,** 93–111.
c. Theory of the method of estimation by minimum chi-square. *Bull. Inst. Inter. Statist.* **35,** 25–32.
d. (with G. Kalianpur) On Fisher's lower bound to asymptotic variance of a consistent estimate. *Sankhyā* **16,** 331–342.

1956

a. On the recovery of interblock information in varietal trials. *Sankhyā* **17,** 105–114.
b. A general class of quasi-factorial and related designs. *Sankhyā* **17,** 165–174.
c. Analysis of dispersion with missing observations. *J. Roy. Statist. Soc.* **B18,** 259–264.
d. (with I. M. Chakravarti) Some small sample tests of significance for a Poisson distribution. *Biometrics* **12,** 264–282.

1957

a. Maximum likelihood estimation for multinomial distribution. *Sankhyā* **18,** 139-148.

1958

a. Quantitative studies in Sociology, need for increased use in India. *Sociology, Social Research, and Social Problems in India.* Asia Publishing House, 1961, 53–74.
b. Some statistical methods for comparison of growth curves. *Biometrics* **14,** 1–17.
c. Maximum likelihood estimation for the multinomial distribution with an infinite number of cells. *Sankhyā* **20,** 211–218.

1959

a. Some problems involving linear hypotheses in multivariate analysis. *Biometrika* **46,** 49–58.
b. Expected values of mean squares in the analysis of incomplete block experiments and some comments based on them. *Sankhyā* **21,** 327–336.
c. Sur une Characterisation de la Distribution Normal Etablie d'apres une Propriete Optimum des Estimations Lineares. *Coll. Inter. due C.N.R.S.* France, **LXXXVII,** 165.

d. (with I. M. Chakravarthy) Tables for some small sample tests of significance for Poisson distributions and 2×3 contingency tables. *Sankhya* **21**, 315–326.

1960

a. Multivariate analysis: An indispensable statistical aid in applied research. *Sankhyā* **22**, 317–338.

b. Experimental designs with restricted randomization. *Bull. Inst. Inter. Statist.* **XXXVII**, 397–404.

1961

a. A study of large sample test criteria through properties of efficient estimates. *Sankhyā A* **23**, 25–40.

b. Asymptotic efficiency and limiting information. *Proc. Fourth Berkeley Symposium on Mathematical Statistics and Probability*, University of California, **1**, 531–546.

c. A study of BIB designs with replications 11 to 16. *Sankhyā* **23**, 117–127.

d. A combinatorial assignment problem. *Nature* **191**, 100.

e. Generation of random permutations of given number of elements using random sampling numbers. *Sankhyā A* **23**, 305–307.

f. Combinatorial arrangements analogous to orthogonal arrays. *Sankhyā A* **23**, 283–286.

g. Some observations on multivariate statistical methods in anthropological research. *Bull. Inst. Inter. Statist.* **XXXVII(4)**, 99–109.

1962

a. Ronald Aylmer Fisher, F.R.S. (an obituary) *Science and Culture* **29**, 80–81.

b. Some observations on anthropometric surveys: *Anthropology today. Essays in memory of D. N. Majumdar*, 135–149. Asia Publishing House.

c. Use of discriminant and allied functions in multivariate analysis. *Sankhyā A* **24**, 149–154.

d. Efficient estimates and optimum inference procedures in large samples (with discussion). *J. Roy. Statist. Soc.* **B 24**, 46–72.

e. Problems of selection with restriction. *J. Roy. Statist. Soc.* **B 24**, 401–405.

f. A note on a generalized inverse of a matrix with applications to problems in mathematical statistics. *J. Roy. Statist. Soc.* **B 24**, 152–158.

g. Apparent anomalies and irregularities in maximum likelihood estimation (with discussion), *Sankhyā B*, **24**, 73–102.

1963

a. Criteria of estimation in large samples. *Sankhyā A* **25**, 189–206.

b. (with V. S. Varadarajan) Discrimination of Gaussian processes. *Sankhyā A* **25**, 303–330.

1964

a. Problems of selection involving programming techniques. *Proceedings of the IBM Scientific Computing Symposium on Statistics*, 29–51.

b. The use and interpretation of principal components analysis in applied research. *Sankhyā A* **26**, 329–358.

c. Sir Ronald Fisher—The architect of multivariate analysis. *Biometrics* **20**, 286–300.

d. (with H. Rubin) On a characterization of the Poisson distribution. *Sankhyā A* **26**, 295–298.

1965

a. Efficiency of an estimator and Fisher's lower-bound to asymptotic variance. *Bull. Inst. Inter. Statistic* XLI(1), 55–63.

b. The theory of least squares when the parameters are stochastic and its applications to the analysis of growth curves. *Biometrika* **52**, 447–458.

c. On discrete distributions arising out of methods of ascertainment. *Sankhyā A* **27**, 311–324.

d. Covariance adjustment and related problems in multivariate analysis. Dayton Symposium on Multivariate Analysis. *Multivariate Analysis* **1**, 87–103. Academic Press Inc., New York.

e. (with A. M. Kagan and Yu. V. Linnik) On a characterization of the normal law based on a property of the sample average. *Sankhyā A* **27**, 405–406.

1966

a. Characterization of the distribution of random variables in linear structural relations. *Sankhyā A* **28**, 251–260.

b. Generalized inverse for matrices and its applications in mathematical statistics—*Festschrift for J. Neyman. Research Papers in Statistics*. 263–299.

c. Discriminant function between composite hypotheses and related problems. *Biometrika* **53**, 315–321.

d. Discrimination among groups and assigning new individuals. *The Role of Methodology of Classification in Psychiatry and Psychopathology*, 229–240, U.S. Department of Health, Education, and Welfare, Public Health Services.

e. On some characterisations of the normal law. *Sankhyā A* **29**, 1–14.

f. (with M. N. Rao) Linked cross sectional study for determining norms and growth curves: A pilot survey on Indian school-going boys. *Sankhyā B* **28**, 231–252 (Partly published under the title, Methods for determining norms and growth rates, A study amongst Indian school-going boys. *Gerentologia* **12**, 200–216).

1967

a. Calculus of generalized inverses of matrices: Part I—general theory. *Sankhyā A* **29**, 317–350.

b. Cyclical generation of linear subspaces in finite geometries, 515–535, Chapter 28 in *Combinatorial Mathematics and Its Applications*. The University of North Carolina Monograph Series No. 4.

c. Least squares theory using an estimated dispersion matrix and its application to measurement of signals. *Proc. Fifth Berkeley Symposium on Mathematical Statistics and Probability* **1**, 355–372, University of California Press.

d. On vector variables with a linear structure and a characterization of the multivariate normal distribution. *Bull. Inst. Inter. Statist.* **XLII(2)**, 1207–1213.

1968

a. A note on a previous lemma in the theory of least squares and some further results. *Sankhyā A* **30**, 259–266.

b. (with C. G. Khatri) Some characterizations of the gamma distribution. *Sankhyā A* **30**, 157–166.

c. (with C. G. Khatri) Solutions to some functional equations and their applications to characterisation of probability distributions. *Sankhyā A* **30**, 167–180.

d. (with B. Ramachandran) Some results on characteristic functions and characterizations of the normal and generalized stable laws. *Sankhyā A* **30**, 125–140.

e. (with S. K. Mitra) Simultaneous reduction of a pair of quadratic forms. *Sankhyā A* **30**, 313–322.

f. (with S. K. Mitra) Some results in estimation and tests of linear hypotheses under the Gauss-Markoff model. *Sankhyā A* **30**, 281–290.

1969

a. A decomposition theorem for vector variables with a linear structure. *Ann. Math. Statist.* **40**, 1845–1849.

b. Some characterizations of the multivariate normal distribution. *Multivariate Analysis* **2**, 322–328. Academic Press Inc., New York.

c. Recent advances in discriminatory analysis. *J. Ind. Soc. Agri. Statist.* **XXI**, 3–15.

d. A multidisciplinary approach for teaching statistics and probability. *Sankhyā B* **31**, 321–340.

e. (with S. K. Mitra) Conditions for optimality and validity of least squares theory. *Ann. Math. Statist.* **40**, 1716–1724.

1970

a. Estimation of heteroscedastic variances in a linear model. *J. Am. Statist. Assoc.* **65**, 161–172.

b. Computers: A great revolution in scientific research. *Proc. Indian National Science Academy* **36**, 123–139.

c. Inference on discriminant function coefficients. *Essays in Probability and Statistics*, Chapter 30, 587–602. Edited by R. C. Bose and others. University of North Carolina and Statistical Publishing Society.

d. (with B. Ramachandran) Solutions of functional equations arising in some regression problems and a characterisation of the Cauchy law. *Sankhāy A* **32**, 1–30.

1971

a. Characterisation of probability laws through linear functions. *Sankhyā A* **33**, 265–270.

b. Some aspects of statistical inference in problems of sampling from finite populations. *Foundations of Statistical Inference*, Holt, Rinehart, and Winston of Canada, 177–202.

c. Estimation of variance and covariance components—Minque Theory. *J. Multivariate Analysis* **1**, 257–275.

d. Minimum variance quadratic unbiased estimation of variance components. *J. Multivariate Analysis* **1**, 445–456.

e. Unified theory of linear estimation. *Sankhyā A* **33**, 370–396 and *Sankhyā A* **34**, 477.

f. Data, analysis, and statistical thinking. *Economic and Social Development, Essays in Honour of C. D. Deshmukh*, Vora and Co. Bombay, 383–392.

g. (with J. K. Ghosh) A note on some translation—parameter families of densities for which the median is an m.l.e. *Sankhyā A* **33**, 91–93.

h. (with S. K. Mitra) Further contributions to the theory of generalized inverse of matrices and its applications. *Sankhyā A* **33**, 289–300.

i. (with S. K. Mitra) Generalized inverse of a matrix and applications. *Sixth Berkeley Symposium on Mathematical Statistics and Probability* **1**, 601–620, University of California Press.

j. Some comments on the logarithmic series distribution. *Statistical Ecology* 1, 131–142. Pennsylvania State University Press.

k. Taxonomy in Anthropology. *Mathematics in Archaeological and Historical Sciences*, 19–29. Edin. Univ. Press.

1972

a. Recent trends of research work in multivariate analysis. *Biometrics* **28**, 3–22.

b. Estimation of variance and covariance components in linear models. *J. Am. Statist. Assoc.* **67**, 112–115.

c. A note on IPM method in the unified theory of linear estimation. *Sankhyā A* **34**, 285–288.

d. Some recent results in linear estimation. *Sankhyā* **B 34**, 369–377.

e. (with P. Bhimasankaram and S. K. Mitra) Determination of a matrix by its subclasses of g-inverse. *Sankhyā A* **34**, 5–8.

f. (with C. G. Khatri) Functional equations and characterisation of probability laws through linear functions of random variables. *J. Multivariate Analysis* **2**, 162–173.

1973

a. Unified theory of least squares. *Communications in Statistics*, **1**, 1–8.

b. Some combinatorial problems of arrays and applications to design of experiments, in *A Survey of Combinatorial Theory*, North Holland, Chapter 29, pp. 349–360.

c. Representation of best linear unbiased estimators in the Gauss-Markoff model with a singular dispersion matrix. *J. Multivariate Analysis* **3,** 276–292.

d. (with S. K. Mitra) Theory and application of constrained inverse matrices. *SIAM J. App. Math.* **24,** 473–488.

e. (with A. M. Kagan and Y. V. Linnik) Extension of Darmois-Skitovic theorem to functions of random variables satisfying an addition theorem. *Communications in Statistics* **1,**

f. (with D. C. Rao and R. Chakravarthy) The generalized Wrights' model. *Genetic structure of Populations,* 55–59. Univ. Press Hawaii.

Author Index

Aitken, A., 44, 78, 220, 221, 312
Aitkinson, F. A., 71, 78
Anderson, S. L., 498, 512
Anderson, T. W., 68, 78, 498, 512, 542, 601
Anscombe, F. J., 426, 428, 442, 485, 512
Arnold, J. C., 322, 380
Ashton, E. H., 579, 601

Bagai, 543
Bahadur, R. R., 140, 141, 153, 366, 380, 425, 442, 470, 512
Banerji, S., 463, 512
Barnard, G., 343, 444, 488, 490, 505, 512
Barnard, M. M., 572, 602
Bartlett, M. S., 551, 556, 559, 573, 602
Basu, D., 505, 512
Berge, C., 53, 78
Bhattacharya, A., 324, 376, 380
Bickle, P. J., 153
Birch, M. W., 363, 380
Birnbaum, A., 444, 505, 512
Blackwell, D., 321, 380, 497, 512
Bock, R. D., 196, 218
Bose, R. C., 220, 224, 312, 542, 597, 599, 602, 603
Bowker, A. K., 542, 602
Box, G. E. P., 286, 312, 498, 512
Burnaby, T. P., 580, 602

Chakravarthy, I. M., 509
Chapman, D. G., 324, 380
Chipman, J. S., 307, 312
Chernoff, H., 441, 442, 453, 464, 470, 482, 497, 504, 513
Cluniess-Ross, 573, 603
Cochran, W. G., 189, 218, 266, 286, 312, 393, 442, 464, 504, 513
Constantine, A. G., 543
Craig, A. T., 189, 210, 218
Cramer, H., 106, 137, 153, 162, 218, 324, 380, 517, 602
Cox, D. R., 505, 513
Cox, G. M., 286, 312, 464, 504, 513

Daniels, H. E., 366, 380
Dantzig, G. B., 455, 486, 513
Darmois, G., 218
Dasgupta, S., 68, 78
David, F. N., 220, 313
Dempster, A. P., 196, 218, 340, 380
Dewess, G., 304, 312
Dharmadhikari, S. W., 106, 153
Dodge, H. F., 473, 513
Doob, J. L., 79, 98, 115, 153, 350, 380, 421, 442
Dugue, D., 366, 380
Dwyer, P. S., 75, 78

Feller, W., 79, 153
Ferguson, T. S., 139, 153
Finney, D. J., 373, 380
Fisher, R. A., 44, 78, 167, 174, 286, 312, 313, 324, 329, 339, 345, 348, 351, 373, 380, 415, 428, 432, 438, 442, 444, 463, 502, 504, 505, 513, 542, 556, 566, 568, 599, 602
Fisz, M., 518, 603
Focke, J., 304, 312
Fortet, R., 154
Fraser, D. A. S., 207, 340, 380, 510, 513
Frechet, M., 518, 602

Gantmacher, F. R., 44, 45, 46, 78
Gart, J. J., 324, 380
Gauss, K. F., 220, 312
Gayen, A. K., 432, 442
Ghosh, J. K., 132, 379
Ghosh, Malay, 599
Gihman, I. I., 422, 442

Girshick, M. A., 497, 512, 542, 602
Glivenko, V. I., 421, 442
Gnanadesikan, R., 573, 603
Gnedenko, B. V., 79, 128, 154
Godambe, V. P., 355, 377, 380, 381, 444, 513
Goldman, A. J., 220, 312
Graybill, F. A., 189, 218, 252, 313
Grenander, U., 79, 154, 518, 602
Guttman, L., 72, 78

Hajek, J., 143, 154
Haldane, J. B. S., 352, 366, 381, 511, 513
Halmos, P. R., 4, 78, 84, 137, 154
Hamburger, H., 106, 154
Hartley, H. O., 313
Healy, M. J. R., 579, 601
Heyde, C. C., 152
Hodges, J. L., Jr., 324, 347, 381, 470, 513
Hoeffding, W., 470, 504, 513
Hoerl, A. E., 306, 312
Hogben, L., 444, 513
Hogg, R. V., 189, 218
Hooke, B. G. E., 272, 280, 313
Hotelling, H., 542, 582, 602
Householder, A. S., 21, 75, 78, 231
Hsu, P. L., 228, 313, 542, 602
Huzurbazar, V. S., 366, 381

James, A. T., 542, 602
Jeffreys, G. M., 154, 336, 381, 444, 513
Jenkins, G. M., 444, 505, 512
John, S., 579, 603
Joos, George, 174, 218

Kac, M., 147, 154
Kagan, A., 79, 218, 379, 527
Kallianpur, G., 347
Katti, S. K., 322, 380
Kempthorne, O., 286, 287, 313, 504, 513
Kennard, R. W., 306, 312
Khatri, C. G., 29, 30, 78, 165, 527, 528, 543, 603
Kiefer, J., 363, 381, 425, 442
Kinchin, A. Ya., 112, 146
Kolmogorov, A. N., 79, 83, 98, 128, 146, 153, 322, 381, 421
Krishnaih, P. R., 543
Kshirsagar, A. M., 593, 603
Kyburgh, H. E., Jr., 444, 514

Lawley, D. V., 587, 603
LeCam, L., 347, 366, 381
Lehmann, E. L., 141, 153, 322, 324, 381, 411, 442, 463, 470, 501, 513, 514
Levi, F. W., 22, 78
Levy, P., 79, 154
Lindley, D. V., 444, 512, 514
Linnik, Yu., V., 79, 154, 218, 219, 286, 313, 379, 463, 464, 514, 527
Lipton, S., 579, 601
Loeve, M., 79, 86, 129, 130, 147, 154

Mahalanobis, P. C., 474, 514, 595, 597, 603
Mann, H. B., 124, 154
Mardia, K. V., 178, 219
Markoff, A. A., 220, 313
Martin, F. B., 220, 313
Matthai, A., 313
Mathai, A. M., 543
Maxwell, A. E., 587, 603
Mayer, J. W., 160, 219
Mayer, M. G., 160, 219
Millikin, G. A., 252, 313
Mitra, S. K., 50, 189, 220, 225, 300, 313, 377, 381, 504, 514, 535, 603
Moore, E. H., 26, 78
Moran, P. A. P., 71, 78
Moses, L. E., 497, 513
Mulholland, H. P., 71, 78
Munroe, M. E., 84, 137, 154

Nair, U. S., 555, 603
Neyman, J., 220, 266, 313, 352, 366, 381, 417, 442, 444, 445, 470, 504, 514
Nievergelt, E., 501, 514
Noether, G. E., 470, 514
Nordskog, A. W., 287, 313

Ogasawara, T., 188, 189, 219
Ogawa, J., 189, 210, 219
Okamoto, M., 125, 154

Parthasarathy, K., 79, 154
Parzen, E., 220, 313
Pathak, P. K., 142, 154, 218, 219
Patil, G. P., 377, 381
Pearson, E. S., 313, 417, 442, 445, 514
Pearson, Karl, 213, 281, 284, 313, 391, 443

Penrose, R., 26, 78
Perlman, M. D., 333, 381
Pillai, R. N., 218, 219
Pillai, S. S., 543
Pitman, E. J. G., 470, 504, 513
Plackett, R. L., 286, 313, 499, 514
Polya, G., 216, 219
Poti, S. J., 456
Pratt, J. W., 136, 154
Prohorov, Yu. V., 518, 603
Puri, M. L., 501, 514

Ramachandran, B., 79, 154
Rao, J. S., 178, 219
Rao, K. S., 542, 603
Rasch, G., 587, 603
Renyi, A., 143, 154
Riffenburgh, R. H., 573, 603
Romig, H. G., 474, 513
Robbins, H., 324, 338, 380, 381, 488, 489, 490, 513
Rosenblatt, M., 470, 513, 518, 602
Roy, J., 377, 381, 556, 603
Roy, S. N., 542, 561, 573, 597, 599, 602, 603

Sakamoto, H., 189, 210, 219
Savage, I. R., 501, 514
Savage, L. J., 95, 154, 444, 470, 501, 504, 514
Scheffe, H., 124, 154, 240, 241, 286, 313, 381, 499, 514
Seigmund, D., 488, 489, 514
Sen, P. K., 501, 514
Sengupta, S., 178, 219
Shaw, D. C., 270
Siegel, S., 501, 515
Skitovic, V. P., 218, 219
Slater, P., 576
Smirnov, N., 421, 422, 443
Smith, C. A. B., 71, 78, 511, 513, 578, 603
Srivastava, J. N., 573, 603
Sprott, D. A., 207, 340, 355, 381, 444, 513
Srivastava, J. N., 573, 603
Stein, C., 324, 343, 381, 486, 515
Student, (W. S. Gosset), 170, 219
Sudakov, V. N., 132, 154

Takahashi, M., 188, 189, 219
Teicher, H., 379, 381
Tukey, J., 249, 251, 313
Turnbull, H. W., 44, 78

Van der Wearden, B. L., 501, 515
Varadarajan, V. S., 128, 154, 518

Wald, A., 124, 128, 154, 366, 381, 417, 443, 455, 456, 474, 495, 497, 500, 515, 579, 603
Walsh, J. E., 501, 515
Watson, G. S., 174, 219, 531, 603
Watterson, G. A., 71, 78
Weiss, L., 425, 443
Welch, W. L., 464, 515
Wijsman, R. A., 542, 603
Wilcoxon, F., 500, 515
Wilks, S. S., 543, 551, 555, 604
Williams, E. J., 174, 219, 286, 313, 573, 604
Winston, C. B., 444, 505, 512
Wishart, J., 291, 313, 573, 604
Wold, H., 106, 153, 517, 518, 604
Wolfowitz, J., 124, 154, 363, 381, 485, 500, 504, 515

Yates, F., 286, 313, 413, 443

Zelen, M., 220, 312
Zyskind, G., 220, 313

Subject Index

For lemmas and theorems named after authors, see under lemmas and theorems respectively.

Analysis of covariance, concomitant variables, 288
univariate, 289
illustrative example, 291
multivariate, 551, 561
Analysis of dispersion, definition, 548
distribution of Wilks' Λ, 555, 556
Gauss-Markoff (multivariate), 543
growth model (multivariate), 561
test for additional information, 551
test for dimensionality, 556
test for structural relationship, 564
test of linear hypotheses, 547–551
Analysis of variance, confidence interval for ratio, 241
covariance adjustment, 289
F-test (Fisher), 238, 239
Gauss-Markoff model, 221
general Gauss-Markoff model, 297
interval estimation, 237, 470
minque (variance components), 65–66, 258, 303
one-way classification, 244
regression. *See* Regression and association
simultaneous confidence intervals, 240
t-test (Student, Gosset), 237, 243
test for nonadditivity, general, 251
Tukey's, 249–251
two-way classification, variance components, 258
with equal numbers, 247, 253
with unequal numbers, 254
Ancillary information, 505

Basis of vector space, definition, 4
Gram-Schmidt orthogonalization, 9
orthogonal basis, 9
orthonormal basis, 9
Basis of vector space, *(Continued)*
Steinitz replacement result, 14
Bayes decision rule, 493
Bivariate normal, characteristic function, 203
density, 202
distribution of correlation coefficient, 206, 208
distribution of means, 205
distribution of regression coefficient, 208
distribution of variances and covariances, 205
properties, 202
Borel field, (σ-field), 83

Central limit theorems, invariance principle, 148
Liapunov, 127
Lindberg - Feller, 128
Lindberg - Levy, 127
multivariate case, 128, 147
Characteristic function, Bochner's theorem, 141
continuity theorem, 119
convergence of, 119
cumulants, 101
definition, 99
examples on, 142–143
expansion of, 100
inversion theorem, 104
moments, 93, 101
non-negative definiteness of, 141
properties of, 100
Characteristic function of, beta, 151
binomial, 102
Cauchy, 151
degenerate, 151
gamma, 151

Characteristic function of, *(Continued)*
  hyperbolic, 151
  Laplace, 151
  multivariate normal, 184, 519, 520
  normal, 103, 159
  Poisson, 102
  rectangular, 151
  triangular, 151
Conditional density, 92, 99
Conditional expectation, 96, 98, 141
Conditional probability, 90
Conditional variance, 97
Confidence sets and invervals, 470
Control charts (O.C. of), 510, 511
Covariance adjustment. *See* Analysis of covariance
Convergence of densities (Scheffe), 124-126
Convergence of distribution functions, definition, 116
  extended Helly-Bray theorem, 118
  Helly lemma, 117
  Helly-Bray theorem, 117
  in law, 116
  Polya's theorem, 120
  theorems involving moments, 121
  weak compactness, 117
Convergence of functions, 134
Convergence of integrals, dominated convergence theorem, 136
  dominated convergence theorem (Pratt's extension), 136
  Fatou's lemma, 135
  monotone convergence theorem, 135
Convergence of random variables, almost sure (strong), 110
  in law, 116
  in probability, 110
  quadratic mean, 110
  rth mean, 148
  theorems on, 122–124, 148
  with probability one, 110
Contingency tables, Dandekar's continuity correction, 414, 415
  test for independence of attributes, 404–415
  Yates' continuity correction, 413, 415
Control charts (O.C. of), 510
Convex sets, definition, 51
  separation theorems, 51, 52
Convolution of distribution functions, 103
Correlation. *See* Regression
Critical difference, 246
Critical region, 445
Cumulants, 101

Dandekar's continuity correction, 414, 415
Decision theory (identification), admissible rules, 492
  Bayes' solution, 492, 493
  complete class (minimal), 495
  minimax rule, 496
  non-randomized rule, 491
  randomized rule, 491
Density, Arc sine, 165
  Bessel function, 217
  beta (central), 151, 165, 167, 168
  beta (noncentral), 172, 217
  binomial, 88, 102, 155
  bivariate Cauchy, 170
  bivariate normal, 201
  Cauchy, 151, 170
  chi-square (central), 166, 181
  chi-square (noncentral), 182
  circular normal, 174, 177
  correlation coefficient (null), 207
  correlation coefficient (non-null), 208
  degenerate, 151
  Dirichlet, 215
  directional data, 175
  equilibrium state of particles, 172
  exponential family, 195
  f (central), 165
  f (noncentral), 172, 216
  F (central), Fisher, 167
  F (noncentral), Fisher, 216
  gamma, 151, 164
  geometric, 88
  Hotelling's T^2 (non-null), 542
  Hotelling's T^2 (null), 542
  Hyperbolic cosine, 151
  hypergeometric, 88
  Laplace, 151, 212
  life testing model, 213
  logarithmic, 88
  logistic, 213
  lognormal, 212
  Makeham's law, 214
  Maxwell, 173
  multinomial, 355
  multiple correlation (central), 273
  multiple correlation (noncentral), 599, 600
  multivariate normal, 184, 194, 527

Density, *(Continued)*
  negative binomial, 88
  normal, 88
  order statistics, 214
  Pareto, 212
  Pearsonian curves, 213
  Pearson P_λ, 168
  Poisson, 88, 102
  Rayleigh, 164
  rectangular, 151
  regression coefficient, 208
  Student's t (central), 170, 183, 197
  Student's t (noncentral), 171
  symmetric normal, 196
  triangular, 151
  Von Mises, 177
  waiting time, 166
  Weibull, 214
  Wishart, 597, 598
  wrapped-up distributions, 178
  z (Fisher), 167
Determinantal equation, 38
Determinants, 22, 31
Discriminatory analysis, discriminant scores, 574
  discrimination of composite hypotheses, 580
  Fisher's discriminant function, 567
  sufficiency of discriminant function, 578
  tests on discriminant functions, 568, 569, 601
Dispersion matrix, 107, 517
Distribution free methods, 499
Distribution function, conditional, 91
  convergence. *See* Convergence of distribution functions
  convolution of, 103
  decomposition of, 85
  definition, 84
  properties of, 85-87
Distributions (sampling, asymptotic), X^2 for contingency tables, 403, 404-412
  X^2 for goodness of fit (Pearson), 391
  deviation in a single cell, 393
  deviation in specified cells, 397
  fractiles, 423-425
  functions of statistics, 385-388
  H-test for homogeneity, 389
  Kolmogorov-Smirnov test, 421
  likelihood ratio tests (Neyman-Pearson), 417
Distributions (sampling, asymptotic), *(Continued)*
  linear functions of frequencies, 383, 384
  quadratic function of frequencies, 384
  scoring test (Rao), 417, 418
  standard errors, 386-388
  Wald's large sample tests, 417, 419
  Wilk's Λ, 556
Distributions (sampling, exact), chi-square (central), 181
  chi-square (noncentral), 182
  correlation, intraclass (non-null), 200
  correlation, intraclass (null), 199
  correlation, multiple (non-null), 599, 600
  correlation, multiple (null), 273
  correlation, product moment (non-null), 208
  correlation, product moment (null), 207
  Fisher's F (null), 167
  Fisher's F (non-null), 216
  Fisher's z, 167
  Hotelling's T^2 (non-null), 542
  Hotelling's T^2 (null), 542
  Mahalanobis' D^2, 542
  P_λ (Pearson), 168
  quadratic forms, 185-193
  regression (bivariate), 208
  sample mean and dispersion, 537
  sample mean and variance, 182
  square of normal variable, 163
  Student's t (non-null), 171
  Student's t (null), 170
  U-statistic (additional information, Rao), 554
  Wishart distribution, 597, 598

Eigen values, 38
Eigen values (restricted), 50
Empirical distribution function, convergence (Glivenko), 421
  definition, 420
  Kolmogorov-Smirnov statistic, 421
  Kolmogorov theorem, 421
  Smirnov theorem, 422
Entropy, 173
Estimation, B A N estimator, 348
  Bayes' estimator, 335
  biased estimator, 332
  C A N estimator, 347
  confidence set, 334, 470
  consistency (Fisher), 345

Estimation, *(Continued)*
consistency (method), 345
consistency (probability), 344
C U A N estimator, 350
efficiency (asymptotic variance), 346
efficiency (first order, Rao), 348
efficiency (Fisher), 348
efficiency (second order, Rao), 352
empirical Bayes estimator (Robbins), 336
fiducial inference (Fisher), 339
general estimation problem, 334
Haldane's discrepancy, 352
Hellinger distance, 352
inadmissibility of sample mean (Stein), 343
information limit to variance (Cramer, Rao), 324-326
information matrix, 331
information measure, 329-331
Kullback-Leibler separator, 352
law of equal ignorance, 336
law of succession, 336
least squares. *See* Linear estimation, Gauss-Markoff
maximum likelihood. *See* Maximum likelihood
method of moments, 351
method of scoring, 367-370
minimax, 341
minimum variance, 317-329
minimum chi-square, 352
pooling of estimators, 389
sufficient statistic. *See* Sufficiency
Expectation, conditional (given statistic), 96
conditional (given subfield), 141
conditional (substitution in), 153
definition and theorems, 92
moments, 93
Exponential family of distributions, 195

Factor analysis, 587
Factorial moments, 102
Fiducial probability, 339
Fisher's F, 167
Fisher's z, 167
Fraser-Sprott equation, 207

g-inverse. *See* Matrix inversion
Gauss-Markoff theory (multivariate), additional information (test for), 551
analysis of dispersion. *See* Analysis of dispersion
Gauss-Markoff theory (multivariate), *(Continued)*
applications, 562-573
covariance analysis, 552
estimation (BLUE etc.), 544-547
growth model, 561
model (general), 543, 544
structural relationship (dimensionality), 556, 564
tests of linear hypotheses, 547-551
Gauss-Markoff theory (univariate), BLE (best linear), 305
BLIMBE (best linear minimum bias), 306
BLUE (best linear unbiased), correlated observations, 229
efficiency of estimators, 233
geometric solution, 228
least squares theory, 223
normal equations, 222, 230
random parameters, 234
simultaneous confidence intervals, 240
simultaneous estimation, 233
singular covariance for observations. *See* Linear estimation-Unified theory
choice of design matrix, 235
Gauss-Markoff model, 221
random parameters, 234
restricted parameters, 231
tests of hypotheses, 237-240
variance of estimator, 226, 230
Gram-Schmidt orthogonalization, 10
Growth model (Pothoff, Roy), 561

Hotelling's T^2, 542
Huzurbazar's conjecture (sufficiency), 132
Hypothesis. *See* Testing of hypotheses

Independence of events, 90
Identification problem, 491
Inequalities (algebraic), Bessel, 10
Cauchy-Schwarz, 54
Frobenius, 31
Hadamard, 56
Hölder, 55
information theory, 58
Jensen, 58
Kantorovich, 74
Lagrange identity, 55
Minkowski, 56
moments, 56

Inequalities (algebraic), *(Continued)*
Sylvester's law, 31
Inequalities (Probability), Berge, 145
Camp-Meidell, 145
Cantelli, 145
Chebyschev, 95
Chebyschev type, 145
Hajek-Renyi, 114, 143
Hölder, 149
Jensen, 149
Kolmogorov, 143
Minkowski, 149
Peek, 145
Infinitely divisible distribution, 149
Information limit to variance (Cramer, Rao), 324–326
Information measure (Fisher), 329–331
Intrinsic accuracy (Fisher), 331
Invariance, principle in estimation, 343
principle for CLT, 148
probability measures, 138
statistical procedure, 139

Law (zero or one), 148
Law of equal ignorance, 336
Law of iterated logarithms, 129, 130, 488
Law of large numbers, Chebychev's theorem (WLLN), 112
definition (WLLN, SLLN), 112
Kinchin's theorem (WLLN), 112, 146
Kolmogorov (SLLN), 114, 115
Kolmogorov (WLLN), 146
other theorems, 146
Law of succession, 336
Least squares. *See* Gauss-Markoff, Linear estimation
Lebegue integral, 133, 134
Lebegue-Stieltjes integral, 132
Lemmas named after authors, Borel-Cantelli, 137
Farka, 53
Fatou, 135
Helly, 117
Kronecker, 148
Neyman-Pearson, 446
Schur, 77
Likelihood, 353
Likelihood ratio (monotone), 461
Likelihood ratio test (Neyman-Pearson), 417, 418
Limit theorems. *See* Central limit theorems, Law of large numbers, Distributions (asymptotic), Convergence of random variables and Convergence of distribution functions
Linear equations, 6
Linear estimation (Unified theory), basic lemma on g-inverse, 294
general Gauss-Markoff model, 297
Gauss-Markoff model. *See* Gauss-Markoff theory
I. P. M. method, 298
test for consistency, 297
test of linear hypotheses, 299
U. L. S. method, 300
variances and covariances, 298, 301

Mahalanobis D^2, 542
Matrix, adjoint, 76
Cayley-Hamilton theorem, 44
circulant, 68
conjugate transpose, 16
derivatives, 71, 72
eigen values and vectors, 38
elementary, 17
g-inverse. *See* Matrix inversion
gramian, 69
hermitian, 17, 42
idempotent, 28
identity (or unit), 15
inverse, 16
inverse, generalized. *See* Matrix inversion
non-negative, 45
normal, 43
null, 15
nullity, 27
null space, 27, 31
orthogonal, 17
partitioned, 17
range space, 27
rank, 16
symmetric, 17
trace, 28, 33, 65
transpose, 15
unit (or identity), 15
unitary, 17
Matrix inversion, computation of g-inverse, 27, 34, 224
duality between A_m^- and A_l^-, 49
g-inverse, 24
least squares inverse, 48

Matrix inversion, *(Continued)*
left inverse, 16
minimum norm inverse, 48
minimum norm least squares inverse, 49
Moore-Penrose inverse, 26, 49
reflexive *g*-inverse, 26
right inverse, 16
Matrix product, Hadamard, 30
new product (Khatri-Rao), 30
Kronecker, 29
Matrix reduction, commuting matrices, 41
diagonal form, 19, 20
echolon form, 18
Frobenius theorem, 46
Hermite canonical form, 18
Hermitian matrix, 42
Householder transformation, 21
Jordan form, 45
Metzler matrix, 46
non-negative matrix, 45
nonsymmetric matrix, 43
normal matrix, 43
pairs of matrices, 41
Perron's theorem, 46
polar form, 70
Q-R-decomposition, 21
rank factorization, 19
triangular form, 20, 22
symmetric matrix, 20, 39
singular value decomposition, 43
spectral decomposition, 40, 45, 46
Maximum likelihood (m.l.), characterization by m.l. estimators, 378
distribution of m.l. estimators, 365
limitation of m.l. method, 355
m.l. equation estimator, 353
m.l. estimation of multinomial, 355, 356-363
m.l. estimator, 353
near m.l. estimator, 353
properties of m.l. estimator, 364
scoring for m.l. estimation, 366
Minimum trace problems, 65-67
Minque theory (variance components), 303
Moment generating function, binomial, 102
definition, 101
normal, 103
Poisson, 102
Moment problem, 106
Multinomial distribution, definition, 355
estimation, 356-363
Multivariate moments, 107
Multivariate normal distribution, bivariate normal, 201
characteristic function, 184, 519
characterization, 525, 526, 532
definition, 184, 518, 522
density (nonsingular dispersion), 184, 194, 517, 527
density (singular dispersion), 527
estimation of parameters, 528
properties of, 519-524
symmetric, 197

Nonparametric inference, distribution free method, 499
Fisher-Yates test, 501
permutation distribution, 503
randomization principle, 501
robustness, 497
sign test, 500
Van der Waerden test, 501
Wald-Wolfowitz test, 500
Wilcoxon test, 500
Normal distribution, characteristic function, 103, 159
circular normal, 174, 177
moment generating function, 101
Normal distribution (characterization), Cramer's theorem, 162
Darmois-Skitovic theorem, 218
De Moivre's theorem, 160
Hagen's hypothesis, 161
Herschell's hypothesis, 158
maximum entropy, 162
Maxwell's hypothesis, 160
other theorems, 218
Polya's theorem, 215, 216

Order statistics, 214, 410
Orthogonal basis of a random variable, 587

Permutation distribution, 503
Principal components, 590
Probability, axioms, 82
conditional, 90
density (p.d.), 87, 89
family of measures, 130
generating function, 101
independence, 90
measure, 83
posterior, 334

Probability, *(Continued)*
prior, 334
space, 84
Probability generating functions, binomial, 102
convergence of, 126
definition, 101
Poisson, 102
Projection operator, definition, 46
explicit expressions, 47, 48
orthogonal, 10, 47

Quadratic forms, classification, 35
distribution of, 185-193
extremum problems, 60-65
reduction of one form, 20, 40
reduction of two forms, 41
transformation, 35

Random variables, conditional distribution, 91
convergence. *See* Convergence of random variables
definition, 84, 87, 89
independence of, 92
Kolmogorov consistency theorem, 108
Randomization principle, 501
Regression and association, best linear predictor, 266, 267
best predictor, 264
canonical correlations, 582
canonical variables, 584
common factors, 585
correlation ratio, 265
covariance analysis, 289
factor analysis, 587
general theory of, 263-270
intraclass correlation, 268
linear predictor, 266, 267
multiple correlation, 266, 268
part correlation, 311
partial correlation coefficient, 268, 269
partial correlation ratio, 268
partial multiple correlation, 268
restricted coefficients, 287
Regression (examples), cranial capacity, 270
equality of coefficients, 281
specification (test of), 284
Robustness, 497

Scoring test (Rao), 417, 418
Second order efficiency (Rao), 352
Sequential analysis, A. S. N. function, 484
control charts (O.C. of), 510
efficiency of S. P. R. T., 478
estimation (Stein), 485
fundamental identity, 482
O.C. function, 484
S. P. R. T. (Wald), 475-478
test with power one, 488
truncation of S. P. R. T., 510
Singular value decomposition, 42
Statistics and subfields, 139
Stieltje's integral, 132
Stirling's approximation, 59
Subfield, 86, 139, 141
Sufficient statistic, bounded completeness, 457
completeness, 321
definition, 130
factorization theorem, 131
Huzurbazar's conjecture, 132
use in estimation (Rao, Blackwell), 321
Sufficient sub-field, 141
Sylvester's law (matrix rank), 31

t-test (Student), 170, 183, 197, 237, 243, 459
Testing of hypotheses (theory), ancillary information, 505
asymptotic efficiency, 464
composite hypothesis, 445, 456
critical function, 450
critical region, 445
distribution free methods, 493
Fisher-Behren's problem, 463
level of significance, 446
Lindley's paradox, 512
locally most powerful test, 453, 454
monotone likelihood ratio, 461
Neyman-Pearson lemma, 446
nonparametric tests. *See* Nonparametric inference
permutation test, 503
Pitman efficiency, 467
probability ratio test, 451
randomized test, 450
robustness, 498
sequential test. *See* Sequential analysis
similar region test, 456
two kinds of errors, 445
unbiased test, 454

Testing of hypotheses (theory), *(Continued)*
uniformly most powerful test, 449
Theorems (named after authors), Bayes, 334
Bochner (n.n.d. of characteristic function), 141
Caley-Hamilton (matrix), 44
Chebychev (W.L.L.N.), 112
Cramer (normal), 162
Darmois-Skitovic, 218, 526
DeMoivre, 160
Fieller, 311
Fisher-Cochran, 185
Frobenius (positive matrix), 46
Fubini, 137
Glivenko, 421
Helly-Bray, 117
Helly-Bray (extended), 118
Kinchin (W.L.L.N.), 112, 146
Kolmogorov (consistency of distributions, 108
Kolmogorov (d.f. of D_n), 421
Kolmogorov (S.L.L.N.), 114, 115
Kolmogorov (sums of random variables), 129
Kolmogorov (W.L.L.N.), 146
Lebesgue dominated convergence, 136
Lebesgue monotone convergence, 135
Liapunov (C.L.T.), 127
Lindberg-Feller (C.L.T.), 128
Lindberg-Levy (C.L.T.), 127
Perron (positive matrix), 46
Poincare separation, 64
Polya (convergence of d.f.), 120
Polya (normal law), 215, 216
Radon-Nikodym, 137
Sakamoto-Craig, 210
Scheffe, 124
Smirnov, 422
Strumian separation, 64

Transformation (linear), 23, 24
Transformation of statistics. binomial proportion ($\sin^{-1}$), 427
correlation coefficient ($\tanh^{-1}$), 432
Poisson variable ($\sqrt{\ }$), 426
Transformation of variables, 156

Variance and covariance components, general model, 302
minque theory, 303
two way data, 258
Vector space, basis (Hamel), 4
definition, 3
direct sum, 11, 40
Gram-Schmidt orthogonalization, 9
inner product, 8, 74
orthogonal basis, 9
orthonormal basis, 9
subspace, 4

Wald's large sample tests, 417, 419
Wald's S.P.R.T., 475
Wilks' Λ criterion, Bartlett's approximation, 556
decomposition, 554
definition, 551
distribution (exact), 555
distribution (asymptotic), 556
variance ratio approximation (Rao), 556
Wishart distribution, 533, 597, 598

Yates' continuity correction, 413, 415

Zero-one law, 148

WILEY SERIES IN PROBABILITY AND STATISTICS

ESTABLISHED BY WALTER A. SHEWHART AND SAMUEL S. WILKS

The ***Wiley Series in Probability and Statistics*** is well established and authoritative. It covers many topics of current research interest in both pure and applied statistics and probability theory. Written by leading statisticians and institutions, the titles span both state-of-the-art developments in the field and classical methods.

Reflecting the wide range of current research in statistics, the series encompasses applied, methodological and theoretical statistics, ranging from applications and new techniques made possible by advances in computerized practice to rigorous treatment of theoretical approaches.

This series provides essential and invaluable reading for all statisticians, whether in academia, industry, government, or research.

ABRAHAM and LEDOLTER · Statistical Methods for Forecasting
AGRESTI · Analysis of Ordinal Categorical Data
AGRESTI · An Introduction to Categorical Data Analysis
AGRESTI · Categorical Data Analysis
ANDĚL · Mathematics of Chance
ANDERSON · An Introduction to Multivariate Statistical Analysis, *Second Edition*
*ANDERSON · The Statistical Analysis of Time Series
ANDERSON, AUQUIER, HAUCK, OAKES, VANDAELE, and WEISBERG · Statistical Methods for Comparative Studies
ANDERSON and LOYNES · The Teaching of Practical Statistics
ARMITAGE and DAVID (editors) · Advances in Biometry
ARNOLD, BALAKRISHNAN, and NAGARAJA · Records
*ARTHANARI and DODGE · Mathematical Programming in Statistics
*BAILEY · The Elements of Stochastic Processes with Applications to the Natural Sciences
BALAKRISHNAN and KOUTRAS · Runs and Scans with Applications
BARNETT · Comparative Statistical Inference, *Third Edition*
BARNETT and LEWIS · Outliers in Statistical Data, *Third Edition*
BARTOSZYNSKI and NIEWIADOMSKA-BUGAJ · Probability and Statistical Inference
BASILEVSKY · Statistical Factor Analysis and Related Methods: Theory and Applications
BASU and RIGDON · Statistical Methods for the Reliability of Repairable Systems
BATES and WATTS · Nonlinear Regression Analysis and Its Applications
BECHHOFER, SANTNER, and GOLDSMAN · Design and Analysis of Experiments for Statistical Selection, Screening, and Multiple Comparisons
BELSLEY · Conditioning Diagnostics: Collinearity and Weak Data in Regression
BELSLEY, KUH, and WELSCH · Regression Diagnostics: Identifying Influential Data and Sources of Collinearity
BENDAT and PIERSOL · Random Data: Analysis and Measurement Procedures, *Third Edition*

*Now available in a lower priced paperback edition in the Wiley Classics Library.

BERRY, CHALONER, and GEWEKE · Bayesian Analysis in Statistics and Econometrics: Essays in Honor of Arnold Zellner
BERNARDO and SMITH · Bayesian Theory
BHAT · Elements of Applied Stochastic Processes, *Second Edition*
BHATTACHARYA and JOHNSON · Statistical Concepts and Methods
BHATTACHARYA and WAYMIRE · Stochastic Processes with Applications
BILLINGSLEY · Convergence of Probability Measures, *Second Edition*
BILLINGSLEY · Probability and Measure, *Third Edition*
BIRKES and DODGE · Alternative Methods of Regression
BLISCHKE AND MURTHY · Reliability: Modeling, Prediction, and Optimization
BLOOMFIELD · Fourier Analysis of Time Series: An Introduction, *Second Edition*
BOLLEN · Structural Equations with Latent Variables
BOROVKOV · Ergodicity and Stability of Stochastic Processes
BOULEAU · Numerical Methods for Stochastic Processes
BOX · Bayesian Inference in Statistical Analysis
BOX · R. A. Fisher, the Life of a Scientist
BOX and DRAPER · Empirical Model-Building and Response Surfaces
*BOX and DRAPER · Evolutionary Operation: A Statistical Method for Process Improvement
BOX, HUNTER, and HUNTER · Statistics for Experimenters: An Introduction to Design, Data Analysis, and Model Building
BOX and LUCEÑO · Statistical Control by Monitoring and Feedback Adjustment
BRANDIMARTE · Numerical Methods in Finance: A MATLAB-Based Introduction
BROWN and HOLLANDER · Statistics: A Biomedical Introduction
BRUNNER, DOMHOF, and LANGER · Nonparametric Analysis of Longitudinal Data in Factorial Experiments
BUCKLEW · Large Deviation Techniques in Decision, Simulation, and Estimation
CAIROLI and DALANG · Sequential Stochastic Optimization
CHATTERJEE and HADI · Sensitivity Analysis in Linear Regression
CHATTERJEE and PRICE · Regression Analysis by Example, *Third Edition*
CHERNICK · Bootstrap Methods: A Practitioner's Guide
CHILÈS and DELFINER · Geostatistics: Modeling Spatial Uncertainty
CHOW and LIU · Design and Analysis of Clinical Trials: Concepts and Methodologies
CLARKE and DISNEY · Probability and Random Processes: A First Course with Applications, *Second Edition*
*COCHRAN and COX · Experimental Designs, *Second Edition*
CONGDON · Bayesian Statistical Modelling
CONOVER · Practical Nonparametric Statistics, *Second Edition*
COOK · Regression Graphics
COOK and WEISBERG · Applied Regression Including Computing and Graphics
COOK and WEISBERG · An Introduction to Regression Graphics
CORNELL · Experiments with Mixtures, Designs, Models, and the Analysis of Mixture Data, *Second Edition*
COVER and THOMAS · Elements of Information Theory
COX · A Handbook of Introductory Statistical Methods
*COX · Planning of Experiments
CRESSIE · Statistics for Spatial Data, *Revised Edition*
CSÖRGŐ and HORVÁTH · Limit Theorems in Change Point Analysis
DANIEL · Applications of Statistics to Industrial Experimentation
DANIEL · Biostatistics: A Foundation for Analysis in the Health Sciences, *Sixth Edition*
*DANIEL · Fitting Equations to Data: Computer Analysis of Multifactor Data, *Second Edition*
DAVID · Order Statistics, *Second Edition*

*Now available in a lower priced paperback edition in the Wiley Classics Library.

*DEGROOT, FIENBERG, and KADANE · Statistics and the Law
DETTE and STUDDEN · The Theory of Canonical Moments with Applications in Statistics, Probability, and Analysis
DEY and MUKERJEE · Fractional Factorial Plans
DILLON and GOLDSTEIN · Multivariate Analysis: Methods and Applications
DODGE · Alternative Methods of Regression
*DODGE and ROMIG · Sampling Inspection Tables, *Second Edition*
*DOOB · Stochastic Processes
DOWDY and WEARDEN · Statistics for Research, *Second Edition*
DRAPER and SMITH · Applied Regression Analysis, *Third Edition*
DRYDEN and MARDIA · Statistical Shape Analysis
DUDEWICZ and MISHRA · Modern Mathematical Statistics
DUNN and CLARK · Applied Statistics: Analysis of Variance and Regression, *Second Edition*
DUNN and CLARK · Basic Statistics: A Primer for the Biomedical Sciences, *Third Edition*
DUPUIS and ELLIS · A Weak Convergence Approach to the Theory of Large Deviations
*ELANDT-JOHNSON and JOHNSON · Survival Models and Data Analysis
ETHIER and KURTZ · Markov Processes: Characterization and Convergence
EVANS, HASTINGS, and PEACOCK · Statistical Distributions, *Third Edition*
FELLER · An Introduction to Probability Theory and Its Applications, Volume I, *Third Edition,* Revised; Volume II, *Second Edition*
FISHER and VAN BELLE · Biostatistics: A Methodology for the Health Sciences
*FLEISS · The Design and Analysis of Clinical Experiments
FLEISS · Statistical Methods for Rates and Proportions, *Second Edition*
FLEMING and HARRINGTON · Counting Processes and Survival Analysis
FULLER · Introduction to Statistical Time Series, *Second Edition*
FULLER · Measurement Error Models
GALLANT · Nonlinear Statistical Models
GHOSH, MUKHOPADHYAY, and SEN · Sequential Estimation
GIFI · Nonlinear Multivariate Analysis
GLASSERMAN and YAO · Monotone Structure in Discrete-Event Systems
GNANADESIKAN · Methods for Statistical Data Analysis of Multivariate Observations, *Second Edition*
GOLDSTEIN and LEWIS · Assessment: Problems, Development, and Statistical Issues
GREENWOOD and NIKULIN · A Guide to Chi-Squared Testing
GROSS and HARRIS · Fundamentals of Queueing Theory, *Third Edition*
*HAHN · Statistical Models in Engineering
HAHN and MEEKER · Statistical Intervals: A Guide for Practitioners
HALD · A History of Probability and Statistics and their Applications Before 1750
HALD · A History of Mathematical Statistics from 1750 to 1930
HAMPEL · Robust Statistics: The Approach Based on Influence Functions
HANNAN and DEISTLER · The Statistical Theory of Linear Systems
HEIBERGER · Computation for the Analysis of Designed Experiments
HEDAYAT and SINHA · Design and Inference in Finite Population Sampling
HELLER · MACSYMA for Statisticians
HINKELMAN and KEMPTHORNE: · Design and Analysis of Experiments, Volume 1: Introduction to Experimental Design
HOAGLIN, MOSTELLER, and TUKEY · Exploratory Approach to Analysis of Variance
HOAGLIN, MOSTELLER, and TUKEY · Exploring Data Tables, Trends and Shapes
*HOAGLIN, MOSTELLER, and TUKEY · Understanding Robust and Exploratory Data Analysis
HOCHBERG and TAMHANE · Multiple Comparison Procedures

*Now available in a lower priced paperback edition in the Wiley Classics Library.

HOCKING · Methods and Applications of Linear Models: Regression and the Analysis of Variables
HOEL · Introduction to Mathematical Statistics, *Fifth Edition*
HOGG and KLUGMAN · Loss Distributions
HOLLANDER and WOLFE · Nonparametric Statistical Methods, *Second Edition*
HOSMER and LEMESHOW · Applied Logistic Regression, *Second Edition*
HOSMER and LEMESHOW · Applied Survival Analysis: Regression Modeling of Time to Event Data
HØYLAND and RAUSAND · System Reliability Theory: Models and Statistical Methods
HUBER · Robust Statistics
HUBERTY · Applied Discriminant Analysis
HUNT and KENNEDY · Financial Derivatives in Theory and Practice
HUSKOVA, BERAN, and DUPAC · Collected Works of Jaroslav Hajek—with Commentary
IMAN and CONOVER · A Modern Approach to Statistics
JACKSON · A User's Guide to Principle Components
JOHN · Statistical Methods in Engineering and Quality Assurance
JOHNSON · Multivariate Statistical Simulation
JOHNSON and BALAKRISHNAN · Advances in the Theory and Practice of Statistics: A Volume in Honor of Samuel Kotz
JUDGE, GRIFFITHS, HILL, LÜTKEPOHL, and LEE · The Theory and Practice of Econometrics, *Second Edition*
JOHNSON and KOTZ · Distributions in Statistics
JOHNSON and KOTZ (editors) · Leading Personalities in Statistical Sciences: From the Seventeenth Century to the Present
JOHNSON, KOTZ, and BALAKRISHNAN · Continuous Univariate Distributions, Volume 1, *Second Edition*
JOHNSON, KOTZ, and BALAKRISHNAN · Continuous Univariate Distributions, Volume 2, *Second Edition*
JOHNSON, KOTZ, and BALAKRISHNAN · Discrete Multivariate Distributions
JOHNSON, KOTZ, and KEMP · Univariate Discrete Distributions, *Second Edition*
JUREČKOVÁ and SEN · Robust Statistical Procedures: Aymptotics and Interrelations
JUREK and MASON · Operator-Limit Distributions in Probability Theory
KADANE · Bayesian Methods and Ethics in a Clinical Trial Design
KADANE AND SCHUM · A Probabilistic Analysis of the Sacco and Vanzetti Evidence
KALBFLEISCH and PRENTICE · The Statistical Analysis of Failure Time Data
KASS and VOS · Geometrical Foundations of Asymptotic Inference
KAUFMAN and ROUSSEEUW · Finding Groups in Data: An Introduction to Cluster Analysis
KENDALL, BARDEN, CARNE, and LE · Shape and Shape Theory
KHURI · Advanced Calculus with Applications in Statistics
KHURI, MATHEW, and SINHA · Statistical Tests for Mixed Linear Models
KLUGMAN, PANJER, and WILLMOT · Loss Models: From Data to Decisions
KLUGMAN, PANJER, and WILLMOT · Solutions Manual to Accompany Loss Models: From Data to Decisions
KOTZ, BALAKRISHNAN, and JOHNSON · Continuous Multivariate Distributions, Volume 1, *Second Edition*
KOTZ and JOHNSON (editors) · Encyclopedia of Statistical Sciences: Volumes 1 to 9 with Index
KOTZ and JOHNSON (editors) · Encyclopedia of Statistical Sciences: Supplement Volume
KOTZ, READ, and BANKS (editors) · Encyclopedia of Statistical Sciences: Update Volume 1
KOTZ, READ, and BANKS (editors) · Encyclopedia of Statistical Sciences: Update Volume 2

*Now available in a lower priced paperback edition in the Wiley Classics Library.

KOVALENKO, KUZNETZOV, and PEGG · Mathematical Theory of Reliability of Time-Dependent Systems with Practical Applications
LACHIN · Biostatistical Methods: The Assessment of Relative Risks
LAD · Operational Subjective Statistical Methods: A Mathematical, Philosophical, and Historical Introduction
LAMPERTI · Probability: A Survey of the Mathematical Theory, *Second Edition*
LANGE, RYAN, BILLARD, BRILLINGER, CONQUEST, and GREENHOUSE · Case Studies in Biometry
LARSON · Introduction to Probability Theory and Statistical Inference, *Third Edition*
LAWLESS · Statistical Models and Methods for Lifetime Data
LAWSON · Statistical Methods in Spatial Epidemiology
LE · Applied Categorical Data Analysis
LE · Applied Survival Analysis
LEE · Statistical Methods for Survival Data Analysis, *Second Edition*
LEPAGE and BILLARD · Exploring the Limits of Bootstrap
LEYLAND and GOLDSTEIN (editors) · Multilevel Modelling of Health Statistics
LINDVALL · Lectures on the Coupling Method
LINHART and ZUCCHINI · Model Selection
LITTLE and RUBIN · Statistical Analysis with Missing Data
LLOYD · The Statistical Analysis of Categorical Data
MAGNUS and NEUDECKER · Matrix Differential Calculus with Applications in Statistics and Econometrics, *Revised Edition*
MALLER and ZHOU · Survival Analysis with Long Term Survivors
MALLOWS · Design, Data, and Analysis by Some Friends of Cuthbert Daniel
MANN, SCHAFER, and SINGPURWALLA · Methods for Statistical Analysis of Reliability and Life Data
MANTON, WOODBURY, and TOLLEY · Statistical Applications Using Fuzzy Sets
MARDIA and JUPP · Directional Statistics
MASON, GUNST, and HESS · Statistical Design and Analysis of Experiments with Applications to Engineering and Science
McCULLOCH and SEARLE · Generalized, Linear, and Mixed Models
McFADDEN · Management of Data in Clinical Trials
McLACHLAN · Discriminant Analysis and Statistical Pattern Recognition
McLACHLAN and KRISHNAN · The EM Algorithm and Extensions
McLACHLAN and PEEL · Finite Mixture Models
McNEIL · Epidemiological Research Methods
MEEKER and ESCOBAR · Statistical Methods for Reliability Data
MEERSCHAERT and SCHEFFLER · Limit Distributions for Sums of Independent Random Vectors: Heavy Tails in Theory and Practice
MYERS, MONTGOMERY, and VINING · Generalized Linear Models. With Applications in Engineering and the Sciences
*MILLER · Survival Analysis, *Second Edition*
MONTGOMERY, PECK, and VINING · Introduction to Linear Regression Analysis, *Third Edition*
MORGENTHALER and TUKEY · Configural Polysampling: A Route to Practical Robustness
MUIRHEAD · Aspects of Multivariate Statistical Theory
MURRAY · X-STAT 2.0 Statistical Experimentation, Design Data Analysis, and Nonlinear Optimization
MYERS and MONTGOMERY · Response Surface Methodology: Process and Product in Optimization Using Designed Experiments
NELSON · Accelerated Testing, Statistical Models, Test Plans, and Data Analyses
NELSON · Applied Life Data Analysis
NEWMAN · Biostatistical Methods in Epidemiology

*Now available in a lower priced paperback edition in the Wiley Classics Library.

OCHI · Applied Probability and Stochastic Processes in Engineering and Physical Sciences
OKABE, BOOTS, SUGIHARA, and CHIU · Spatial Tesselations: Concepts and Applications of Voronoi Diagrams, *Second Edition*
OLIVER and SMITH · Influence Diagrams, Belief Nets and Decision Analysis
PANKRATZ · Forecasting with Dynamic Regression Models
PANKRATZ · Forecasting with Univariate Box-Jenkins Models: Concepts and Cases
*PARZEN · Modern Probability Theory and Its Applications
PEÑA, TIAO, and TSAY · A Course in Time Series Analysis
PIANTADOSI · Clinical Trials: A Methodologic Perspective
PORT · Theoretical Probability for Applications
POURAHMADI · Foundations of Time Series Analysis and Prediction Theory
PRESS · Bayesian Statistics: Principles, Models, and Applications
PRESS and TANUR · The Subjectivity of Scientists and the Bayesian Approach
PUKELSHEIM · Optimal Experimental Design
PURI, VILAPLANA, and WERTZ · New Perspectives in Theoretical and Applied Statistics
PUTERMAN · Markov Decision Processes: Discrete Stochastic Dynamic Programming
*RAO · Linear Statistical Inference and Its Applications, *Second Edition*
RENCHER · Linear Models in Statistics
RENCHER · Methods of Multivariate Analysis
RENCHER · Multivariate Statistical Inference with Applications
RIPLEY · Spatial Statistics
RIPLEY · Stochastic Simulation
ROBINSON · Practical Strategies for Experimenting
ROHATGI and SALEH · An Introduction to Probability and Statistics, *Second Edition*
ROLSKI, SCHMIDLI, SCHMIDT, and TEUGELS · Stochastic Processes for Insurance and Finance
ROSS · Introduction to Probability and Statistics for Engineers and Scientists
ROUSSEEUW and LEROY · Robust Regression and Outlier Detection
RUBIN · Multiple Imputation for Nonresponse in Surveys
RUBINSTEIN · Simulation and the Monte Carlo Method
RUBINSTEIN and MELAMED · Modern Simulation and Modeling
RYAN · Modern Regression Methods
RYAN · Statistical Methods for Quality Improvement, *Second Edition*
SALTELLI, CHAN, and SCOTT (editors) · Sensitivity Analysis
*SCHEFFE · The Analysis of Variance
SCHIMEK · Smoothing and Regression: Approaches, Computation, and Application
SCHOTT · Matrix Analysis for Statistics
SCHUSS · Theory and Applications of Stochastic Differential Equations
SCOTT · Multivariate Density Estimation: Theory, Practice, and Visualization
*SEARLE · Linear Models
SEARLE · Linear Models for Unbalanced Data
SEARLE · Matrix Algebra Useful for Statistics
SEARLE, CASELLA, and McCULLOCH · Variance Components
SEARLE and WILLETT · Matrix Algebra for Applied Economics
SEBER · Linear Regression Analysis
SEBER · Multivariate Observations
SEBER and WILD · Nonlinear Regression
SENNOTT · Stochastic Dynamic Programming and the Control of Queueing Systems
*SERFLING · Approximation Theorems of Mathematical Statistics
SHAFER and VOVK · Probability and Finance: It's Only a Game!
SMALL and McLEISH · Hilbert Space Methods in Probability and Statistical Inference
STAPLETON · Linear Statistical Models
STAUDTE and SHEATHER · Robust Estimation and Testing

*Now available in a lower priced paperback edition in the Wiley Classics Library.

STOYAN, KENDALL, and MECKE · Stochastic Geometry and Its Applications, *Second Edition*
STOYAN and STOYAN · Fractals, Random Shapes and Point Fields: Methods of Geometrical Statistics
STYAN · The Collected Papers of T. W. Anderson: 1943–1985
SUTTON, ABRAMS, JONES, SHELDON, and SONG · Methods for Meta-Analysis in Medical Research
TANAKA · Time Series Analysis: Nonstationary and Noninvertible Distribution Theory
THOMPSON · Empirical Model Building
THOMPSON · Sampling
THOMPSON · Simulation: A Modeler's Approach
THOMPSON and SEBER · Adaptive Sampling
TIAO, BISGAARD, HILL, PEÑA, and STIGLER (editors) · Box on Quality and Discovery: with Design, Control, and Robustness
TIERNEY · LISP-STAT: An Object-Oriented Environment for Statistical Computing and Dynamic Graphics
UPTON and FINGLETON · Spatial Data Analysis by Example, Volume II: Categorical and Directional Data
VIDAKOVIC · Statistical Modeling by Wavelets
WEISBERG · Applied Linear Regression, *Second Edition*
WELSH · Aspects of Statistical Inference
WESTFALL and YOUNG · Resampling-Based Multiple Testing: Examples and Methods for p-Value Adjustment
WHITTAKER · Graphical Models in Applied Multivariate Statistics
WINKER · Optimization Heuristics in Economics: Applications of Threshold Accepting
WONNACOTT and WONNACOTT · Econometrics, *Second Edition*
WOODING · Planning Pharmaceutical Clinical Trials: Basic Statistical Principles
WOOLSON · Statistical Methods for the Analysis of Biomedical Data
WU and HAMADA · Experiments: Planning, Analysis, and Parameter Design Optimization
YANG · The Construction Theory of Denumerable Markov Processes
*ZELLNER · An Introduction to Bayesian Inference in Econometrics

*Now available in a lower priced paperback edition in the Wiley Classics Library.

9 780471 218753